# NAME REACTIONS
# AND REAGENTS
# IN ORGANIC SYNTHESIS

# NAME REACTIONS
# AND REAGENTS
# IN ORGANIC SYNTHESIS
## Second Edition

**Bradford P. Mundy**
Prof. of Chemistry, Emeritus
Colby College
Waterville, ME

**Michael G. Ellerd**
Maxim Technologies
Bozeman, MT

**Frank G. Favaloro, Jr.**
Helicon Therapeutics
Farmingdale, NY

**WILEY-INTERSCIENCE**

A JOHN WILEY & SONS, INC., PUBLICATION

Published by John Wiley & Sons, Inc., Hoboken, New Jersey.
Published simultaneously in Canada.

For general information on our products and services contact our Customer Care Department within the U.S. at 877-762-2974, outside the U.S. at 317-572-3993 or fax 317-572-4002.

Wiley also publishes its books in a variety of electronic formats. Some content that appears in print, however, may not be available in electronic format.

*Library of Congress Cataloging-in-Publication Data is available.*
ISBN 0-471-22854-0

Printed in the United States of America

10 9 8 7 6 5 4 3 2 1

# Preface

It has been a long haul. The start for this revision came almost the same way that the original edition started. For the first edition it was Mike Ellerd, then an undergraduate at Montana State, who organized my crude Name Reaction handouts so well that others encouraged the conversion into a book. At Colby College, Frank Favaloro did the same thing, making "study sheets" and adding to the list of Name Reactions. He graduated in 1996 and I started reformatting and expanding. With encouragement from Darla Henderson, this became a project. By then Frank had finished graduate school and was enthusiastic about participating. I had also retired from formal teaching and found much more time for creative work. The three of us started to work in earnest!

This edition differs substantially from the first by the inclusion of many modern Name Reactions instead of sticking exclusively with the old, tried and true. There are many reactions not covered; indeed, we ultimately eliminated those that had little contemporary use. We generally applied a "rule of thumb" that a newer name had to be cited by multiple authors. Therefore there are some relatively new protocols that have not stood the test of time; however the breadth of recent use warranted inclusion. As for reagents, we have focused on both Name Reagents and those whose acronyms are often used in place of the actual name. We have noted the common use of these forms in current literature.

First and foremost, this is a book to be used. Feel free to write in the text . . . use any available blank space to add your own notes. Transform this into *your* book of Name Reactions! It is intended to serve as a starting point. Within a two page format for reactions and one page for reagents, the reader will find a basic, generalized definition / formula, a mechanism that conveys a possible course from starting material to product, notes which describe a few of the major highlights of the reaction or which points the reader to related reactions (by name or similarity) and recent examples of use. We have tried to convey the current mechanistic thinking with special care to show intermediate steps, point out proton exchanges, and sometimes suggest transition states, but without going through kinetics, isotope effects, etc.

Wherever appropriate, we have included references to selected secondary sources. They contain more detailed discussions on the topics introduced in this book. In all cases, we recommend use of the primary literature. The examples in the following pages are but a small taste of the detail, variation, scope and experimental detail available. Our choices reflect our personal interests; there is no "better or worse" implied! We tried to use current examples from journals that seem to be most commonly accessible, both in paper form and electronically, to student and professional alike. When recent references were difficult to come by, we made use of the abstracts and reaction-search engine of *SciFinder* (American Chemical Society). In these cases, we supplied a number [AN year: XXXX] that will allow ready access to the abstract. To the authors of the works we have chosen to describe, we hold the most sincere gratitude and we hope we have faithfully represented your work.

Colby College
Waterville, ME
Feb 1, 2005

# ACKNOWLEDGMENTS

As always, completion of a project requires more than just the work of the authors. Without the consideration, support and patience of spouses: Margaret (Brad), Mary (Mike) and Michelle (Frank), this probably could not have been completed.

Special thanks goes to the chemistry community for their endless development of new methods for creating C-C and C-heteroatom bonds. It has been an enlightening experience to chronicle the explosion of new "named" reactions and protocols. We have not lost view of the obvious new participation of the world chemical community.

Each of us can thank mentors and special people that have given us encouragement:

**Brad:**
I still owe much to my formal mentors:
Richard F. Smith who first provided the excitement of chemistry, A.Paul Krapcho, graduate mentor and friend, and the late Henry Rapoport, postdoctoral advisor.

I thank my colleagues from Colby College, Dasan Thamattoor and Jeff Katz, for their help in reading parts of this manuscript. And, of course my former graduate and undergraduate students . . . two of the latter are now coauthors, who were the reason for my continued interest in the academic life. Special thanks goes to Prof.Tom Poon (Claremont McKenna, Pitzer, & Scripps Colleges) for a great two years as a Dreyfus Fellow with me at Colby. He taught me much, and worked closely with Frank Favaloro.

I would like to thank several Colby staff that made my working easier: Susan W. Cole of the Science Library could always be depended on to solve any library problem that developed in the absolutely great electronic resources of Colby College, and patiently put up with my many requests, piled up books and journals and general use of the library. The Colby College ITS staff was extremely good-natured and helpful for computer questions. Their help was greatly appreciated.

**Mike:**
My appreciation goes out to all of my professors at Montana State, who,years ago sparked my interest in chemistry, and to those who still today keep that interest very much alive.

**Frank:**
I would like to thank all of those who not only taught me organic chemistry, but also to be excited for the art it contains: Gordon W. Gribble, Tadashi Honda, Thomas Spencer, Peter Jacobi, David Lemal, Thomas Poon, Philip Previte and, most importantly, Brad Mundy. Thank you to the many friends and co-workers who provided support, advice and the occasional reference: Erin Pelkey, Janeta Popovici-Müller, Tara Kishbaugh, Jeanese Badenock, Alison Rinderspacher and Chaoyang Dai.

Of course a project with a publisher requires interaction. Darla Henderson, Amy Byers, Camille Carter and Dean Gonzalez were the people who kept the ball rolling and the project in focus.

Colby College
Waterville, ME

Feb 1, 2005

# CONTENTS

# ACRONYMS AND ABBREVIATIONS

| Acronym | Name | |
|---------|------|---|
| Ac | Acetyl | Me, C=O structure |
| Acac | Acetylacetonate | Me–C(O)–CH$_2$–C(O)–Me structure |
| AcOH (HOAc) | Acetic acid | Me –COOH |
| AIBN | 2,2'-Azobisisobutyronitrile | NC–C(Me)$_2$–N=N–C(Me)$_2$–CN structure |
| ACN | 1,1´-Azobis-1-cyclohexanenitrile | structure |
| *9-BBN* | 9-Borobicyclo[3.3.1]nonane | structure |
| *BINAP* | 2,2'-Bis(Diphenylphosphino)-1,1'-binaphththyl | PPh$_2$ / PPh$_2$ structure |
| *BINOL* | 1,1'-bi-2,2'-naphthol | OH / OH structure |
| BITIP | Binol/Titanium isopropoxide | Ti(iPrO)$_4$ / BINOL |
| *BMDA* | Bromomagnesium Diisopropylamide | (iPr)$_2$N$^{\ominus}$ MgBr$^{\oplus}$ structure |
| *BMS* | Borane Dimethylsulfide | BH$_3$-Me$_2$S |

| | | |
|---|---|---|
| **BMS** | Borane Dimethylsulfide | $BH_3\text{-}Me_2S$ |
| Bn- | Benzyl | |
| **Boc-** (t-Boc) | t-Butoxycarbonylchloride | |
| **BOM-** | Benzyloxymethyl- | |
| Bs | Brosylate | |
| **Bu₃SnH** | tri-ⁿbutylstannane | $^{n}Bu)_3SnH$ |
| Bz | Benzoyl | |
| **CAN** | Ceric ammonium nitrate | $Ce(NH_4)_2(NO_3)_6$ |
| **CAS** | Ceric ammonium sulfate | $Ce(NH_4)_4(SO_4)_4$ |
| **Cbz-** | Carbobenzyloxy | |
| **CDI** | 1,1'-Carbonyldiimidazole | |
| Cetyl | Hexadeca- | $C_{16}H_{33}\text{-}$ |
| **cod** | Cyclooctadiene | |
| cp | Cyclopentadienyl | |
| cp* | Tetramethylcyclopentadienyl | |
| **CSA** | Camphorsulfonic Acid | |
| **DABCO** **TED** | 1,4-Diazabicylo[2.2.2]octane, TED, triethylenediamine | |

| | | |
|---|---|---|
| **_DAST_** | Diethylamino)sulfur trifluoride | Et—N—SF₃ (Et, Et on N) |
| **_DBN_** | 1,5-Diazabicyclo[4.3.0]non-5-ene | |
| **_DBU_** | 1,5-Diazabicyclo[5.4.0]undec-7-ene | |
| **_DCC_** | Dicyclohexylcarbodiimide | cyclohexyl—N=C=N—cyclohexyl |
| **_DDQ_** | 2,3-Dichloro-5,6-dicyano-1,4-benzoquinone | |
| **_DDO_** | Dimethyldioxirane | |
| **_DEAD_** | Diethyl Azodicarboxylate | EtOOC—N=N—COOEt |
| DEIPS | Diethylisopropylsilyl | i-Pr—Si—$\xi$ (Et, Et) |
| DET | Dietkyl tartrate | EtOOC-CH-CH-COOEt with OH and HO; in R-, S-, and meso forms |
| **_DIBAL_** **_DIBAL-H_** | Disobutylaluminum hydride | |
| **_DIEA_** **_DIPEA_** | Diisopropylethylamine **_Hunig's base_** | |
| **_DIPT_** | Diisopropyl tartrate | iPrOOC-CH-CH-COOiPr with OH and HO; in R-, S-, and meso forms |
| **_Diglyme_** | Diethylene glycol dimethyl ether | MeO⌒⌒O⌒⌒OMe |

| | | |
|---|---|---|
| **DMAP** | 4-(Dimethylamino)pyridine | Me N Me structure (pyridine ring with N,N-dimethylamino group) |
| **DME** | 1,2-Dimethoxyethane<br>Glyme | MeO—CH₂CH₂—OMe structure |
| DMIPS | Dimethylisopropylsilyl | i-Pr—Si(Me)₂— structure |
| DMF | Dimethylformamide | H—C(=O)—N(Me)Me structure |
| DMP | Dimethylpyrazole | dimethylpyrazole ring structure |
| **DMPU** | N,N'-Dimethylpropyleneurea | cyclic urea structure with Me on both N |
| DMS | Dimethylsulfide | Me—S—Me |
| **DMSO** | Dimethylsulfoxide | Me—S(=O)—Me |
| DNP | 2,4-dinitrophenyl | 2,4-dinitrophenyl ring structure (O₂N groups) |
| **dppe** | 1,2-Bis(diphenylphosphino)ethane<br>(DIPHOS) | Ph—P(Ph)—CH₂CH₂—P(Ph)—Ph structure |
| **dppp** | 1,2-Bis(diphenylphosphino)propane | Ph—P(Ph)—CH₂CH₂CH₂—P(Ph)—Ph structure |
| ee | enantiomeric excess<br>= % major enantiomer - % minor enantiomer | |
| **Fmoc** | 9-Fluorenylmethoxycarbonyl | fluorenylmethoxycarbonyl structure |

| HCTU | 2-(6-Chloro-1H-benzotriazole-1-yl)-1,1,3,3-tetramethyluronium hexafluorophosphate | |
|------|-----------------------------------------------------------------------------------|---|
| ***HMPT*** ***HMPA*** | Hexamethylphosphoric triamide | |
| HMTA | Hexamethylenetetramine | |
| HTIB | Hydroxy(tosyloxy)-iodobenzene | |
| Im | Imidazoyl | |
| ***Icp₂BH*** | Diisopinocampheylborane | |
| ***LTA*** | Lead tetraacetate | |
| ***LTMP*** ***LiTMP*** | Lithium 2,2,6,6-tetramethylpiperidide | |
| ***MAD*** | Methylaluminum bis(2,6-di-t-butyl-4-methylphenoxide) | |
| MCPBA | m-Chlorperoxybenzoic acid | |
| MeCN | Acetonitrile | |
| ***MEM-*** | 2-Methoxyethoxymethyl | |
| Ms | Mesyl , Methanesulfonyl | |

| MTM | Methylthiomethyl | $\mathsf{\xi\!-\!S\!-\!Me}$ |
|---|---|---|
| MVK | Methyl Vinyl Ketone | Me–C(=O)–CH=CH₂ |
| **_NBS_** | N-Bromosuccinimide | succinimide with N–Br |
| **_NCS_** | N-Chlorosuccinimide | succinimide with N–Cl |
| **_NMM_** | 4-Methylmorpholine | morpholine with N–Me |
| **_NMO_** | N-Methylmorpoline-N-oxide | morpholine N-oxide with Me |
| NMP | N-Methylpyrrolidone | pyrrolidone with N–Me |
| PCC | Pyridinium chlorochromate Corey's Reagent | pyridinium with O=Cr(–Cl)(–O⁻) |
| PDC | Pyridinium dichromate | (pyridinium)₂ Cr₂O₇²⁻ |
| **_Pd(dba)₂_** | Bis(dibenzylideneacetone)palladium (0) | dibenzylideneacetone structure |
| PMB | p-Methoxybenzyl | ξ–CH₂–C₆H₄–OMe |
| PNB | para-Nitrobenzoyl | ξ–C(=O)–C₆H₄–NO₂ |
| **_PPA_** | Polyphosphoric Acid | Unspecified mixture with High concentration of P₂O₅ |
| **_PTT_** (PTAB) | Phenyltrimethylammonium tribromide Phenyltrimethylammonium perbromide | Ph–N⁺(Me)₃ Br₃⁻ |

| | | |
|---|---|---|
| ***PPTS*** | Pyridinium para-toluenesulfonate | |
| ***PTSA*** | p-Toluenesulfonic acid; Tosic acid | Me—⟨⟩—SO$_3$H |
| Pv | Pivaloyl | |
| Py | Pyridine | |
| ***RAMP*** | (R)-1-Amino-2-Methoxymethylpyrrolidine | |
| ***SAMP*** | (S)-1-Amino-2-Methoxymethylpyrrolidine Ender's Reagent | |
| SEM | 2-Trimethylsilylethoxy-methoxy | |
| ***SMEAH*** | Sodium Bis(2-methoxyethoxy)aluminum Hydride | |
| ***TBAF*** | Tetrabutylammonium fluoride | |
| TBDPS | *tert*-Butyldiphenylsilyl | |
| ***TBHP*** | *t*-Butyl hydroperoxide | |
| TBS TBDMS | *tert*-Butyldimethylsilyl | |
| TEA | Triethylamine | |
| ***TEBA*** ***TEBAC*** | Benzyltriethylammonium chloride | |
| ***TEMPO*** | 2,2,6,6-Tetramethylpiperidin-1-oxyl | |

| TES | Triethylsilyl | Et<br>\<br>Et—Si—§<br>/<br>Et |
|---|---|---|
| Tf | Triflate | O<br>‖<br>§—S—CF$_3$<br>‖<br>O |
| THF | Tetrahydrofuran | |
| THP | Tetrahydropyranyl | |
| TIPS | Triisopropylsilyl | i-Pr<br>\<br>i-Pr—Si—§<br>/<br>i-Pr |
| TMEDA | N,N,N',N'-Tetramethylethylenediamine | Me—N      N—Me<br>\|              \|<br>Me          Me |
| **_TPAP_** | Tetra-n-Propylammonium Perruthenate | Pr$_4$N$^+$RuO$_4^-$ |
| TPP | Triphenyl phosphine | Ph<br>\|<br>Ph—P<br>\<br>Ph |
| TMS | Trimethylsilyl | Me<br>\<br>Me—Si—§<br>/<br>Me |
| **_TMSOTf_** | Trimethylsilyltrifluoro-methanesulfonate | TMS$^{-O}\diagdown$SO$_2$CF$_3$ |
| TPS | Triphenylsilyl | Ph<br>\<br>Ph—Si—§<br>/<br>Ph |
| Trt | Trityl | Ph<br>\|<br>Ph——§<br>\|<br>Ph |
| Ts-<br>Tos- | Tosyl<br>p-toluenesulfonyl | O<br>‖<br>§—S—⟨◯⟩—Me<br>‖<br>O |

# NAME REACTIONS

In this section we provide a summary of Name Reactions. The format is slightly modified from our previous book, but maintains the essential features:

**Reaction:**
   Summary reaction.

**Proposed Mechanism:**
   Currently accepted mechanisms. We have tried to be complete in showing steps, intermediates and the necessary curly arrow notations.

**Notes:**
   Additional comments and references from key sources.

**Examples:**
   Current examples if possible.

When a term is underlined, (for example, ***Aldol Condensation***) it means that the concept can be found under an independent heading in the book.

**General Bibliography:**
B. P. Mundy, M. G. Ellerd, *Name Reactions and Reagents in Organic Synthesis*, John Wiley and sons, Inc., New York, 1988;
M. B. Smith, J. March in *March's Advanced Organic Chemistry*, 5$^{th}$ ed., John Wiley and Sons, Inc., New York, 2001;
T. Laue, A. Plagens, *Named Organic Reactions*, John Wiley and Sons, Inc., New York, 1998;
V. K. Ahluwalia, R. K. Parashar, *Organic Reaction Mechanisms*, Alpha Science International Ltd., Pangbourne, U.K., 2002;
J. J. Li, *Name Reactions*, Springer, Berlin, 2002;
*Comprehensive Organic Synthesis*, B. M. Trost, editor-in-chief, Pergamon Press, Oxford, 1991;
M. B. East, D. J. Ager, *Desk Reference for Organic Chemists*, Krieger Publishing Company, Malabar, FL, 1995;
M. Orchin, F. Kaplan, R. S. Macomber, R. M. Wilson, H. Zimmer, *The Vocabulary of Organic Chemistry*, John Wiley and Sons, Inc., New York, 1980;
A. Hassner, C. Stumer, *Organic Syntheses Based on Name Reactions and Unnamed Reactions*, Pergamon, Oxford, 1994;
*The Merck Index*, Merck & CO., Inc., Whitehouse Station, N. J. (now in the 13$^{th}$ Edition) Each edition has an updated list of Named Reactions.
See also: http://themerckindex.cambridgesoft.com/TheMerckIndex/NameReactions/TOC.asp

Other URL's to Name Reaction Websites:
www.monomerchem.com/display4.html
www.chempensoftware.com/organicreactions.htm
www.organic-chemistry.org/namedreactions/
http://orgchem.chem.uconn.edu/namereact/named.html

Some references are provided with a SciFinder (American Chemical Society) number so that one can access the abstract if needed.

# Acetoacetic Ester Synthesis

## The Reaction:

$$\text{(structure)} \xrightarrow[\substack{\text{3. Base} \\ \text{4. R''-X}}]{\substack{\text{1. Base} \\ \text{2. R'-X}}} \text{(structure)} \xrightarrow[\text{decarboxylation}]{\text{hydrolysis}} \text{(structure)}$$

## Proposed Mechanism:

The methylene protons are the most acidic by influence from both carbonyls.

X can be Cl, Br, I, OTs, etc.

$$\xrightarrow[\text{2. R''-X}]{\text{1. Base}} \text{(structure)} \xrightarrow[\text{2. H}^+]{\text{1. HO}^-, \text{H}_2\text{O}} \text{(structure)} \xrightarrow[-\text{CO}_2]{\Delta}$$

Alkylation can be done a second time (with a different R) if desired.

Ester hydrolysis/saponification, then with heat, the β-keto acid decarboxylates to give an enol.

$$\text{(structure)} \xrightleftharpoons[\text{source}]{\text{proton}} \text{(structure)}$$

keto-enol tautomerism

## Notes:

Acetoacetic Ester can be prepared by the condensation of ethyl acetate, called the *Acetoacetic Ester Condensation Reaction*, a ***Claisen Condensation***:

$$\text{(structure)} \xrightarrow{\text{Base}} \text{(structure)}$$

J. K. H. Inglis and K. C. Roberts
*Organic Syntheses* CV1, 235

See M. B. Smith, J. March in *March's Advanced Organic Chemistry*, 5th ed., John Wiley and Sons, Inc., New York, 2001, p 549; and C. R. Hauser, B. E. Hudson, Jr., *Organic Reactions* 1, 9

***Weiler Modification***: By using very strong bases, a dianion can be formed that will preferentially alkylate at the methyl group:

$$\text{(structure)} \xrightarrow[\text{THF, 30 min}]{\text{NaH, } n\text{-BuLi}} \left[ \text{(structure)} \right] \xrightarrow{\text{Br}} \text{(structure)}$$

83%

S. N. Huckin, L. Weiler *Journal of the American Chemical Society* **1974**, 96, 1082

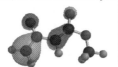

Simple AM1 calculation on Me ester shows the *HOMO* corresponding to the reactive intermediate

## Examples:

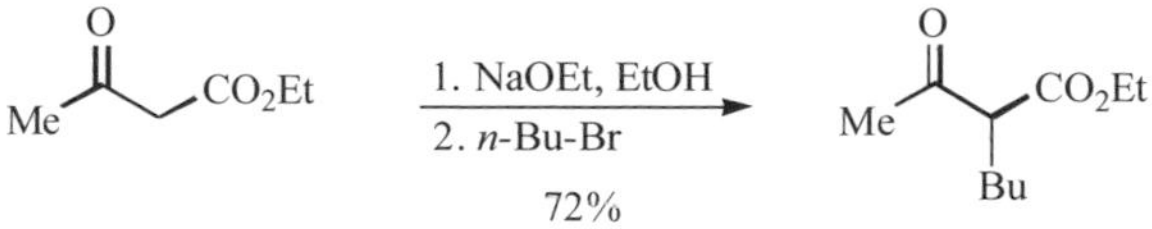

72%

C. S. Marvel, F. D. Hager, *Organic Syntheses* **1941**, <u>1</u>, 248

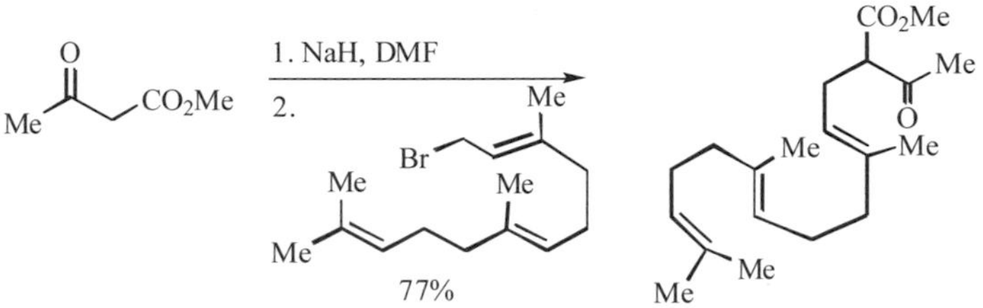

77%

K. A. Parker, L. Resnick, *Journal of Organic Chemistry* **1995**, <u>60</u>, 5726

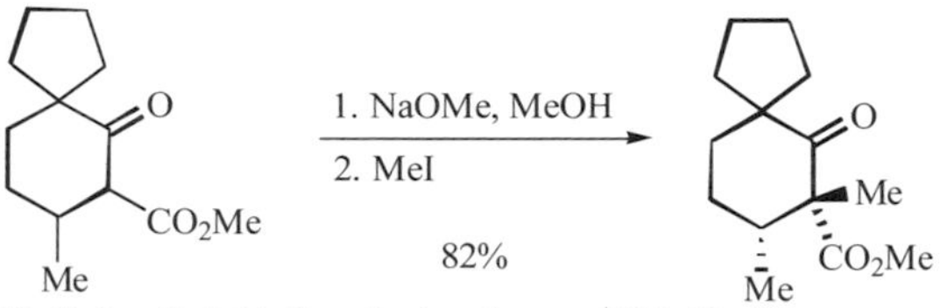

82%

Y.-Q. Lu, C.-J. Li, *Tetrahedron Letters* **1996**, <u>37</u>, 471

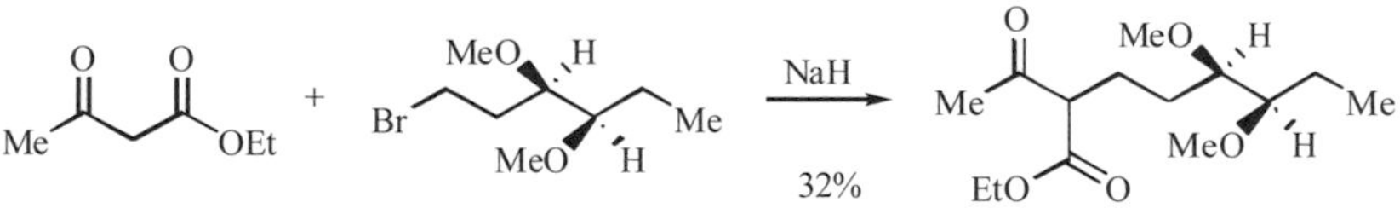

32%

K. Mori, *Tetrahedron* **1974**, <u>30</u>, 4223

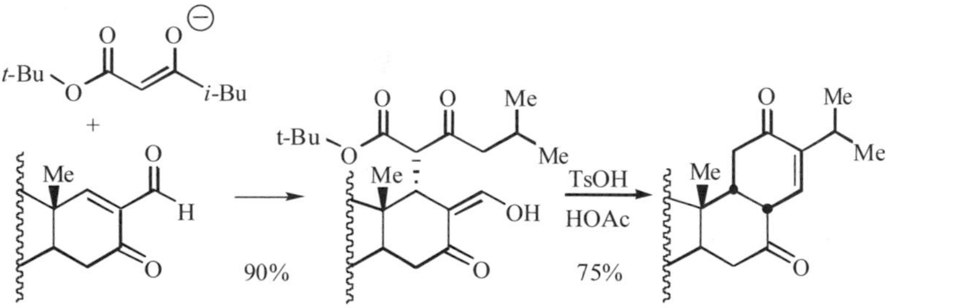

90%    75%

W. L. Meyer, M. J. Brannon, C. da G. Burgos, T. E. Goodwin, R. W. Howard, *Journal of Organic Chemistry* **1985**, <u>50</u>, 438

# Acyloin Condensation

## The Reaction:

## Proposed Mechanism:

An electron adds to the LUMO of the ester.

Two of these radical anions react.

Alkoxide leaves to give a 1,2 dione that further reacts with electrons in solution.

## Notes:

M. B. Smith, J. March in *March's Advanced Organic Chemistry*, 5th ed., John Wiley and Sons, Inc., New York, 2001, p 1562; T. Laue, A. Plagens, *Named Organic Reactions*, John Wiley and Sons, Inc., New York, 1998, pp. 1-3; S. M. McElvain, *Organic Reactions*, **4**, 4; J. P. Schaefer, J. J. Bloomfield, *Organic Reactions*, **4**, 15; J. J. Bloomsfield, J. M. Owsley, J. M. Nelke, *Organic Reactions* **23**, 2

The ***Rühlmann modification (Bouveault-Blanc Condensation or Rühlmann Reaction)*** traps the dienolate as a TMS derivative. This protocol generally results in improved yields.

This reaction is better than either the ***Dieckmann*** or ***Thorpe-Zeigler*** reactions for preparing large rings.

## Examples:

N. L. Allinger, *Organic Syntheses* **1963**, <u>4</u>, 840

E. Butkus, A. Ilinskasa, S. Stoniusa, R. Rozenbergasa, M. urbanováb, V. Setnikac, P. Bouc, K. Volkac, *Tetrahedron: Asymmetry* **2002**, <u>13</u>, 633

Na-K, toluene
TMSCl
taken to next step w/o purification

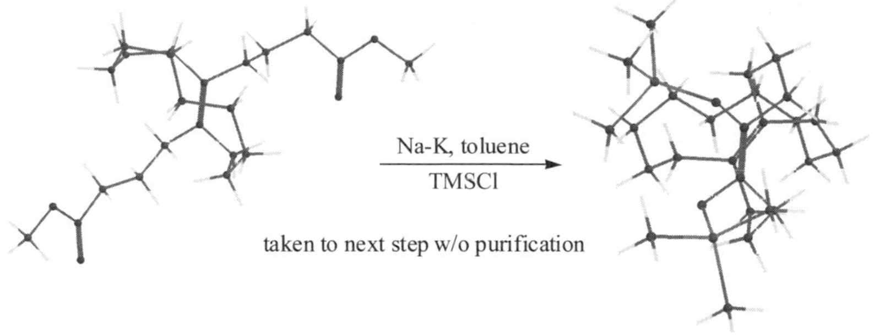

J. A. Marshall, J. C. Peterson, L. Lebioda, *Journal of the American Chemical Society* **1984**, 106, 6006

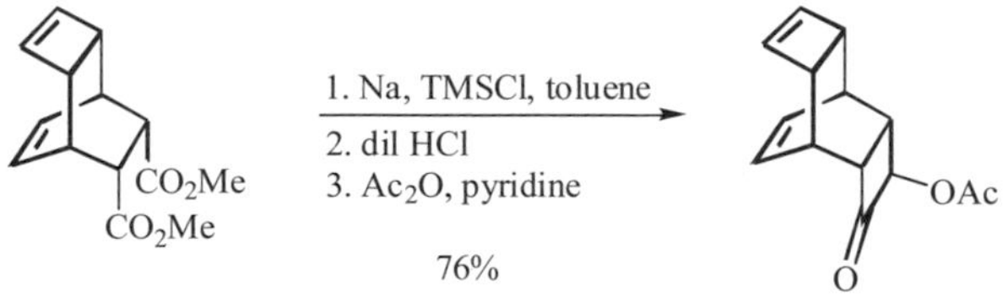

CO2Me
CO2Me
1. Na, TMSCl, toluene
2. dil HCl
3. Ac2O, pyridine
76%
OAc
O

G. Mehta, R. Vidya, *Journal of Organic Chemistry* **2001**, 66, 6913

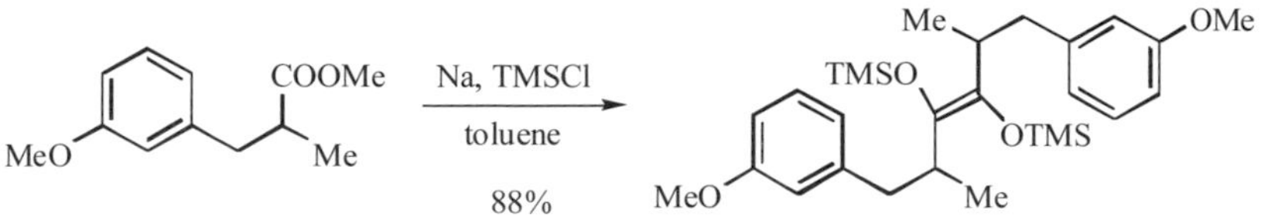

COOMe
MeO
Me
Na, TMSCl
toluene
88%
Me
OMe
TMSO
OTMS
MeO
Me

M. J. Meyers, J. Sun, K. E. Carlson, B. S. Katzenellenbogen, J. A. Katzenellenbogen, *Journal of Medicinal Chemistry* **1999**, 42, 2456

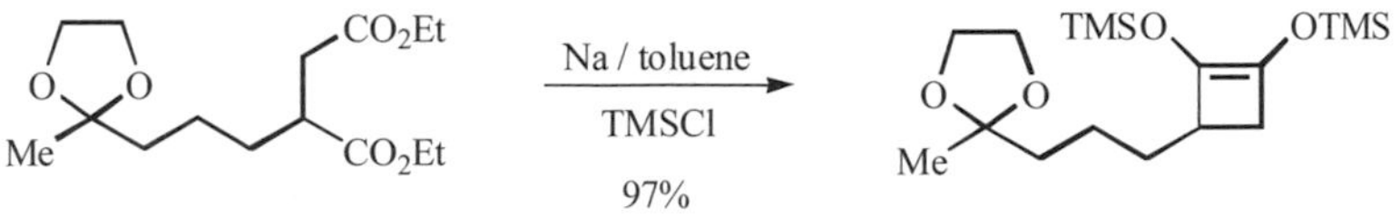

O
O
Me
CO2Et
CO2Et
Na / toluene
TMSCl
97%
TMSO
OTMS
O
O
Me

A. N. Blanchard, D. J. Burnell, *Tetrahedron Letters* **2001**, 42, 4779

# Acyloin Rearrangement

## The Reaction:

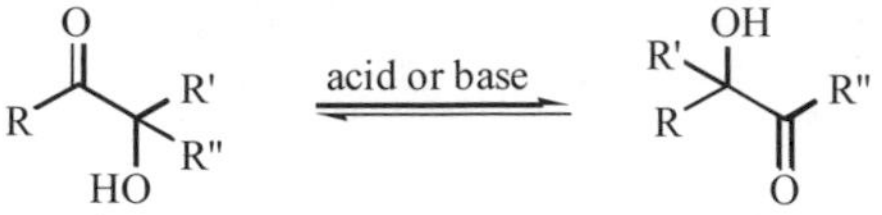

$$\text{acid or base}$$

## Proposed Mechanism:
In acid:

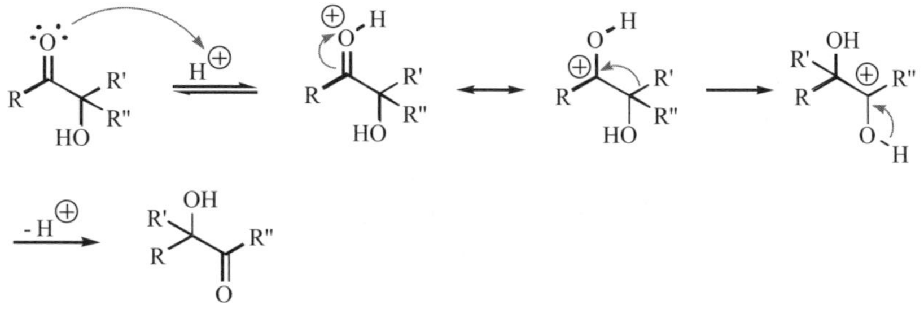

In base:

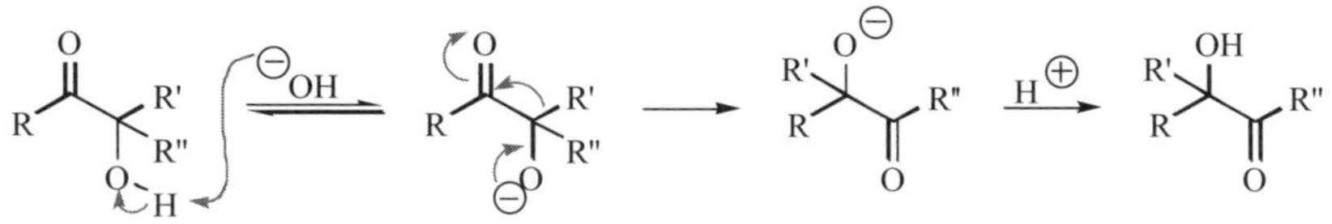

## Examples:

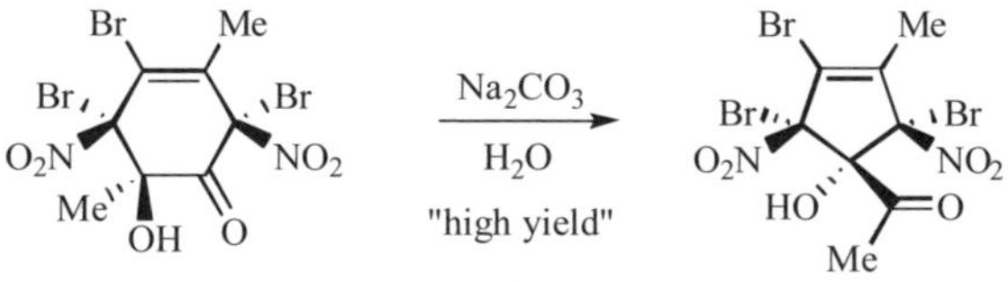

$$\xrightarrow[\text{H}_2\text{O}]{\text{Na}_2\text{CO}_3}$$

"high yield"

P. A. Bates, E. J. Ditzel, M. P. Hartshorn, H. T. Ing, K. E. Richards, W. T. Robinson, *Tetrahedron Letters* **1981**, <u>22</u>, 2325

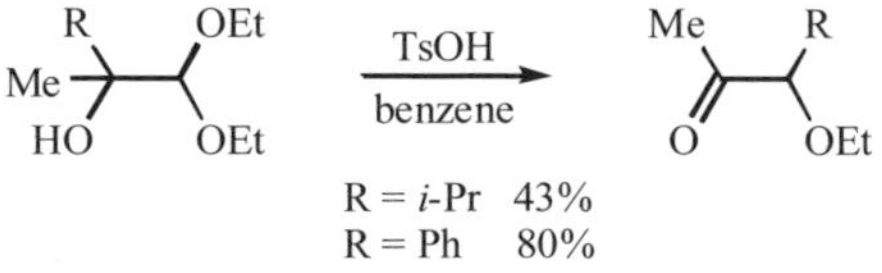

$$\xrightarrow[\text{benzene}]{\text{TsOH}}$$

R = *i*-Pr   43%
R = Ph    80%

T. Sate, T. Nagata, K. Maeda, S. Ohtsuka, *Tetrahedron Letters* **1994**, <u>35</u>, 5027 a mixture of acyl esters

M. Rentzea, E. Hecker, *Tetrahedron Letters* **1982**, <u>23</u>, 1785

J. Liu, L. N. Mander, A. C. Willis, *Tetrahedron* **1998**, <u>54</u>, 11637

# Adamantane Rearrangement (Schleyer Adamantization)

## The Reaction:

## Proposed Mechanism:

P. von R. Schleyer, P. Grubmcller, W. F. Maier, O. Vostrowsky, *Tetrahedron Letters* **1980**, <u>21</u>, 921

M. Farcasiu, E. W. Hagaman, E. Wenkert, P. von R. Schleyer *Tetrahedron Letters* **1981**, <u>22</u>, 1501

E. M. Engler, M. Farcasiu, A. Sevin, J. M. Cense, P. V. R. Schleyer, *Journal of the American Chemical Society* **1973**, <u>95</u>, 5769

M. A. McKervey, *Tetrahedron* **1980**, <u>36</u>, 971 provides a useful review:

This reaction consists of a series of deprotonations, protonations, hydride transfers and ***Wagner-Meerwein rearrangements***. There are postulated to be 2897 possible routes between starting material and product! A few of the steps have been tested experimentally; most of the data are computational. The following structural features seem to be supported:

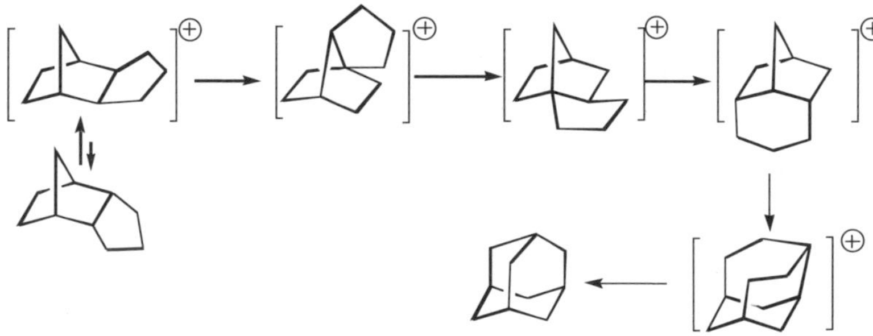

## Notes:

Tricyclic molecules having 10 carbon atoms are converted to adamantane with Lewis acids. Additional carbon atoms become alkyl appendages:

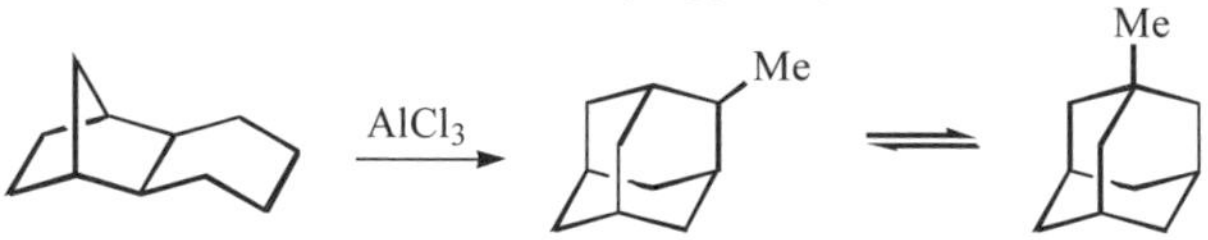

M. A. McKervey, *Tetrahedron* **1980**, <u>36</u>, 971

## Examples:

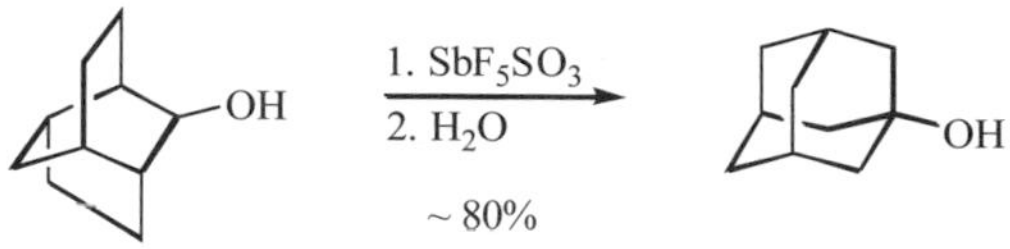

~ 80%

H. W. Whitlock, Jr., M. W. Siefken, *Journal of the American Chemical Society* **1968**, 90, 4929

Verification of the first steps:

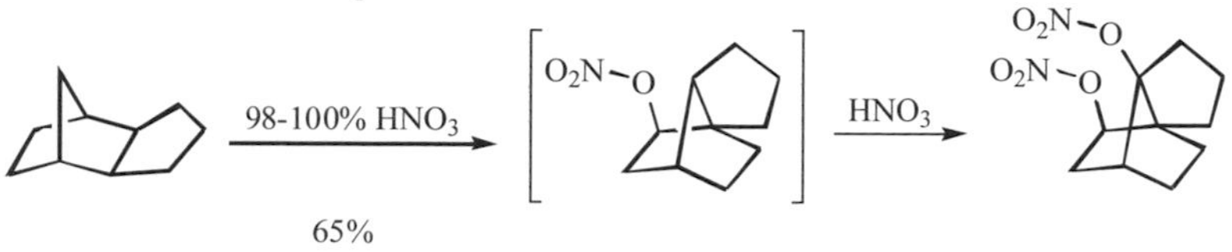

65%

P. A. Krasutsky, I. R. Likhotvorik, A. L. Litvyn, A. G. Yurchenko, D. Van Engen *Tetrahedron Letters* **1990**, 31, 3973

# Aldehyde Syntheses

## *Arens-van Dorp Cinnamaldehyde Synthesis*

## *Bodroux-Chichibabin Aldehyde Synthesis*

## *Bouveault Aldehyde Synthesis*

# DMSO-based Oxidations

### *Albright-Goldman Oxidation / Albright-Goldman Reagent*

## *Corey-Kim Oxidation / Corey-Kim Reagent*

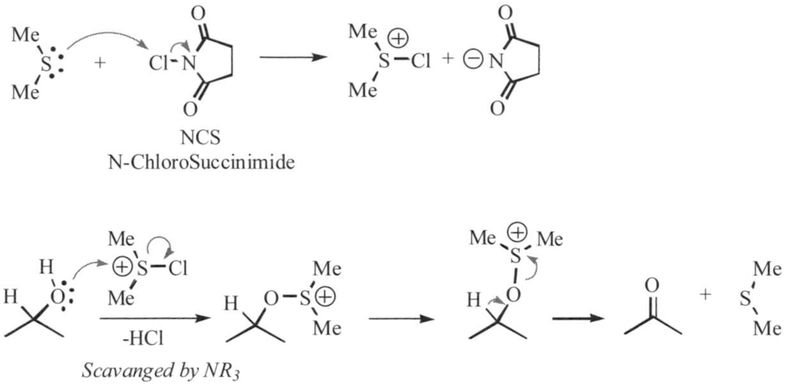

NCS
N-ChloroSuccinimide

*Scavanged by NR₃*

### Kornblum Aldehyde Synthesis

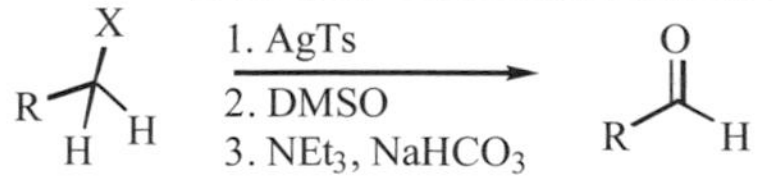

1. AgTs
2. DMSO
3. NEt₃, NaHCO₃

X = I, Br, OTs

## *Onodera Oxidation*

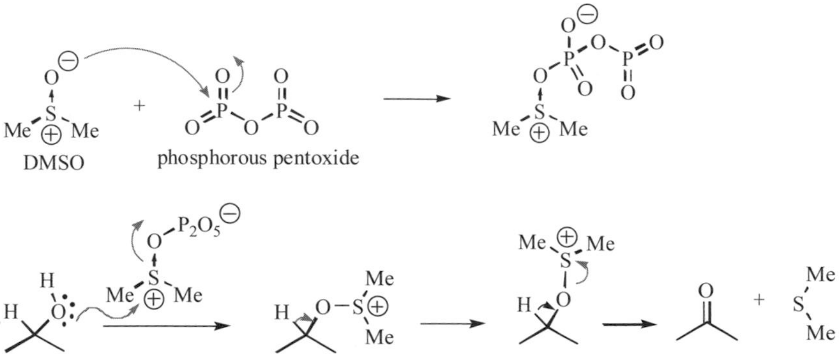

DMSO        phosphorous pentoxide

### Pfitzner-Moffatt Oxidation

DCC, HX
DMSO

also for ketones

### Swern Oxidation

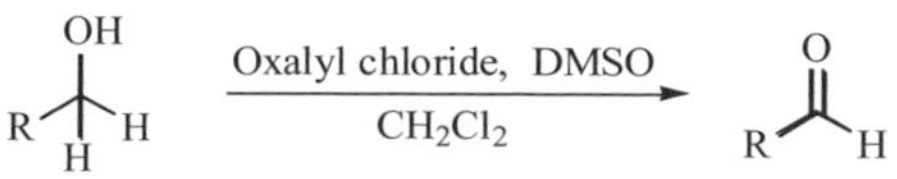

Oxalyl chloride, DMSO
CH₂Cl₂

also for ketones

### *Dess-Martin Oxidation*

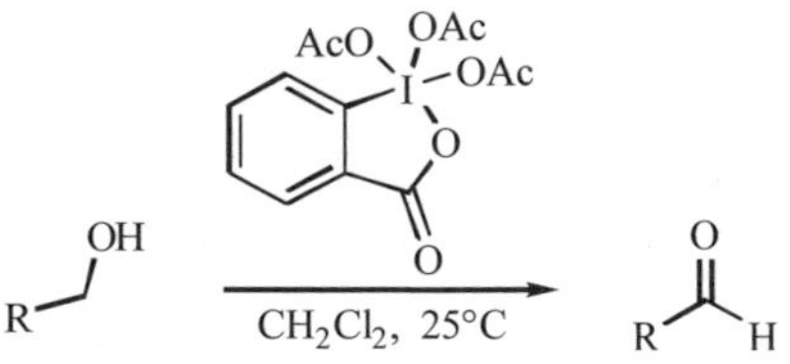

also for ketones

### *Duff Reaction*

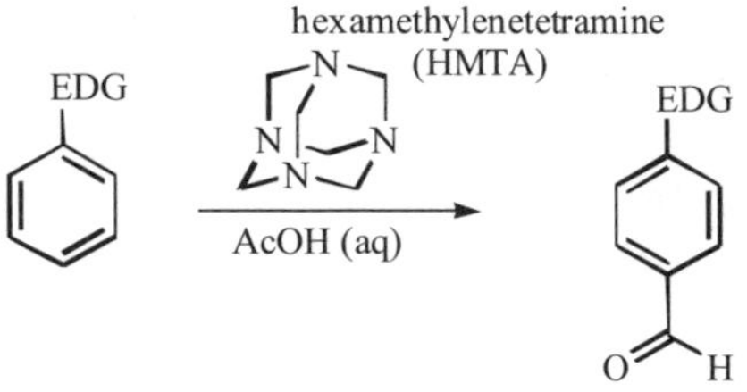

hexamethylenetetramine (HMTA)

### *Étard Reaction*

+ HOCrCl + HCl

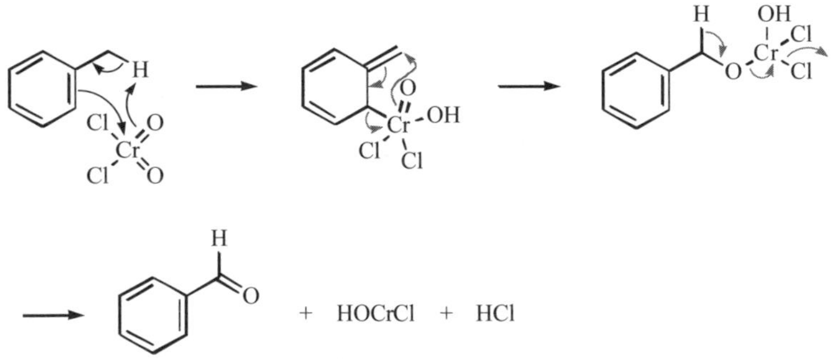

### *Fukuyama Reduction*

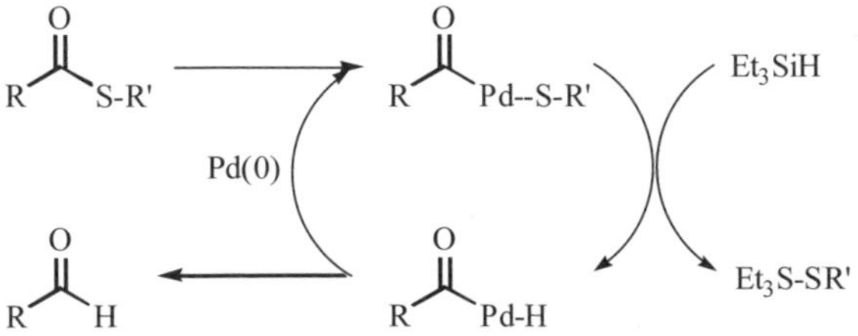

M. Kimura, M. Seki, *Tetrahedron Letters* **2004**, <u>45</u>, 3219

## *Ganem Oxidation*

TMNO

S_N2 displacement
helped by DMSO

## *Gattermann Reaction (Gatterman Aldehyde Synthesis) / Gattermann Reagent*

$$\xrightarrow[\text{H}_2\text{O}]{\text{Zn(CN)}_2\text{ , HCl}}$$

G = alkyl, OR

## *Gatterman-Koch Reaction (see under <u>Gatterman Reaction</u>)*

There seems to be agreement that the product-forming part of the mechanism is:

However, the details of the formation of the formyl cation seem to be less assured.

$$\xrightarrow[\text{Cu}^{\oplus}]{\text{HCl, AlCl}_3}$$

See S. Raugei, M. L. Klein, *Journal of Physical Chemistry B*, **2001**, <u>105</u>, 8213 for pertinent references to experiment, and their computational study of the formyl cation.

## *Grundmann Aldehyde Synthesis*

$$\xrightarrow[i\text{PrOH}]{\text{Al(O}i\text{Pr)}_3}$$

*<u>Meerwein-Ponndorf-Verley Reduction</u>*

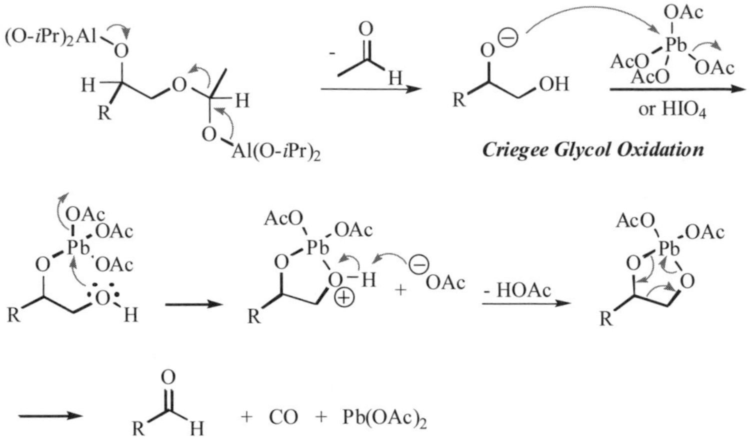

$Criegee\ Glycol\ Oxidation$

## Hass-Bender Reaction

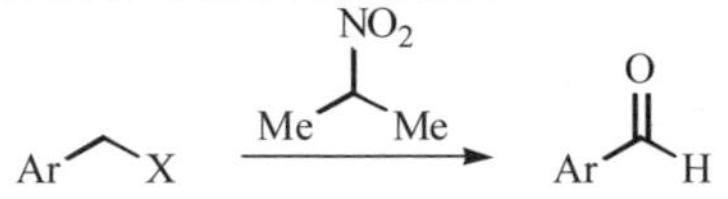

## Kröhnke Aldehyde Synthesis

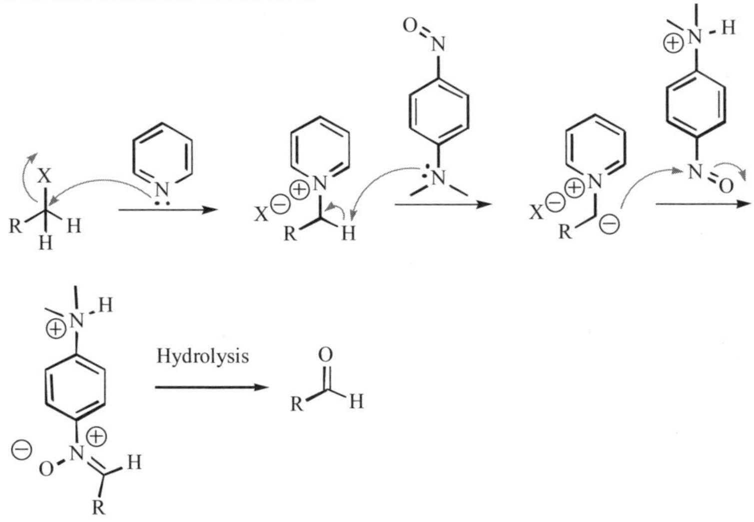

Hydrolysis

## McFadyen-Stevens Aldehyde Synthesis

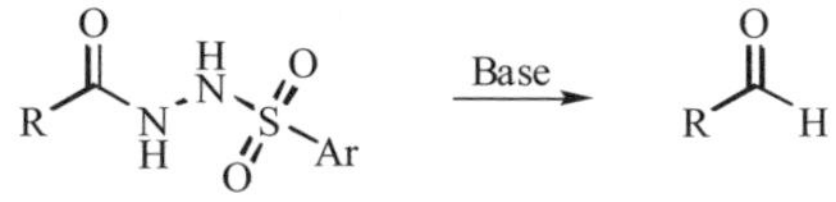

Base

R = Ar or alkyl with no α-protons

## *Meyers Aldehyde Synthesis / Meyers Reagents*

## *Polonovski Reaction*

## *Reimer-Tiemann Reaction*

## *Reissert Reaction (Grosheintz-Fischer-Reissert Aldehyde Synthesis)*

## __Rosenmund Reduction__

## __Sommelet Reaction__

## *Sonn-Muller Method*

## *Stephen Reduction (Stephen Aldehyde Synthesis)*

## *Vilsmeier-Haack Reaction*

nucleophilic aromatic
compounds only

## *Wacker Oxidation Reaction*

# Alder-Rickert Reaction

## The Reaction:

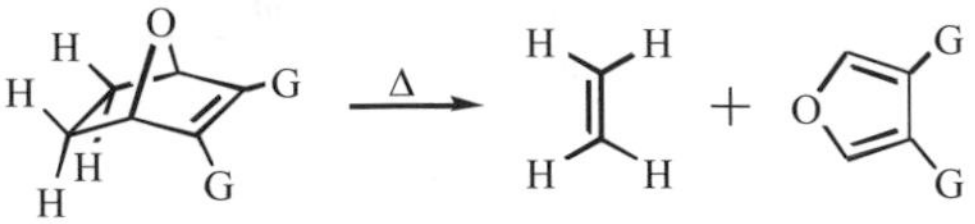

## Proposed Mechanism:

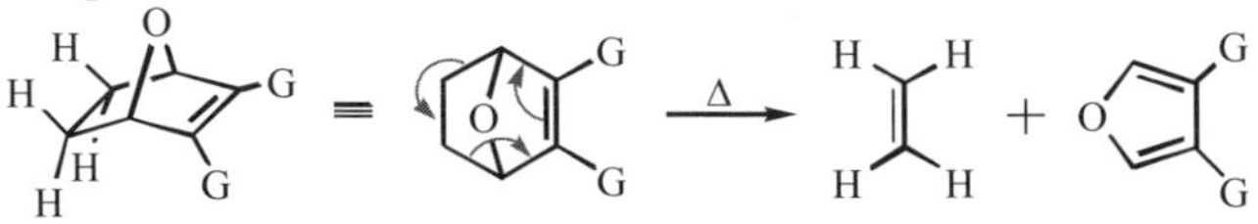

This reaction is a reverse ***Diels-Alder Reaction***. The orbital considerations controlling the "backward: reaction are the same as the "forward" reaction.

## Notes:

It seems accepted that almost any "***retro-Diels-Alder***" reaction can be included in the grouping, "***Alder-Rickert Reaction***".

## Examples:

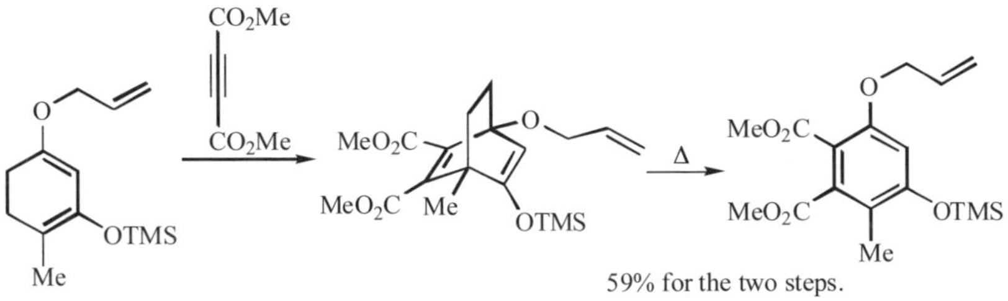

59% for the two steps.

J. W. Patterson, *Tetrahedron* **1993**, <u>49</u>, 4789

R. N Warrener, J.-M. Wang, K. D. V. Weerasuria, R. A. Russell, *Tetrahedron Letters* **1990**, <u>31</u>, 7069

D. W. Landry, *Tetrahedron* **1983**, <u>39</u>, 2761

D. Schomburg,  M. Thielmann, E. Winterfeldt, *Tetrahedron Letters* **1985**, <u>26</u>, 1705

M. E. Jung, L. J. Street, *Journal of the American Chemical Society* **1984**, <u>106</u>, 8327

# Aldol Type Reactions

## The Reaction:
This reaction has become an extremely important tool in the reaction toolbox of organic chemists. Because of the variety of approaches to the aldol products, this summary section is prepared.

Most synthetically useful approaches use a preformed enolate as one of the reactants.

Strong Base $\xrightarrow{\text{Ketone added to base}}$

"Kinetic enolate"

With a weaker base and / or slow addition of base to the ketone, an equilibrium will be established and a *"thermodynamic enolate"* will predominate.

"Thermodynamic enolate"

The most useful approach is when the enolate can be trapped and used in a configurationally stable form.

A generic analysis of enolate addition to an aldehyde:

**Z-enolate**

not

*Unfavorable steric interaction*

A similar exercise can be provided for the *E*-enolate.

### *Zimmerman-Traxler model*
An analysis of the steric effects in a chair-transition state for the reaction:

A *directed aldol reaction* requires that one partner provides a preformed enolate (or chemically equivalent reactive species) and is then added to the second carbonyl-containing molecule.

When one of the reactants is chiral, asymmetric induction can provide enantioselective products:

## Cram's Rule and Related Views on Asymmetric Induction

This rule was developed to rationalize the steric course of addition to carbonyl compounds.[1] The conformations of the molecules are shown in their *Newman structures*, and a preferred conformation is selected in which the *largest group,* **L**, is situated *anti* to the carbonyl oxygen. This conformation assumes a model having a *large* oxygen, sometimes referred to as the "big O" model.[2] Examination of steric hindrance to nucleophile trajectory determined the major product.[3] We might point out, at the start, that Reetz has recently reported that "how" the reaction is carried out; for example "slow" vs. "fast" mixing, can dramatically alter product ratios.[4]

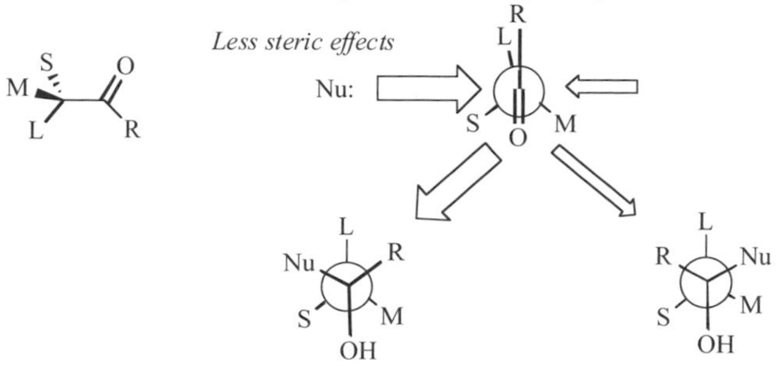

**Major product**        **Minor product**

In cases where the alpha-carbon is chiral, attack at the carbonyl carbon introduces a new stereogenic center. The two carbonyl faces are *diastereotopic* and attack at the *re* and *se* faces are different

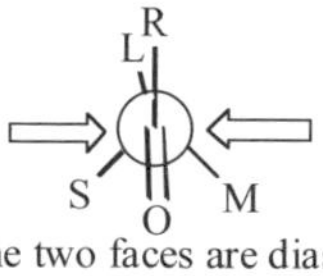

The two faces are diastereotopic

A modification of the *Cram model*, in which the medium sized group, **M**, eclipsed the carbonyl oxygen, was developed by Karabatsos[5]; however, it generally predicted the same product as the *Cram model*. In this model, which assumes two major conformations, the major product is that which is derived from attack at the less hindered side of the more stable conformer.

-----------------------------------------------------------------------------------------------------

1. a. See J. D. Morrison, H. S. Mosher, *Asymmetric Organic Reactions*, Prentice-Hall, Englewood Cliffs, 1971, Chapter 3, for a somewhat dated, but excellent account of this concept.

  b Cram's first work, (D. J. Cram, F. A. Abd Elhafez, *Journal of the American Chemical Society* **1952**, 74, 5828) set the stage for intense studies that have spanned 50 years.

2. The original thought included the notion that there was a large steric bulk associated with the oxygen by nature of metal complexing.

3. Application of the *Curtin-Hammett Principle* would suggest that the different ground state conformers have minimal influence on the product composition. It is the difference in activation energies for the two different isomers that controls the reaction, and the diastereomeric transition states would be attained from either ground state conformation.

4. M. T. Reetz, S. Stanchev, H. Haning, *Tetrahedron* **1992**, 48, 6813

5. a G. J. Karabatsos *Journal of the American Chemical Society* **1967**, 89, 1367;
  b. G. J. Karabatsos, D. J. Fenoglio, *Topics in Stereochemistry* **1970**, 5, 167

Preferred

## Felkin–Cherest-Anh Rule

Like ***Cram's Rule***, the ***Felkin-Cherest-Anh model***, developed by Felkin and coworkers[6], is an attempt to understand and predict the stereochemistry of addition to a carbonyl group. This model requires a "small O" interpretation in which the largest group is oriented *anti* to the attacking nucleophile's trajectory. One should note that the ***Felkin-Cherest-Anh model*** neglects the interaction of the carbonyl oxygen. In this approach, the *R/S* or *R/M* interactions dominate.

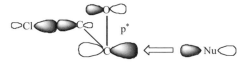

This is the important interaction that must be minimized. Note that in this approach the carbonyl substituent plays an important role.

Calculations in this model are based on an orbital interaction as described below. It should also be noted that the trajectory of delivery of nucleophile to the carbonyl carbon is defined by an angle of about 109°.

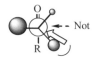

***Bürgi- Dunitz trajectory***

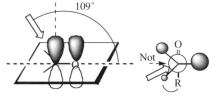

109°

Not

Not

**Preferred conformation.** Less interaction between the small group and the R-group. We also note that this model "feels" the influence of increasing size of R.

We see in this coformer an increased interaction between the medium group and R. Also, there is more interaction with the nucleophile.

This model often leads to the same conclusions obtained from the other models. It does, however, recognize the nonpassive role of the **R**-group in ketones. In this model one would predict an *increase* of stereodifferentiation as the size of **R-** increases. This has been found experimentally.

For aldehydes the transition state model will be:

---------------------------------------------------------------------------------------------------

6. M. Cherest, H. Felkin, N. Prudent, *Tetrahedron Letters*, **1968**, <u>9</u>, 2199

A useful orbital approach by Cieplak[7] has suggested that the nucleophile will attack the carbonyl *anti* to the best donor ligand.

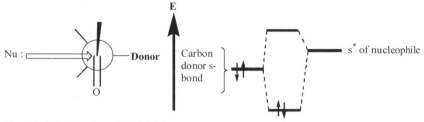

### Cases for Modification of the Models

Sometimes the Lewis acid that coordinates with the carbonyl oxygen is sufficiently bulky that it seriously influences the stereochemistry of attack. Sometimes these reaction products, which seem opposite of the expected ***Cram Rule*** analysis, are termed ***"anti-Cram" products***. Compare the "normal" situation with the influence of a sterically bulky Lewis acid:

This gives the
Cram product

This gives the
"anti-Cram" product

### *Dipolar Model*

There is evidence to suggest that competing dipole effects will alter the preferred conformation. Thus, for example, halogens will prefer a conformation in which the dipoles are *anti* to one another. This is often described as the ***Cornforth model***.[8] In this model the highly polarized group will take the place of the L-group of the ***Cram model***.

Preferred direction
of attack.

---

7. a. A. S. Cieplak, B. D. Yait, C. R. Johnson, *Journal of the American Chemical Society* **1989** <u>111</u>, 8447

   b.   A. S. Cieplak, *Journal of the American Chemical Society* **1981**, <u>103</u>, 4548

8. J. W. Cornforth, R. H. Cornforth, K. K. Methew, *Journal of the Chemical Society* **1959**, 112

***Chelation Control***[9]

Neighboring heteroatoms can provide a site for complexing.

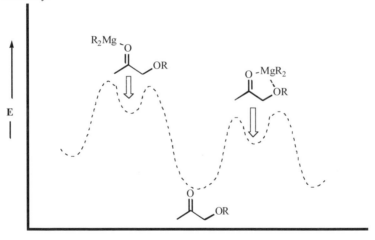

Het = heteroatom
M = metal

Product stereochemistries can be greatly influenced by these chelation control effects. This was first observed by Cram.[10] There are many controversies about this topic, and the issue remains a topic of investigative interest.[11] Without kinetic data, it has been suggested that it is impossible to distinguish the following two mechanistic types:[12]

Chelate $\rightleftharpoons$ Ketone + MgMe$_2$ $\longrightarrow$ Product    (Non-chelation)

Ketone + MgMe$_2$ $\rightleftharpoons$ Chelate $\longrightarrow$ Product    (Chelation)

Rate enhancement should be a requirement for chelation control because if chelation is the source of stereoselectivity it necessarily follows that the chelation transition state should be of a lower energy pathway.[13]

These concepts are seen on the energy diagram below. It should be noted that an interesting conclusion from this analysis is that *increased selectivity* is associated with *increased reactivity*. This might be considered to run counter to a number of other analyses of reactivity and selectivity.

----------------------------------------------------------------------------------------------------

9. M. T. Reetz, *Accounts of Chemical Research* **1993**, <u>26</u>, 462

10. D. J. Cram, K. R. Kopecky, *Journal of the American Chemical Society* **1959**, <u>81</u>, 2748

11. a. W. C. Still, J. H. McDonald, *Tetrahedron Letters,* **1984**, 1031
    b. M. T. Reetz, *Angewandte Chemie, International Edition in English,* **1984**, <u>23</u>, 556
    c. G. E. Keck, D. E. Abbott, *Tetrahedron Letters* **1984**, <u>25</u>, 1883
    d. S. V. Frye, E. L. Eliel, *Journal of the American Chemical Society* **1988**, <u>110</u>, 484

12. J. Laemmle, E. C. Ashby, H. M. Neumann, *Journal of the American Chemical Society* **1971**, <u>93</u>, 5120

13. X. Chen, E. R. Hortelano, E. L. Eliel, S. V. Frye *Journal of the American Chemical Society* **1992**, <u>114</u>, 1778

The *a priori* prediction of which functional groups will provide complexation are not always obvious. Keck[14] demonstrated some dramatic differences in oxygen chelation resulting from minor differences in substitution.

Me                              Me

Ph
With SnCl₄
*Stong chelate*              *No chelate*

A potentially useful extension of the **Cram's rule** is the asymmetric induction provided by a remote ester (**Prelog's rule**):

R'MgX
⟶

## Reactions based on the Aldol Reaction:

*Claisen-Schmidt:*

*Henry Reaction*

*Knoevenagel Reaction*

no α protons        a methylene with two
               electron withdrawing groups

-------------------------------------------------------------------------------------------------

14. G. E. Keck, S. Castellino, *Tetrahedron Letters* **1987**, <u>28</u>, 281

# Aldol Condensation

## The Reaction:

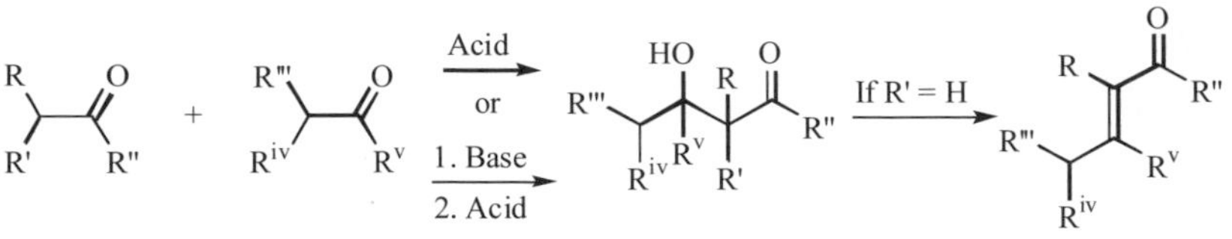

## Proposed Mechanism:
*Acid Catalyzed*

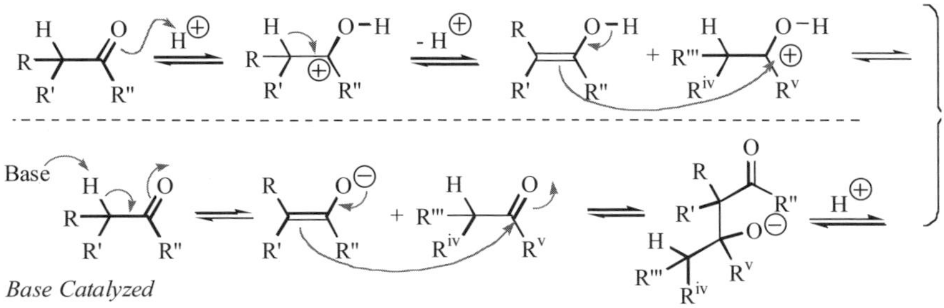

*Base Catalyzed*

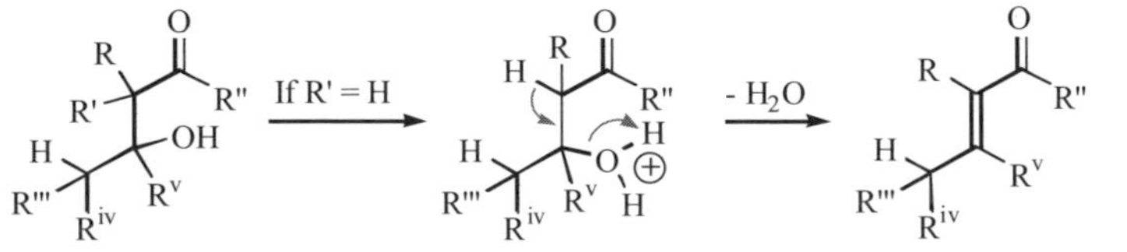

If R' = H, dehydration is possible to give the α, β unsaturated ketone.

Dehydration is often irreversible and a driving force.

## Notes:
If the starting materials are not the same, the reaction is known as a ***"mixed" aldol condensation***.

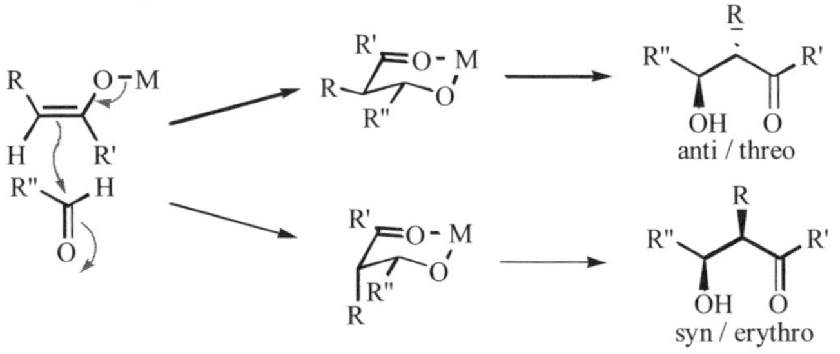

anti / threo

syn / erythro

M. B. Smith, J. March in *March's Advanced Organic Chemistry*, 5th ed., John Wiley and Sons, Inc., New York, 2001, pp 1218-1213; T. Laue, A. Plagens, *Named Organic Reactions*, John Wiley and Sons, Inc., New York, 1998, pp. 4-10; A. T. Nielsen, W. J. Houlihan, *Organic Reactions* **16** (full volume); T. Mukaiyama, *Organic Reactions*, **28**, 3; C. J. Cowden, I. Patterson, *Organic Reactions* **51**, 1.

## Examples:

M. Haussermann, *Helvetica Chimica Acta* **1951**, 34, 1482 (Reported in A. T. Nielsen, W. J. Houlihan, *Organic Reactions* **16**, page 8).

P. M. McCurry, Jr., R. K. Singh, *Journal of Organic Chemistry* **1974**, 39, 2316

E. J. Corey, S. Nozoe, *Journal of the American Chemical Society* **1965**, 87, 5728

M. T. Crimmins, K. Chaudhary, *Organic Letters* **2000**, 2, 775

D. Zuev, L. A. Paquette, *Organic Letters* **2000**, 2, 679

P. M. Pihko, A. Erkkila, *Tetrahedron Letters* **2003**, 44, 7607

# Algar-Flynn-Oyamada Reaction

## The Reaction:

$$\xrightarrow[\text{HO}^{\ominus}]{\text{H}_2\text{O}_2}$$

## Proposed Mechanism:

$$\text{H}_2\text{O}_2 \ + \ \text{HO}^{\ominus} \longrightarrow \ ^{\ominus}\text{OOH} \ + \ \text{H}_2\text{O}$$

$$\xrightarrow{\text{HO}^{\ominus}}$$

Enone Epoxidation Route:

$$\xrightarrow{^{\ominus}\text{OOH}}$$

$$\xrightarrow{\text{Oxidation}}_{" -\text{H}_2 "}$$

_**Michael**_ Route

T. M. Gormley, W. I. O'Sullivan, _Tetrahedron_ **1973**, _29_, 369
See: M. Bennett, A. J. Burke, W. I. O'Sullivan, _Tetrahedron_ **1996**, _52_, 7178 for a detailed analysis of the role of the epoxide intermediate.

## Notes:

Sometimes an "arone" can be formed.

$$\xrightarrow{\text{Br}_2}$$

## Auwers Synthesis

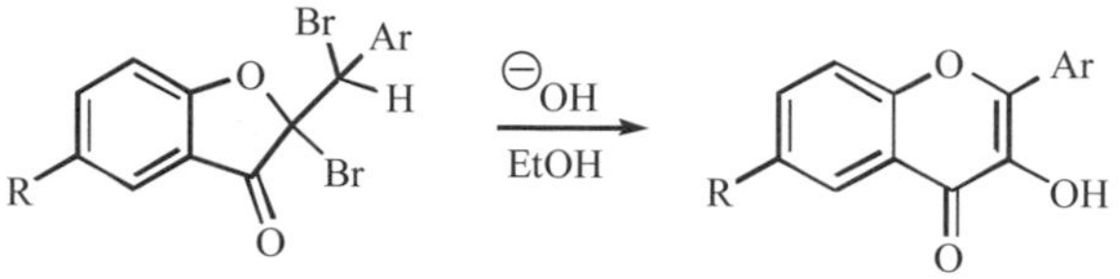

## The Rasoda Reaction:

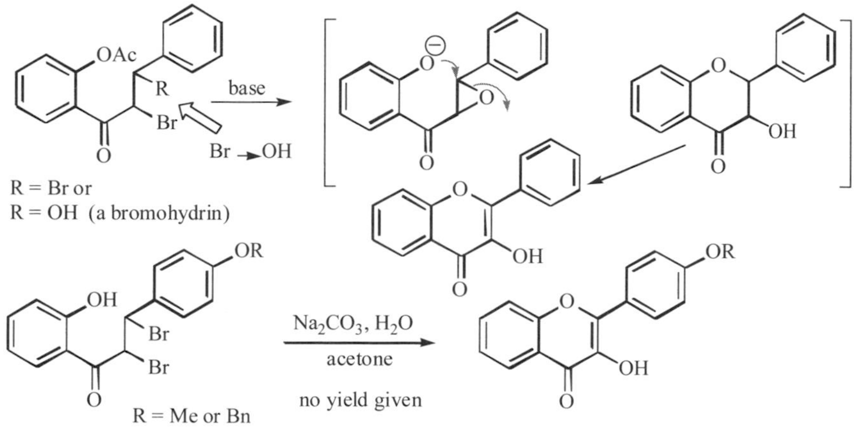

R = Br or
R = OH  (a bromohydrin)

R = Me or Bn

$Na_2CO_3$, $H_2O$

acetone

no yield given

M. G. Marathey, *Journal of Organic Chemistry* **1955**, 20, 563

## Examples:

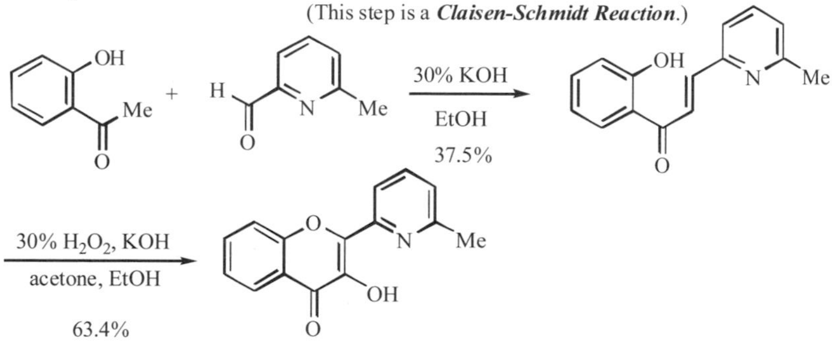

(This step is a *Claisen-Schmidt Reaction*.)

30% KOH

EtOH

37.5%

30% $H_2O_2$, KOH

acetone, EtOH

63.4%

K. B. Raut, S. H. Wender, *Journal of Organic Chemistry* **1960**, 25, 50

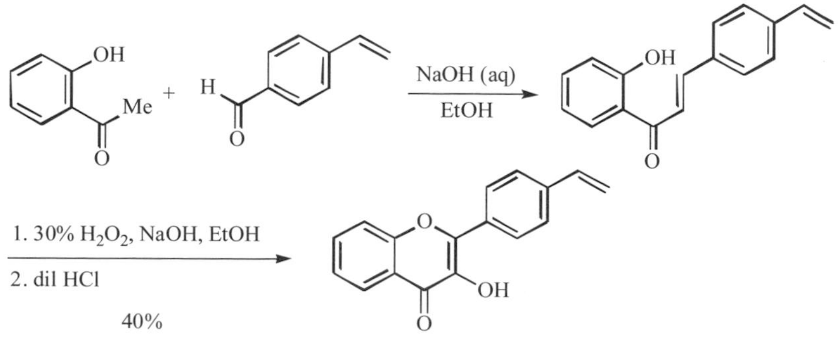

NaOH (aq)

EtOH

1. 30% $H_2O_2$, NaOH, EtOH

2. dil HCl

40%

J. R. Dharia, K. F. Johnson, J. B. Schlenof, *Macromolecules* **1994**, 27, 5167

# Alkyne Coupling

## The Reaction:

$$2 \quad R{-}{\equiv}{-}H \quad \longrightarrow \quad R{-}{\equiv}{-}{\equiv}{-}R$$

## General Discussion:

See P. Siemsen, R. C. Livingston, F. Diederich, *Angewandte Chemie International Edition in English* **2000**, *39*, 2632 and K. Sonogashira, *Comprehensive Organic Synthesis*, Vol 3, Chapter 2.5

The earliest of the alkyne coupling reactions is that of Glaser, who had noted:

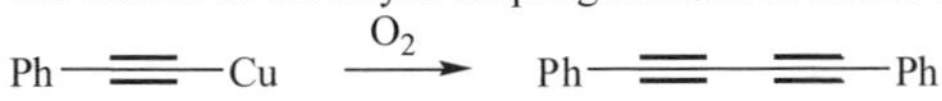

$$Ph{-}{\equiv}{-}Cu \xrightarrow{O_2} Ph{-}{\equiv}{-}{\equiv}{-}Ph$$

In much of the early work, the copper acetylides were prepared from the reaction of a terminal alkyne with Cu(I) salts.

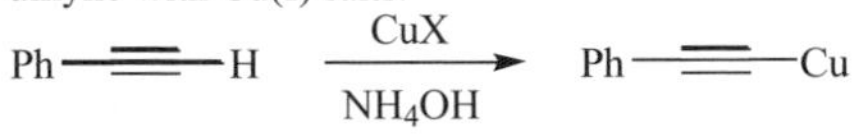

$$Ph{-}{\equiv}{-}H \xrightarrow[\text{NH}_4\text{OH}]{\text{CuX}} Ph{-}{\equiv}{-}Cu$$

The reaction was of limited use due to the explosive nature of copper acetylides.

In the ***Hay modification*** of the ***Glaser reaction***, it was noted that the reaction could be modified to avoid isolation of the acetylide:

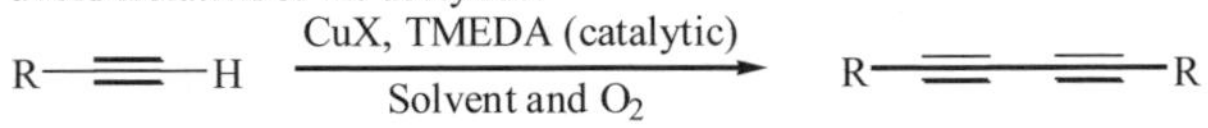

$$R{-}{\equiv}{-}H \xrightarrow[\text{Solvent and O}_2]{\text{CuX, TMEDA (catalytic)}} R{-}{\equiv}{-}{\equiv}{-}R$$

A. S. Hay, *Journal of Organic Chemistry* **1962**, *27*, 3320

In this process the TMEDA-Cu complex readily binds to the alkyne. Various interpretations of the binding are possible:

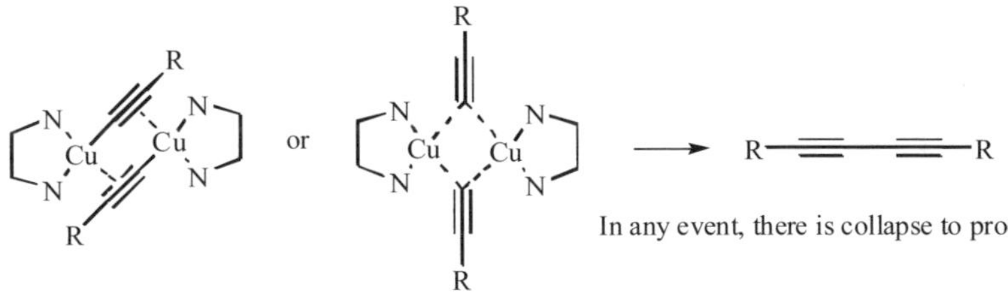

In any event, there is collapse to product.

There is evidence that the role of oxidant is to convert Cu(I) to Cu(II). It may be:

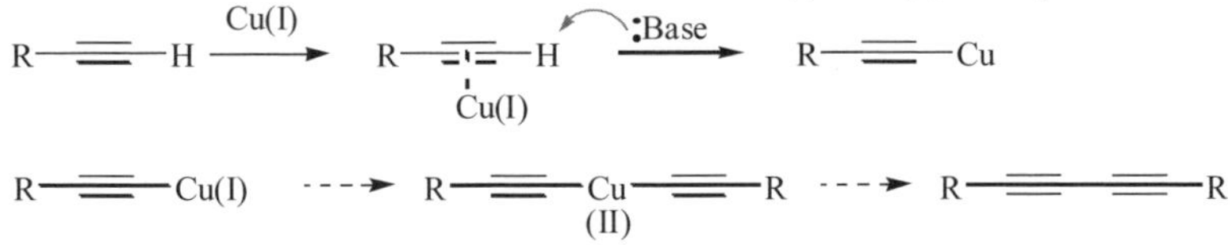

An older view of these reactions involved formation of a radical:

$$2 \ R\!\!-\!\!\equiv\!\!-Cu \ + \ 2 \ Cu(II) \longrightarrow 2 \ R\!\!-\!\!\equiv\!\!\cdot \ + \ 4 \ Cu(I)$$

$$R\!\!-\!\!\equiv\!\!-\!\!\equiv\!\!-R$$

This would require that hetero-coupling of two different alkynes give a statistical product mix. This is not observed.

A computational study of the ***Glaser reaction*** provides additional mechanistic insight.
L. Fomina, B. Vazquez, E. Tkatchouk, S. Fomine, *Tetrahedron* **2002**, <u>58</u>, 6741

Selected intermediates are shown:

There is much to learn about the details of these reactions. In different sections the following reactions will be described:

**Cadiot-Chodkiewicz Coupling**

$$R\!\!-\!\!\equiv\!\!-H \ + \ X\!\!-\!\!\equiv\!\!-R' \longrightarrow R\!\!-\!\!\equiv\!\!-\!\!\equiv\!\!-R'$$

**Castro-Stephens Coupling**

$$R\!\!-\!\!\equiv\!\!-H \ + \ Ar\!-\!X \xrightarrow[\text{CuCl, } \Delta]{\text{base}} R\!\!-\!\!\equiv\!\!-Ar$$

**Eglinton Reaction**

$$2 \ R\!\!-\!\!\equiv\!\!-H \xrightarrow[\text{pyridine}]{CuX_2} R\!\!-\!\!\equiv\!\!-\!\!\equiv\!\!-R$$

**Glaser Coupling**

$$R\!\!-\!\!\equiv\!\!-H \xrightarrow[\text{Solvent and } O_2]{\text{CuX, TMEDA (catalytic)}} R\!\!-\!\!\equiv\!\!-\!\!\equiv\!\!-R$$

**Sonogashira Coupling**

$$R\!-\!X \ + \ H\!\!-\!\!\equiv\!\!-R' \xrightarrow[\text{CuI, NEt}_3]{PdCl_2(PPh_3)_2} R\!\!-\!\!\equiv\!\!-R'$$

# Allan-Robinson Reaction

## The Reaction:

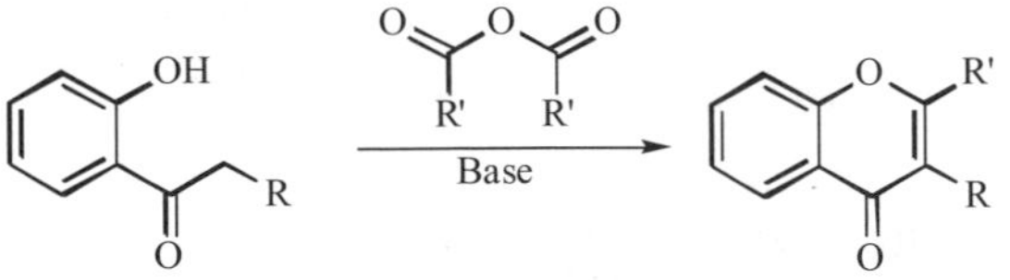

## Proposed Mechanism:

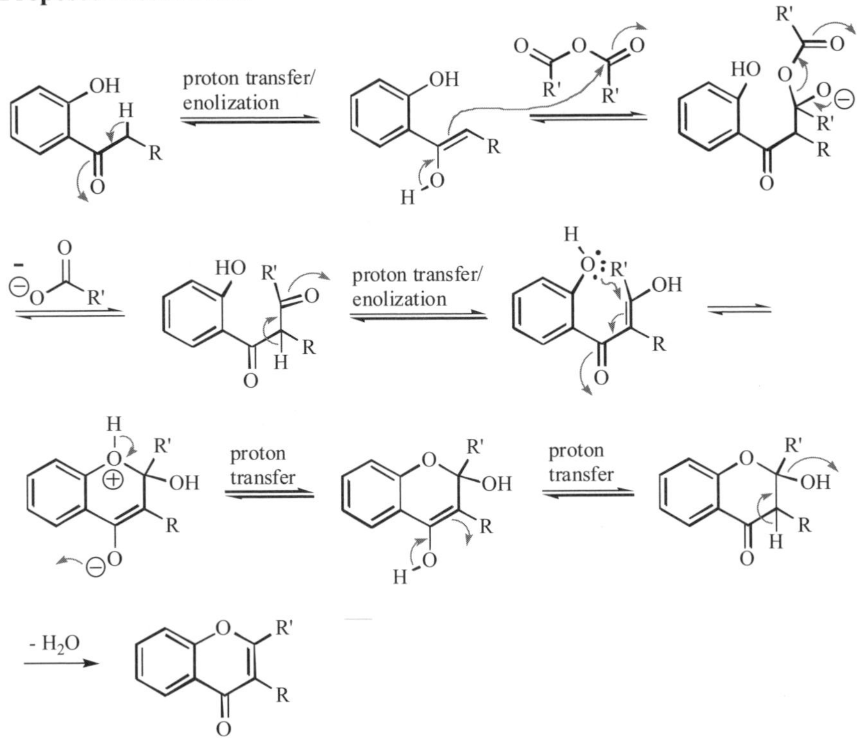

## Notes:

The rate determining step is dependent on both the concentration of enolacetate and acetate ion. T. Szell, D. M. Zorandy, K. Menyharth, *Tetrahedron* **1969**, <u>25</u>, 715

In the related ***Kostanecki Reaction***, the same reagents give a different product. In that case, the attacking species is the phenol oxygen, rather than the enol tautomer of the ketone.

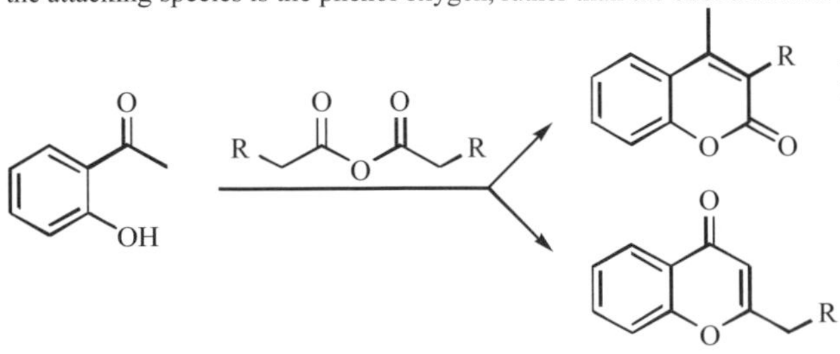

***Kostanecki Acylation***
Product
   = Coumarin

***Allan-Robinson
Reaction*** Product
   = Chromone

## Examples:

T. Szell, *Journal of the Chemical Society*, <u>C</u>, **1967**, 2041 (AN 1968:2779)

C. Riva, C. De Toma, L. Donadel, C. Boi, R. Pennini, G. Motta, A. Leosardi, *Synthesis* **1997**, 195

No yield given, product synthesis to confirm structure of an isolated compound.

G. Berti, O. Liv, D. Segnini, I. Cavero, *Tetrahedron* **1967**, <u>23</u>, 2295

# Amine Preparations

See R. E. Gawley, *Organic Reactions* **1988**, <u>35</u>, 1

### *Delépine Reaction*

### *Fischer-Hepp and Related Rearrangements*
#### *Fischer-Hepp Rearrangement*

HCl is the preferred acid.

### *Orton Rearrangement*

### *Hofmann-Martius Rearrangement*

### *Reilly-Hickinbottom Rearrangement*

Similar to *Hoffmann-Martius Rearrangement* except that it uses Lewis acids and the amine rather than protic acid and the amine salt.

M = Co, Cd, Zn

## *Forster Reaction (Forster-Decker Method)*

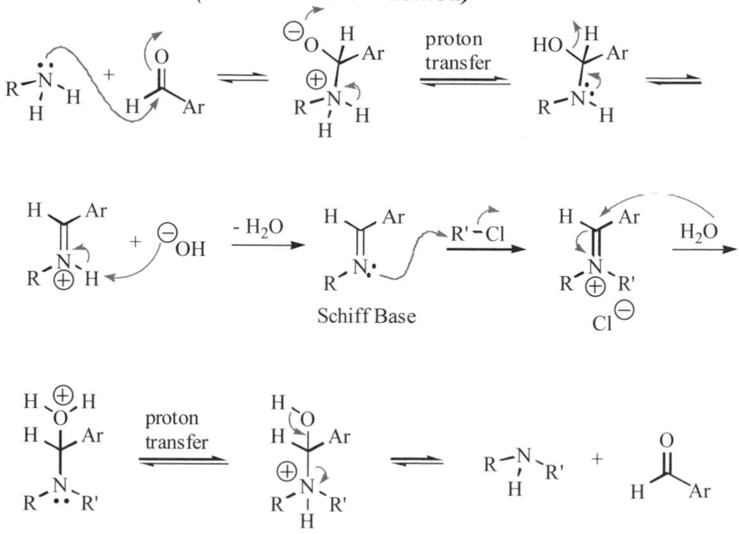
proton
transfer
- H2O
Schiff Base
H2O
proton
transfer

## *Fukuyama Amine Synthesis*

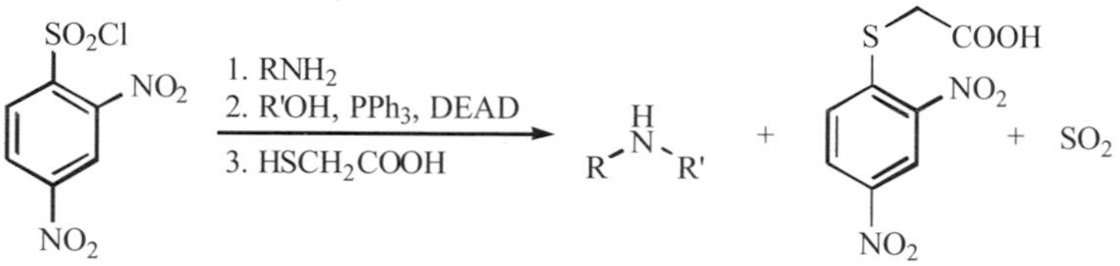
1. RNH2
2. R'OH, PPh3, DEAD
3. HSCH2COOH

**via:**

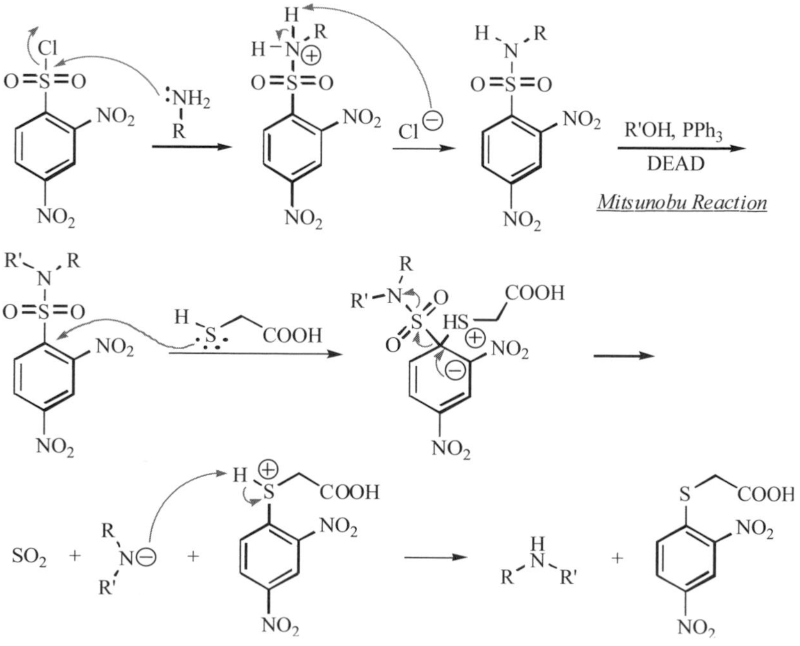
R'OH, PPh3
DEAD
Mitsunobu Reaction

## *Gabriel Synthesis*

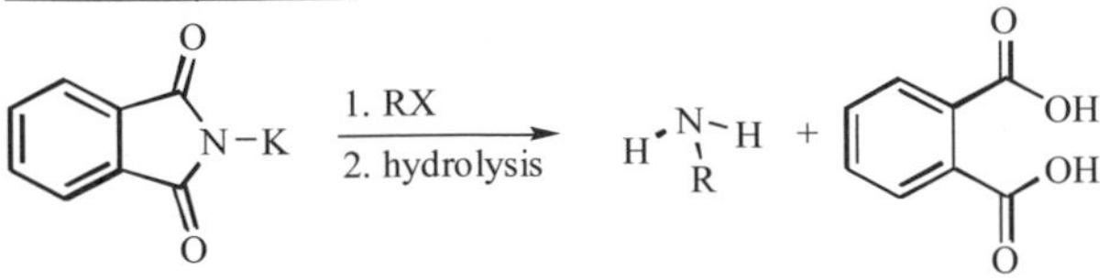

**via:**

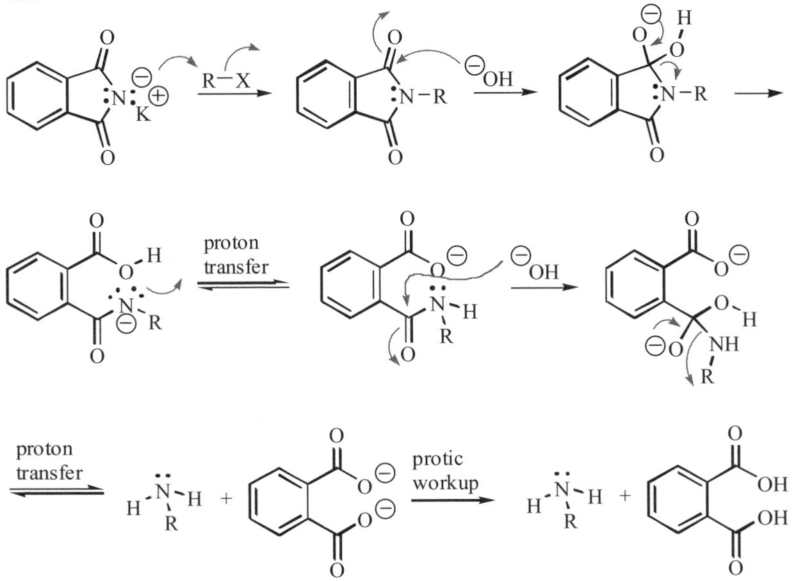

## *Gabriel-Colman Rearrangement*

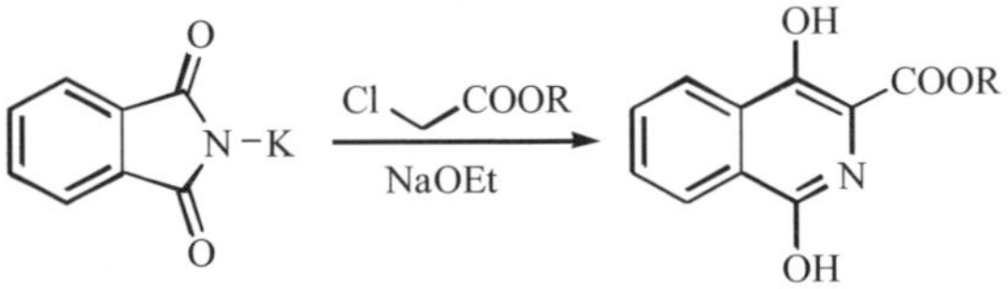

## *Gabriel-Cromwell Reaction*

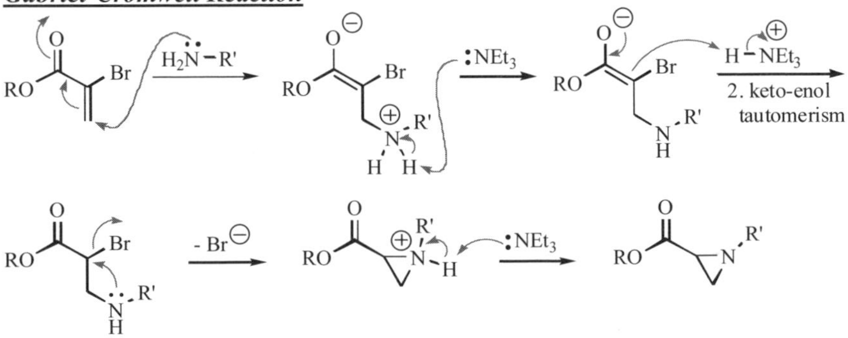

## *Gabriel (-Marckwald) Ethylenimine Method*

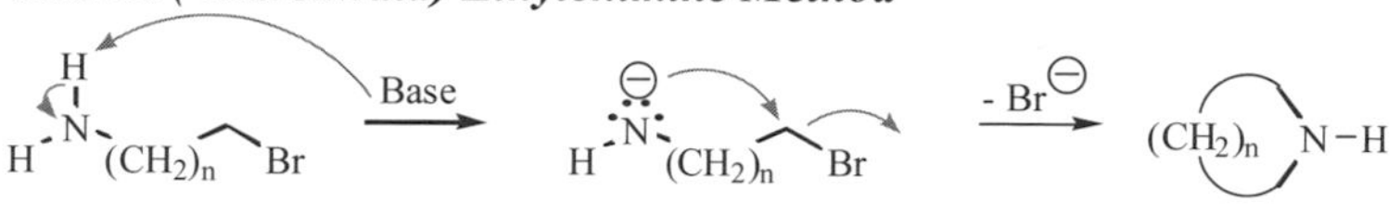

## *Schweizer Allyl Amine Synthesis*

A combination of *Gabriel* and *Wittig* chemistry

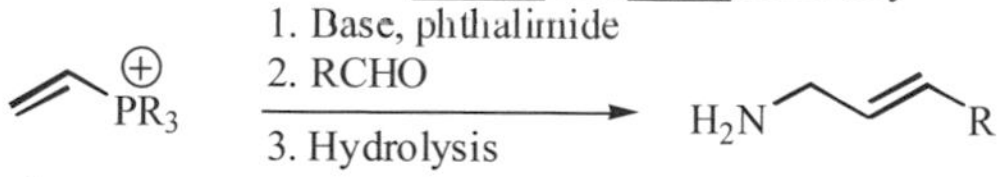

1. Base, phthalimide
2. RCHO
3. Hydrolysis

**via:**

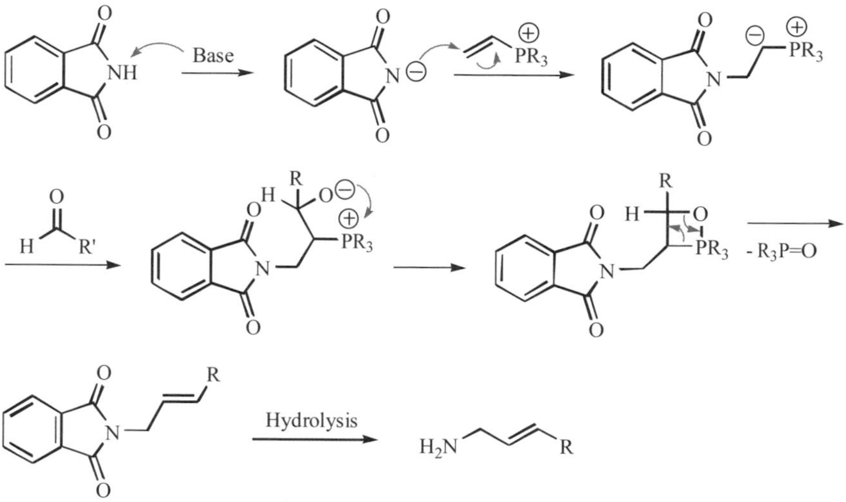

## *Voight Amination / Reaction*

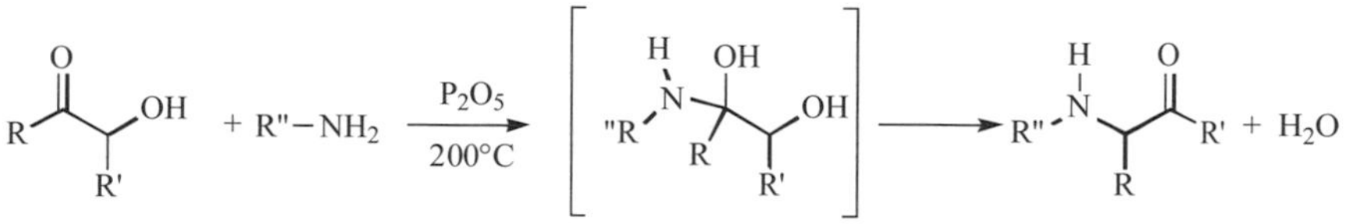

# Andersen Sulfoxide Synthesis

## The Reaction:

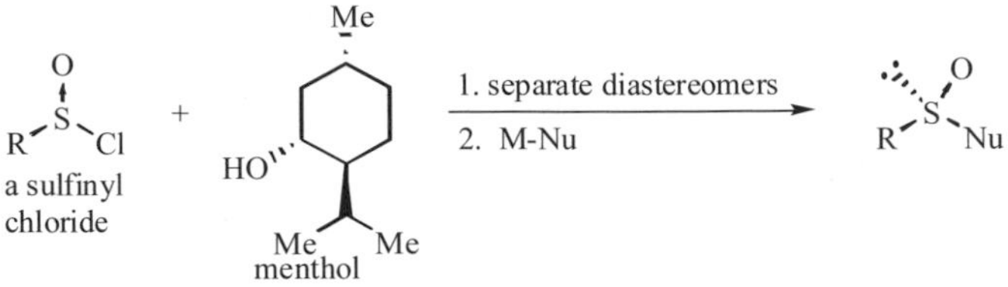

## Proposed Mechanism:

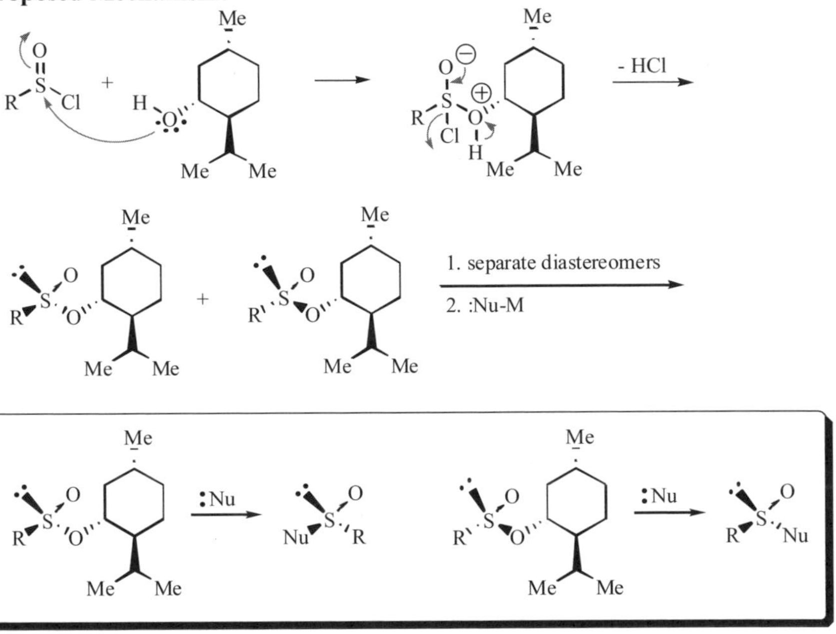

K. K. Andersen, *Tetrahedron Letters* **1962**, <u>3</u>, 93

## Notes:

Other chiral auxiliaries have been used besides menthol.

### Sulfoxide Designations:

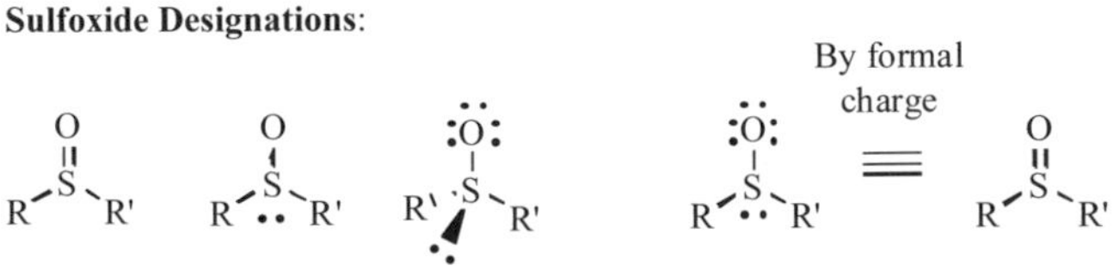

## Examples:

1. cholesterol
2. RMgX

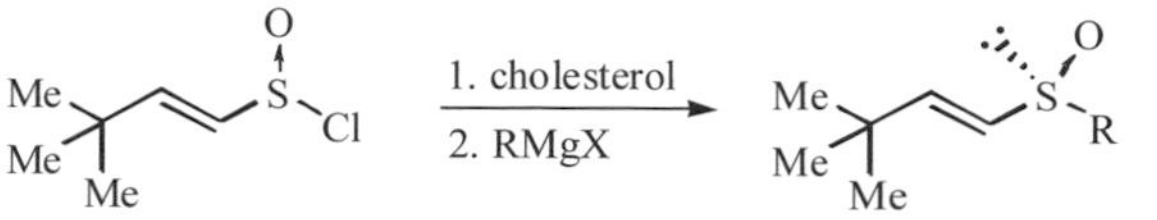

R = $n$-Bu: yield = 61%, 83% ee
R = Ph-$CH_2$: yield = 78%, 91% ee

R. R. Strickler, A. L. Schwan, *Tetrahedron: Asymmetry* **1999**, 10, 4065

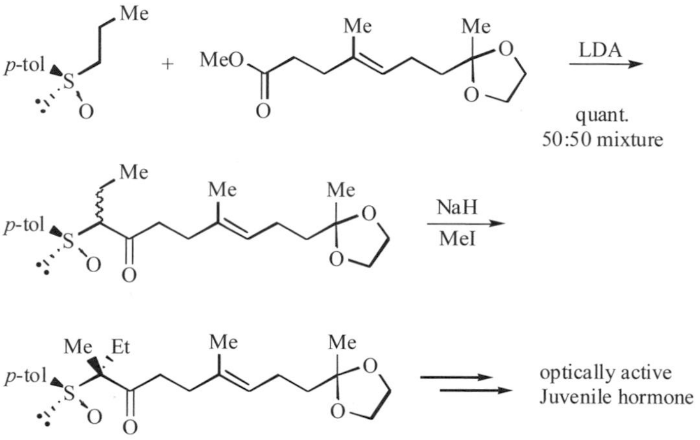

LDA
quant.
50:50 mixture
NaH
MeI
optically active
Juvenile hormone

9:1 diastereoselectivity

H. Kosugi, O. Kanno, H. Uda, *Tetrahedron: Asymmetry* **1994**, 5, 1139

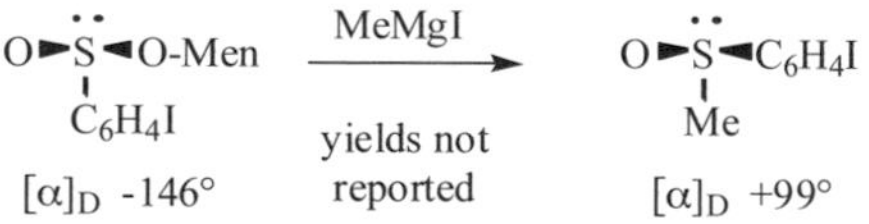

MeMgI
yields not
reported
[α]D -146°
[α]D +99°

P. Bickart, M. Axelrod, J. Jacobs, K. Mislow, *Journal of the American Chemical Society* **1967**, 89, 697

# Appel Reaction

## The Reaction:

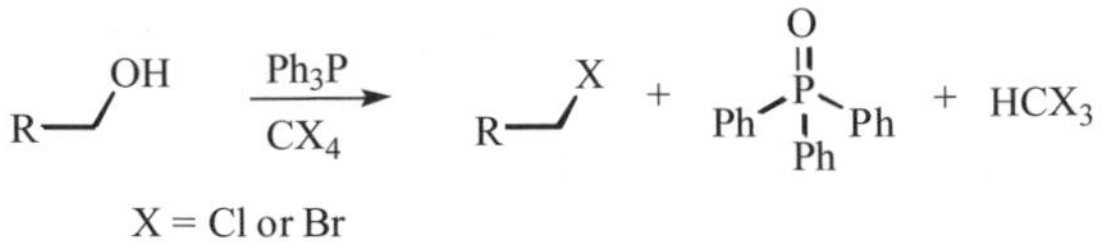

X = Cl or Br

## Proposed Mechanism:

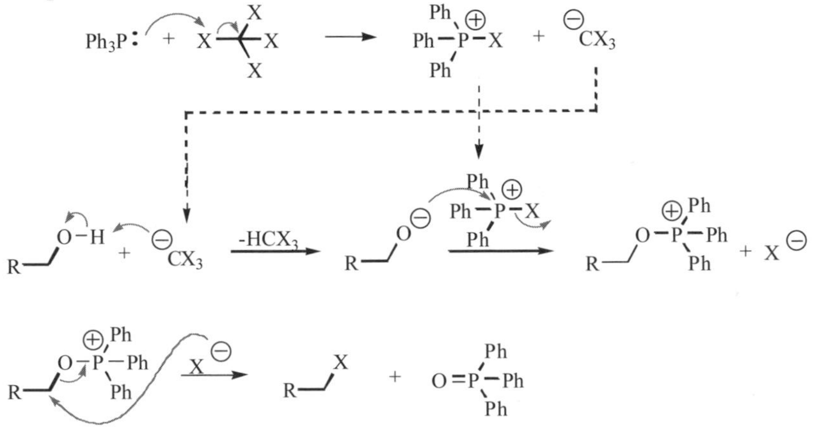

## Notes:

There are two processes called the *Appel Reaction*. Although similar, the second is concerned with reactions of phosphorous:

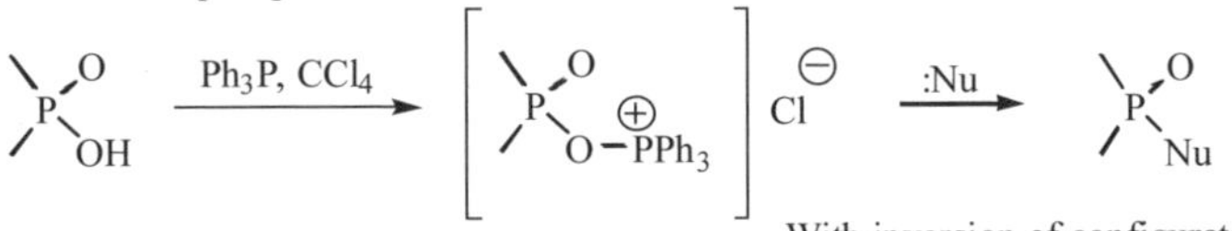

With inversion of configuration around P.:

J. Baraniak, W. J. Stec, *Tetrahedron Letters* **1985**, 26, 4379

See also: J. Beres, W. G. Bentrude, L. Parkanji, A. Kalman, A. E. Sopchik, *Journal of Organic Chemistry* **1985**, 50, 1271

## Examples:

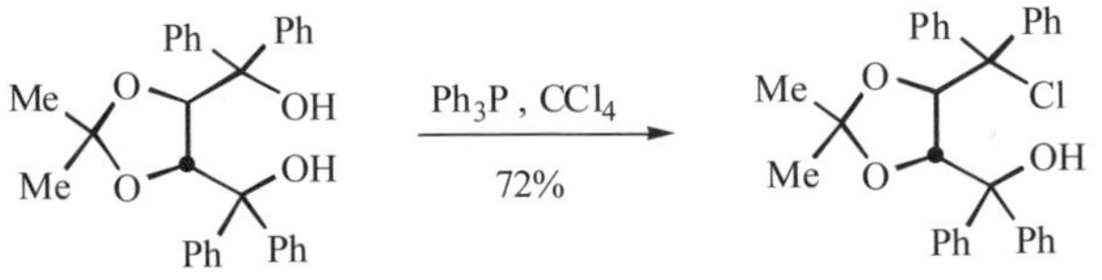

D. Seebach, A. Pichota, A. K. Beck, A. B. Pinkerton, T. Litz, J. Karjalainen, V. Gramlich, *Organic Letters* **1999**, 1, 55

M. Dubber, T. K. Lindhorst, *Organic Letters* **2001**, <u>3</u>, 4019

ODMT (ODMTr) is the 4,4'-dimethoxytrityl group, a common –OH protecting group for the carbohydrate moieties in syntheses of polynucleotides.

4,4'-dimethoxytrityl chloride [40615-36-9]

B. Nawrot, O. Michalak, M. Nowak, A. Okruszek, M. Dera, W. J. Stee, *Tetrahedron Letters* **2002**, <u>43</u>, 5397

# Arbuzov Reaction (Michaelis-Arbuzov Reaction)

## The Reaction:

## Proposed Mechanism:

## Notes:

T. Laue, A. Plagens, *Named Organic Reactions*, John Wiley and Sons, Inc., New York, 1998, p. 12.;
M. B. Smith, J. March in *March's Advanced Organic Chemistry*, 5$^{th}$ ed., John Wiley and Sons, Inc.,
New York, 2001, p. 1234.

### *The Photo-Arbuzov Reaction:*

### *Michaelis-Becker Reaction (Michaelis Reaction)*
#### The Reaction:

#### Proposed Mechanism:

### *Kabachnik-Fields Reaction*

H.-J. Cristan, A. Herve, D. Virieux, *Tetrahedron* **2004**, <u>60</u>, 877

## Examples:

M. S. Landis, N. J. Turro, W. Bhanthumnavin, W. G. Bentrude, *Journal of Organometallic Chemistry* **2002**, <u>646</u>, 239

P.-Y. Renard, P. Vayron, C. Mioskowski, *Organic Letters* **2003**, <u>5</u>, 1661

92%

S.-S Chou, D.-J. Sun, J.-Y. Huang, P.-K. Yang, H.-C. Lin, *Tetrahedron Letters* **1996**, <u>37</u>, 7279

94%

R. W. Driesen, M. Blouin, *Journal of Organic Chemistry* **1996**, <u>61</u>, 7202

33%

S. Fortin, F. Dupont, P. Deslongchamps, *Journal of Organic Chemistry* **2002**, <u>67</u>, 5437

Described as a tandem ***Staudinger-Arbuzov Reaction***:

R = Me, 78%
R = Ph, 66%

M. M. Sá, G. P. Silveira, A. J. Bortoluzzi, A. Padwa, *Tetrahedron* **2003**, <u>59</u>, 5441

"high yields"

I. Pergament, M. Srebnik, *Organic Letters* **2001**, <u>3</u>, 217

88%

H.-P. Guan, Y.-L. Qui, M. B. Ksebati, E. A. Kern, J. Zemlicka, *Tetrahedron* **2002**, <u>58</u>, 6047

# Arndt-Eistert Homologation Reaction

## The Reaction:

## Proposed Mechanism:

Elimination of nitrogen gives a carbene, followed by migration of the R group.

The rearrangement of the diazo ketone is known as the ***Wolff Rearrangement***.

Alternatively, R migration and $N_2$ elimination may be concerted, avoiding the formation of a carbene.

keto-enol tautomerism

## Notes:

See: *Diazomethane*

M. B. Smith, J. March in *March's Advanced Organic Chemistry*, 5th ed., John Wiley and Sons, Inc., New York, 2001, pp 1405-1407; T. Laue, A. Plagens, *Named Organic Reactions*, John Wiley and Sons, Inc., New York, 1998, pp. 13-15; W. E. Bachmann, W. S. Struve, *Organic Reactions* **1**, 2

The ***Kowalski Ester Homologation*** provides a similar conversion (C. Kowalski, M. S. Haque, *Journal of Organic Chemistry* **1985**, <u>50</u>, 5140)

See also: P. Chen, P. T. W. Cheng, S. H. Spergel, R. Zahler, X. Wang, J. Thottathil, J. C. Barrish, R. P. Polniaszek, *Tetrahedron Letters* **1997**, <u>38</u>, 3175

## *Nierenstein Reaction*

## Examples:

T. Hudlicky, J. P. Sheth, *Tetrahedron Letters* **1979**, <u>29</u>, 2667

J. M. Jimenez, R. M. Ortuno, *Tetrahedron: Asymmetry* **1996**, <u>7</u>, 3203

A number of examples to show that this method is more mild than the ***Arndt-Eistert*** reaction

D. Gray, C. Concello´, T. Gallagher, *Journal of Organikc Chemistry* **2004**, <u>69</u>, 4849

N. J. Garg, R. Sarpong, B. M. Stoltz, *Journal of the American Chemical Society* **2002**, <u>124</u>, 13179

R. A. Ancliff, A. T. Russell, A. J. Sanderson *Tetrahedron: Asymmetry* **1997**, <u>8</u>, 3379

# Aza-Cope Rearrangement

## The Reaction:

## Proposed Mechanism:

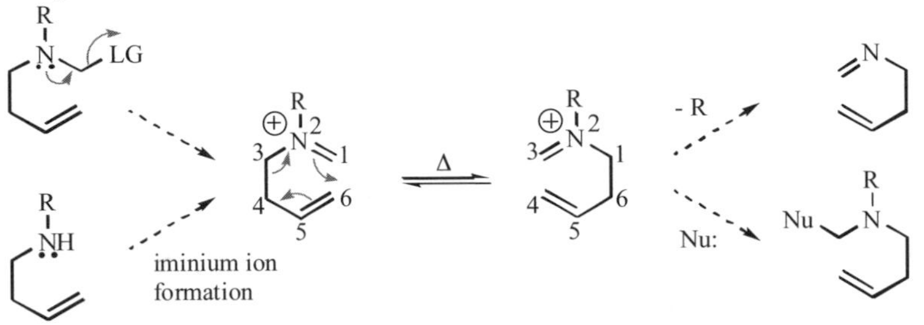

iminium ion
formation

## Notes:

M. B. Smith, J. March in *March's Advanced Organic Chemistry*, 5$^{th}$ ed., John Wiley and Sons, Inc., New York, 2001, p. 1445.

iminium ion formation

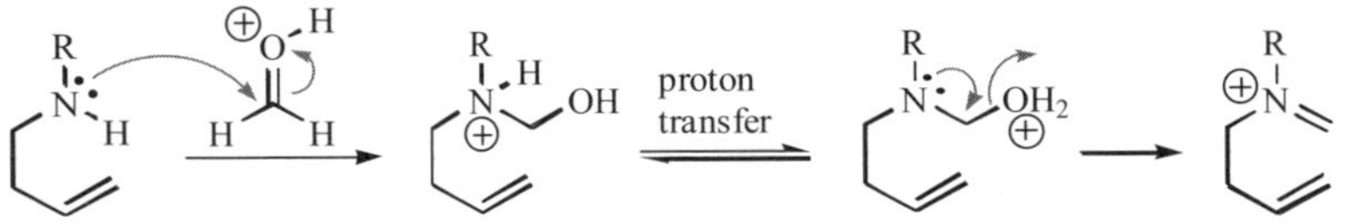

proton
transfer

The *Azo-Cope Rearrangement*:

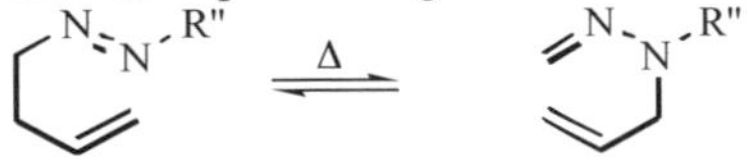

## Examples:

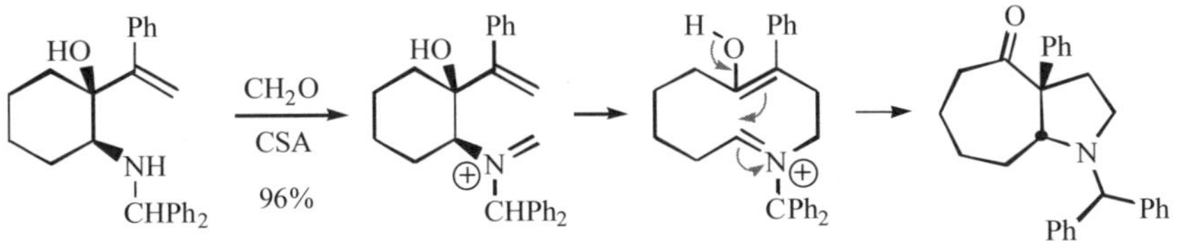

CH$_2$O
CSA
96%

L. E. Overman, E. J. Jacobsen, R. J. Doedens, *Journal of Organic Chemistry* **1983**, <u>48</u>, 3393

98%

K. Shishido, K. Hiroya, K. Fukumoto, T. Kametani, *Tetrahedron Letters* **1986**, <u>27</u>, 1167

H. Ent, H. De Koning, W. N. Speckamp *Journal of Organic Chemistry* **1986**, <u>51</u>, 1687

M. Bruggemann, A. I. McDonald, L. E. Overman, M. D. Rosen, L. Schwink, J. P. Scott, *Journal of the American Chemical Society* **2003**, <u>125</u>, 15284

No yield given for this step, catalyzed by tosic acid in benzene.

K. M. Brummond, J. Lu, *Organic Letters* **2001**, <u>3</u>, 1347

# Baeyer-Villiger Reaction

## The Reaction:

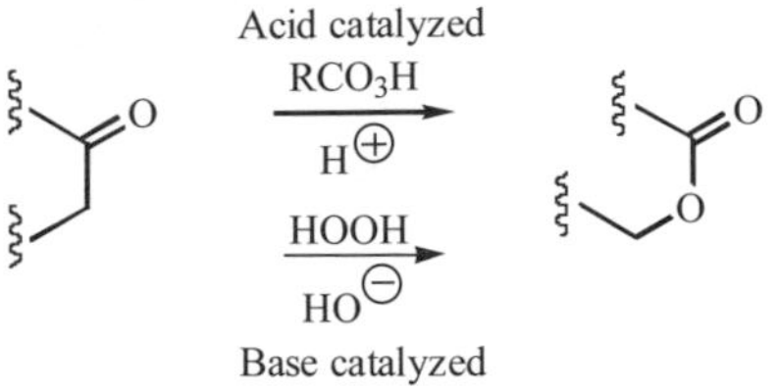

Acid catalyzed

Base catalyzed

## Proposed Mechanism:

Acid catalyzed:

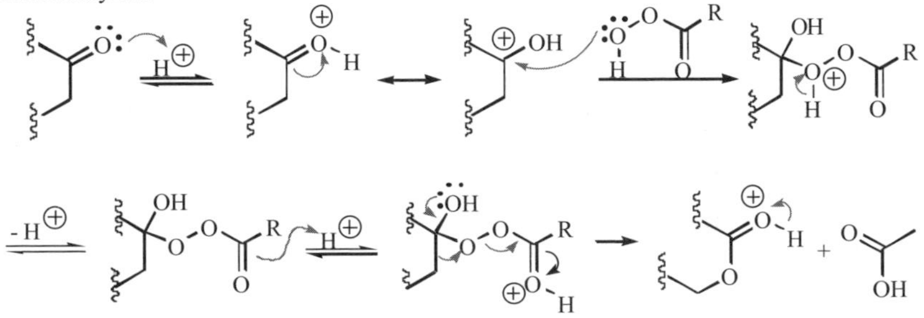

Base catalyzed:

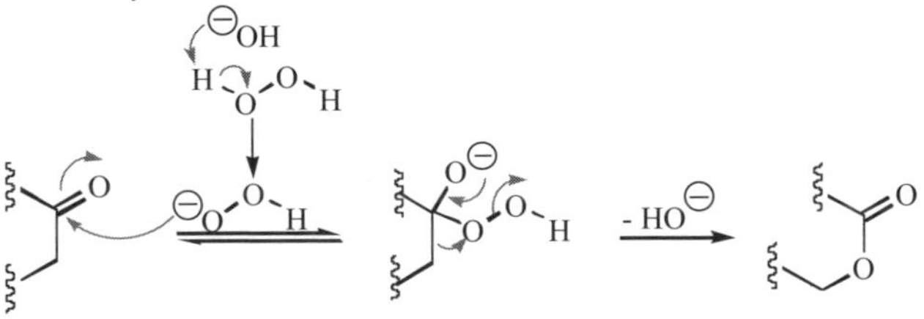

## Notes:

M. B. Smith, J. March in *March's Advanced Organic Chemistry*, 5th ed., John Wiley and Sons, Inc., New York, 2001, pp 1417-1418; T. Laue, A. Plagens, *Named Organic Reactions*, John Wiley and Sons, Inc., New York, 1998, pp. 16-19; C. H. Hassall, *Organic Reactions* **9**, 3; G. R. Krow, *Organic Reactions* **43**, 3.

**Migratory Apptitude**:  3° > 2°> Ph-CH$_2$-> Ph- > 1°> Me > H

Hydrolysis or reduction of the lactone ring provided by reaction with cyclic ketones provides a useful strategy for construction of ring systems:

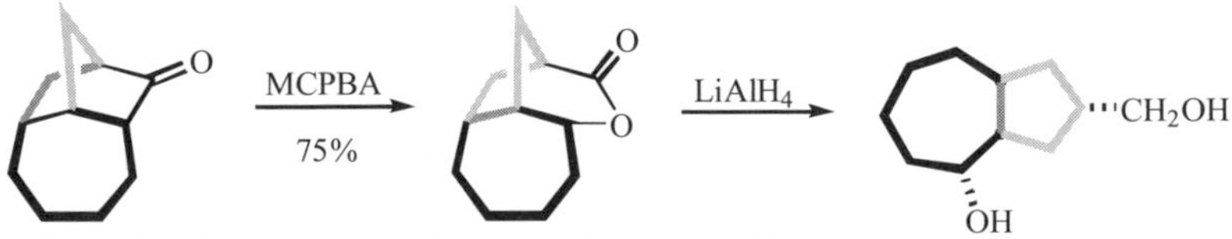

Y. Chen, J. K. Snyder, *Tetrahedron Letters* **1997**, <u>38</u>, 1477

MCPBA → Hydrolysis → n-BuI / HMPA

—COOn-Bu

A. E. Greene, C. Le Drian, P. Crabbe, *Journal of the American Chemical Society* **1980**, <u>102</u>, 7584

## Examples:

Note the retention of stereochemistry after the oxygen insertion. This is a general observation.

*m*-CPBA, Li$_2$CO$_3$
CH$_2$Cl$_2$, rt, 5h
80%

N. Haddad, Z. Abramovich, I. Ruhman *Tetrahedron Letters* **1996**, <u>37</u>, 3521

Baeyer-Villiger
75%

F. W. J. Demnitz, R. A. Parhael, *Synthesis* **1996**, <u>11</u>, 1305

H$_2$O$_2$, (CF$_3$CO)$_2$O
CHCl$_3$

42.8%   +   25.2%

B. Voigt, J. Schmidt, G. Adam, *Tetrahedron* **1996**, <u>52</u>, 1997

MCPBA
69%

G. Magnusson, *Tetrahedron Letters* **1977**, <u>18</u>, 2713

# Baker-Venkataraman Rearrangement

**The Reaction:**

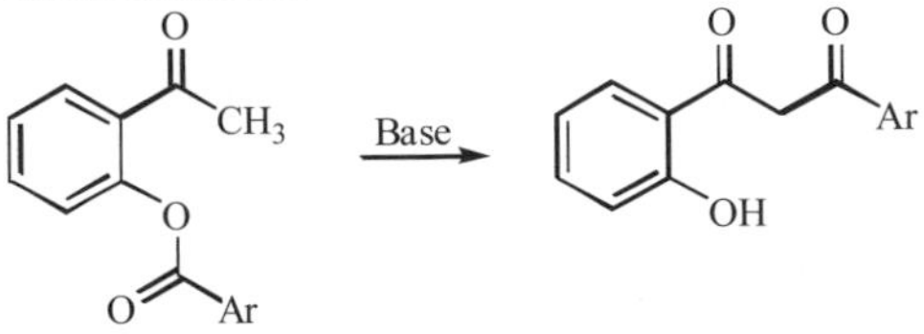

**Proposed Mechanism:**

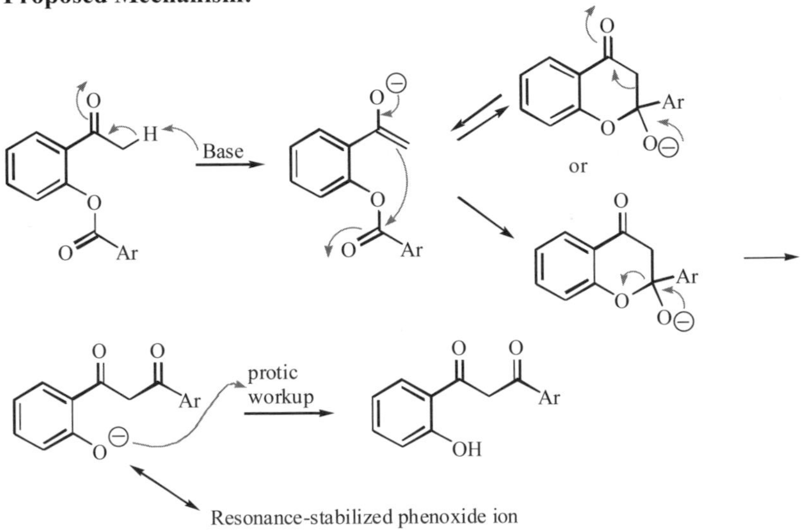

Resonance-stabilized phenoxide ion

See: T. Szell, G. Balaspiri, T. Balaspiri, *Tetrahedron* **1969**, 25, 707

**Notes:**

These β-diketones are useful intermediates for the synthesis of flavones and chromones:

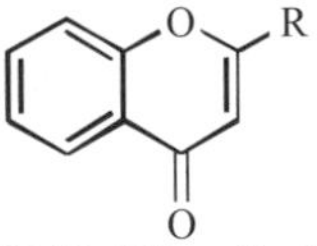

R = Ph: Flavone; R = Me: Chromone

V. K. Ahluwalia, R. K. Parashar, *Organic Reaction Mechanisms*, Alpha Science International Ltd., Pangbourne, U.K., 2002, pp. 277-278

**Examples:**

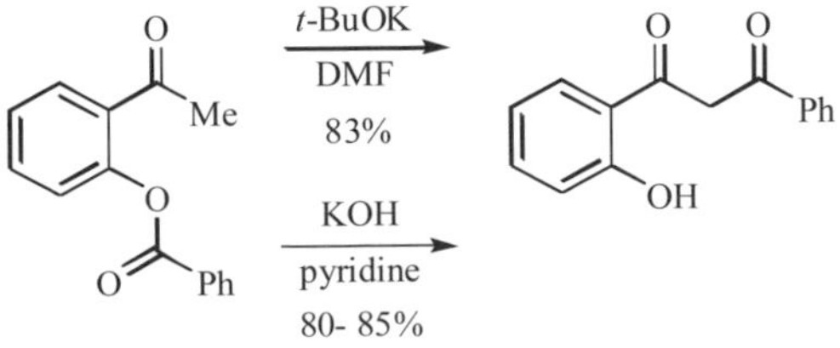

A. Nishinaga, H. Ando, K. Maruyama, T. Mashino, *Synthesis* **1992**, 839
T. S. Wheeler, *Organic Syntheses*, CV4, 478

A ring closure that is often associated with the reaction is called the **Baker-Venkataraman Reaction**.

N. Thasana, S. Ruchirawat, *Tetrahedron Letters* **2002**, 43, 4515

S. J. Cutler, F. M. El-Kabbani, C. Keane, S. L. Fisher-Shore, C. D. Blanton, *Heterocycles* 1990, 31, 651 (AN 1990:552089)

P. F. Devitt, A. Timoney, M. A. Vickars, *Journal of Organic Chemistry* **1961**, 26, 4941

A. V. Kalinin, A. J. M. da Silva, C. C. Lopes, R. S. C. Lopes, V. Snieckus *Tetrahedron Letters* **1998**, 39, 4995

# Balz-Schiemann Reaction (Schiemann Reaction)

## The Reaction:

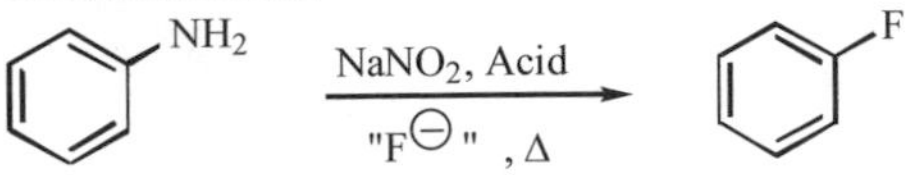

## Proposed Mechanism:

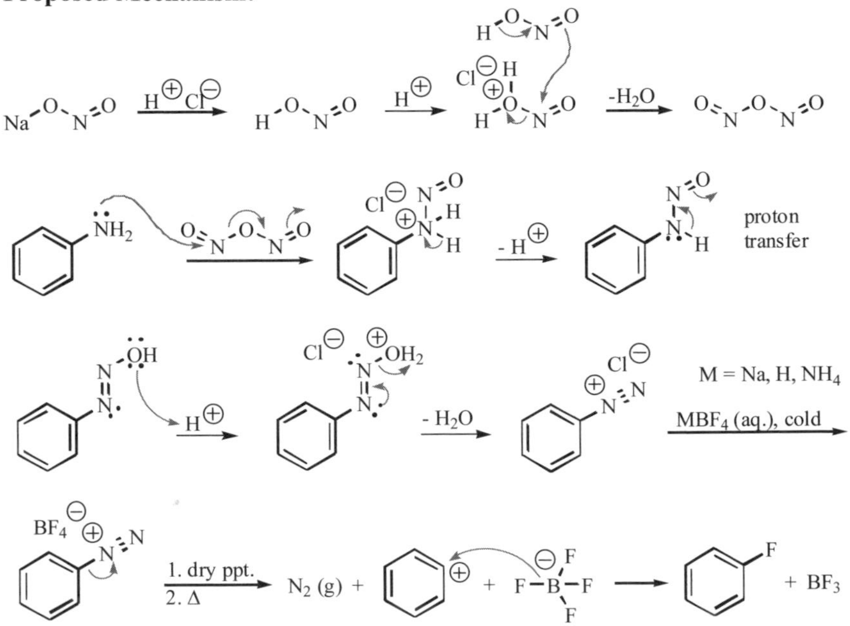

M = Na, H, NH₄

salt precipitates

## Notes:

The original work:

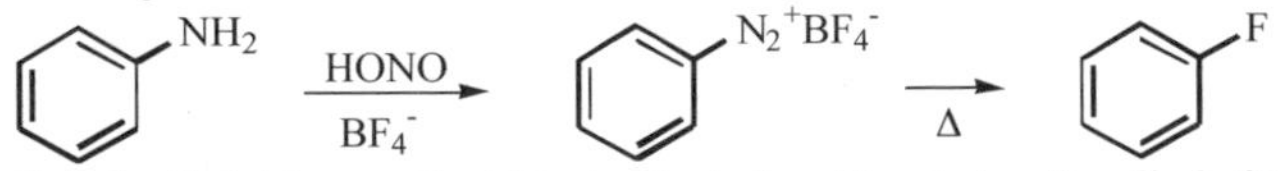

G. Balz, G. Schiemann, *Berichte der Deutschen Chemischen Gesellschaft* **1927**, 60, 1186

T. Laue, A. Plagens, *Named Organic Reactions*, John Wiley and Sons, Inc., New York, 1998, pp. 237-238; M. B. Smith, J. March in *March's Advanced Organic Chemistry*, 5th ed., John Wiley and Sons, Inc., New York, 2001, p. 875; A. Roe, *Organic Reactions* 5, 4

Reaction is often incorporated into the **_Sandmeyer Reactions_** series,

Procedural improvement to avoid isolation of the (toxic) intermediate:
D. J. Milner, P. G. McMunn, J. S. Moilliet, *Journal of Fluorine Chemistry* **1992**, 58, 317 and D. J. Milner, *Journal of Fluorine Chemistry* **1991**, 54, 382

Reaction improvement by using ionic liquid salts:
K. K. Laali, V. J. Gettwert, *Journal of Fluorine Chemistry* **2001**, 107, 31

## Examples:

A. Kiryanov, A. Seed, P. Sampson, *Tetrahedron Letters* **2001**, 42, 8797

A modified *Balz-Schiemann Reaction*:

63%

F. Dolle, L. Dolci, H. Valette, F. Hinnen, F. Vaufrey, H. Guenther, C. Fuseau, C. Coulon, M. Buttalender, C. Crouzel *Journal of Medicinal Chemistry* **1999**, 42, 2251

M. Argentini, C. Wiese, R. Weinreich, *Journal of Fluorine Chemistry* **1994**, 68, 141

H. Hart, J. F. Janssen, *Journal of Organic Chemistry* **1970**, 35, 3637

# Bamberger Rearrangement

**The Reaction:**

**Proposed Mechanism:**

See discussion in: N. Haga, Y. Endo, K.-i. Kataoka, K. Yamaguchi, K. Shudo, *Journal of the American Chemical Society* **1992**, 114, 9795

## Notes:

M. B. Smith, J. March in *March's Advanced Organic Chemistry*, 5th ed., John Wiley and Sons, Inc., New York, 2001, p. 878; V. K. Ahluwalia, R. K. Parashar, *Organic Reaction Mechanisms*, Alpha Science International Ltd., Pangbourne, U.K., 2002, p. 449

By addition of azide ion to the reaction, the intermediate can be competitively trapped:

J. C. Fishbein, R. A. McClelland, *Journal of the American Chemical Society* **1987**, 109, 2824

A related reaction:

N. Haga, Y. Endo, K.-i. Kataoka, K. Yamaguchi, K. Shudo, *Journal of the American Chemical Society* **1992**, 114, 9795

## Examples:

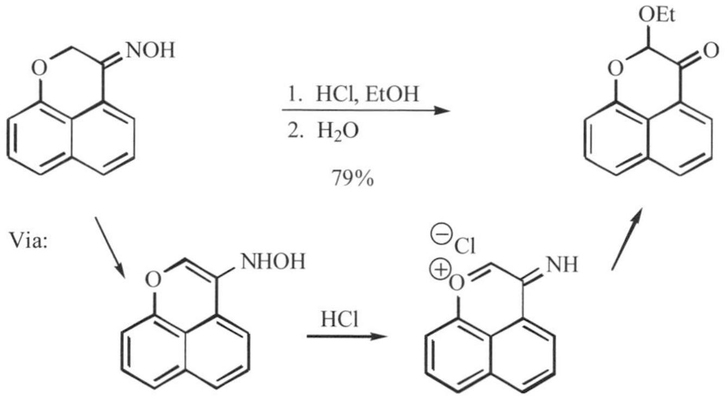

1. HCl, EtOH
2. H₂O

79%

Via:

HCl

J. C. Jardy, M. Venet, *Tetrahedron Letters* **1982**, <u>23</u>, 1255

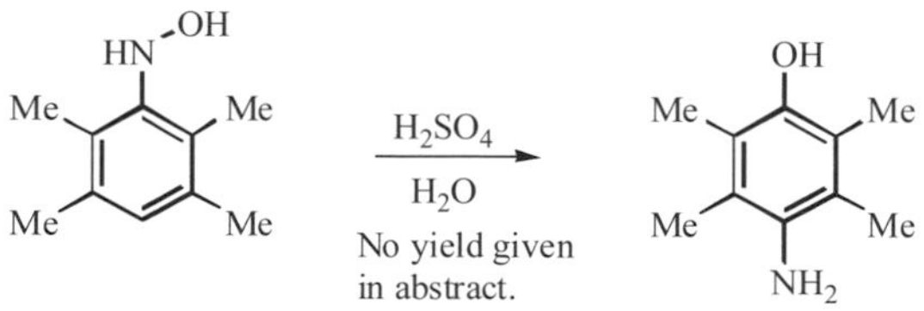

H₂SO₄

H₂O

No yield given
in abstract.

G. G. Barclay, J. P. Candlin, W. Lawrie, P. L. Paulson, *Journal of Chemical Research Synopses*
**1992**, 245

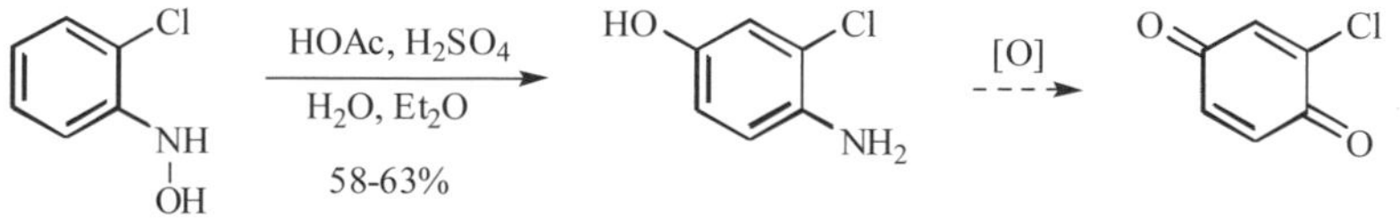

HOAc, H₂SO₄

H₂O, Et₂O

58-63%

[O]

R. E. Harman, *Organic Syntheses* <u>CV4</u>, 148

# Bamford-Stevens Reaction

## The Reaction:

H
H
N
N
Ts
Base
H

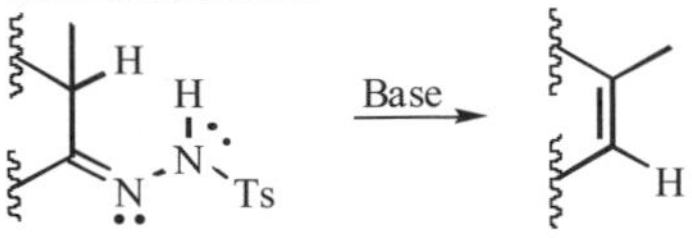

## Proposed Mechanism:

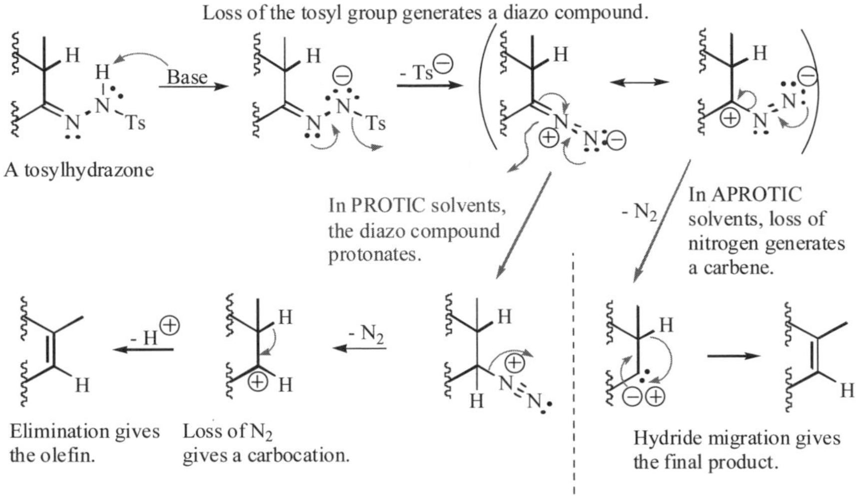

Loss of the tosyl group generates a diazo compound.
Base
- Ts⊖
A tosylhydrazone
In PROTIC solvents, the diazo compound protonates.
- N2
In APROTIC solvents, loss of nitrogen generates a carbene.
- H⊕
- N2
Elimination gives the olefin.
Loss of N2 gives a carbocation.
Hydride migration gives the final product.

## Notes:

M. B. Smith, J. March in *March's Advanced Organic Chemistry*, 5th ed., John Wiley and Sons, Inc., New York, 2001, p. 1335; T. Laue, A. Plagens, *Named Organic Reactions*, John Wiley and Sons, Inc., New York, 1998, pp. 19-22; R. H. Shapiro, *Organic Reactions* **23**, 3

In the related ***Shapiro reaction***, two equivalents of an alkyl lithium are used and the less substituted alkene is formed.

Bamford-Stevens
Na
ethylene glycol
Ts
H2N−N
H
O
1
Me
N
N
Ts
H
1. 2 RLi
2. H2O
Shapiro

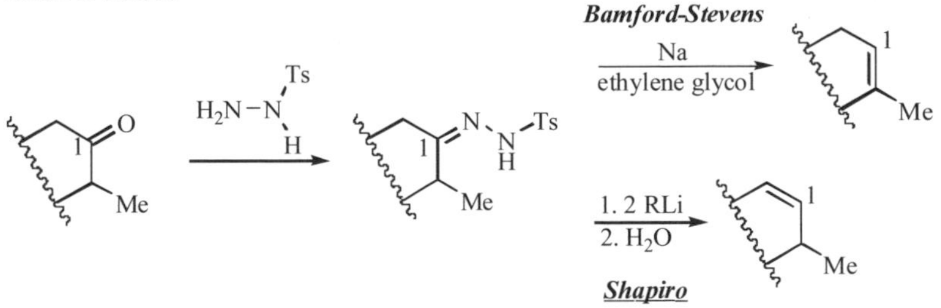

## Examples:

P. A. Grieco, T. Oguri, C.-L. J. Wang, E. Williams, *Journal of Organic Chemistry* **1977**, 42, 4113

C. Marchioro, G. Pentassuglia, A. Perboni, D. Donati, *Journal of the Chemical Society Perkin Transactions 1* **1997**, 463

S. J Hecker, C. H. Heathcock, *Journal of the American Chemical Society* **1986**, 108, 4586

A general method for the homologation of aldehydes to benzylic ketones makes use of the *Bamford-Stevens* approach, via intermediate aryldiazomethanes:

S. R. Angle, M. L. Neitzel, *Journal of Organic Chemistry* **2000**, 65, 6458

# Barbier (Coupling) Reaction

## The Reaction:

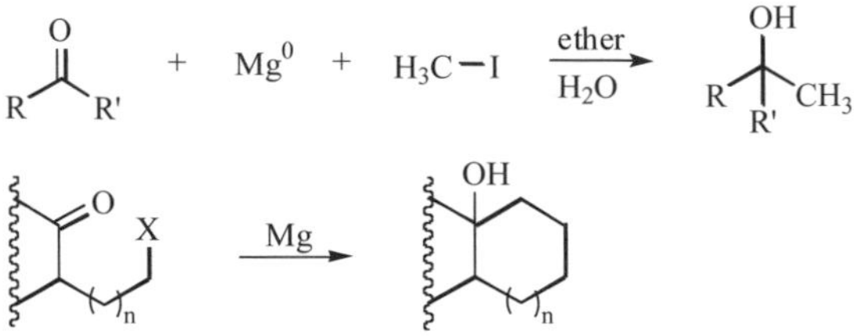

## Proposed Mechanism:

Resembles an internal **_Grignard reaction_**:

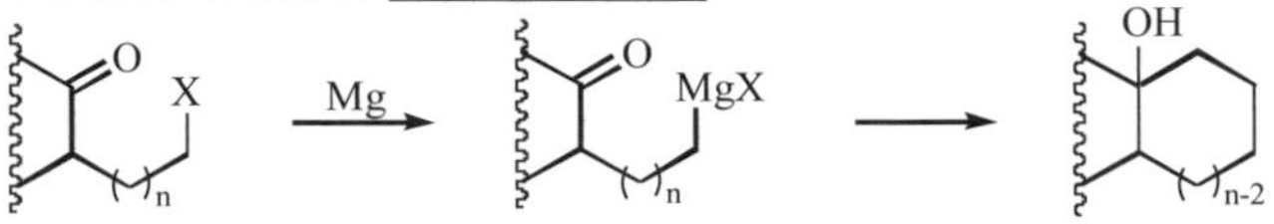

## Notes:

M. B. Smith, J. March in *March's Advanced Organic Chemistry*, 5th ed., John Wiley and Sons, Inc., New York, 2001, p. 1205

This reaction was used before it was noted that adding the halide to magnesium prior to the addition the carbonyl gave a better reaction.  See the **_Grignard Reaction_**.

Other metals may be used.

A variety of reactions of a carbonyl and an organohalogen compound are classified as **Barbier and Barbier-type**.

## Examples:

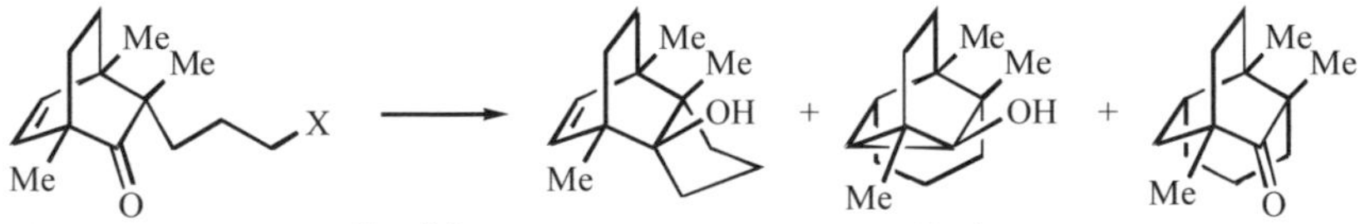

| X  | Conditions                              | Ratio |    |    |
|----|-----------------------------------------|-------|----|----|
| Br | Li powder, ultrasound, Et$_2$O          | 34    | 50 | -  |
| Br | Mg turning, HgCl$_2$, ultrasound, THF   | 10    | 12 | 37 |
| I  | BuLi, THF                               | 71    | -  | -  |

W. Zhang, P. Dowd, *Tetrahedron Letters* **1993**, <u>34</u>, 2095

C. A. Molander, J. B. Etter, L. S. Harring, P.-J. Thorel, *Journal of the American Chemical Society* **1991**, 113, 3889

For a review of diiodosamarium chemistry (including *Barbier Reactions*) see:
H. Kagan, *Tetrahedron* **2003**, 59, 10351

**Mechanistically:**

C. C. K. Keh, C. Wei, C.-J, Li, *Journal of the American Chemical Society* **2003**, 125, 4062

LiDBB = Lithium 4,4'-dit-butylbiphenyl =

J. Shin, O. Gerasimov, D. H. Thompson, *Journal of Organic Chemistry* **2002**, 67, 6503

X = O, Y = S 68% (cis/trans = 95/5)   X = S, Y = S 72% (cis/trans = 95/5)
X = O, Y = O 71% (cis/trans = 56/44)   X = S, Y = O 43% (cis/trans = 52/48)

A. S.-Y. Lee, Y.-T. Chang, S.-H. Wang, S.-F. Chu, *Tetrahedron Letters* **2002**, 43, 8489

# Barbier-Wieland Degradation (Barbier-Locquin Degradation)

**The Reaction:**

1. EtOH, H⊕
2. excess Ph-MgX
3. Ac$_2$O or Δ or H⊕
4. CrO$_3$ or NaIO$_4$-RuO$_4$

A procedure for decreasing a chain length by one carbon.

**Proposed Mechanism:**

proton transfer

EtOH

-H$_2$O       PhMgX       -OEt       PhMgX

-H$_2$O

CrO$_3$

**Notes:**

M. B. Smith, J. March in *March's Advanced Organic Chemistry*, 5$^{th}$ ed., John Wiley and Sons, Inc., New York, 2001, p. 1526.

A variation of this procedure, the ***Meystre-Miescher-Wettstein Degradation (Miescher Degradation)*** removes three carbons from the chain:

PhMgX       - H$_2$O

NBS       - HBr

CrO$_3$

*Gallagher-Hollander Degradation*

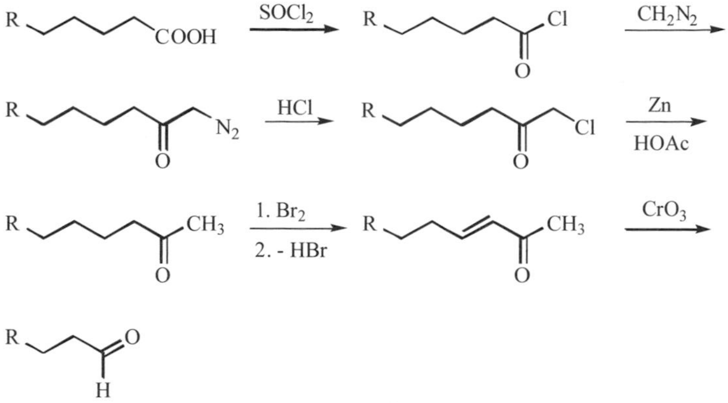

R～～COOH $\xrightarrow{\text{SOCl}_2}$ R～～C(=O)Cl $\xrightarrow{\text{CH}_2\text{N}_2}$

R～～C(=O)～N$_2$ $\xrightarrow{\text{HCl}}$ R～～C(=O)～Cl $\xrightarrow[\text{HOAc}]{\text{Zn}}$

R～～C(=O)CH$_3$ $\xrightarrow[\text{2. - HBr}]{\text{1. Br}_2}$ R～～C(=O)CH$_3$ $\xrightarrow{\text{CrO}_3}$

R～～CHO

*Krafft Degradation*

R～～C(=O)-O-M-O-C(=O)～～R $\xrightarrow{\text{Ac}_2\text{O}}$ R～～C(=O)CH$_3$

$\xrightarrow{[\text{O}]}$ R～～COOH

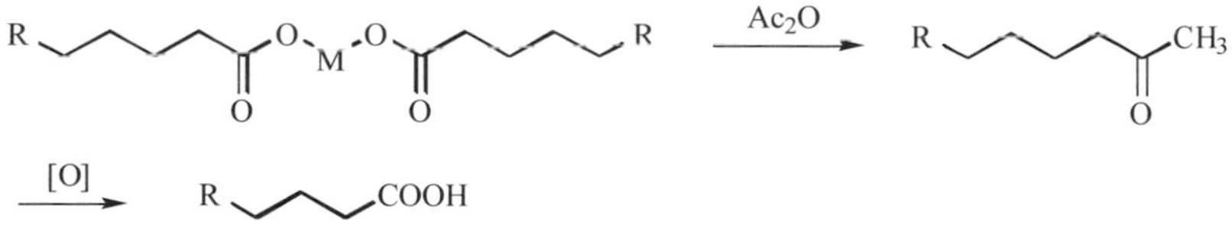

## Bartoli Indole Synthesis

### The Reaction:

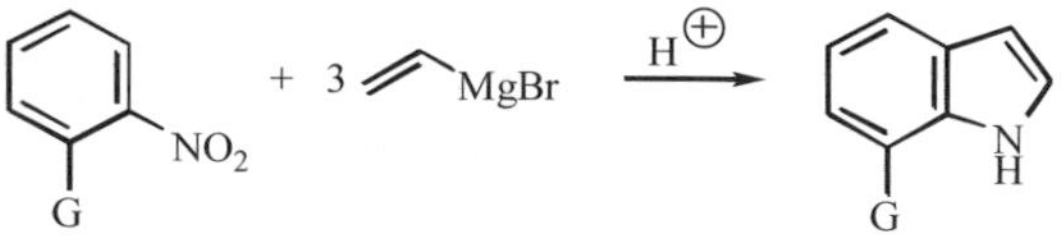

### Proposed Mechanism:

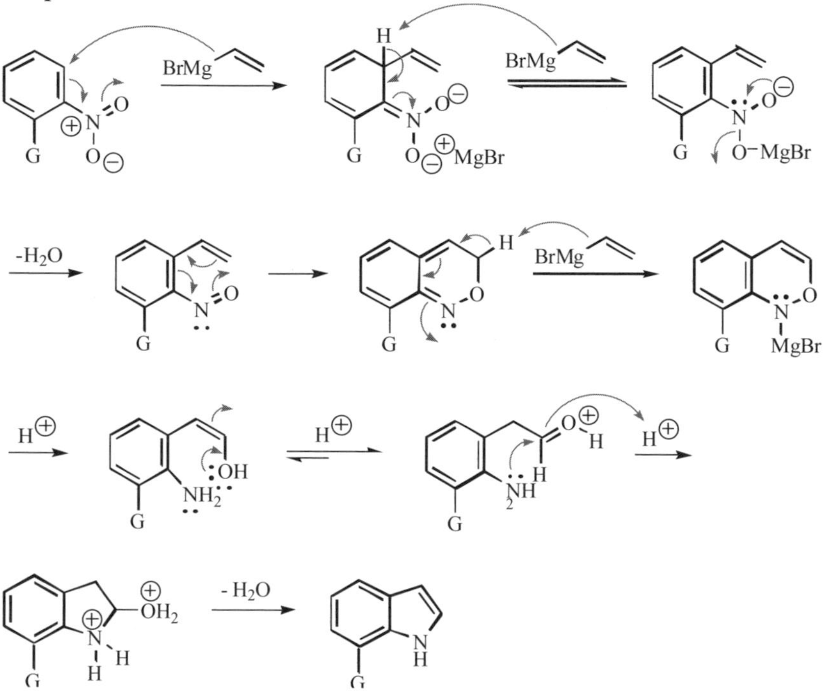

### Notes:

G. Bartoli, G. Palmieri, M. Bosco, R. Dalpozzo, *Tetrahedron Letters* **1989**, <u>30</u>, 2129;
G. Bartoli, M. Bosco, R. Dalpozzo, *Tetrahedron Letters* **1985**, <u>26</u>, 115

The reaction works only with the ortho position of the nitrobenzene occupied.

## Examples:

Me
NO$_2$   1. $\diagup$ MgBr   →   Me   H / N indole

2. NH$_4$Cl

67%

G. Bartoli, G. Palmieri, M. Bosco, R. Dalpozzo, *Tetrahedron Letters* **1989**, <u>30</u>, 2129

CH$_2$Br          Me          CH$_2$Br

          $=\!\!\!<$          Me

          MgBr          N / H

NO$_2$          61%          Br

Br

A. Dobbs, *Journal of Organic Chemistry* **2001**, <u>66</u>, 638

(resin bead)–O–C(=O)– aryl –Cl          Me   MgBr          MeO$_2$C

NO$_2$          1. Me $=$ Me          Me

          2. acid          Me   NH

          37%          Me

K. Knepper, S. Brase, *Organic Letters* **2003**, <u>5</u>, 2829

# Barton Decarboxylation

**The Reaction:**

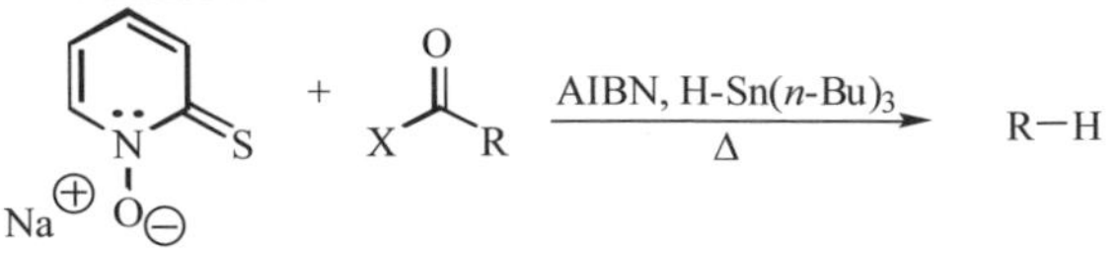

**Proposed Mechanism:**

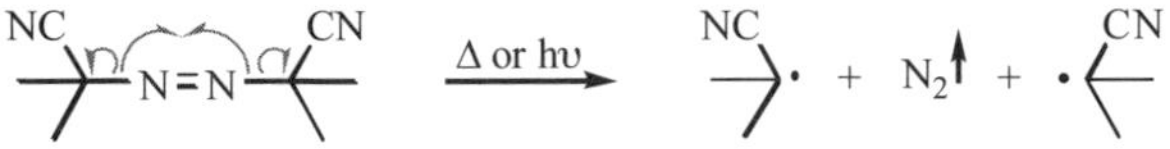

AIBN = Azo-*bis*-isobutyronitrile

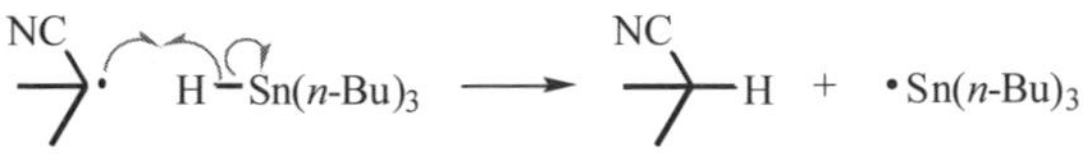

The decarboxylation step:

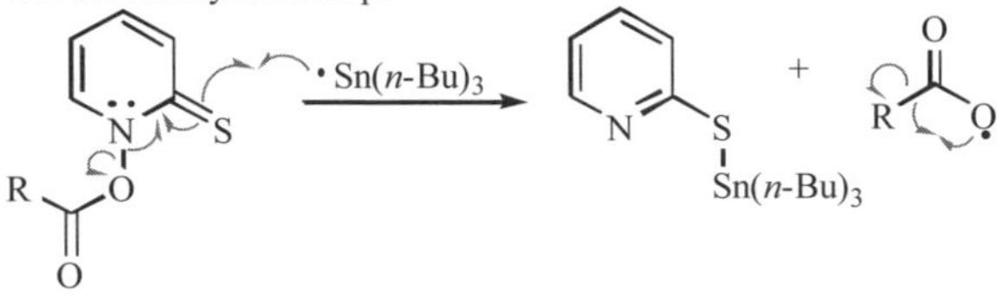

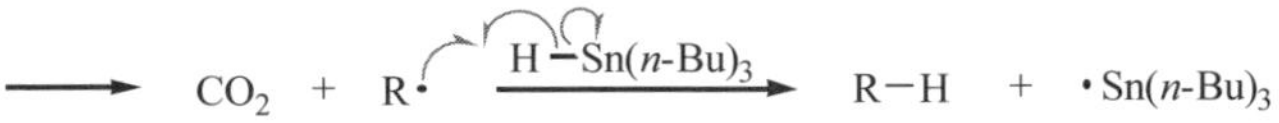

$$\longrightarrow \quad CO_2 \; + \; R\cdot \xrightarrow{\; H-Sn(n\text{-}Bu)_3 \;} \quad R-H \; + \; \cdot Sn(n\text{-}Bu)_3$$

**Notes:**

Starting material preparation:

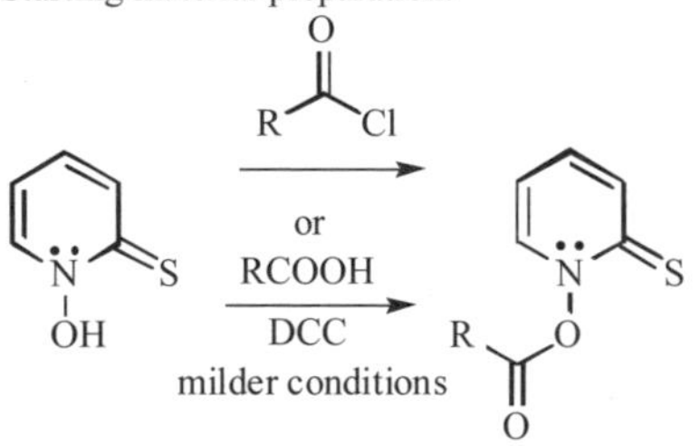

Rather than direct reaction of the the the acid chloride with oxygen, the following takes place:

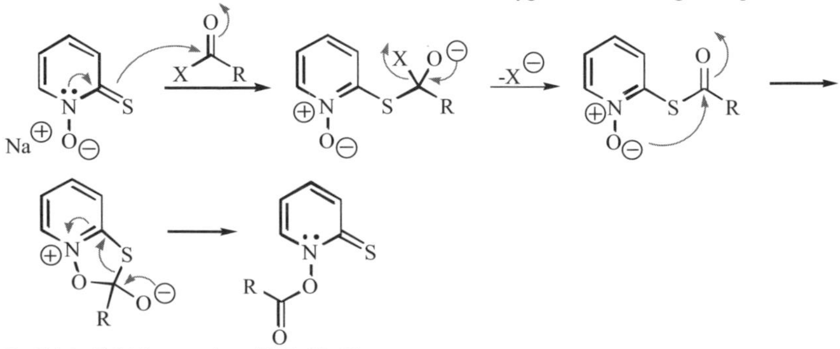

D. Crich *Aldrichimica Acta* **1987**, <u>20</u>, 35

## Examples:

J. T. Starr, G. Koch, E. M. Carreira, *Journal of the American Chemical Society* **2000**, <u>122</u>, 8793

D. H. R. Barton, Y. Herve, P. Potier, J. Thierry, *Tetrahedron*, **1988**, <u>44</u>, 5479

S. F. Martin, K. X. Chen, C. T. Eary, *Organic Letters* **1999**, <u>1</u>, 79

E. Bacque, F. Pautrat, S. Z. Zand, *Organic Letters* **2003**, <u>5</u>, 325

# Barton Reaction (Barton Nitrite Photolysis Reaction)

## The Reaction:

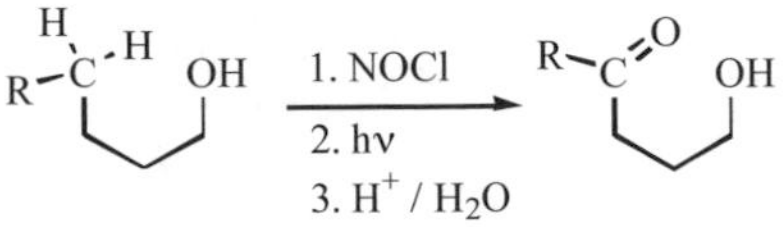

## Proposed Mechanism:

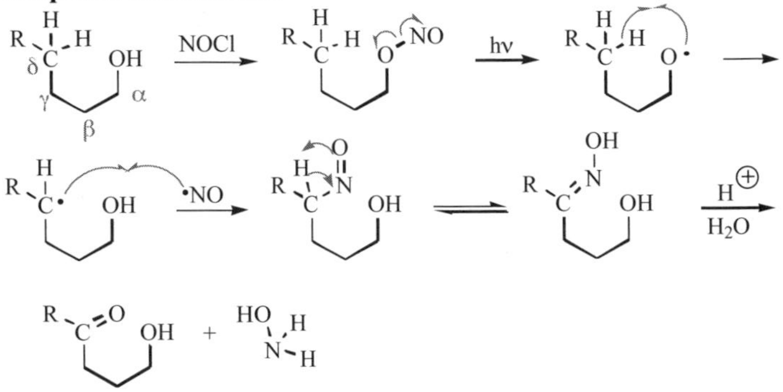

## Notes:

**Hydrolysis mechanism**

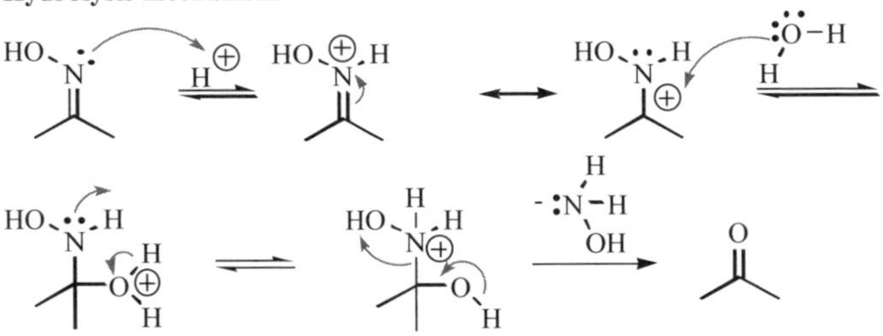

This reaction is a useful method for functionalizing a remote position (the δ–position).

Also by Barton is the

## Barton-Kellogg Reaction (Barton Olefin Synthesis)

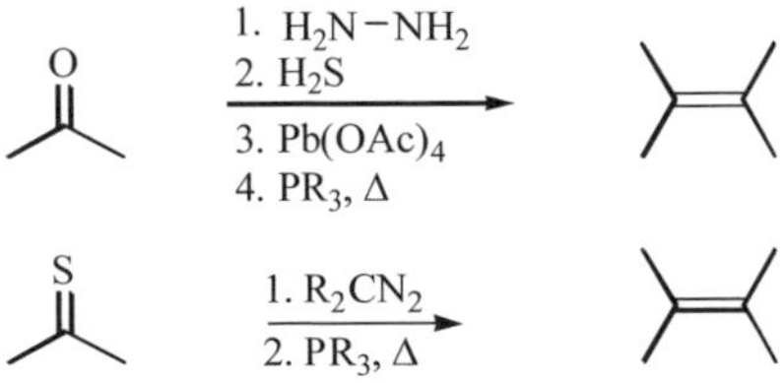

1. $H_2N-NH_2$
2. $H_2S$
3. $Pb(OAc)_4$
4. $PR_3, \Delta$

1. $R_2CN_2$
2. $PR_3, \Delta$

## Examples:

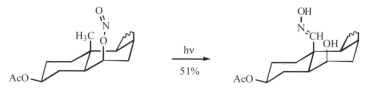

D. H. R. Barton, I. M. Beaton, L. E. Geller, M. M. Pechet, *Journal of the American Chemical Society* **1960**, <u>82</u>, 2640

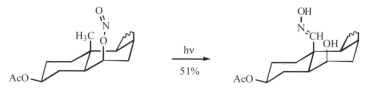

Not purified          51% yield of hemiacetal

P. D. Hobbs, P. D. Magnus *Journal of the American Chemical Society* **1976**, <u>98</u>, 4594

# Barton-McCombie Reaction (Barton-Deoxygenation)

## The Reaction:

$$\underset{R'}{\overset{S}{\diagdown}}\underset{O}{\diagdown}\underset{R'}{\overset{R}{\diagup}} \quad \xrightarrow[\Delta]{\text{AIBN, H-Sn}(n\text{-Bu})_3} \quad \underset{H}{\overset{R}{\diagdown}}R'$$

R' = H, CH₃, SCH₃ (= xanthate, see ***Chugaev Reaction***), Ph, OPh, imidazolyl

## Proposed Mechanism:

**AIBN** = Azo-*bis*-isobutyronitrile

## Notes:

M. B. Smith, J. March in *March's Advanced Organic Chemistry*, 5ᵗʰ ed., John Wiley and Sons, Inc., New York, 2001, p. 527.

For a discussion of mechanism: D. Crich, *Tetrahedron Letters* **1988**, *29*, 5805

Depending on the substrate, different thiocarbonyl compounds have been used:

*primary alcohols with xanthates*

imidoyl chloride methochloride
(from phosgene and an amide)

*secondary alcohols with thiocarbonyl esters*

*tertiary alcohols with thioformates*

## Examples:

G. L. Lange, C. Gottardo, *Tetrahedron Letters* **1994**, <u>35</u>, 8513

M. T. Crimmins, J. M. Pace, P. G. Nantermet, A. S. Kim-Meade, J. B. Thomas, S. H. Watterson, A. S. Wagman, *Journal of the American Chemical Society* **1999**, <u>121</u>, 10249

J. R. Williams, D. Chai, J. D. Bloxton, II, H. Gong, W. R. Solvibile, *Tetrahedron* **2003**, <u>59</u>, 3183

K. Paulvannan, J. R. Stille, *Tetrahedron Letters* **1993**, <u>34</u>, 6673

# Barton-Zard Pyrrole Synthesis

## The Reaction:

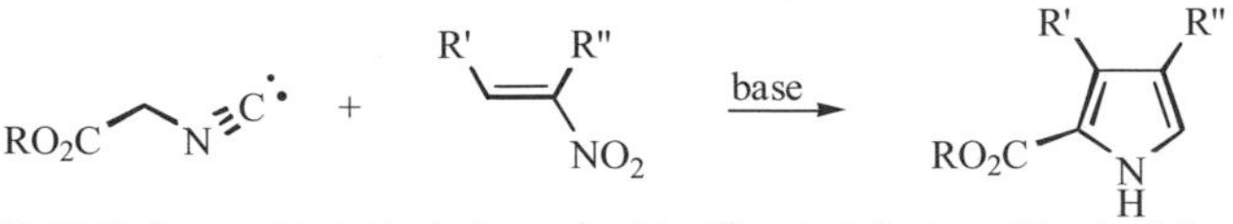

D. H. R. Barton, S. Z. Zard, *Journal of the Chemical Society, Chemical Communications* **1985**, 1098
D. H. R. Barton, J. Kervagoret, S. Zard, *Tetrahedron* **1990**, <u>46</u>, 7587

## Proposed Mechanism:

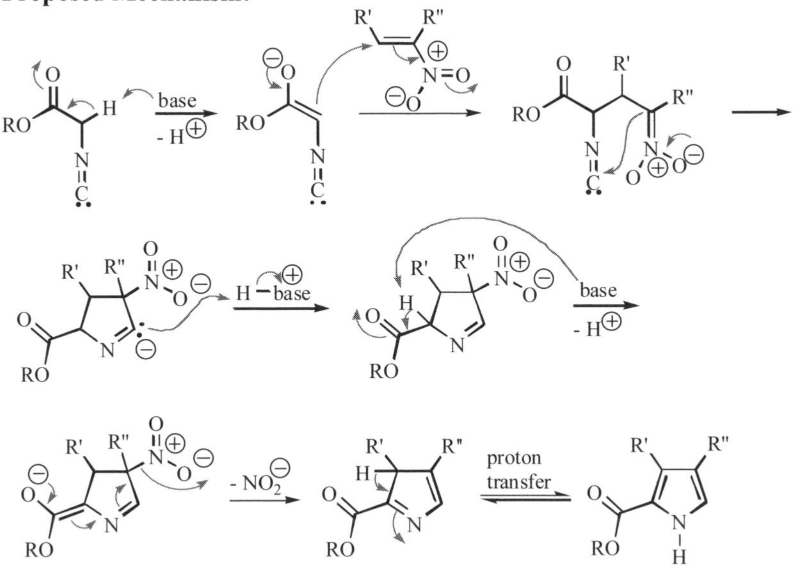

D. H. R. Barton, J. Kervagoret, S. Zard, *Tetrahedron* **1990**, <u>46</u>, 7587

## Notes:

One possible starting material preparation: A ***Henry Reaction*** followed by trapping with Ac$_2$O and elimination of the resultant acetate.

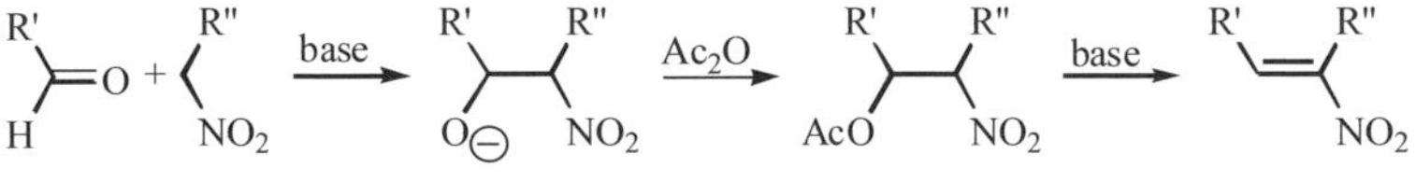

## Examples:

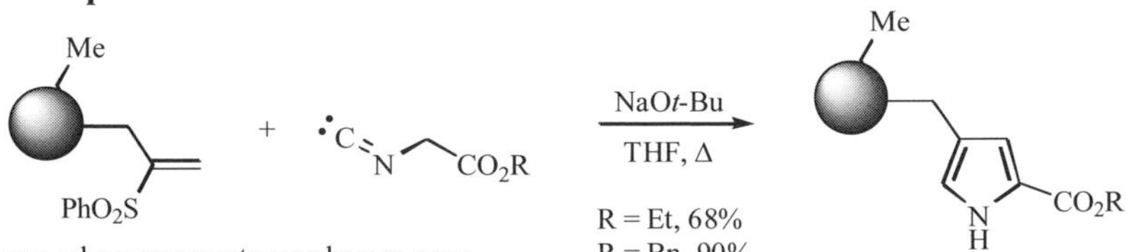

grey sphere represents a carborane cage
(see reference for a better picture)

R = Et, 68%
R = Bn, 90%

S. Chayer, L. Jaquinod, K. M. Smith, M. G. H. Vicente *Tetrahedron Letters* **2001**, <u>42</u>, 7759 part of a 7 step procedure of overall 32% yield

D. Lee, T. M. Swager *Journal of the American Chemical Society* **2003**, <u>125</u>, 6870

72%

J. Bergman, S. Rehn *Tetrahedron* **2002**, <u>58</u>, 9179

X = N, R = Et, 53%
X = CH, R = Et, 86%			X = N, R = *t*-Bu, 83%
X = CH, R = *t*-Bu, 65%		X = N, R = Bz, 73% (refluxing THF / *i*-PrOH)

T. D. Lash, B. H. Novak, Y. Lin *Tetrahedron Letters* **1994**, <u>35</u>, 2493

R' = H, R" = Ph, 78%		R' = Me, R" = Ph, 75%
R' = H, R" = Py, 66%		R' = Me, R" = Py, 74%

S. H. Hwang, M. J. Kurth *Tetrahedron Letters* **2002**, <u>43</u>, 53

A rearranged / abnormal **Barton-Zard Pyrrole** product is observed when the protecting group on nitrogen is phenyl sulfonyl. However, when R = Bn, $CO_2Et$ or 2-pyridyl, the expected pyrrolo[3,4,*b*]indole is obtained.

R = $SO_2Ph$, 85%

R = Bn, 31%
R = $CO_2Et$, 85%
R = 2-pyridyl, 91%

E. T. Pelkey, L. Chang, G. W. Gribble *Chemical Communications* **1996**, 1909
E. T. Pelkey, G. W. Gribble *Chemical Communications* **1997**, 1873

# Baudisch Reaction

## The Reaction:

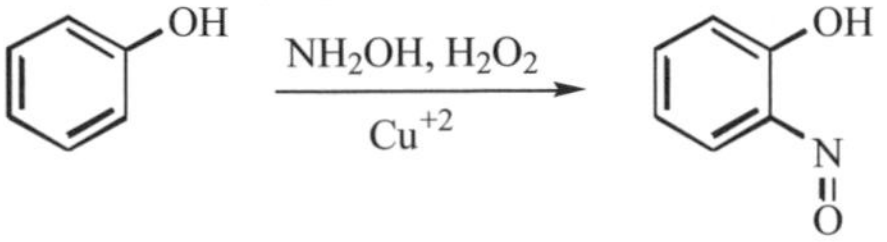

## Proposed Mechanism:
There is much not known about the details of this reaction.

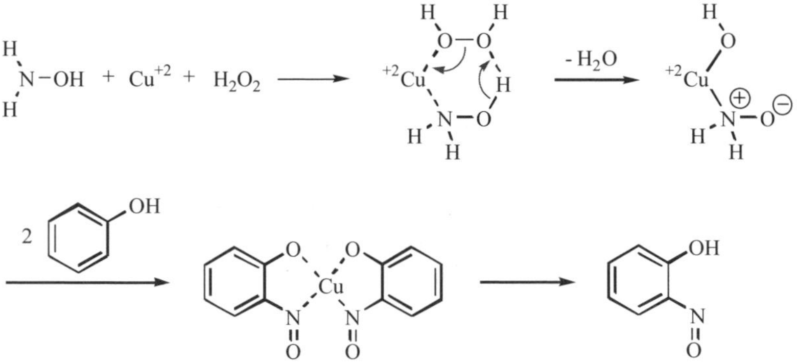

## Notes:

For studies on the mechanism:  See K. Maruyama, I. Tanimoto, R. Goto, *Tetrahedron Letters* **1966**, <u>47</u>, 5889

**Examples:**

R. J. Maleski, M. Kluge, D. Sicker, *Synthetic Communications* **1995**, <u>25</u>, 2327 (AN 1995-63432)

M. C. Cone, C. R. Melville, J. R. Carney, M. P. Gore, S. J. Gould, *Tetrahedron* **1995**, <u>51</u>, 3095

# Baylis-Hillman Reaction (Morita-Baylis-Hillman)

## The Reaction:

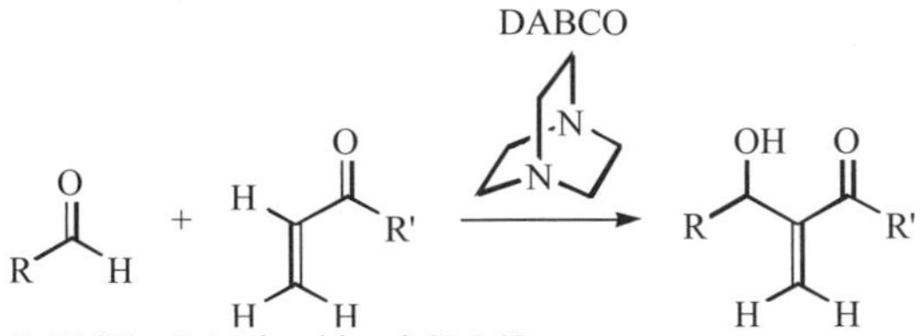

*DABCO* = 1,4-Diazabicyclo[2.2.2]octane

## Proposed Mechanism:

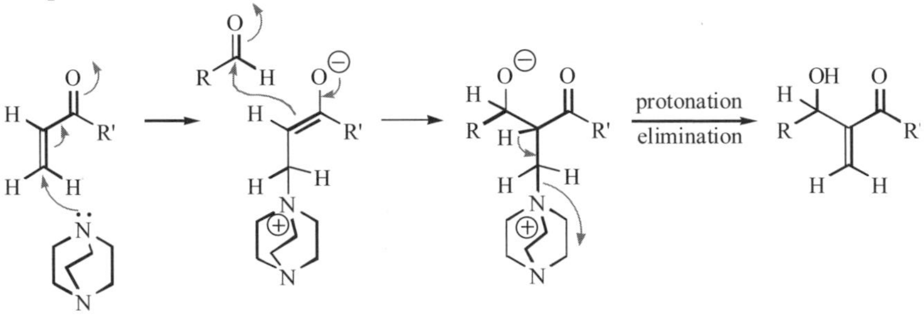

Attack of the tertiary amine generates the
enolate which will attack the aldehyde.

Protonation of the alkoxide and elimination
of the amine gives the final product.

## Notes:

M. B. Smith, J. March in *March's Advanced Organic Chemistry*, 5th ed., John Wiley and Sons, Inc.,
New York, 2001, p. 1212; E. Ciganek, *Organic Reactions* **51**, 2; D. Basavaiah, A. J. Rao, T.
Satyanarayana, *Chemical Reviews* **2003**, 103, 811

The *Rauhut-Currier Reaction* is a similar reaction involving two enone coupling partners:

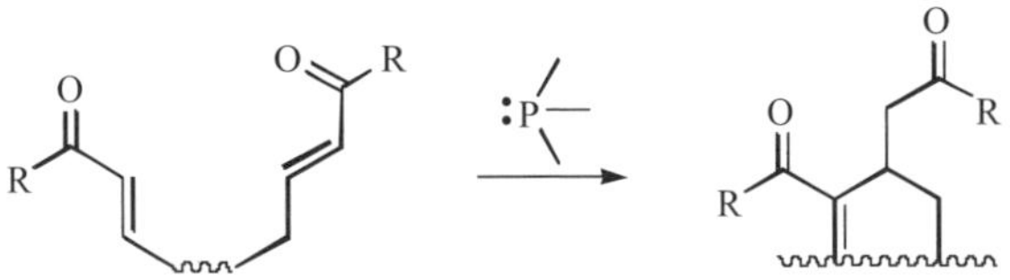

## Proposed Mechanism:

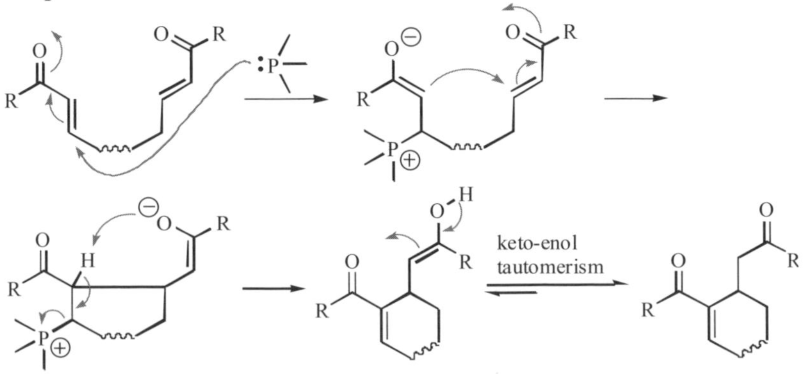

## Examples:

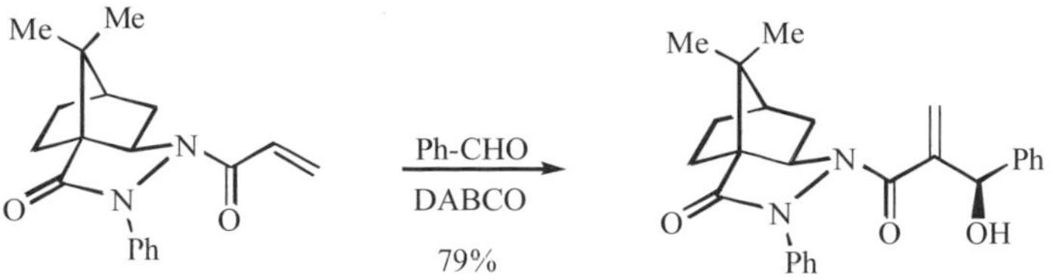

79%

K.-S. Yang, K. Chen, *Organic Letters* **2000**, 2, 729

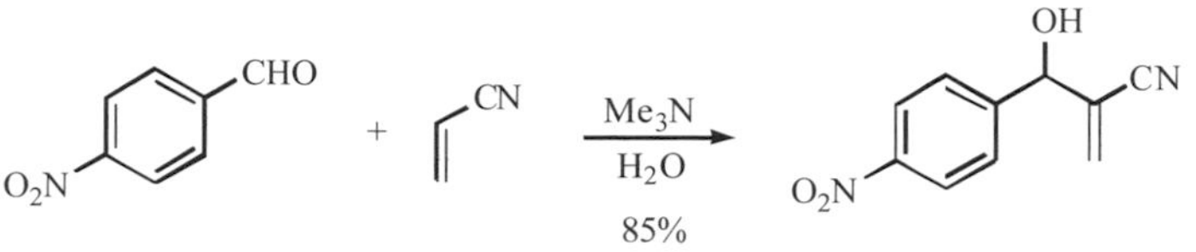

85%

K,-S. Yang, K. Chen, *Organic Letters* **2000**, 2, 729

40%

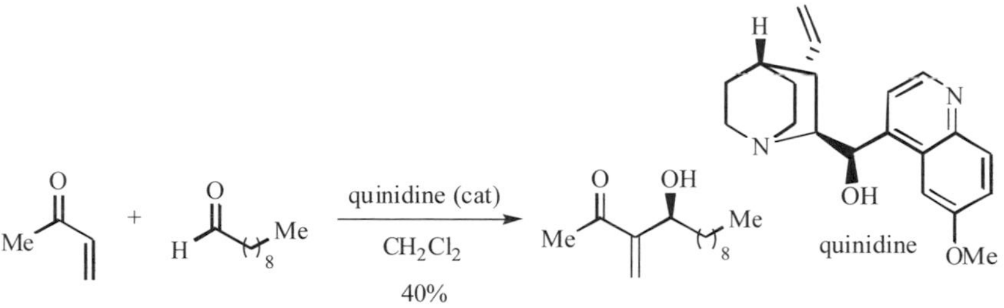

I. E. Mariko, P. R. Giles, N. J. Hindley, *Tetrahedron* **1997**, 53, 1015

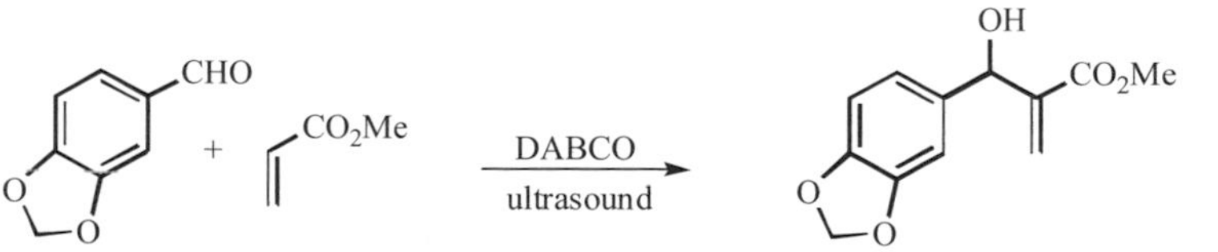

F. Coelho, W. P. Almeida, D. Veronese, C. R. Mateus, E. C. S. Lopes, R. C. Rossi, G. P. C. Silveira, C. H. Pavam, *Tetrahedron* **2002**, 58, 7437

### Examples for the *Rauhut-Currier Reaction*:

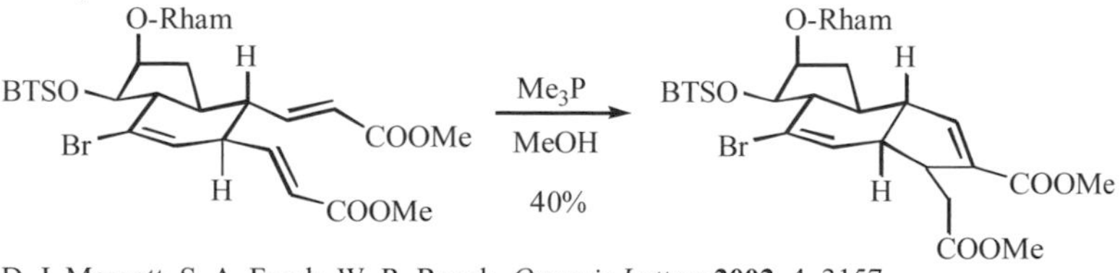

40%

D. J. Mergott, S. A. Frank, W. R. Roush, *Organic Letters* **2002**, 4, 3157

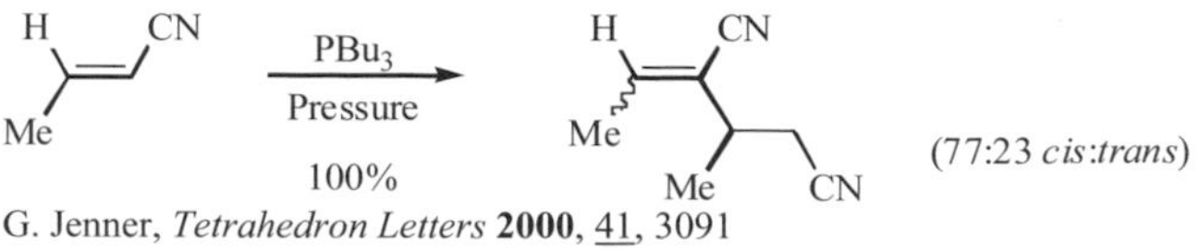

100%

(77:23 *cis:trans*)

G. Jenner, *Tetrahedron Letters* **2000**, 41, 3091

# Béchamp Reduction

## The Reaction:

$$O_2N\text{-}C_6H_5 + 2Fe + 6HCl \longrightarrow H_2N\text{-}C_6H_5 + 2H_2O + 2FeCl_3$$

## Proposed Mechanism:

$$Ar\text{-}NO_2 + 6e^{\ominus} + 6 H^{\oplus} \longrightarrow Ar\text{-}NH_2 + 2 H_2O$$

| 2 Fe | 2 Fe+3 |
|------|--------|
| 6 H+ | 2 H$_2$O |

The reaction is a metal-catalyzed oxidation-reduction process.

## Notes:

C. S. Hamilton, J. F. Morgan, *Organic Reactions* **2**, 10

### Other new approaches to the reduction:

1,1'-Dialkyl-4,4-bipyridinium halides (viologens) are useful electron-transfer catalysts.

R: *para* -CH$_3$, 83%
R: *para* -CH$_2$CH$_2$N$_3$, 92%

C. Yu, B. Liu, L. Hu, *Journal of Organic Chemistry* **2001**, 6̲6̲, 919

92%

R. W. Fitch, F. A. Luzzio, *Tetrahedron Letters* **1994**, 3̲5̲, 6013

$$R-NO_2 \xrightarrow[\substack{H_2O,\ rt,\ \text{)))} \\ \text{high yields}}]{\text{In wire, acidic}} R-NH_2$$

Y. Se.Cho, B. K. Jun, S. Kim, J. H. Cha, A. N. Pae, H. Y. K., M. C. Ho, S.-Y. Han
*Bulletin of the Korean Chemical Society* **2003**, 2̲4̲, 653 (AN 2003:513428)

## Examples:

S. Mukhopadhyay, G. K. Gandi, S. B. Chandalia, *Organic Process Research & Development* **1999**, 3, 201

L. Wang, P. Li, Z. Wu, J. Yan, M. Wang, Y. Ding, *Synthesis* **2003**, 2001

1. Me$_2$N-CH(OMe)$_2$

2. NaIO$_4$, THF, H$_2$O
3. Fe, AcOH, HCl (cat), EtOH, H$_2$O

73%

E. C Riesgo, X. Jin, R. P. Thummel, *Journal of Organic Chemistry* **1996**, 61, 3017

Bechamp

reduction

A. Courtin, *Helvetica Chemica Acta* **1980**, 63, 2280 ( AN 1981: 406876)

Fe, MeOH

HCl
89%

M. W. Zettler, *Encyclopedia of Reagents for Organic Synthesis*, John Wiley and Sons, Inc., L. A. Paquette, Ed., New York, 1995, **4**, 2871

# Beckmann Fragmentation

## The Reaction:

no α protons
or a stable
carbocation
eg (Ar₂C⁺H)

R, OH on the oxime, with R', R", R''' substituents → R—≡N + Cl on the carbocation fragment

an oxime

$PCl_5$

## Proposed Mechanism:

A better leaving group is made when
oxygen bonds to the Lewis Acid

R—≡N⁺—C(R")(R")(R') nitrilium ion → R—≡N + R'—C⁺(R")(R''') $\xrightarrow{PCl_5}$ R'—C(Cl)(R")(R''')

nitrilium ion

Because a stable carbocation can be formed, the nitrile is liberated before it can be trapped by water
as in the usual ***Beckmann Rearrangement***.

## Notes:

Starting material preparation:

R, R'—C=O $\xrightarrow{H_2N-OH}$ R, R'—C=N—OH

an oxime

The -OH group is generally anti to the larger R

Hydroxylamine-O-sulfonic acid
(HOSA)

R, R'—C=O $\xrightarrow{H_2N-O-SO_3H}$ R, R'—C=N—SO₃H

Provides an intermediate with a reactive leaving group already incorporated.

## Examples:

G. Rosini, M. Greir, E. Marcotta, M. Petrini, R. Ballini, *Tetrahedron* **1986**, <u>42</u>, 6027

H. Nishiyama, K. Sakuta, N. Osaka, H. Arai, M. Matsumoto, K. Itoh, *Tetrahedron* **1988**, <u>44</u>, 2413

M. G. Rosenberg, U. Haslinger, U. H. Brinker, *Journal of Organic Chemistry* **2002**, <u>67</u>, 450

J. D. White, J. Kim, N. E. Drapela, *Journal of the American Chemical Society* **2000**, <u>122</u>, 8665

# Beckmann Rearrangement

## The Reaction:

*Can be a ring enlargement.*

The reaction can also be carried out with $PCl_5$, PPA, $P_2O_5$ or TsCl.

## Proposed Mechanism:

nitrilium ion

## Notes:

Starting Material Preparation:

an oxime

The -OH group is generally anti to the larger R.

T. Laue, A. Plagens, *Named Organic Reactions*, John Wiley and Sons, Inc., New York, 1998, pp. 22-24; M. B. Smith, J. March in *March's Advanced Organic Chemistry*, 5th ed., John Wiley and Sons, Inc., New York, 2001, pp 1349, 1381, 1384, 1415-1416; L. G. Donaruma, W. Z. Heldt, *Organic Reactions* **11**, 1; R. E. Gawley, *Organic Reactions* **35**, 1

The rearrangement of amidoximes to derivatives of urea is called the ***Tiemann Rearrangement***.

## Examples:

72%

J. A. Robl, E. Dieber-McMaster, R. Sulsky, *Tetrahedron Letters* **1996**, <u>37</u>, 8985

30%                                          53%

M. Han, D. F. Covey, *Journal of Organic Chemistry* **1996**, <u>61</u>, 7614

95%

O. Muraoka, B.-Z. Zheng, K. Okumura, G. Tanabe, T. Momose, C. H. Eugster, *Journal of the Chemical Society, Perkin Transactions 1* **1996**, <u>13</u>, 1567

N. Komatsu, S. Simizu, T. Sugita, *Synthetic Communications* **1992**, <u>22</u>, 277 (AN 1992:151543)

79%

P. W. Jeffs, G. Molina, N. A. Cortese, P. R. Hauck, J. Wolfram, *Journal of Organic Chemistry* **1982**, <u>47</u>, 3876

In this example, the intermediate carbocation is trapped by Me₃Al rather than water:

54%

Y. Matsmura, K. Maruoka, H. Yamamoto, *Tetrahedron Letters* **1982**, <u>23</u>, 1929

# Beirut Reaction

## The Reaction:

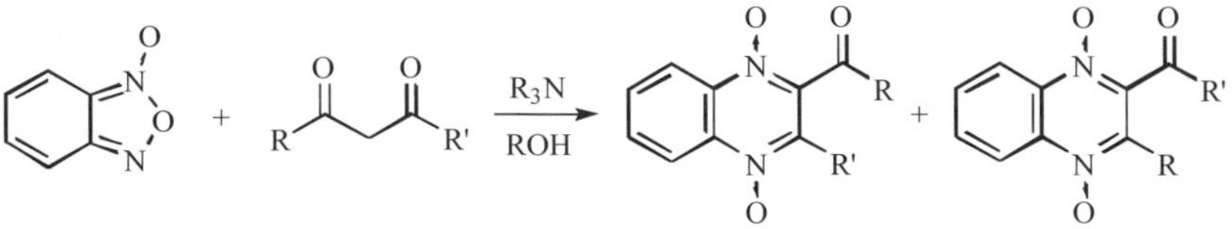

## Proposed Mechanism:

L. Turker, E. Dura, *Journal of Molecular Structure (Theochem)* **2002**, <u>593</u>, 143

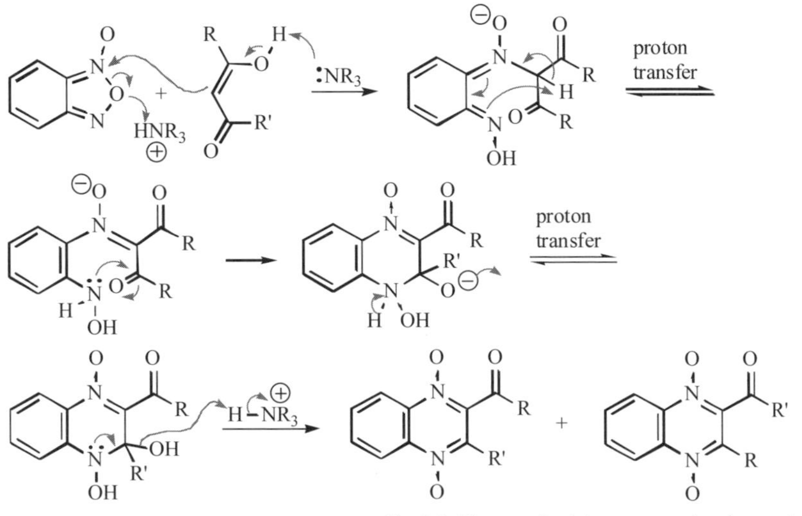

proton transfer

proton transfer

For 1,3-diketones both isomers can be observed.

## A second mechanistic interpretation:

C. H. Issidorides, M. J. Haddadin, *Journal of Organic Chemistry* **1996**, <u>31</u>, 4067

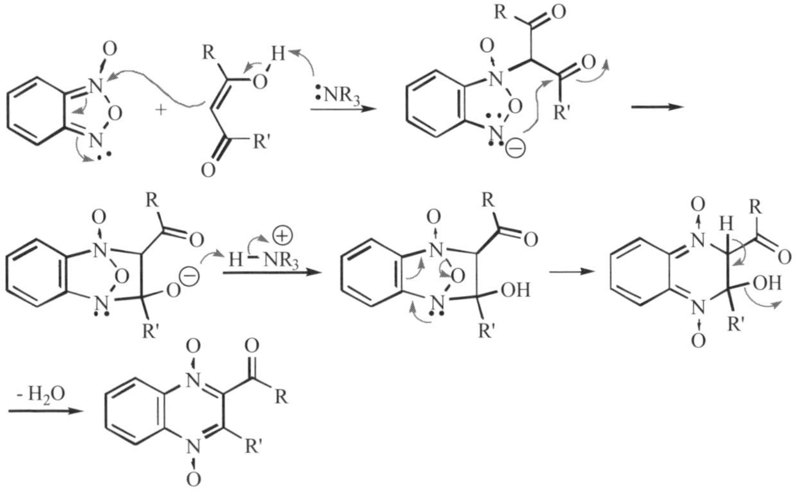

- H₂O

## Notes:

The reaction provides access to a number of quinoxaline-1,4-dioxide derivatives, by reaction of the benzofurazan oxide with 1,3-diketones, β-ketoesters, enals, enamines, phenols and α,β-unsaturated ketones.

## Examples:

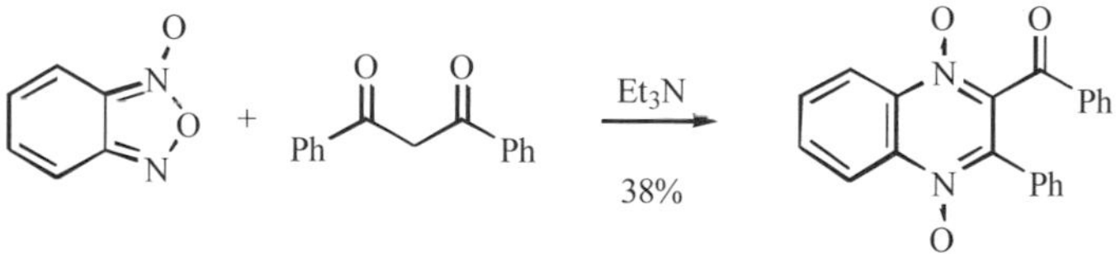

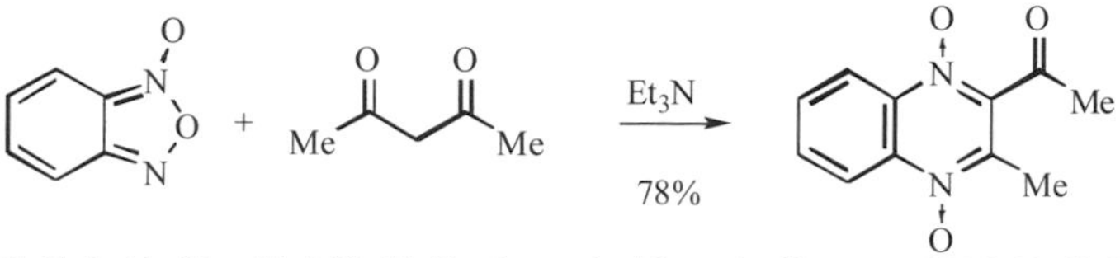

38%

78%

C. H. Issidorides, M. J. Haddadin, *Journal of Organic Chemistry* **1966**, <u>31</u>, 4067

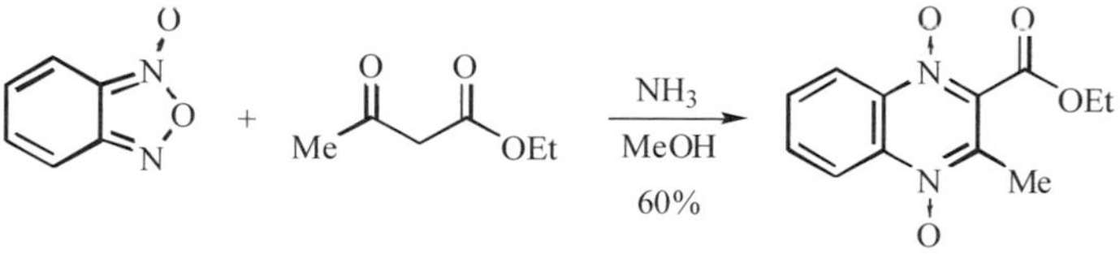

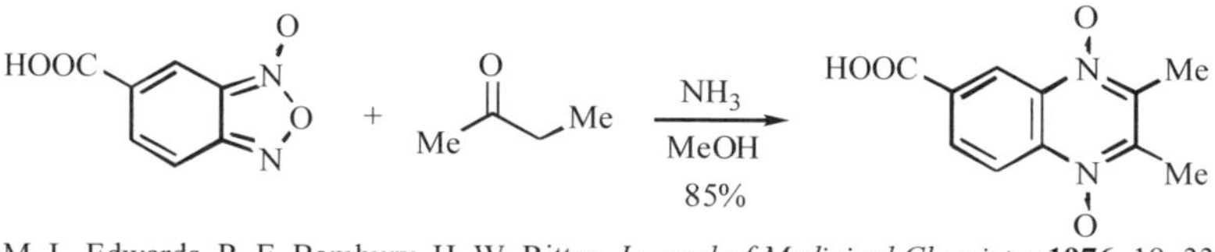

60%

M. L. Edwards, R. E. Bambury, H. W. Ritter, *Journal of Medicinal Chemistry* **1976**, <u>19</u>, 330

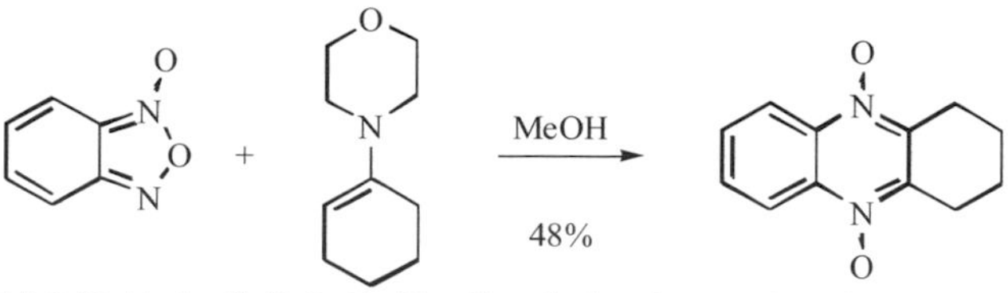

48%

M. J. Haddadin, C. H. Issidorides, *Tetrahedron Letters* **1965**, <u>6</u>, 3253

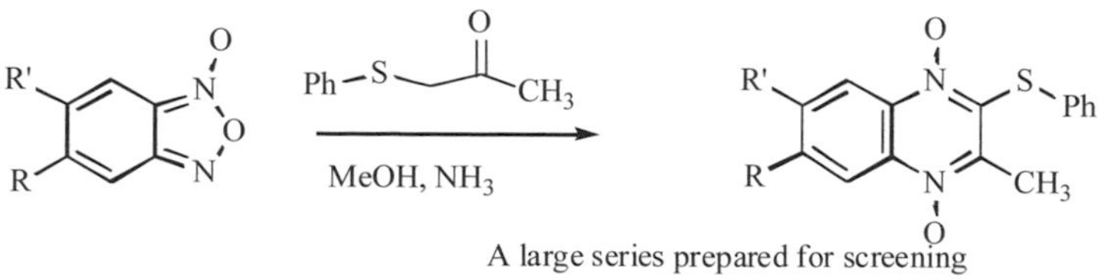

A large series prepared for screening

A. Carta, G. Paglietti, M. E. R. Nikookar, P. Sanna, L. Sechi, S. Zanetti, *European Journal of Medicinal Chemistry* **2002**, <u>37</u>. 355

# Benzidine Rearrangement (Zinin Benzidine Rearrangement)

**The Reaction:**

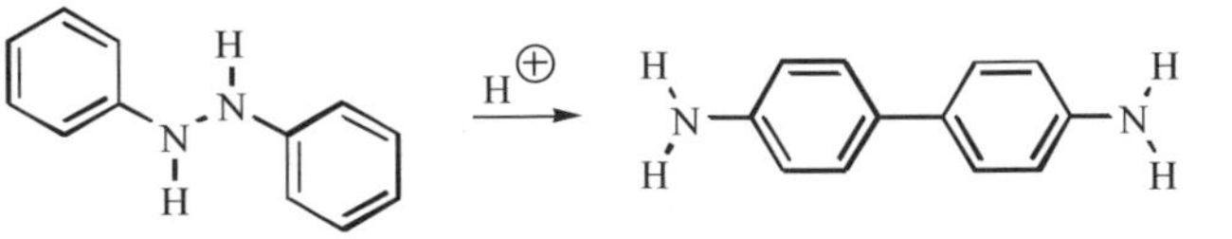

**Proposed Mechanism:**

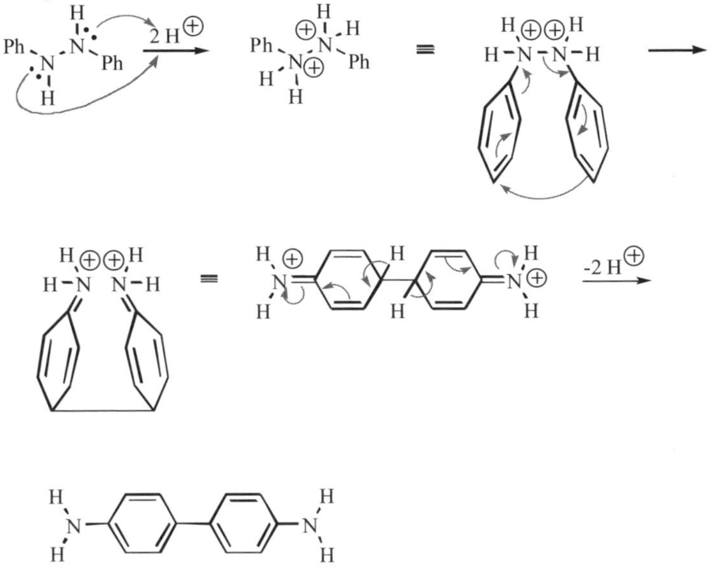

**Notes:**

M. B. Smith, J. March in *March's Advanced Organic Chemistry*, 5th ed., John Wiley and Sons, Inc., New York, 2001, pp. 1455-1456; T. Laue, A. Plagens, *Named Organic Reactions*, John Wiley and Sons, Inc., New York, 1998, pp. 24-26.

Byproducts sometimes include *semidines*:

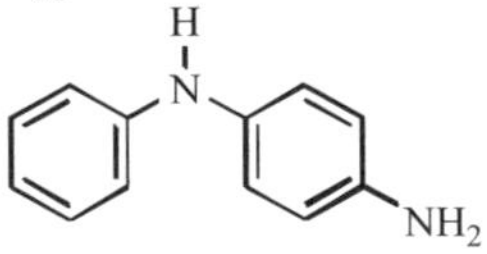

## Examples:

This work included kinetic details.

H. J. Shine, K. H. Park, M. L. Brownawell, J. S. Filippo, Jr., *Journal of the American Chemical Society* **1984**, <u>106</u>, 7077

A. Burawoy, C. E. Vellins *Journal of the Chemical Society* **1954**, 90

H. R. Snyder, C. Weaver, C. D. Marshall, *Journal of the American Chemical Society* **1949**, <u>71</u>, 289

M. Nojima, T. Ando, N. Tokura, *Journal of the Chemical Society, Perkin Transaction 1* **1976**, <u>14</u>, 1504

The following is an example of the **Sheradsky Rearrangement**. It is an oxygenated analog of the **Benzidine Rearrangement**.

T. Sheradsky, S. Auramovki-Grisaru, *Journal of Heterocyclic Chemistry* **1980**, <u>17</u>, 189 (AN 1980:407167)

# Benzilic Acid Rearrangement

## The Reaction:

Typically the R groups are aromatic, but if not, must be devoid of α-hydrogens.

## Proposed Mechanism:

no α hydrogens

## Notes:

M. B. Smith, J. March in *March's Advanced Organic Chemistry*, 5<sup>th</sup> ed., John Wiley and Sons, Inc., New York, 2001, p. 1403; T. Laue, A. Plagens, *Named Organic Reactions*, John Wiley and Sons, Inc., New York, 1998, pp. 26-27.

This reaction can be a ring contraction:

Reported in: T. Laue, A. Plagens, *Named Organic Reactions,* John Wiley and Sons, Inc., New York, 1998, p 27 (A. Schaltegger, P. Bigler, *Helvetica Chemica Acta*, **1986**, <u>69</u>, 1666)

## Examples:

1. NaOH,
2. H ⊕

82%

S. Deb, R. Chakraborti, U. R. Ghatak, *Synthetic Communications* **1993**, <u>23</u>, 913 (AN 1993:494997)

J. M. Robinson, E. T. Flynn, T. L. McMahan, S. L. Simpson, J. C. Trisler, K. B. Conn, *Journal of Organic Chemistry* **1991**, 56, 6709

V. Georgian, N. Kundu, *Tetrahedron* **1963**, 19, 1037

E. Campaigne, R. C. Bourgeois, *Journal of the American Chemical Society* **1953**, 75, 2702

# Benzoin Condensation

## The Reaction:

No protons on the α carbon. The group is typically aromatic.

## Proposed Mechanism:

## Notes:

M. B. Smith, J. March in *March's Advanced Organic Chemistry*, 5th ed., John Wiley and Sons, Inc., New York, 2001, p 1240, 1243; T. Laue, A. Plagens, *Named Organic Reactions*, John Wiley and Sons, Inc., New York, 1998, pp. 27-29; W. S. Ide, J. S. Buck, *Organic Reactions* **4**, 5.

Use of thiazolium ion catalysis allows the ***benzoin condensation*** of aldehydes with α-protons.

71-74%

H. Stetter, H. Kuhlmann, *Organic Syntheses*, <u>CV 7</u>, 95

In these reactions a thiazolium salt forms an ion that participates much like cyanide:

## Stetter Reaction (Stetter 1,4-Dicarbonyl Synthesis)

H. Stetter, H. Kuhlmann, W. Haese, *Organic Syntheses*, <u>CV 8</u>, 620

## Examples:

G. Sumrell, J. I. Stevens, G. Goheen, *Journal of Organic Chemistry* **1957**, 22, 39

A. S. Demir, P. Ayhan, A. C. Igdir, A. N. Duygu, *Tetrahedron* **2004** 60 6509

Y. Tachibana, N. Kihara, T. Takata, *Journal of the American Chemical Society* **2004**, 126, 3438

ThDP = Thiamine diphosphate =

BFD = Benzoylformate decarboxylase

P. Dünkelmann, D. Kolter-Jung, A. Nitsche, A. S. Demir, P. Siegert, B. Lingen, M. Baumann, M. Pohl, M. Müller, *Journal of the American Chemical Society* **2002**, 124, 12084

# Bergman Cyclization

## The Reaction:

R
R'
Δ
2 H•
R
R'

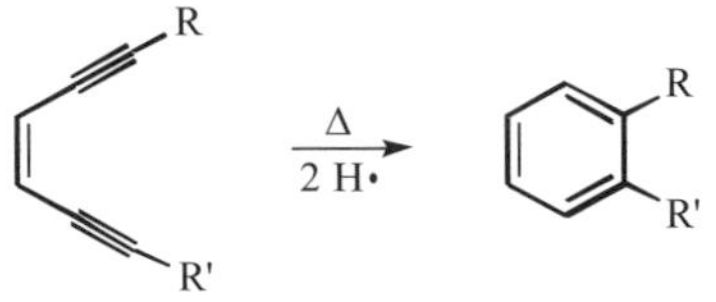

The source is frequently 1,4-dihydrobenzene

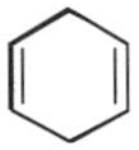

## Proposed Mechanism:

Δ
2 H•

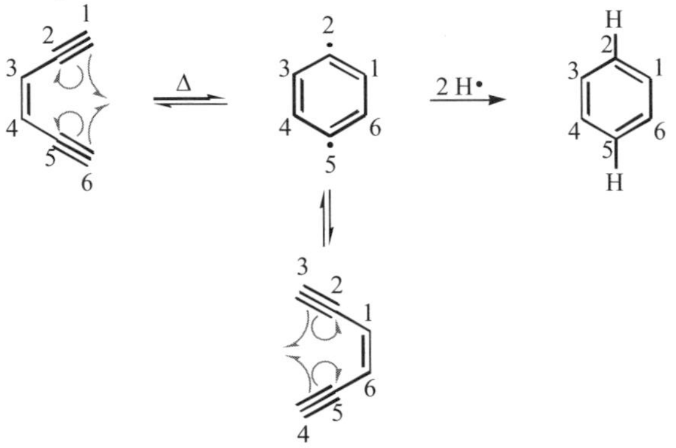

## Notes:

M. B. Smith, J. March in *March's Advanced Organic Chemistry*, 5th ed., John Wiley and Sons, Inc., New York, 2001, p. 1432; T. Laue, A. Plagens, *Named Organic Reactions*, John Wiley and Sons, Inc., New York, 1998, pp. 29-33.

The ***Myers-Saito Cyclization*** is a similar reaction with a different substrate:

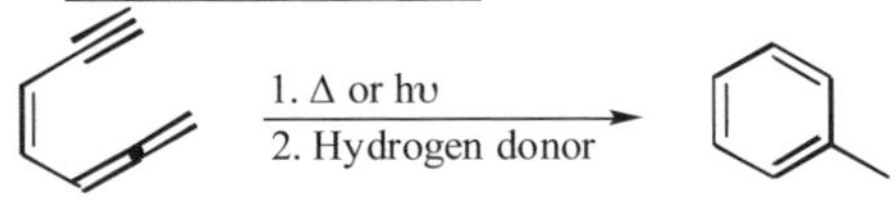

1. Δ or hυ
2. Hydrogen donor

allenyl enyne

## Examples:

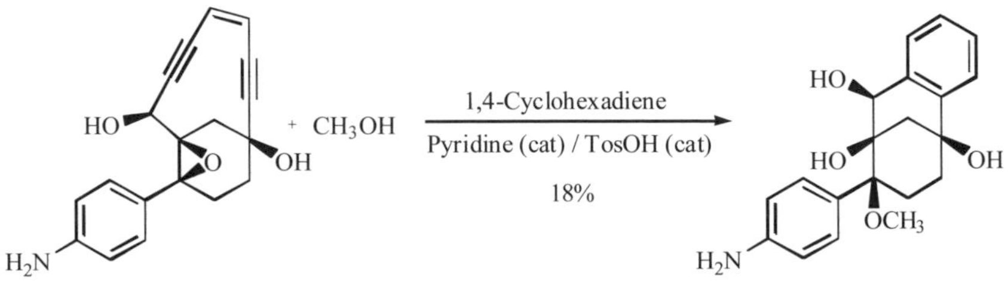

HO
O
OH
H2N
+ CH3OH
1,4-Cyclohexadiene
Pyridine (cat) / TosOH (cat)
18%
HO
HO
OH
OCH3
H2N

H. Mastalerz, T. W. Doyle, J. F. Kadow, D. M. Vyas, *Tetrahedron Letters* **1996**, 37, 8683

K. Iida, M. Kirama, *Journal of the American Chemical Society* **1995**, 117, 8875

D. P. Magnus, R. T. Lewis, J. C. Huffman, *Journal of the American Chemical Society* **1988**, 110, 6921

M. F. Semmelhack, T. Neu, F. Foubelo, *Tetrahedron Letters* **1992**, 33, 3277

T. Brandstetter, M. E. Maier, *Tetrahedron* **1994**, 50, 1435

M. M. McPhee, S. M. Kerwin, *Journal of Organic Chemistry* **1996**, 61, 9385

# Biginelli Reaction (Biginelli Pyrimidone Synthesis)

## The Reaction:

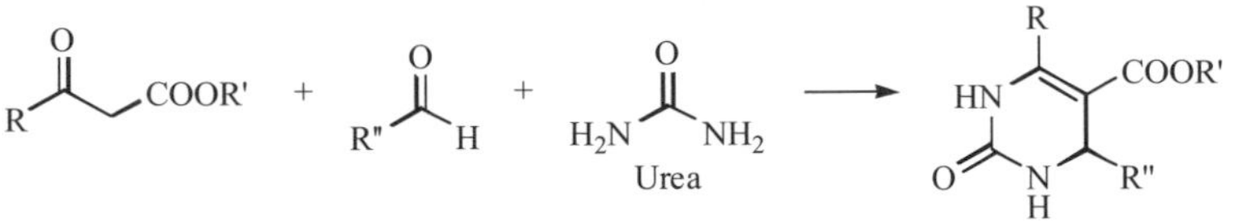

## Proposed Mechanism:
G. Jenner, *Tetrahedron Letters* **2004**, <u>45</u>, 6195
C. O. Kappe, *Journal of Organic Chemistry* **1997**, <u>62</u>, 7201

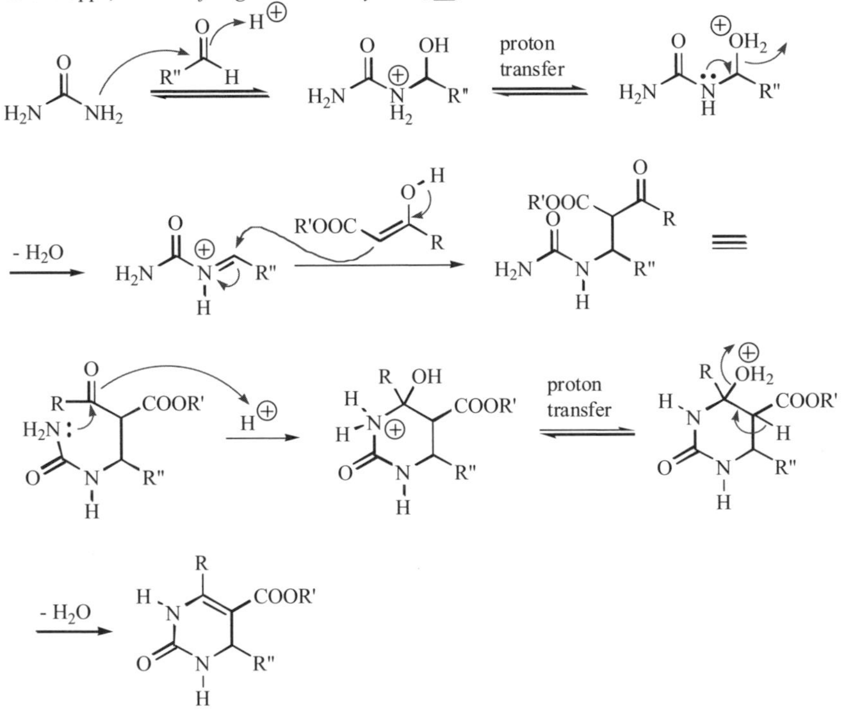

## Notes:
C. O. Kappe, A. Stadler, *Organic Reactions* **63**, 1

## Examples:

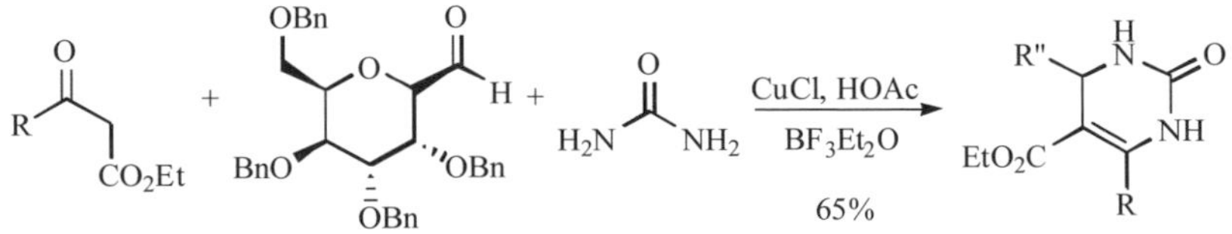

A. Dondoni, A. Mass, S. Sabbatini, V. Bertolasi, *Journal of Organic Chemistry* **2002**, <u>67</u>, 6979

$Cl_3C$-CO-CH$_2$-COOEt + $H_2N$-CO-$NH_2$ + Ph-CHO $\xrightarrow{INBr_3, EtOH}$ product

70%

M. A. P. Martins, M. V. M. Teixeira, W. Cunico, E. Scapin,a, R. Mayer, C. M. P. Pereira, N. Zanatta, H.G. Bonacorso, C.Peppeb Y.-F. Yuan, *Tetrahedron Letters* **2004** , 45, 8991

Me-CHO + Ph-CO-CH$_2$-CO-OEt + $H_2N$-CO-$NH_2$ $\xrightarrow{EtOH, HCl}$

78%

H. E. Zaugg, W. B. Martin, *Organic Reactions* **1965**, 14, 130

An α-ketoacid approach

M. M. Abelman, S. C. Smith, D. R. James, *Tetrahedron Letters* **2003**, 44, 4559

$H_3C$-CO-CH$_2$-COOEt + $H_2N$-CO-$NH_2$ + ArCHO $\xrightarrow{I_2, toluene}$

92%

R. S. Bhosale, S. V. Bhosale, S. V. Bhosale, T. Wang, P. K. Zubaidha, *Tetrahedron Letters* **2004**, 45, 9111

HOOC-CO-CH$_2$-CO-OH + Ph-CHO + $H_2N$-CO-$NH_2$ $\longrightarrow$

J. C. Bussolari, P. A. McDonnel *Journal of Organic Chemistry* **2000**, 65, 6777

# Birch Reduction

## The Reaction:

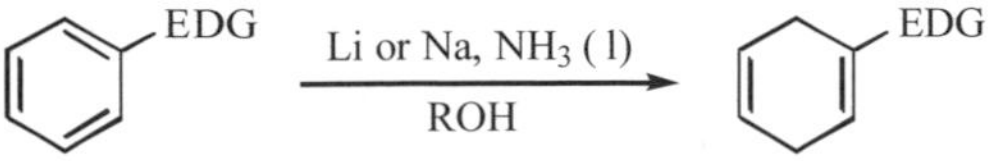

EDG can be
-R, -OR

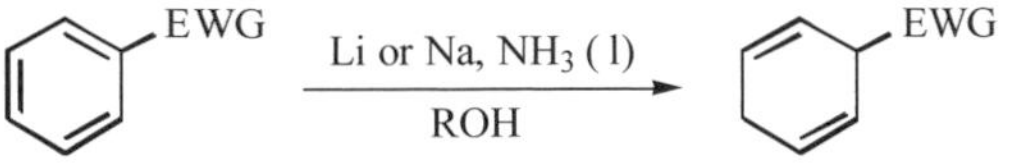

EWG can be
-COOH

## Proposed Mechanism:

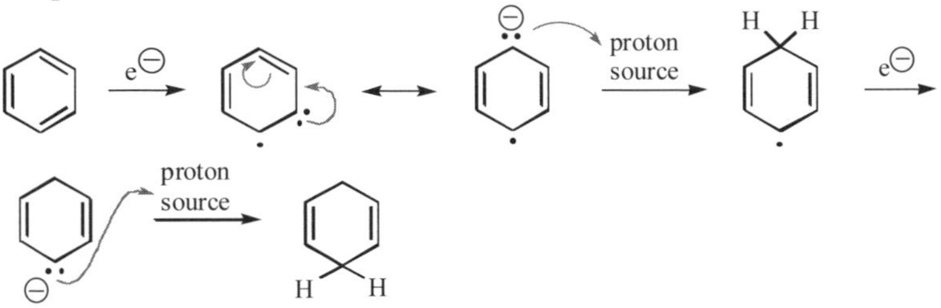

## Notes:

M. B. Smith, J. March in *March's Advanced Organic Chemistry*, 5th ed., John Wiley and Sons, Inc., New York, 2001, p. 1010; T. Laue, A. Plagens, *Named Organic Reactions*, John Wiley and Sons, Inc., New York, 1998, pp. 33-35; D. Caine, *Organic Reactions* **23**,1; P. W. Rabideau, Z. Marcinow, *Organic Reactions* **42**,1.

Na / $NH_3$ is more prone to Fe-catalyzed conversion to $NaNH_2$. The ***Wilds Modification*** (A. L. Wilds, N. A. Nelson, *Journal of the American Chemical Society* **1953**, 75, 5360) uses Li, and is less likely to be converted to amide ion. It is often helpful to distill the liquid ammonia before use.

## Other susceptible functional groups:

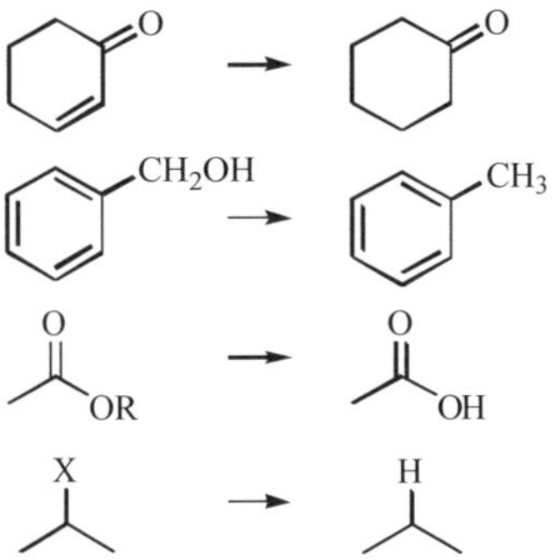

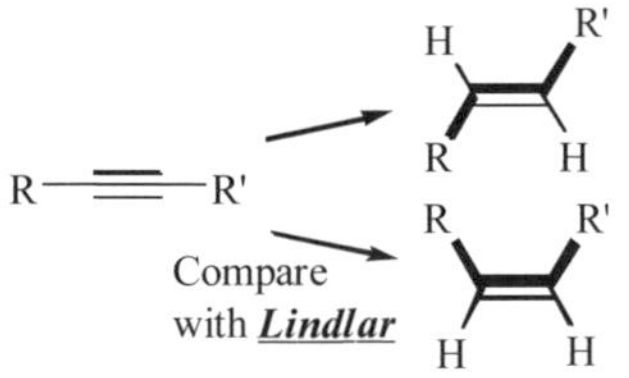

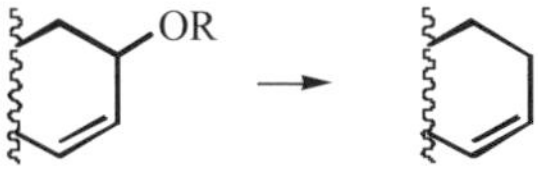

R = OAc, H, $POCl_2$

Compare with ***Lindlar***

***Benkeser Reduction***

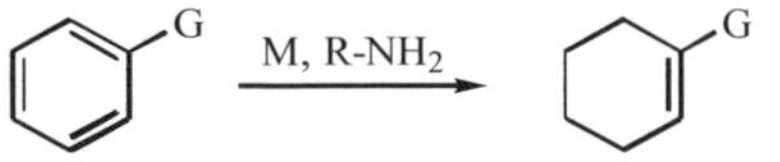

$R\text{-}NH_2$ = various amines    eg. $Et_2NH$ and $Me_2NH$
M = Na. Li, Ca

**Henbest Reductino**

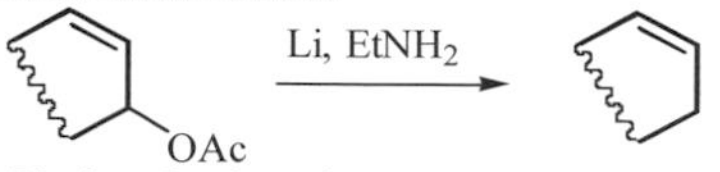

**Birch reduction** of enones provides for a number of useful synthetic applications:

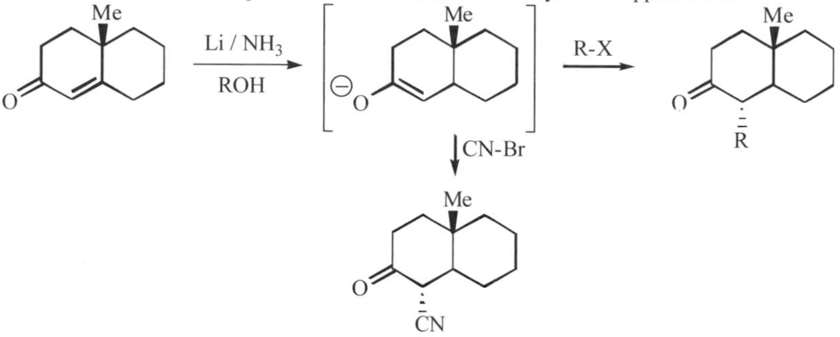

**Examples:**

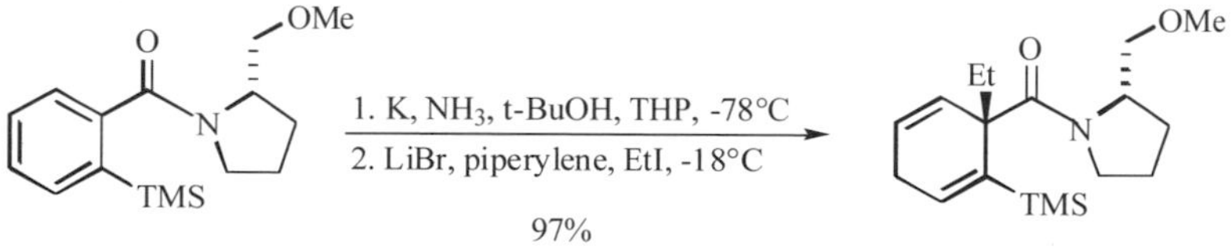

97%

A. G. Schultz, L. Pettus, *Journal of Organic Chemistry* **1997**, <u>62</u>, 6855

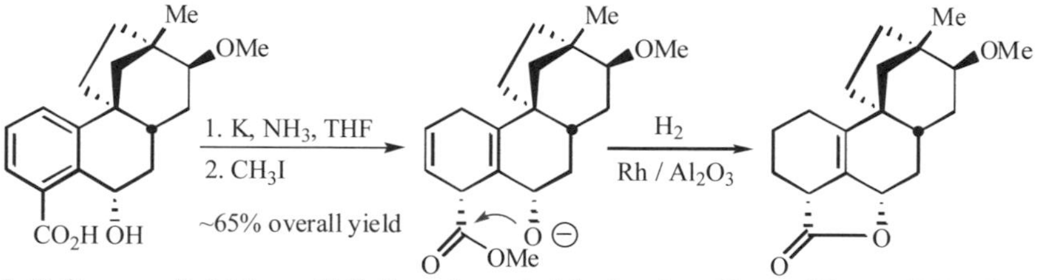

~65% overall yield

L. E. Overman, D. J. Riccan, V. D. Tran, *Journal of the American Chemical Society* **1997**, <u>119</u>, 12031

With a weaker organic acid used in workup, the enol ether can be selectively converted to the ketone without conjugation of the resulting enone.

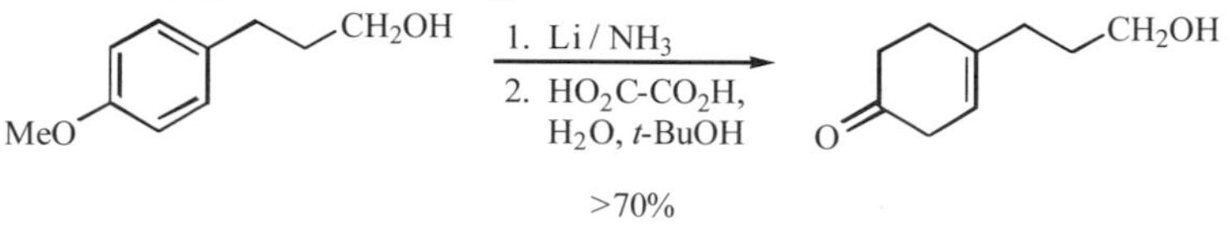

>70%

E. J. Corey, N. W. Boaz, *Tetrahedron Letters* **1985**, <u>26</u>, 6015

The intermediate anion can be captured by electrophiles other than protons.

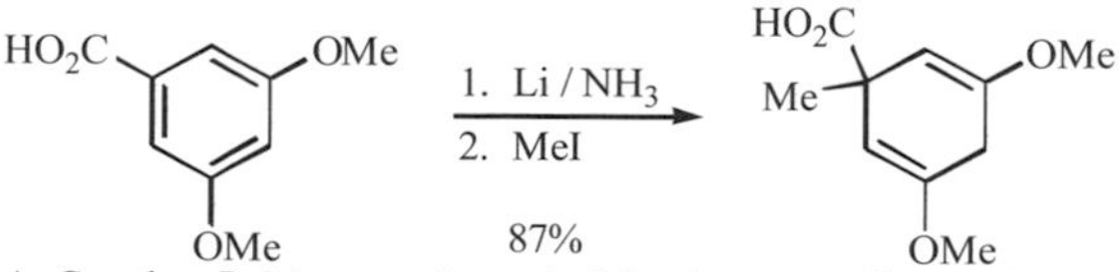

87%

A. Gopalan, P. Mangus, *Journal of the American Chemical Society* **1980**, <u>102</u>, 1756

# Bischler-Napieralski Reaction

## The Reaction:

Other catalysts are possible. (e.g. $ZnCl_2$ and $PO_5$)

## Proposed Mechanism:

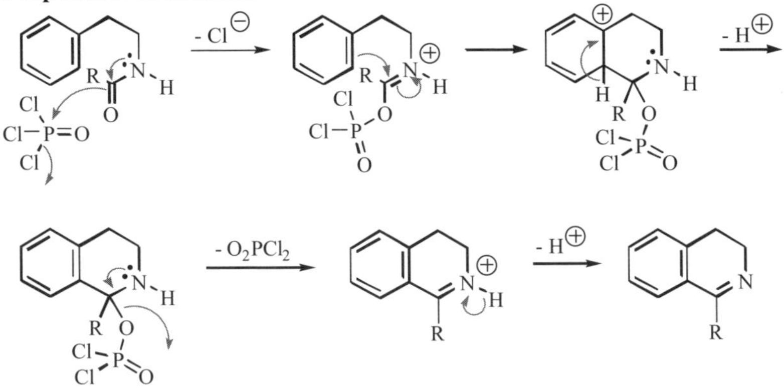

## Notes:

M. B. Smith, J. March in *March's Advanced Organic Chemistry*, 5th ed., John Wiley and Sons, Inc., New York, 2001, p. 721; W. M. Whaley, T. R. Govindachari, *Organic Reactions* **6**, 12

## Starting Material Preparation:

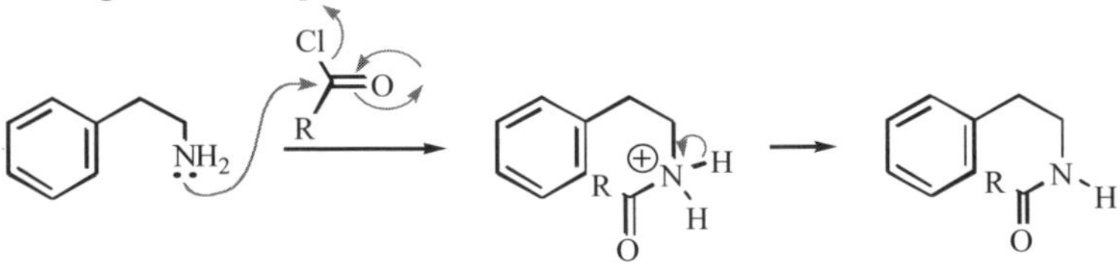

Anhydrides can also be used.

## Examples:

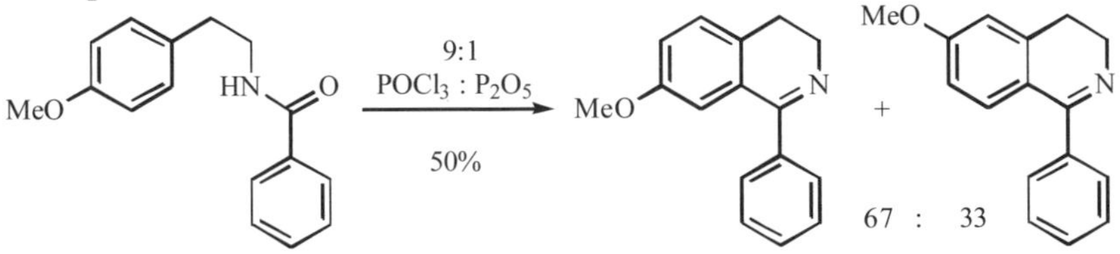

67 : 33

S. Doi, N. Shirai, Y. Sato, *Journal of the Chemical Society, Perkin Transactions 1* **1997**, <u>15</u>, 2217

E. E. Van Tamelen, C. Placeway, G. P. Schiemenz, I. G. Wright, *Journal of the American Chemical Society* **1969**, <u>91</u>, 7359

C. V. Denyer, J. Bunyan, D. M. Loakes, J. Tucker, J. Gillam, *Tetrahedron* **1995**, <u>51</u>, 5057

S. Jeganathan, M. Srinivasan, *Synthesis* **1980**, 1021

A. Brossi, L. A. Dolan, S. Teitel, *Organic Synthesis* **1977**, <u>56</u>, 3

C. S. Hilger, B. Fugmann, W. Steglich, *Tetrahedron Letters* **1985**, <u>26</u>, 5975

# Blaise Reaction

## The Reaction:

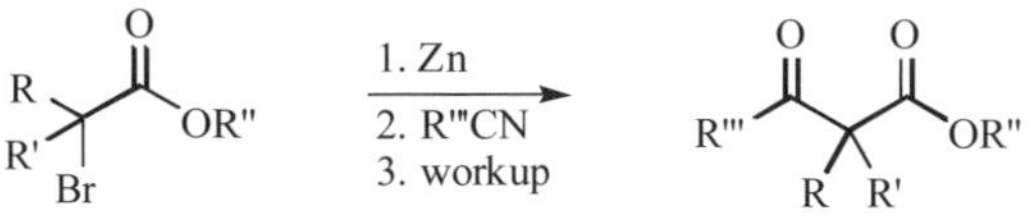

## Proposed Mechanism:

The starting reagent is the ***Reformatsky***-type. In this case, instead of adding to a carbonyl group, addition is to the nitrile. See ***Reformatsky Reaction*** for a discussion of reagent.

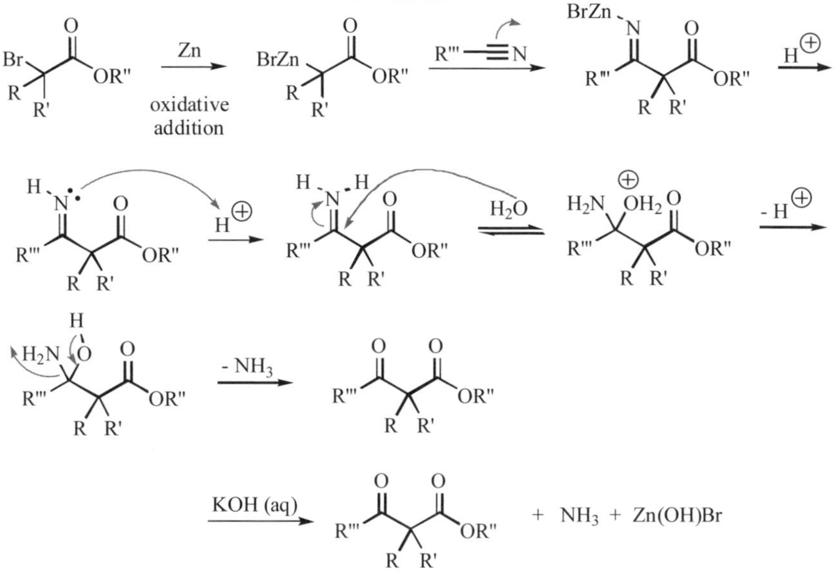

## Notes:

M. B. Smith, J. March in *March's Advanced Organic Chemistry*, 5th ed., John Wiley and Sons, Inc., New York, 2001, p. 1213.

See ***Reformatsky Reaction*** for further comments on the organo-zinc reagent.

It is possible to arrest the workup to provide the enamine product:

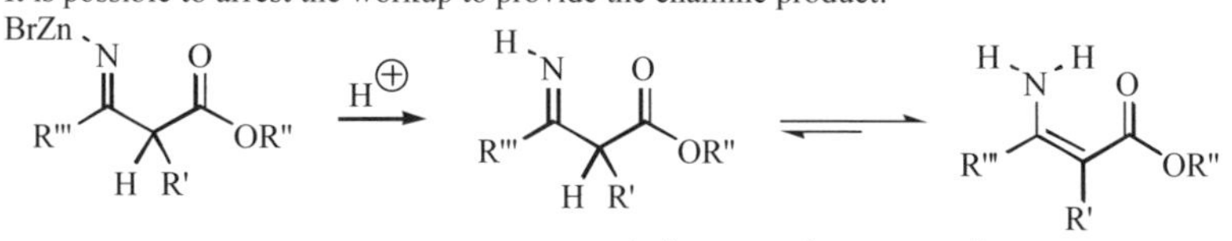

imine - enamine tautomerism

## Examples:

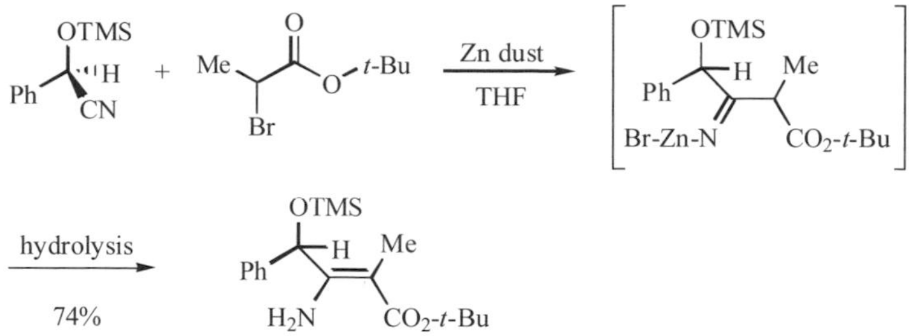

J. Syed, S. Forster, F. Effenberger, *Tetrahedron: Asymmetry* **1998**, <u>9</u>, 805

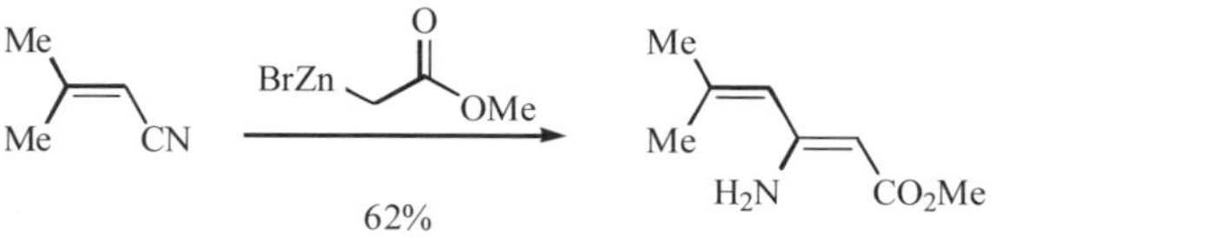

M. Mauduit, C. Kouklovsky, Y. Langlois, C. Riche, *Organic Letters* **2000**, <u>2</u>, 1053

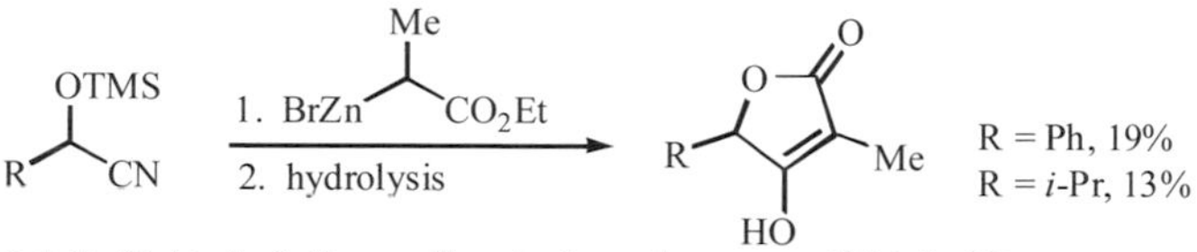

R = Ph, 19%
R = *i*-Pr, 13%

J. J. Duffield, A. C. Regan, *Tetrahedron: Asymmetry* **1996**, <u>7</u>, 663

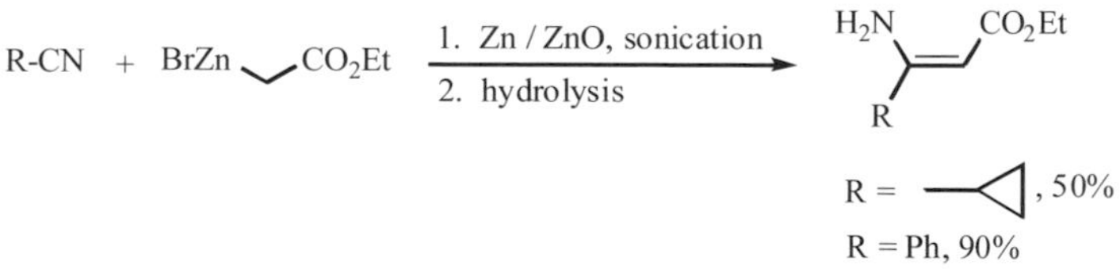

R = ◁ , 50%
R = Ph, 90%

A. S.-Y. Lee, R.-Y. Cheng, O.-G. Pan, *Tetrahedron Letters* **1997**, <u>38</u>, 443

# Blanc Chloromethylation Reaction

## The Reaction:

Ar−H + [formaldehyde, O with H, H] + HCl $\xrightarrow{ZnCl_2}$ [product: Ar attached to CH with Cl, H, H]

## Proposed Mechanism:

The HCl and $ZnCl_2$ form the reactive reagent with formaldehyde:

[structure: H, H C=O + HCl + $ZnCl_2$] $\rightleftharpoons$ [H, H C=$\overset{+}{O}$H] + $\ominus ZnCl_3$

[mechanism scheme with benzene reacting with protonated formaldehyde, forming benzyl intermediate, −H$^{\oplus}$, then further steps]

[second row of mechanism scheme: benzyl oxocarbenium + Cl$^{\ominus}$ → benzyl chloride with Cl]

## Notes:

See: **_Formaldehyde_**. The use of aqueous formaldehyde sometimes gives better yield than using paraformaldehyde.

See the similar **_Quelet Reaction_**:

[reaction scheme: R'—O substituted benzene + R—CHO, HCl/ZnCl₂ → para product with R–CH–Cl group + H₂O]

If the *p*-position is filled, the substitution will go to an open *ortho* position.

## Examples:

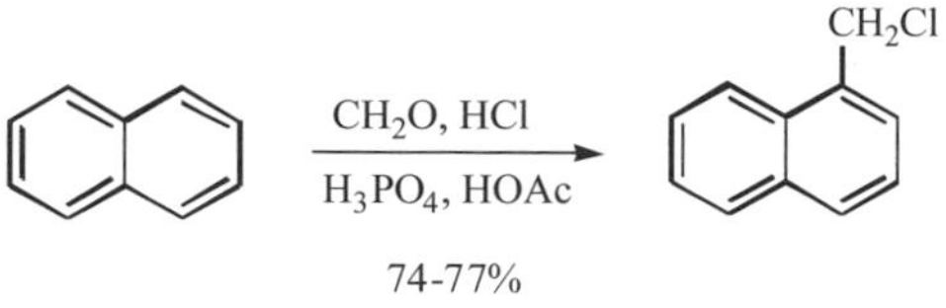

CH$_2$O, HCl
H$_3$PO$_4$, HOAc

74-77%

O. Grummitt, A. Buc, *Organic Syntheses* CV 3, 195

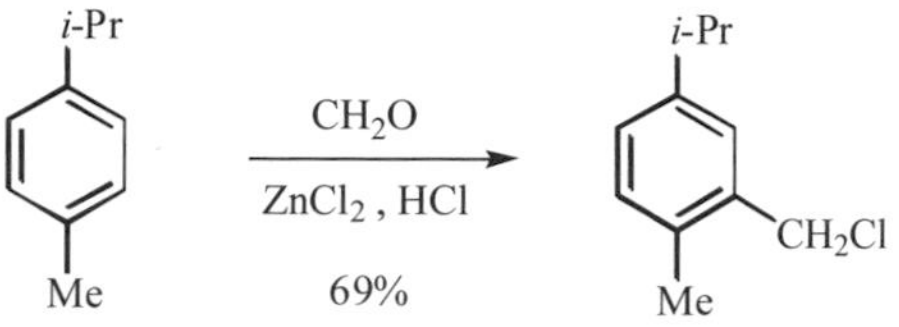

*i*-Pr

CH$_2$O
ZnCl$_2$ , HCl

69%

Me

W. G. Whittleston, *Journal of the American Chemical Society* **1937**, <u>59</u>, 825

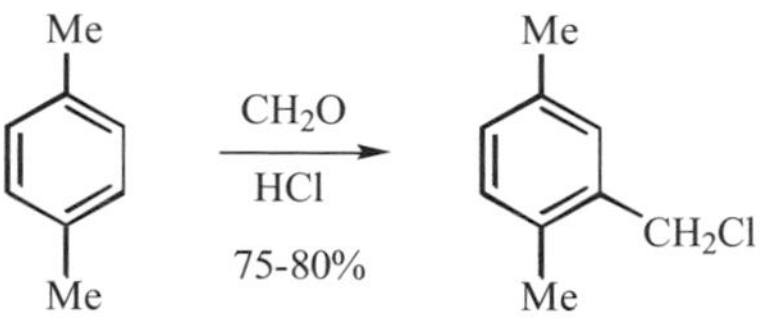

Me

CH$_2$O
HCl

75-80%

Me

J. V. Braun, J. Nelles, *Journal of the American Chemical Society* **1951**, <u>73</u>, 766

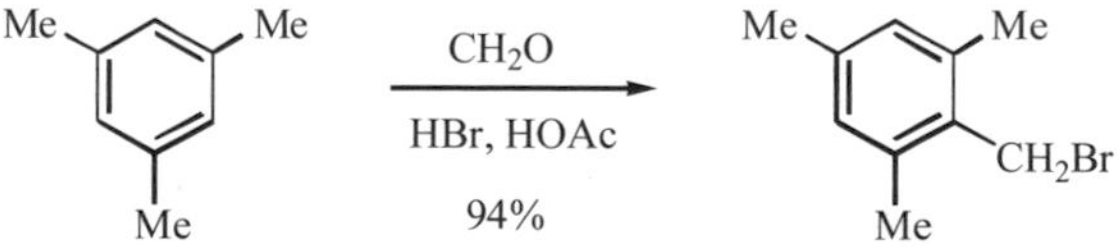

Me            Me

CH$_2$O
HBr, HOAc

94%

Me

A. W. Van der Made R. H Van Der Made, *Journal of Organic Chemistry* **1993**, <u>58</u>, 1262

# Blanc Cyclization / Reaction (Blanc Rule)

## The Reaction:

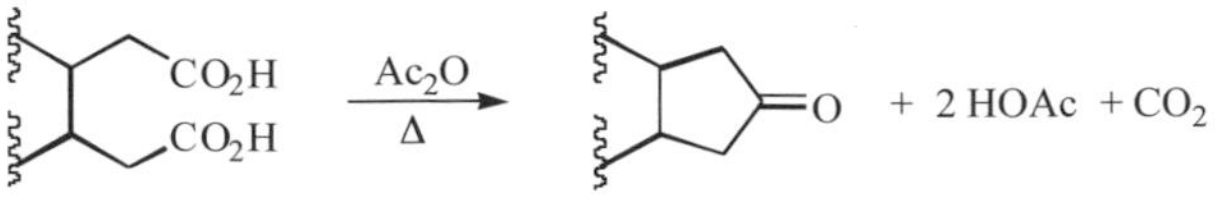

This reaction works for 1,6 dicarboxylic acids or larger.

## Proposed Mechanism:

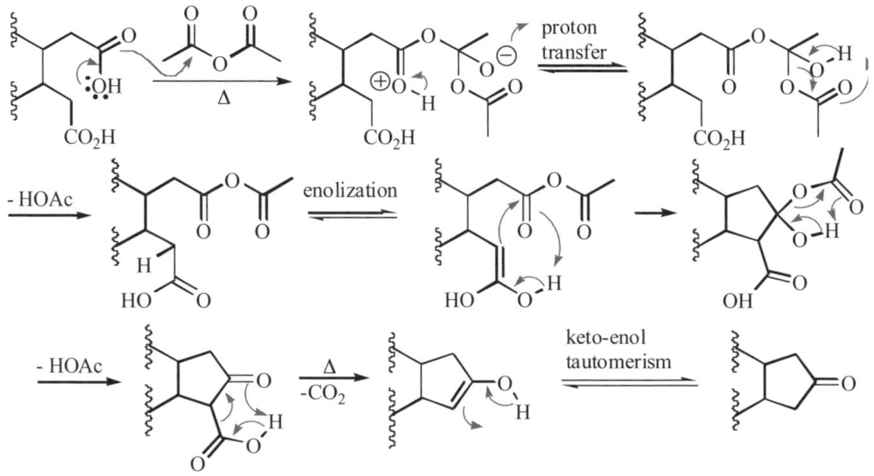

## Notes:

For 1,4 or 1,5 diacids the following is observed:

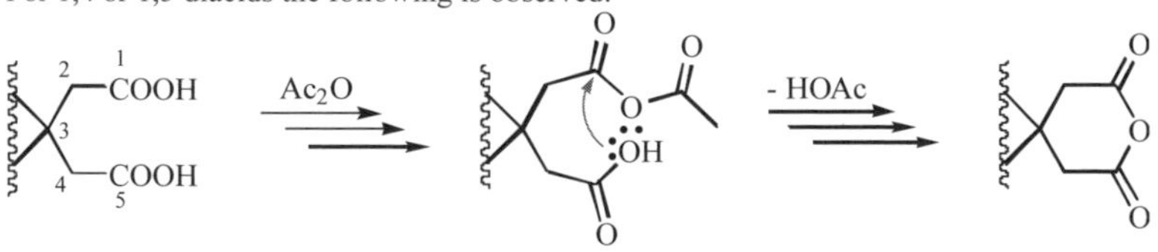

The **Ruzicka Cyclization (or Ruzicka Large Ring Synthesis)** is a similar reaction in which cyclic ketones are formed from salts (Ca or Ba for smaller rings and Th or Ce for larger rings) of diacids:

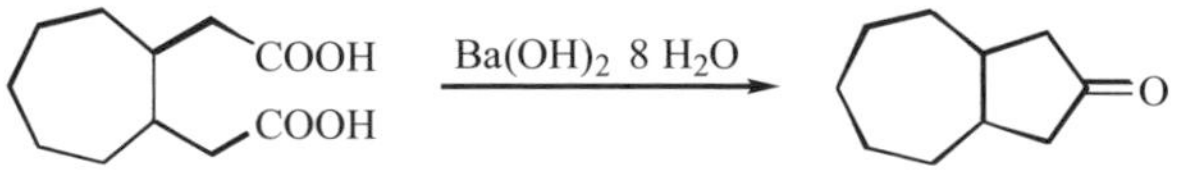

P. A. Plattner, A. Furst, K. Jirasek, *Helvetica Chimica Acta* **1944**, _29_, 730 (AN 1946:23967)

## Examples:

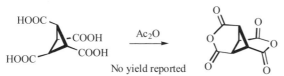

H.-J. Gais, G. Bülow, A. Zatorski, M. Jentsch, P. Maidonis, *Journal of Organic Chemistry* **1989**, <u>54</u>, 5115

L. Crombie, J. E. H. Hancock, R. P. Linstead, *Journal of the Chemical Society* **1953**, 3496

W. E. Backmann, N. C. Deno, *Journal of the American Chemical Society* **1949**, <u>71</u>, 3540

J. Dressel, K. L. Chasey, L. A. Paquette, *Journal of the American Chemical Society* **1988**, <u>110</u>, 5479

**Examples of the *Ruzicka Cyclization*:**

W. S. Johnson, D. K. Banerjee, W. P. Schneider, C. D. Gutsche, W. E. Shelberg, L. J. Chinn, *Journal of the American Chemical Society* **1952**, <u>74</u>, 2832

H. E. Baumgarten, D. C. Gleason, *Journal of Organic Chemistry* **1951**, <u>16</u>, 1658

# Boord Olefin Synthesis

## The Reaction:

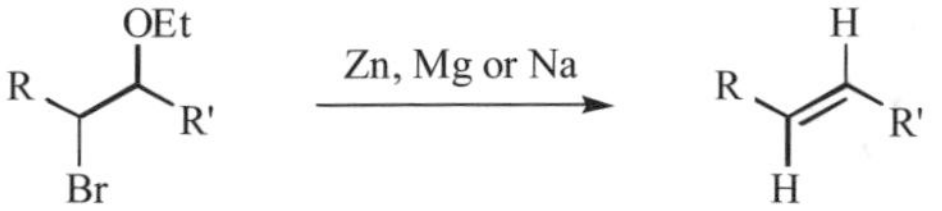

## Proposed Mechanism:

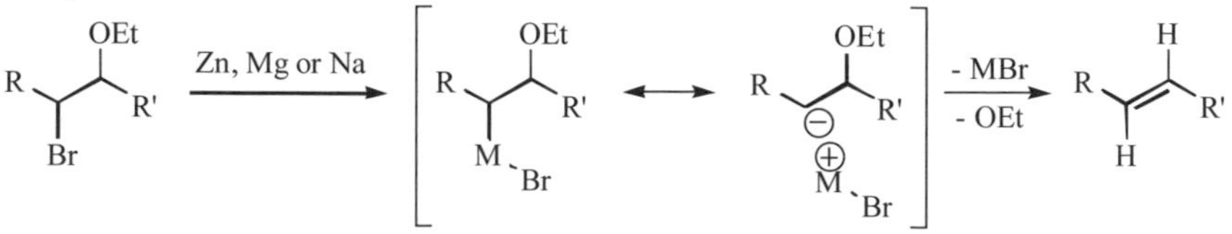

The E1-CB mechanism associated with this reaction results in similar product yields independent of stereochemistry.

## Notes:

M. B. Smith, J. March in *March's Advanced Organic Chemistry*, 5$^{th}$ ed., John Wiley and Sons, Inc., New York, 2001, p. 1344.

## Starting Material Preparation:

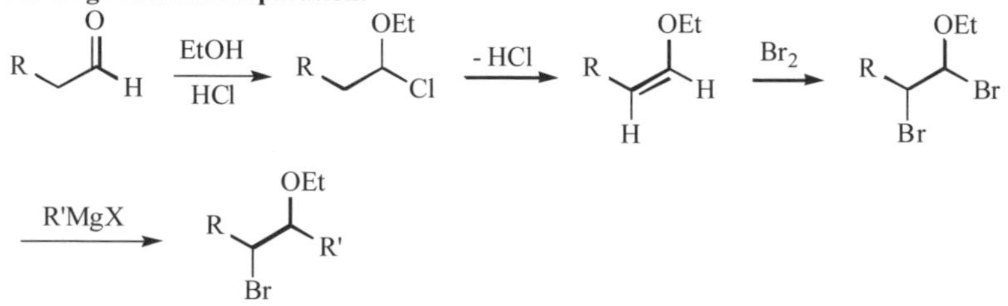

## Examples:

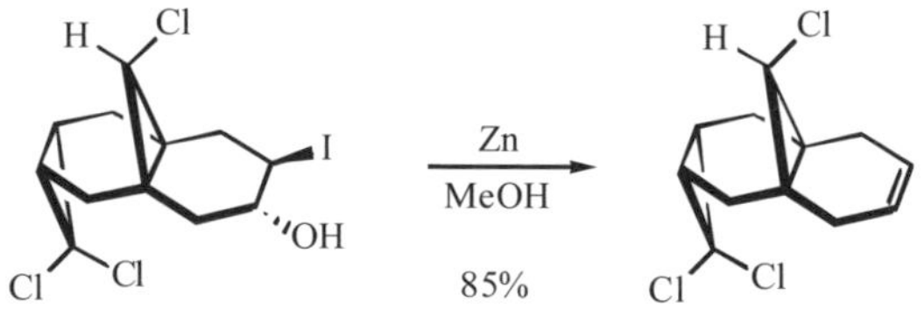

B. Halton, S. G. G. Russell, *Journal of Organic Chemistry* **1991**, <u>56</u>, 5553

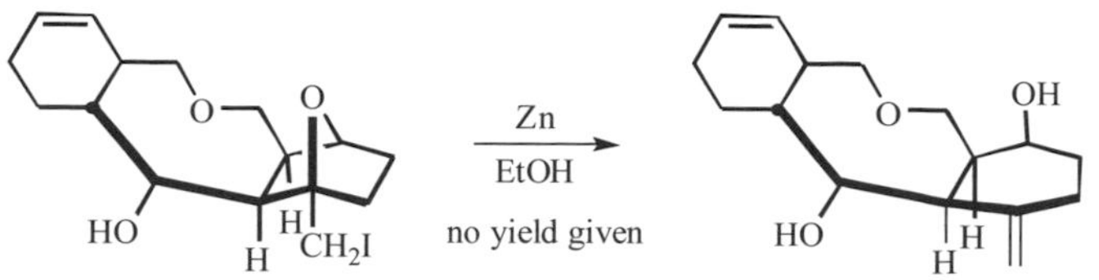

$$\xrightarrow[\text{EtOH}]{\text{Zn}}$$

no yield given

J. S. Yadav, R. Renduchintala, L. Samala, *Tetrahedron Letters* **1994**, 35, 3617

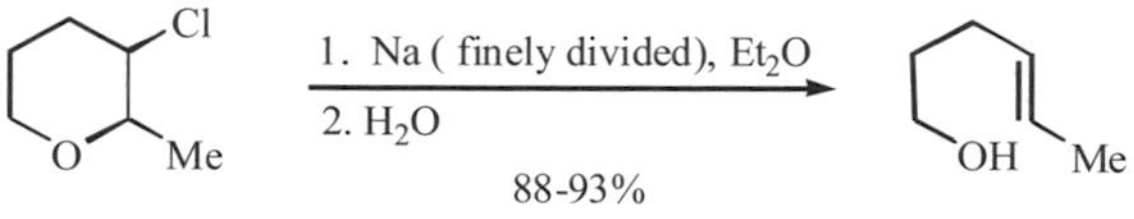

1. Na ( finely divided), Et$_2$O
2. H$_2$O

88-93%

R. Paul, O. Riobé, M. Maumy, *Organic Syntheses* CV 6, 675

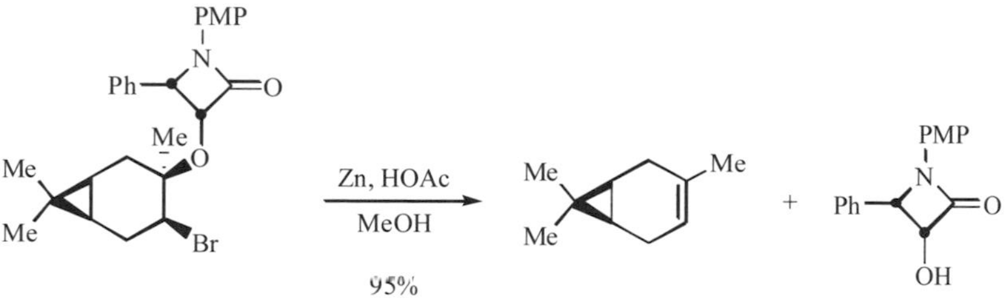

$$\xrightarrow[\text{MeOH}]{\text{Zn, HOAc}}$$

95%

S. N. Joshi, A. R. A. S. Deshmukh, B. M. Bhawal, *Tetrahedron: Asymmetry* **2000**, 11, 1477

# Borche Reduction

## The Reaction:

H−N
LiCNBH3
O
R
R'
N
R
R'

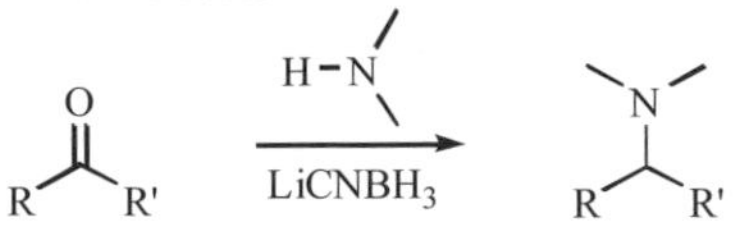

## Proposed Mechanism:

N
H
LiCNBH3
"H⊖"

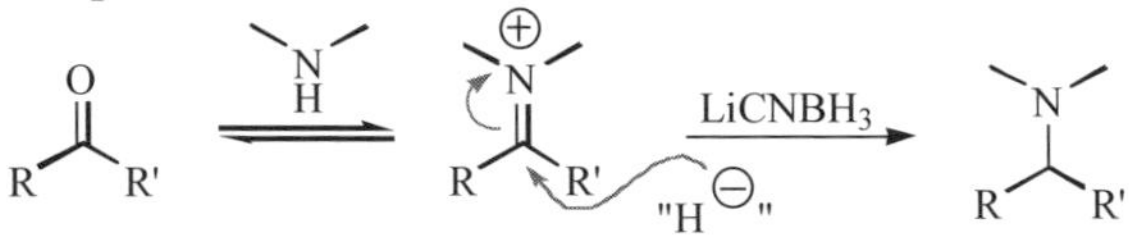

## Notes:

The success of this procedure rests on the much greater reactivity of

>>

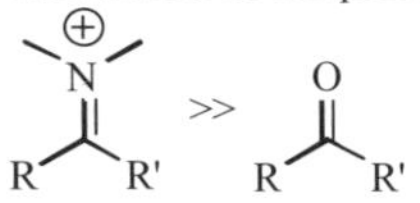

to the reducing agent

## Examples:

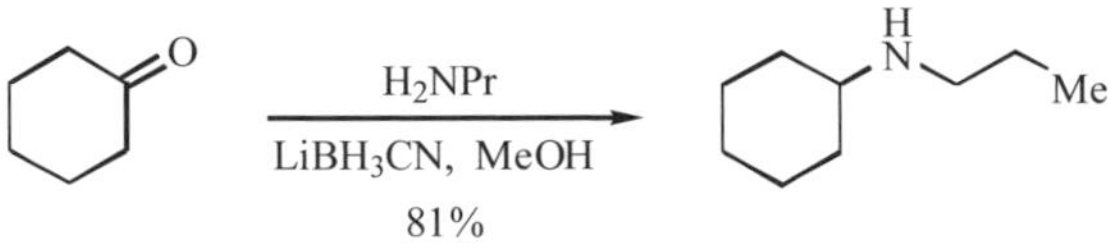

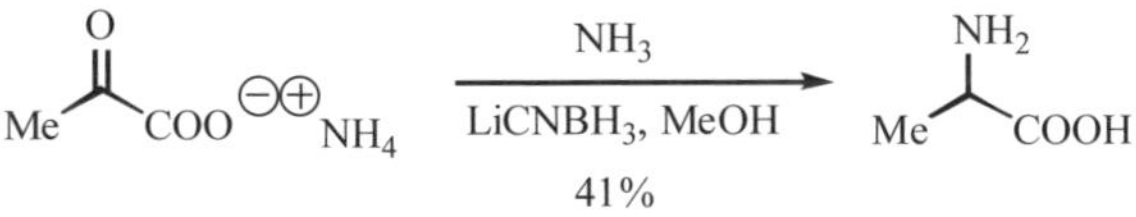

H2NPr
LiBH3CN, MeOH
81%

NH3
LiCNBH3, MeOH
41%
Me
COO
NH4
NH2
COOH

R. F. Borch, H. D. Durst, *Journal of the American Chemical Society* **1969**, 91, 3996

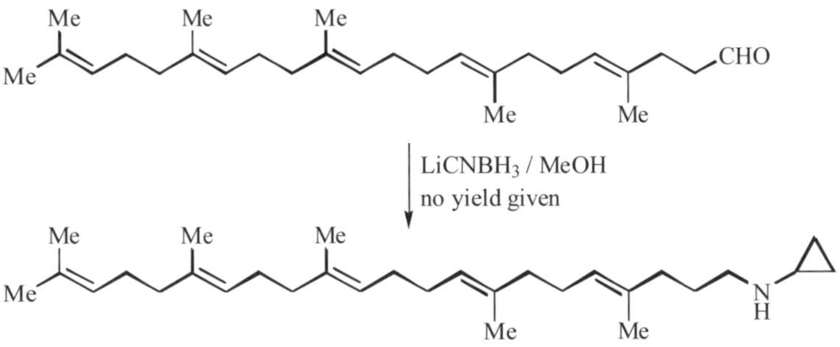

Me
CHO
LiCNBH3 / MeOH
no yield given
N
H

S. E. Sen, G. D. Prestwich, *Journal of the American Chemical Society* **1989**, 111, 8761

A. S. Kende, T. J. Bentley, R. A. Mader, D. Ridge, *Journal of the American Chemical Society* **1974**, 96, 4332

To see the use of the method for generating a library, with a modification of the reducing agent:
R. A. Tommasi, L. W. Whaley, H. R. Marepalli, *Journal of Combinatorial Chemistry* **2000**, 2, 447

A. J. Frontier, S. Raghaven, S. J. Danishefsky, *Journal of the American Chemical Society* **1997**, 119, 6686

K. M. Werner, J. M. de los Santos, S. M. Weinreb, M. Shang, *Journal of Organic Chemistry* **1999**, 64, 686

# Borsche-Drechsel Cyclization

## The Reaction:

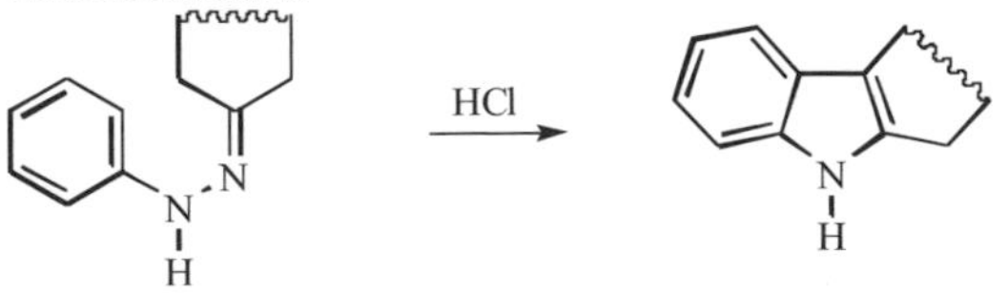

HCl

## Proposed Mechanism:

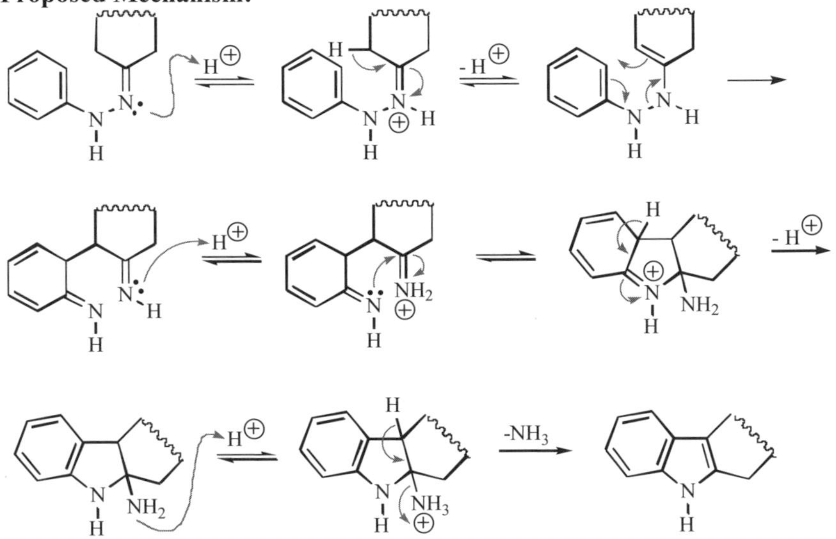

## Notes:

See this general concept in the ***Fischer Indole Synthesis*** and the ***Bucherer Carbazole Synthesis***.

M. B. Smith, J. March in *March's Advanced Organic Chemistry*, 5th ed., John Wiley and Sons, Inc., New York, 2001, pp 865-66; T. Laue, A. Plagens, *Named Organic Reactions*, John Wiley and Sons, Inc., New York, 1998, p. 37

## *Bucherer Carbazole Synthesis*
### The Reaction:

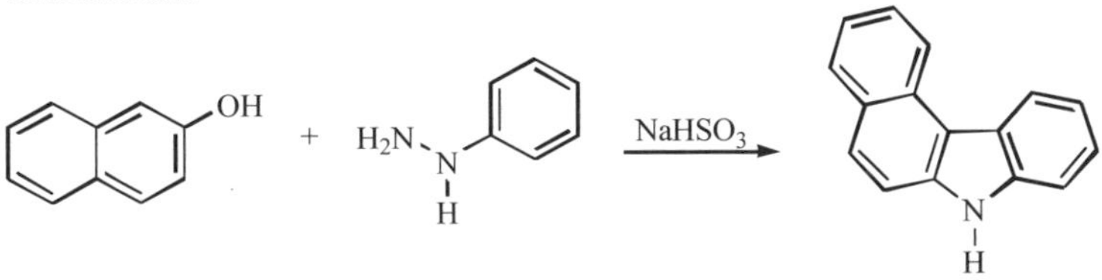

OH

+   $H_2N$   NaHSO$_3$

### Proposed Mechanism:

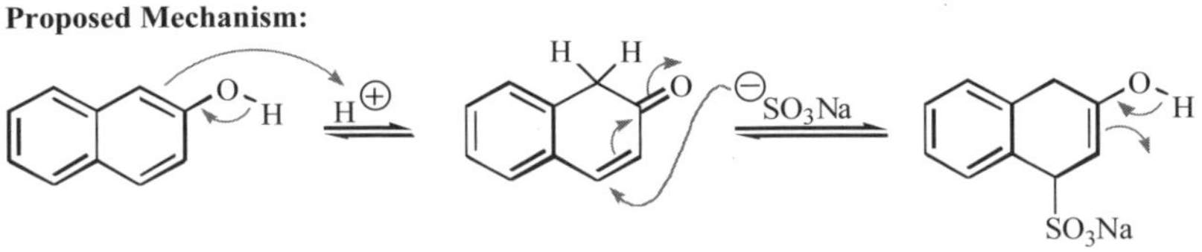

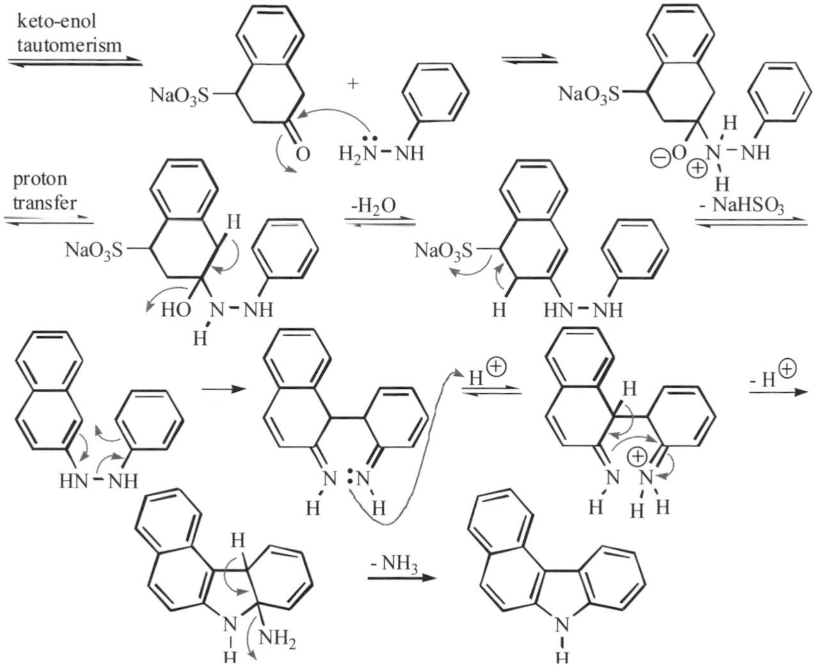

keto-enol
tautomerism

proton
transfer

-H₂O

- NaHSO₃

+ H⊕

- H⊕

- NH₃

**Examples of *Borsche-Drechsel Cyclization***

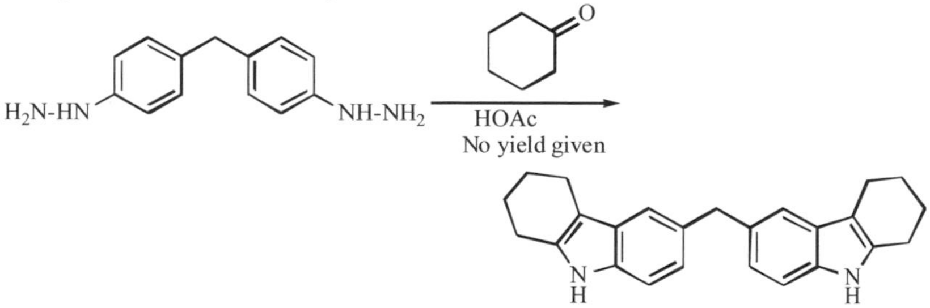

HOAc
No yield given

P. Bruck, *Journal of Organic Chemistry* **1970**, <u>35</u>, 2222

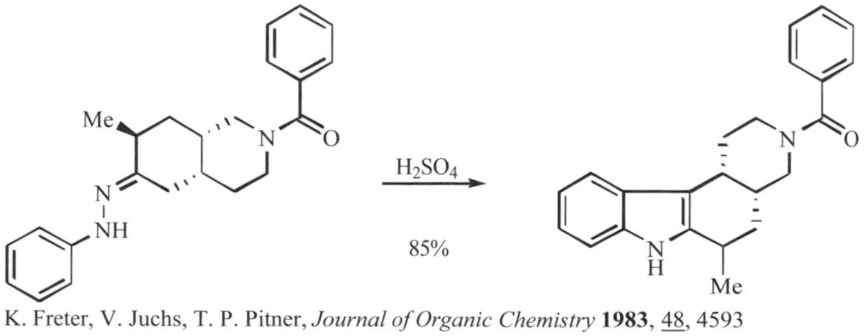

H₂SO₄

85%

K. Freter, V. Juchs, T. P. Pitner, *Journal of Organic Chemistry* **1983**, <u>48</u>, 4593

# Bouveault Aldehyde Synthesis / Reaction

## The Reaction:

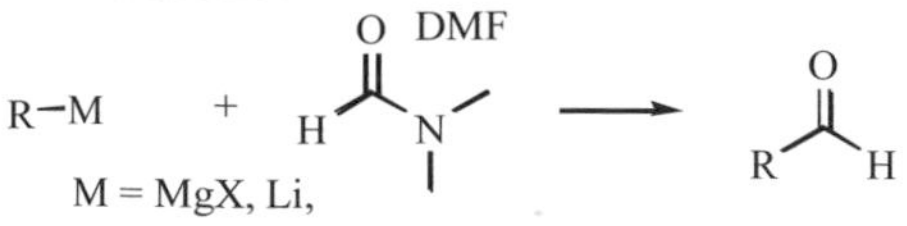

M = MgX, Li,

## Proposed Mechanism:

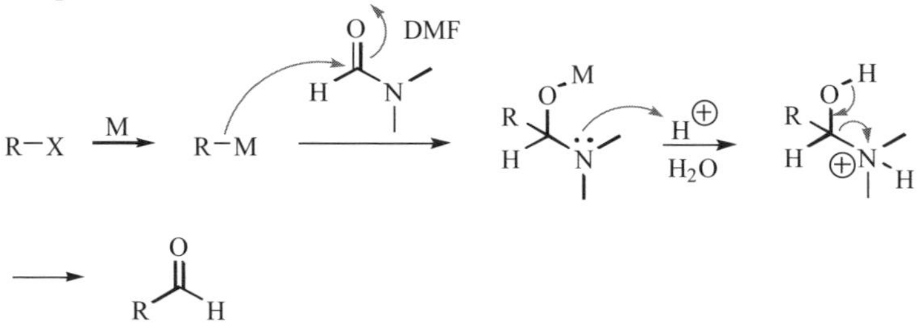

*Comin modification*:
D. L. Comins, J. D. Brown, *Journal of Organic Chemistry* **1984**, <u>49</u>, 1078

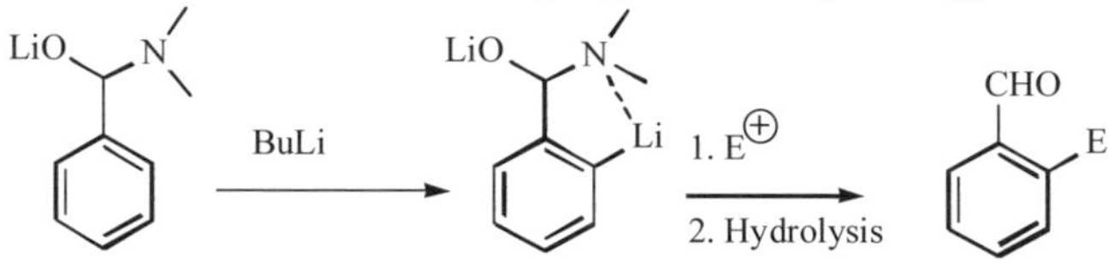

## Notes:

Based on the mechanism, one might suggest an extension to ketone synthesis. This is not a useful reaction. (M. B. Smith, J. March in *March's Advanced Organic Chemistry*, 5th ed., John Wiley and Sons, Inc., New York, 2001, p 1215).

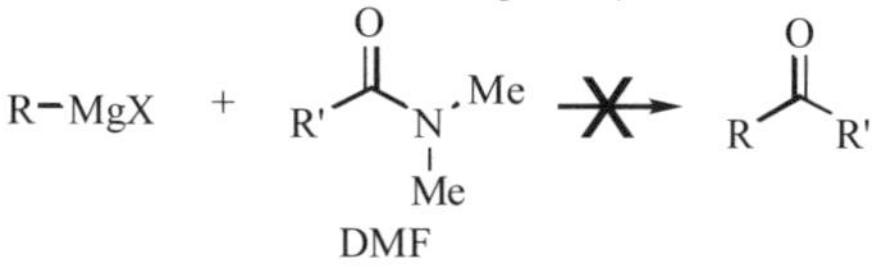

## *Bodroux-Chichibabin Aldehyde Synthesis*

### The Reaction:

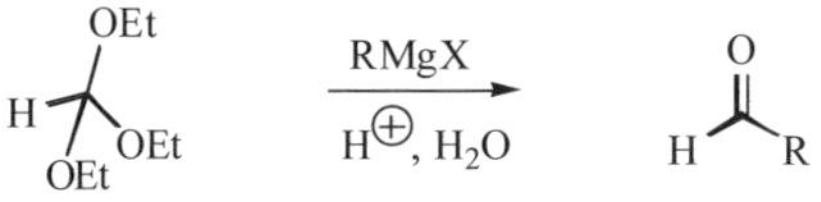

**Proposed Mechanism:**

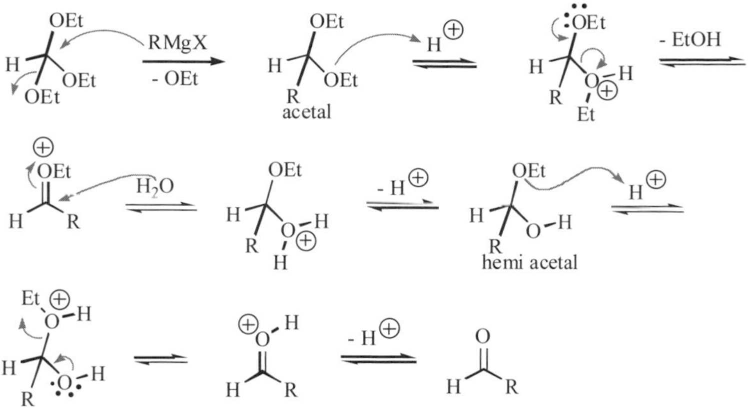

**Examples:**

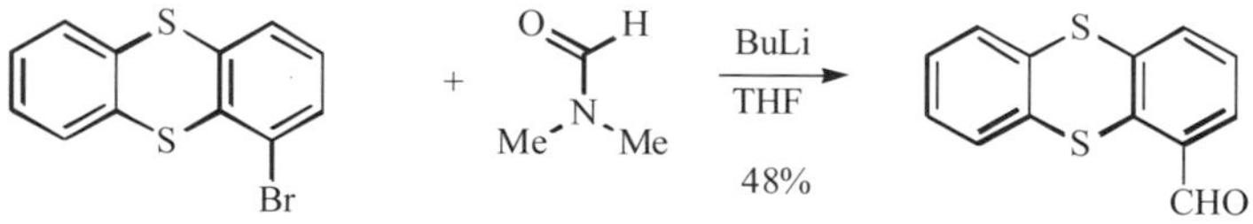

J. M. Lovell, J. A. Joule, *Journal of the Chemical Society, Perkin Transactions 1* **1996**, 2391

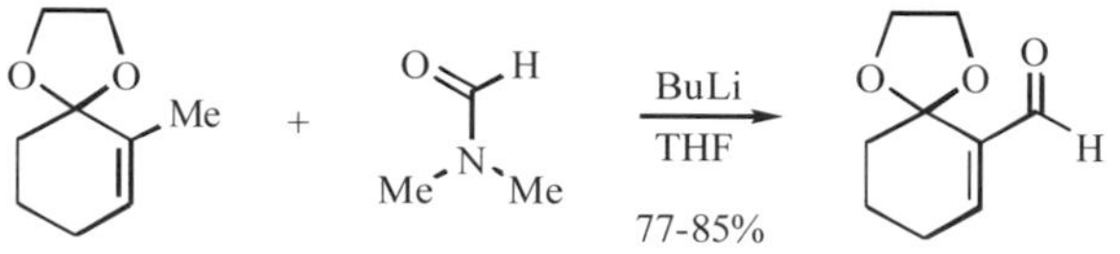

G. J. Bodwell, Z. Pi, *Tetrahedron Letters* **1997**, <u>38</u>, 309

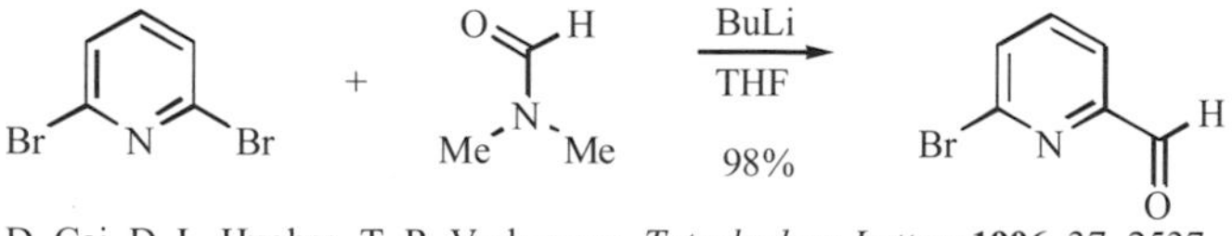

D. Cai, D. L. Hughes, T. R. Verhoeven, *Tetrahedron Letters* **1996**, <u>37</u>, 2537

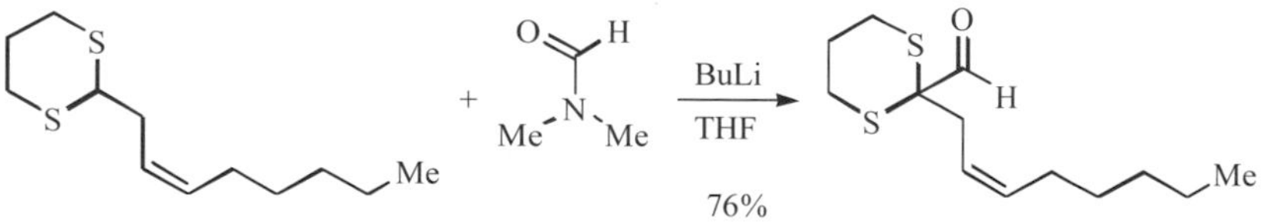

S. P. Khanapure, S. Manna, J. Rokach, R. C. Murphy, P. Wheelan, W. S. Powell, *Journal of Organic Chemistry* **1995**, <u>60</u>, 1806

# Bouveault-Blanc Reduction

## The Reaction:

$$R \overset{O}{\underset{OR'}{\bigvee}} \xrightarrow[\text{EtOH}]{\text{Na}} R \overset{OH}{\underset{H}{\bigvee}} H$$

## Proposed Mechanism:

*Dissolving metal reduction of carbonyl group*

## Notes:

M. B. Smith, J. March in *March's Advanced Organic Chemistry*, 5th ed., John Wiley and Sons, Inc., New York, 2001, pp 1191, 1551.

The related reduction of simple carbonyl groups with Na / alcohol (see boxed portion of mechanism) has been largely replaced by hydride reduction. However, the method has some advantages:
1. "thermodynamic products" are favored;
2. Oximes are converted to amines.

For direct conversion of an acid to an aldehyde in the presence of a ketone:

85-93%

T. Fujisawa, T. Sato, *Organic Syntheses* CV 8, 498

## Examples:

It is appropriate here to mention that in the following example LiAlH$_4$ would reduce both the acid and the ester; diborane would preferentially reduce the acid; and DIBAH would convert the ester to an aldehyde.

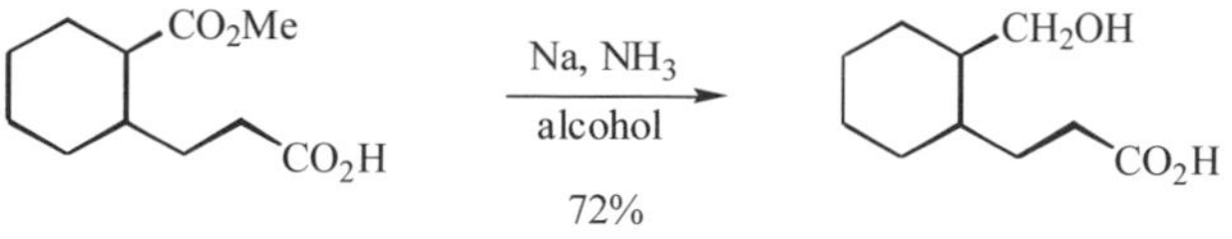

$$\xrightarrow[\text{alcohol}]{\text{Na, NH}_3}$$

72%

L. A. Paquette, N. A. Nelson, *Journal of Organic Chemistry* **1962**, 27, 2272

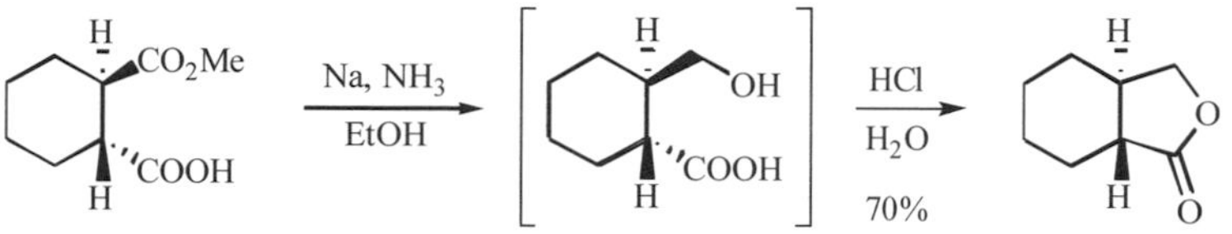

$$\xrightarrow[\text{EtOH}]{\text{Na, NH}_3}$$

$$\xrightarrow[\text{H}_2\text{O}]{\text{HCl}}$$

70%

R. M. Borzilleri, S. M. Weinreb, M. Parvez, *Journal of the American Chemical Society* **1995**, 117, 10905

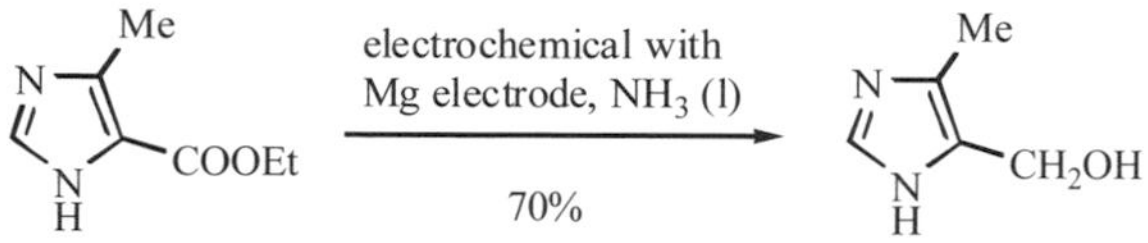

electrochemical with
Mg electrode, NH$_3$ (l)

70%

J. Chaussard, C. Combellas, A. Thiebault, *Tetrahedron Letters* **1987**, 28,1173

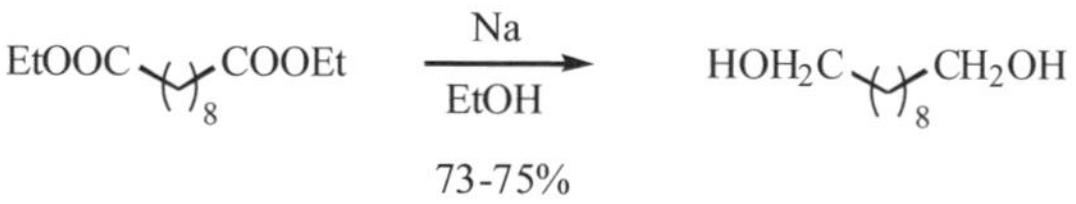

EtOOC $\underset{8}{\diagdown\!\!\!\bigwedge}$ COOEt $\xrightarrow[\text{EtOH}]{\text{Na}}$ HOH$_2$C $\underset{8}{\diagdown\!\!\!\bigwedge}$ CH$_2$OH

73-75%

R. H. Manske, *Organic Syntheses* CV 2, 154

# Boyland-Sims Oxidation

## The Reaction:

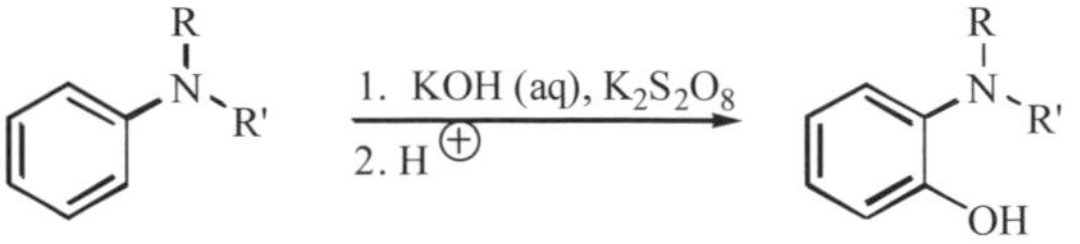

## Proposed Mechanism:

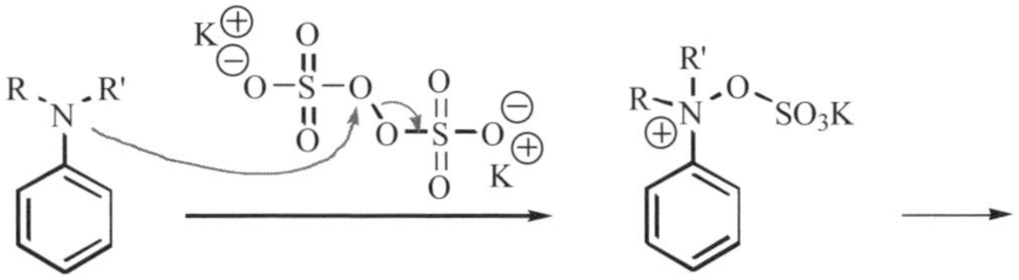

E. J. Behrman, *Journal of Organic Chemistry* **1992**, <u>57</u>, 2266

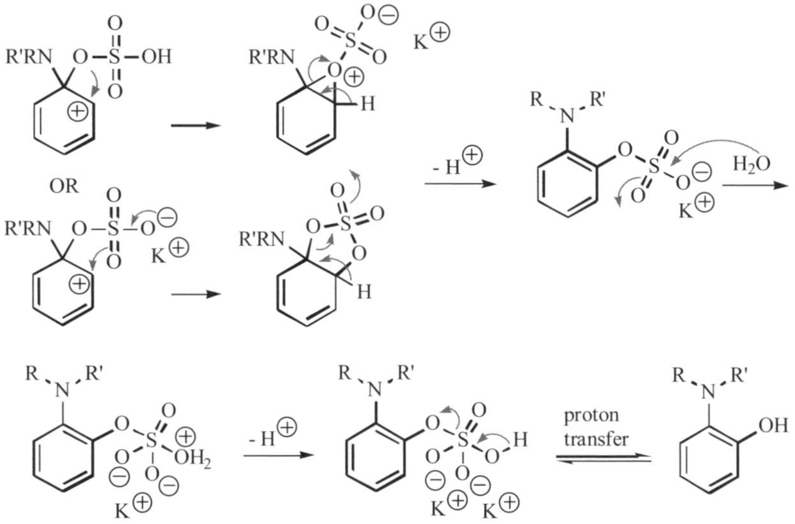

## Notes:

For comments on this and the ***Elbs Reaction***, see: M. B. Smith, J. March in *March's Advanced Organic Chemistry*, 5th ed., John Wiley and Sons, Inc., New York, 2001, p 724; E. J. Behrman, *Organic Reactions* **35**, 2

See the similarity in the:
# Elbs Persulfate Oxidation (Elbs Reaction)

## The Reaction:

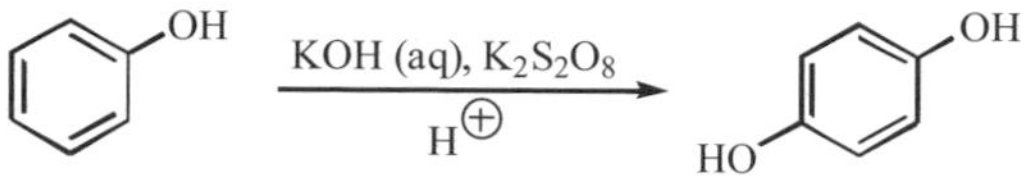

## Proposed Mechanism:

T. Laue, A. Plagens, *Named Organic Reactions*, John Wiley and Sons, Inc., New York, 1998, pp. 92-93; V. K. Ahluwalia, R. K. Parashar, *Organic Reaction Mechanisms*, Alpha Science International, Ltd., Pangbourne, U.K., 2002, pp 320-321.

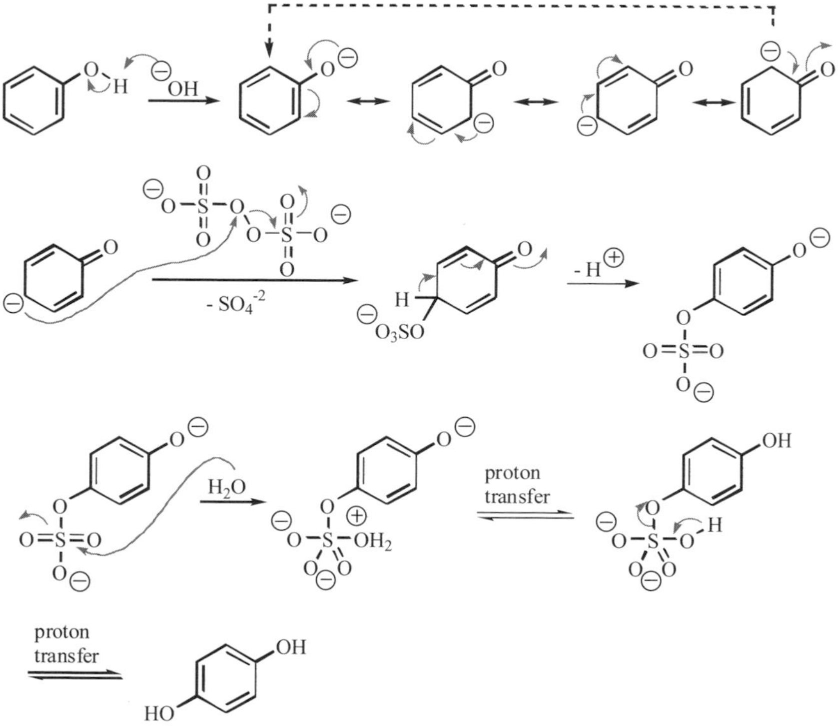

## Notes:

Oxidation usually occurs at the *para* position. If the *para* position is occupied, the *ortho* position is the next likely site for reaction.

# Bradsher Reaction

## The Reaction:

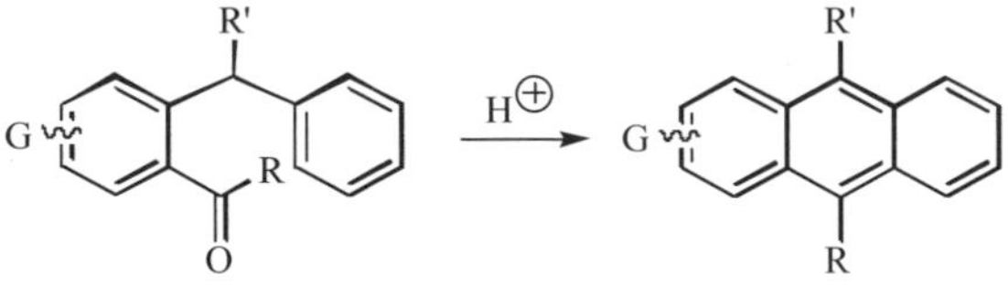

## Proposed Mechanism:

M. B. Smith, J. March in *March's Advanced Organic Chemistry*, 5th ed., John Wiley and Sons, Inc., New York, 2001, p 720

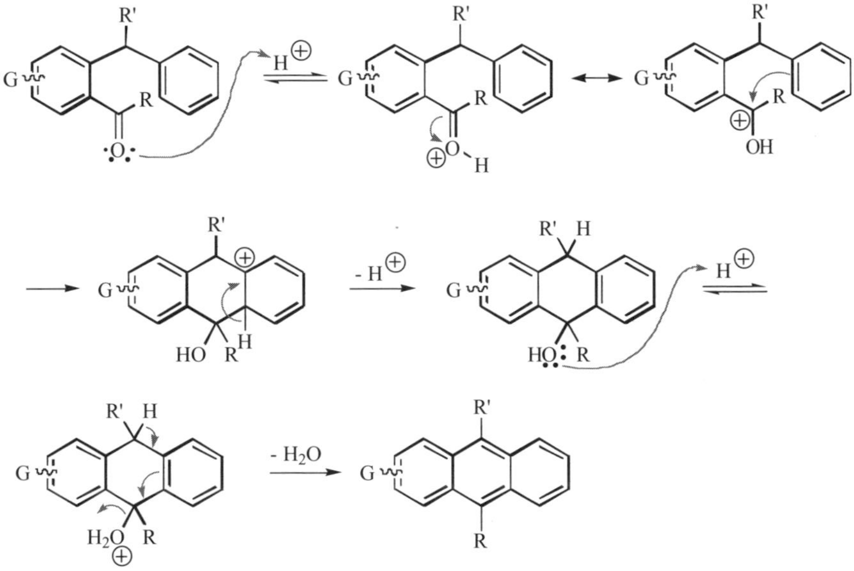

## Notes:

This reaction also works with heterocyclic derivatives, where G = O, S, or Se.

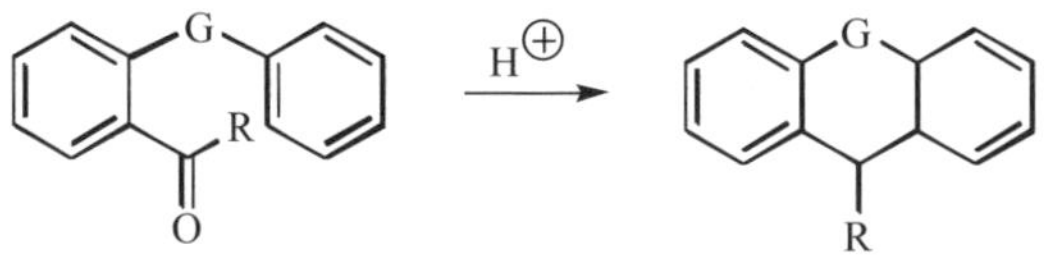

See: ***Polycarbocyclic Syntheses***

## Examples:

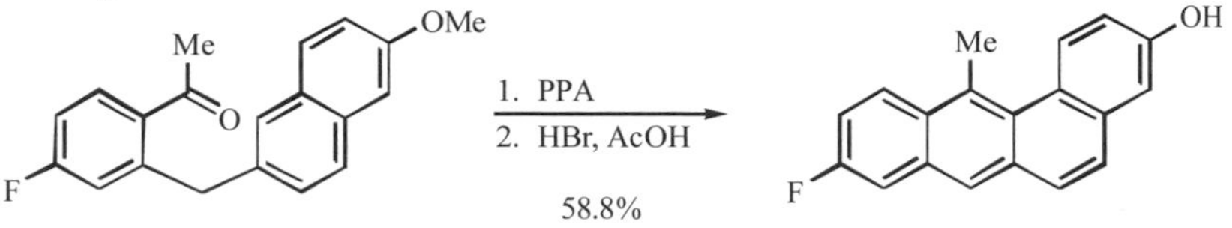

1. PPA
2. HBr, AcOH

58.8%

R. G. Harvey, C. Cortez, Tetrahedron **1997**, <u>53</u>, 7101

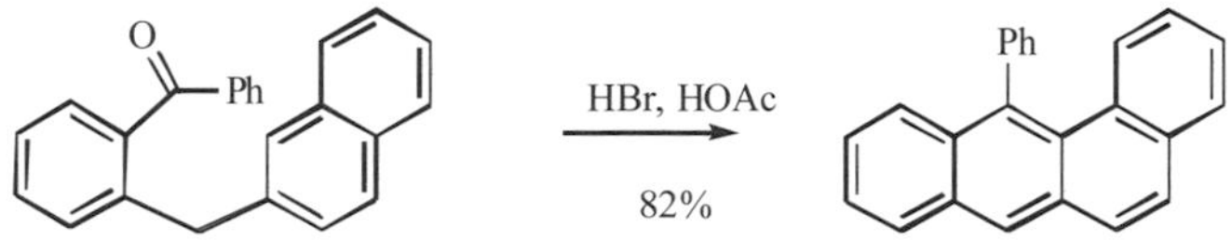

HBr, HOAc

82%

C. K Bradsher, *Journal of the American Chemical Society* **1940**, <u>62</u>, 486

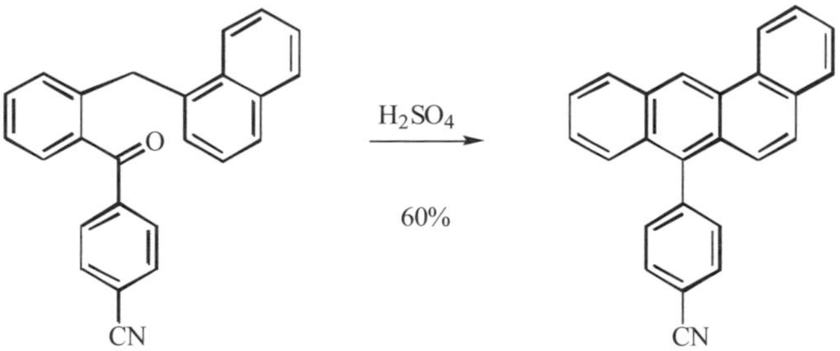

$H_2SO_4$

60%

F. A. Vingiello, R. K. Stevens, *Journal of the American Chemical Society* **1958**, <u>80</u>, 5256

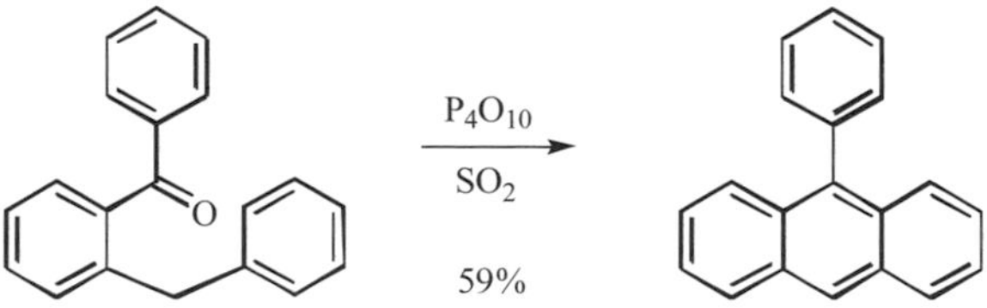

$P_4O_{10}$

$SO_2$

59%

C. K. Bradsher, E. F. Sinclair, *Journal of Organic Chemistry* **1957**, <u>22</u>, 79

# Brook Rearrangement

## The Reaction:

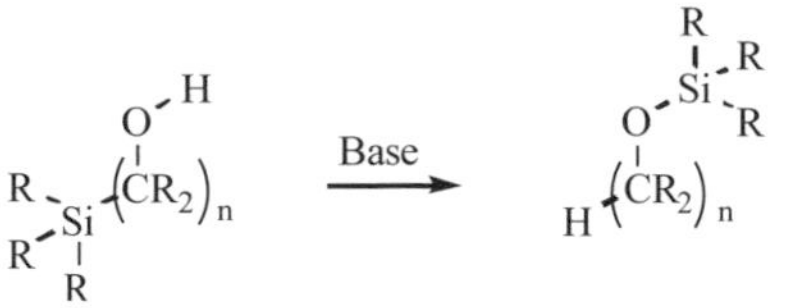

R R group structure with Base arrow converting to silyl ether product

n = 1 – 3, where for 3 – 5 membered transition states can be formed.

## Proposed Mechanism:

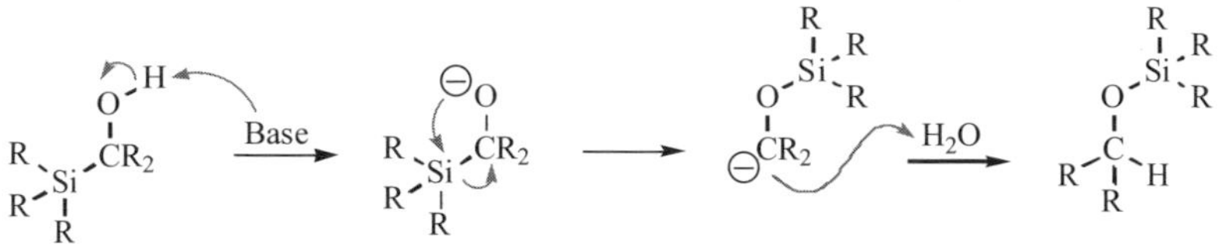

Mechanism scheme showing silyl alcohol with Base, then anion intermediate, then Brook rearrangement with $H_2O$ to product.

## Examples:

$Me_3Si$ structure, $HO$, $Ph$, $(CF_2)_5CF_3$ with Excess $NH_3$ / $Et_2O$ arrow to $Me_3Si-O$, H, $Ph$, $(CF_2)_5CF_3$

95%

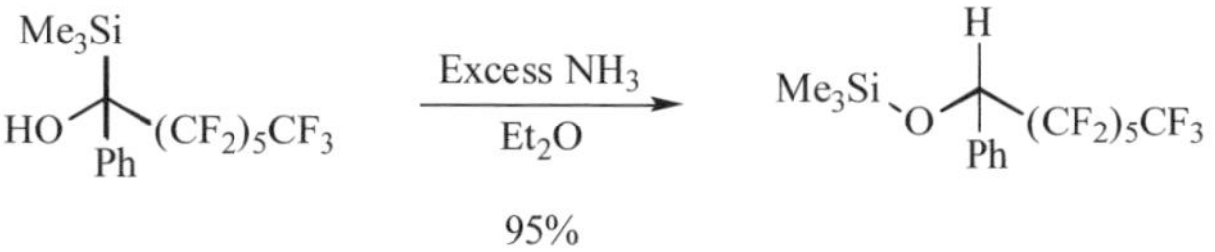

B. Dondy, P. Doussot, C. Portella, *Tetrahedron Lett*ers **1994**, <u>35</u>, 409

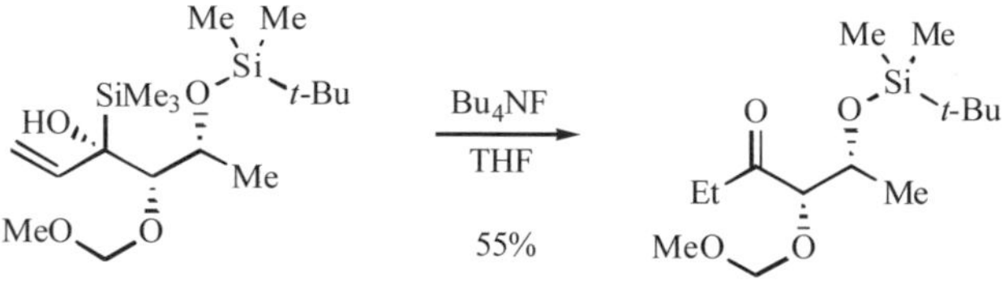

Structure with Me Me, Si, O, SiMe₃, t-Bu, HO, Me, MeO, O with $Bu_4NF$ / THF arrow

55%

Product with Me Me, Si, t-Bu, O, Et, Me, MeO, O

P. F. Cirillo, J. S. Panek, *Journal of Organic Chemistry* **1990**, <u>55</u>, 6071

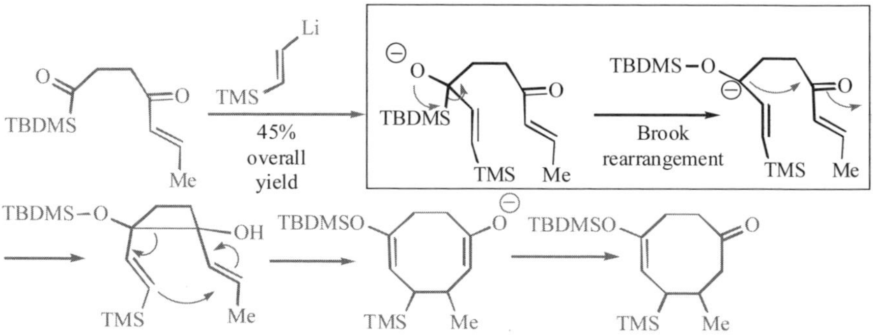

Reaction scheme with TBDMS, O, TMS, Li, Me, 45% overall yield, then mechanism steps showing TBDMS, TMS, Me, Brook rearrangement, TBDMS–O, and final products with TBDMSO, TMS, Me.

K. Takeda, H. Haraguchi, Y. Okamoto, *Organic Letters* **2003**, <u>5</u>, 3705

7%                                        51%

M. Koreeda, S. Koo, *Tetrahedron Letters* **1990**, <u>31</u>, 831

An example of a ***Homo-Brook Rearrangement***

Yields were varied, depending on base used.
R. Ducray, N. Cramer, M. A. Ciufolini, *Tetrahedron Letters* **2001**, <u>42</u>, 9175

L. A. Calvo, A. M. Gonza´lez-Nogal, A. Gonza´lez-Ortega, M. C. San˜udo, *Tetrahedron Letters* 2001, 42, 8981

# Brown's Hydroboration

See: T. Laue, A. Plagens, *Named Organic Reactions*, John Wiley and Sons, Inc., New York, 1998, pp. 157-160.

## The Reaction:

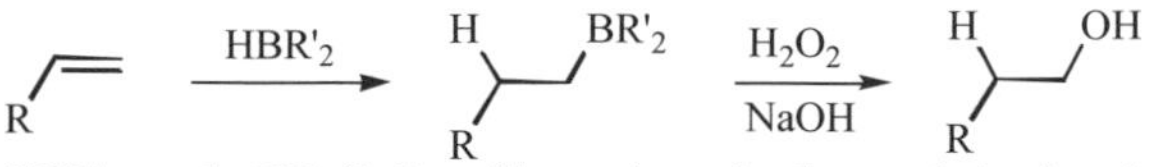

HBR'$_2$ may be BH$_3$ (B$_2$H$_6$ = diborane) or other borane derivatives (see below).

## Proposed Mechanism:

In the presence of ethers (and especially dimethyl sulfide) diborane can dissociate into a complexed borane:

Z = O, S, generally from Et$_2$O, THF, and DMS

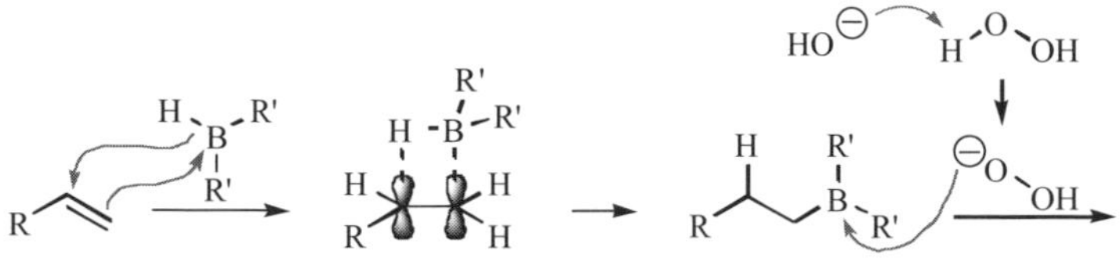

Borane attacks from the least hindered face. *Stereochemical control.*

Boron adds as an electrophile and hydride as the nucleophile in a cis-fashion. *Regiochemical control.*

Boron hydrolysis begins with the attack of peroxide.

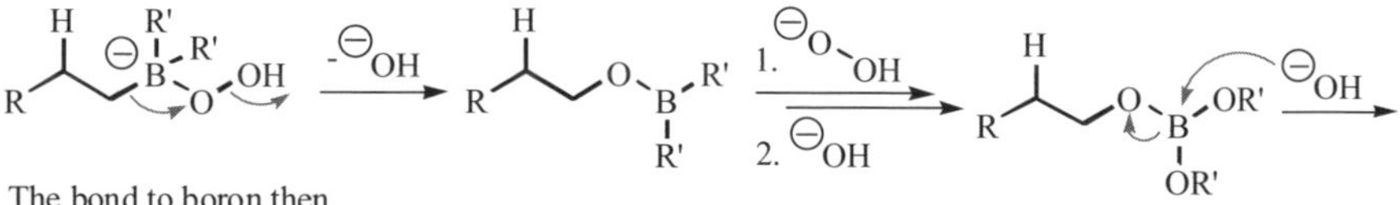

The bond to boron then migrates to oxygen.

Two more addition / migrations take place

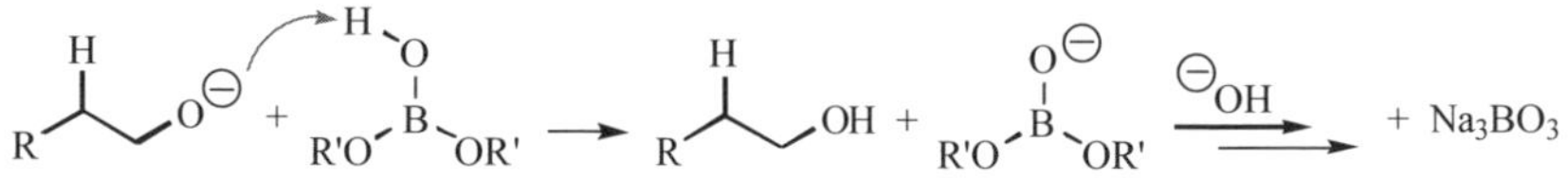

The result is a cis, anti-Markownikoff addition of water. The C-B bond is converted into a C-OH bond *with retention of stereochemistry.*

## Notes:

With hindered alkenes, it is more difficult to add three alkenes to borane. This becomes the basis for unique, borane derivatives. See ***Hydroboration Reagents***.

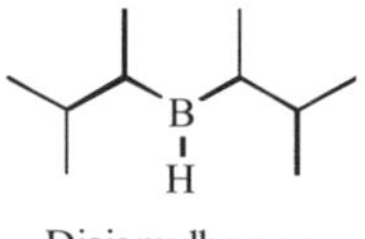

Disiamylborane
Sia$_2$BH

Thexylborane

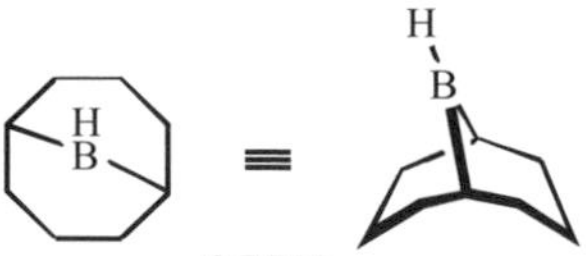

9-BBN
9-borabicyclo[3.3.1]nonane

## Examples:

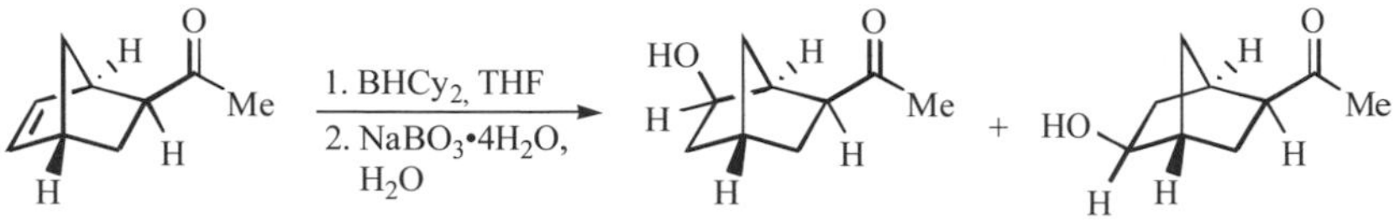

G. W. Kabalka, S. Yu, N.-S. Li, *Tetrahedron Letters* **1997**, <u>38</u>, 5455

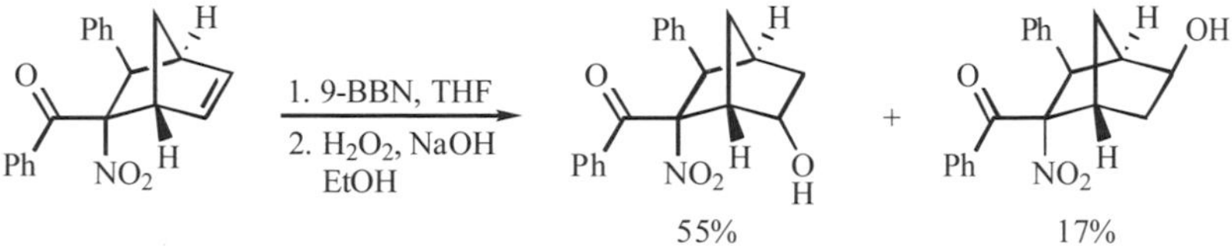

55%                    17%

D. L. Gober, R. A. Lerner, B. F. Cravatt, *Journal of Organic Chemistry* **1994**, <u>59</u>, 5078

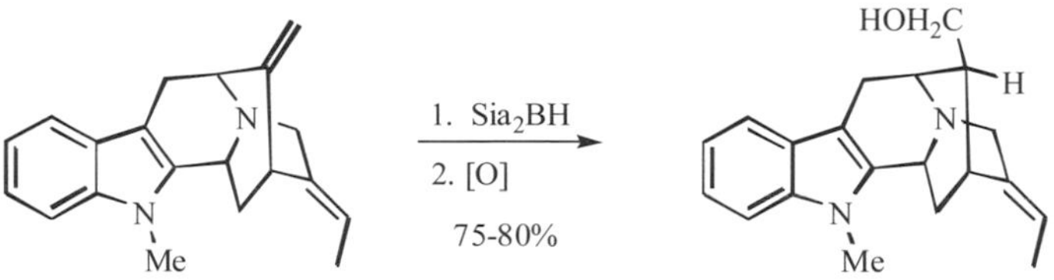

1. Sia₂BH

2. [O]

75-80%

D. L. Gober, R. A. Lerner, B. F. Cravatt, *Journal of Organic Chemistry* **1994**, <u>59</u>, 5078

1. 9-BBN

2. [O]

99%

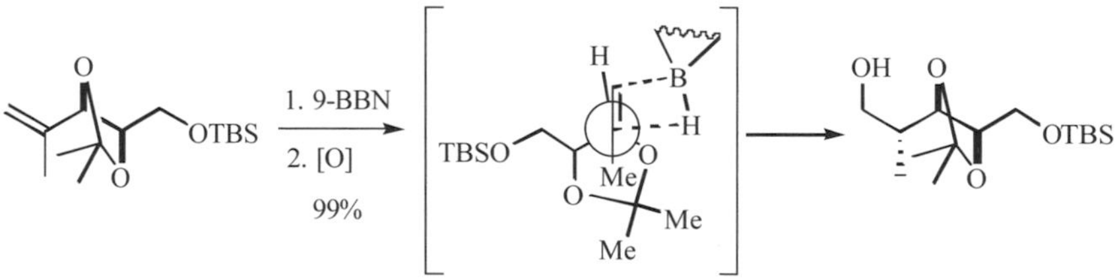

K. Suenaga, K. Araki, T. Sengoku, D. Uemura, *Organic Letters* **2001**, <u>3</u>, 527

# Bucherer-Bergs Reaction

## The Reaction:

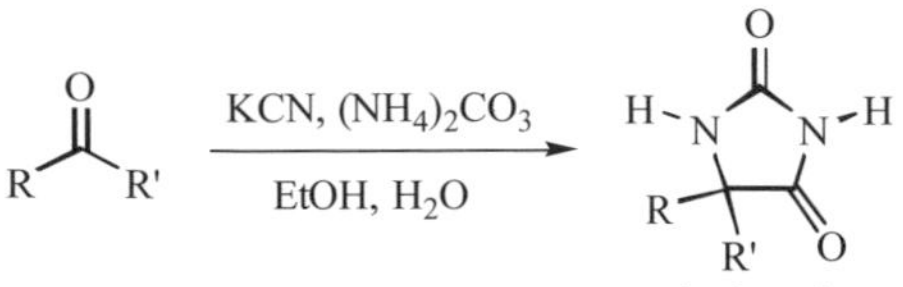

hydantoin

Hydrolysis of the hydantoin provides an approach to amino acids:

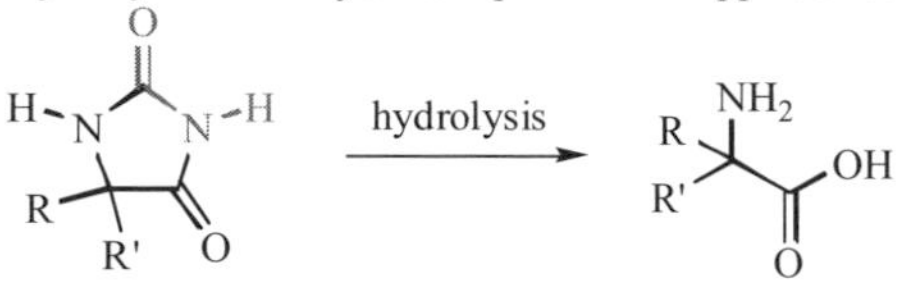

## Proposed Mechanism:

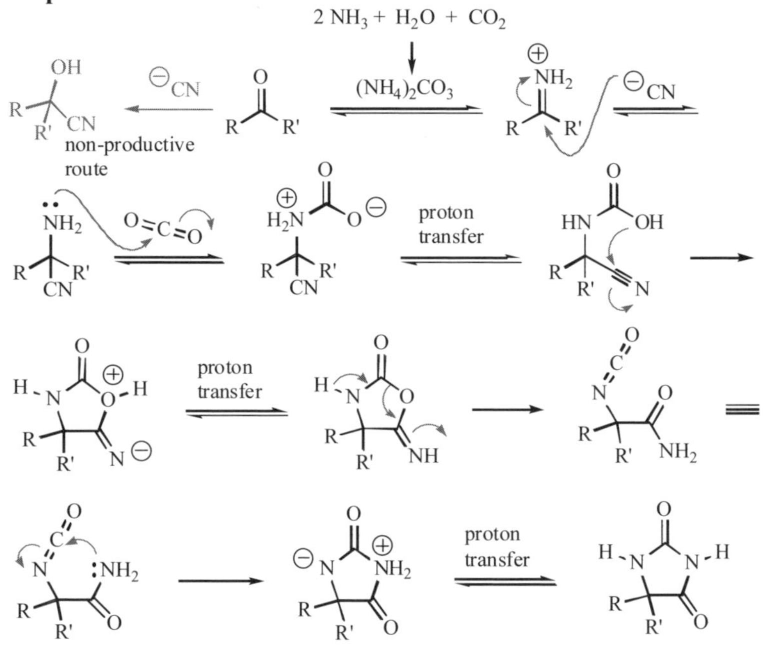

$$2 NH_3 + H_2O + CO_2$$

F. L. Chubb, J. T. Edward, S. C. Wong, *Journal of Organic Chemistry* **1980**, <u>45</u>, 2315
An intermediate α-aminonitrile carbamate is found. A. Rousset, M. Lasperas, J. Taillades, A. Commeyras, *Tetrahedron* **1980**, <u>36</u>, 2649

## Examples:

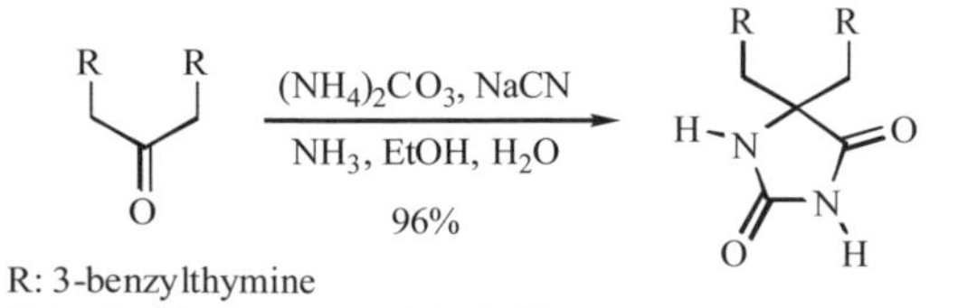

R: 3-benzylthymine
F. L. Chubb, J. T. Edward, S. C. Wong, *Journal of Organic Chemistry* **1980**, <u>45</u>, 2315

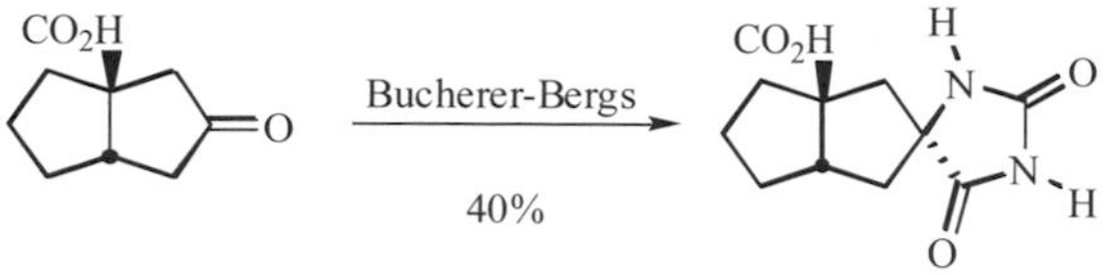

Bucherer-Bergs

40%

J. Ezquerra, B. Yruretaguyena, C. Avendano, E. de la Cuesta, R. Gonzalez, L. Prieto, C. Peqregal, M. Espada, W. Prowse, *Tetrahedron* **1995**, 51, 3271

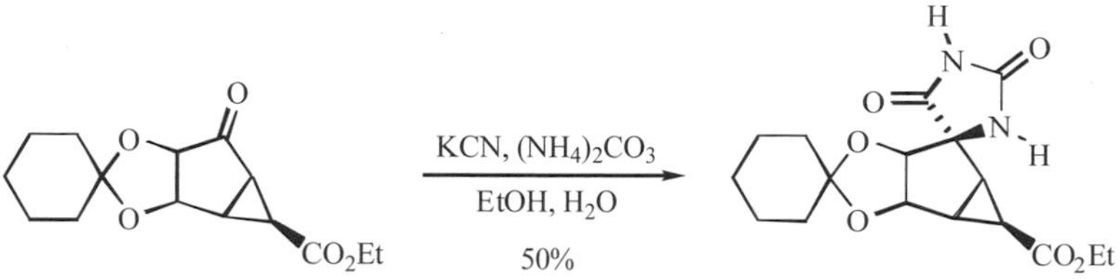

KCN, (NH$_4$)$_2$CO$_3$

EtOH, H$_2$O

50%

C. Dominguez, J. Ezquerra, S. R. Baker, S. Burrelly, L. Prieto, M. Espada, C. Pedregal, *Tetrahedron Letters* **1998**, 39, 9305

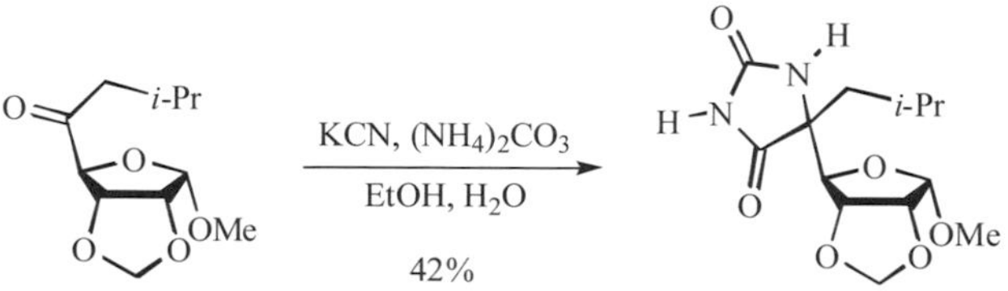

KCN, (NH$_4$)$_2$CO$_3$

EtOH, H$_2$O

42%

B. Steiner, J. Micova, M. Koos, V. Langer, D. Gyepesova, *Carbohydrate Research* **2003**, 338, 1349

# Bucherer Reaction

## The Reaction:

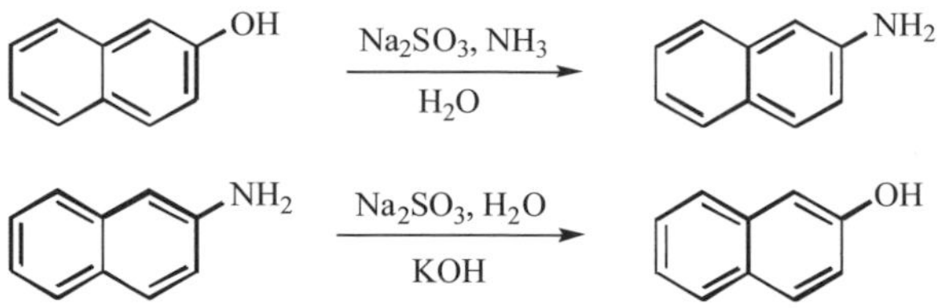

This reaction can be carried out in either direction with modest modifications of reaction conditions.

## Proposed Mechanism:

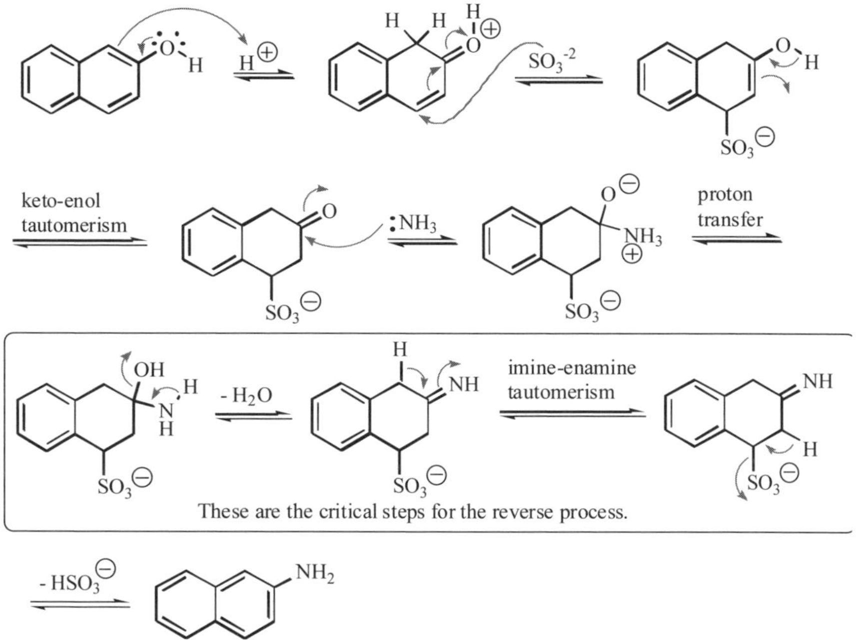

keto-enol
tautomerism

proton
transfer

imine-enamine
tautomerism

These are the critical steps for the reverse process.

## Notes:

M. B. Smith, J. March in *March's Advanced Organic Chemistry*, 5th ed., John Wiley and Sons, Inc., New York, 2001, pp 861, 865; T. Laue, A. Plagens, *Named Organic Reactions*, John Wiley and Sons, Inc., New York, 1998, pp. 37-39; N. L. Drake, *Organic Reactons* **1**, 5.

# Examples:

R. S. Coleman, M. A. Mortensen, *Tetrahedron Letters* **2003**, <u>44</u>, 1215

S. Vyskocil, M. Smrcina, M. Lorenc, I. Tislerova, R. D. Brooks, J. J. Kulagowski, V. Langer, L. J. Farrugia, P. Kocovsky, *Journal of Organic Chemistry* **2001**, <u>66</u>, 1359

L. F. Fieser, E. B. Hershberg, L. Long, Jr., M. S. Newman, *Journal of the American Chemical Society* **1937**, <u>59</u>, 475

K. Korber, W. Tang, X. Hu, X. Zhang, *Tetrahedron Letters* **2002**, <u>43</u>, 7163

# Buchwald-Hartwig Reactions

## The Reaction:

$$Ar-X \quad + \quad \underset{H}{\overset{R \diagdown_{N} \diagup R'}{|}} \quad \xrightarrow[\text{NaO}t\text{-Bu, tol, } \Delta]{\text{Pd(0)}} \quad Ar-N\overset{R}{\underset{R'}{<}}$$

## Proposed Mechanism:

Buchwald calls the replacement of halide by nitrogen metathesis.

Sources for mechanism:

J. P. Wolfe, S. Wagaw, S. L. Buchwald, *Accounts of Chemical Research* **1998**, <u>31</u>, 805

J. Louie, J. F. Hartwig, *Tetrahedron Letters* **1995**, <u>36</u>, 3609

B. H. Yang, S. L. Buchwald, *Journal of Organometallic Chemistry* **1999**, <u>576</u>, 125

## Notes:

Representative catalysts:

$$Pd_2(dba)_3 \quad \xrightarrow{PPh_3} \quad 2 \; Pd(PPh_3)_4$$

dba
dibenzylidineacetone

Pd(dppf)Cl$_2$ =
[1,1'-Bis(diphenylphosphino)ferrocene]dichloropalladium(II)

A useful Review describes the use of BINAP for the conversion:

$$\text{Ar-Br} + \text{R'NH}_2 \xrightarrow[\text{BINAP, NaO}t\text{-Bu}]{\text{0.25 mol\% Pd}_2(\text{dba})_3} \text{Ar-NHR'}$$

J. P. Wolfe, S. Wagaw, S. L. Buchwald, *Accounts of Chemical Research* **1998**, <u>31</u>, 805

## Examples:

When Ar =

72%                      83%

DMT = Dimethoxytrityl, a useful protecting group

L. C. J. Gillet, O. D. Scharer, *Organic Letters* **2002**, <u>4</u>, 4205

See ***Verkade's Base***

96%

92%

S. Urgaonkar, J.-H. Xu, J. G. Verkade, *Journal of Organic Chemistry* **2003**, <u>68</u>, 8416

A number of ligands and resins were examined.
K. Weigand, S. Pelka, *Organic Letters* **2002**, <u>4</u>, 4689

# Cadiot-Chodkiewicz Coupling

## The Reaction:

$$H\!\!=\!\!=\!\!-R'$$
$$+$$  $\xrightarrow[\text{CuCl}]{\text{base}}$  $R\!-\!\!=\!\!=\!\!=\!\!=\!\!-R'$
$$R\!-\!\!=\!\!=\!\!-X$$

## Proposed Mechanism:

$R'\!-\!\!=\!\!=\!\!-H$ $\xrightarrow{\text{Cu(I)}}$ $R'\!-\!\!=\!\!=\!\!-H$ $\xrightarrow[\text{CuCl}]{\text{:N}\!-}$ $R'\!-\!\!=\!\!=\!\!-Cu$ $\xrightarrow[\text{oxidative addition}]{R\!-\!\!=\!\!=\!\!-X}$
$$\underset{\overset{|}{Cu}}{}$$

$R\!-\!\!=\!\!=\!\!-X$ $\longrightarrow$ $R\!-\!\!=\!\!=\!\!-Cu\!-\!\!=\!\!=\!\!-R'$ $\longrightarrow$ $R\!-\!\!=\!\!=\!\!=\!\!=\!\!-R'$
$$\underset{Cu\!-\!\!=\!\!=\!\!-R'}{} \qquad \underset{X}{} \qquad \text{reductive elimnation}$$

## Notes:

See *Alkyne Coupling*,
$CuCl = Cu_2Cl_2$

M. B. Smith, J. March in *March's Advanced Organic Chemistry*, 5th ed., John Wiley and Sons, Inc., New York, 2001, p. 937.

*Strauss Coupling*:
In the absence of oxygen or oxidizing atmosphere:

$R\!-\!\!=\!\!=\!\!-Cu$ $\xrightarrow[\text{non-oxidative conditions}]{\text{HOAc}}$ $R\!-\!\!=\!\!<$

## Examples:

| | |
|---|---|
| A | $NH_2OH\ HCl$, $BuNH_2$, CuCl, EtOH |
| | 15% |
| B | CuI, $Pd_2(dba)_3$, $Pr_2NH$, |

Reported

77%

D. Elbaum, T. B. Nguyen, W. L. Jorgensen, S. L. Schreiber, *Tetrahedron* **1994**, _50_, 1503

U. Fritzsche, S. Hunig, *Tetrahedron Letters* **1972**, <u>13</u>, 4831

H. A. Stansbury, W. R. Proops, *Journal of Organic Chemistry* **1962**, <u>27</u>, 320

A. S. Hay, *Journal of Organic Chemistry* **1962**, <u>27</u>, 3320

J.-P. Gotteland, I. Brunel, F. Gendre, J. Desire, A. Delhon, D. Junquero, P. Oms, S. Halazy, *Journal of Medicinal Chemistry* **1995** <u>38</u>, 3207

# Cannizzaro Reaction / Aldehyde Disproportionation

## The Reaction:

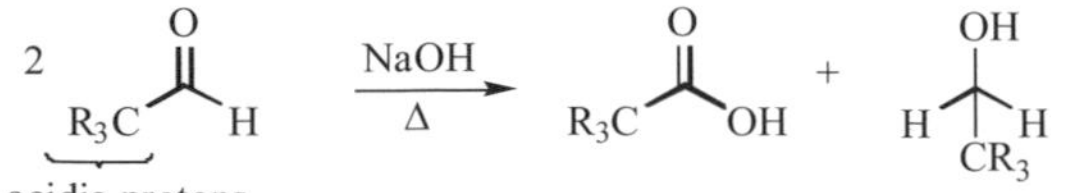

$$2 \quad \underset{R_3C}{\overset{O}{\underset{\phantom{x}}{\parallel}}}H \quad \xrightarrow[\Delta]{NaOH} \quad \underset{R_3C}{\overset{O}{\parallel}}OH \quad + \quad \underset{\underset{CR_3}{\overset{|}{H}}}{\overset{OH}{\underset{|}{C}}}H$$

no acidic protons
typically CR$_3$ = Ar

## Proposed Mechanism:

Hydride transfer to 2nd aldehyde.

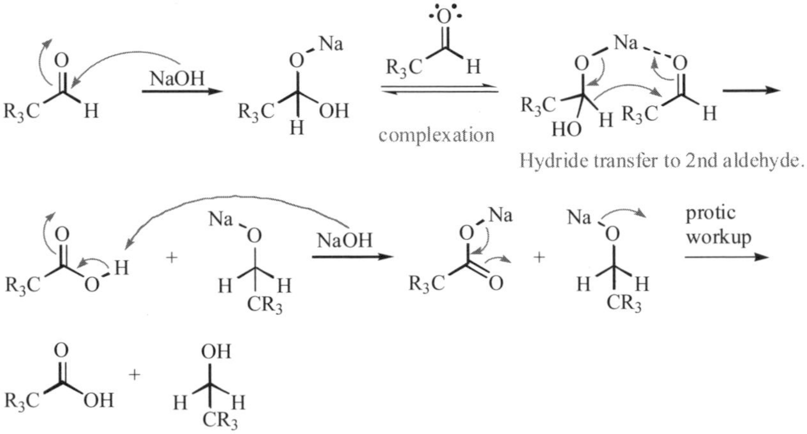

## Notes:

T. Laue, A. Plagens, *Named Organic Reactions*, John Wiley and Sons, Inc., New York, 1998, pp 40-42; M. B. Smith, J. March in *March's Advanced Organic Chemistry*, 5$^{th}$ ed., John Wiley and Sons, Inc., New York, 2001, pp 1564-1565; T. A. Geissman, *Organic Reactions* **2**, 3.

## Tollens Reaction

The reaction is an ***aldol condensation*** followed by a ***Cannizzaro reaction***.

### The Reaction:

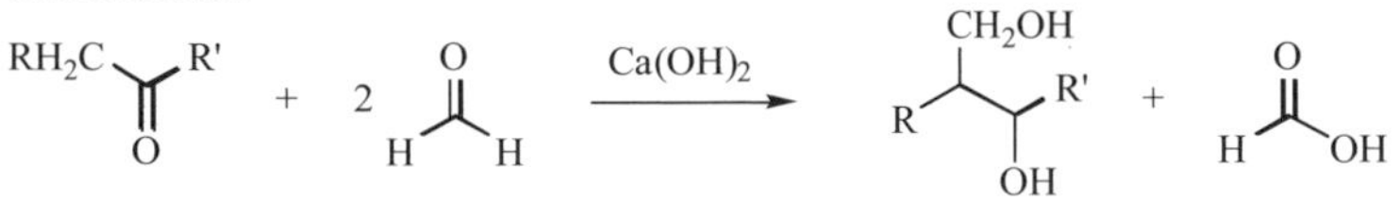

### Proposed Mechanism:

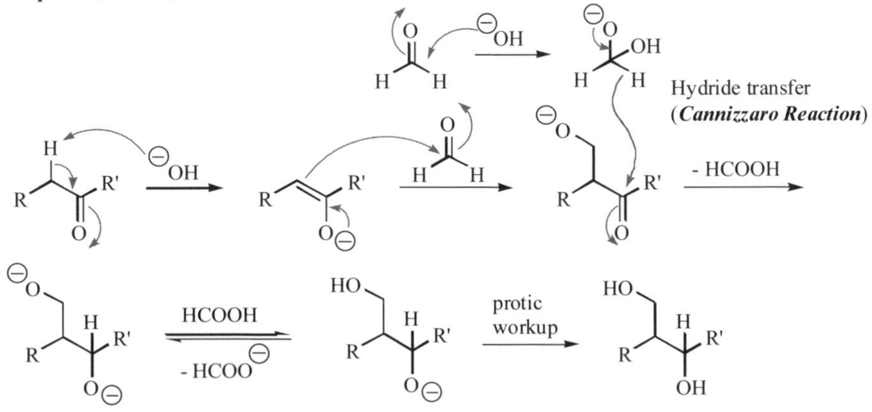

# Examples:

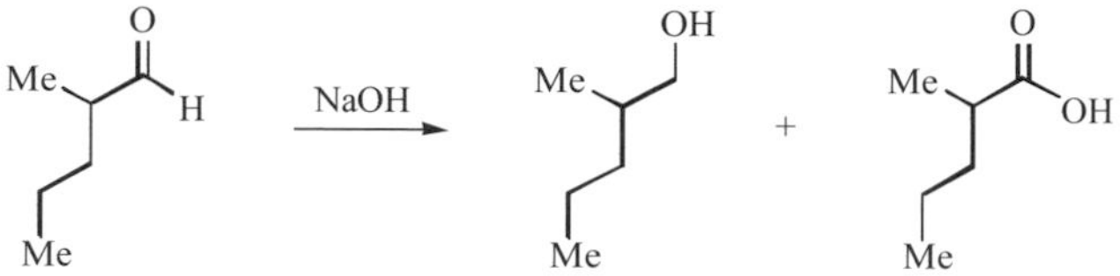

M. Hausermann, *Helvetica Chemica Acta* **1951**, <u>34</u>, 1211 (AN 1952:2485)

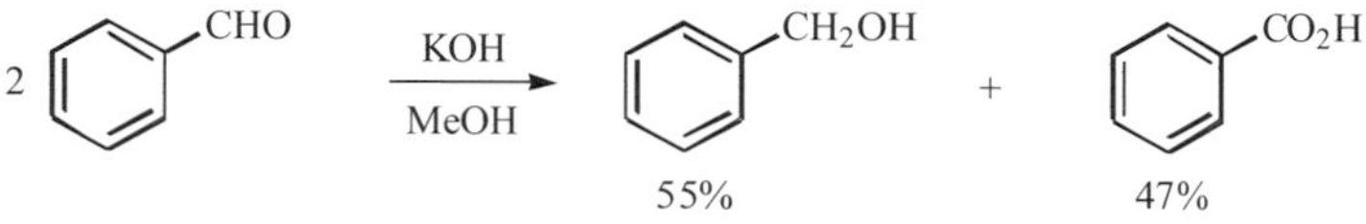

55%                                      47%

C. G. Swain, A. L. Powell, W. A. Sheppard, C. R. Morgan, *Journal of the American Chemical Society* **1979**, <u>101</u>, 3576

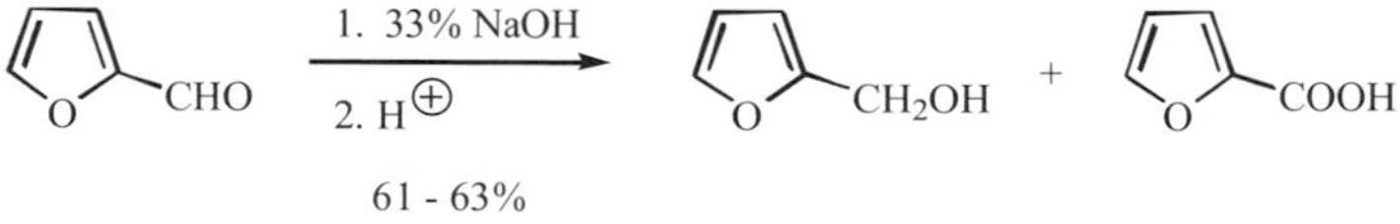

61 - 63%

W. C. Wilson, *Organic Syntheses*, <u>CV1</u>, 276

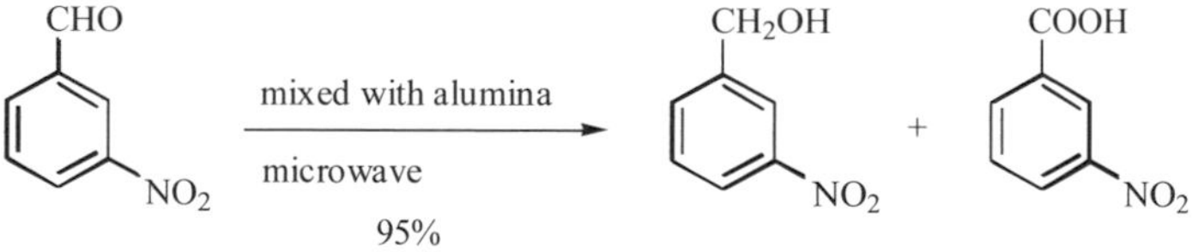

95%

A. Pourjavadi, B. Soleimanzadeh, G. B. Marandi, *Reactive & Functional Polymers* **2002**, <u>51</u>, 49

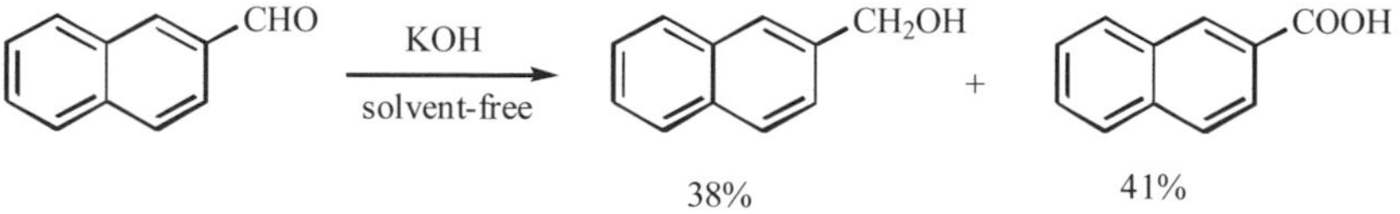

38%                                      41%

K. Yoshizawa, S. Toyota, F. Toda, *Tetrahedron Letters* **2001** <u>42</u>, 7983

# Cargill Rearrangement

## The Reaction:

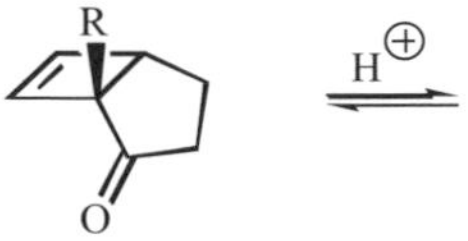

## Proposed Mechanism:

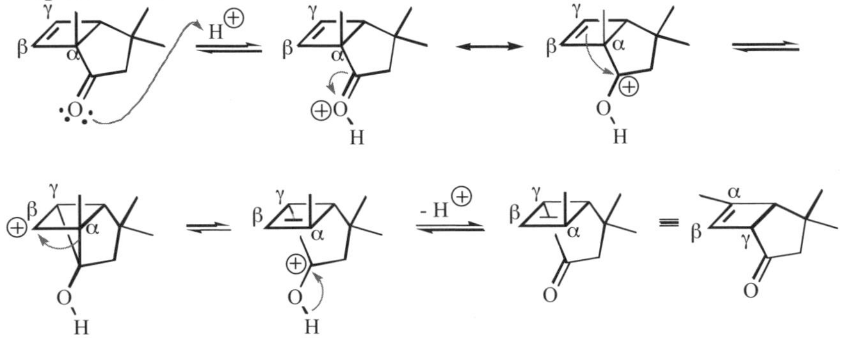

R. L. Cargill, T. E. Jackson, N. P. Peet, D. M. Pond, *Accounts of Chemical Research* **1974**, <u>7</u>, 106

## Notes:

The name of this rearrangement is applied to acid conditions on β-γ unsaturated ketones where either (or both) the carbonyl and/or alkene bond is in a strained environment.

In later works, the name seems to include highly congested and strained multi-ring systems.

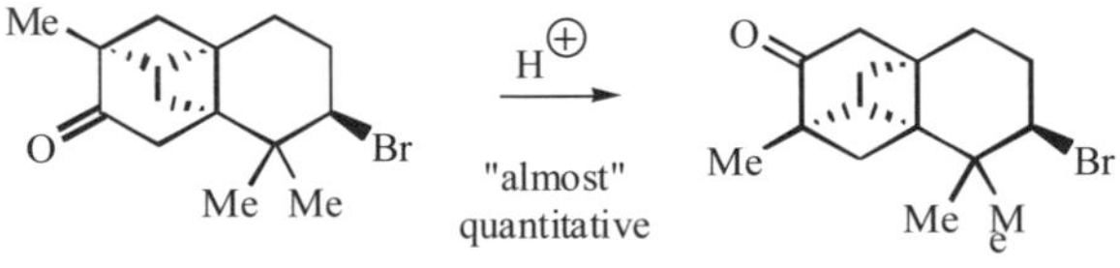

S. N. Fedorov, O. S. Radchenko, L. K. Shubina, A. I. Kalinovsky, A. V. Gerasimenko, D. Y. Popov, V. A. Stonik *Journal of the American Chemical Society* **2001**, <u>123</u>, 504

## Examples:

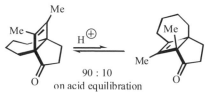

90 : 10
on acid equilibration

R. L. Cargill, T. E. Jackson, N. P. Peet, D. M. Pond, *Accounts of Chemical Research* **1974**, <u>7</u>, 106

93%

A. B. Smith, III, P. J. Jerris, *Journal of the American Chemical Society* **1981**, <u>103</u>, 194

R. L. Cargill, D. M. Ponds, S. O. LeGrande *Journal of Organic Chemistry* **1970**, <u>35</u>, 359

# Carroll Rearrangement (Kimel-Cope Rearrangement)

## The Reaction:

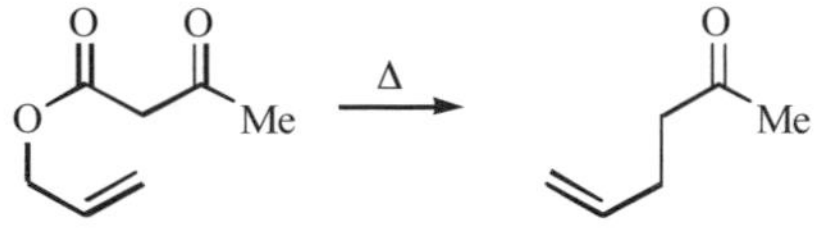

## Proposed Mechanism:

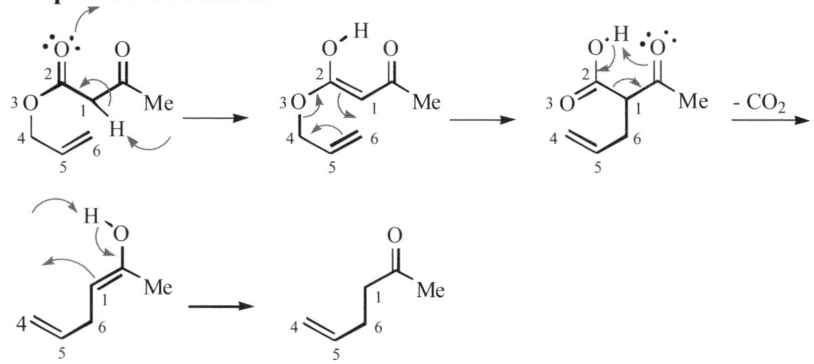

## Notes:

See *Claisen, Cope and Related Rearrangements*.

Enol formation provides the necessary "1,5-diene" for the "*Cope-like*" rearrangement. If one provides an enolate anion, the ionic nature of the reaction provides the expected acceleration of rate.

*Anion-accelerated Carroll Rearrangement:*

1. LDA, THF
2. toluene, Δ

60%

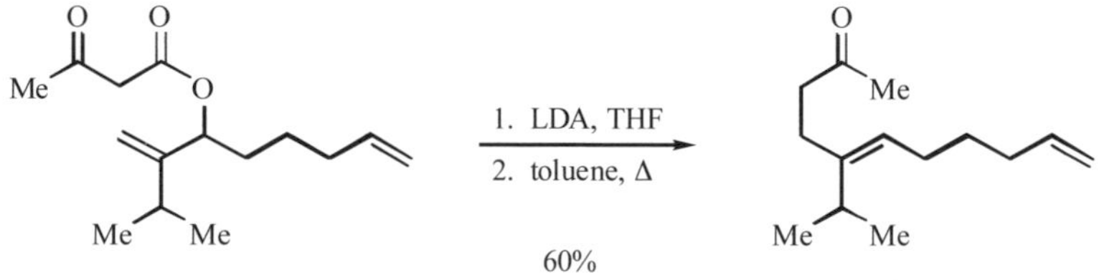

B. Shi, N. A. Hawrylik, B. B. Snider, *Journal of Organic Chemistry* **2003**, <u>68</u>, 1030

## Examples:

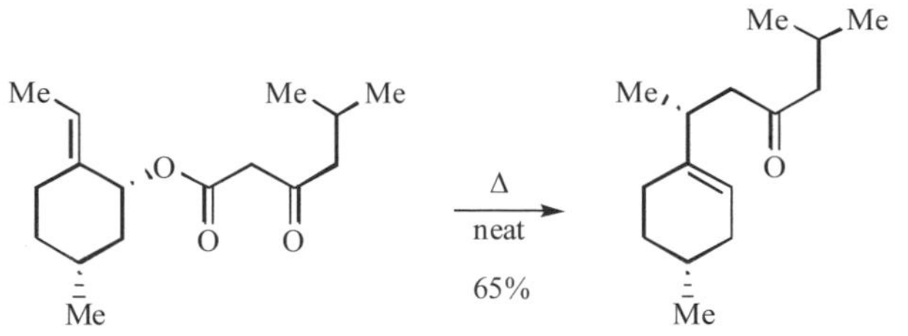

M. Koreeda, L. Brown, *Journal of Organic Chemistry* **1983**, <u>48</u>, 2122

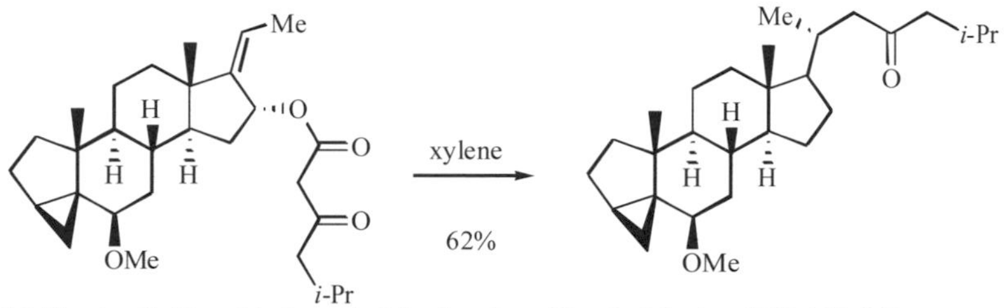

M. Tanabe, K. Hayashi, *Journal of the American Chemical Society* **1980**, <u>102</u>, 862

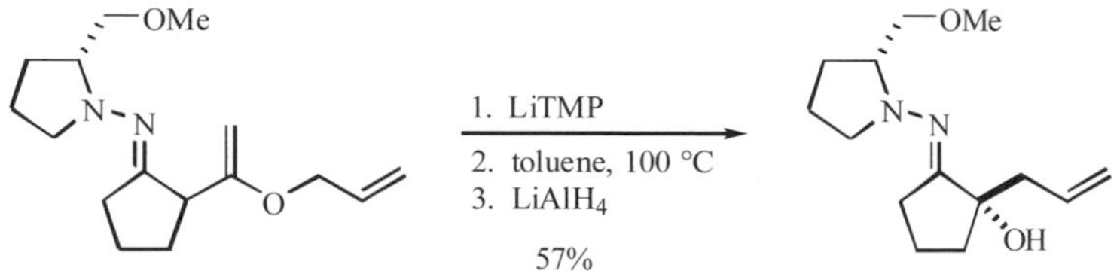

57%

D. Enders, M. Knopp, *Tetrahedron* **1996**, <u>52</u>, 5805

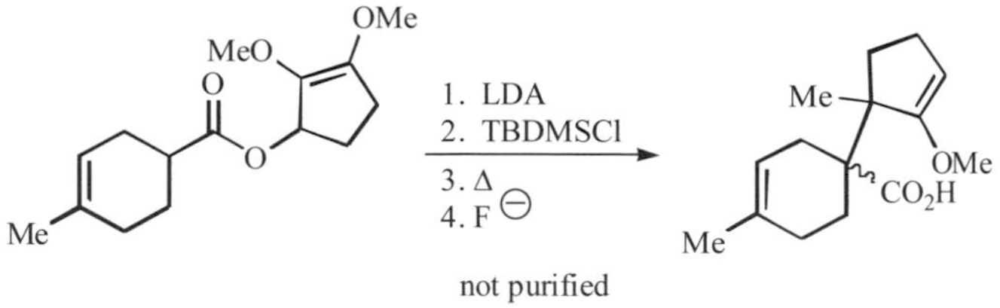

not purified

G. A. Kraus, P. J. Thomas, *Journal of Organic Chemistry* **1986**, <u>51</u>, 503

# Castro-Stephens Coupling

**The Reaction:**

$$Cu(I) \quad H-C\equiv C-R$$

$$Ar-X \quad + \quad Cu-C\equiv C-R \quad \xrightarrow[\Delta]{pyridine} \quad Ar-C\equiv C-R$$

**Proposed Mechanism:**

See *Alkyne Coupling*

$$H\!\!=\!\!=\!\!R \quad \xrightarrow{CuX} \quad H\!\!=\!\!=\!\!R \quad \xrightarrow{Et_3N} \quad Cu\!\!=\!\!=\!\!R \quad \xrightarrow{Ar-X}$$

$$\underset{X}{\overset{Cu}{|}}$$

$$\underset{X}{\overset{|}{Ar-Cu}}\!\!=\!\!=\!\!R \quad \longrightarrow \quad Ar\!\!=\!\!=\!\!R \quad + \quad CuX$$

**Notes:**

It appears that common use of the Name now extends to halides other than Ar-X. Thus, for example:

J. D. White, R. G. Carter, K. F. Sundermann, M. Wartmann, *Journal of the American Chemical Society,* **2001,** 123, 5407

**Examples:**

M. A. Ciufolini, J. W. Mitchell, F. Roschangar, *Tetrahedron Letters* **1996,** 37, 8281

Although the authors call this reaction a *Castro-Stephens reaction,* they point out that it is really a *Linstrumelle modification of the Sonogashira protocol.*

D. Guillerm, G. Linstrumelle, *Tetrahedron Letters* **1985,** 26, 3811

J. Mulzer, M. Berger, *Tetrahedron Letters*, **1998**, <u>39</u>, 803

R. Garg, R. S. Coleman, *Organic Letters*, **2001**, <u>3</u>, 3487

G. A. Krause, K. Frazier, *Tetrahedron Letters*, **1978**, <u>19</u>, 3195

# Chan Alkyne Reduction

## The Reaction:

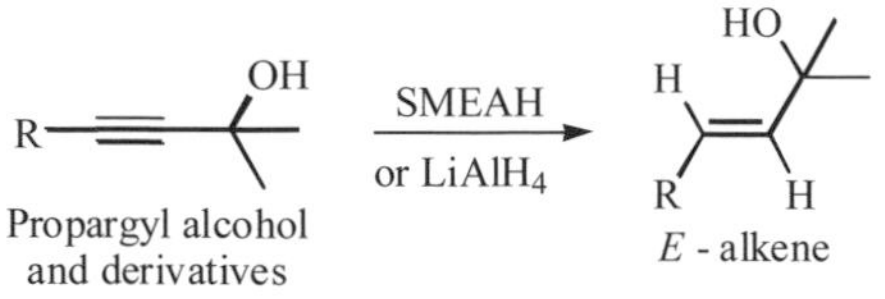

Propargyl alcohol
and derivatives

*E* - alkene

## Proposed Mechanism:

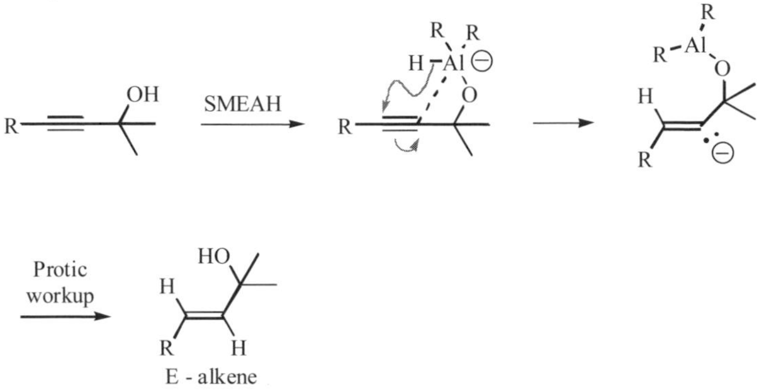

*E* - alkene

## Notes:

Reagent preparation:

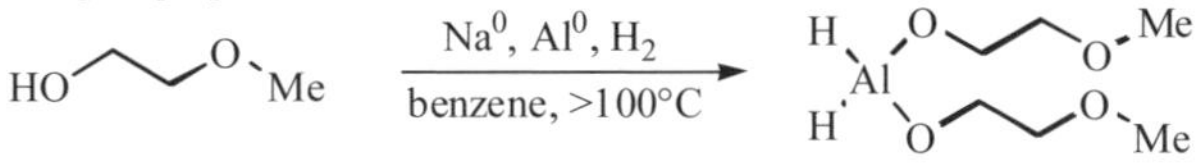

### *SMEAH*

Sodium bis(2-methoxymthoxy)aluminum Hydride

The reaction bears similarity to the **Whiting Reaction**:

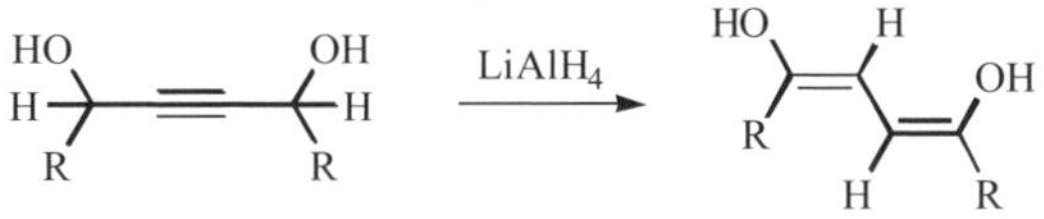

## Examples:

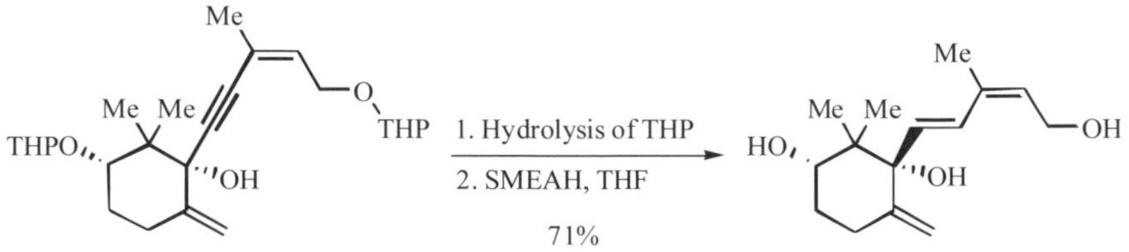

1. Hydrolysis of THP
2. SMEAH, THF

71%

H. Yamamoto, T. Oritani, *Phytochemistry,* **1995**, <u>40</u>, 1033

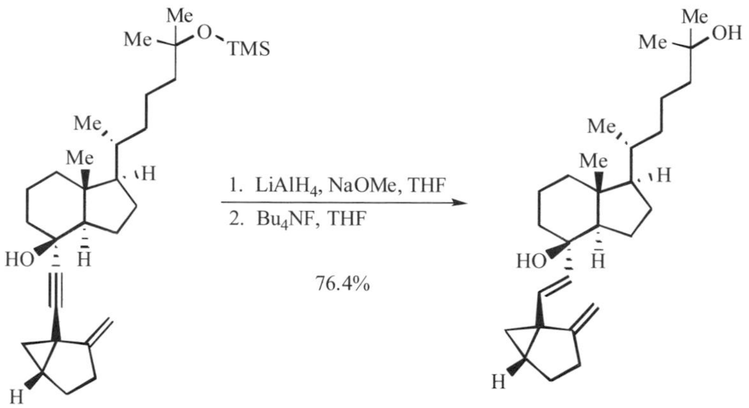

1. LiAlH$_4$, NaOMe, THF
2. Bu$_4$NF, THF

76.4%

M. M. Kabat, J. Kiegiel, N. Cohen, K. Toth, P. M. Wovkulich, M. R. Uskokovic, *Journal of Organic Chemistry*, **1996**, <u>61</u>, 118

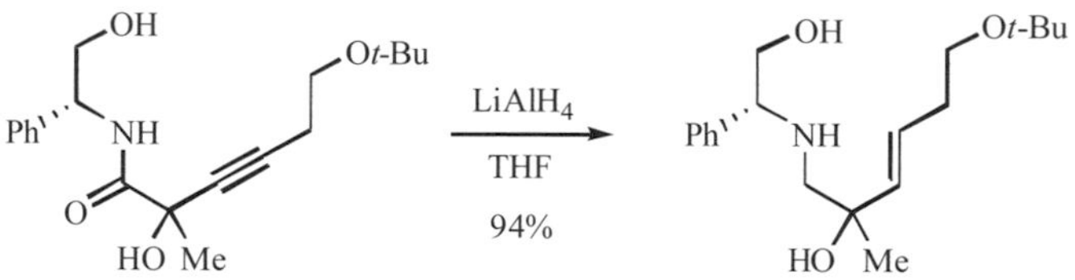

LiAlH$_4$

THF

94%

C. Agami, M. Cases, F. Couty, *Journal of Organic Chemistry*, **1994**, <u>59</u>, 7937

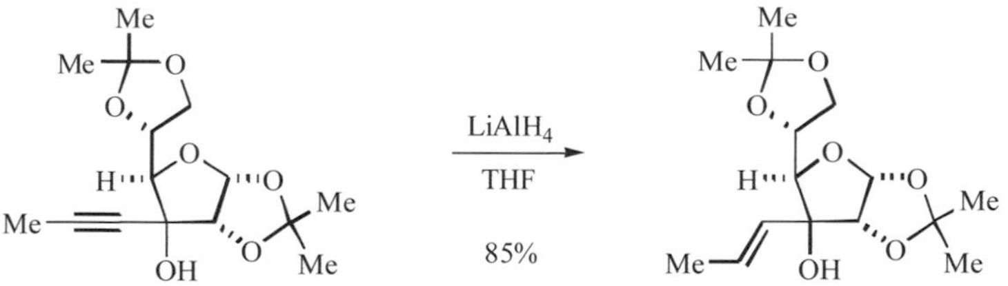

LiAlH$_4$

THF

85%

T. Eguchi, T. Koudate, K. Kakinuma, *Tetrahedron*, **1993**, <u>49</u>, 4527

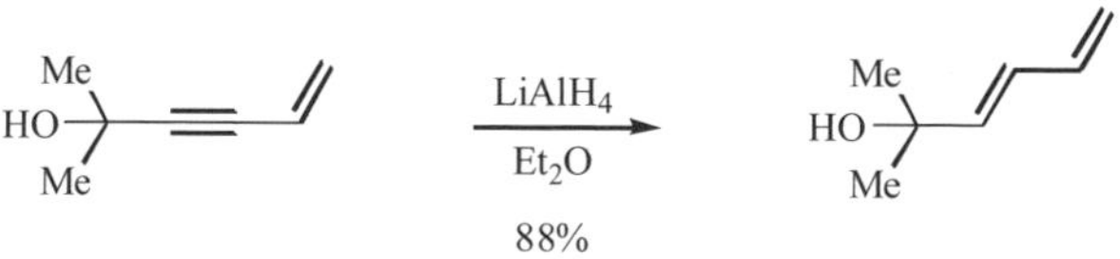

LiAlH$_4$

Et$_2$O

88%

E. B. Bates, E. R. H. Jones, M. C. Whiting, *Journal of the Chemical Society*, **1954**, 1854

# Chan-Lam Coupling

## The Reaction:

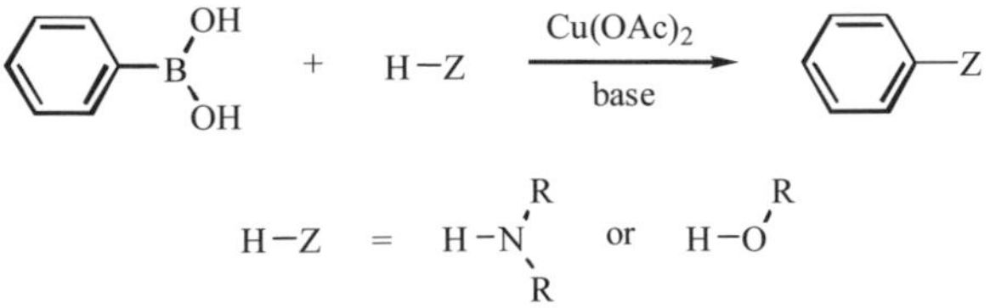

## Proposed Mechanism:

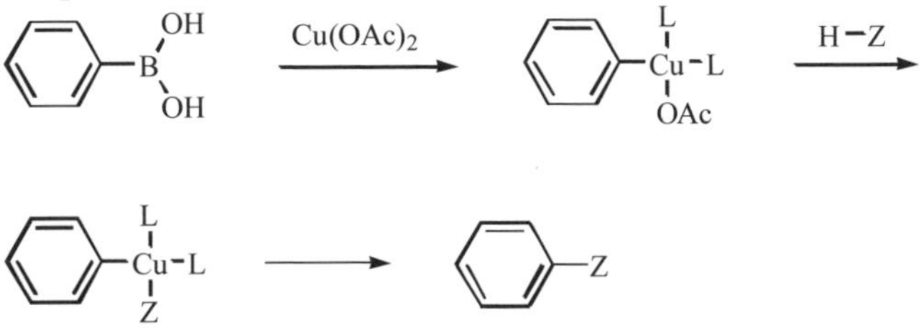

## Notes:

Evans describes the possibility of a common intermediate for the *Ullman* and *Chan-Lam reactions*.
D. A. Evans, J. Katz, T. R. West, *Tetrahedron Letters* **1998**, 39, 2937

## Examples:

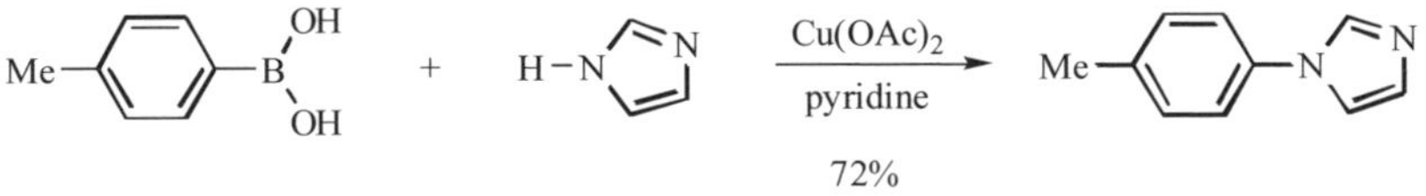

72%

P. Y. S. Lam, C. G. Clark, S. Saubern, J. Adams, M. P. Winters, D. M. T. Chan, A. Combs, *Tetrahedron Letters* **1998**, 39, 2941

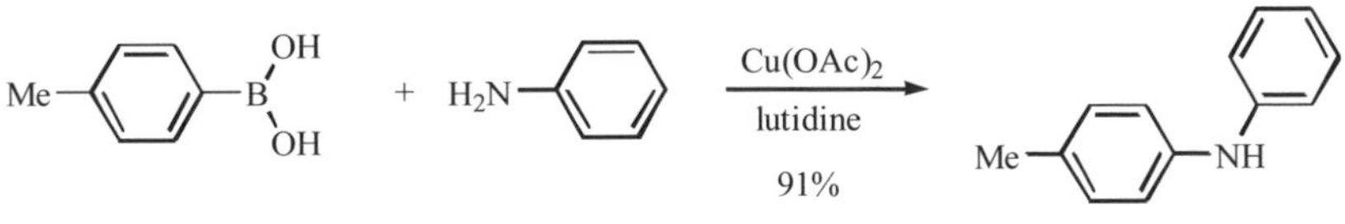

91%

J. C. Antilla, S. L. Buchwald, *Organic Letters* **2001**, 3, 2077

J. C. Antilla, S. L. Buchwald, *Organic Letters* **2001**, <u>3</u>, 2077

D. M. T. Chan, K. L. Monaca, R.-P. Wanag, M. P. Winters, *Tetrahedron Letters* **1998**, <u>39</u>, 2933

D. A. Evans, J. Katz, T. R. West, *Tetrahedron Letters* **1998**, <u>39</u>, 2937

# Chapman Rearrangement

## The Reaction:

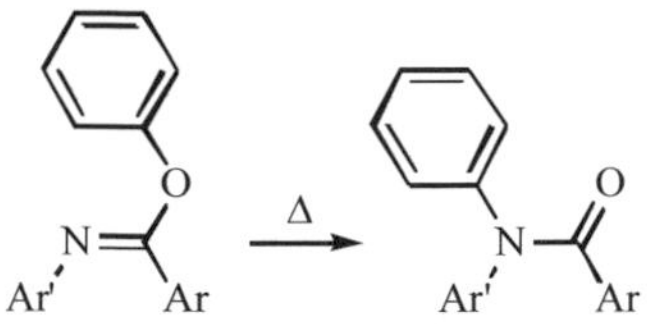

## Proposed Mechanism:

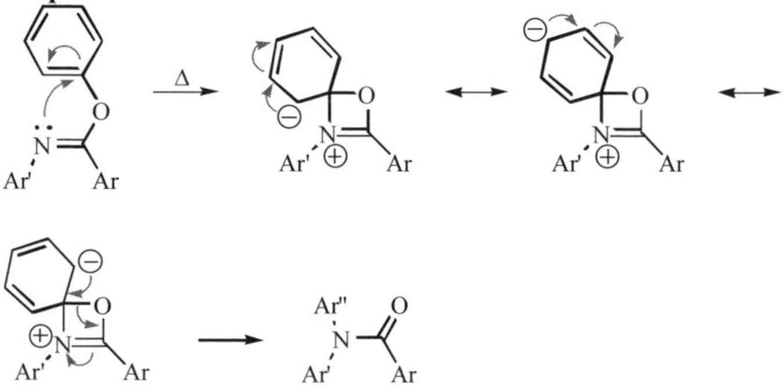

## Notes:

Similar to the *Newman-Kwart Rearrangement*:

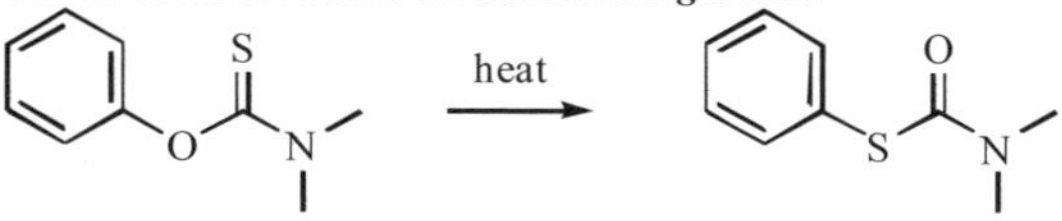

## Examples:

J. D. McCullough, Jr., D. Y. Curtin, I. C. Paul, *Journal of the American Chemical Society* **1972**, <u>94</u>, 874

D. M. Hall, E. E. Turner, *Journal of the Chemical Society* **1945**, 694

The methyl migration is rationalized.

T. Kuroda, F. Zuzuki, *Tetrahedron Letters* **1992**, <u>33</u>, 2027

W. G. Dauben, R. L. Hodgson, *Journal of the American Chemical Society* **1950**, <u>72</u>, 3479

# Chichibabin Reaction

## The Reaction:

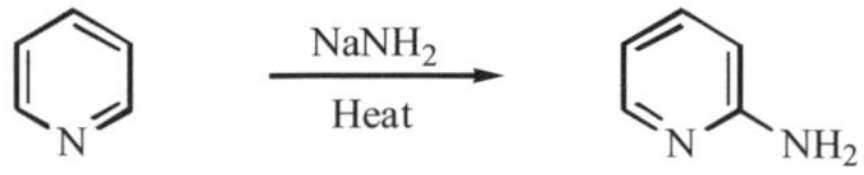

## Proposed Mechanism:

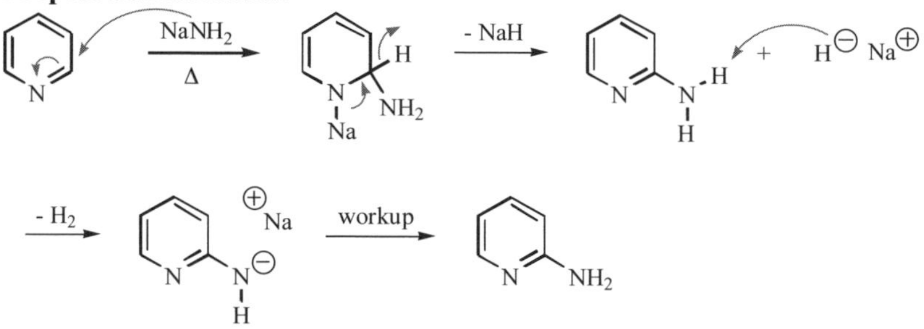

## Notes:

See: M. B. Smith, J. March in *March's Advanced Organic Chemistry*, 5th ed., John Wiley and Sons, Inc., New York, 2001, p. 873.

This reaction will work with other nitrogen containing heterocycles such as quinoline.

$RNH^{\ominus}$ and $R_2N^{\ominus}$ will also work

Attack at the 2- or 4- positions can leave negative charge on the ring nitrogen.

A related reaction: *Ziegler alkylation*:

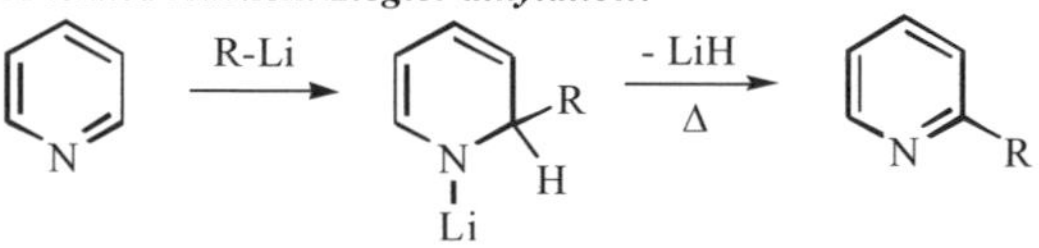

In the pyrimidine series an $S_N$(ANRORC) mechanism has been proposed:

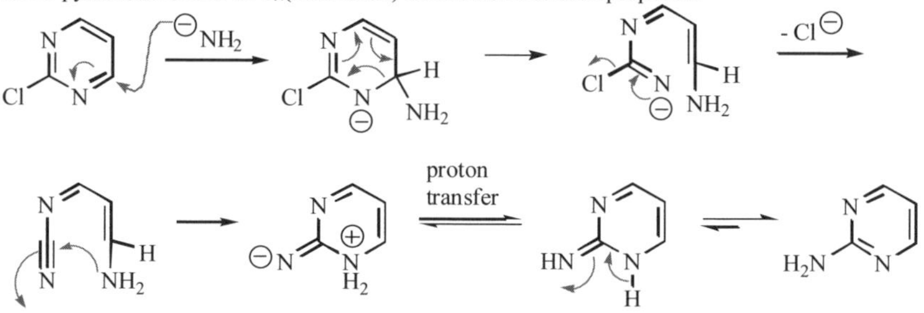

H. C. van der Plas, *Accounts of Chemical Research* **1978**, 11, 462

## Examples:

M. Palucki, D. L. Hughes, N. Yasuda, C. Yang, P. J. Reider, *Tetrahedron Letters* **2001**, 42, 6811

N. J. Kos, H. C. van der Plas, B. van Veldhuizen, *Journal of Organic Chemistry* **1979**, 44, 3140

Y. Kobayashi, I. Kumadaki, S. Taguchi, Y. Hanzawa, *Tetrahedron Letters* **1970**, 11, 3901

F. W. Bergstrom, H. G. Sturz, H. W. Tracy, *Journal of Organic Chemistry* **1945**, 11, 239

# Chugaev Reaction

## The Reaction:

xanthate                    less substituted alkene

## Proposed Mechanism:

less substituted alkene

A *cis* elimination, often providing the least substituted alkene.

## Notes:

T. Laue, A. Plagens, *Named Organic Reactions*, John Wiley and Sons, Inc., New York, 1998, pp. 42-44; M. B. Smith, J. March in *March's Advanced Organic Chemistry*, 5th ed., John Wiley and Sons, Inc., New York, 2001, p. 1330; H. R. Nace, *Organic Reactions* **12**, 2.

Starting Material Preparation:

See the related ***Grieco-Sharpless Elimination***.

## Examples:

74%

T. M. Meulemans, G. A. Stork, F. Z. Macaev, B. J. M. Jansen, A. deGroot, *Journal of Organic Chemistry* **1999**, <u>64</u>, 9178

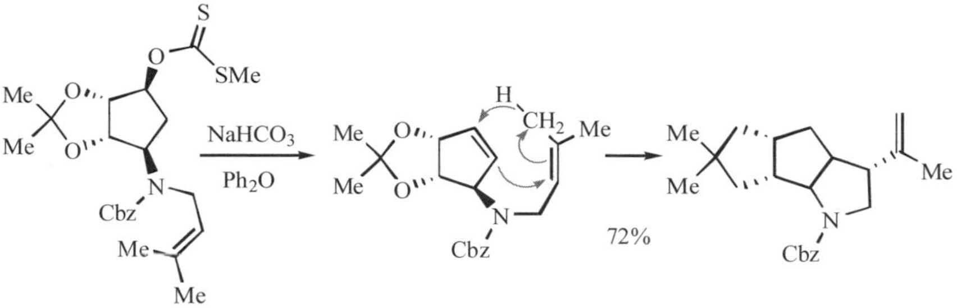

H. Nakagawa, T. Sugahara, K. Ogasawara, *Organic Letters* **2000**, *2*, 3181

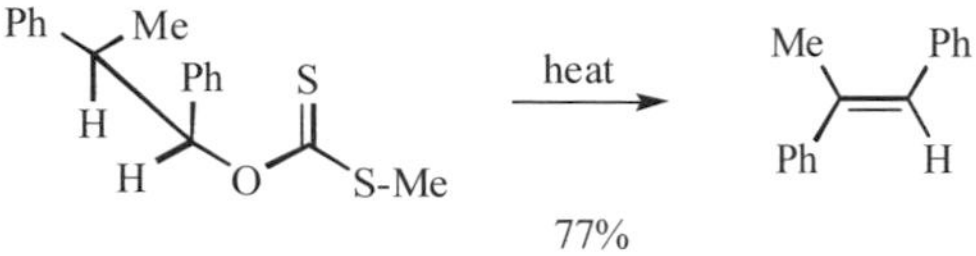

77%

D. J. Cram and F. A. A. Elhafez, *Journal of the American Chemical Society* **1952**, *74*, 5828

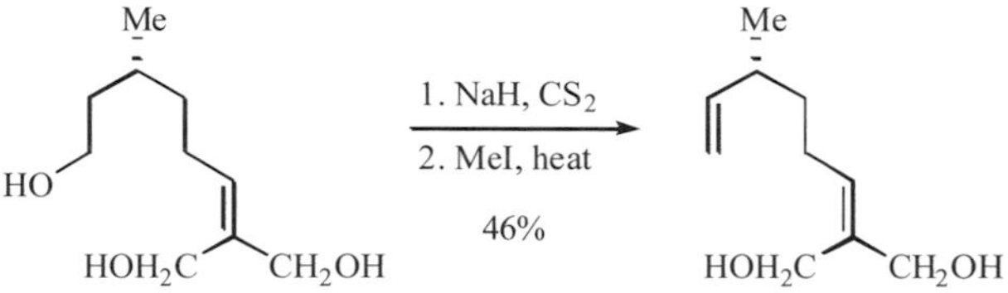

1. NaH, CS₂

2. MeI, heat

46%

G. Cernigliano, P. Kocienski, *Journal of Organic Chemistry* **1977**, *42*, 3622

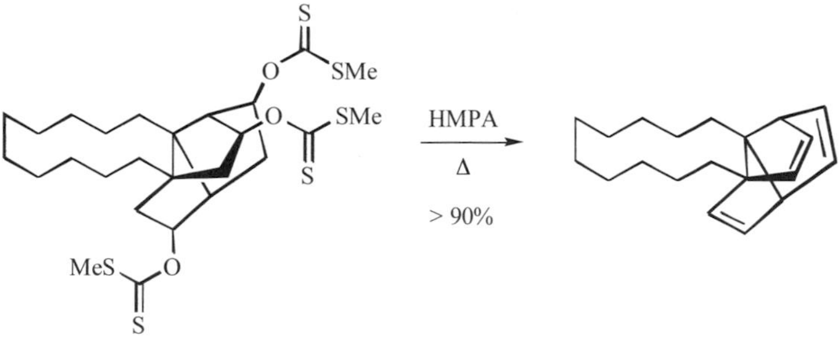

HMPA

Δ

> 90%

X. Fu, C. M. Cook, *Tetrahedron Letters* **1990**, *31*, 3409

# Ciamician-Dennstedt Rearrangement

## The Reaction:

Dichlorocarbene
OR
Cl
N
H

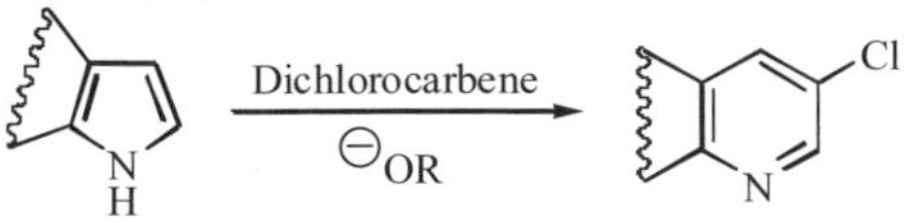

## Proposed Mechanism:

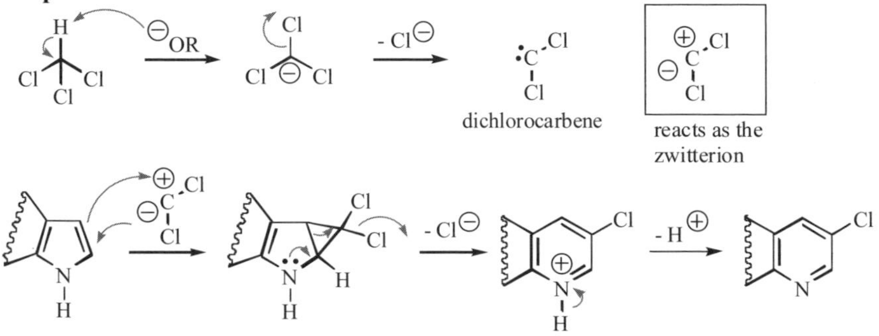

OR
- Cl
dichlorocarbene
reacts as the zwitterion
- H

## Notes:

empty p-orbital
Two electrons

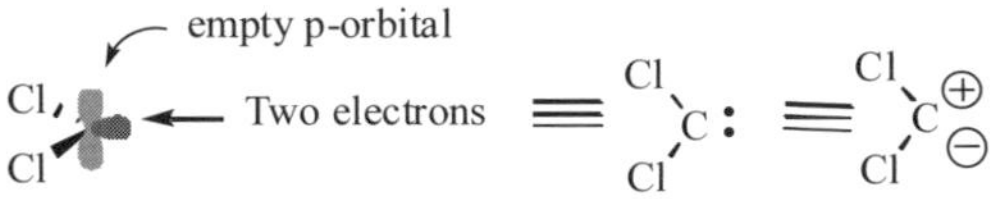

The ***Skattebol Rearrangement*** is another dihalocarbene-based rearrangement:

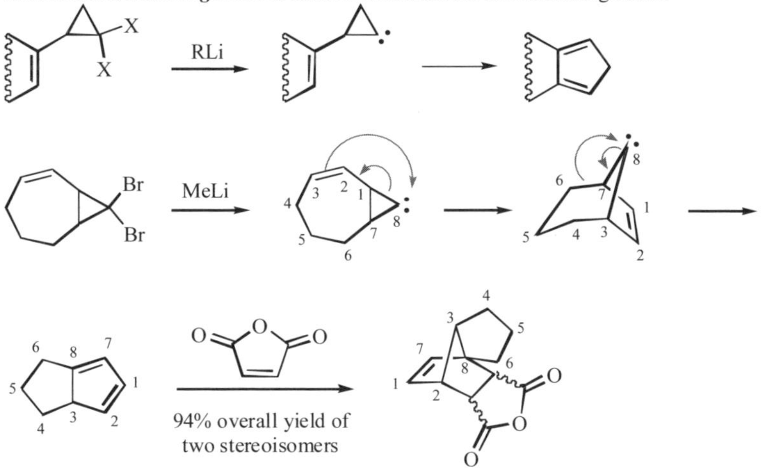

RLi
MeLi
94% overall yield of
two stereoisomers

K. H. Holm, L. Skattebol, *Tetrahedron Letters* **1977**, 18, 2347

L. A. Paquette, M. Gugelchuk, M. L. McLaughlin, *Journal of Organic Chemistry* **1987**, 52, 4372

## Examples:

F. DeAngelis, A. Inesi, M. Feroci, R. Nicoletti, *Journal of Organic Chemistry* **1995**, 60, 445

K. C. Joshi, R. Jain, S. Arora, *Journal of the Indian Chemical Society* **1993**, 70, 567 (AN 1994:508439)

E. R. Alexander, A. B. Herrick, T. M. Roder, *Journal of the American Chemical Society* **1950**, 72, 2760

*a dichlorocarbene precursor*

W. E. Parham, R. W. Davenport, J. B. Biasotti, *Tetrahedron Letters* **1969**, 10, 557

Ph-Hg-CX₃ are known as **Seyferth Reagents**.

# Claisen Condensation (ester attacking ester)

## The Reaction:

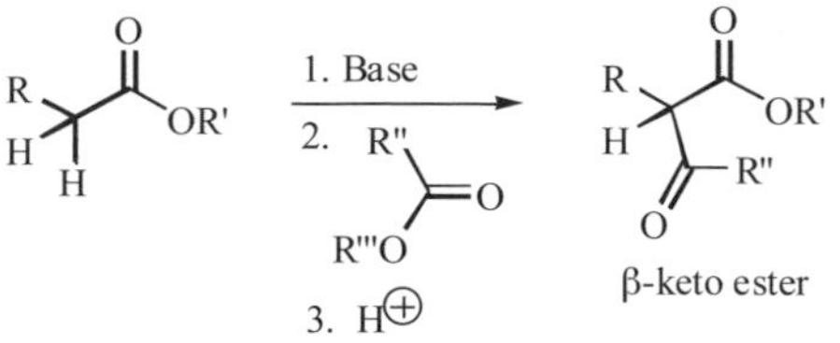

β-keto ester

## Proposed Mechanism:

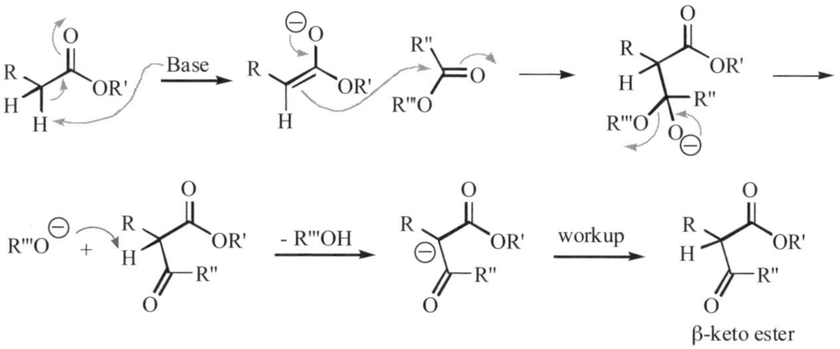

β-keto ester

## Notes:

R" is usually a group that cannot form an enolate, such as a phenyl ring.

It is important to note that an equivalent of base must be used for this reaction; unlike the *Aldol Condensation*, this cannot be used catalytically.

See: T. Laue, A. Plagens, *Named Organic Reactions*, John Wiley and Sons, Inc., New York, 1998, pp. 45-48.

### Aldehyde or ketone attacking ester

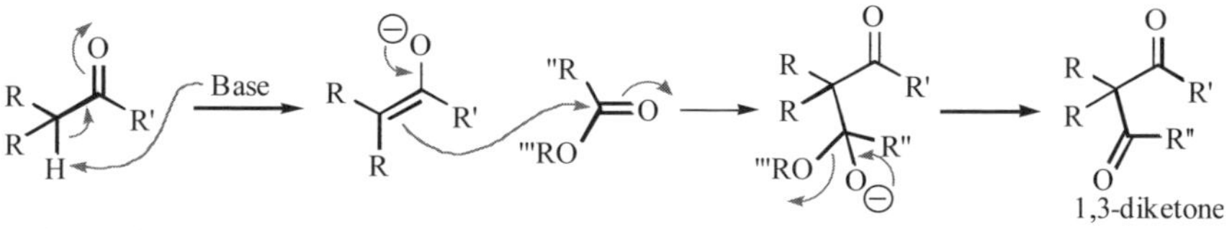

1,3-diketone

R" is usually a group that cannot form an enolate.

## Examples:

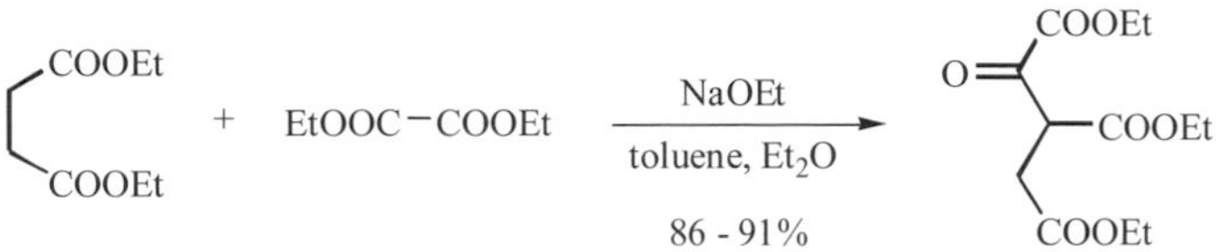

NaOEt

toluene, Et$_2$O

86 - 91%

L. Friedman, E. Kosower, *Organic Syntheses,* CV3, 510

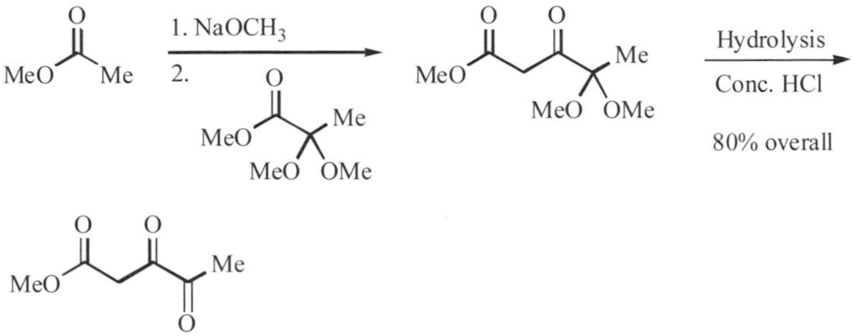

1. NaOCH$_3$

2.

Hydrolysis

Conc. HCl

80% overall

A. G. Cameron, A. T. Hewson, M. I. Osammur *Tetrahedron Letters* **1984**, 25, 2267

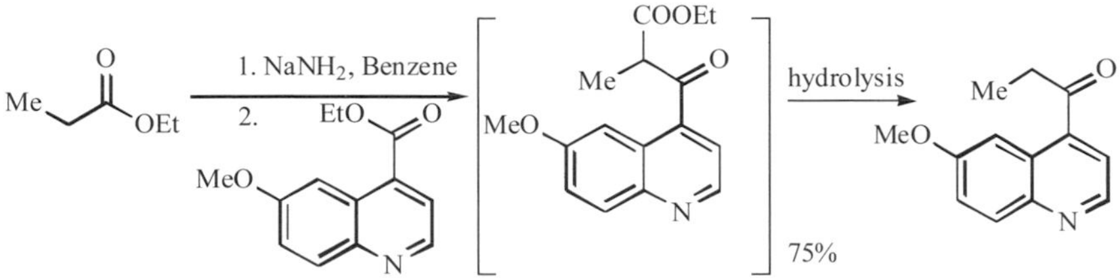

1. NaNH$_2$, Benzene

2.

hydrolysis

75%

J. W. Cornforth, R. H. Cornforth, *Journal of the Chemical Society* **1948**, 93

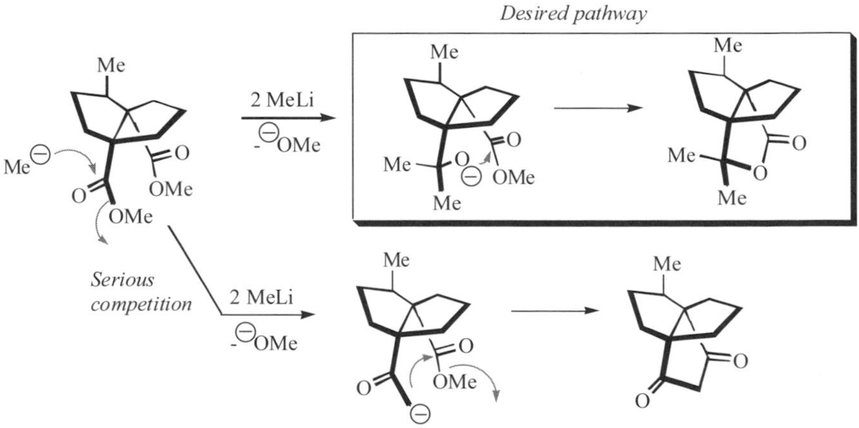

*Desired pathway*

2 MeLi

$-^{\ominus}$OMe

*Serious competition*

2 MeLi

$-^{\ominus}$OMe

B. P. Mundy, D. Wilkening, K. B. Lipkowitz *Journal of Organic Chemistry* **1985**, 50, 5727

# Claisen, Cope and Related Rearrangements

## Claisen-Type Reactions:

## Cope-Type Reactions:

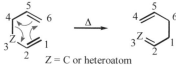

## Notes:

All reactions are classified as [3,3]-sigmatropic reactions. They are orbital symmetry regulated processes.

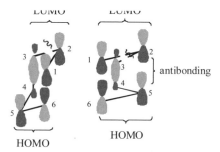

Z = C or heteroatom

By Frontier Molecular Orbital theory several important predictions can be made:
1. There will be a preferred chair transition-state for the reaction:

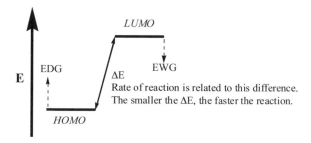

In the boat transition-state there is an *antibonding* interaction between C-2 and C-5.

2. Reactions will be accelerated by charged intermediates:

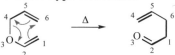

Rate of reaction is related to this difference.
The smaller the $\Delta E$, the faster the reaction.

## Claisen and Related Rearrangements

This reaction has developed a number of related protocols and variations. Some have earned sufficient recognition to act as "stand alone" reactions and will be found under their own headings in this monograph.

### Carroll Rearrangement (Kimel-Cope Rearrangement)

β-keto esters can rearrange to give β-keto acids which will decarboxylate to give δ,γ-unsaturated ketones.

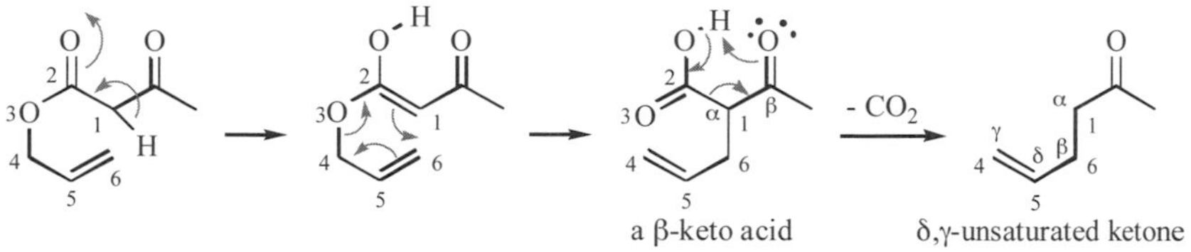

a β-keto acid                    δ,γ-unsaturated ketone

### Claisen-Arnold Reaction:

Enolate formation with an allyl ester can give 2-substituted carboxylic acids.

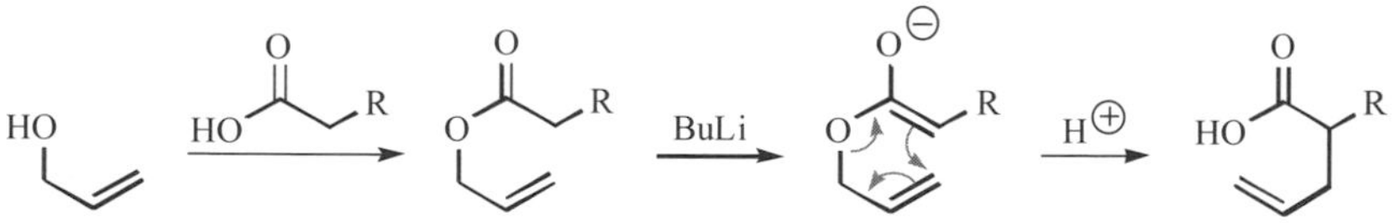

### Claisen-Eschenmoser Reaction (Eschenmoser-Claisen Rearrangement):

Amides are produced after rearrangement with heating.

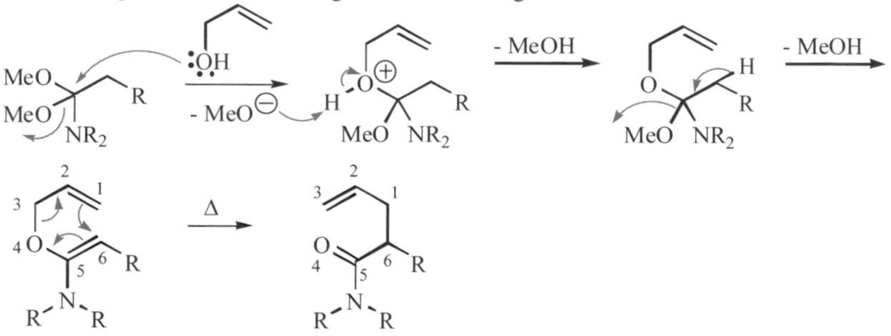

### Claisen-Ireland Reaction (Ireland-Claisen Rearrangement):

Formation of a silyl enol ether will generate an allyl vinyl ether which after rearrangement can be desilylated to give a carboxylic acid.

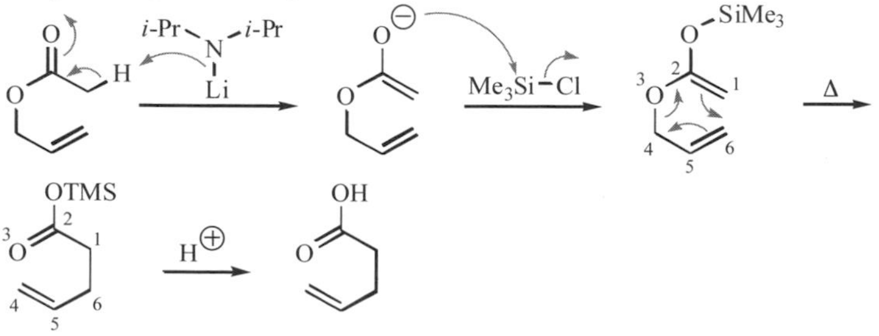

**Dauben-Dietsche Rearrangement**:

**Denmark Rearrangement**:
The *Claisen rearrangement* of phenyl sulfonyl substituted allyl vinyl ethers.

**Johnson-Claisen Rearrangement (Johnson Orthoester Rearrangement)**
Elimination of two equivalents of ROH after condensation between an allyl alcohol and an
orthoester will provide the allyl vinyl ether which undergoes the [3,3]-sigmatropic rearrangement.

**Marbet-Saucy Reaction / Variation**
A vinyl ether and allyl alcohol will react under acidic conditions to give the allyl vinyl ether which
then rearranges under the reaction conditions.

# Cope and Related Reactions

## *Aza-Cope Rearrangement*

## *Azo-Cope Rearrangement*

## *Cope Rearrangement*

![Cope Rearrangement scheme showing cyclohexene ring numbered 1-6 rearranging under Δ]

## *Oxy-Cope Rearrangement*

![Oxy-Cope Rearrangement scheme: HO-substituted diene to enol via Δ then keto-enol to aldehyde]

# Claisen Rearrangement (allyl phenyl ethers)

## The Reaction:

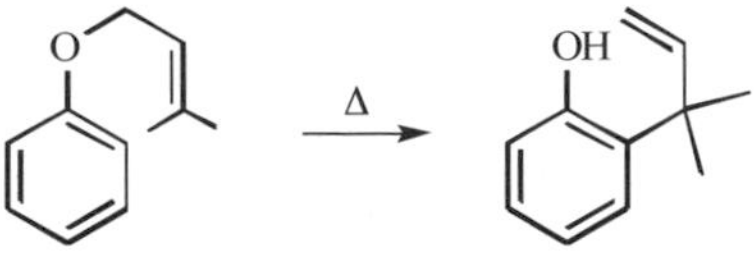

## Proposed Mechanism: a [3.3]-sigmatropic rearrangement

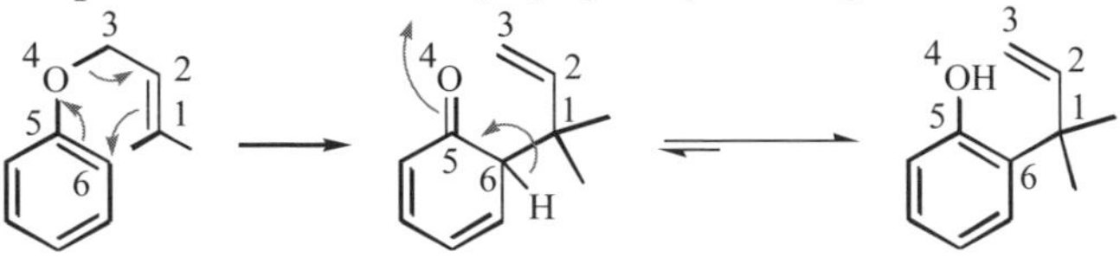

keto-enol tautomerism

Note that the pi-system of the allyl group is readily aligned over the aromatic ring, providing a 1,5-diene motif for the sigmatropic rearrangement.

## Notes:

If the ortho-positions are blocked (only one shown, below), rearrangement continues:

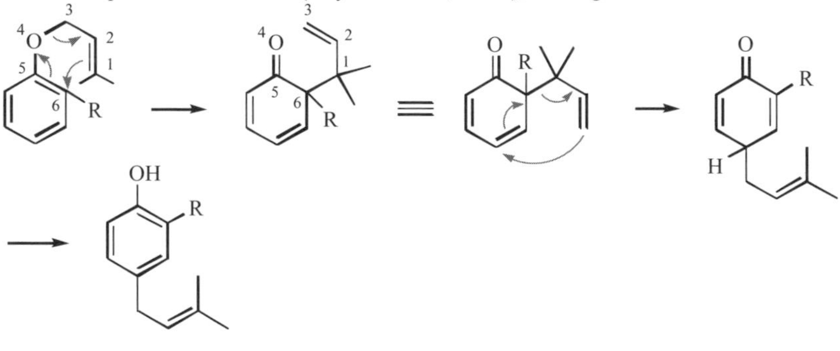

Simple [1,3]-rearrangements have been observed:

Florisil®

50%

F. X. Talams, D. B. Smith, A. Cervantes, F. Franco, S. T. Cutler, D. G. Loughhead, D. J. Morgans, Jr., R. J. Weikert *Tetrahedron Letters* **1997**, _38_, 4725

## Examples:

D. B. Smith, T. R. Elworthy, D. J. Morgns, Jr., J. T. Lenson, J. W. Patterson, A. Vasquez, A. M. Waltos, *Tetrahedron Letters* **1996**, 37, 21

**\*Reagent used for the enantioselective rearrangement**:
(S,S)-2-Bromo-4,5-diphenyl-1,3-tosyl-1,3-diaza-2-borolidine
H. Ito, A. Sato, T. Taguchi, *Tetrahedron Letters* **1997**, 38, 4815

J. Barluenga, R. Sanz, F. J. Fananas, *Tetrahedron Letters* **1997**, 38, 6103

S. Lambrecht, H. J. Schaefer, R. Froehlich, M. Grehl, *Synlett*, **1996**, 283 (AN 1996:201759)

# Claisen Rearrangement (allyl vinyl ethers)

## The Reaction:

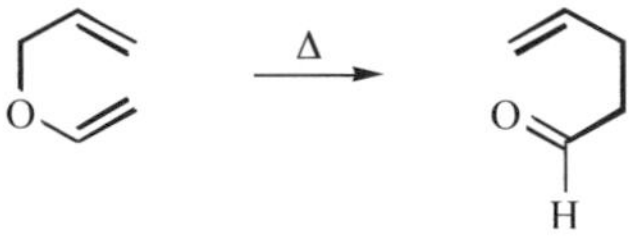

## Proposed Mechanism:

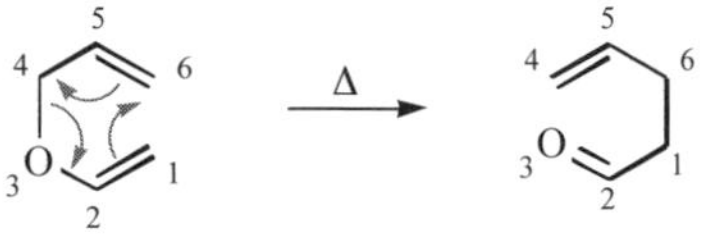

## Notes:

See: ***Claisen, Cope and Related Rearrangements***

*Ficini-Ynamine-Claisen Rearrangement*

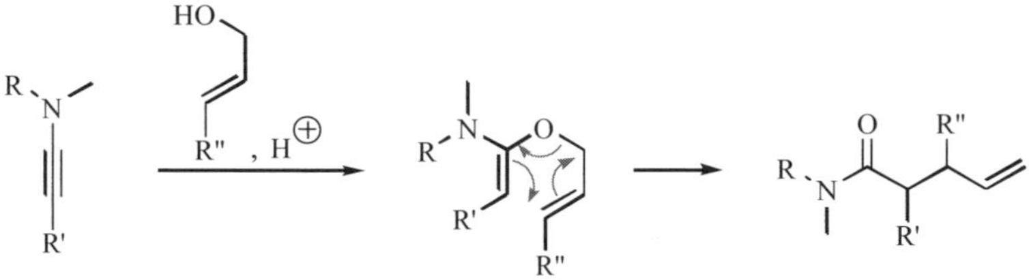

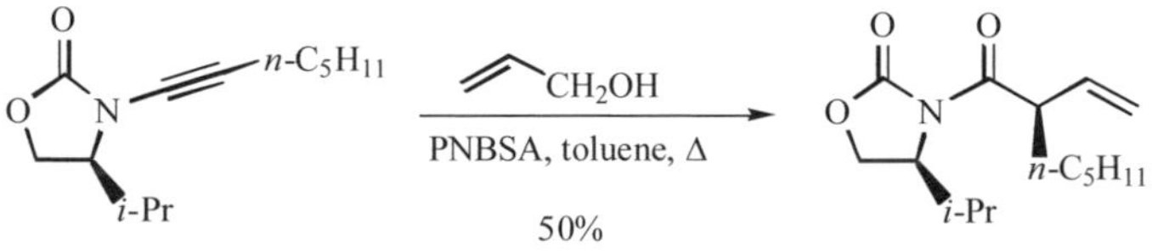

J. A. Mulder, R. P. Hsung, M. O. Frederick, M. R. Tracey, C. A. Zificsak, *Organic Letters* **2002**, <u>4</u>, 1383

## Examples:

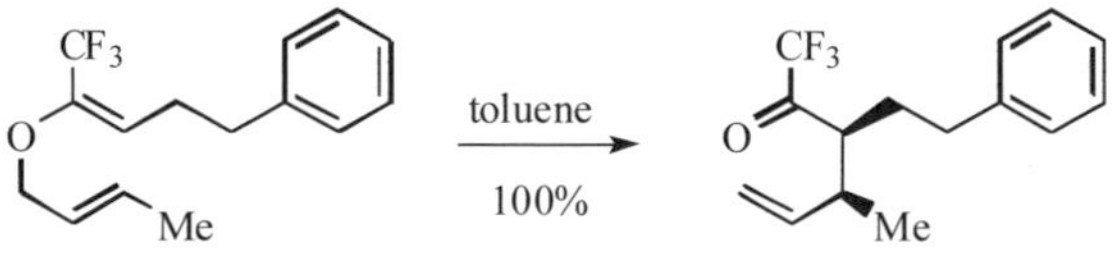

J.-P. Begue, D. Bonnet-Delpon, S.-W. Wu, A. M'Bida, T. Shintani, T. Nakai, *Tetrahedron Letters*, **1994**, <u>35</u>, 2907

L. A. Paquette, S.-Q. Sun, D. Friedrich, P. G. Savage, *Journal of the American Chemical Society* **1997**, 119, 8438

C. M. G. Philippo, V. Nha Huu, L. A. Paquette, *Journal of the American Chemical Society* **1991**, 113, 2762

Y. Masuyama, Y. Nimura, Y. Jurusu, *Tetrahedron Letters* **1992**, 33, 6477

β-ketoacid

E. Marotta, P. Righi, G. Rosini, *Organic Letters* **2000**, 2, 4145

T. V. Ovaska, S. E. Reisman, M. A. Flynn, *Organic Letters* **2001**, 3, 115

# Clemmensen Reduction

## The Reaction:

(zinc amalgam)

$$\overset{O}{\underset{}{\bigwedge}} \xrightarrow[\text{HCl}]{\text{Zn(Hg)}} \overset{H}{\underset{H}{\bigwedge}}$$

Aldehydes and Ketones

## Proposed Mechanism:

This mechanism is not yet resolved. There are a number of possibilities:

1. 

2. 

## Notes:

T. Laue, A. Plagens, *Named Organic Reactions*, John Wiley and Sons, Inc., New York, 1998, pp. 52-53; M. B. Smith, J. March in *March's Advanced Organic Chemistry*, 5th ed., John Wiley and Sons, Inc., New York, 2001, p. 1547; E. L. Martin, *Organic Reactions* **1**, 7; E. Vedejs, *Organic Reactions* **22**, 3.

Other mechanistic interpretations:
Zn-carbene: J. Burdon, R. C. Price, *Chemical Communications* **1986**, 893
One electron transfer: M. L. DiVona, V. Ruanati, *Journal of Organic Chemistry* **1991**, 56, 4269

## Examples:

$$\xrightarrow[\text{HCl}]{\text{Zn-Hg}}$$

30%          +          45%

F. J. C. Martins, L. Fourie, H. J. Venter, P. L. Wessels *Tetrahedron* **1990**, 46, 623

$$\xrightarrow[\text{Et}_2\text{O}]{\text{Zn, HCl}}$$

40%

I. Elphimoff-Felkin, P. Sarda, *Tetrahedron Letters* **1983**, 24, 4425

Zn-Hg, HCl
toluene

14%

+

27%

Probably via:

+

44%

S. K. Talapatra, S. Chakrabarti, A. K. Mallik, B. Talapatra, *Tetrahedron* **1990**, <u>46</u>, 6047

Zn
HCl

Yield
not reported

J. B. Thomas, K. M. Gigstad, S. E. Fix, J. P. Burgess, J. B. Cooper, S. W. Mascarella, B. E. Cantrell,
D. M. Zimmerman, F. I. Carroll, *Tetrahedron Letters* **1999**, <u>40</u>, 403

Zn, HOAc
ultra-sound

90%

J. A. R. Salvador, M. L. SaeMelo, A. S. Campos Neves *Tetrahedron Letters* **1993**, <u>34</u>, 361

Zn-Hg
HCl

+

51%                                              6%

H₂, Pd-C

70%

K. M. Werner, J. M. de los Santos, S. M. Weinreb, M. Shang, *Journal of Organic Chemistry* **1999**,
<u>64</u>, 686

# Collman Carbonylation Reaction

## The Reaction

## Proposed Mechanism:

J. P. Collman, *Accounts of Chemical Research* **1975**, $\underline{8}$, 342

## Notes:

Reagent preparation:

K- Selectride                                    *Collman's Reagent*

For an alternative approach:  A. Schoenberg, R. F. Heck, *Journal of the American Chemical Society* **1974**, $\underline{96}$, 7761

$$R-X \ + \ CO \ + \ H_2 \ \xrightarrow[\text{R}_3\text{N}]{\text{PdX}_2(\text{PPh}_3)_2} \ \underset{R}{\overset{O}{\|}}{\text{C}}-H \ + \ HX$$

## Examples:

D. Bankston, F. Fang, E. Huie, S. Xie, *Journal of Organic Chemistry* **1999**, $\underline{64}$, 3461

K. Yoshida, H. Kuwata, *Journal of the Chemical Society, Perkin Transactions 1* **1996**, 1873

R. G. Finke, T. N. Sorrel, *Organic Synthesis* <u>CV6</u>, 807

J. A. Gladysz, W. Tam, *Journal of Organic Chemistry* **1978**, 43, 2279

# Combes Quinoline Synthesis

## The Reaction:

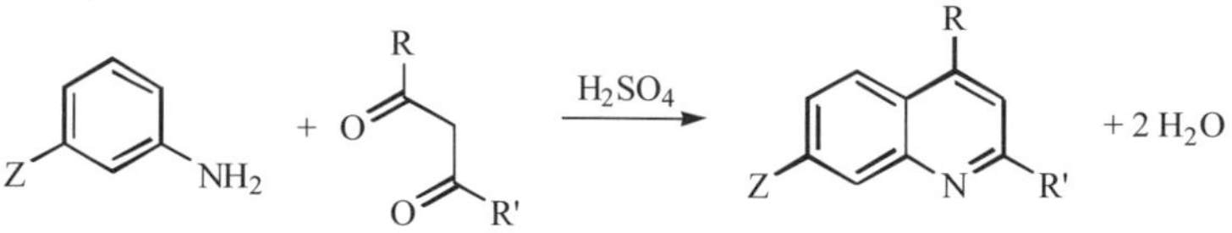

$$+ 2 H_2O$$

## Proposed Mechanism:

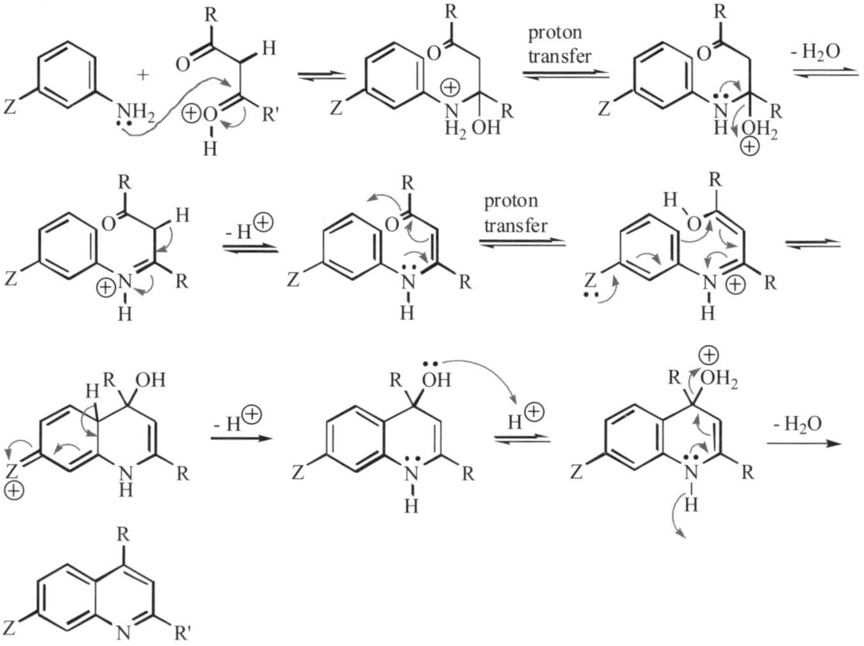

## Notes:

The rate of cyclization is enhanced when Z = EDG. G. R. Newkome, W. W. Paudler, *Contemporary Heterocyclic Chemistry*, John Wiley and Sons, New York, 1982, p. 203

J. L. Born, *Journal of Organic Chemistry* **1972**, <u>37</u>, 3952 provides mechanistic insight based on deuterium incorporation.

## Examples:

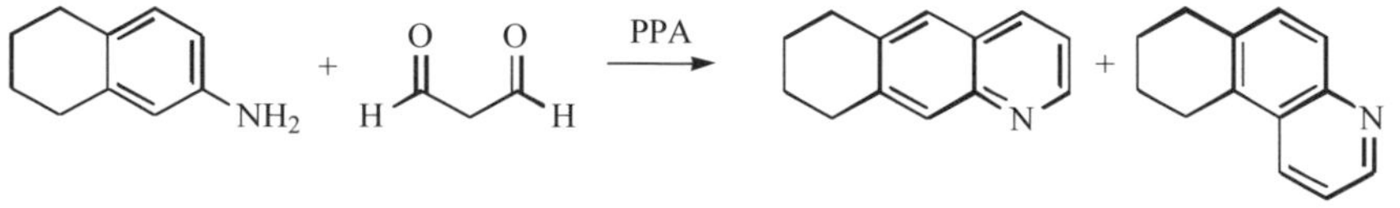

L. Takeuchi, M. Ushida, Y. Hamada, T. Yuzure, H. Suezawa, M. Hirota, *Heterocycles* **1995**, <u>41</u>, 2221 (AN 1996:201759)

A. Baba, N. Kawamura, H. Makino, Y. Ohta, S. Taketome, T. Sohda, *Journal of Medicinal Chemistry* **1996**, <u>39</u>, 5176

J. L. Born, *Journal of Organic Chemistry* **1972**, <u>37</u>, 3952

J.-C. Perche, G. Saint-Ruf, N. P. Buu-Hoi, *Journal of the Chemical Society Perkin Transactions 1* **1972**, 260

# Conrad-Limpach Reaction

## The Reaction:

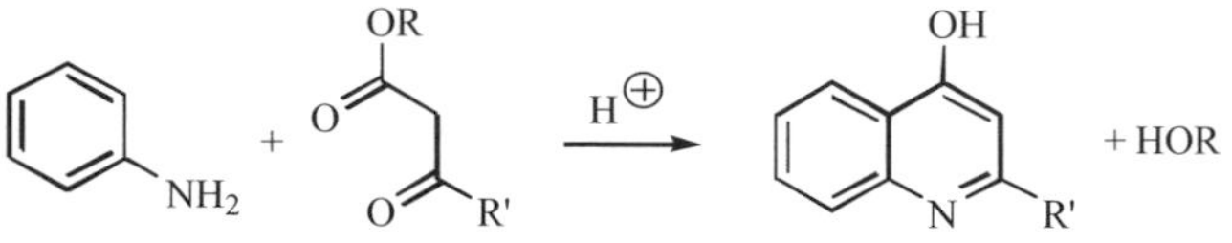

## Proposed Mechanism:

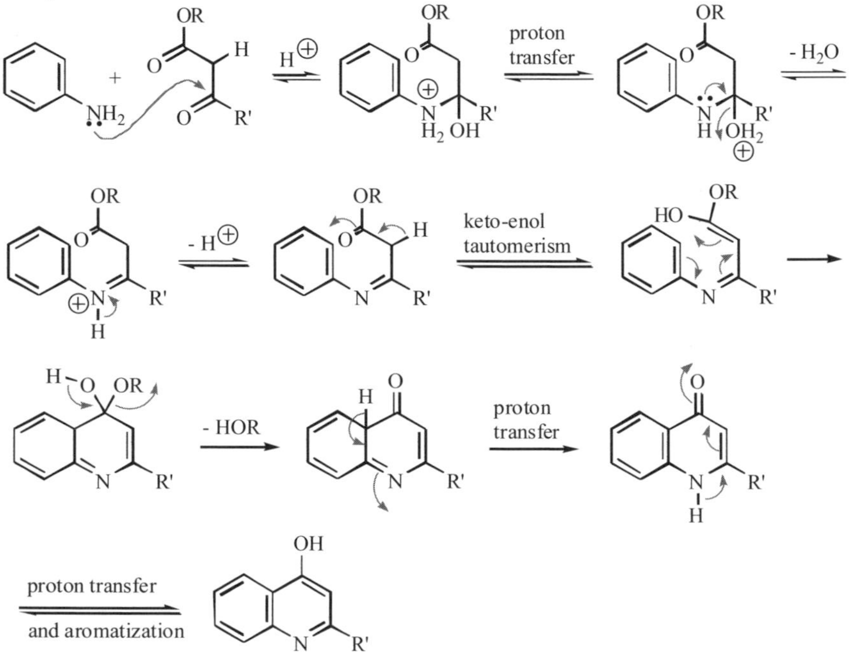

## Notes:

See the very similar ***Knorr Quinoline Reaction***, conducted with the same reagents at elevated temperatures.

## Examples:

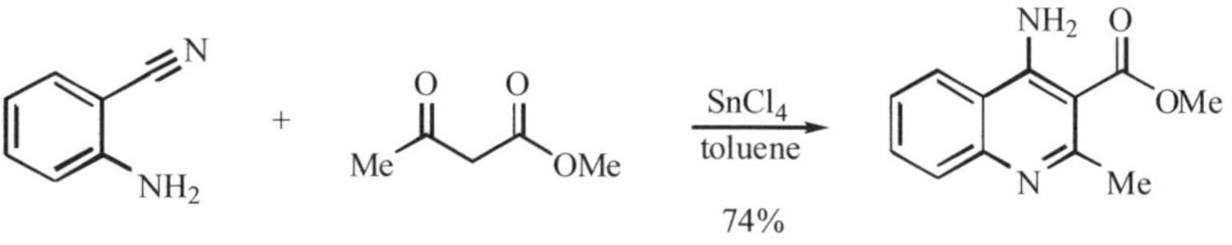

74%

A. C. Veronese, R. Callegari, D. F. Morelli, *Tetrahedron* **1995**, 51, 12227

D. R. Sliskovic, J. A. Picard, W. H. Roark, B. D. Roth, E. Ferguson, B. R. Krause, R. S. Newton, C. Sekerke, M. K. Shaw, *Journal of Medicinal Chemistry* **1991**, 34, 267

L. A. Bastiaansen, J. A. M. V. Schijndel, H. M. Buck, *Organic Preparations and Procedures International* **1988**, 20, 102 (AN 1998:510222)

B. Staskun, S. S. Isrealstam, *Journal of Organic Chemistry* **1961**, 26, 3191

C. E. Kaslow, M. M. Marsh, *Journal of Organic Chemistry* **1947**, 12, 456

# Cope Elimination (Reaction)

## The Reaction:

## Proposed Mechanism:

## Notes:

See: M. B. Smith, J. March in *March's Advanced Organic Chemistry*, 5$^{th}$ ed., John Wiley and Sons, Inc., New York, 2001, pp 1322-1326; T. Laue, A. Plagens, *Named Organic Reactions*, John Wiley and Sons, Inc., New York, 1998, pp. 54-56; A. C. Cope, E. R. Trumbull, *Organic Reactions* **11**, 5.

Starting Material Preparation:                                                    Alternate Starting Materials:

## Examples:

K.-H. Lee, S.-H. Kim, H. Furukawa, C. Piantadosi, E.-S. Huang, *Journal of Medicinal Chemistry* **1975**, <u>18</u>, 59

I. A. O'Neil, E. Cleator, V. E. Ramos, A. P. Chorlton, D.J. Topolczay, *Tetrahedron Letters* **2004**, <u>45</u>, 3655

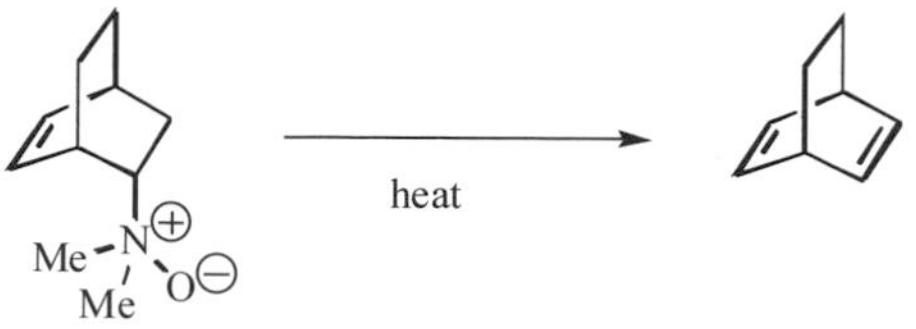

C. A. Grob, H. Kny, A. Gagneux, *Helvetica Chimica Acta* **1957**, <u>40</u>, 130 (AN 1958:20977)

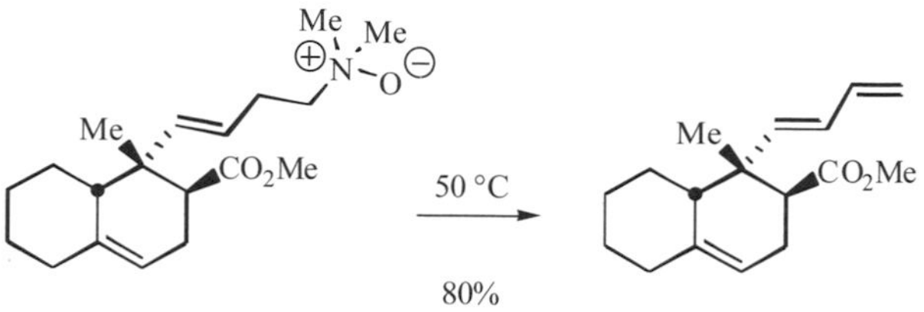

80%

E. J. Corey, M. C. Desai, *Tetrahedron Letters* **1985**, 5747

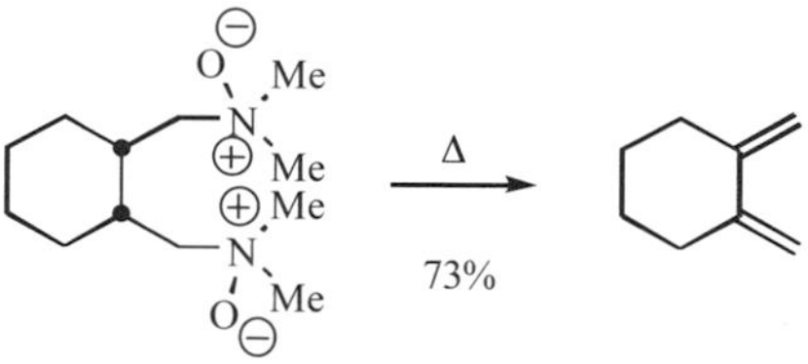

73%

L. D. Quin, J. Leimert, E. D. Middlemass, R. W. Miller, A. T. McPhail, *Journal of Organic Chemistry* **1979** <u>44</u>, 3496

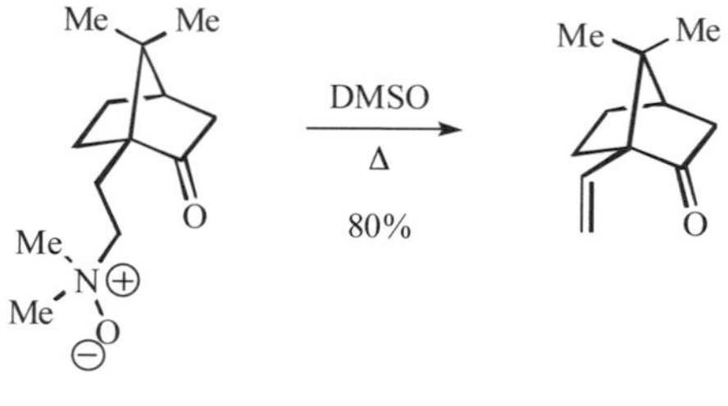

80%

A. G. Martinez, E. T. Vilar, A. G. Fraile, S. de la Moya Cereo, B. L. Maroto, *Tetrahedron: Asymmetry* **2002**, <u>13</u>, 17

# Cope Rearrangement

## The Reaction:

## Proposed Mechanism:

The preferred chair transition state geometry by FMO theory:

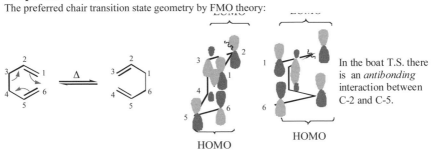

In the boat T.S. there is an *antibonding* interaction between C-2 and C-5.

T. Laue, A. Plagens, *Named Organic Reactions*, John Wiley and Sons, Inc., New York, 1998, pp. 56-59.

## Notes:

See ***Claisen, Cope and Related Rearrangements***.

An interesting study of competing pathways:

P. A. Wender, R. J. Ternansky, S. McN. Sieburth, *Tetrahedron Letters* **1985**, <u>26</u>, 4319

# Examples:

H. M. L. Davies, B. D. Doan, *Tetrahedron Letters* **1996**, <u>37</u>, 3967

G. Jomme, F. Orsini, M. Resmini, M. Sisti, *Tetrahedron Letters* **1991**, <u>32</u>, 6969

N. Kato, H. Takeshita, S. Tanaka, H. Kataoka, *Journal of the Chemical Society Perkin: Transactions 1* **1989**, 1833

Major isomer of several products formed

A. M. Adio, C. Paul, P. Kloth, W. A. König, *Phytochemistry* **2002**, <u>65</u>, 199

# Corey-Bakshi-Shibata (CBS Reduction)

## The Reaction:

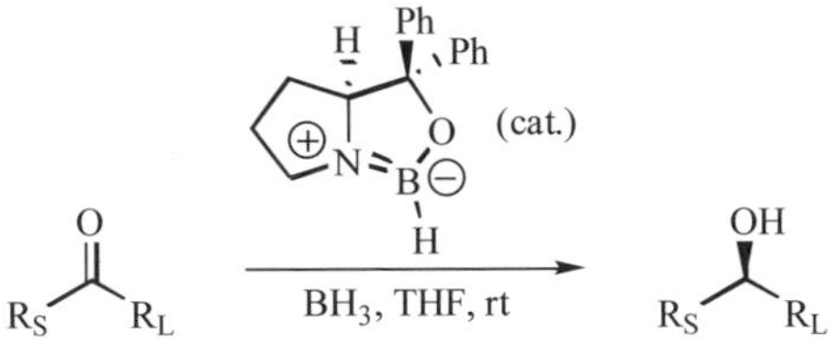

## Proposed Mechanism:

The face syn to the ring proton is more available to attack

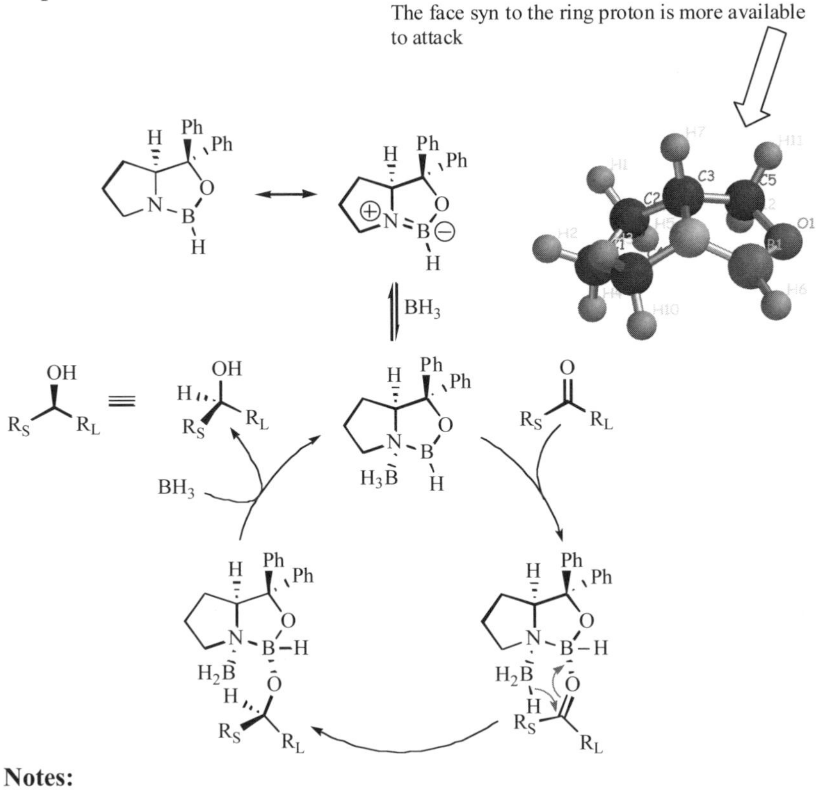

## Notes:

For a discussion of the reaction, see D. J. Mathre, I. Shinkai, *Encyclopedia of Reagents for Organic Synthesis,* John Wiley and Sons, Inc., New York, 1995, **7**, 4767

Enatioselectivity for the reduction improves as the size differential of the groups on the carbonyl increases.

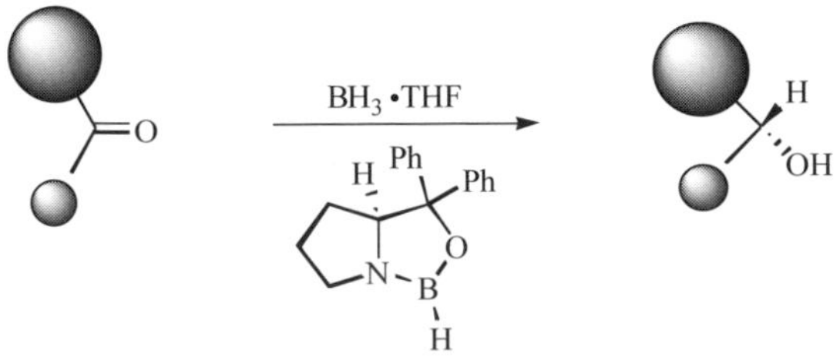

## Examples:

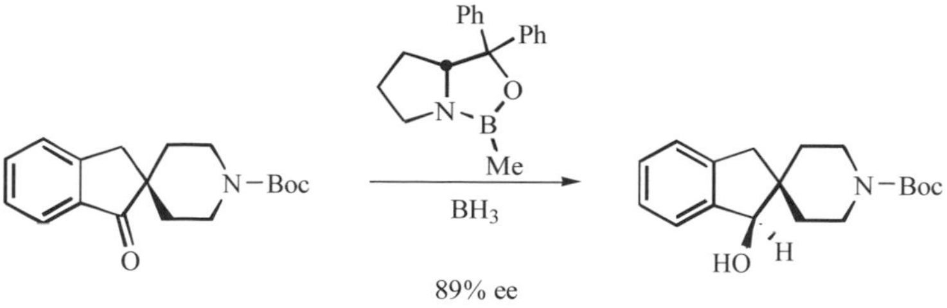

89% ee

T. Takemoto, K. Nakajima, Y. Iio, M. Tamura, T. Nishi, *Tetrahedron Asymmetry* **1999**, <u>10</u>, 1787

major isomer

J. Mulzer, M. Berger, *Tetrahedron Letters* **1998**, <u>39</u>, 803

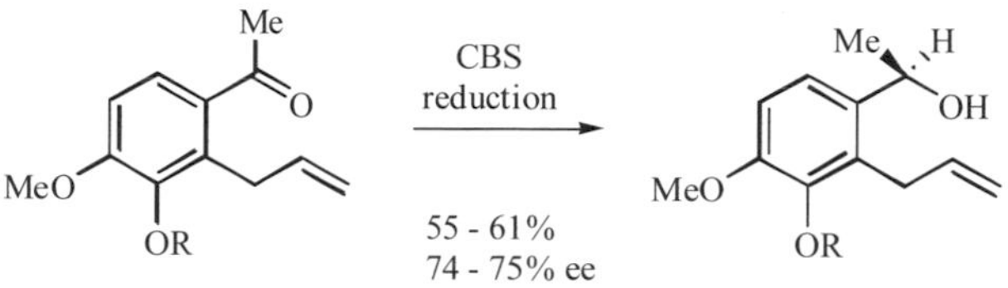

55 - 61%
74 - 75% ee

C. B. de Koning, R.-G. F. Giles, I. R. Green, N. M. Jahed, *Tetrahedron Letters* **2002**, <u>43</u>, 4199

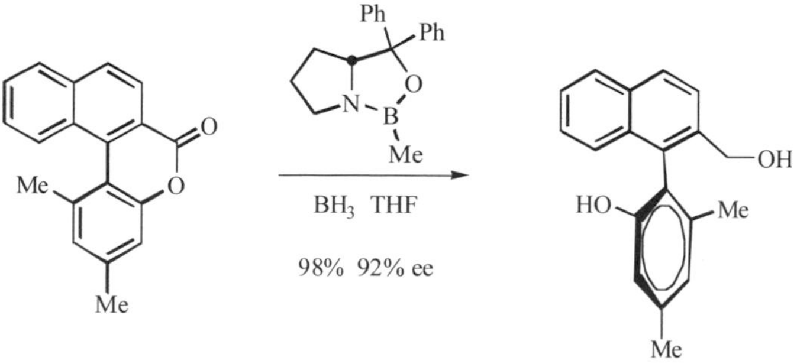

BH₃ THF

98% 92% ee

G. Bringmann, M. Breuning, P. Henschel, J. Hinrichs, *Organic Syntheses*, <u>79</u>, 72

# Corey-Chaykovsky Reaction

## The Reaction:

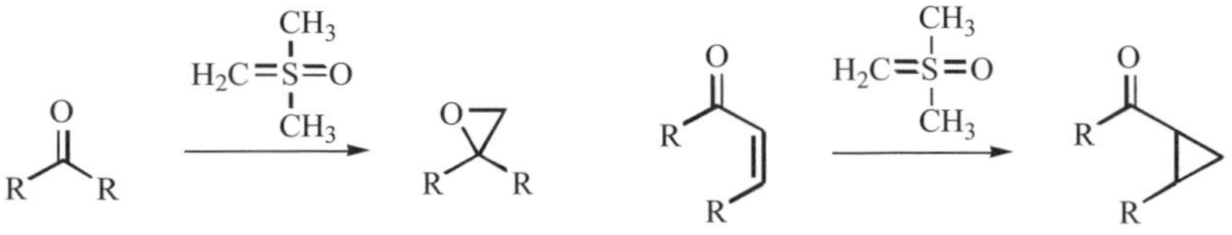

## Proposed Mechanism:

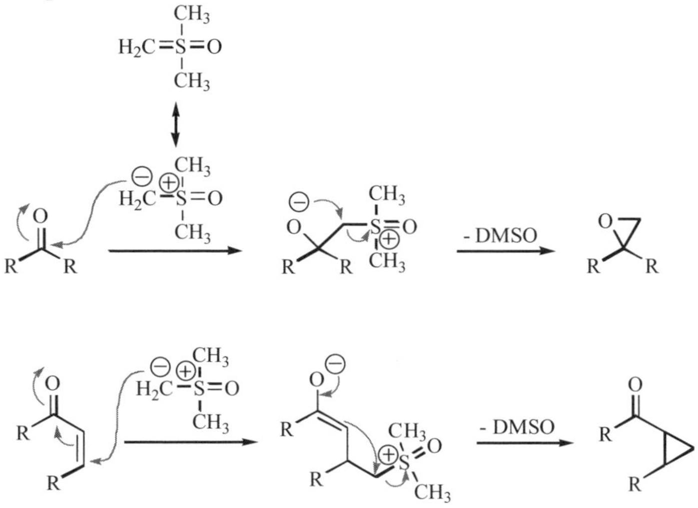

## Notes:

See: *Corey-Chaykovsky Reagent*.

A chiral approach: See V. K. Aggarwal C. L. Winn, *Accounts of Chemical Research* **2004**, _37_, 611

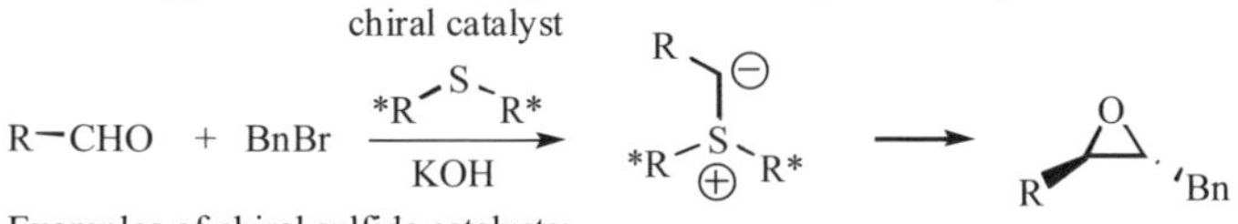

chiral catalyst

$$R-CHO \quad + \quad BnBr \xrightarrow[\text{KOH}]{*R\diagdown{S}\diagup{R*}} $$

Examples of chiral sulfide catalysts:

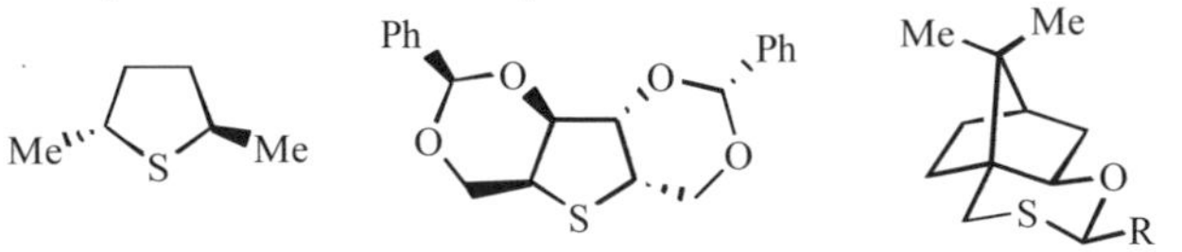

## Examples:

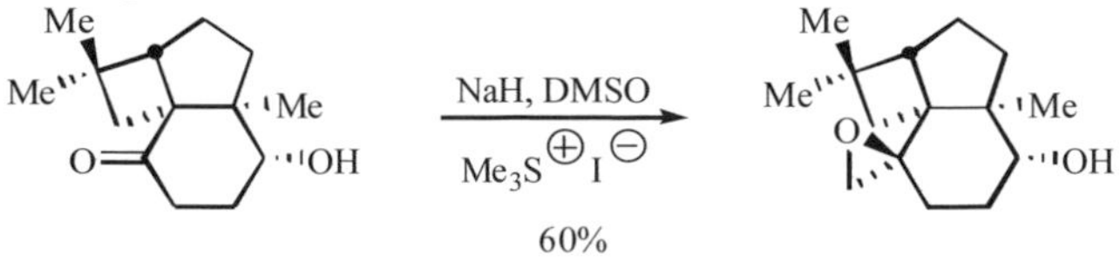

60%

C. F. D. Amigo, I. G. Collado, J. R. Hanson, R. Hernandez-Galan, P. B. Hitchcock, A. J. Macias-Sanchez, D. J. Mobbs, *Journal of Organic Chemistry* **2001**, _66_, 4327

K. Hantawong, W. S. Murphy, N. Russell, D. R. Boyd, *Tetrahedron Letters* **1984**, <u>25</u>, 999

A novel-in-situ preparation of an epoxidation reaction (***Simmons-Smith Epoxidation***):

This reaction is useful for base-sensitive aldehydes; no epimerization.

V. K. Aggarwal, M. P. Coogan, R. A. Stenson, R. V. H. Jones, R. Fieldhouse, J. Blacker, *European Journal of Organic Chemistry* **2002**, 319

A polymer-bound reagent:

M. K. W. Choi, P. H. Toy, *Tetrahedron* **2004**, <u>60</u>, 2875

# Corey-Fuchs Reaction

## The Reaction:

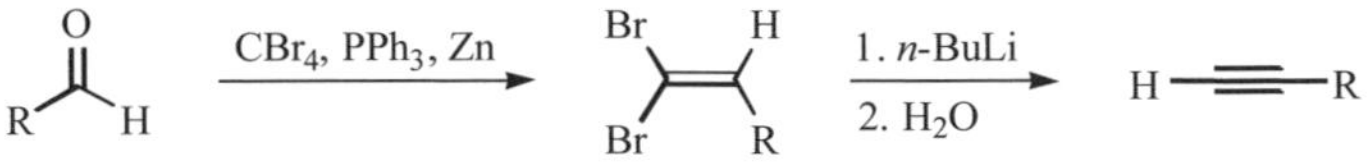

E. J. Corey, P. L. Fuchs, *Tetrahedron Letters* **1972**, *36*, 3769

## Proposed Mechanism:

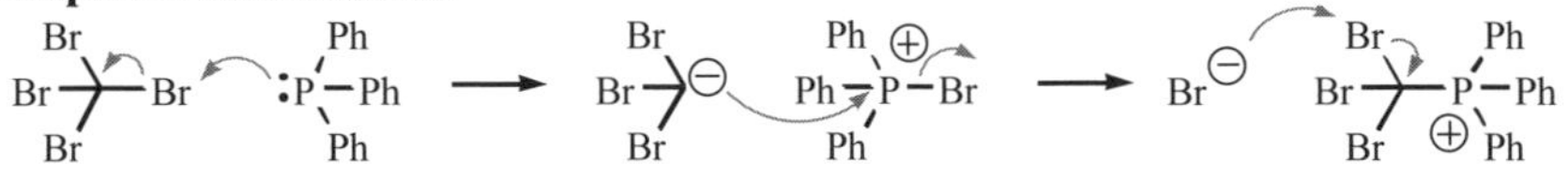

$$\left( \begin{array}{c} \text{The bromine produced in this} \\ \text{reaction is decomposed with zinc:} \end{array} \quad Br_2 \quad + \quad Zn \quad \longrightarrow \quad ZnBr_2 \right)$$

phosphorous ylide      betaine      oxaphosphetane

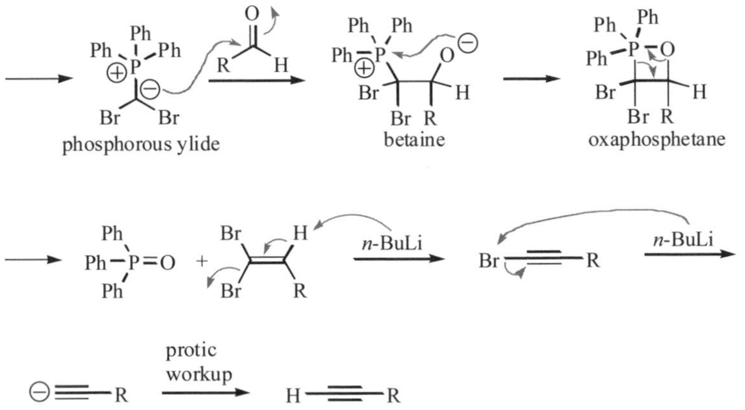

## Notes:

This is an alkyne-analog of the ***Wittig reaction***. Since organocopper chemistry finds utility in converting the C-Br bond to other alkyl groups, we show only the first part of the transformation in most examples.

See: ***Fritsch-Buttenberg-Wiechell Rearrangement.***

The ***Seyferth Protocol:*** is another method for the conversion:

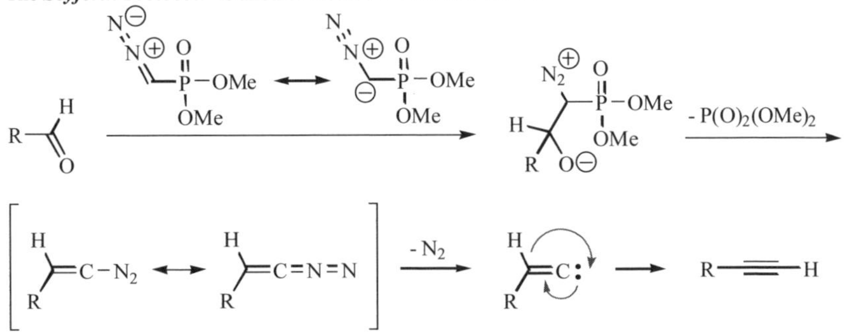

# Examples:

Ph₃P, CBr₄

85%

Y. Mizuno, R. Mori, H. Irie, *Journal of the Chemical Society, Perkin Transactions 1* **1982**, 2849

1. CBr₄, Ph₃P

2. BuLi, MeI

75%

A. B. Smith, S. S.-Y. Chen, F. C. Nelson, J. M. Reichert, B. A. Salvatore, *Journal of the American Chemical Society* **1995**, 117, 12013

1. CBr₄, Ph₃P

2. BuLi, ClCOOEt

It was reported that these were just two of nice steps with average yield of 90%
W. C. Still, J. C. Barrish, *Journal of the American Chemical Society* **1983**, 105, 2487

CBr₄, Zn, Ph₃P

CH₂Cl₂

W. H. Okamura, G.-D. Zhu, D. K. Hill, R. J. Thomas, K. Ringe, D. B. Borchardt, A. W. Norman, L. J. Mueller, *Journal of Organic Chemistry* **2002**, 67, 1637

Ph₃P

CBr₄

LDA

K. H. H. Fabian, A. H. M. Elwahy, K. Hafner, *Tetrahedron Letters* **2000**, 41, 2855

# Corey-Gilman-Ganem Oxidation

## The Reaction:

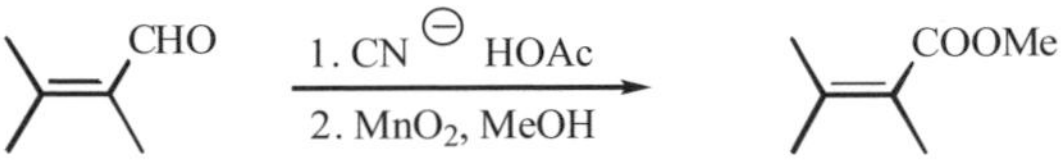

## Proposed Mechanism:

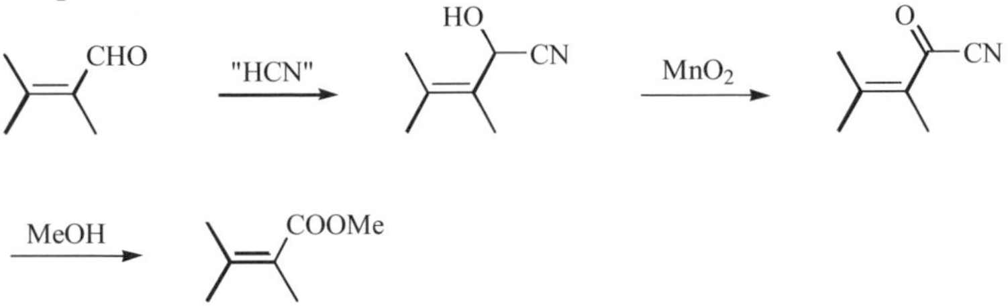

E. J. Corey, N. W. Gilman, B. E. Ganem, *Journal of the American Chemical Society* **1968**, 90, 5616

## Examples:

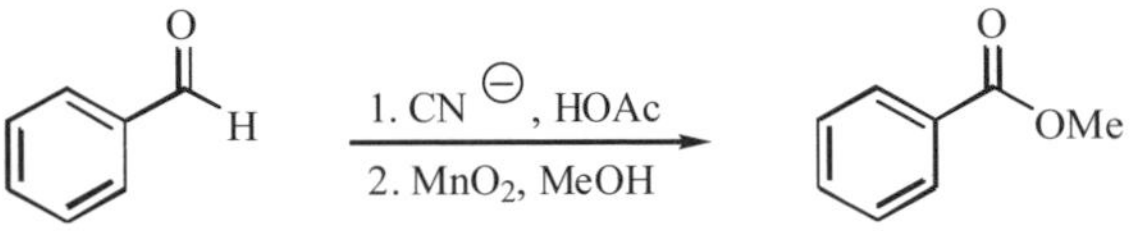

>95%

E. J. Corey, N. W. Gilman, B. E. Ganem, *Journal of the American Chemical Society* **1968**, 90, 5616

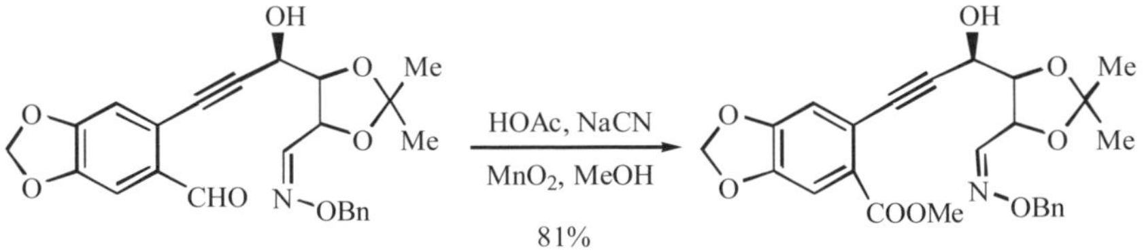

81%

G. E. Keck, T. T. Wagner, J. F. D. Rodriguez, *Journal of the American Chemical Society* **1999**, 121, 5176

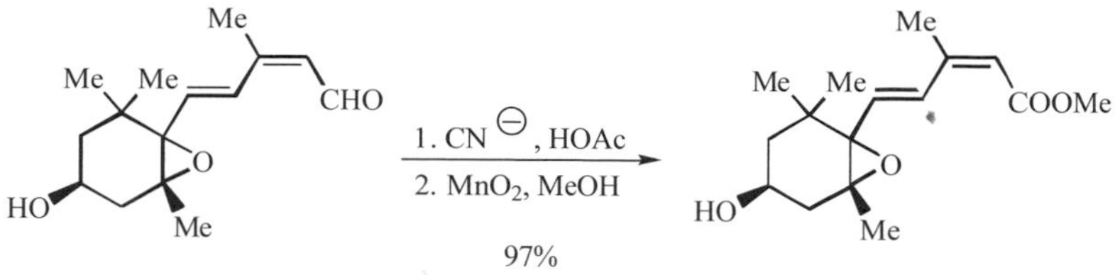

1. CN⊖, HOAc
2. MnO₂, MeOH

97%

H. Yamamoto, T. Oritani, *Tetrahedron Letters* **1995**, <u>36</u>, 5797

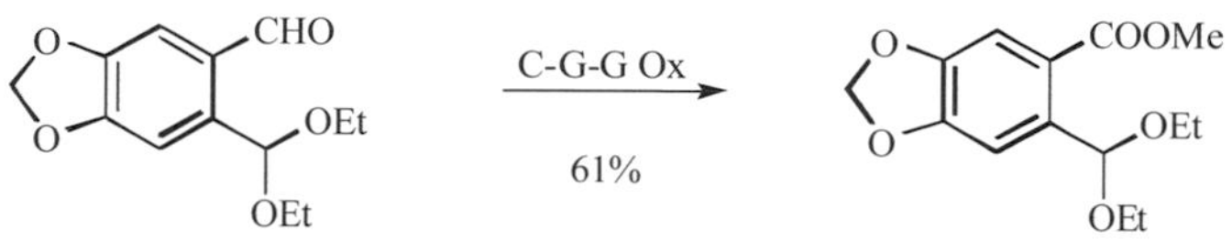

C-G-G Ox

61%

D. L. Boger, S. E. Wolkenberg, *Journal of Organic Chemistry* **2000**, <u>65</u>, 9120

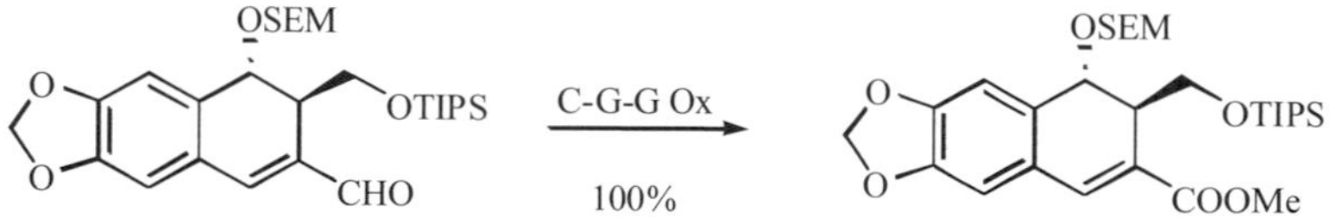

C-G-G Ox

100%

D. B. Berkowitz, S. Choi, J.-H. Maeng, *Journal of Organic Chemistry* **2000**, <u>65</u>, 847

# Corey-House-Posner-Whitesides Reaction

## The Reaction:

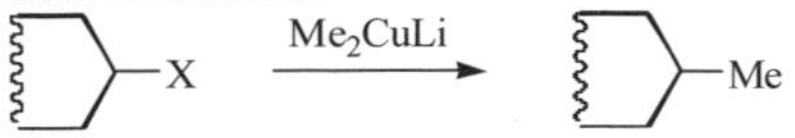

X = halogen or tosylate

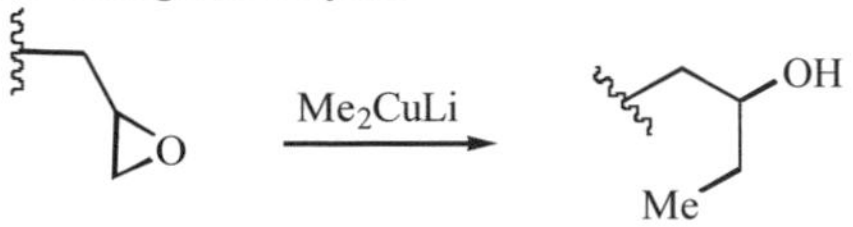

## Proposed Mechanism:

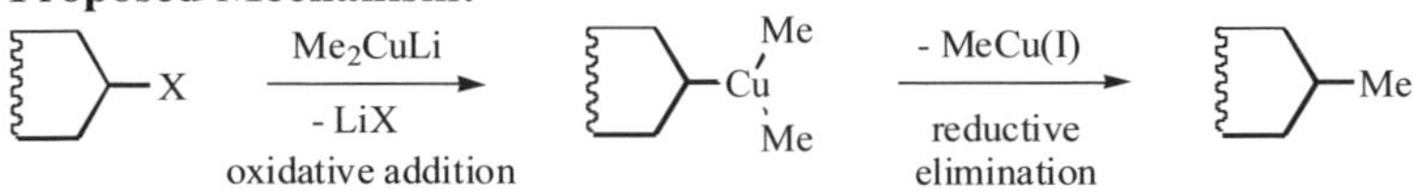

See: N. Yoshikai, N. Eiichi, *Journal of the American Chemical Society* **2004**, <u>126</u>, 12264 for a mechanistic discussion of the reaction at sp$_2$ centers.

## Notes:

The reagent is most likely a more complex species.

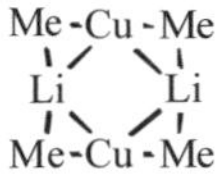

When X = tosylate there is evidence of a direct S$_N$2 displacement with inversion of stereochemistry.

The general reaction of R$_2$CuLi is often included in this class.

## Examples:

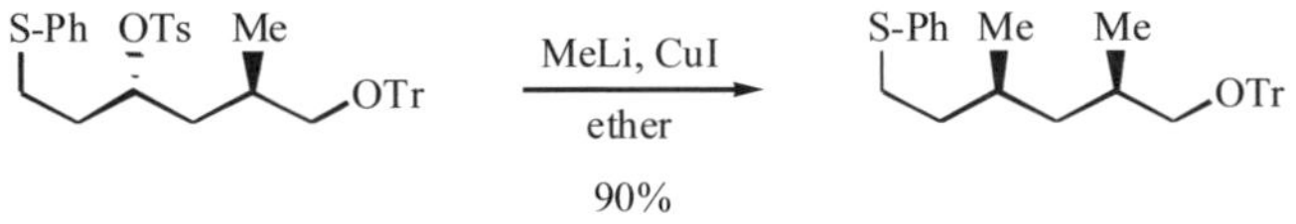

S. Hanessian, N. G. Cooke, B. DeHoff, Y. Sakito *Journal of the American Chemical Society* **1990**, <u>112</u>, 5276

| R = | Me | 83% |
| R = | Bu | 72% |
| R= | Allyl | 98% |

S. Hanessian, B. Thavonekham, B. DeHoff, *Journal of Organic Chemistry* **1989**, <u>54</u>, 5831

B. D. Johnstone, A. C. Oelschlager, *Journal of Organic Chemistry* **1982**, <u>47</u>, 5384

T. Wakamatsu, H. Nakamura, M. Taniguchi, Y. Ban, *Tetrahedron Letters* **1986**, <u>27</u>, 6071

D. Caine, A. S. Frobese, V. C. Ukachukwu, *Journal of Organic Chemistry* **1983**, <u>48</u>, 740

# Corey-Link Reaction

## The Reaction:

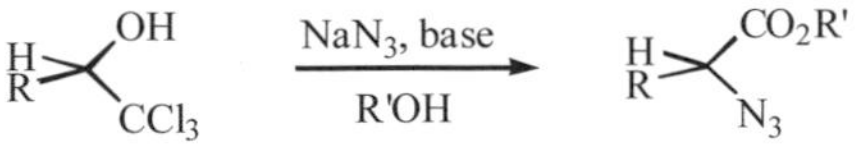

## Proposed Mechanism:

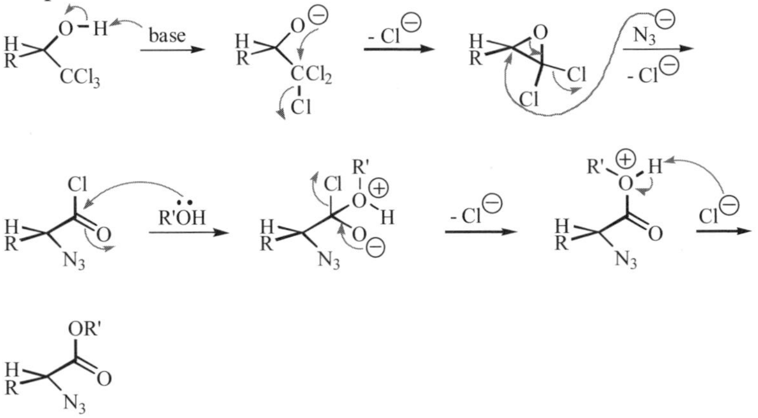

## Notes:

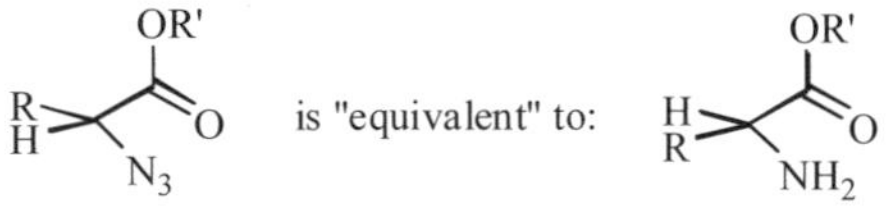

is "equivalent" to:

See a similar mechanistic reaction in the **Bargellini Reaction**:

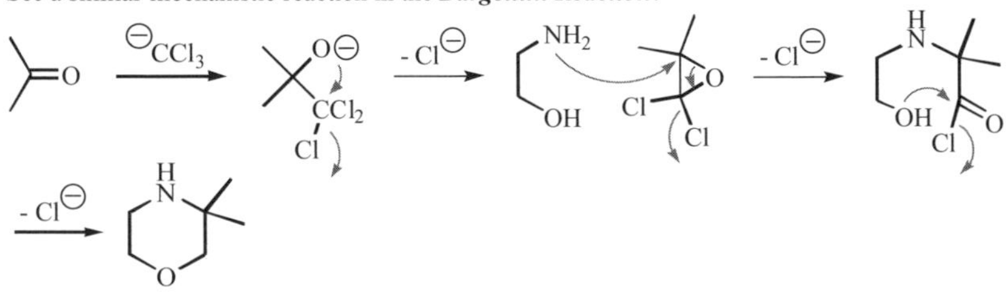

Starting material preparation:

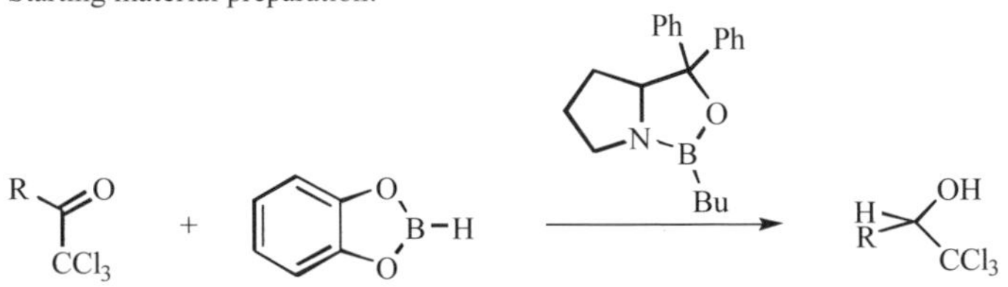

## Examples:

TBDPSO—OH O⤬
Cl₃C—¦-O¦O¦

NaN₃, 18-Crown-5
—————————————
DBU, MeOH

68%

TBDPSO—⤬ O⤬
MeO₂C—O¦O¦
N₃

M. H. Sorensen, C. Nielsen, P. Nielsen, *Journal of Organic Chemistry* **2001**, 66, 4878

EtO₂C H H
F
O

LiHDMS
—————————
HCCl₃

96%

EtO₂C H H
OH
F
Cl₃C

NaN₃
—————————
DBU, EtOH

98%

EtO₂C H H
CO₂Et
F
N₃

C. Pedregal, W. Prowse, *Bioorganic and Medicinal Chemistry Letters* **2002**, 10, 433

OH
CCl₃

NaOH, NaN₃
—————————
DME, H₂O

N₃
CO₂H

H₂, Pd/C
—————————

70%,  0% ee

NH₂
CO₂H

when DBU is the base used, 62%, 99% ee

C. Mellin-Morliere, D. J. Aitken, S. D. Bull, S. G. Davies, H.-P. Husson, *Tetrahedron: Asymmetry* **2001**, 12, 149

CO₂Me
HO—CCl₃

DBU, NaN₃
—————————
MeOH

CO₂Me
N₃—CO₂Me

⟹

CO₂H
H₂N—CO₂H

S. R. Baker, T. C. Hancox, *Tetrahedron Letters* **1999**, 40, 781

# Corey-Nicoloau Macrocyclization

## The Reaction:

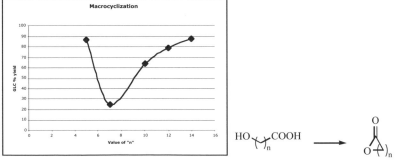

## Proposed Mechanism:

This thiopyridyl ester uniquely activates the carbonyl
as well as the hydroxyl group by proton exchange.

## Notes:
See: ***Macrolactonization Methods***

E. J. Corey, K. C. Nicolaou, *Journal of the American Chemical Society* **1974**, 96, 5614

## Examples:

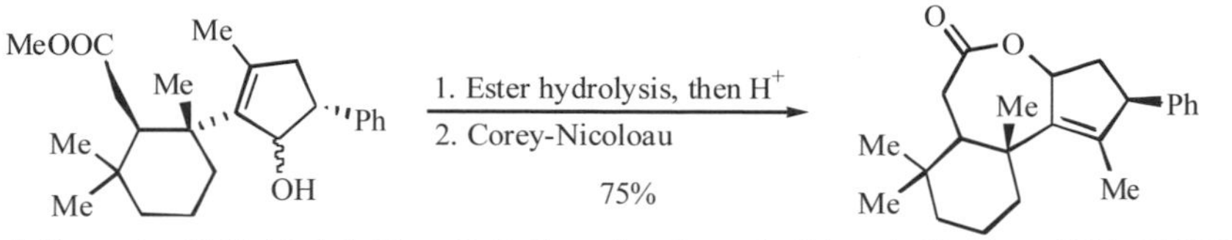

1. Ester hydrolysis, then H⁺
2. Corey-Nicoloau

75%

A. Fernandez, E. M. M. de la Nava, R. R. Gonzalez, *Journal of Organic Chemistry*, **2001**, <u>66</u>, 7632

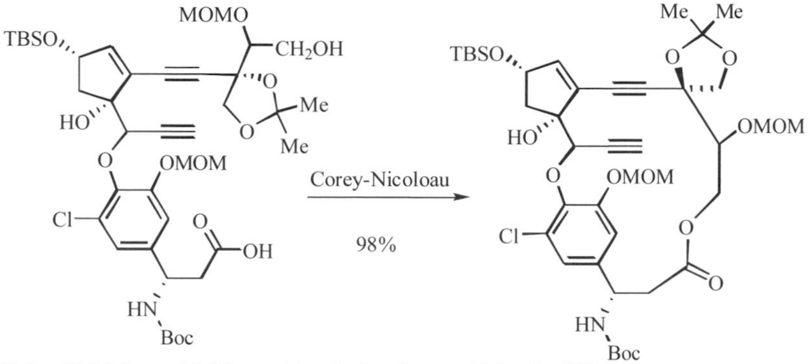

Corey-Nicoloau

98%

T. Sasaki, M. Inoue, M. Hirama, *Tetrahedron Letters*, **2001**, <u>42</u>, 5299

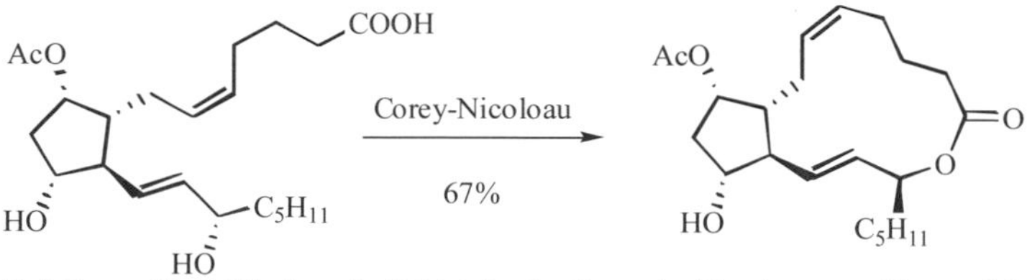

Corey-Nicoloau

67%

E. J. Corey, K. C. Nicoloau, L. S. Melvin, Jr., *Journal of the American Chemical Society* **1975**, <u>97</u>, 653

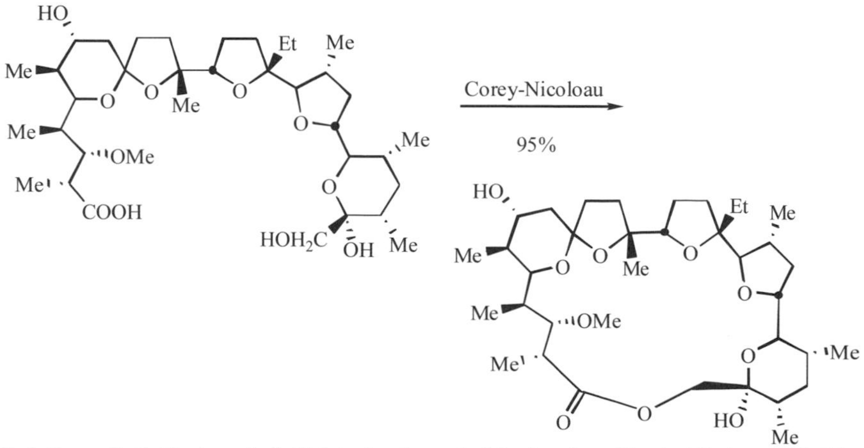

Corey-Nicoloau

95%

E. J. Corey, K. C. Nicoloau, L. S. Melvin, Jr., *Journal of the American Chemical Society*, **1975**, <u>97</u>, 653

# Corey-Seebach Reaction

## The Reaction:

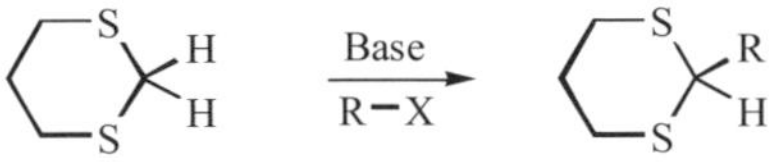

Alkylation can take place twice.

## Proposed Mechanism:

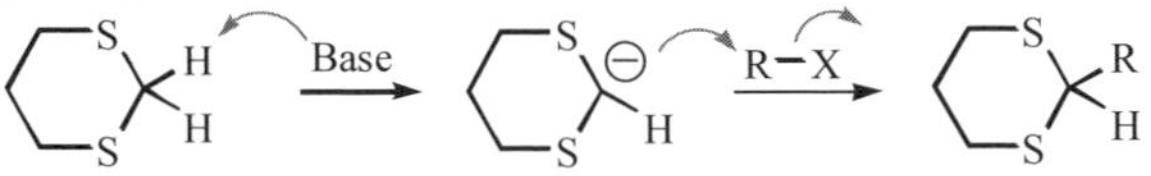

Adjacent sulfur atoms allow the
methylene protons to be acidic.

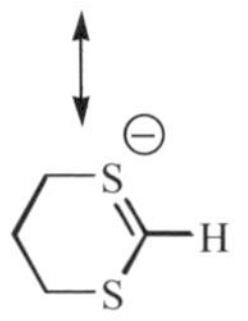

## Notes:
Although the five-ring analog is easily prepared, it is not useful for the transformation due to the
tendency for base-induced decomposition:

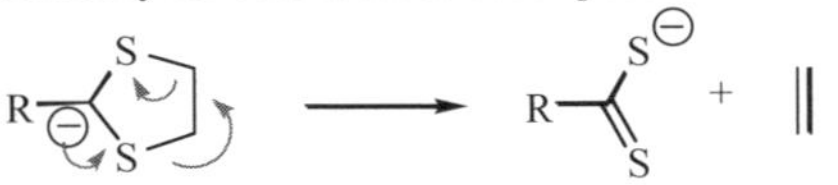

See: A. B. Smith, III, C. M. Adams, *Accounts of Chemical Research* **2004**, *37*, 365 for an excellent
tour through the authors' use of the dithiane-based chemistry for the construction of complex
molecules.

## Examples:

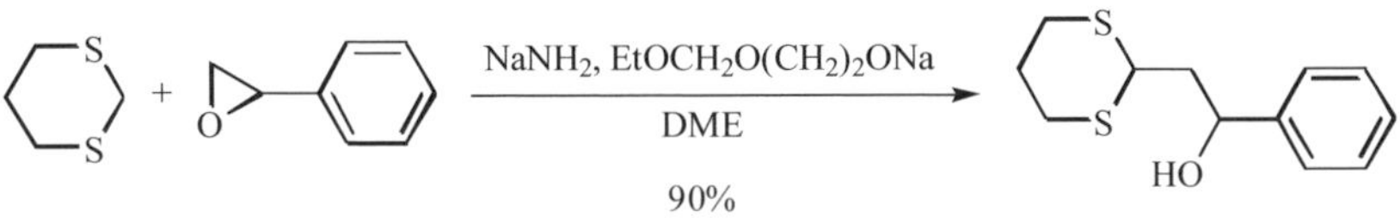

90%

P. Gros, P. Hansen, P. Caubere, *Tetrahedron* **1996**, _52_, 15147

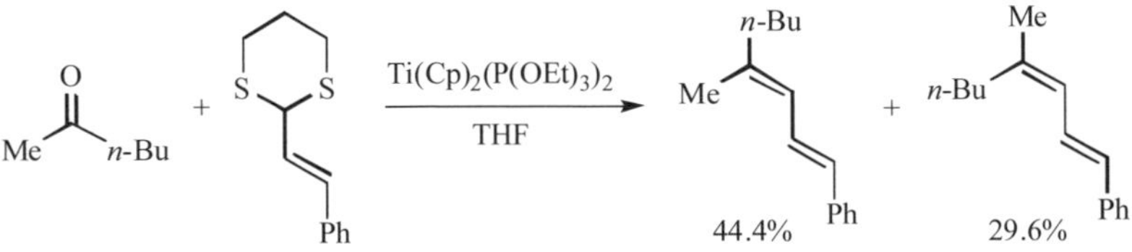

44.4%          29.6%

Y. Horikawa, M. Watanabe, T. Fujiwara, T. Takeda, *Journal of the American Chemical Society*
**1997**, _119_, 1127

A. S. Pilcher, P. Deshong, *Journal of Organic Chemistry* **1996**, <u>61</u>, 6901

E. Juaristi, B. Gordillo, L. Valle, *Tetrahedron* **1986**, <u>42</u>, 1963

M. H. B. Stowell, R. S. Rock, D. C. Rees, S. I. Chan, *Tetrahedron Letters* **1996**, <u>37</u>, 307

# Corey-Winter Olefination

## The Reaction:

## Proposed Mechanism:

See: T. Laue, A. Plagens, *Named Organic Reactions*, John Wiley and Sons, Inc., New York, 1998, pp. 59-61; M. B. Smith, J. March in *March's Advanced Organic Chemistry*, 5th ed., John Wiley and Sons, Inc., New York, 2001, p. 1340; E. Block, *Organic Reactions* **30**, 2

## Notes:

Instead of using $Cl_2C=S$:

carbonyldiimidazole    1,1'-thiocarbonyldiimidazole
                              TCDI

A modification is the *Corey-Hopkins olefination*. See *Tetrahedron Letters*, **1982**, <u>23</u>, 1979:

J. Dressel, K. L. Chasey, L. A. Paquette, *Journal of the American Chemical Society* **1988**, <u>110</u>, 5479

## Examples:

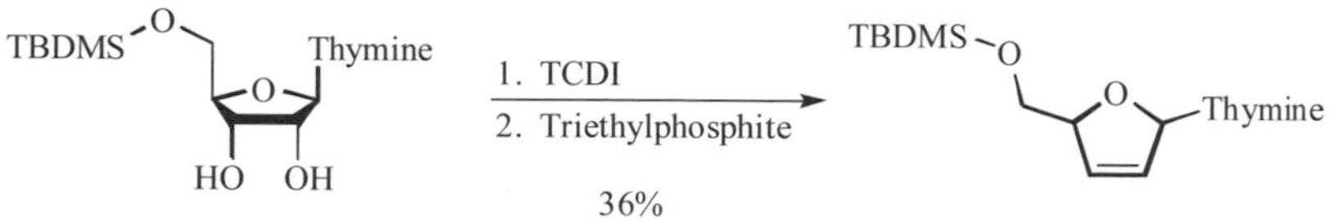

TBDMS$^-$O Thymine

HO   OH

1. TCDI
2. Triethylphosphite

36%

TBDMS$_-$O

O Thymine

Y. Saito, T. A. Zevaco, L. A. Agrofoglio, *Tetrahedron* **2002**, 58, 9593

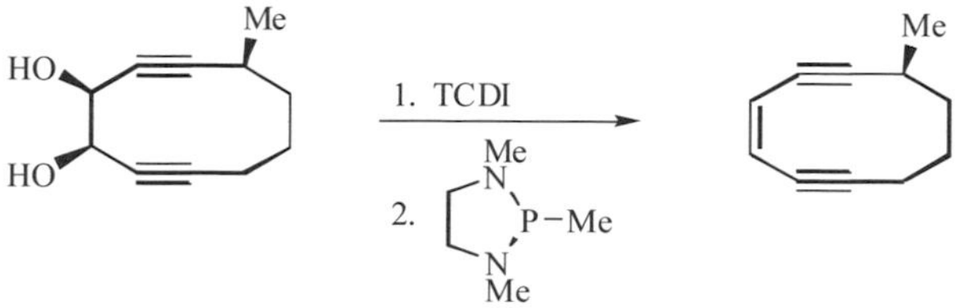

HO
HO
Me

1. TCDI

2. 
Me
N
P–Me
N
Me

Me

M. F. Semmelhack, J. Gallagher, *Tetrahedron Letters* **1993**, 34, 4121

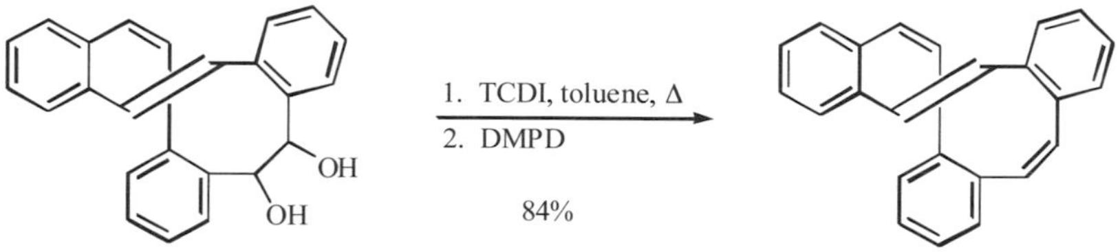

OH
OH

1. TCDI, toluene, Δ
2. DMPD

84%

Y. Kuwatani, T. Yoshida, A. Kusaka, M. Iyoda, *Tetrahedron Letter* **2000**, 41, 359

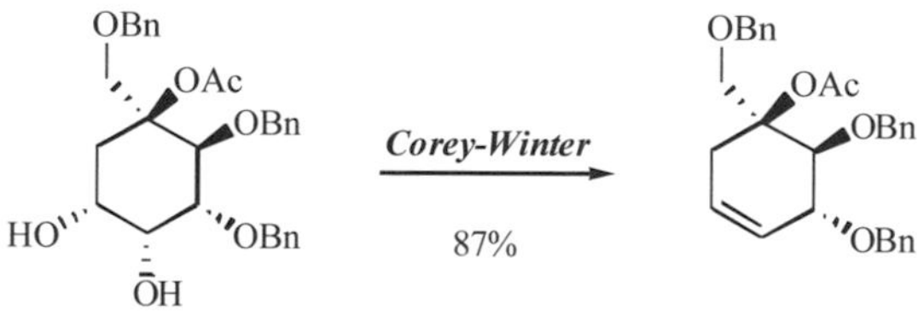

OBn
OAc
OBn

HO'''
OH

OBn

*Corey-Winter*

87%

OBn
OAc
OBn

OBn

S. H.-L. Koc, C. C. Lee, T. K. M. Shing, *Journal of Organic Chemistry* **2001**, 66, 7184

# Cornforth Rearrangement

## The Reaction:

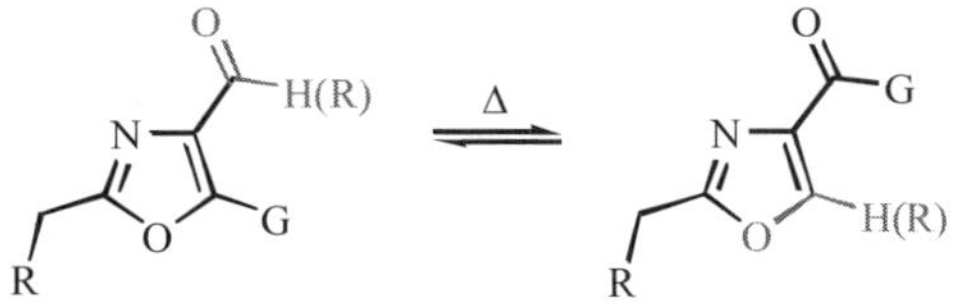

## Proposed Mechanism:

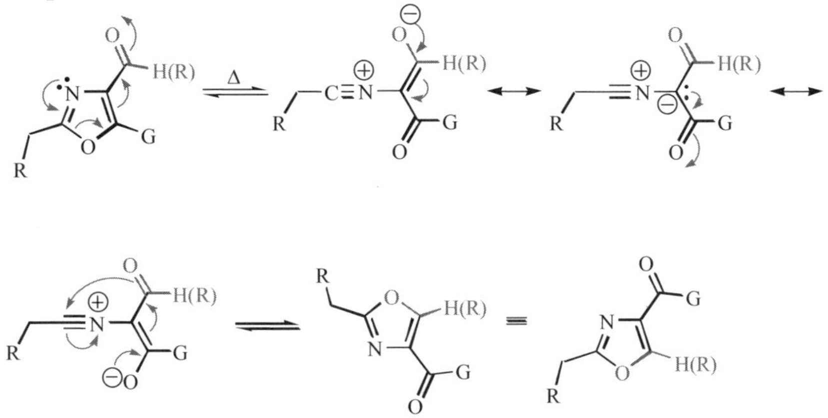

## Notes:

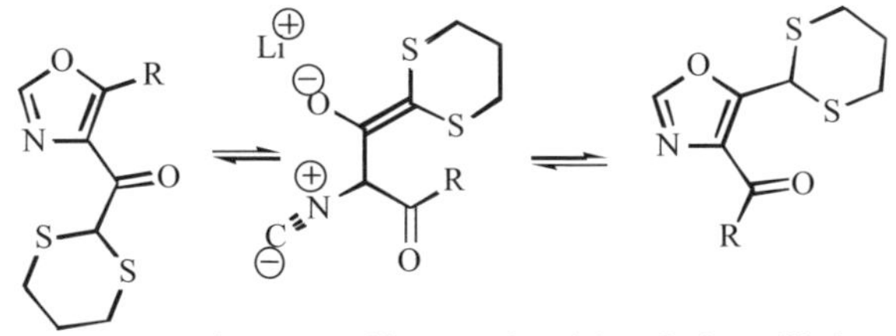

First reported base-catalyzed Cornforth equilibrium.

D. R. Williams, E. L. McClymont, *Tetrahedron Letters* **1993**, <u>34</u>, 7705

## Examples:

G. L'Abbe, A.-M. Llisiu, W. Dehaen, S. Toppett, *Journal of the Chemical Society Perkin Transactions 1* **1993**, 2259

I. J. Turchi, C. A. Maryanoff, *Synthesis* **1983**, 837

M. J. S. Dewar, I. K. Turchi, *Journal of Organic Chemistry* **1975**, <u>40</u>, 1521

# Curtius Rearrangement / Reaction / Degradation

## The Reaction:

## Proposed Mechanism:

To this point has been called the
***Curtius Rearrangement / Reaction***

stops here if have R'          *Carbamic acid*

OR with no nitrene...

## Notes:

T. Laue, A. Plagens, *Named Organic Reactions*, John Wiley and Sons, Inc., New York, 1998, pp.
61-63; M. B. Smith, J. March in *March's Advanced Organic Chemistry*, 5$^{th}$ ed., John Wiley and
Sons, Inc., New York, 2001, pp 1412-1413; P. A. S. Smith, *Organic Reactions* **3**, 9.

Alternate Starting Material:

As with the ***Hofmann***, ***Schmidt*** and ***Lossen Rearrrangements***, there is a common isocyanate
intermediate.

# Examples:

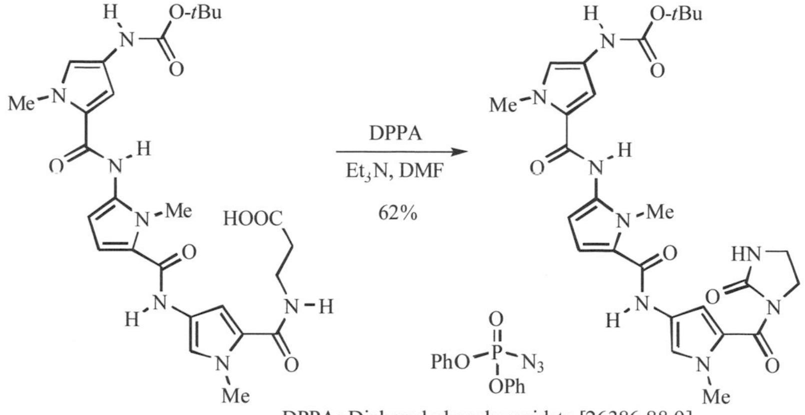

DPPA: Diphenyl phosphorazidate [26386-88-9]

I. Beria, M. Nesi, *Tetrahedron Letters* **2002**, <u>43</u>, 7323

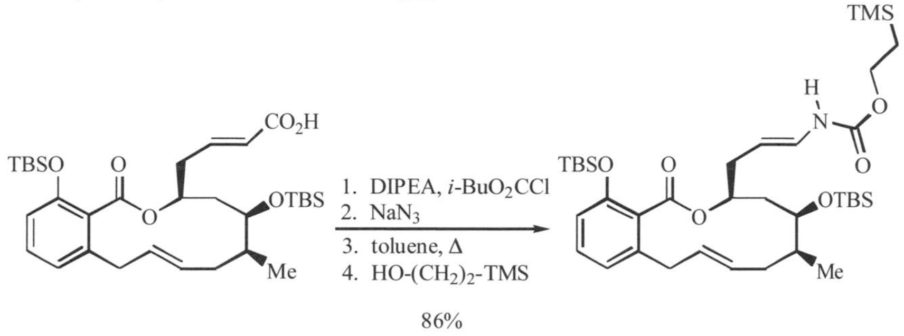

1. DIPEA, *i*-BuO$_2$CCl
2. NaN$_3$
3. toluene, Δ
4. HO-(CH$_2$)$_2$-TMS

86%

A. B. Smith, III, J. Zheng, *Tetrahedron* **2002**, <u>43</u>, 7323

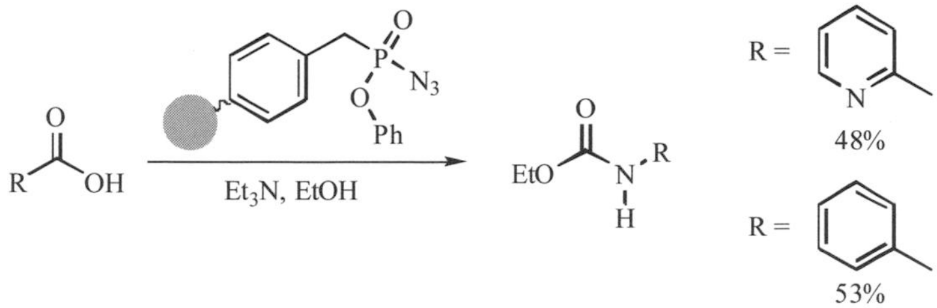

Et$_3$N, EtOH

R = 48%

R = 53%

Y. Lu, R. T. Taylor, *Tetrahedron Letters* **2003**, <u>44</u> 9267

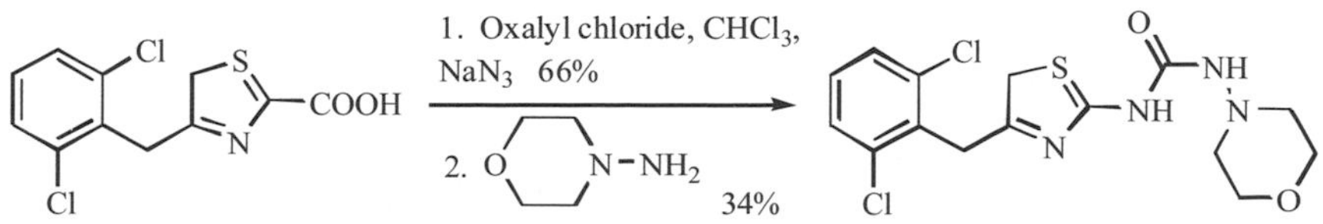

1. Oxalyl chloride, CHCl$_3$, NaN$_3$  66%

2. 34%

S. D. Larsen, C. F. Stachew, P. M. Clare, J. W. Cubbage, K. L. Leach *Bioorganic & Medicinal Chemistry Letters* **2003**, <u>13</u>, 3491

# Dakin Reaction

## The Reaction:

The reaction requires an -OH or -NH$_2$ in the *ortho* or *para* position.
R = H or Me

## Proposed Mechanism:
(The ketone oxygen is bolded to demonstrate labeling studies.)

## Notes:
M. B. Smith, J. March in *March's Advanced Organic Chemistry*, 5$^{th}$ ed., John Wiley and Sons, Inc., New York, 2001, p. 1528; V. K. Ahluwalia, R. K. Parashar, *Organic Reaction Mechanisms*, Alpha Science International Ltd., Pangbourne, U.K., 2002, pp. 432-433

Peroxy acids (***Baeyer-Villiger Reaction***) can provide the same products.

## Hydroperoxide Rearrangement

Phenol and acetone

# Examples:

G. W. Kabalka, N. K. Reddy, C. Narayana, *Tetrahedron Letters* **1992**, <u>33</u>, 865

R. S. Varma, K. P. Naiker, *Organic Letters* **1999**, <u>1</u>, 189

M. E. Jung, T. I. Lazarova, *Journal of Organic Chemistry* **1997**, <u>62</u>, 1553

P. Wipf, S. M. Lynch, *Organic Letters*, **2003**, <u>5</u>, 1155

G. W. Kabalka, N. K. Reddy, C. Narayana, *Tetrahedron Letters* **1992**, <u>33</u>, 865

# Danheiser Annulation

## The Reactions:

A number of examples are presented

## Proposed Mechanisms:

R. L. Danheiser, S. K. Gee, *Journal of Organic Chemistry* **1984**, _49_, 1672

## Examples:

82%

R. L. Danheiser, D. M. Fink, Y.-M. Tsai, *Organic Synthesis* **1988**, _66_, 8

R. L. Danheiser, D. J. Carini, A. Basak, *Journal of the American Chemical Society* **1981**, <u>103</u>, 1604

R. L. Danheiser, D. J. Carini, D. M. Fink, A. Basak *Tetrahedron*, **1983**, <u>39</u>, 935

A. B. Smith, III, C. M. Adams, S. A. Kozmin, D. V. Paone, *Journal of the American Chemical Society* **2001**, <u>123</u>, 5925

# Darzens Condensation (Darzens-Claisen Reaction, Darzens Glycidic Ester Condensation)

## The Reaction:

X = halogen

glycidic ester = an α, β-epoxy ester

## Proposed Mechanism:

## Notes:

T. Laue, A. Plagens, *Named Organic Reactions*, John Wiley and Sons, Inc., New York, 1998, pp. 71-72; M. B. Smith, J. March in *March's Advanced Organic Chemistry*, 5th ed., John Wiley and Sons, Inc., New York, 2001, p. 1230; M. S. Newman, B. J. Magerlein, *Organic Reactions* **5**, 10.

One possible extension:

saponification

In a reaction with enones:

## Examples:

83-95%

R. Hunt, L. Chinn, W. Johnson, *Organic Syntheses* <u>CV4</u>, 459

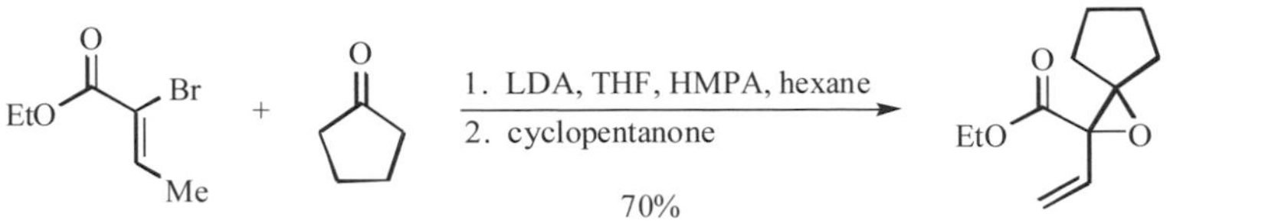

EtO
Br
Me
1. LDA, THF, HMPA, hexane
2. cyclopentanone
70%
EtO
O

F. E. Anderson, H. Luna, T. Hudlicky, L. Radesca, *Journal of Organic Chemistry* **1986**, 51, 4746

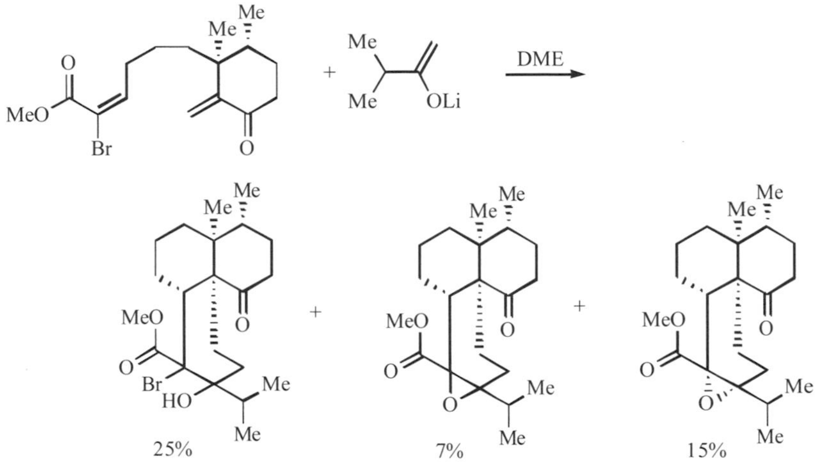

MeO
Br
Me
Me
OLi
DME
HO
25%
7%
15%

S. Danishefsky, S. Chackalamannil, P. Harrison, M. Silvestri, P. Cole, *Journal of the American Chemical Society* **1985**, 107, 2474

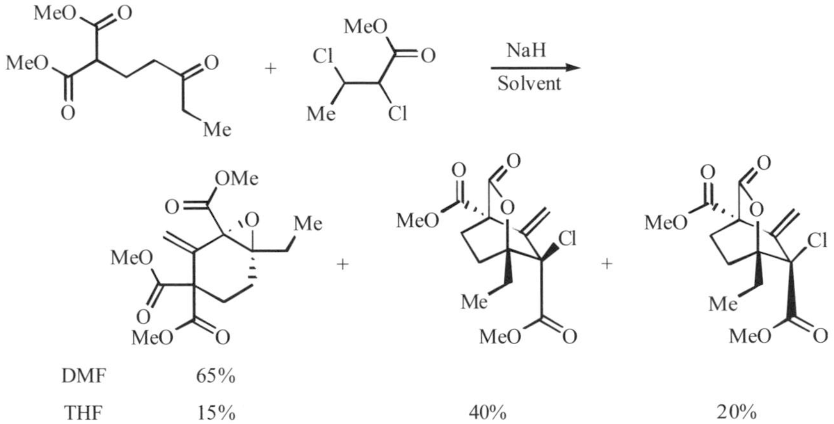

MeO
Cl
Me
NaH
Solvent
OMe
DMF
65%
THF
15%
40%
20%

J. G. Bauman, R. C. Hawley, H. Rapoport, *Journal of Organic Chemistry* **1984**, 49, 3791

# de Mayo Reaction

## The Reaction:

## Proposed Mechanism:

The orientation of the alkene is dictated
by the identity of the R groups.

*retro-Aldol*

keto-enol
tautomerism

## Notes:

The reaction can be a ring expansion:

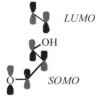

*LUMO*

*SOMO*

## Examples:

58%

T. M. Qaevilllon, A. C. Weedon, *Tetrahedron Letters* **1996**, <u>37</u>, 3939

R. Kaczmarek, S. Blechert, *Tetrahedron Letters* **1986**, *27*, 2845

46%

D. E. Minter, C. D. Winslow, *Journal of Organic Chemistry* **2004**, *69*, 603

# Dess-Martin Oxidation

## The Reaction:

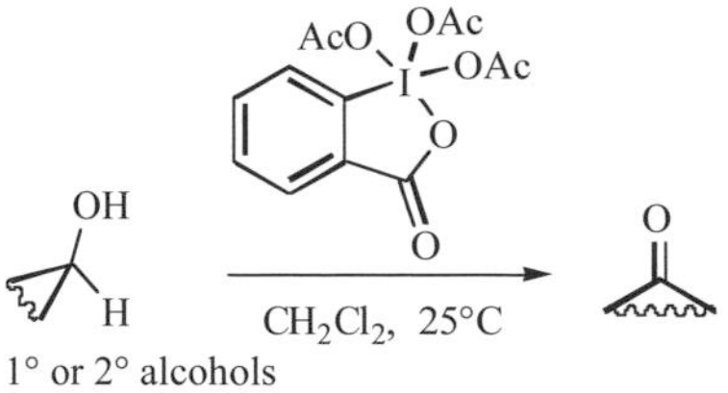

1° or 2° alcohols

## Proposed Mechanism:

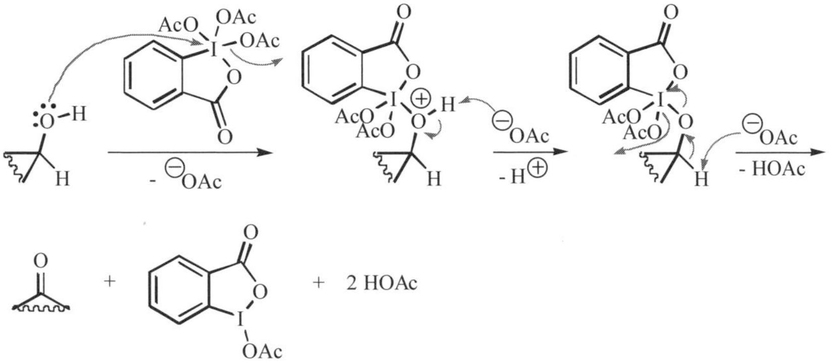

## Notes:

M. B. Smith, J. March in *March's Advanced Organic Chemistry*, 5th ed., John Wiley and Sons, Inc., New York, 2001, p. 1516; F. A. Luzzio, *Organic Reactions* **53**, 1.

This reaction is noted for its mildness.

**Reagent Preparation**:  D. B. Dess, J. C. Martin, *Journal of Organic Chemistry* **1983** <u>48</u>, 4155

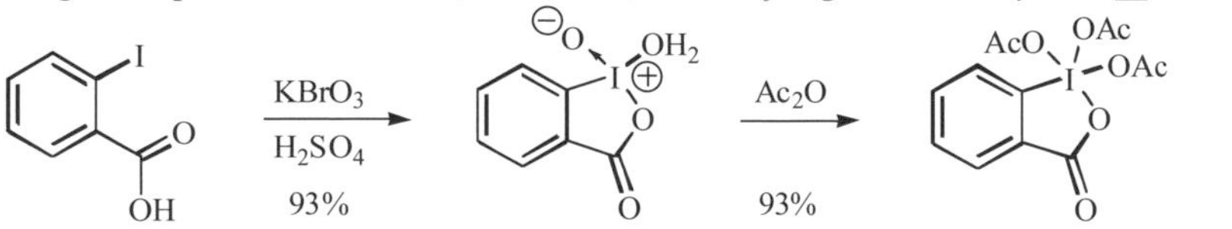

*2-Iodoxybenzoic acid*
*(IBX )*

*Dess-Martin Periodinane*
*Reagent*

*2-Iodoxybenzoic acid (IBX)* is also a useful oxidizing reagent.  Insoluble in most solvents (except DMSO) it can be used with other cosolvent mixtures.

See ***Dess-Martin Reagent*** and ***IBX Reagent***.

## Examples:

M. Mori, T. Tomita, Y. Kita, T. Kitamura, *Tetrahedron Letters* **2004**, <u>45</u>, 4397

See: ***Bmim***
J. S. Yadav, B. V. S. Reddy, A. K. Basak, A. Venkat Narsaiah, *Tetrahedron* **2004**, <u>60</u>, 2131

K. Suenaga, K. Araki, T. Sengoku, D. Uemura, *Organic Letters* **2001**, <u>3</u>, 527

Y.-F. Lu, A. G. Fallis, *Tetrahedron Letters* **1993**, <u>34</u>, 3367

P. A. Wender, S. G. Hegde, R. D. Hubbard, L. Zhang, *Journal of the American Chemical Society* **2002**, <u>124</u>, 4956

# Dieckmann Condensation / Cyclization / Reaction

## The Reaction:

## Proposed Mechanism:

The product remains an
enolate until protic workup.

## Notes:

V. K. Ahluwalia, R. K. Parashar, *Organic Reaction Mechanisms*, Alpha Science International Ltd.,
Pangbourne, U.K., 2002, pp. 308-309; M. B. Smith, J. March in *March's Advanced Organic
Chemistry*, 5th ed., John Wiley and Sons, Inc., New York, 2001, pp 569-570; C. R. Hauser, B. E.
Hudson, Jr., *Organic Reactions* **1**, 9; J. P. Schaefer, J. J. Bloomfield, *Organic Reactions*, **15**, 1.

### Regiochemistry Issues:

P. Compain, J. Gore, J.-M. Vatele, *Synthetic Communications* **1995**, <u>25</u>, 3075 (AN 1995:752767)

### Kinetic vs. Thermodynamic Control:

F. Duus, *Tetrahedron* **1981**, <u>57</u>, 2633

## Examples:

COOEt ⟍ ⟋ —Me

1. NaH, Cat EtOH, benzene
2. Hydrolysis
77%

COOEt ⟍ ⟋ —Me

O ⟍ ⟋ —Me

D. P. Provencal, J. W. Leahy, *Journal of Organic Chemistry* **1994**, <u>59</u>, 5496

EtOOC

EtOOC

,,Me

N      O

Me

1. Na / Ph-H / EtOH

2. NaCl/ DMSO / H₂O

Me

O ⟍ ⟋ ,,Me

N      O

Me

60%

O ⟍ ⟋ ,,Me

N      O

Me

9%

W. Zhu, D. Ma, *Organic Letters* **2003**, <u>5</u>, 5063

Me

COOMe

,,,COOMe

Me    Me

NaN(TMS)₂

THF

99%

Me  O

COOMe

Me    Me

L. A. Paquette, H.-L. Wang, *Journal of Organic Chemistry* **1996**, <u>61</u>, 5352

COOEt

—COOEt

Na, Toluene

cat. EtOH
followed by weak
acid in workup: 74 - 81%

O

—COOEt

In the absence of solvent
*t*BuONa        74%
EtONa           61%

P. S. Pinkney, *Organic Syntheses*, <u>CV2</u>, 116
F. Toda, T. Suzuki, S. Higa, *Journal of the Chemical Society: Perkin Transactions 1* **1988**, 3207

# Diels-Alder Reaction

## The Reaction:

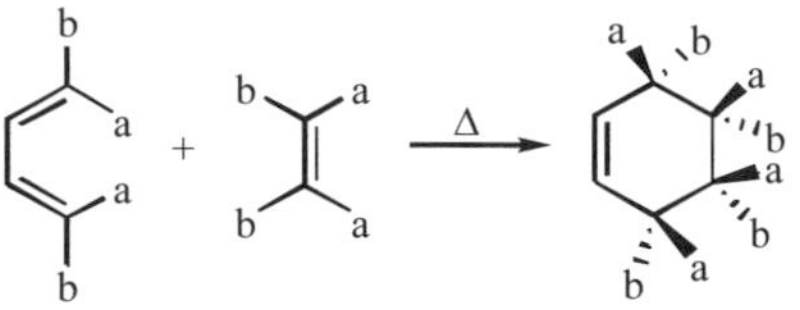

## Proposed Mechanism:

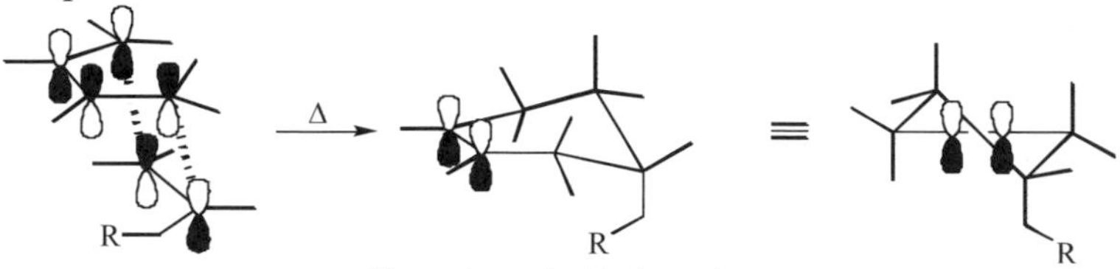

The endo product is favored.

This is a [4+2] concerted reaction. Bonds are broken and formed simultaneously, however, the arrows shown below are usually used to illustrate the mechanism.

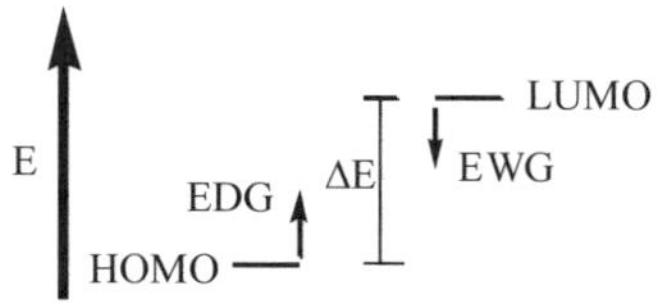

The smaller the $\Delta E$, the faster the reaction. Therefore, electron donating groups on the diene and/or electron withdrawing groups on the dieneophile will accelerate the reaction.

Although FMO theory shows the reaction to be a ground-state process, photochemical reactions with a trans-ring juncture product have been observed. This has been attributed to an excited state isomerization of the ene-portion followed by a ground state *Diels-Alder reaction*:

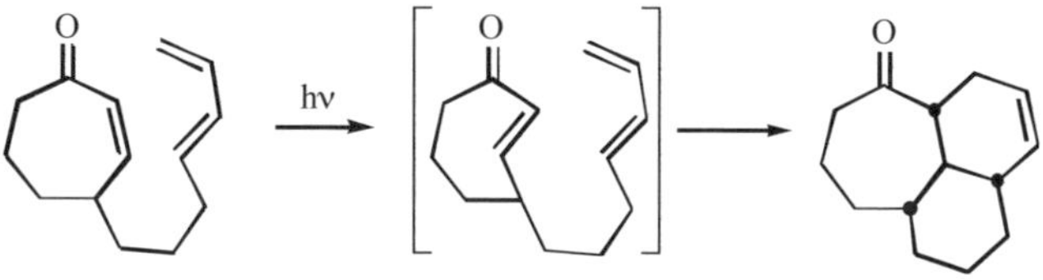

H. Dorr, V. H. Rawal, *Journal of the American Chemical Society* **1999**, <u>121</u>, 10229

T. Laue, A. Plagens, *Named Organic Reactions*, John Wiley and Sons, Inc., New York, 1998, pp. 78-85; M. B. Smith, J. March in *March's Advanced Organic Chemistry*, 5th ed., John Wiley and Sons, Inc., New York, 2001, pp 1062-1075; M. C. Klotzel, *Organic Reactions* **4**, 1; E. Ciganek, *Organic Reactions* **32**, 1.

## Notes:

The reverse reaction can also be performed, particularly if one or both fragments are stable.

### Boger Heterocycle Synthesis

D. L. Boger, J. S. Panek, *Journal of Organic Chemistry* **1981**, <u>46</u>, 2179

The **Diels-Alder Reaction** can be used in a number of creative ways. For example, extremely reactive dienes can be generated by thermal cycloreversion reactions:

65% *cis*, 28% *trans*

H. Pellissier, M. Santelli, *Tetrahedron* **1996**, <u>52</u>, 9093

benzene

C. W. Jefford, G. Bernardinelli, Y. Wang, D. C. Spellmeyer, A. Buda, K. N. Houk, *Journal of the American Chemical Society* **1992**, <u>114</u>, 1157

The driving force for the **Diels-Alder reaction** is, in part, due to the rearomatization process. Fluoride-induced elimination creates the same opportunity:

CsF

Y. Ito, M. Nakatsuka, T. Saeguso *Journal of the American Chemical Society* **1982**, <u>104</u>, 7609

Extrusion of SO$_2$ provides a unique opportunity for diene preparation:
Because of the sulfoxide influence, substitution at these positions quite easy.

J. Leonard, A. B. Hague, G. Harms, M. F. Jones, *Tetrahedron Letters* **1999**, <u>40</u>, 8141

T. Heiner, S. I. Kozhushkov, M. Noltemeyer, T. Hauman, R. Boese, A. DeMeijere, *Tetrahedron* **1996**, <u>52</u>, 12185

M. Toyota, T. Wada, Y. Nishikawa, K. Yanai, K. Fukumoto, C. Kabuto, *Tetrahedron* **1995**, <u>51</u>, 6927

B. M. Trost, M. Lautens, *Journal of the American Chemical Society* **1983**, <u>105</u>, 3345

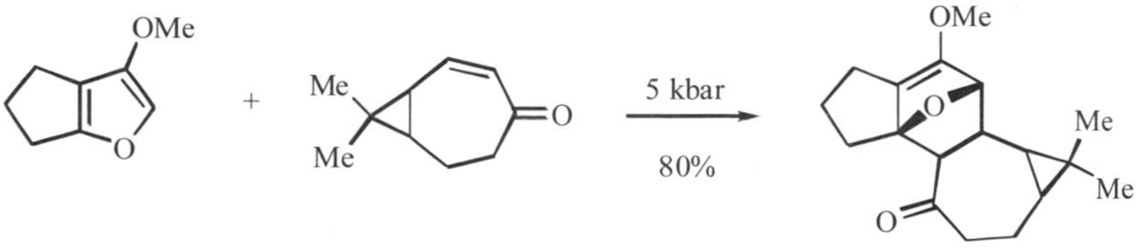

A. B. Smith, III, N. J. Liverton, N. J. Hrib, H. Sivaramakrishnan, K. Winzenberg, *Journal of Organic Chemistry* **1985**, <u>50</u>, 3239

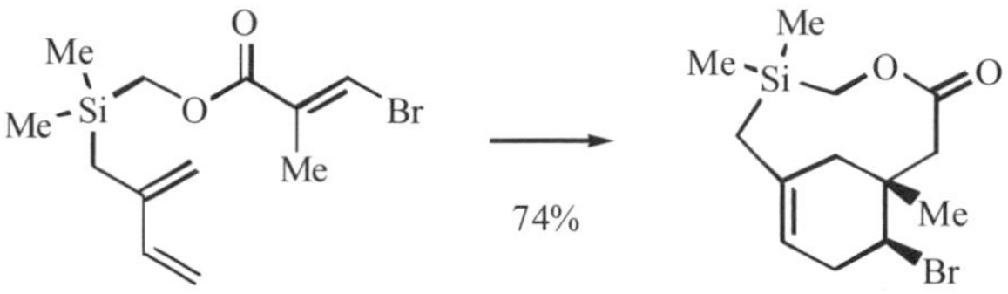

J. M. Whitney, J. S. Parner, K. J. Shea, *Journal of Organic Chemistry* **1997**, <u>62</u>, 8962

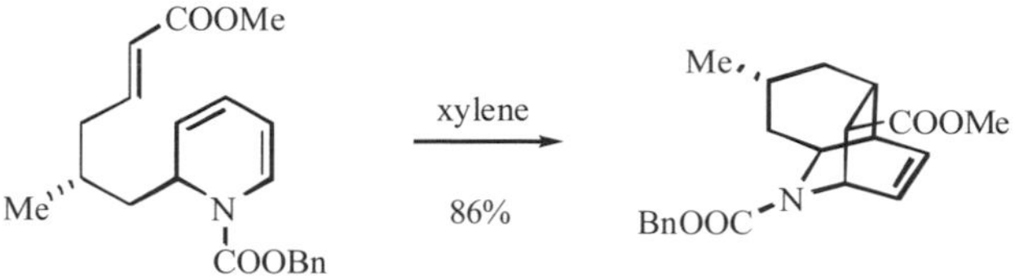

D. L. Comins, C. A. Brooks, R. S. Al-awar, R. R. Goehring, *Organic Letters* **1999**, <u>1</u>, 229

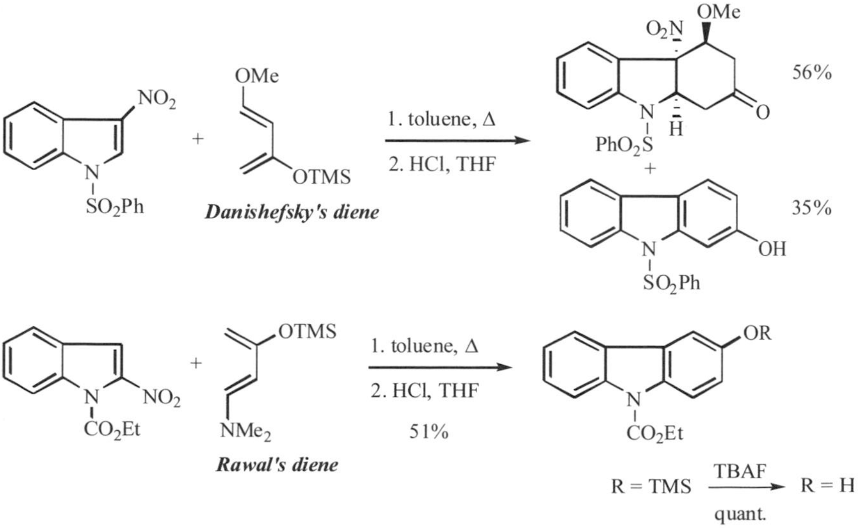

*Danishefsky's diene*

*Rawal's diene*

T. L. S. Kishbaugh, G. W. Gribble *Tetrahedron Letters* **2001**, <u>42</u>, 4783

# Dienone-Phenol Rearrangement

## The Reaction:

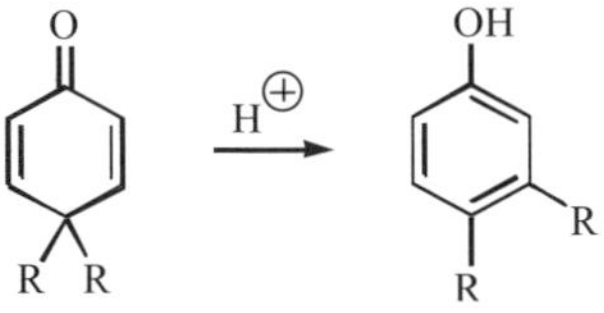

## Proposed Mechanism:

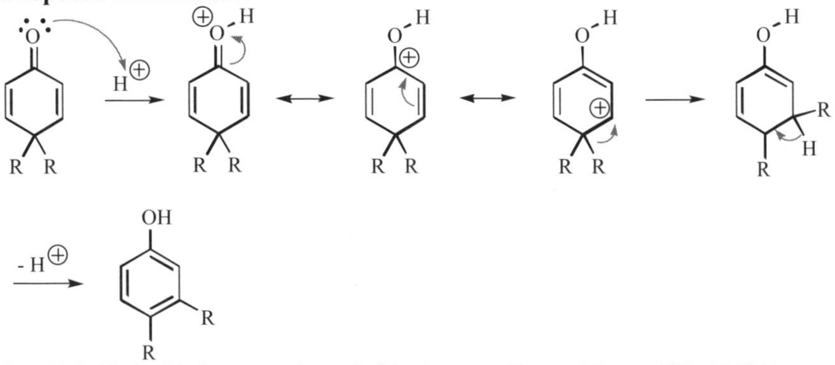

See: V. P. Vitullo, N. Grossman, *Journal of the American Chemical Society* **1972**, <u>94</u> 3844

## Notes:

If the R-groups are different, product mixtures to be expected.

### *Dienol-Benzene Rearrangement*

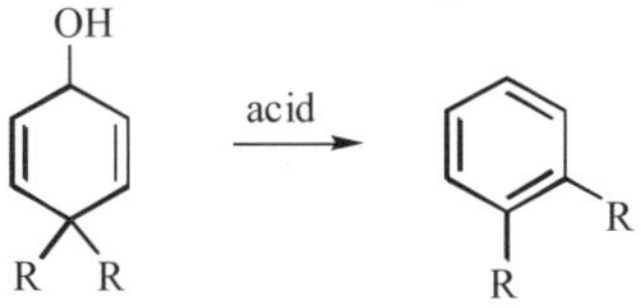

A dienone preparation:

### Zincke-Suhl Reaction

### The Reaction:

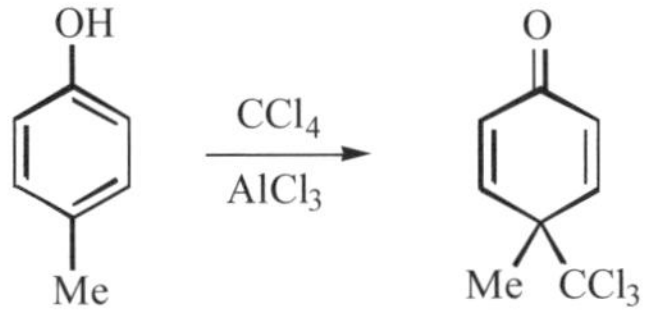

## Examples:

A. Planas, J. Tomas, J.-J. Bonet, *Tetrahedron Letters* **1978**, 28, 471

A. Sandoval, L. Miramontes, G. Rosenkranz, C. Djerassi, *Journal of the American Chemical Society* **1951**, 73, 990

A. J. Waring, *Tetrahedron Letters* **1975**, 12, 172

P. J. Kropp, *Tetrahedron Letters* **1963**, 4, 1671

S. Kodama, H. Takita, T. Kajimoto, K. Nishide, M. Node, *Tetrahedron* **2004**, 60, 4901

# Doebner Reaction (Beyer Synthesis, Beyer Method for Quinolines)

## The Reaction:

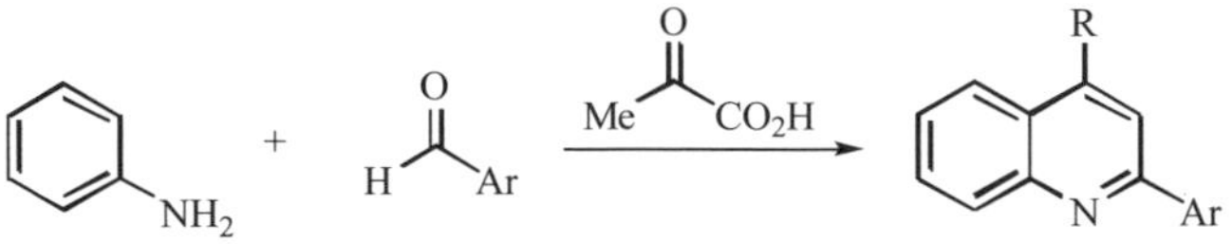

## Proposed Mechanism:

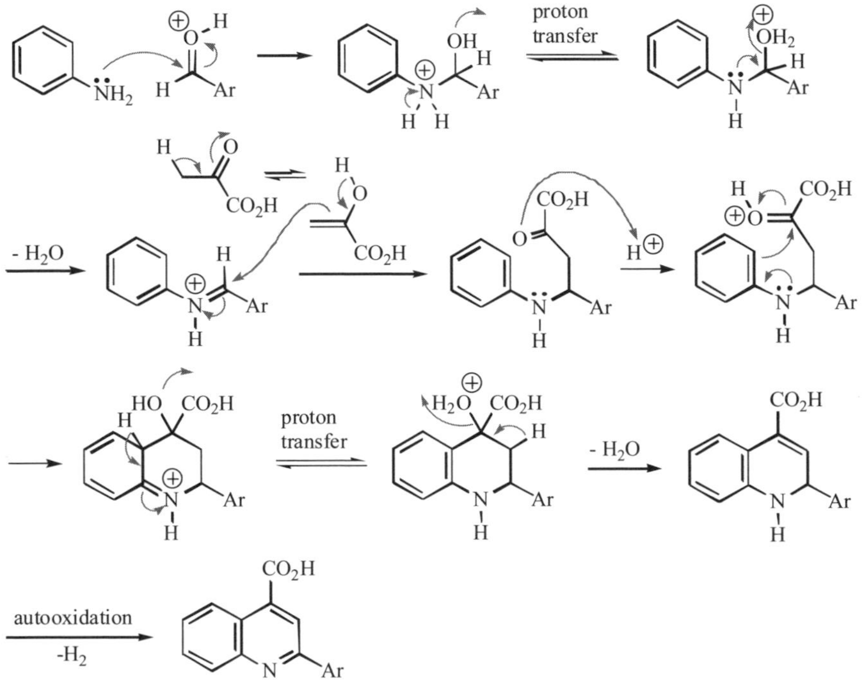

## Notes:
## Doebner-von Miller Reaction
**The Reaction:** V. K. Ahluwalia, R. K. Parashar, *Organic Reaction Mechanisms*, Alpha Science International Ltd., Pangbourne, U.K., 2002, p. 314

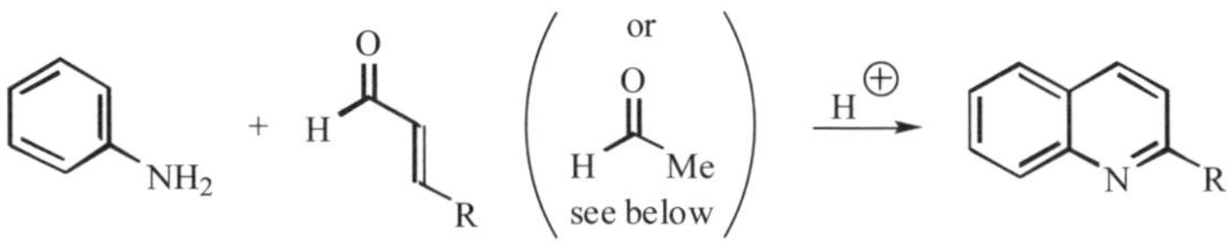

## Proposed Mechanism:

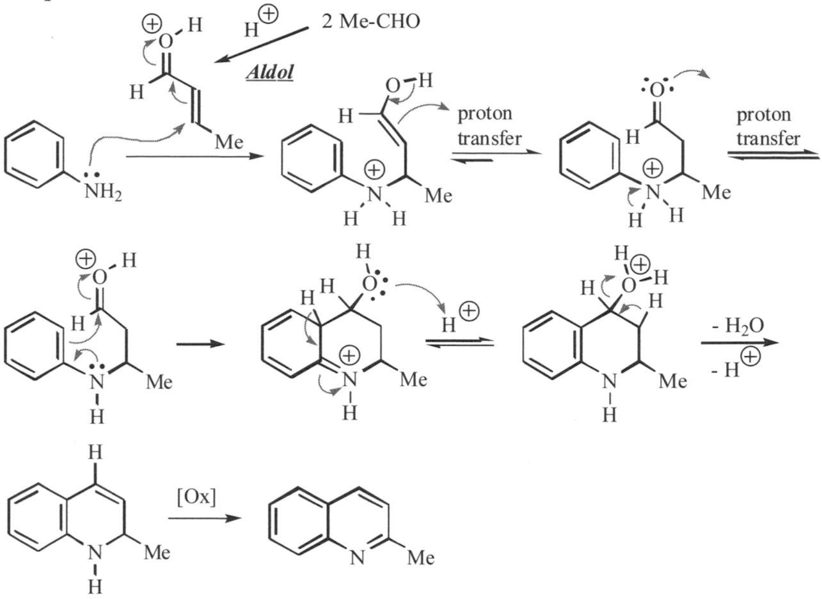

## Examples:

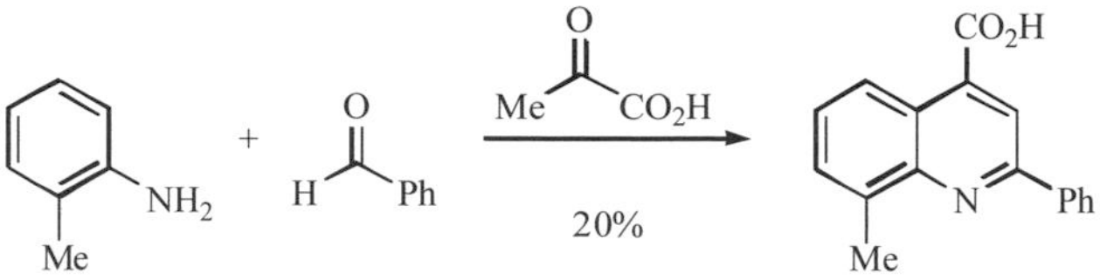

G. J. Atwell, B. C. Baguley, W. A. Denny, *Journal of Medicinal Chemistry* **1989**, <u>32</u>, 396

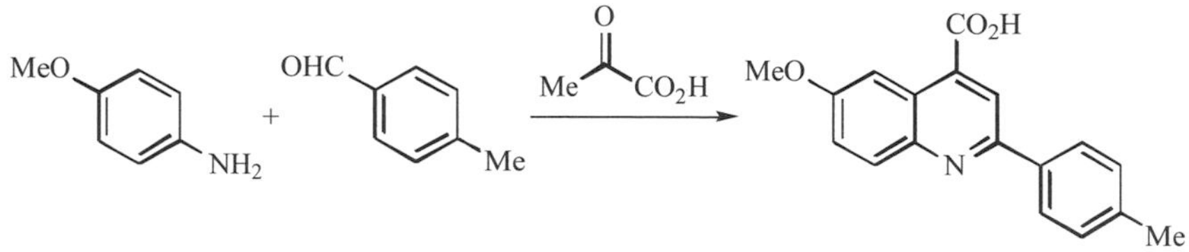

G. A. Epling, K. Y. Lin *Journal of Heterocyclic Chemistry* **1987**, <u>24</u>, 853 (AN 1998: 55860)
See also: G. SA. Epling, A. A. Provatas, *Chemical Communications* **2002**, 1036

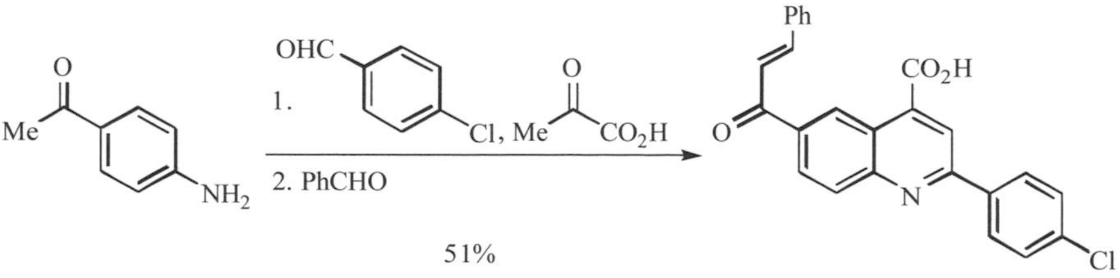

51%

D. J. Bhatt, G. C. Kamdar, A. R. Parikh *Journal of the Indian Chemical Society* **1984**, <u>61</u>, 816
(AN 1985:453938)

# Dondoni Homologation

## The Reaction:

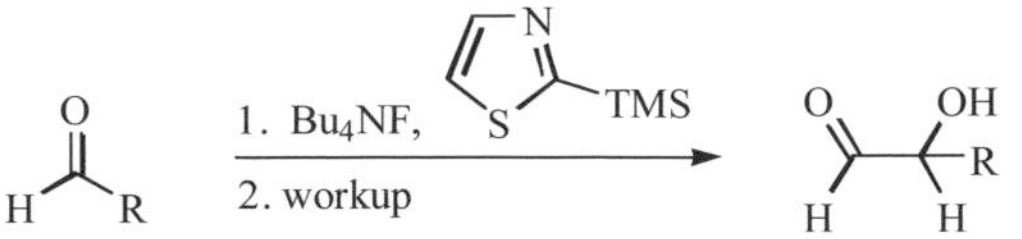

## Proposed Mechanism:

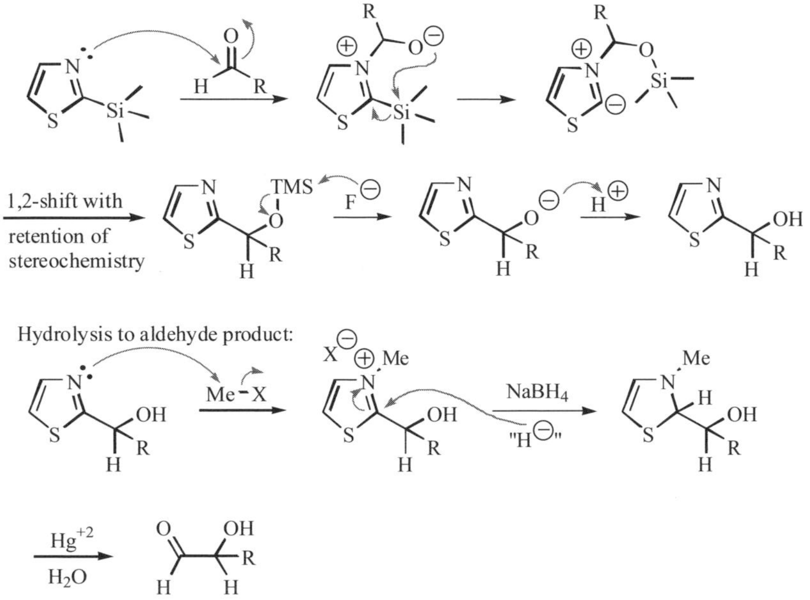

**1,2-shift with retention of stereochemistry**

Hydrolysis to aldehyde product:

A. Dondoni,.G. Fantin, M. Fogagnolo, A. Medici, P. Pedrini, *Journal of Organic Chemistry* **1989**, <u>54</u>, 693

A. Dondoni, G. Fantin, M. Fogagnolo, A. Medici, P. Pedrini, *Tetrahedron Letters* **1985**, <u>26</u>, 5477

## Notes:

Rationalization of stereochemistry for addition given by:

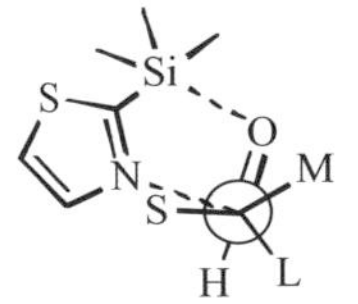

## Examples:

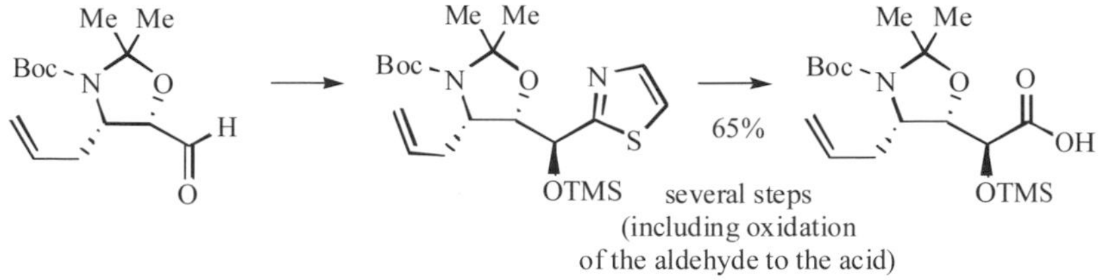

Me Me
Boc
N
O
H
O
OTMS
S
65%
several steps
(including oxidation
of the aldehyde to the acid)
OH

A. K. Ghosh, A. Bischoff, J. Cappiello, *Organic Letters* **2001**, 3, 2677

Me
I
CHO
Dondoni
Homologation
OBn
N
S
Me Me

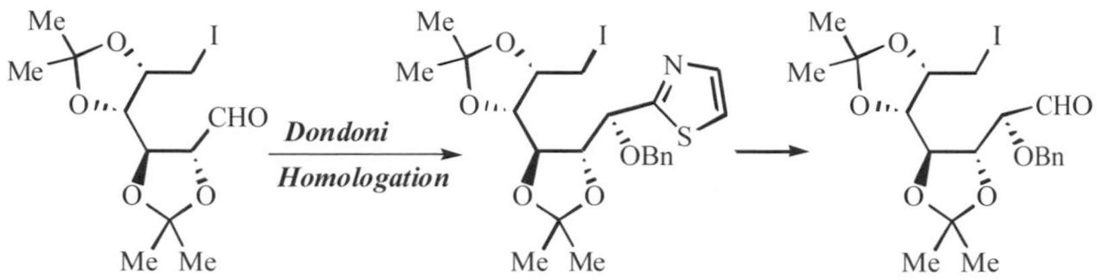

J. Marco-Contelles, E. de Opazo, *Journal of Organic Chemistry* **2002**, 67, 3705

O
H
Boc
N
O
Me
Dondoni
Homologation
OBn

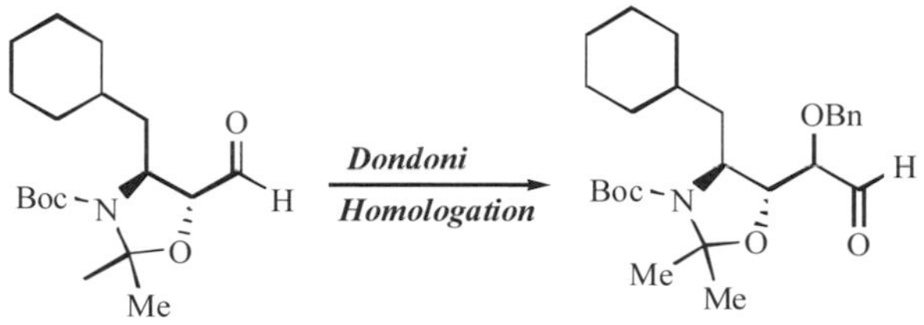

A. Wagner, M. Mollath, *Tetrahedron Letters* **1993**, 34, 619

Me
O
CHO
Dondoni
Homologation
N
S
OH

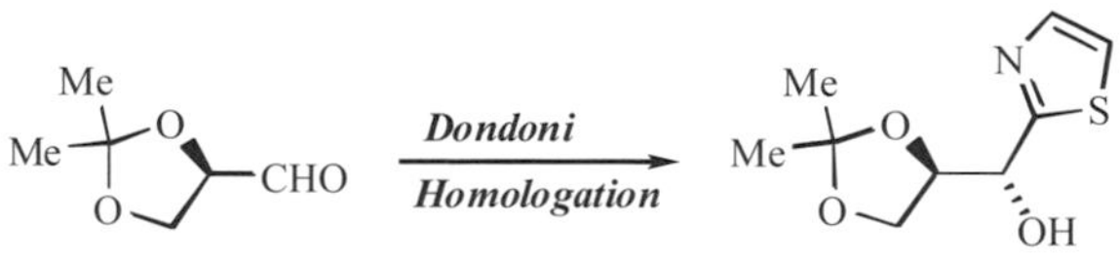

A. Dondoni, G. Fantin, M. Fogagnolo, A. Medici, P. Pedrini, *Journal of Organic Chemistry* **1989**, 54, 693

# Dötz Reaction

## The Reaction:

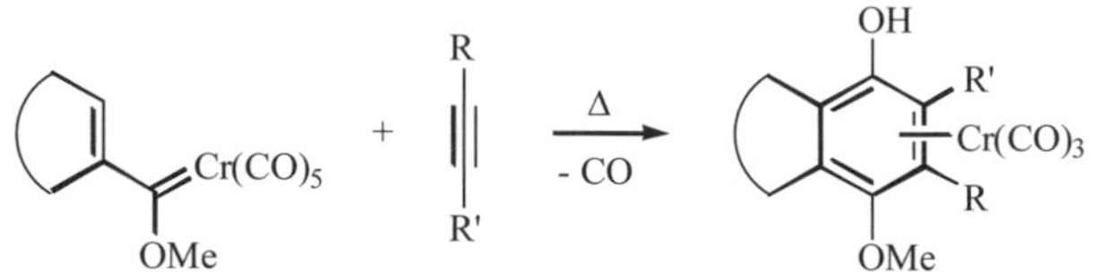

vinyl or aromatic alkoxy pentacarbonyl      and sometimes the regioisomer
chromium carbene complex                    for unsymmetrical alkynes

## Proposed Mechanism:

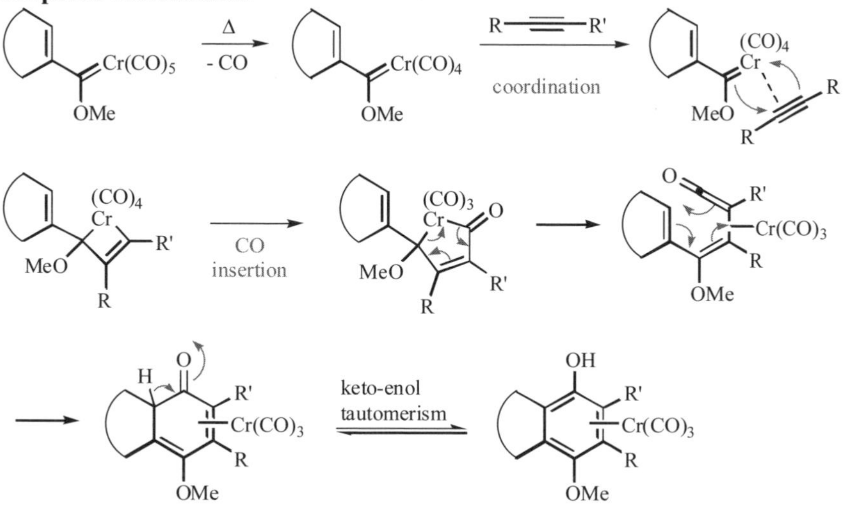

## Notes:

See: T. Laue, A. Plagens, *Named Organic Reactions*, John Wiley and Sons, Inc., New York, 1998, pp. 88-91. These authors note that there is poor regioselectivity for non-symmetrical alkynes.

Often the product is directly oxidized:

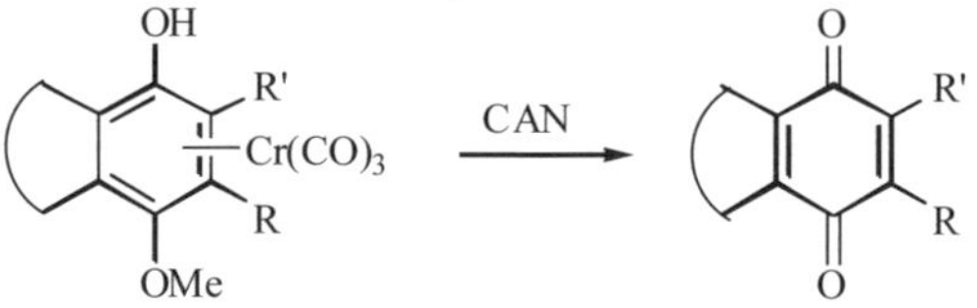

The starting material can be prepared:

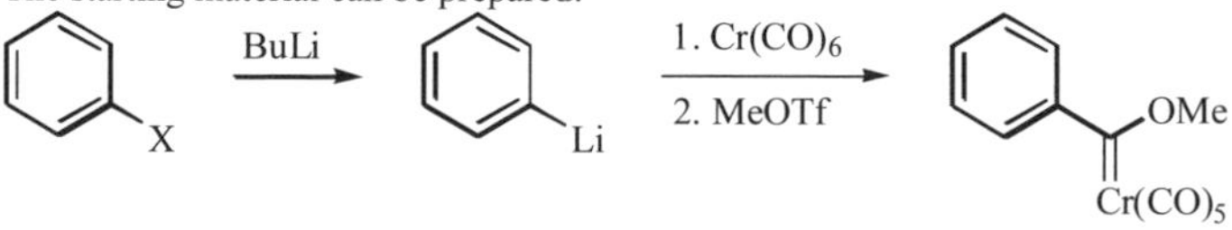

## Examples:

S. R. Pulley, B. Czako, *Tetrahedron Letters* **2004**, <u>45</u>, 5511

J. C. Anderson, J. W. Cran, N. P. King *Tetrahedron Letters* **2002**, <u>43</u>, 3849

35 - 40%

W. R. Roush, R. J. Neitz, *Journal of Organic Chemistry* **2004**, <u>69</u>, 4906

# Dowd-Beckwith Ring Expansion

## The Reaction:

## Proposed Mechanism:

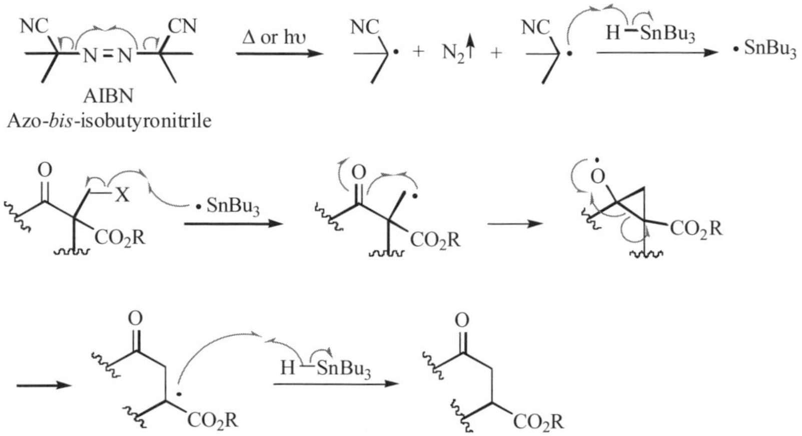

AIBN
Azo-*bis*-isobutyronitrile

P. Dowd, S.-C. Choi, *Tetrahedron* **1989**, <u>45</u>, 77

## Examples:

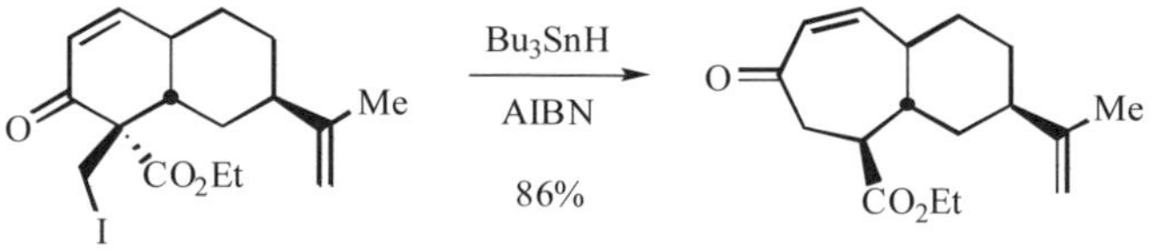

86%

M. G. Banwell, J. M. Cameron, *Tetrahedron Letters* **1996**, <u>37</u>, 525

C. Wang, X. Gu, M. S. Yu, D. P. Curran, *Tetrahedron Letters* **1998**, <u>54</u>, 8355

75%

P. Dowd, S.-C. Choi, *Tetrahedron* **1989**, <u>45</u>, 77

M. T. Crimmins, Z. Wang, L. A. McKerlie, *Journal of the American Chemical Society* **1998**, <u>120</u>, 1747

# Doyle-Kirmse Reaction

## The Reaction:

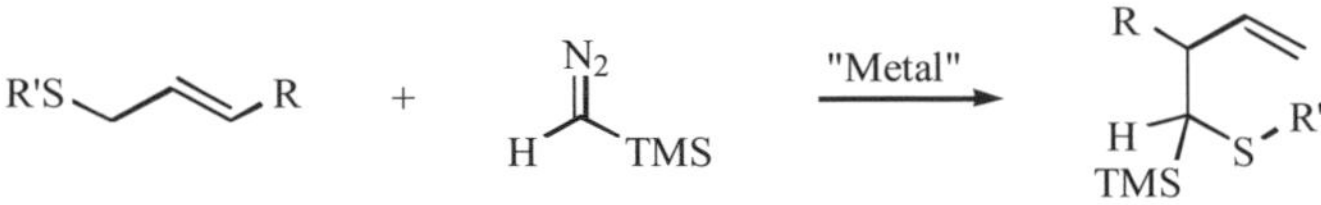

## Proposed Mechanism:

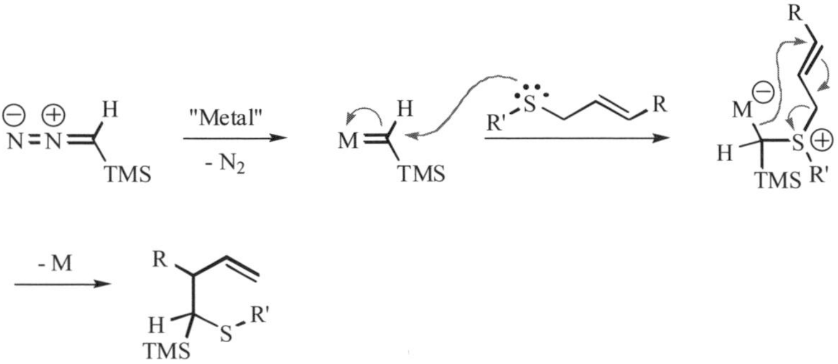

M. P. Doyle, W. H. Tamblyn, V. Bagher, *Journal of Organic Chemistry* **1981**, <u>40</u>, 5094

## Notes:

Influence of metal catalyst:

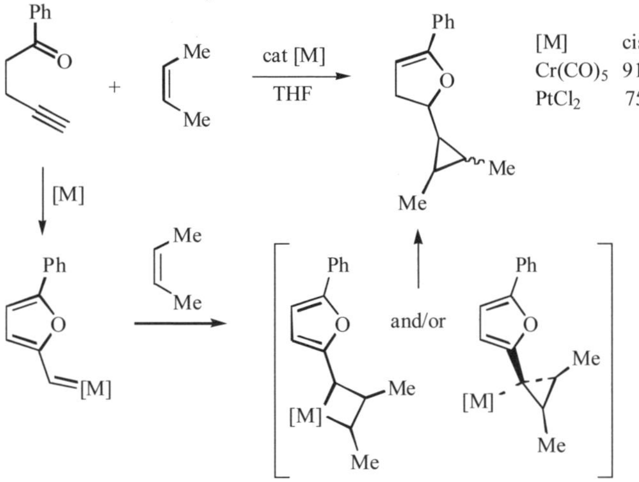

| [M] | cis | trans |
|-----|-----|-------|
| Cr(CO)₅ | 91 | 9 |
| PtCl₂ | 75 | 25 |

| [M] | Yield |
|-----|-------|
| Cr(CO)₅ | 14 |
| PtCl₂ | 26 |

K. Miki, T. Yokoi, F. Nishino, Y. Kato, Y. Washitake, K. Ohe, S. Uemura, *Journal of Organic Chemistry* **2004**, <u>69</u>, 1557

## Examples:

PhS~~Me + H–C(N₂)–TMS —(Fe⁺²)→ (88%) → Me₂C(CH=CH₂)–CH(TMS)(SPh)

PhS~~Ph + H–C(N₂)–TMS —(Fe⁺²)→ (94%) → Ph–CH(CH=CH₂)–CH(TMS)(SPh)

D. S. Carter, D. L. Van Vranken, *Organic Letters* **2000**, <u>2</u>, 1303

Cyclohexene–CHO / C≡C–CO₂Me —[Rh(OAc)₂]₂→ [ bicyclic furan with MeO₂C=[Rh] ] —(allyl–SMe)→ (91%) → furan product MeO₂C / SMe

Y. Kato, K. Miki, F. Nishino, K. Ohe, S. Uemura, *Organic Letters* **2003**, <u>5</u>, 2619

Me / terpene–S-Et + H–C(N₂)–TMS, FeCl₂, dppe —(Cl-CH₂CH₂-Cl)→ (89%) → Me / vinyl–S-Et / TMS product

J. B. Perales, N. F. Makino, D. L. Van Vranken, *Journal of Organic Chemistry* **2002**, <u>67</u>, 6711

# Duff Reaction

## Reaction:

hexamethylenetetramine
(HMTA)

EDG = OH or NR$_2$

minor

major

+

## Proposed Mechanism:

:EDG

dehydrogentation

$-H^{\oplus}$

$H_2O$

proton
transfer

+

## Notes:

V. K. Ahluwalia, R. K. Parashar, *Organic Reaction Mechanisms*, Alpha Science International Ltd.,
Pangbourne, U.K., 2002, pp. 315-316; M. B. Smith, J. March in *March's Advanced Organic
Chemistry*, 5$^{th}$ ed., John Wiley and Sons, Inc., New York, 2001, p. 717.
A detailed kinetic and product (intermediate) study (Y. Ogata, A. Kawaqsaki, F. Sugiura,
*Tetrahedron* **1968**, <u>24</u>, 5001) describes the rapid decomposition of HMTA:

$H_2C=NH_2$
$\oplus$

which can add to the phenol and also serves as an
oxidizing agent

$H_2C=NH_2$
$\oplus$

$H_2C=NH_2$
$\oplus$

hyd

Product

## Examples:

R. A. Johnson, R. R. Gorman, R. J. Wnuk, N. J. Drittenden, J. W. Aikem, *Journal of Medicinal Chemistry* **1993**, <u>36</u>, 3202

M. A. Weidner-Wells, S. A. Fraga-Spano, *Synthetic Communications* **1996**, <u>26</u>, 2775

C. F. Allen, G. W. Leubner, *Organic Syntheses* <u>CV4</u>, 866

T. Ikemoto, T. Kawamoto, H. Wada, T. Ishida, T. Ito, Y. Isogami, Y. Miyano, Y. Mizuno, K. Tomimatsu, K. Hamamura, M. Takatani, M. Wakimasu, *Tetrahedron* **2002**, <u>58</u>, 489

J. F. Larrow, E. N. Jacobson, Y. Gao, Y. Hong, X. Nie, C. M. Zepp, *Journal of Organic Chemistry* **1994**, <u>59</u>, 1939

H.-M. He, M. Cushman, *Bioorganic & Medicinal Chemistry Letters* **1994**, <u>4</u>, 1725

# Eglinton Reaction

## The Reaction:

$$2 \quad R\!\!-\!\!\equiv\!\!-H \quad \xrightarrow[\text{pyridine}]{\text{CuX}_2} \quad R\!\!-\!\!\equiv\!\!-\!\!\equiv\!\!-R$$

terminal alkyne

X = OAc, Cl

## Proposed Mechanism:

See *Alkyne Coupling* for a general discussion.

## Notes:

M. B. Smith, J. March in *March's Advanced Organic Chemistry*, 5th ed., John Wiley and Sons, Inc., New York, 2001, p 927.

## Examples:

$$\xrightarrow[65\ °C]{\text{CuCl, CuCl}_2}$$

15 - 20%
(no product at higher temperatures)

S. Hoger, K. Bonrad, L. Karcher, A.-D. Meckenstock, *Journal of Organic Chemistry* **2000**, <u>65</u>, 1588

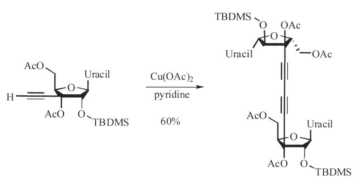

M. H. Haley, M. L. Bell, S. C. Brand, D. B. Kimball, J. J. Pak, W. B. Wan, *Tetrahedron Letters* **1997**, <u>38</u>, 7483

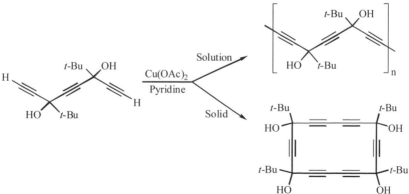

F. Jung, A. Berger, J.-F. Biellmann, *Organic Letters* **2003**, <u>5</u>, 383

The product composition can differ depending on whether the reaction is carried out in solid or solution:

F. Toda, Y. Tokumaru, *Chemistry Letters* **1990**, 987
(Reported in P. Siemsen, R. C. Livingston, F. Diederich, *Angewandte Chemie International Edition in English* **2000**, <u>39</u>, 2632)

# Ene Reaction (Alder-Ene Reaction)

## The Reaction:

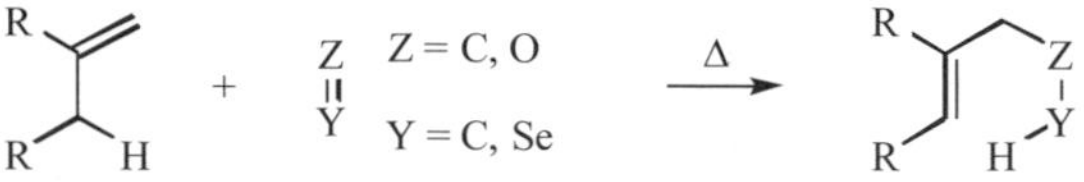

R⌁ + Z‖Y  Z = C, O   Y = C, Se  $\xrightarrow{\Delta}$   product

## Proposed Mechanism:

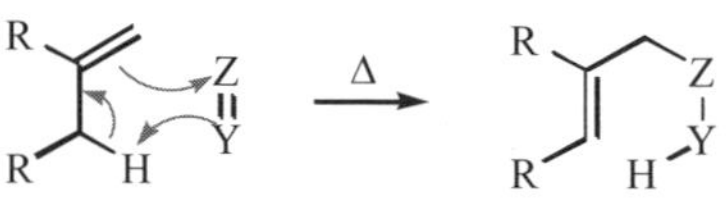

From FMO Theory:

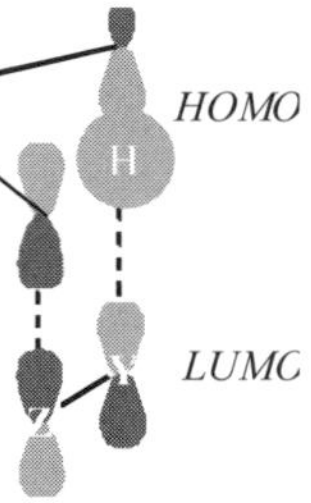

*HOMO*

*LUMO*

## Notes:

See: T. Laue, A. Plagens, *Named Organic Reactions*, John Wiley and Sons, Inc., New York, 1998, pp. 93-97.

Other reactions following the "ene" mechanistic interpretation:

Reactions of singlet oxygen (***Schenck ene reaction***),

SeO$_2$ oxidations,

***Etard Reaction***,

***Conia Cyclization***:

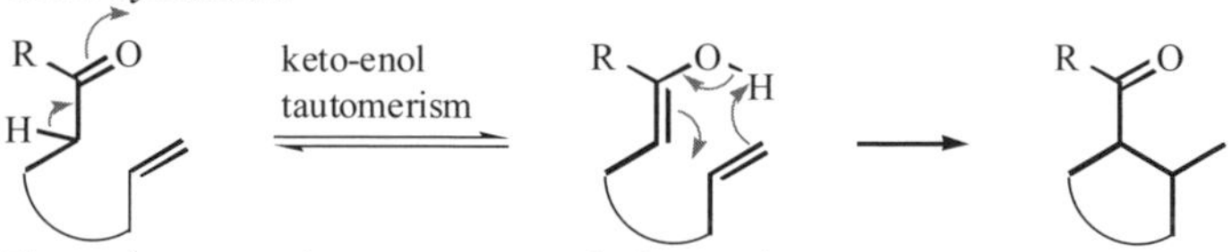

R⌁O, H   $\xrightleftharpoons[\text{tautomerism}]{\text{keto-enol}}$   enol   $\longrightarrow$   product

The enol serves as the ene component in the reaction.

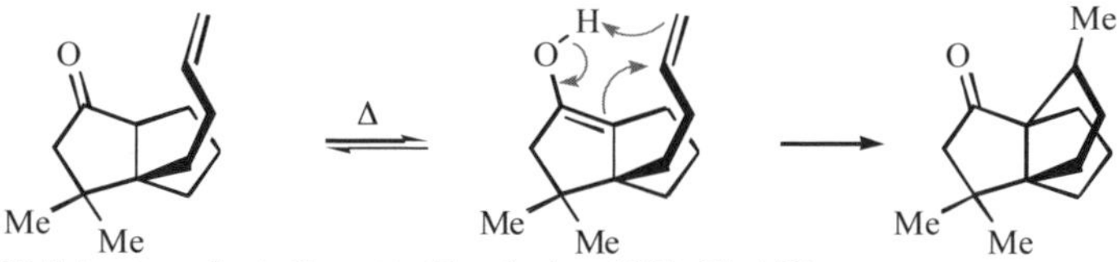

$\xrightleftharpoons{\Delta}$

H. Schostarez, L. A. Paquette, *Tetrahedron* **1981**, <u>37</u>, 4431

## Examples:

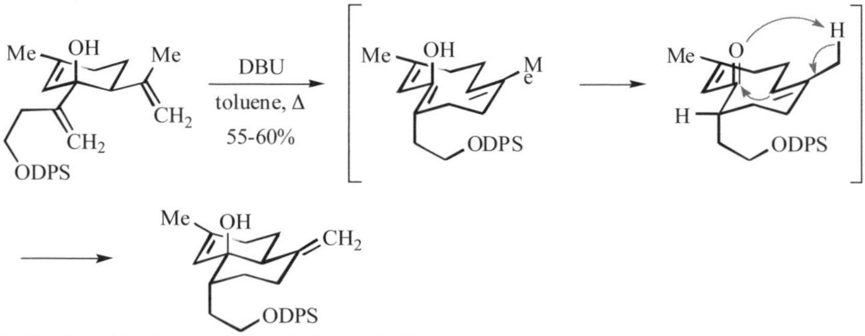

Me, OH, Me, CH$_2$, CH$_2$, ODPS   $\xrightarrow[\substack{\text{toluene, }\Delta \\ 55\text{-}60\%}]{\text{DBU}}$   [ Me, OH, Me, ODPS   $\longrightarrow$   Me, O, H, ODPS ]

$\longrightarrow$   Me, OH, CH$_2$, ODPS

L. Barriault, D. H. Deon, *Organic Letters* **2001**, <u>3</u>, 1925

The ***magnesium-ene reaction*** provides a unique approach to remote functional groups:

W. Oppolzer, A. Nakao, *Tetrahedron Letters* **1986**, <u>27</u>, 5471

***Carbonyl-ene Reaction***:

D. A. Evans, S. W. Tregay, C. S. Burgey, N. A. Paras, T. Vojkovsky, *Journal of the American Chemical Society* **2000**, <u>122</u>, 7936

J. Henninger, K. Polborn, H. Mayr, *Journal of Organic Chemistry* **2000**, <u>65</u>, 3569

# Eschenmoser Coupling Reaction

## The Reaction:

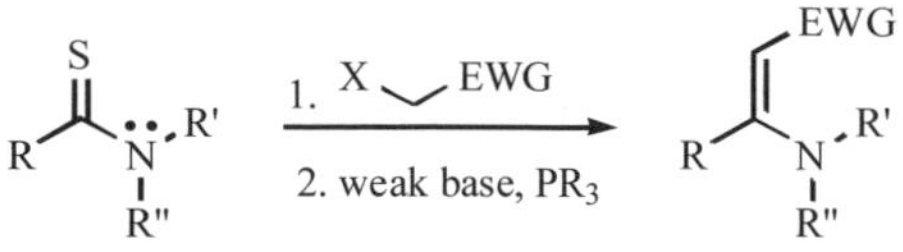

## Proposed Mechanism:

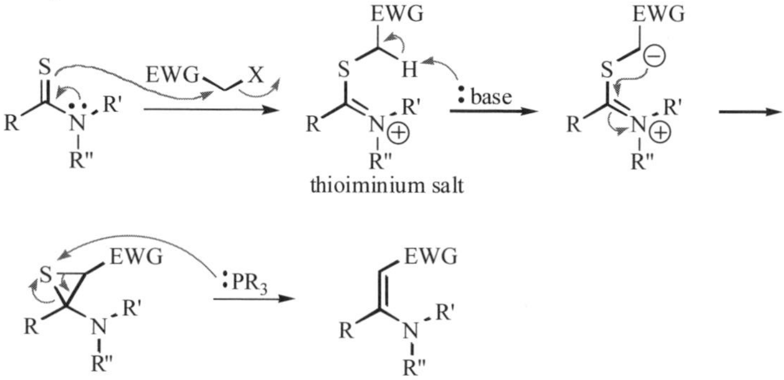

thioiminium salt

## Notes:

Possible starting material preparation:

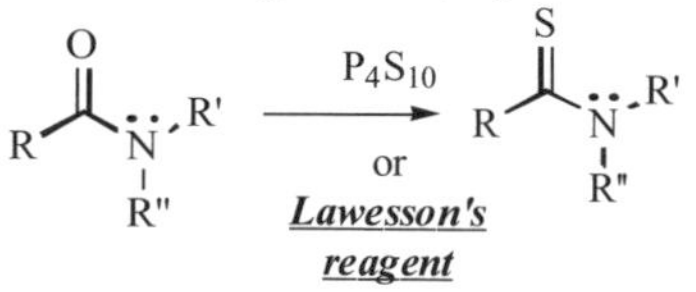

$P_4S_{10}$

or

*Lawesson's reagent*

## Examples:

DBU

60%

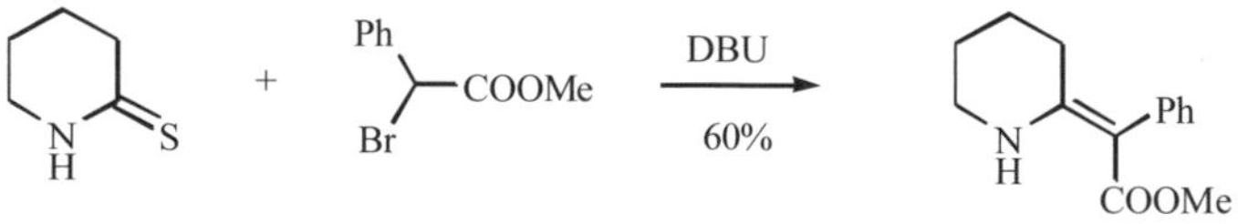

D. Russowsky, B. A. da Silveira Neto, *Tetrahedron Letters* **2004**, 45, 1437

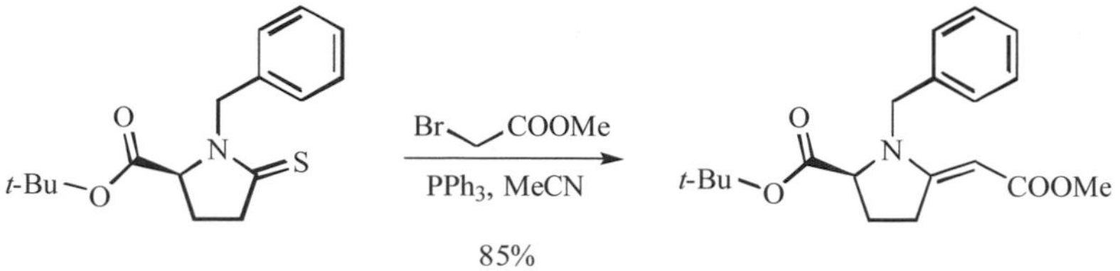

85%

J. S. Petersen, G. Fels, H. Rapoport, *Journal of the American Chemical Society* **1984**, <u>106</u>, 4359

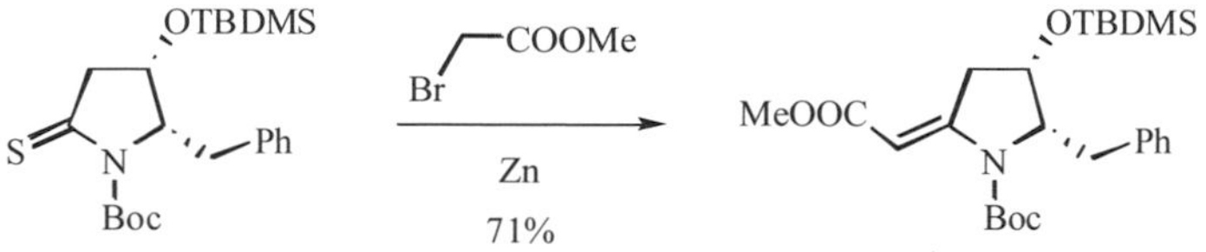

71%

H. K. Lee, J. Kim, C. S. Pak, *Tetrahedron Letters* **1999**, <u>40</u>, 2173

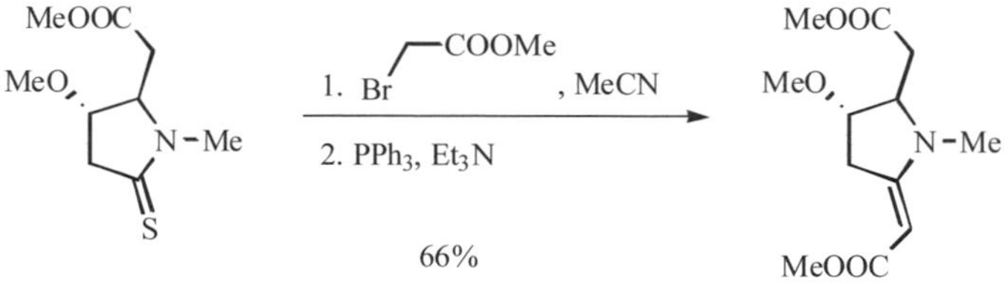

66%

D. J. Hart, L.-Q. Sun, A. P. Kozikowski, *Tetrahedron Letters* **1995**, <u>36</u>, 7787

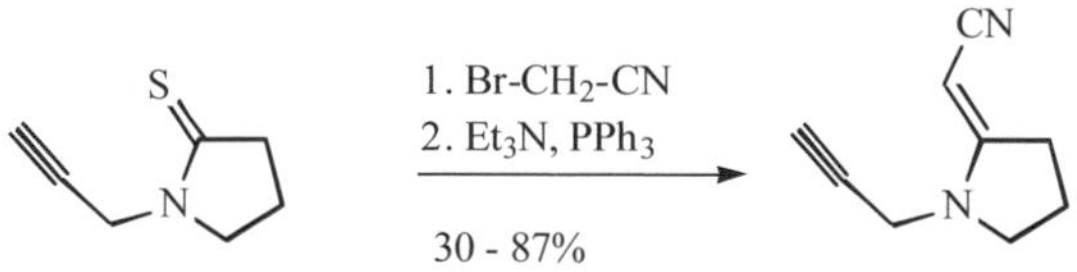

1. Br-CH$_2$-CN
2. Et$_3$N, PPh$_3$

30 - 87%

D. Gravenstock, I. G. Pierson, *Tetrahedron Letters* **2000**, <u>41</u>, 3497

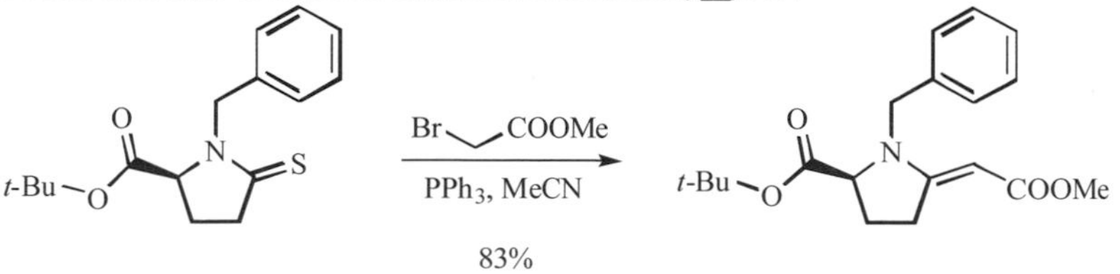

83%

J. A. Campbell, H. Rapoport, *Journal of Organic Chemistry* **1996**, <u>61</u>, 6313

# Eschenmoser Fragmentation (Eschenmoser-Tanabe Fragmentation / Ring Cleavage)

## The Reaction:

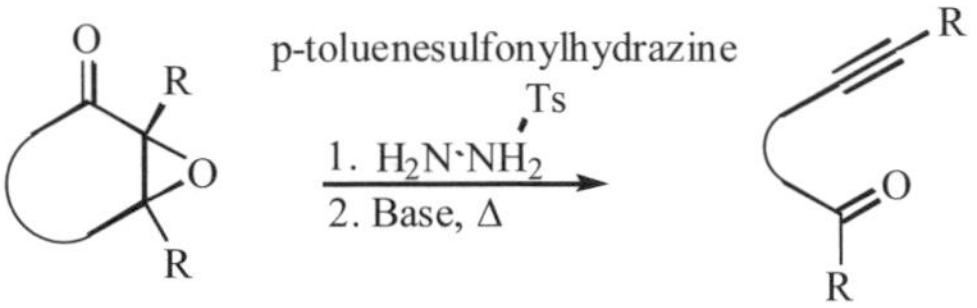

p-toluenesulfonylhydrazine

1. H₂N·NH₂
2. Base, Δ

## Proposed Mechanism:

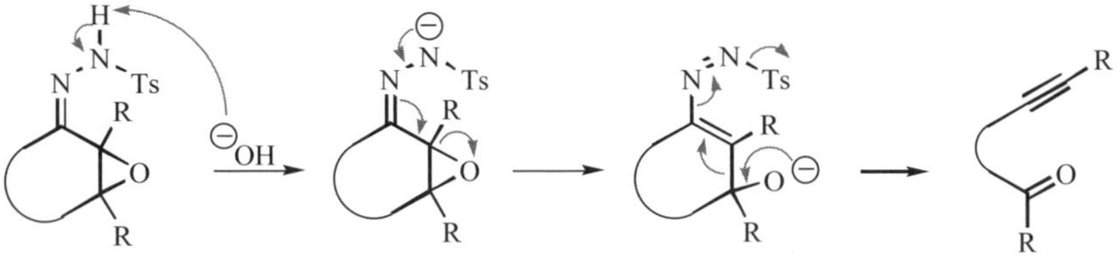

## Notes:

See: M. B. Smith, J. March in *March's Advanced Organic Chemistry*, 5$^{th}$ ed., John Wiley and Sons, Inc., New York, 2001, p. 1347. The authors note that aldehydes can be formed if 2,4-dinitrophenylhydrazine is used.

## Examples:

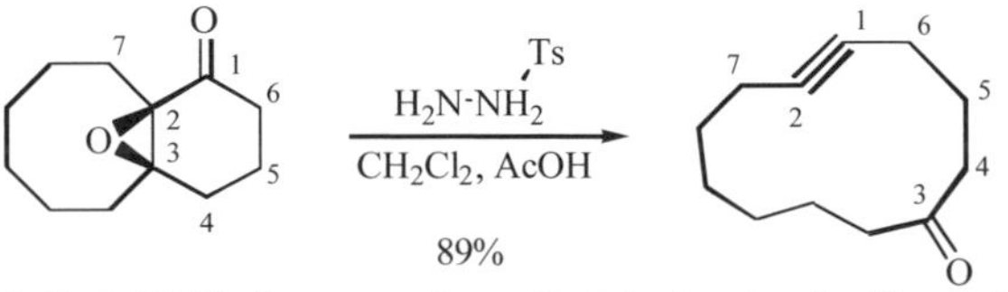

Ts
H₂N·NH₂
CH₂Cl₂, AcOH

89%

P. Kraft, W. Tochtermann, *Justus Liebig's Annalen der Chemie* **1994**, 1161 (AN 1994:700495)

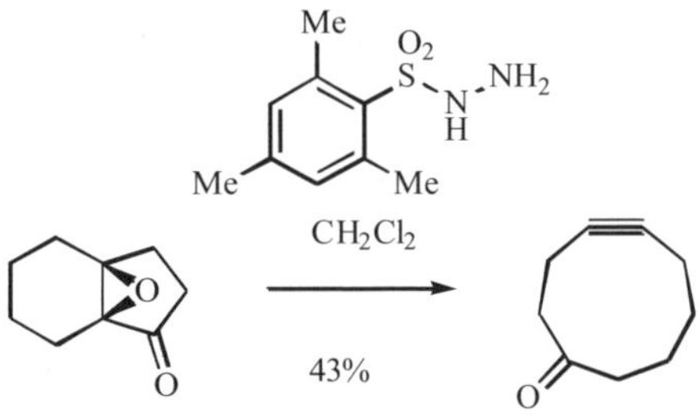

CH₂Cl₂

43%

C. B. Reese, H. P. Sander, *Synthesis* **1981**, 276

E. J. Corey, H. S. Sachdev, *Journal of Organic Chemistry* **1975**, <u>40</u>, 579

R = Pr, *i*-Bu, *i*-Pr,

W. Dai, J. A. Katzenellenbogen, *Journal of Organic Chemistry* **1993**, <u>58</u>, 1900

L. N. Mander, M. M. McLachlan, *Journal of the American Chemical Society* **2003**, <u>125</u>, 2400

# Eschweiler-Clarke (Clark) Methylation

## The Reaction:

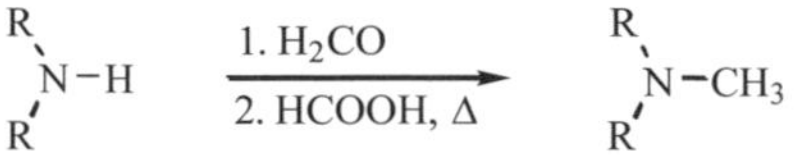

$$
\begin{array}{c}
R \\
\diagdown \\
N-H \\
\diagup \\
R'
\end{array}
\quad
\xrightarrow[\text{2. HCOOH, } \Delta]{\text{1. H}_2\text{CO}}
\quad
\begin{array}{c}
R \\
\diagdown \\
N-CH_3 \\
\diagup \\
R'
\end{array}
$$

## Proposed Mechanism:

immonium ion

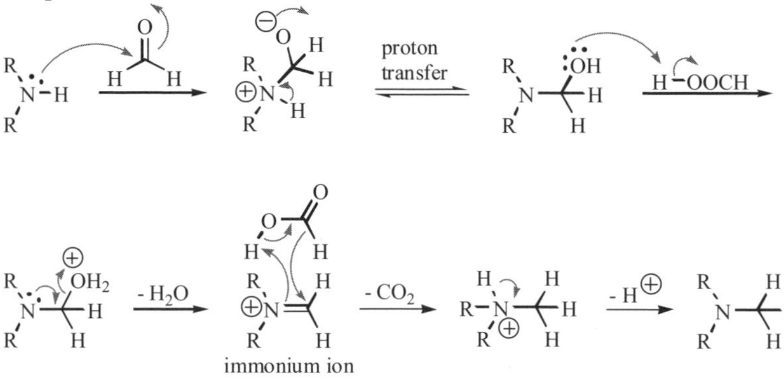

## Notes:

This reaction specifically refers to the case when a primary or secondary amine is reductively methylated with formaldehyde and formic acid.

See also the **Wallach Reaction** and the **Leuckart(-Wallach) Reaction**.

## Examples:

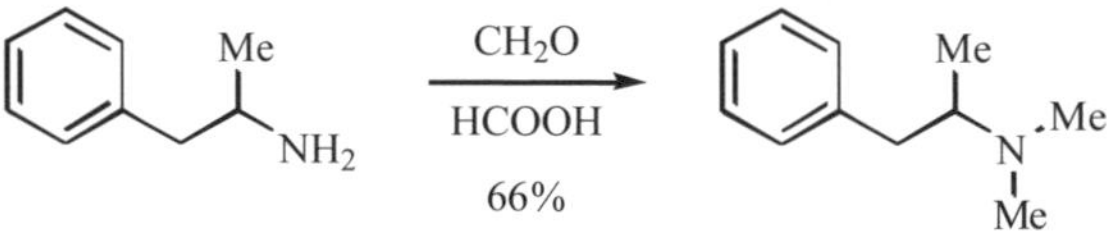

66%

W. E. Parham, W. T. Hunter, R. Hanson, T. Lahr, *Journal of the American Chemical Society* **1952**, <u>74</u>, 5646

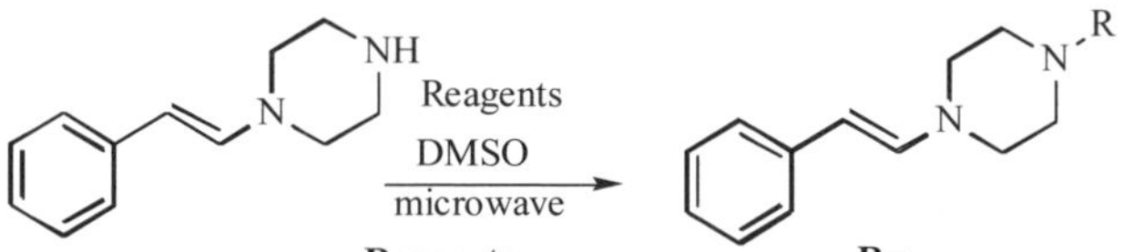

| Reagents | R= |
|---|---|
| HCHO, DCOOD | -CH$_2$D |
| DCHO, HCOOH | -CHD$_2$ |
| DCDO, DCOOD | -CD$_3$ |

J. R. Harding, J. R. Jones, S.-Y. Lu, R. Wood, *Tetrahedron Letters* **2002**, <u>43</u>, 9487

G. Lakshmaiah, T. Kawabata, M. Shang, K. Fuji, *Journal of Organic Chemistry* **1999**, <u>64</u>, 1699

P. Sahakitpichan, S. Ruchirawat, *Tetrahedron Letters* **2003**, <u>44</u>, 5239

K. Watanabe, T. Wakabayashi, *Journal of Organic Chemistry* **1980**, <u>45</u>, 357

# Evans Chiral *N*-Acyloxazolidinone Methodology

## The Reaction:

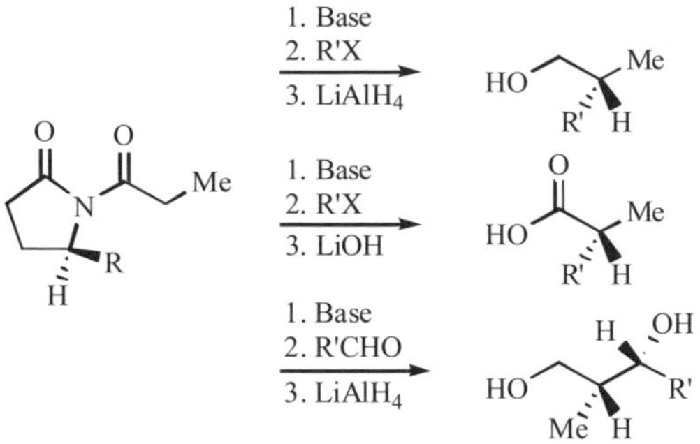

## Proposed Mechanism:

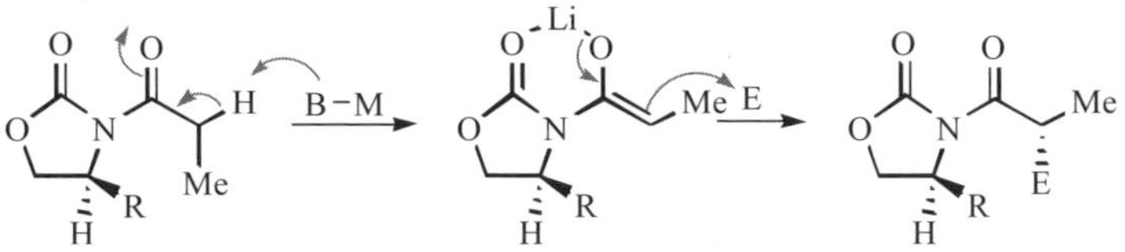

## Notes:

R is often *i*-Pr or Bn.
Bases (B-M) include LDA, Li or NaHMDS, amines with Lewis acids, etc.
E = electrophile (halide, aldehyde).

Removal of chiral auxiliary:

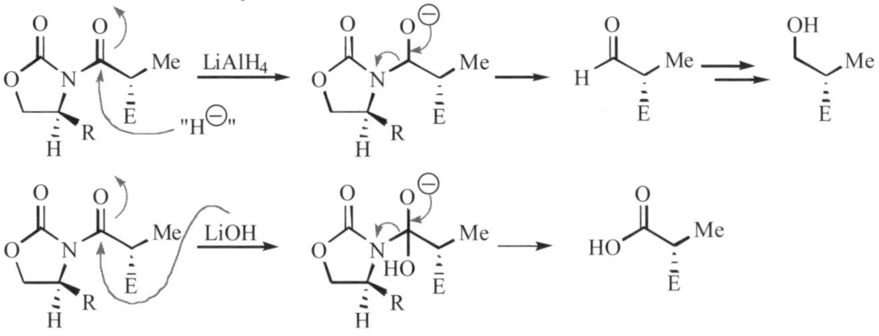

Z-enolates produced. Steric effects on direction of enolate alkylation are quite obvious:

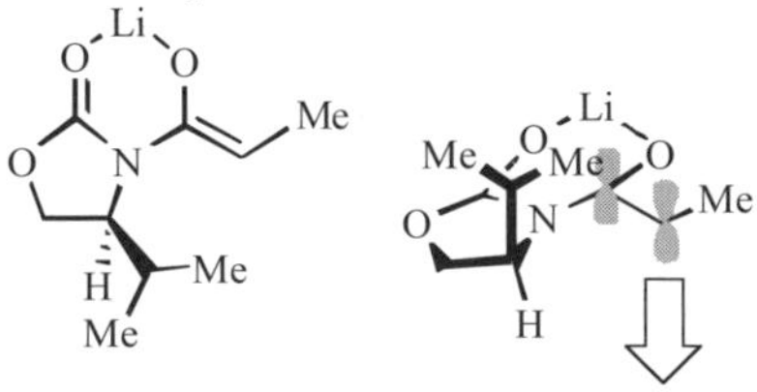

## Examples:

75%

A. N. Hulme, E. M. Rosser, *Organic Letters* **2002**, *4*, 265

M. R. Pitts, J. Mulzer, *Tetrahedron Letters* **2002**, *43*, 8471

A. K. Ghosch, J.-H. Kim, *Tetrahedron Letters* **2003**, *44*, 7659

A. K. Mandal, *Organic Letters* **2002**, *4*, 2043

Generally high yields
Protecting Groups = Bn-, Et₃Si-, Me-, MOM-.

M. T. Crimmins, K. A. Emmitte, J. D. Katz, *Organic Letters* **2000**, *2*, 2165

# Evans-Tischenko Reaction

## The Reaction:

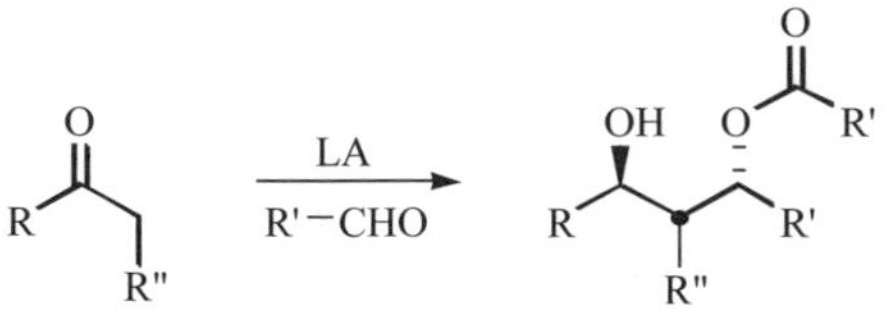

## Proposed Mechanism:

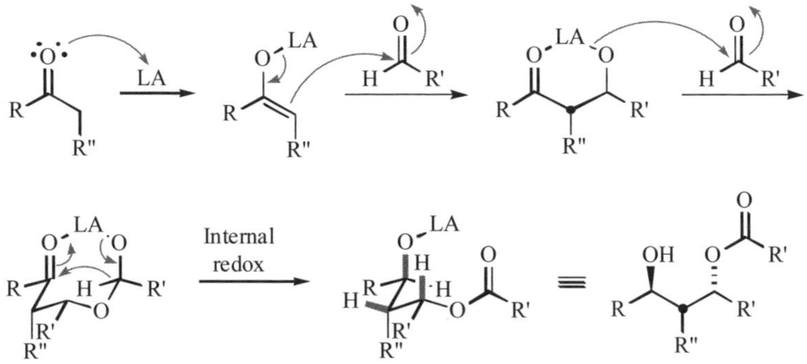

Note syn-relationships
(highlighted bonds)

## Notes:

The **Evans-Tischenko Reaction** generally requires a β-hydroxyketone (developed from an **Aldol reaction**) to react with an aldehyde. The resulting glycol monoester will be characterized as having high anti-selectivity.

## Examples:

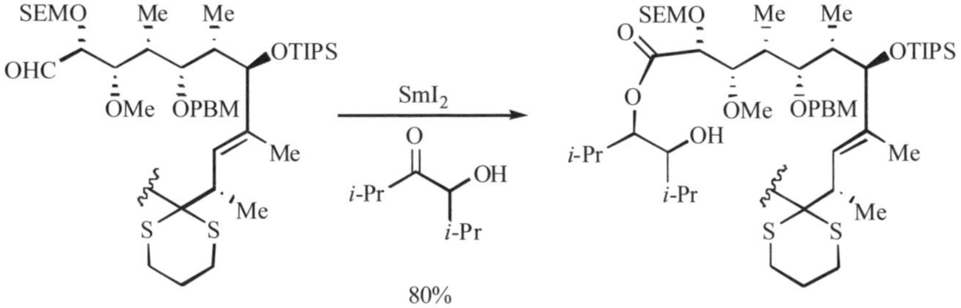

80%

A. B. Smith, III, C. M. Adams, S. A. L. Barbosa, A. P. O. Degnan, *Journal of the American Chemical Society* **2003**, <u>125</u>, 350

L. Lu, H.-Y. Chang, J.-M. Fang, *Journal of Organic Chemistry* **1999**, <u>64</u>, 843

I. Paterson, M. E. DiFrancesco, T. Kuhn, *Organic Letters* **2003**, <u>5</u>, 599

L. Lu, H.-Y. Chang, J.-M. Fang, *Journal of Organic Chemistry* **1999**, <u>64</u>, 843

# Favorskii Rearrangement (Favorsky)

## The Reaction:

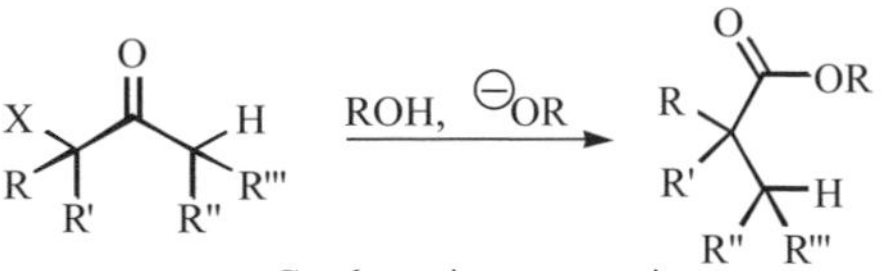

Can be a ring contraction.

## Proposed Mechanism:

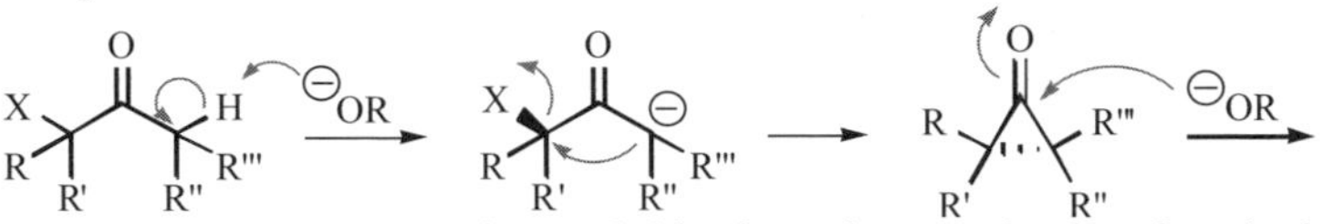

The new C-C is of opposite stereochemistry from the departing X-group.

G. Stork, J. J. Borowitz, *Journal of the American Chemical Society* **1960**, *82*, 4370.

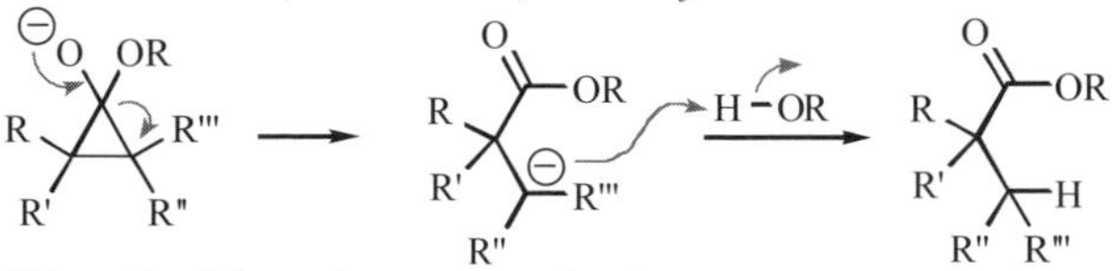

Either side of the cyclopropane can break.
The more stabilized carbanion will be favored.

## Notes:

T. Laue, A. Plagens, *Named Organic Reactions*, John Wiley and Sons, Inc., New York, 1998, pp. 100-103; M. B. Smith, J. March in *March's Advanced Organic Chemistry*, 5th ed., John Wiley and Sons, Inc., New York, 2001, pp 1403-1405; A. S. Kende, *Organic Reactions* **11**, 4.

Alkoxide or hydroxide bases can be used, giving esters (as shown above) or acids, respectively.

**An alternative rationale:**

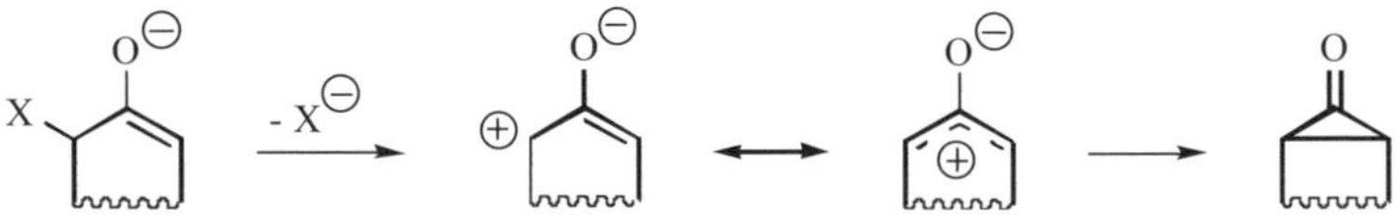

The oxyallyl intermediate has been trapped:

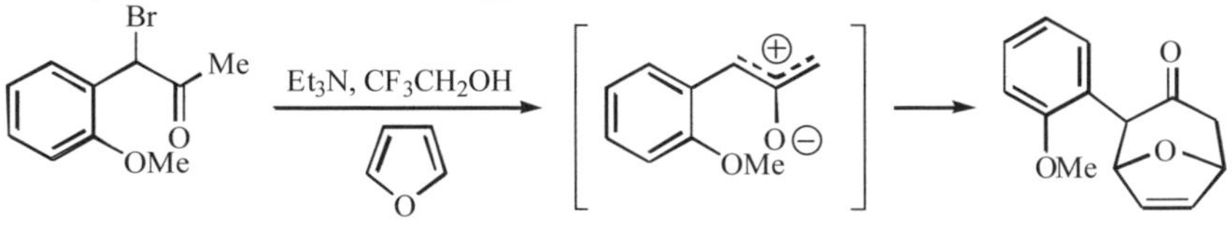

73%

M. W. Finch, J. Mann, P. D. Wilde, *Tetrahedron* **1987**, *43*, 5431

The rearrangement was essential for the design and execution of the first preparation of the cubane skeleton:

Isolated as dimethyl ester
in overall yield of 30%

P. E. Eaton, T. W. Cole, Jr., *Journal of the American Chemical Society* **1964**, <u>86</u>, 962

## Examples:

R. Xu, G. Chu, D. Bai, *Tetrahedron Letters* **1996**, <u>37</u>, 1463

>56%

J. D. White, J. Kim, N. E. Drapela, *Journal of the American Chemical Society* **2000**, <u>122</u>, 8665

39% overall

***Quasi Favorski Rearrangement:***

KH

76%

- Cl⊖

M. Harmata, P. Rashatasakhon, *Organic Letters* **2001**, *3*, 2533

THF

70%

G. A. Kraus, J. Shi, *Journal of Organic Chemistry* **1991**, <u>56</u>, 4147

EtOAc

80%

J. M. Llera, B. Fraser-Reid, *Journal of Organic Chemistry* **1989**, <u>54</u>, 5544

# Ferrier Rearrangement

## The Reaction:

Several reaction types appear to be covered under this heading:

1.

2.

3. *A Petasis-Ferrier rearrangement*

N. A. Petasis, S.-P. Lu, *Tetrahedron Letters* **1996**, *37*, 141

## Proposed Mechanism:

1.

2.

3.

## Examples:

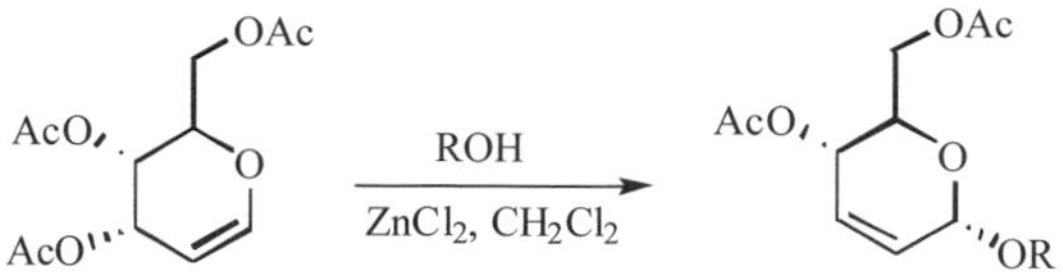

| ROH | % Yield |
|-----|---------|

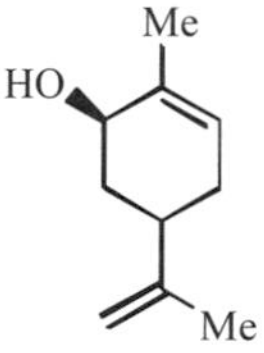

70%

Ph-CH₂OH 95%

B. K. Bettadaiah, P. Srinivas, *Tetrahedron Letters* **2003**, 44, 7275

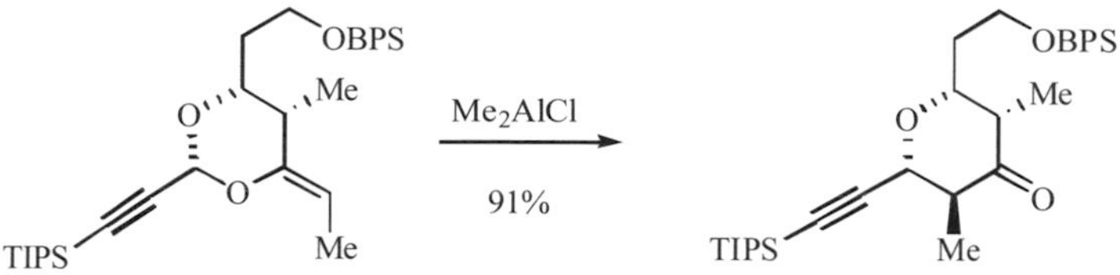

91%

A. B. Smith, III, K. P. Minbiule, P. R. Verhoest, T. J. Beauchamp, *Organic Letters* **1999**, 1, 913

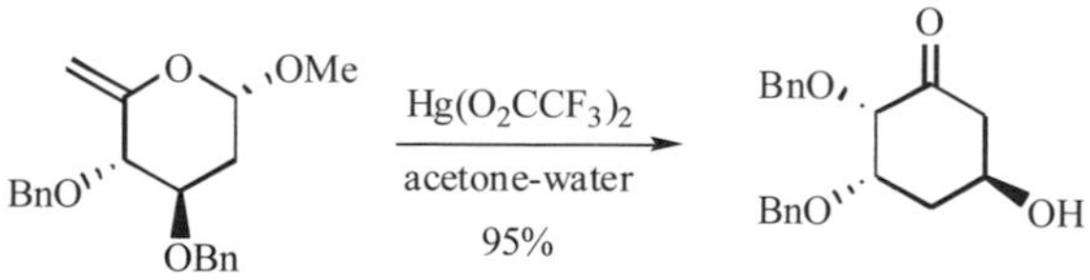

95%

H. Fuwa, Y. Okamura, H. Natsugari, *Tetrahedron* **2004**, 60, 5341

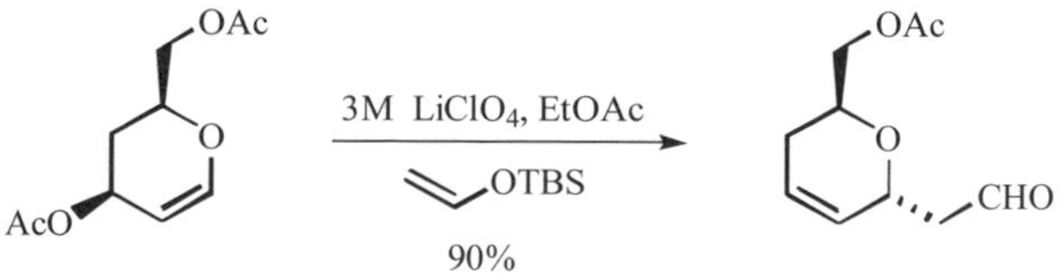

90%

P. A. Grieco, J. D. Speake, *Tetrahedron Letters* **1998**, 39, 1275

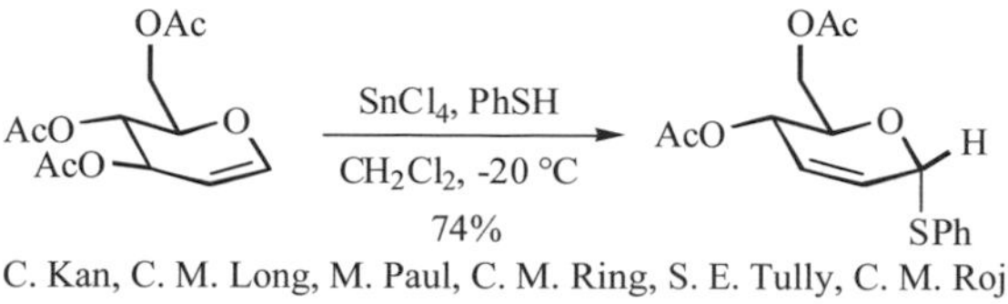

74%

C. Kan, C. M. Long, M. Paul, C. M. Ring, S. E. Tully, C. M. Rojas, *Organic Letters* **2001**, 3, 381

# Finkelstein Reaction

## The Reaction:

$$R-X \ + \ Na-X' \ \xrightarrow{\text{acetone}} \ R-X' \ + \ Na-X \text{ (s)}$$

X = Br or Cl
X' = I or F

## Proposed Mechanism:

See: T. Laue, A. Plagens, *Named Organic Reactions*, John Wiley and Sons, Inc., New York, 1998, pp. 102-103; M. B. Smith, J. March in *March's Advanced Organic Chemistry*, 5th ed., John Wiley and Sons, Inc., New York, 2001, p. 517

This is a classical $S_N2$ reaction, therefore $1° > 2° > 3°$.

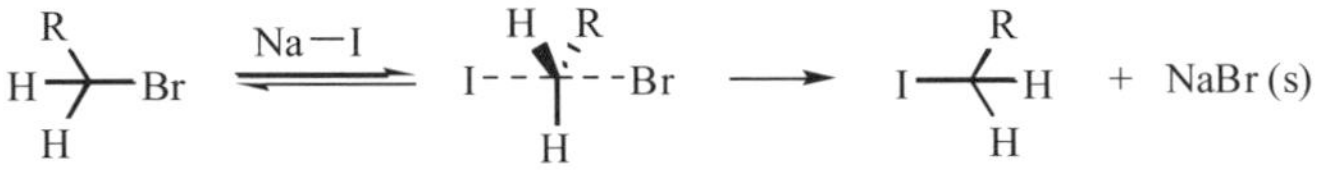

$3°$ chlorides -> $3°$ iodides with NaI in $CS_2$ and $ZnCl_2$ catalyst.

NaBr and NaCl are not soluble in acetone, driving the equilibrium.

## Notes:

Although the reaction was considered to involve simple displacement on alkyl halides, an interpretation now appears to include other systems:

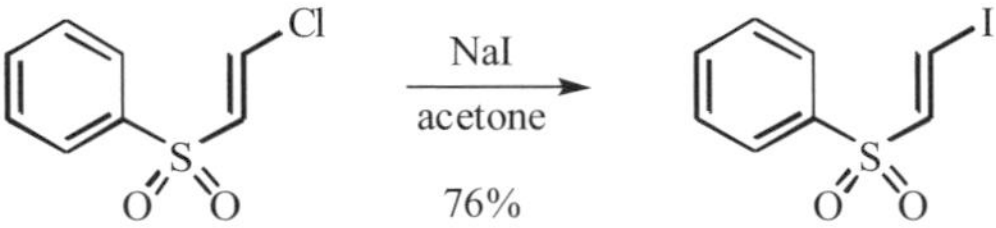

76%

T. Zoller, D. Uguen, A. DeCian, J. Fischer, *Tetrahedron Letters* **1998**, _39_, 8089

An "*Aromatic Finkelstein*" reaction:

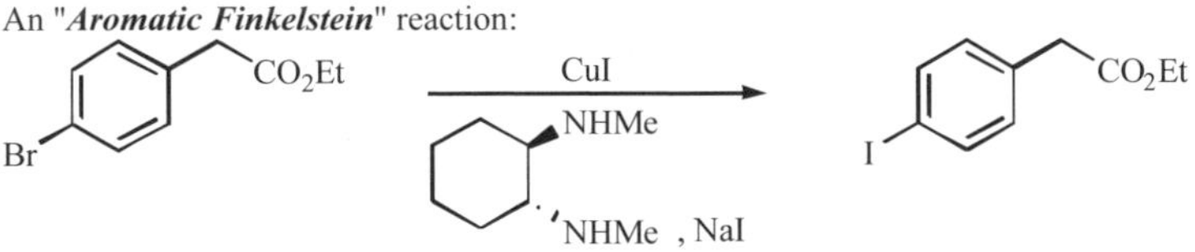

A. Klapars, S. L. Buchwald, *Journal of the American Chemical Society* **2002**, _124_, 14844

## Examples:

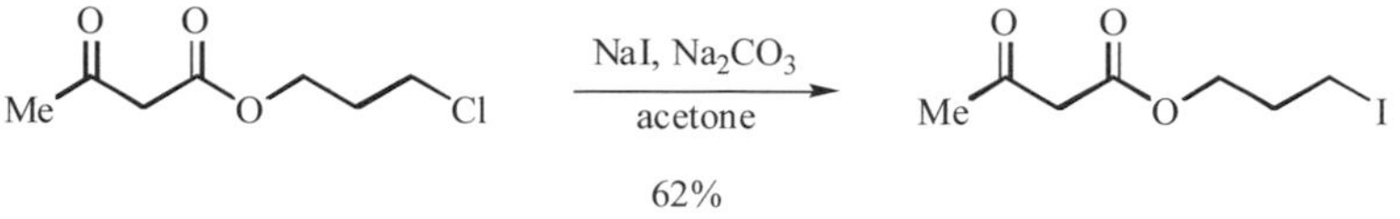

NaI, $Na_2CO_3$
acetone
62%

J. Christoffers, H. Oertling, P. Fischer, W. Frey, *Tetrahedron* **2003**, 59, 3769

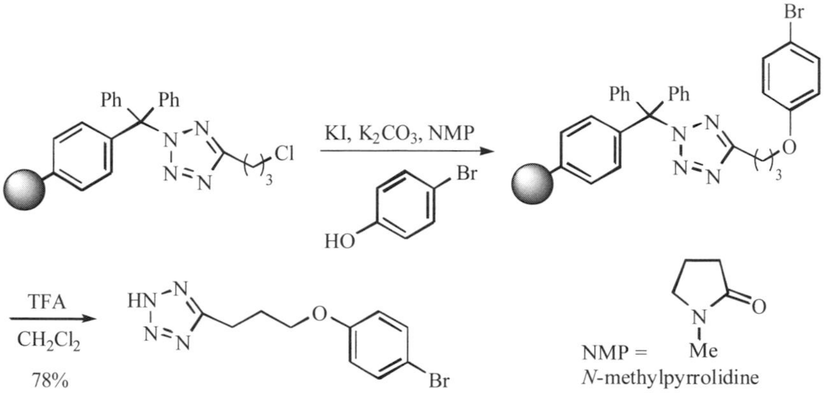

KI, $K_2CO_3$, NMP
TFA
$CH_2Cl_2$
78%
NMP =
*N*-methylpyrrolidine

D. P. Matthews, J. E. Green, A. J. Shuker, *Journal of Combinatorial Chemistry* **2000**, 2, 19

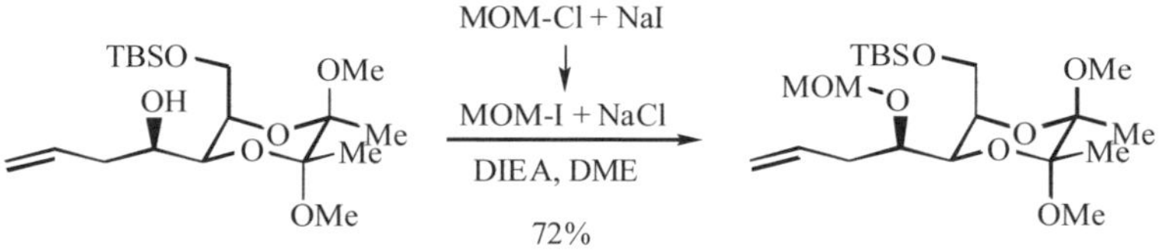

MOM-Cl + NaI
MOM-I + NaCl
DIEA, DME
72%

DIEA = ***Hunig's Base*** = diisopropylethylamine
D. J. Dixon, A. C. Foster, S. V. Ley, *Organic Letters* **2000**, 2, 123

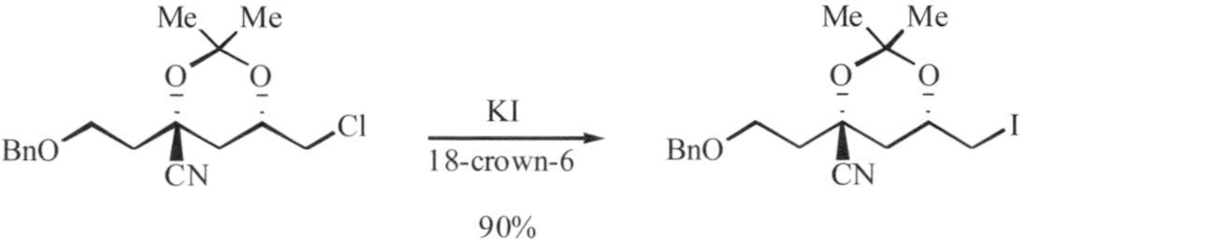

KI
18-crown-6
90%

T. I. Richardson, S. D. Rychnovsky, *Journal of the American Chemical Society* **1997**, 119, 12360

# Fischer Indole Synthesis

## The Reaction:

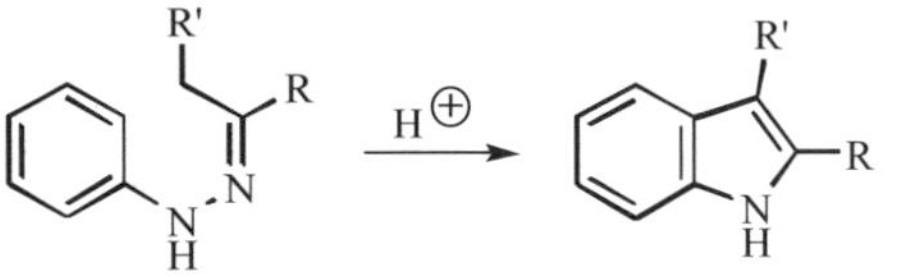

## Proposed Mechanism:

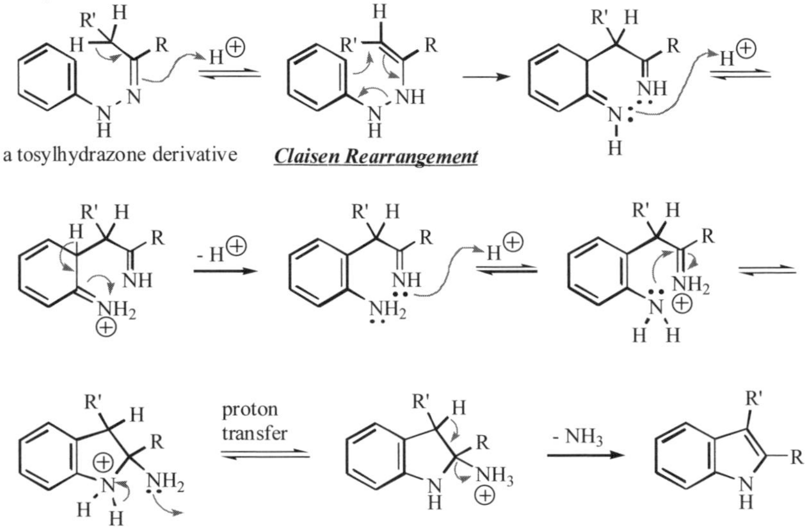

a tosylhydrazone derivative        *Claisen Rearrangement*

See: T. Laue, A. Plagens, *Named Organic Reactions*, John Wiley and Sons, Inc., New York, 1998, pp. 103-106; M. B. Smith, J. March in *March's Advanced Organic Chemistry*, 5th ed., John Wiley and Sons, Inc., New York, 2001, pp 1452-1453

## Notes:

A variety of Lewis acids can be used.

The *Barry Reaction*:

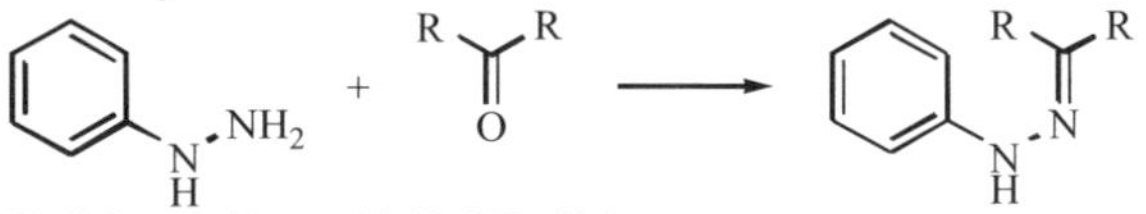

V. C. Barry, *Nature* **1948**, <u>152</u>, 537

*Abramovitch-Shapiro Protocol*

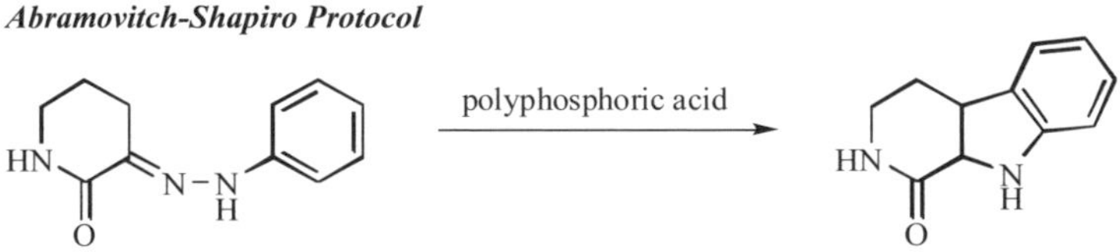

polyphosphoric acid

## Examples:

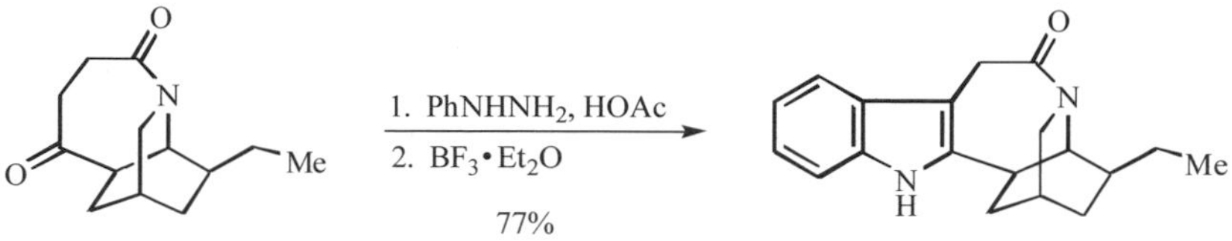

1. PhNHNH₂, HOAc
2. BF₃·Et₂O

77%

J. D. White, Y. Choi, *Organic Letters,* **2000**, 2, 2373

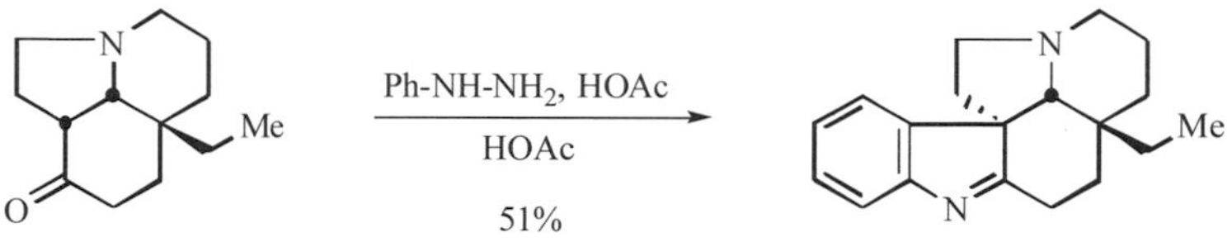

Ph-NH-NH₂, HOAc

HOAc

51%

R. Iyengar, K. Schildknegt, J. Aube, *Organic Letters* **2000**, 2, 1625

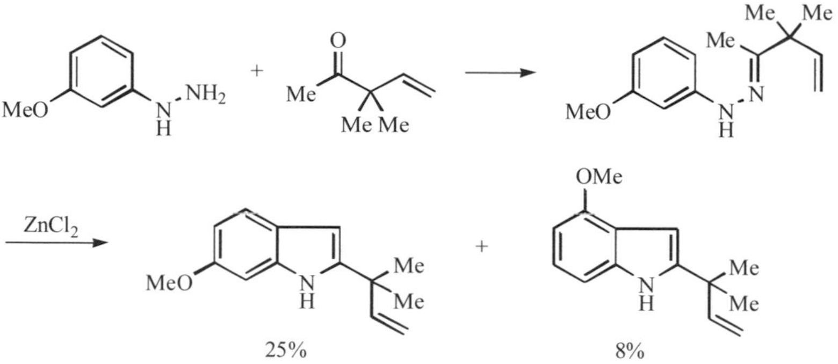

ZnCl₂

25%        +        8%

R. J. Cox, R. M. Williams, *Tetrahedron Letters* **2002**, 43, 2149

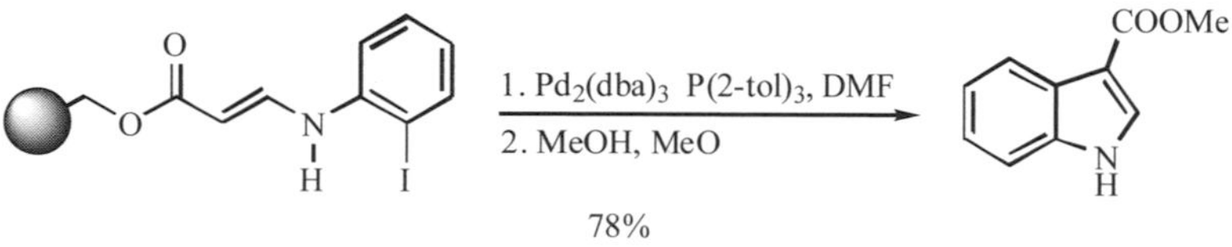

1. Pd₂(dba)₃ P(2-tol)₃, DMF
2. MeOH, MeO

78%

K. Yamazaki, Y. Kondo, *Journal of Combinatorial Chemistry* **2002**, 4, 191

# Fischer-Hepp and Related Rearrangements

## The Reaction:

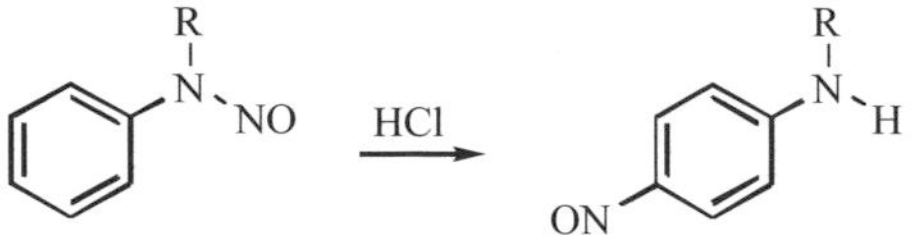

HCl is the preferred acid.

## Proposed Mechanism:

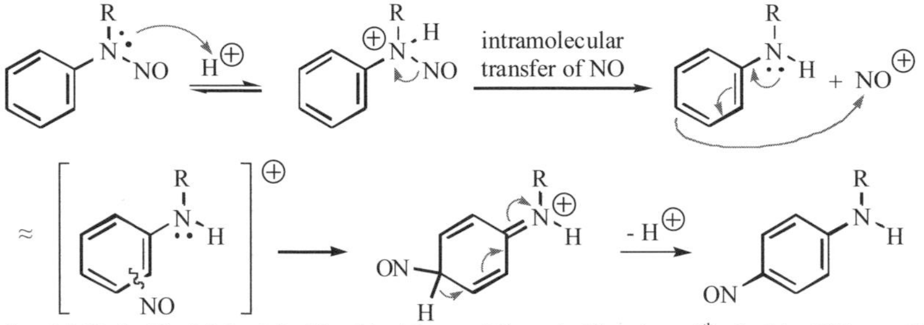

See: M. B. Smith, J. March in *March's Advanced Organic Chemistry*, 5th ed., John Wiley and Sons, Inc., New York, 2001, pp. 728-730 for the **Fischer-Hepp Rearrangement** as well as the *Orton and Hofmann-Martius Rearrangements*.
D. L. H. Williams, *Tetrahedron* **1975**, 31, 1343

See similarity to:

# Orton Rearrangement
## The Reaction:

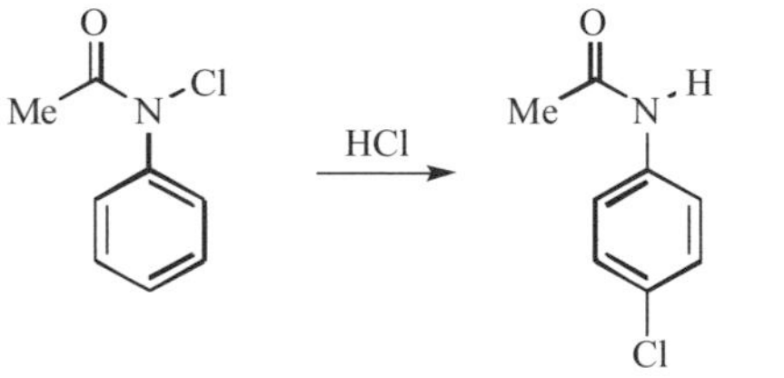

D. L. H. Williams, *Tetrahedron* **1975**, 31, 1343 comments on the similarities.

## Proposed Mechanism:

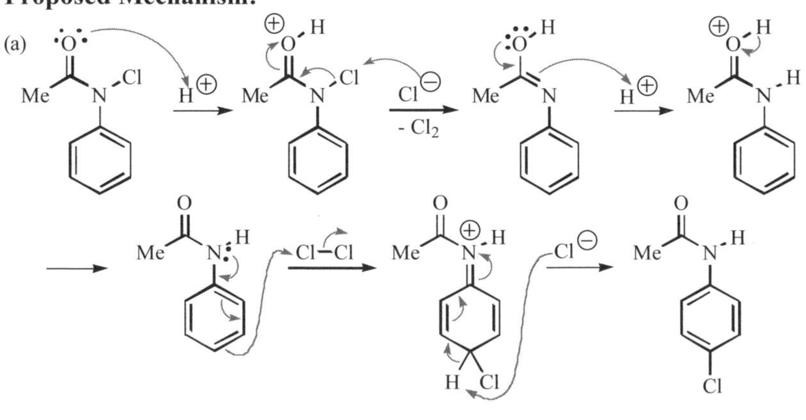

(b)

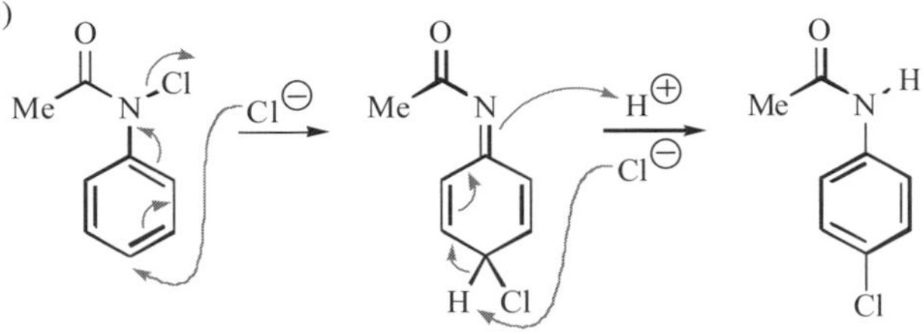

# Hofmann-Martius Rearrangement

## The Reaction:

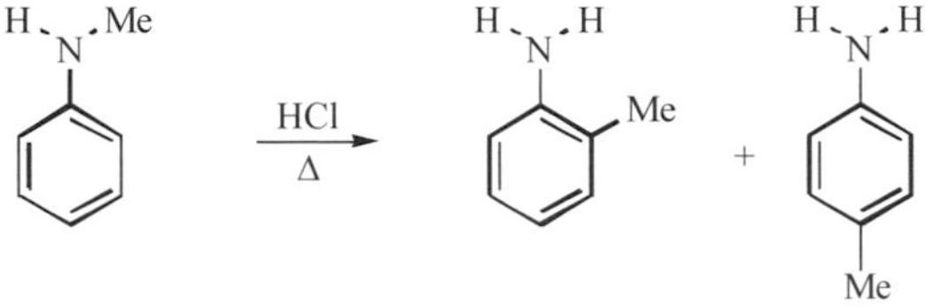

## Proposed Mechanism:

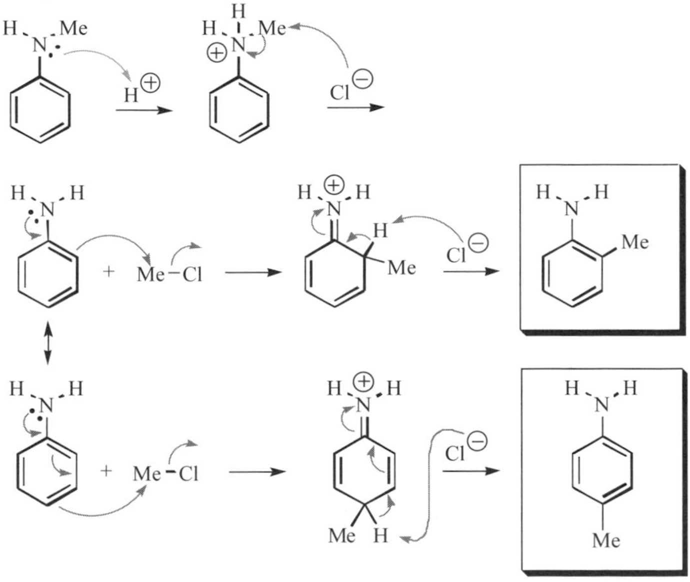

*Reilly-Hickinbottom Rearrangement* - uses Lewis acids and the amine rather than protic acid and the amine salt.

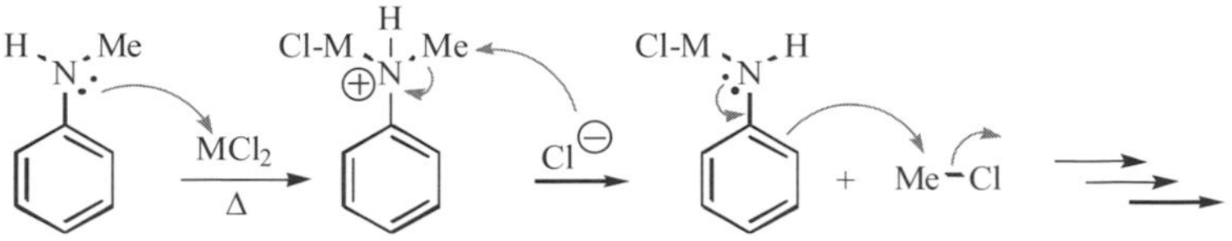

M = Co, Cd, Zn

# Fischer Esterification / Fischer-Speier Esterification

## The Reaction:

## Proposed Mechanism:

**Notes:** R. Bruchner, *Advanced Organic Chemistry, Reaction Mechanisms*, Academic Press, San Diego, CA, 2002, p. 247

This is an equilibrium process, and the extent of reaction is controlled by forcing the equilibrium to the right. Trapping of water, or forcing by having one reactant (clearly the most available!) in excess are methods often employed.

See: K. G. Kabza, B. R. Chapados, J. E. Gestwicki, J. L. McGrath, *Journal of Organic Chemistry* **2000**, <u>65</u>, 121 for an interesting overview in the context of microwave use.
This process is common for undergraduate laboratory experiments.

## Examples:

66-70%

S. Zen, M. Koyama, S. Koto, *Organic Syntheses* <u>CV 6</u>, 797

75%

A. N. Hulme, E. M. Rosser, *Organic Letters* **2002**, <u>4</u>, 265

# Steglich Esterification

Useful method for sterically-hindered esterification reactions.

## Reaction:

See: B. Neises, W. Steglich, *Organic Syntheses* <u>CV 7</u>, 93

## Mechanism:

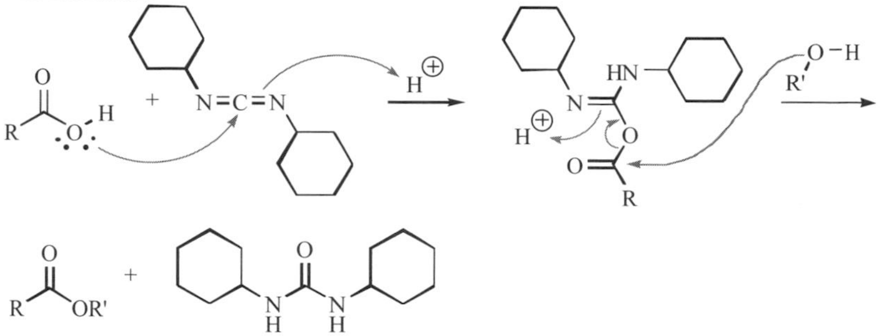

## Examples:

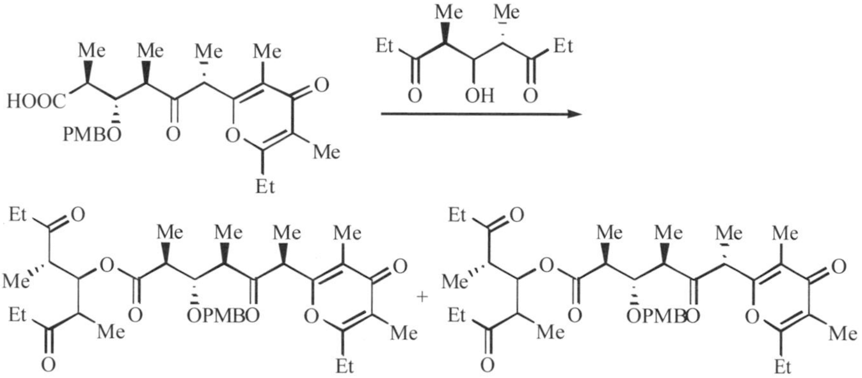

| Conditions | | Ratio | |
|---|---|---|---|
| DCC, DMAP, DMAP-HCl | 2.5 | : | 1 |
| DMAP, Et$_3$N, 2,4,6-Cl$_3$(C$_6$H$_2$)COC | 1 | : | 10 |

I. Paterson, D. Y.-K. Chen, J. L. Acena, A. S. Franklin, *Organic Letters* **2000**, <u>2</u>, 1513

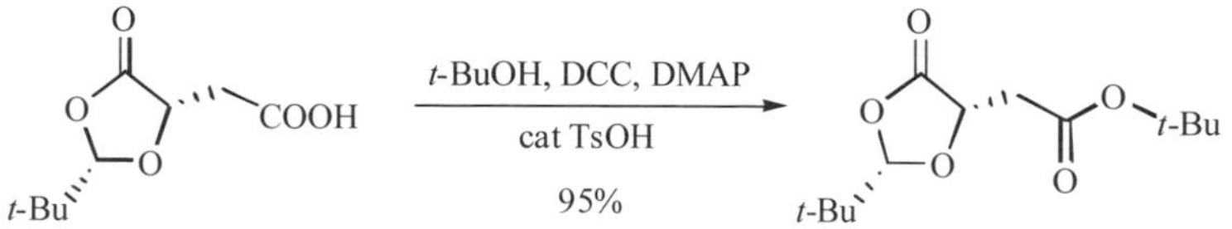

M. Sefkow, *Journal of Organic Chemistry* **2001**, <u>66</u>, 2343

# Fleming Oxidation

## The Reaction:

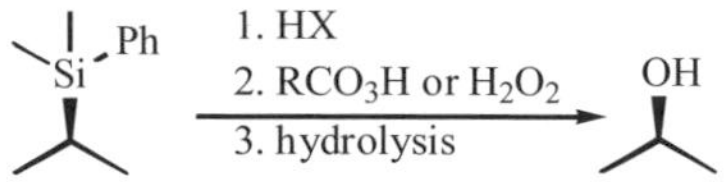

## Proposed Mechanism:

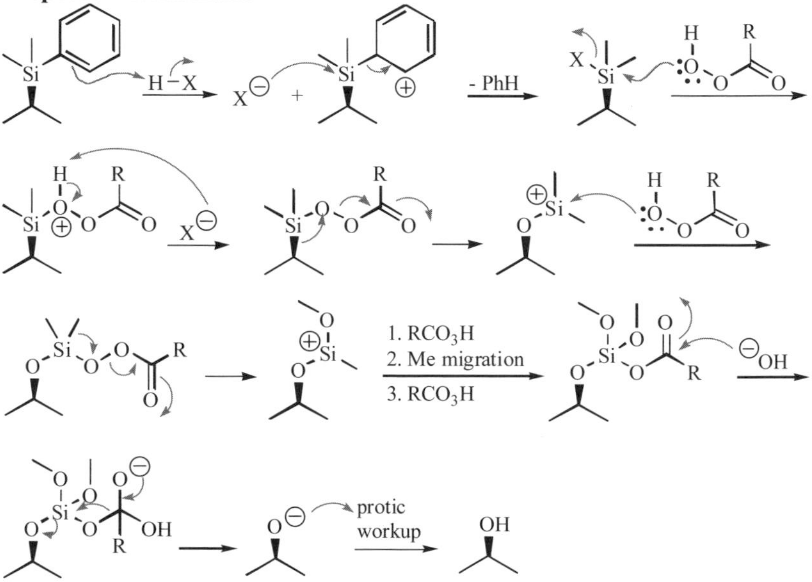

## Notes:

This reaction, depending on the details of the reaction, is known as the *Fleming Oxidation*, the *Fleming-Tamao Oxidation*, the *Tamao-Fleming Oxidation*, or the *Tamao-Kumada Oxidation*.

## Fleming-Tamao-Kumada Oxidation

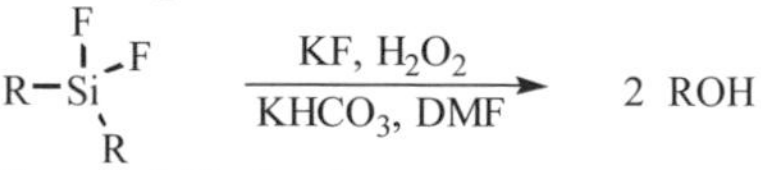

## Proposed Mechanism:

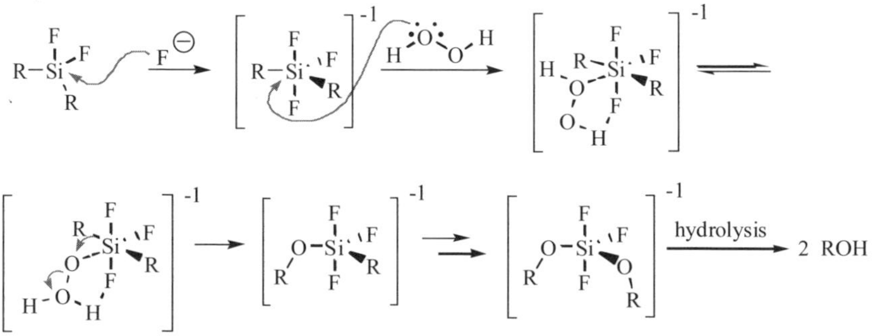

## Examples:

Me
Me—Si—Ph

HO,,, ⟨ring⟩ ,,OH

HO'' ''OBn

$\xrightarrow{\text{1. HF, pyridine}}_{\text{2. Hg(OAc)}_2,\ \text{AcOOH}}$

OH
HO,, ⟨ring⟩ ,,OH

HO'' ''OBn

51%

J. N. Heo, E. B. Holson, W. R. Roush, *Organic Letters* **2003**, 5, 1697

Me
Me—Si—Ph  OH
        H

⟨bicyclic ring with N⟩

$\xrightarrow{\text{1. AcOH, TFA}}_{\text{2. Hg(CF}_3\text{COO)}_2,\ \text{AcOOH}}$

OH   OH
   H

⟨bicyclic ring with N⟩

81%

J. A. Vanecko, F. G. West, *Organic Letters* **2002**, 4, 2813

Me—⟨ring⟩  H  O
      Me   Me
        Si—O*i*-Bu
      Me  Me

$\xrightarrow{\text{1. TBAF}}_{\text{2. H}_2\text{O}_2,\ \text{KF}}$

56%

Me—⟨ring⟩  H  O
      Me   Me
             OH

M. K. O'Brien, A. J. Pearson, A. A. Pinkerton, W. Schmidt, K. Willamn, *Journal of the American Chemical Society* **1989**, 111, 1499

TsO,, ⟨ring⟩ H O—N—O
TBSO—
   *i*-Pr—Si—O
        *i*-Pr

$\xrightarrow{\text{H}_2\text{O}_2,\ \text{KHCO}_3}$

90%

TsO,, ⟨ring⟩ H O—N—O
TBSO—
        HO   OH

Note: When the silicon unit is part of an oxasilacycloalkane, F$^{\ominus}$ is not needed.

S. E. Denmark, J. J. Cottell, *Journal of Organic Chemistry* **2001**, 66, 4276

Me
Me—Si—O
Br—
       ⟨ring⟩  Me
              Me
TBDMSO  Me

$\xrightarrow{\text{Bu}_3\text{SnH}}_{\text{AIBN}}$

Me
Me—Si—O
   ⟨ring⟩  Me
          Me
        Me
        ÓTBDMS

$\xrightarrow{\text{KH}}_{\text{H}_2\text{O}_2}$

OH   OH
   ⟨ring⟩  Me
          Me
        Me
        ÓTBDMS

M. R. Elliot, A.-L. Dhimane, M. Malacria, *Journal of the American Chemical Society* **1997**, 119, 3427

# Frater-Seebach Alkylation

## The Reaction:

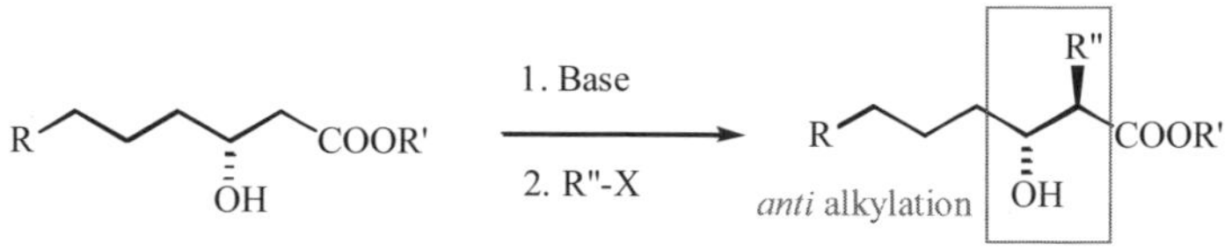

## Proposed Mechanism:

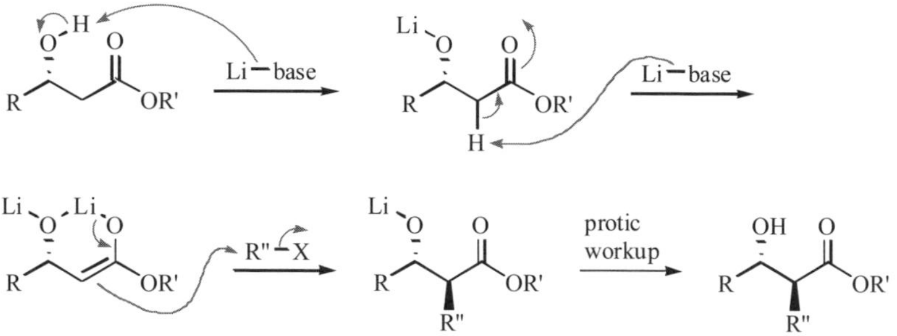

## Notes:

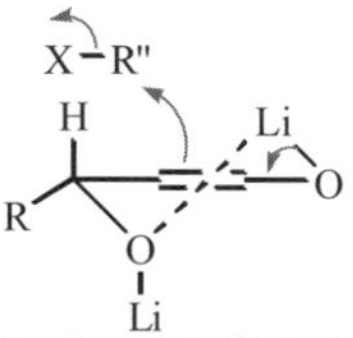

Preferential alkylation is *anti* to the hydroxyl group.

## Examples:

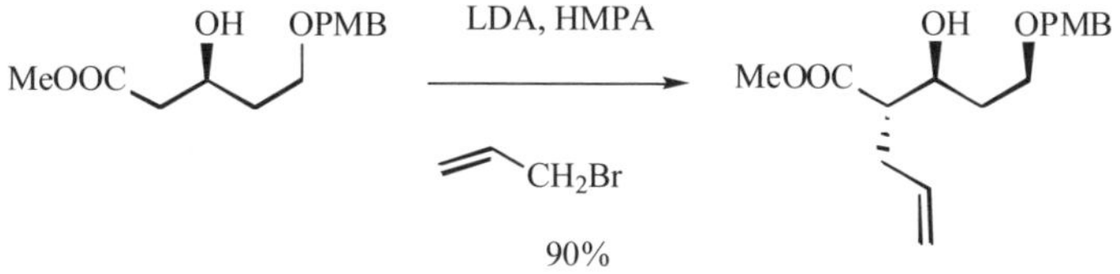

90%

M. Nakatsuka, J. A. Ragan, T. Sammakia, D. B. Smith, D. E. Vehling, S. L. Schreiber, *Journal of the American Chemical Society* **1990**, <u>112</u>, 5583

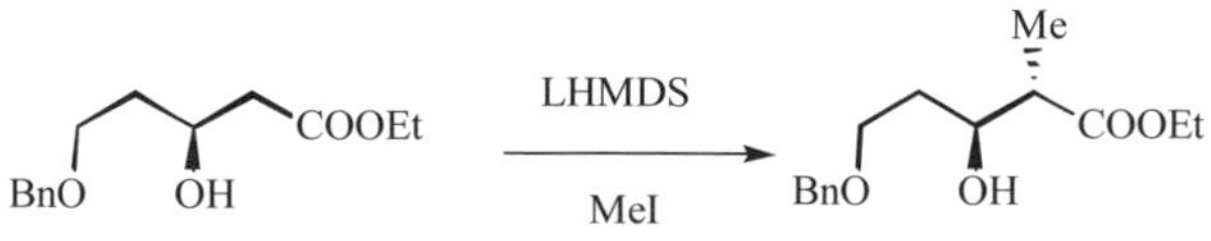

80%

S. D. Rychnovsky, C. Rodriguez, *Journal of Organic Chemistry* **1992**, <u>57</u>, 4793

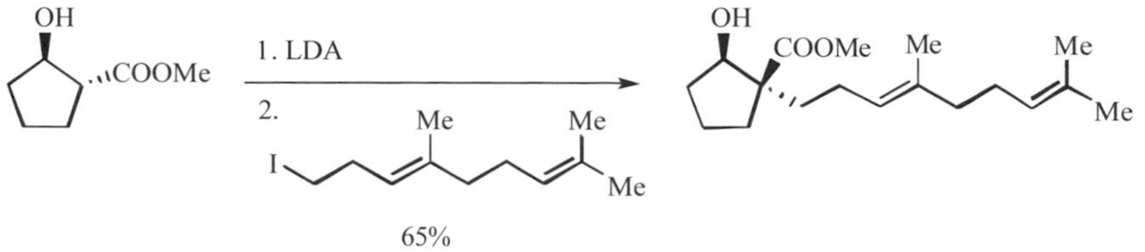

65%

C. H. Heathcock, C. J. Kath, R. B. Ruggeri, *Journal of Organic Chemistry* **1995**, <u>60</u>, 112

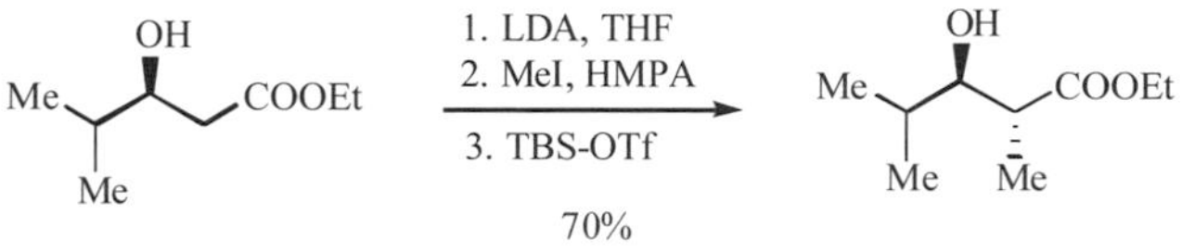

70%

W. R. Roush, T. D. Bannister, M. D. Wendt, J. A. Jablonowski, K. A. Scheidt, *Journal of Organic Chemistry* **2002**, <u>67</u>, 4275

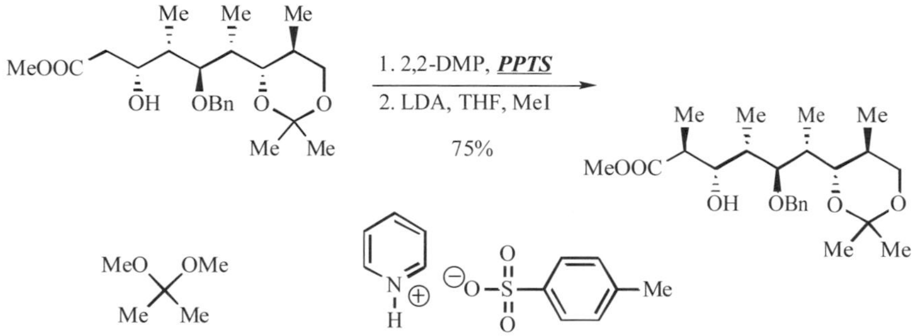

75%

2,2-DMP = 2,2-dimethoxypropane  ***PPTS = pyridinium p-toluenesulfonate***
[77-76-9]

K. W. Hunt, P. A. Grieco, *Organic Letters* **2001**, <u>3</u>, 481

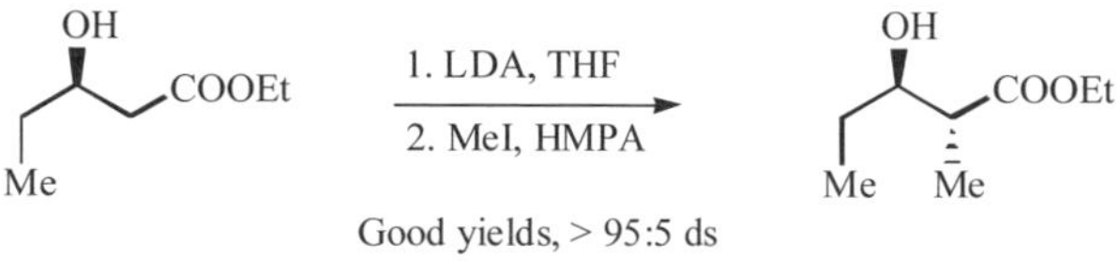

Good yields, > 95:5 ds

W. R. Roush, J. S. Newcom, *Organic Letters* **2002**, <u>4</u>, 4739

# Friedel-Crafts Acylation

**The Reaction:**

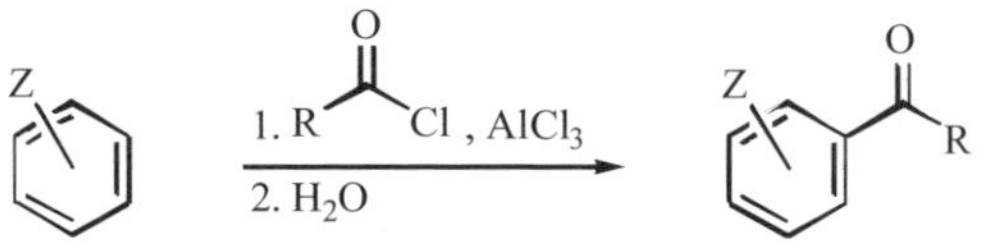

**Proposed Mechanism:**

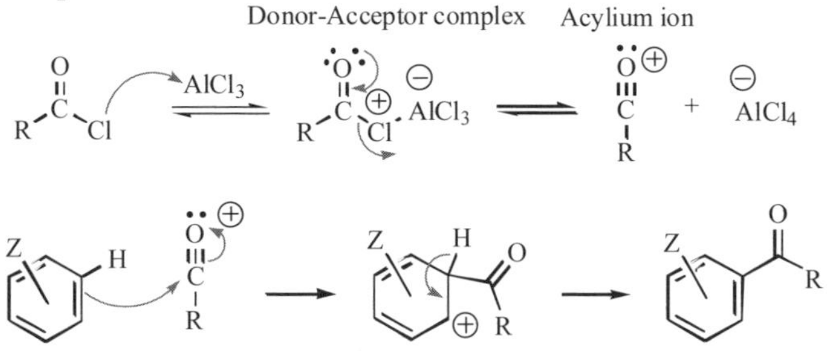

**Notes:**
Unlike the carbocation intermediate on the ***Friedel-Crafts alkylation***, this **acylium ion** intermediate will not undergo rearrangement.
See: T. Laue, A. Plagens, *Named Organic Reactions*, John Wiley and Sons, Inc., New York, 1998, pp. 106-109.

**Nencki Reaction:** Acylation of polyphenolic compounds.
**Example:**

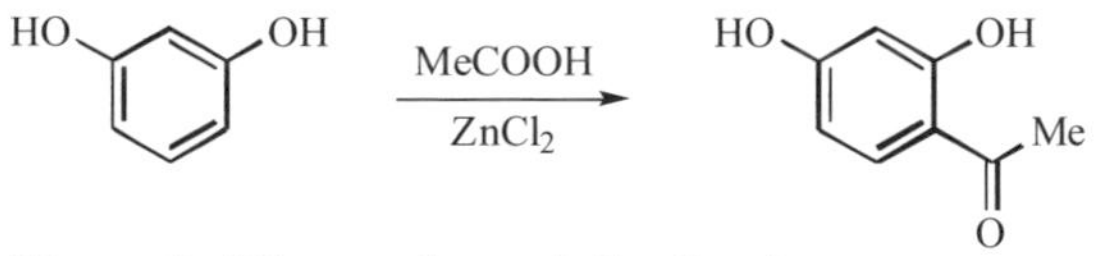

# Haworth (Phenanthrene) Synthesis
**The Reaction:**

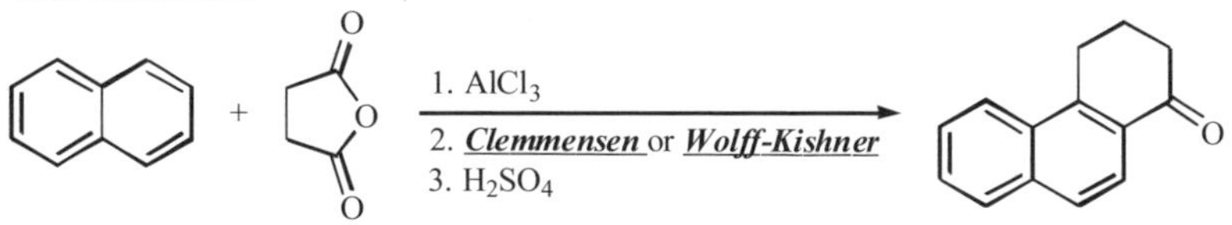

**Proposed Mechanism:**

succinic anhydride

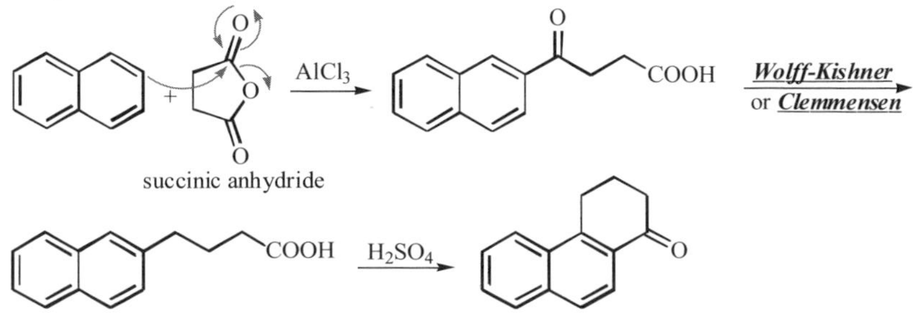

**Examples:**

95%

D.-M. Cui, C. Zhang, M. Kawamura, S. Shimada, *Tetrahedron Letters* **2004**, <u>45</u>, 1741

aluminum dodecatungstophosphate
($AlPW_{12}O_{40}$),
stable and non-hygroscopic

96%

H. Firouzabadi, N. Iranpoor, F. Nowrouzi, *Tetrahedron Letters* **2003**, <u>44</u>, 5343

F. Bevacqua, A. Basso, R. Gitto, M. Bradley, A. Chimirri, *Tetrahedron Letters* **2001**, <u>42</u> 7683

| LA | Reagent | R= | Yield |
|----|---------|-----|-------|
| $AlCl_3$ | (acetic anhydride) | $CH_3$ | 78% |
| $SnCl_4$ | Me—CN | $CH_3$ | 96% |
| $SnCl_4$ | (chloroacetyl chloride) | $ClCH_2$ | 80% |

O. Ottoni, A. de V. F. Neder, A. K. B. Dias, R. P. A. Cruz, L. B. Aquino, *Organic Letters* **2001**, <u>3</u>, 1005

# Friedel-Crafts Alkyation

## The Reaction:

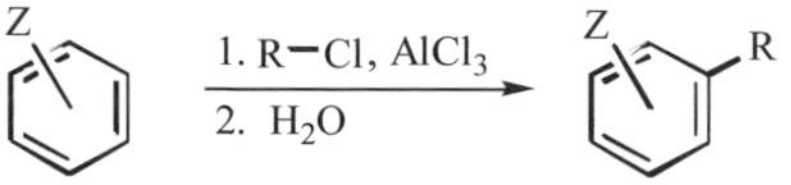

Z

1. R—Cl, AlCl$_3$

2. H$_2$O

→ R

## Proposed Mechanism:

R—Cl + AlCl$_3$ ⇌ R$^{\oplus}$ + $\overset{\ominus}{AlCl_4}$

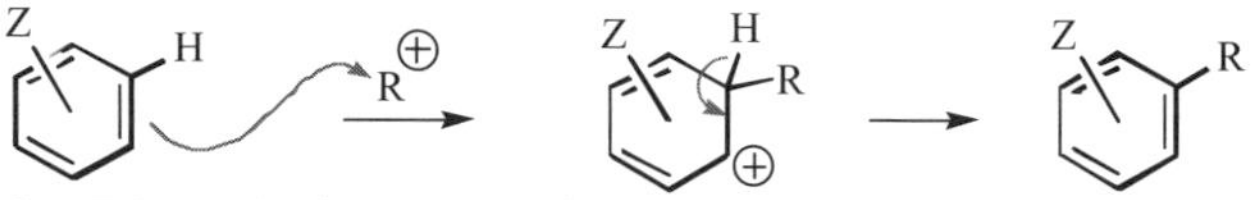

See: T. Laue, A. Plagens, *Named Organic Reactions*, John Wiley and Sons, Inc., New York, 1998, pp. 110-113; C. A. Price, *Organic Reactions* **3**, 1; E. Berliner, *Organic Reactions* **5**, 5.

Alcohols and alkenes are also useful alkylating agents. Some classify the latter reactions as:

## Darzens-Nenitzescu Alkylation

## The Reaction:

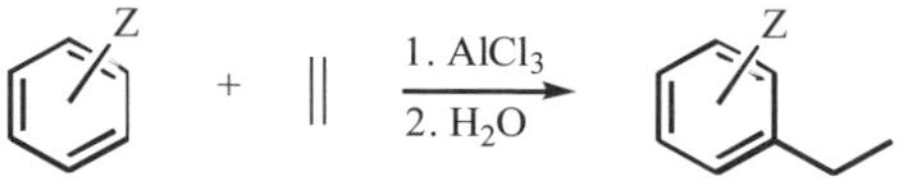

Z

+ ||

1. AlCl$_3$

2. H$_2$O

→ Z

## Proposed Mechanism:

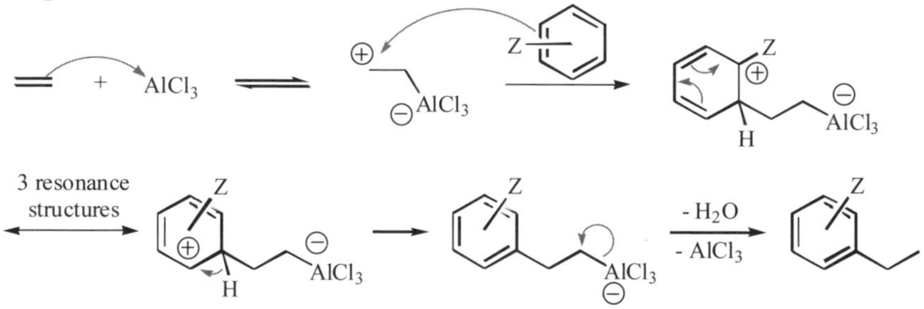

3 resonance structures

- H$_2$O
- AlCl$_3$

## Notes:

A variety of protic and Lewis acids have been used.

**Examples:**

87%

93% ee

N. A. Paras, D. W. C. MacMillan, *Journal of the American Chemical Society* **2001**, <u>123</u>, 4370

Use of CO as a catalyst:

$$R\text{-}Cl + CO \longrightarrow \left[ \underset{R}{\overset{O}{\diagdown}}\overset{\bullet\bullet}{\underset{Cl}{\diagup}} \rightleftharpoons R^{\oplus} + O\equiv C + Cl^{\ominus} \right] \xrightarrow{Ar\text{-}H} Ar\text{-}R$$

alkoxy halo carbene

$$\xrightarrow[\text{CO}]{\text{Ph-H}}$$

86%          $E/Z = 69:31$

S. Ogoshi, H. Nakashima, K. Shimonaka, H. Kurosawa, *Journal of the American Chemical Society* **2001**, <u>123</u>, 8626

1. CF$_3$SO$_3$H
2. H$_2$O

100%

N. Dennis, B. E. D. Ibrahim, A. R. Katritzky, *Synthesis* **1976**, 105

$$\xrightarrow[\text{NaCl}]{\text{AlCl}_3}$$

90%

$$\xrightarrow{\text{AlCl}_3}$$

D. B. Bruce, A. J. S. Sorrie, R. H. Thomson, *Journal of the Chemical Society* **1953**, 2403

# Friedlander Quinoline Synthesis

## The Reaction:

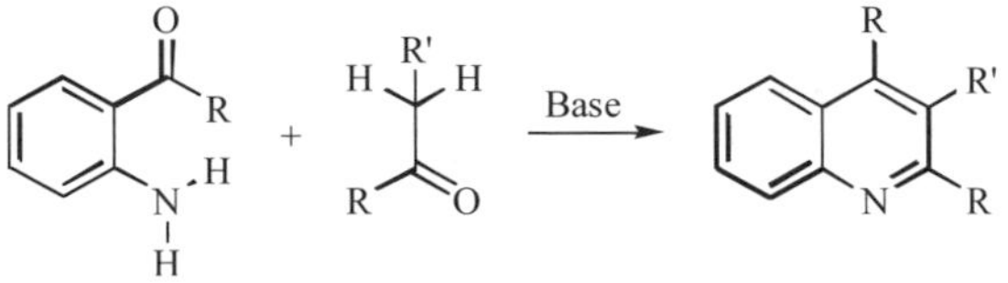

## Proposed Mechanism:

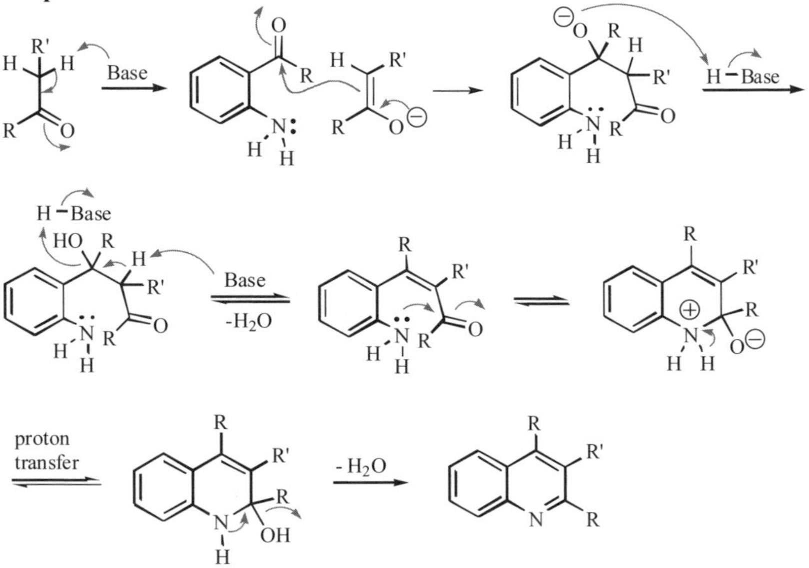

*The Imine-enamine Approach*

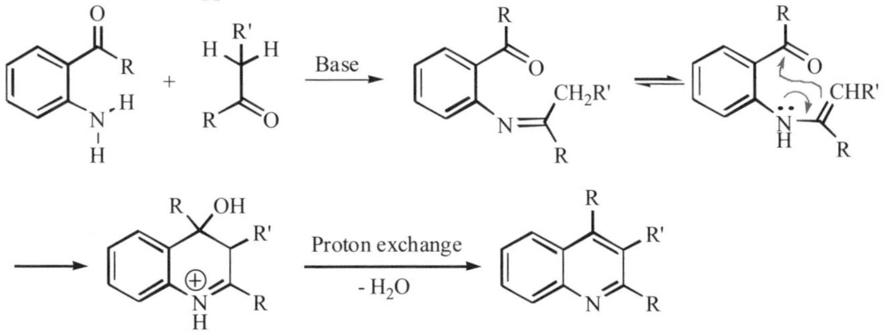

## Notes:

See: T. Laue, A. Plagens, *Named Organic Reactions*, John Wiley and Sons, Inc., New York, 1998, pp. 114-116

Intermediates from both interpretations have been isolated.

The **_PfitzingerReaction_** and the **_Niementowski Quinoline Synthesis_** are based on this concept:

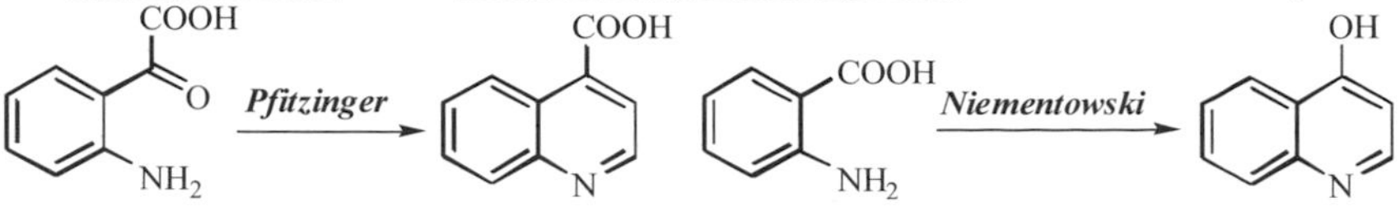

## Examples:

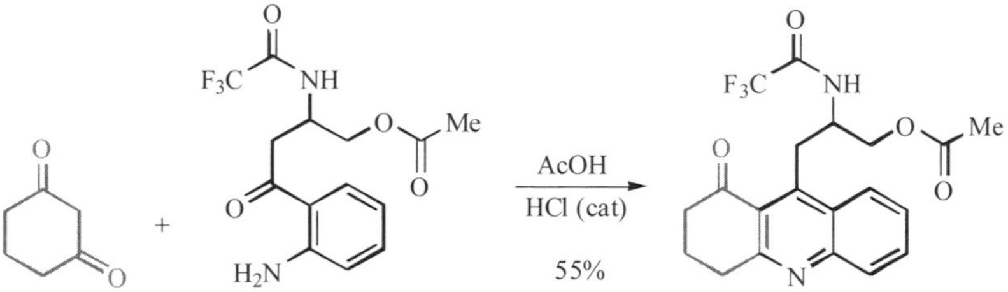

G. Gellerman, A. Rudi, Y. Kashman, *Tetrahedron* **1994**, _50_, 12959

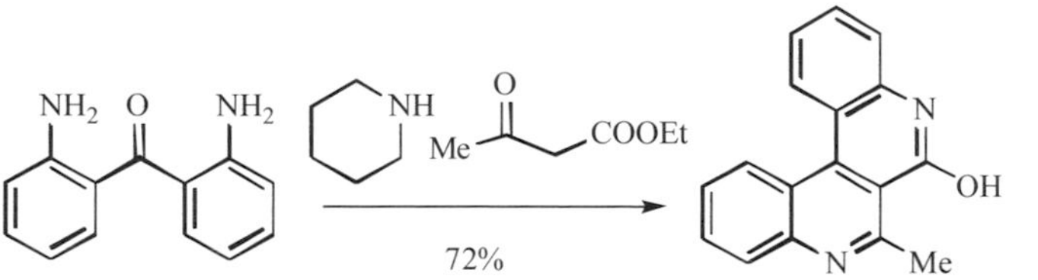

R. T. Parfitt, *Journal of Medicinal Chemistry* **1966**, _9_, 161

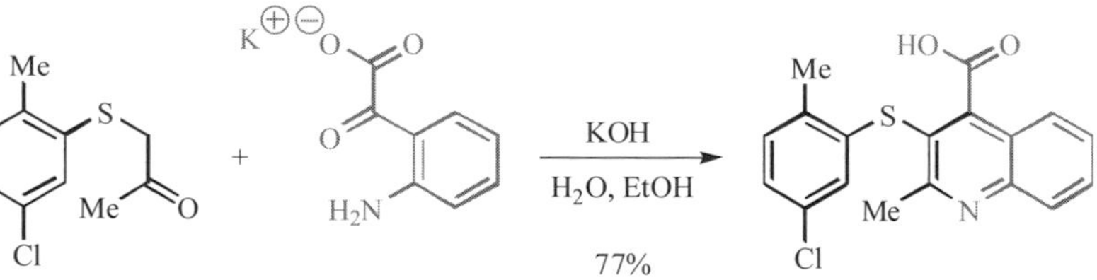

M. Croisy-Delcey, C. Huel, A. Croisy, E. Bisagni, *Heterocycles* **1991**, _32_, 1933 (AN 1992:128706)

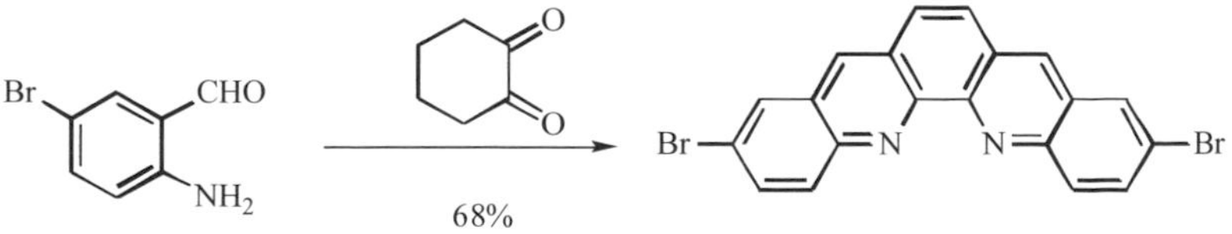

Y.-Z Hu, G. Zhang, R P. Thummel *Organic Letters* **2003**, _5_, 2251

# Fries Rearrangement / Photo Fries Rearrangement

## The Reaction:

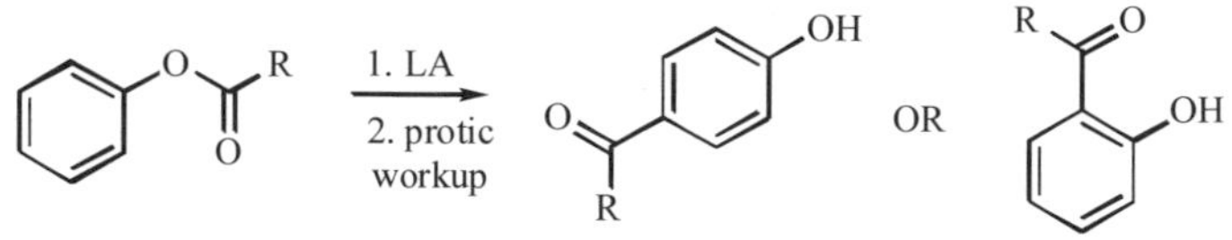

## Proposed Mechanism:

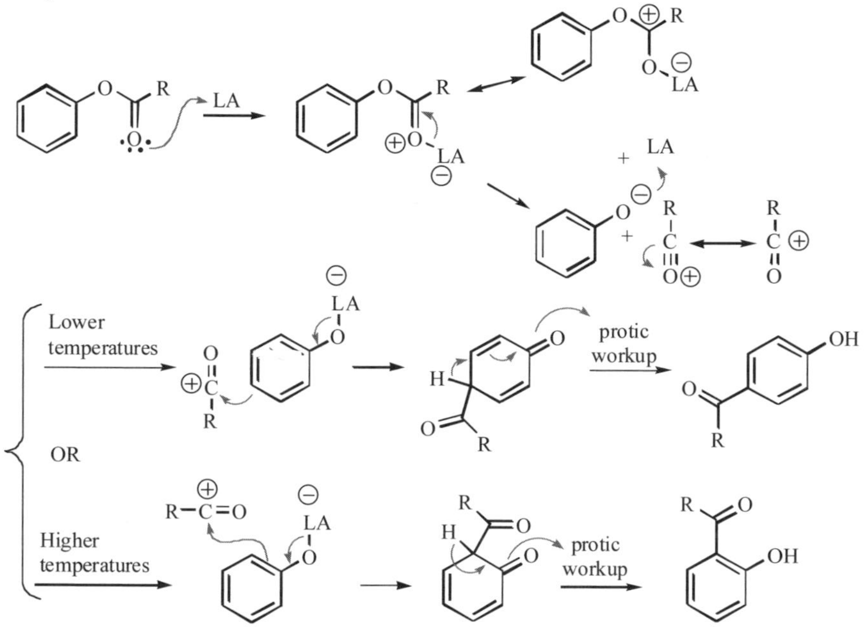

## Notes:

See:T. Laue, A. Plagens, *Named Organic Reactions*, John Wiley and Sons, Inc., New York, 1998, pp. 116-119; A. H. Blatt, *Organic Reactions* **1**, 11.

## *The Photo-Fries Reaction*:

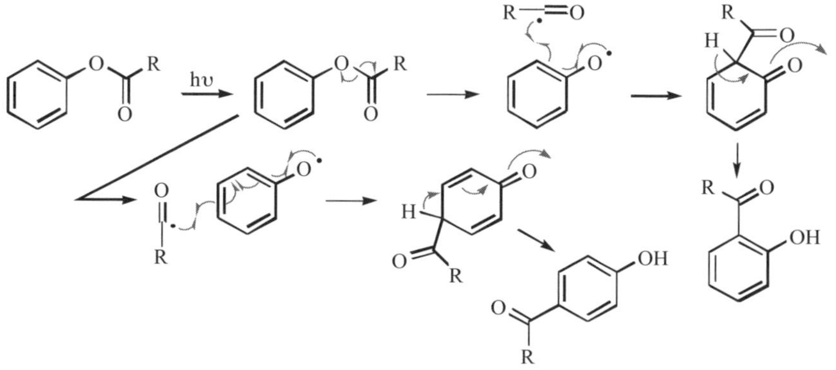

## Examples:

Both Cu(OTf)$_2$ and Y(OTf)$_3$ gave the product in 90% yield.

O. Mouhtady, H. Gaspard-Iloughmane, N. Roques, C. LeRoux, *Tetrahedron Letters* **2003**, <u>44</u>, 6379

47%

32%

D. C. Harrowven, R. F. Dainty, *Tetrahedron Letters* **1996**, <u>37</u>, 7659

Montmorillonite clays are layered silicates; montmorillonite K-10 is a specially manufactured acidic catalyst (Montmorillonite K10, [1318-93-0] A. Cornélis, P. Laszlo, M. W. Zettler in *eEROS Encyclopedia of Reagents for Organic Synthesis*, L. A. Paquette, Ed., John Wiley and Sons, Inc., online reference available at http://www.intersciene.wiley.com)

K-10 clay

microwave
78%

S. A. Khanum, V. T. D, S. Shashikantha, A. Firdouseb, *Bioorganic & Medicinal Chemistry Letters* **2004**, <u>4</u>, 5351

51.3%          28.5%          15.2%

R. Saua, G. Torres, M. Valpuesta, *Tetrahedron Letters* **1995**, <u>36</u>, 1311

# Fritsch-Buttenberg-Wiechell Rearrangement

## The Reaction:

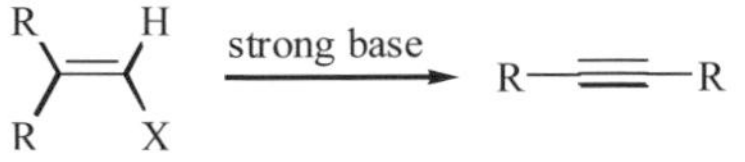

$$\underset{R}{\overset{R}{>}}=\underset{X}{\overset{H}{<}} \xrightarrow{\text{strong base}} R-\!\!\!\equiv\!\!\!-R$$

## Proposed Mechanism:

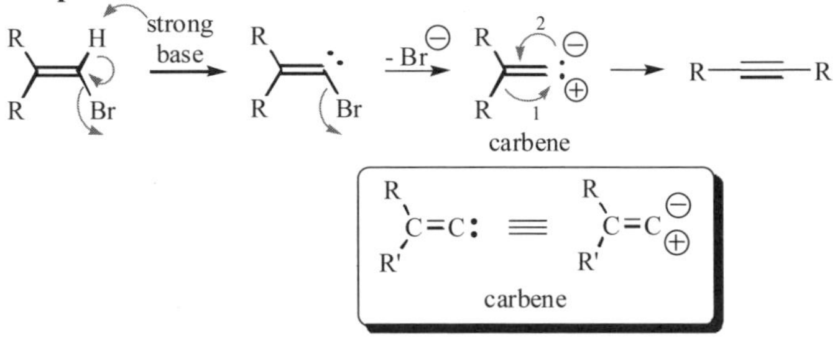

## Notes:

See the similarities in:

**Arens-van Dorp Synthesis - Isler Modification**

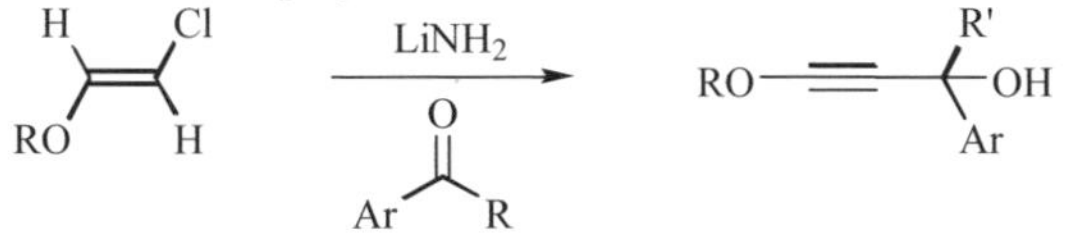

The original **Arens-van Dorp Synthesis** (see below) used ethoxyacetylene, which is difficult to make, as the starting material. The advantage of the **Isler Modification** is the use of β-chlorovinyl ether which will generate lithium ethoxyacetylene in situ.

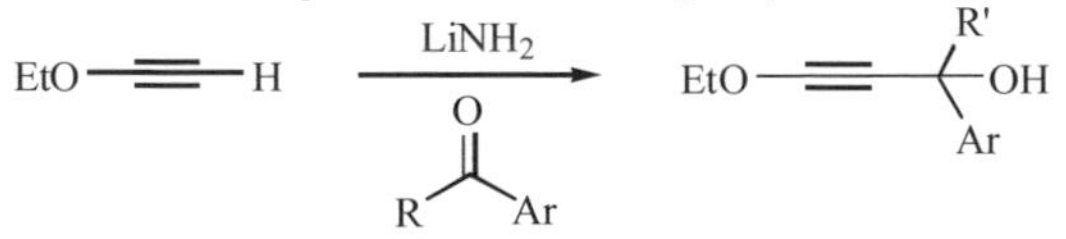

## Examples:

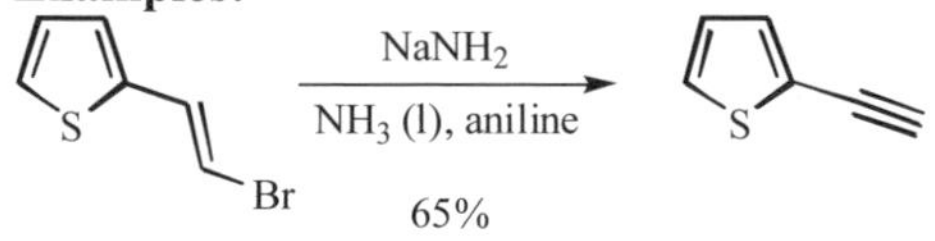

A. Vaitiekunas, F. F. Nord, *Journal of Organic Chemistry* **1954**, <u>19</u>, 902

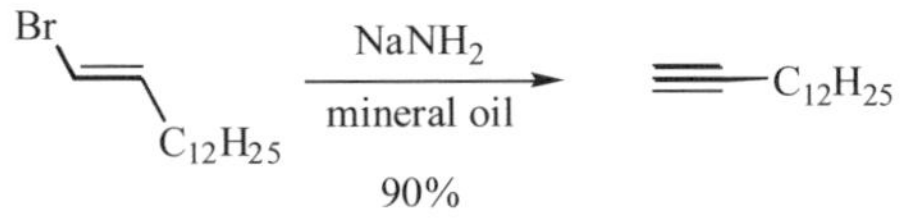

W. G. Knipprath, R. A. Stein, *Lipids* **1966**, <u>1</u>, 81

I. Creton, H. Rezaei, I. Marek, J. F. Normant, *Tetrahedron Letters* **1999**, <u>40</u>, 1899

P. Pianetti, P. Rollin, J. R. Poughny, *Tetrahedron Letters* **1987**, <u>27</u>, 5853

E. Ottow, R. Rohde, W. Schwede, R. Wiechert, *Tetrahedron Letters* **1993**, <u>34</u>, 5253

T. Suzuki, T. Sonoda, S. Kobayashi, H. Taniguchi, *Journal of the Chemical Society, Chemical Communications* **1976**, 180

# Gabriel Synthesis

## The Reaction:

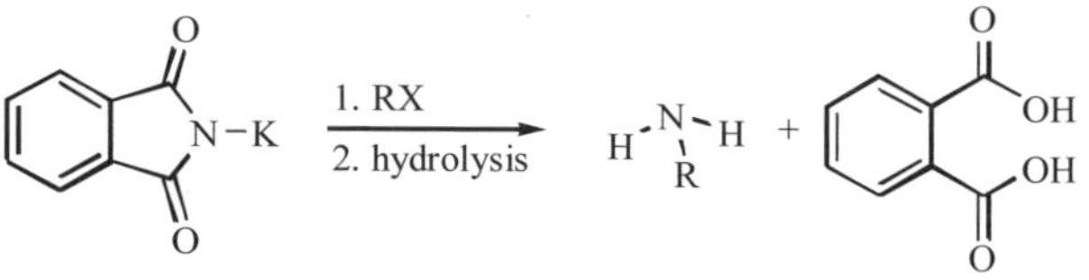

## Proposed Mechanism:

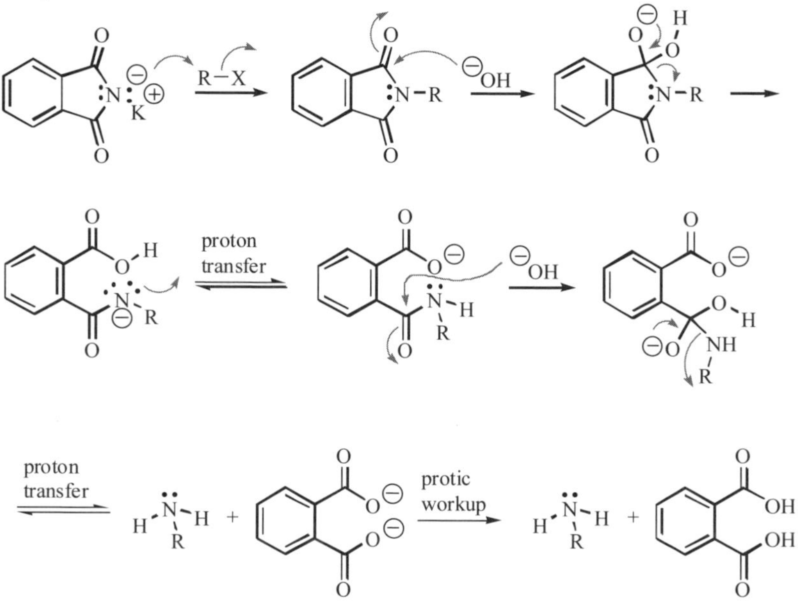

## Notes:

The **Manske Modification**: Using hydrazine to release the primary amine.

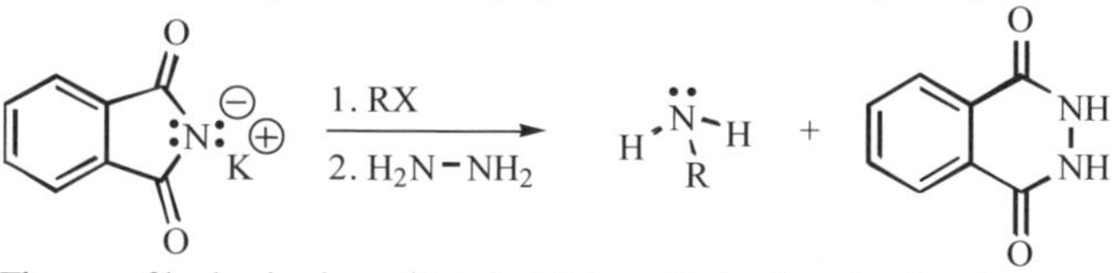

The use of hydrazine is useful in that it is gentle to other functional groups.

See T. Laue, A. Plagens, *Named Organic Reactions*, John Wiley and Sons, Inc., New York, 1998, pp. 120-122; M. B. Smith, J. March in *March's Advanced Organic Chemistry*, 5th ed., John Wiley and Sons, Inc., New York, 2001, pp 500, 513, 864.

# Gabriel-Colman Rearrangement

## The Reaction:

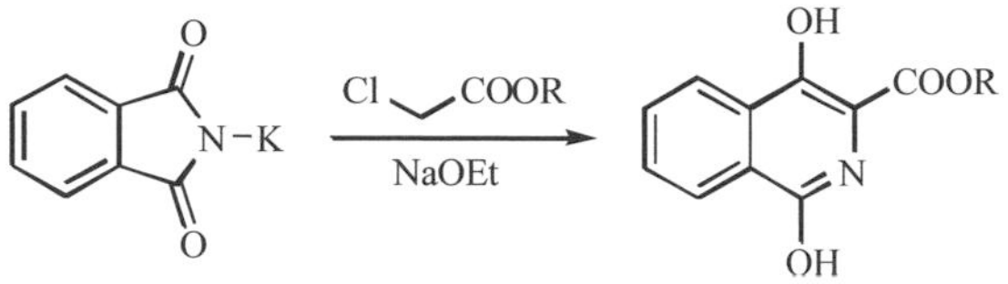

## Proposed Mechanism:

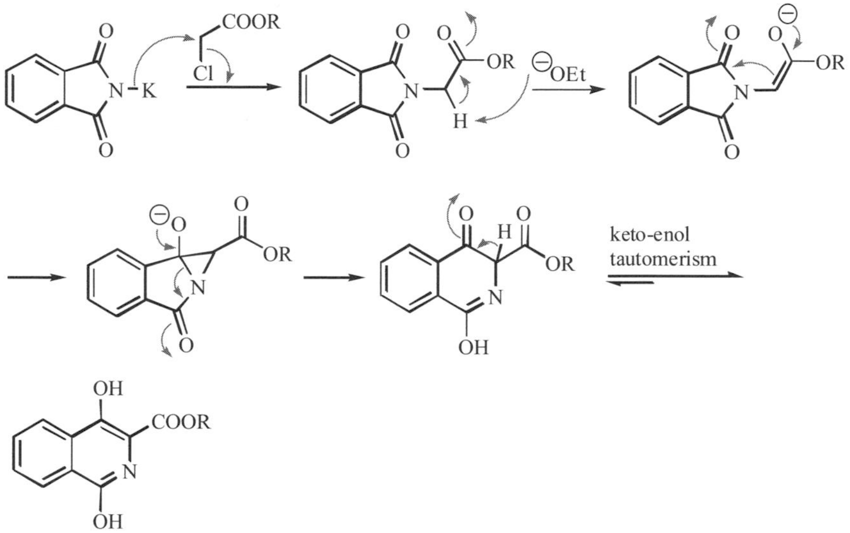

## Examples:

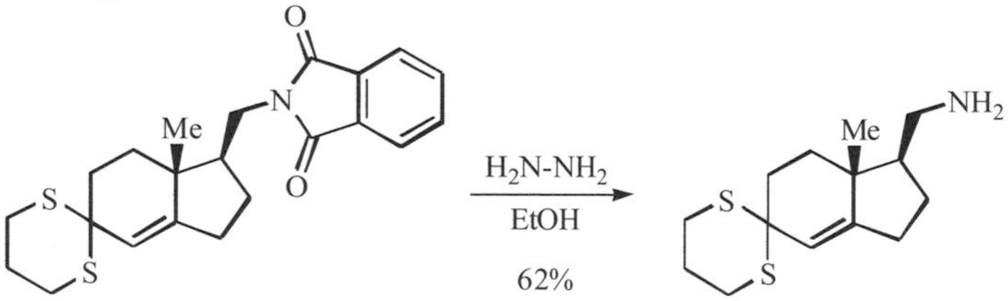

62%

L. G. Sevillano, C. P. Melero, E. Caballero, F. Tome, L. G. Lelievre, K. Geering, G. Crambert, R. Carron, M. Medarde, A. San Feliciano, *Journal of Medicinal Chemistry* **2002**, <u>45</u>, 127

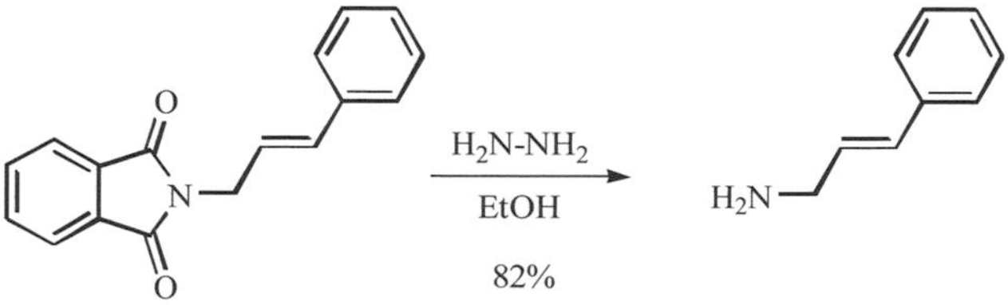

82%

C. Serino, N. Stehle, Y. S. Park, S. Florio, P. Beak, *Journal of Organic Chemistry* **1999**, <u>64</u>, 1160

# Gabriel-Cromwell Reaction

## The Reaction:

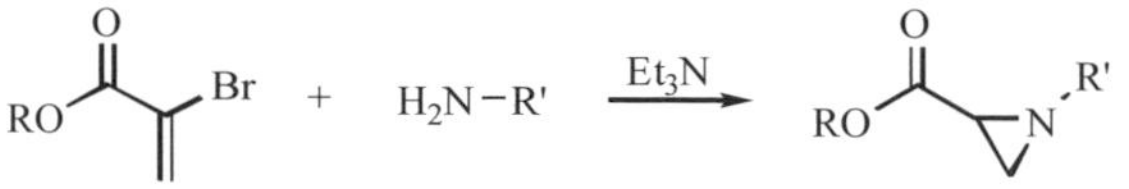

## Proposed Mechanism:

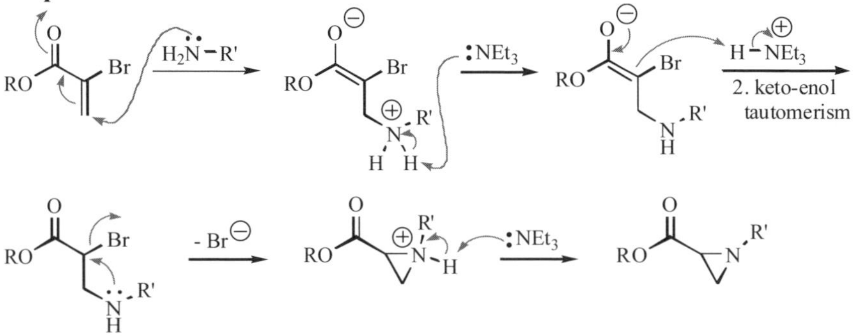

## Notes:

General reactions on solid phase are described in a review:

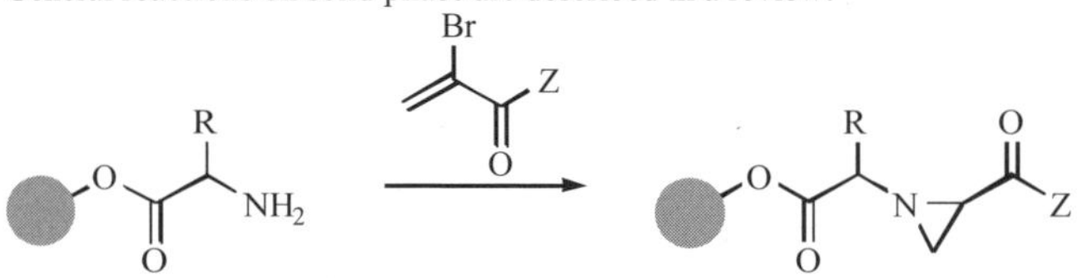

R. G. Franzen, *Journal of Combinatorial Chemistry* **2000**, *2*, 195

## Examples:

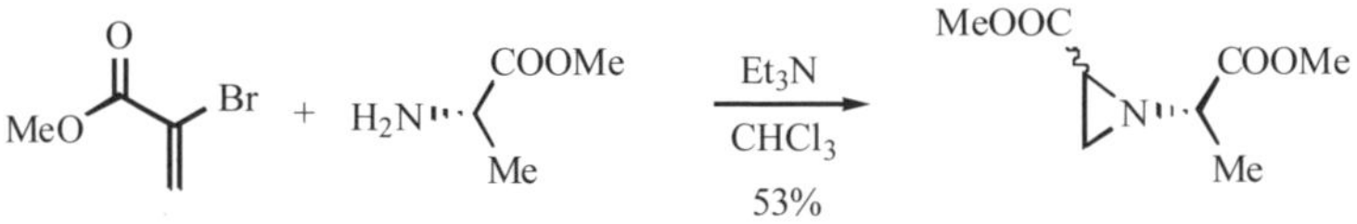

S. N. Filigheddu, M. Taddei, *Tetrahedron Letters* **1998**, *39*, 3857

L. R. Comstock, S. R. Rajski, *Tetrahedron* **2002**, 58, 6019

M. T. Barros, C. D. Maycock, M. R. Ventura, *Tetrahedron Letters* **2002**, 43, 4329

G. Cardillo, L. Gentilucci, C. Tomasini, M. P. V. Castjon-Bordas, *Tetrahedron Asymmetry* **1998**, 39, 3857

# Garegg-Samuelsson Reaction

## The Reaction:

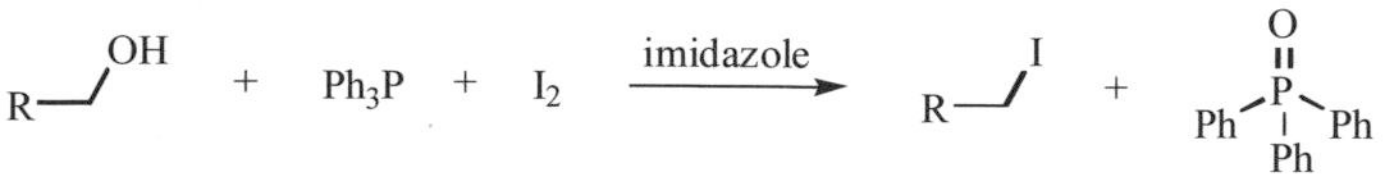

## Proposed Mechanism:

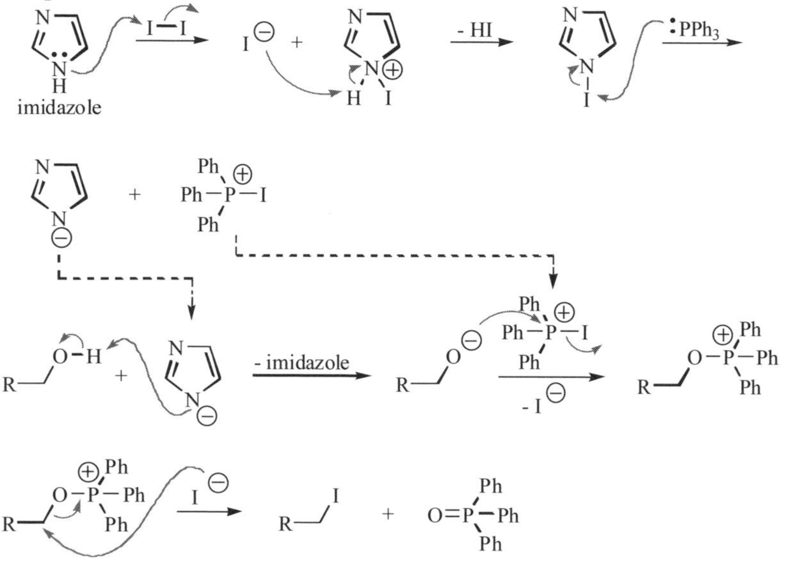

## Notes:

1. The reaction will also work with bromine.
2. General reactivity:  1° > 2° alcohols.
3. Secondary alcohols often experience inversion of configuration.
4. 1,2-diols in this reaction result in alkene formation [*Garegg-Samuelsson Olefination*]

## Examples:

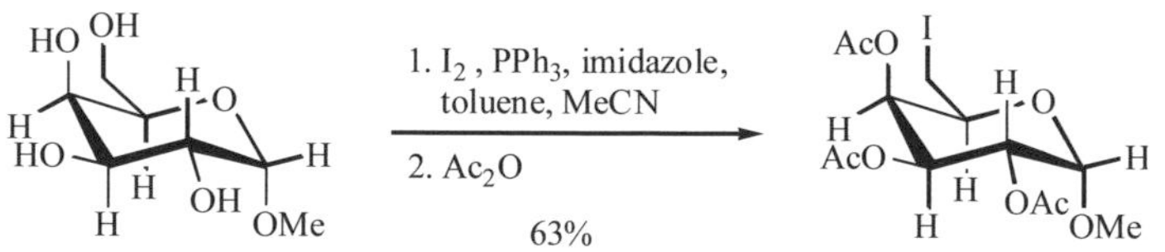

1. I$_2$ , PPh$_3$, imidazole, toluene, MeCN

2. Ac$_2$O

63%

P. J. Garegg, R. Johansson, C. Ortega, B. Samuelsson, *Journal of the Chemical Society, Perkin Transaction I,* **1982**, 681

Z. Pakulski, A. Zamojski, *Carbohydrate Research* **1990**, <u>205</u>, 410

Pht = phthalyl

J. T. Starr, G. Koch, E. M. Carreira, *Journal of the American Chemical Society* **2000**, <u>122</u>, 8793

G. Anikumar, H. Nambu, Y. Kita, *Organic Process and Research Development* **2002**, <u>6</u>, 190

J. M. G. Fernandez, A. Gadelle, J. Defaye, *Carbohydrate Research* **1994**, <u>265</u>, 249

I. Izquierdo, M. T. Plaza, F. Franco, *Tetrahedron Asymmetry* **2002**, <u>13</u>, 1503

# Garst-Spencer Furan Annulation

## The Reaction:

R—C(=O)—CH$_2$—R'

1. NaOMe, EtOCHO
2. n-BuSH, p-TsOH, Δ
3. H$_2$C=SMe$_2$
4. Hg$^{+2}$

→ furan with R and R'

## Proposed Mechanism:

R—C(=O)—CH$_2$—R'

NaOMe / EtO—CHO (O)

→ R—C(=O)—C(R')=CH—OH

n-BuSH, p-TsOH / Δ

→ R—C(=O)—C(R')=CH—S n-Bu

H$_2$C=S(Me)(Me)

→ epoxide with R, C(R')=CH—S n-Bu

→ dihydrofuran with R, R', S n-Bu

Hg$^{+2}$

→ furan with R, R'

## Examples:

An improvement by Kurosu, Marcin and Kishi:

R—C(=O)—CH$_2$—R'

NaOMe / EtO—CHO (O)

→ R—C(=O)—C(R')=CH—OH

p-MePhSH, p-TsOH / benzene, Δ

→ R—C(=O)—C(R')=CH—S p-MePh

Me$_3$SI, NaHMDS / Et$_2$O

→ epoxide with R, C(R')=CH—S p-MePh

Δ / toluene

→ dihydrofuran with R, R', S p-MePh

I$_2$ / toluene

→ furan with R, R'

1. Me$_3$SI, NaHMDS, Et$_2$O
2. Δ, toluene
3. I$_2$, toluene

70%

M. Kurosu, L. R. Marcin, Y. Kishi, *Tetrahedron Letters* **1989**, <u>39</u>, 8929
see also M. Kurosu, L. R. Marcin, T. J. Grinsteiner, Y. Kishi *Journal of the American Chemical Society* **1998**, <u>120</u>, 6627

1. NaH, EtOCHO, MeOH (cat.) then HCl
2. *p*-TsOH, pyr, 0 °C then *n*-BuSH

3. DMSO, Me₂S=CH₂, THF
4. HgSO₄, Et₂O

57%

P. A. Zoretic, M. Wang, Y. Zhang, Z. Shen, *Journal of Organic Chemistry* **1996**, <u>61</u>, 1806

Phase-transfer conditions are used in this example:

1. Me₃S⁺MeSO₄⁻, 50% NaOH (aq),
   CH₂Cl₂, Δ

2. 2N HCl, THF

59% (69% rec. st. mat.)

M. E. Price, N. E. Schore *Journal of Organic Chemistry* **1989**, <u>54</u>, 2777

1. HCO₂Et, NaOMe, benzene 67%
2. *n*-BuSH, *p*-TsOH, benzene 100%

3. Me₃S⁺MeSO₄⁻, 50% NaOH (aq),
   CH₂Cl₂, Δ
4. 2N HCl, THF 70%

M. E. Price, N. E. Schore *Journal of Organic Chemistry* **1989**, <u>54</u>, 5662

1. H₂C=SMe₂

2. 6M HCl / MeOH

84%

L. C. Garver, E. E. van Tamelen, *Journal of the American Chemical Society* **1982**, <u>104</u>, 867

# Gattermann Aldehyde Synthesis (Gattermann-Adams Formylation)

## The Reaction:

G = alkyl, OR

## Proposed Mechanism:

$$Zn(CN)_2 \; + \; 2\; HCl \longrightarrow 2\; HCN \; + \; ZnCl_2 \xrightarrow{\;HCl\;} H-C\equiv\overset{\oplus}{N}-H \quad Cl^{\ominus}$$

iminium salt

α amino alcohol
(not isolated)

## Notes:
T. Laue, A. Plagens, *Named Organic Reactions*, John Wiley and Sons, Inc., New York, 1998, pp. 123-124; M. B. Smith, J. March in *March's Advanced Organic Chemistry*, 5th ed., John Wiley and Sons, Inc., New York, 2001, p. 715; W. E. Truce, *Organic Reactions* **9**, 2.

The reaction works with alkylbenzenes, phenols, their ethers, and many heterocycles, it doesn't work with aromatic amines.

In strong acid solutions, there is evidence of a dication intermediate:

Y. Sato, M. Yato, T. Ohwada, S. Saito, K. Shudo, *Journal of the American Chemical Society* **1995**, 117, 3037

The original conditions, called the **Gattermann Reaction / Formylation**, were to add HCN, HCl and ZnCl₂ (known as **Adam's Catalyst**) directly. Use of **Adam's catalyst** avoids using gaseous HCN.

## Gattermann Method

## Examples:

HO COOMe → Zn(CN)₂ / HCl, AlCl₃ / 65% → HO COOMe CHO (structures)

A. V. Rama Rao, N. Sreenivasan, D. Reddeppa Reddy, V. H. Deshpande, *Tetrahedron Letters* **1987**, 28, 455

HO Me → Zn(CN)₂ / HCl, AlCl₃ / 95% → HO Me CHO (structures)

G. Solladie, A. Rubio, M. C. Carreno, J. L. G. Ruano, *Tetrahedron Asymmetry* **1990**, 1, 187

## Gattermann-Koch Reaction

**The Reaction:**

benzene + CO → HCl, AlCl₃ / CuCl → benzaldehyde (G substituted)

**Proposed Mechanism:**
The details of the formation of the formyl cation seem to be less assured.

See S. Raugei, M. L. Klein, *Journal of Physical Chemistry B* **2001** 105, 8212 for pertinent references to experiment, and their computational study of the formyl cation. The work shows the uncertainty associated with acid concentration, counter ion and relative O- vs-C protonation possibilities. We provide a very simplistic possibility:

(mechanism scheme)

The probable role of copper is to facilitate the transport of carbon monoxide, with which it binds. See: N. N. Crouse, *Organic Reactions* **5**, 6.

There seems to be agreement that the product-forming part of the mechanism is:

$:C{\equiv}O: + H^{\oplus} \rightleftharpoons$ (mechanism scheme)

This reaction is limited to benzene and alkylbenzenes.

# General Coupling Reactions

A number of coupling reactions have already been summarized under the title *__Alkyne Coupling__*.
This section summarizes a number of other common coupling protocols.

## The Reaction:
These important synthesis reactions have a common mechanistic theme:

Organic Substrate (S) $\xrightarrow{\text{M}}$ $L{>}M{<}^L_S$ $\xrightarrow{\text{S}}$ $L{>}M{<}^S_L$ $\Longrightarrow$

$L{>}M{<}^S_S$ $\longrightarrow$ S-S'

The difference in the reactions is in: M (Ni, Cu, Pd).
In these reactions there are many variables: Ligand (type and number), Metal (oxidation state and
identity), X (halogen, triflate, etc), M'.

## Proposed Mechanism:
In a catalytic cycle the common trend is:

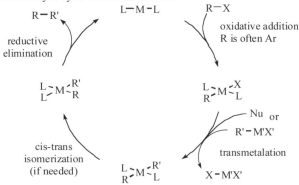

This step generally has some nucleophile
replace a ligand. In the cases where the
nucleophile is derived from an organometallic
reagent it is called a *__transmetallation__*.

| Named Reaction | Metal | Nucleophile |
|---|---|---|
| *__Buchwald-Hartwig__* | Pd | $R_2NH$ or ROH |
| *__Fukuyama__* | Pd | R-Zn-I * |
| *__Heck__* | Pd | alkene |
| *__Kumada__* | Pd or Ni | R-MgX |
| *__Negishi__* | Pd or Ni | R-ZnX |
| *__Sonogashira__* | Pd | R-≡-Cu |
| *__Stille__* | Pd | $R\text{-}SnBu_3$ |
| *__Suzuki__* | Pd | $R\text{-}B(OH)_2$ |

Under the specific heading one can find details and examples specific for the reaction.

**Common Ligands:**

Ph—P(Ph)Ph                Ph—P(Ph)—P(Ph)—Ph              Ph—P(Ph)—P(Ph)—Ph

Triphenylphosphine              dppm                              dppe

The ligands can be particularly useful in preparing the organometallic-substrate for reductive elimination.

$$\underset{L}{\overset{L}{M}}\overset{Y}{\underset{Z}{}} \longrightarrow \overset{L}{\underset{L}{M}} \qquad Y-Z$$

In this generic example, the oxidation state of the metal is reduced by two, while the product is released in the reductive elimination step. In many reactions the oxidative addition does not provide the proper stereochemistry for elimination; an isomerization must occur.

$$Z-Y \;+\; M\underset{L}{\overset{L}{}} \quad\xrightarrow[\text{addition}]{\text{Oxidative}}\quad \underset{L}{\overset{Y}{M}}\overset{L}{\underset{Z}{}} \quad\underset{}{\overset{L}{\rightleftarrows}}\quad \underset{L}{\overset{Y}{M}}\overset{L}{\underset{Z}{}} \quad\xrightarrow{-L}\quad \underset{L}{\overset{L}{M}}\overset{Y}{\underset{Z}{}}$$

# Glaser (Oxidative) Coupling (Reaction)

## The Reaction:

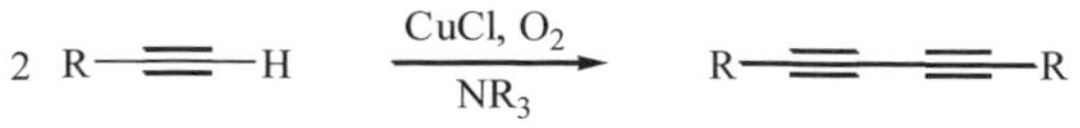

$$2 \ R\!-\!\!\equiv\!\!-H \quad \xrightarrow[\text{NR}_3]{\text{CuCl, O}_2} \quad R\!-\!\!\equiv\!\!-\!\!\equiv\!\!-R$$

## Proposed Mechanism:

Mechanistic details are not well understood. It is known that the reaction has been successfully accomplished with a variety of solvents, amines and copper salts.

There seems to be consensus that a copper-alkyne bond is homolytically broken by oxygen (often just air) and the radicals thus formed, couple:

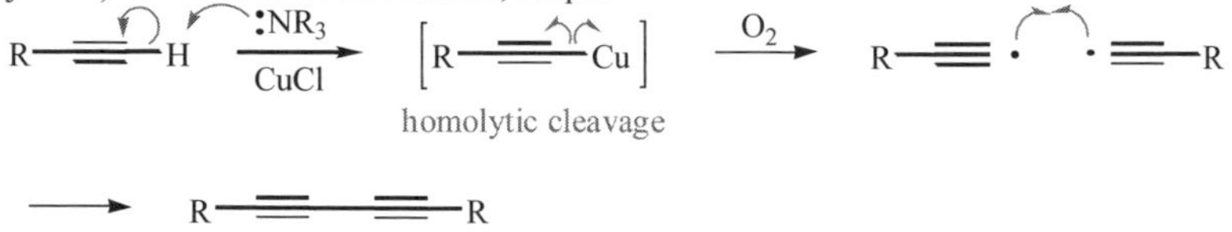

$$R\!-\!\!\equiv\!\!-H \quad \xrightarrow[\text{CuCl}]{:\text{NR}_3} \quad \left[ R\!-\!\!\equiv\!\!-Cu \right] \quad \xrightarrow{O_2} \quad R\!-\!\!\equiv\!\!\cdot \quad \cdot\!\!\equiv\!\!-R$$

homolytic cleavage

$$\longrightarrow \quad R\!-\!\!\equiv\!\!-\!\!\equiv\!\!-R$$

A more detailed interpretation is available:

P. Siemsen, R. C. Livingston, F. Diederich, *Angewandte Chemie International Edition* **2002**, <u>39</u>, 2632

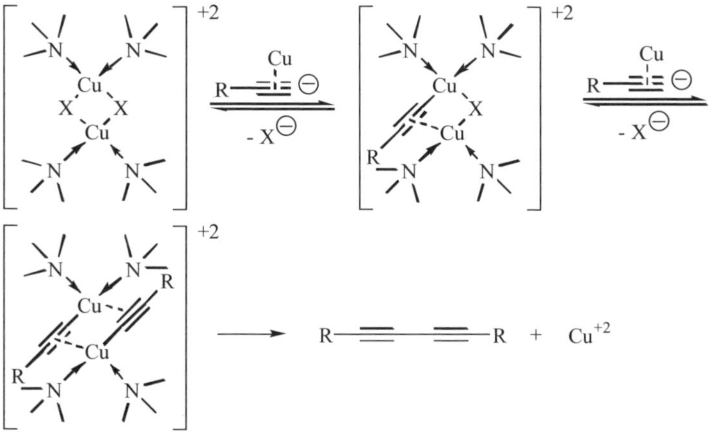

## Notes:

See: T. Laue, A. Plagens, *Named Organic Reactions*, John Wiley and Sons, Inc., New York, 1998, pp. 125-127; M. B. Smith, J. March in *March's Advanced Organic Chemistry*, 5th ed., John Wiley and Sons, Inc., New York, 2001, p. 927

$O_2$ was an early choice for oxidant; however, other oxidants have been used.

## Examples:

The reaction can be carried out without amines, using sodium acetate in super-critical $CO_2$.

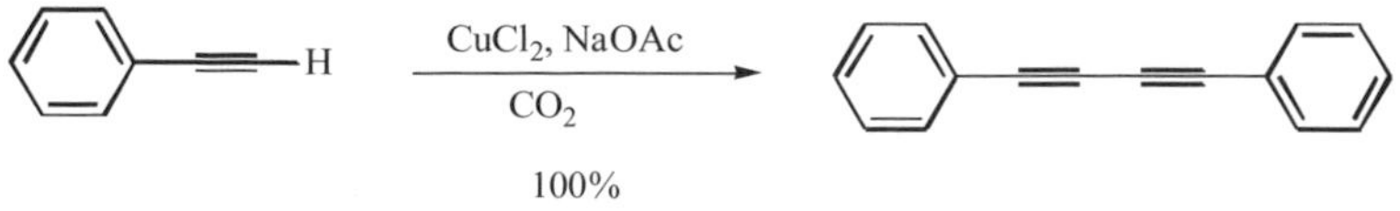

100%

J. Li, H. Jiang, *Chemical Communications* **1999**, 2369

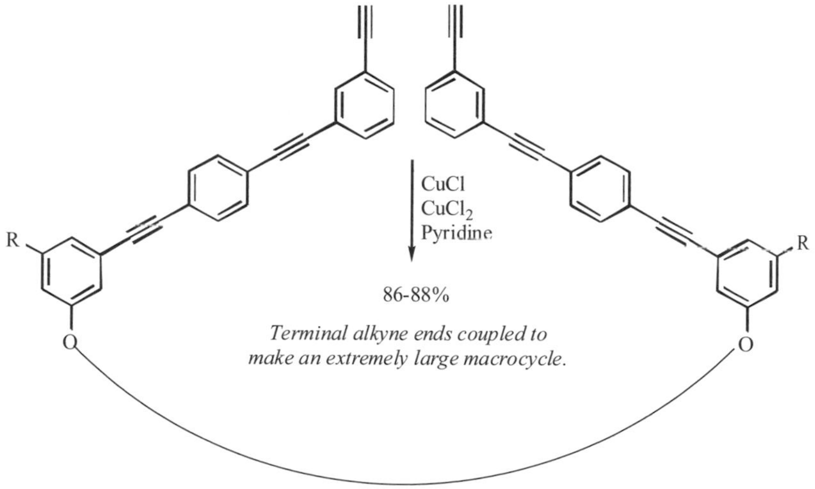

CuCl
$CuCl_2$
Pyridine
86-88%
*Terminal alkyne ends coupled to make an extremely large macrocycle.*
R
R
O
O

Large Carbon Span

S. Hoger, A.-D. Meckenstock, H. Pellen, *Journal of Organic Chemistry* **1997**, 62, 4556

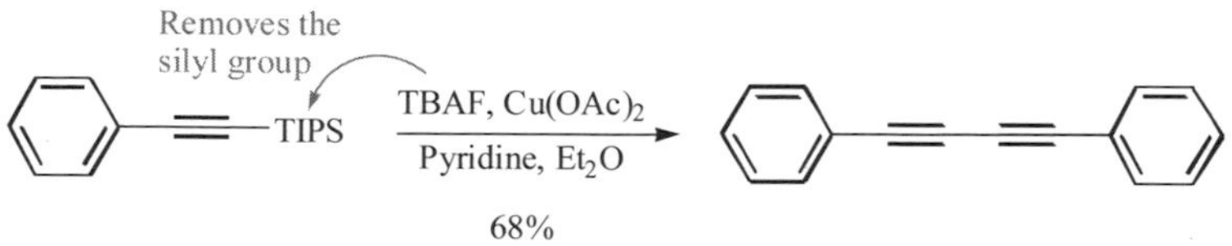

Removes the silyl group
TIPS
TBAF, $Cu(OAc)_2$
Pyridine, $Et_2O$
68%

M. A. Heuft, S. K. Collins, G. P. A. Yap, A. G. Fallis, *Organic Letters* **2001**, 3, 2883

BnO
$(CH_2)_6$
$(CH_2)_6$
OBn
O
O
$O-(CH_2)_5$
$(H_2C)_5 O$
CuCl, $CuCl_2$
pryidine, air
75%

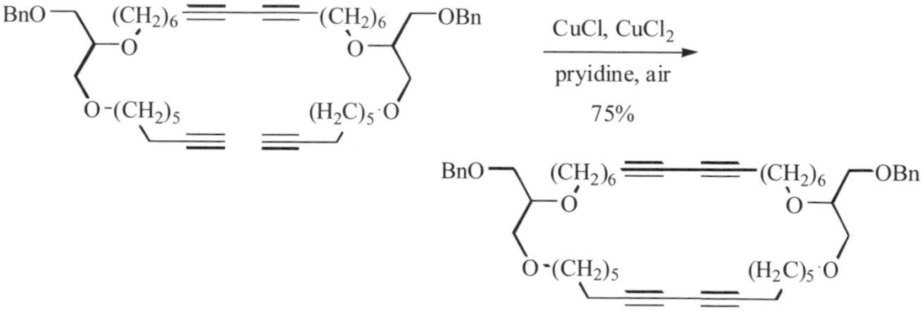

A. P. Patwardhan, D. H. Thompson, *Organic Letters* **1999**, 1, 241

# Gould-Jacobs Reaction

## The Reaction:

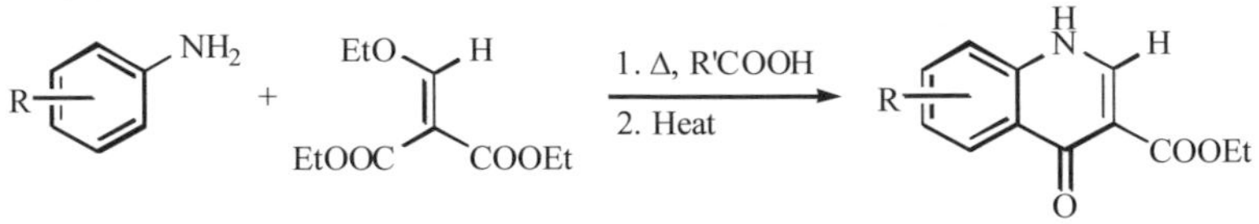

## Proposed Mechanism:

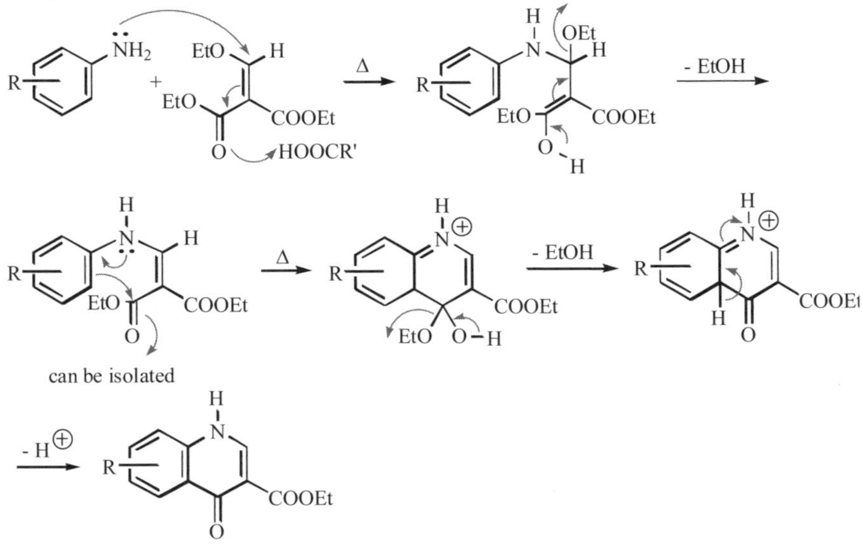

can be isolated

## Notes:

Hydrolysis readily removes the ester and aromatizes the second ring:

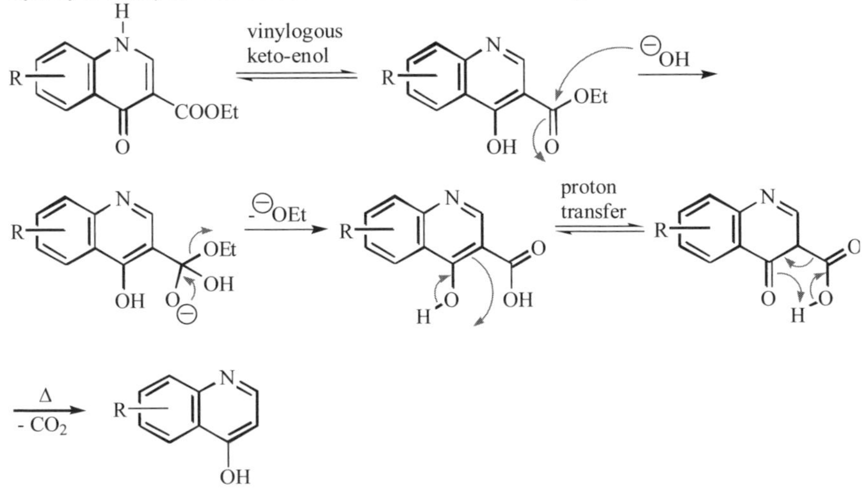

A different approach to similar molecules:

N. D. Heindel, I. S. Bechara, T. F. Lemke, V. B. Fish, *Journal of Organic Chemistry* **1967**, <u>32</u>, 4155

## Examples:

J. N. Kim, K. Y. Lee, H. S. Kim, T. Y. Kim, *Organic Letters* **2000**, <u>2</u>, 343

J. C. Carretero, J. L. Garcfa Ruano, M. Vicioso, *Tetrahedron* **1992**, <u>48</u>, 7373

# Gribble Indole Reduction

## The Reaction:

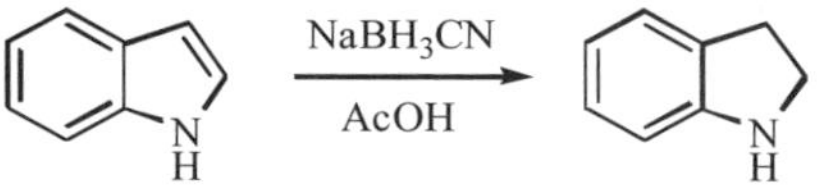

G. W. Gribble, P. D. Lord, J. Skotnicki, S. E. Dietz, J. T. Eaton, J. L. Johnson, *Journal of the American Chemical Society* **1974**, <u>96</u>, 7812

## Proposed Mechanism:

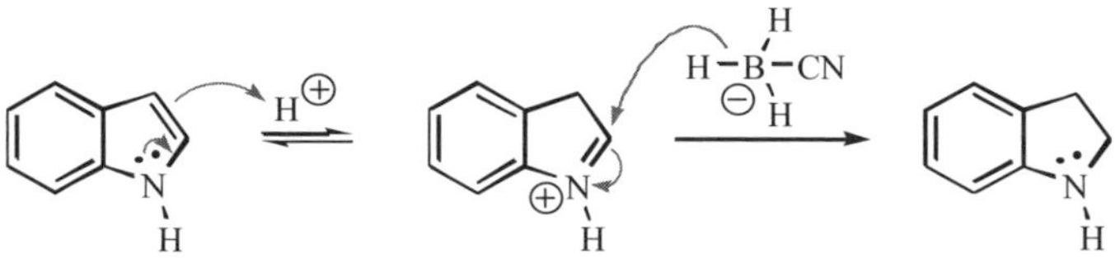

## Notes:

G. W. Gribble, *Chemical Society Reviews* **1998**, <u>27</u>, 395

## Examples:

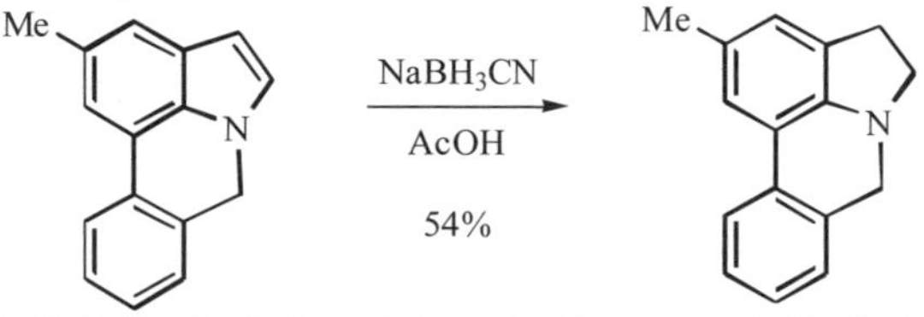

54%

J. C. Torres, A. C. Pinto, S. J. Garden *Tetrahedron* **2004**, <u>60</u>, 9889

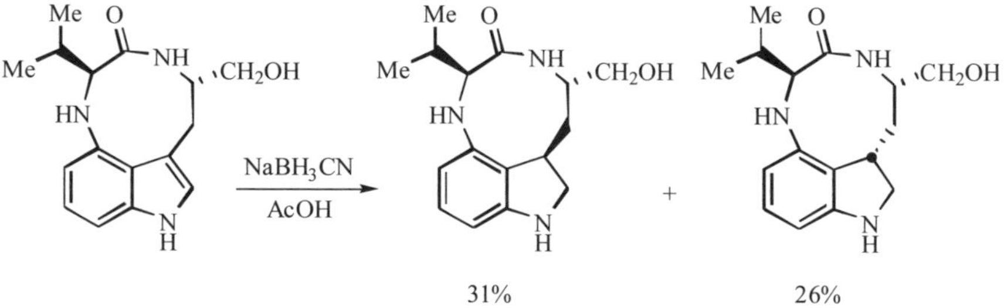

31%                    26%

Y. Nakagawa, K. Irie, Y. Komiya, H. Ohigashia, K. Tsuda, *Tetrahedron* **2004**, <u>60</u> 7077

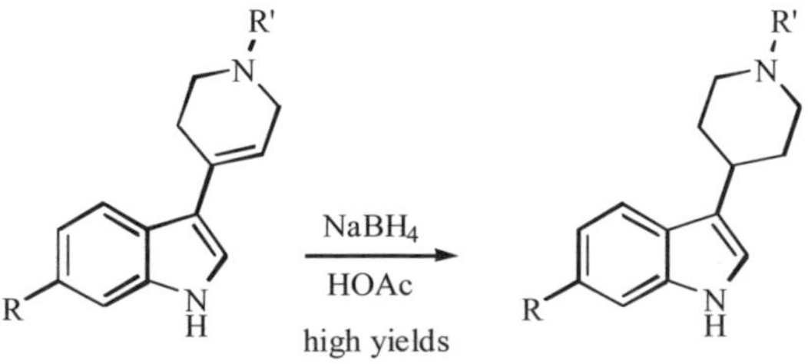

high yields

A. Borghese, L. Antoine, G. Stephenson, *Tetrahedron Letters* **2002**, <u>43</u>, 8087

# Gribble Reduction of Diaryl Ketones and Methanols

## The Reaction:

G. W. Gribble, M. Leese, B. E. Evans, *Synthesis* **1977**, 172
G. W. Gribble, W. J. Kelley, S. E. Emery, *Synthesis* **1978**, 763

## Proposed Mechanism:

$$NaBH_4 + CF_3CO_2H \longrightarrow (CF_3CO_2)_3BH$$

(TFA = trifluoroacetic acid = $CF_3CO_2H$)

## Examples:

G. Bringmann, T. Pabst, P. Henschel, M. Michel, *Tetrahedron* **2001**, *57*, 1269

calix[4]arene
(abbreviated as Ar)

J. Wang, C. D. Gutsche, *Journal of Organic Chemistry* **2002**, *67*, 4423

| Acid | yield |
|------|-------|
| $CF_3CO_2H$ | 70 - 81% |
| $CF_3SO_3H$ | 94 - 99% |

G. A. Olah, A.-h. Wu, O. Farooq, *Journal of Organic Chemistry* **1988**, *53*, 5143

# Grieco-Sharpless Elimination

## The Reaction:

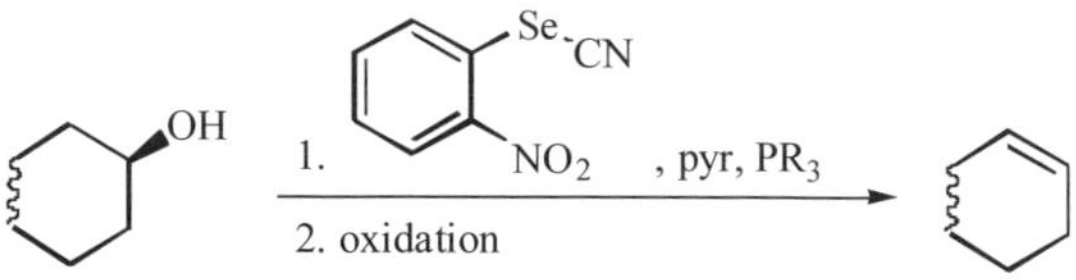

## Proposed Mechanism:

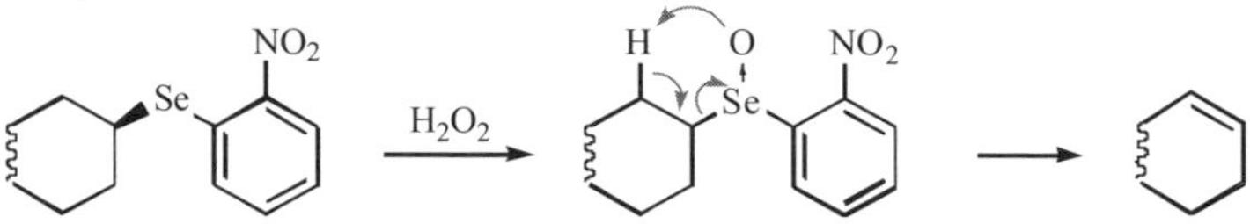

## Notes:

Selenium-based reactions are generally very mild, often with the advantage of preparing the selenium derivative and carrying out the oxidative elimination as a one-pot sequence.

The five-membered transition state for selenoxide elimination is more restrictive than the six-membered transition state of the *Chugaev Elimination*.

## *Chugaev:*

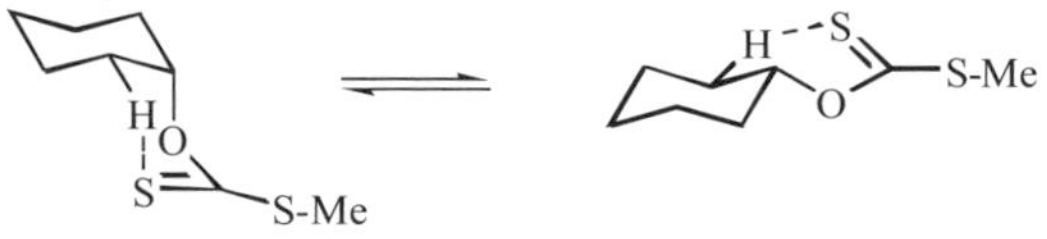

*syn* elimination                          *anti* elimination

## *Selenoxide:*

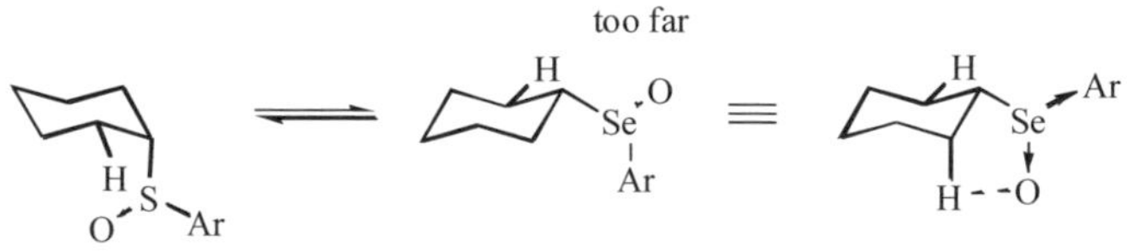

*syn* elimination                          *syn* elimination

## Examples:

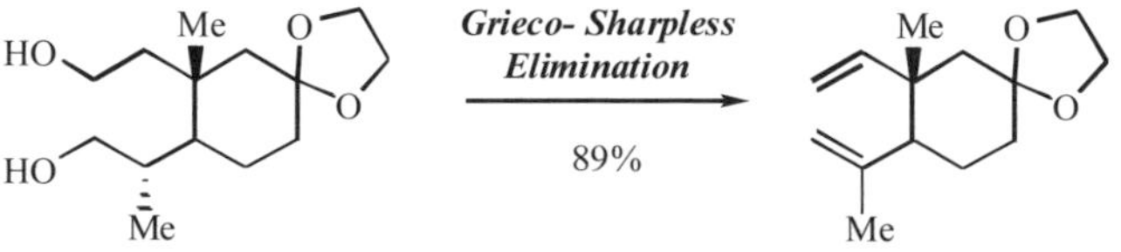

G. Majetich, P. A. Grieco, M. Nishizawa, *Journal of Organic Chemistry* **1977**, _42_, 2327

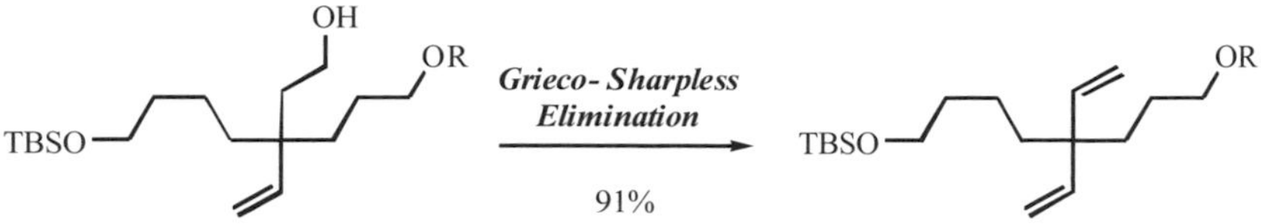

OH
OR
TBSO
Grieco- Sharpless
Elimination
91%
TBSO
OR

Y. Fukuda, M. Shindo, K. Shishido, *Organic Letters* **2003**, 5, 749

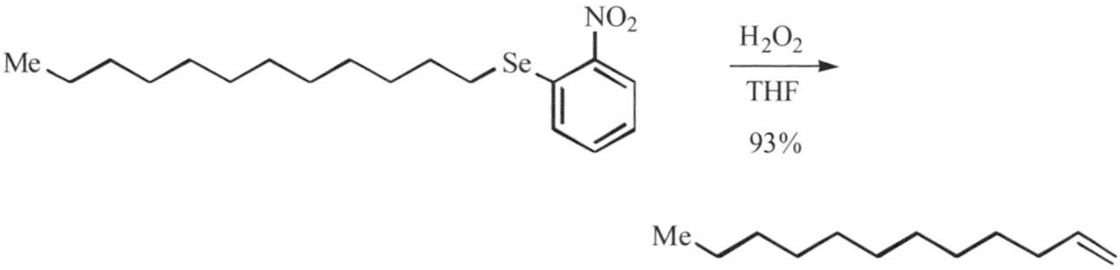

NO2
Me
Se
H2O2
THF
93%
Me

K. B. Sharpless, M. Y. Young, *Journal of Organic Chemistry* **1975**, 40, 947

# Grignard Reactions

## The Reaction:

R−X + Mg $\longrightarrow$ R−MgX   The **Grignard reagent**

formaldehyde $\xrightarrow{\quad R-MgX \quad}$ 1° alcohols

aldehydes $\longrightarrow$ 2° alcohols

ketones $\longrightarrow$ 3° alcohols

esters $\longrightarrow$ 3° alcohols

$CO_2$ $\longrightarrow$ carboxylic acid

nitrile $\longrightarrow$ ketone

Chloramine $\longrightarrow$ Amine

NC-CN $\longrightarrow$ Nitrile

$HC(OR)_3$ $\longrightarrow$ Aldehyde

## Proposed Mechanism:

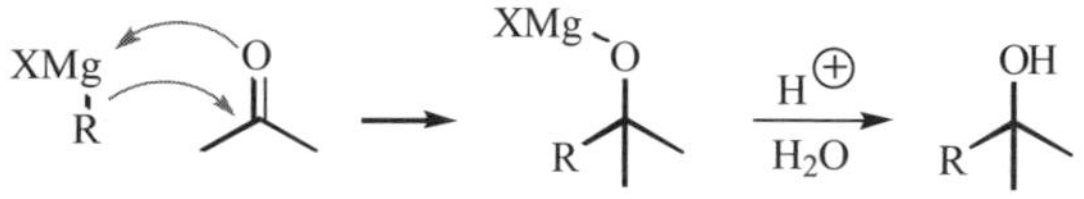

- - - - - - - - - - - - - - - - - - - - - - - - - - - - - - - - - - - - - - -

nitrile $\longrightarrow$ ketone

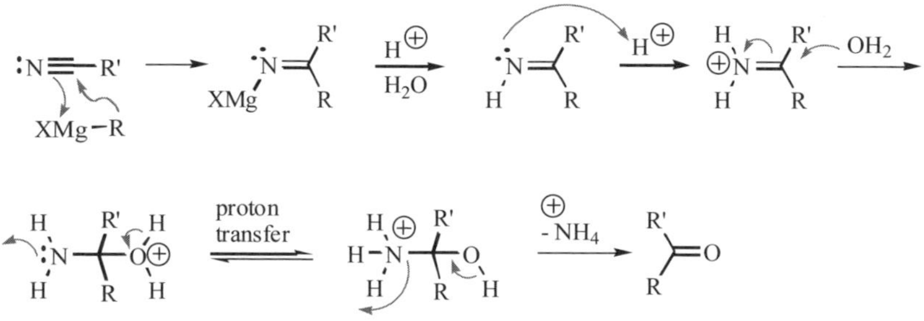

## Notes:

T. Laue, A. Plagens, *Named Organic Reactions*, John Wiley and Sons, Inc., New York, 1998, pp. 132-138; M. B. Smith, J. March in *March's Advanced Organic Chemistry*, 5<sup>th</sup> ed., John Wiley and Sons, Inc., New York, 2001, pp 1205-1209.

**Schlenk equilibrium** - The "**Grignard reagent**" is really an equilibrium.
Shown below are some of the structures:

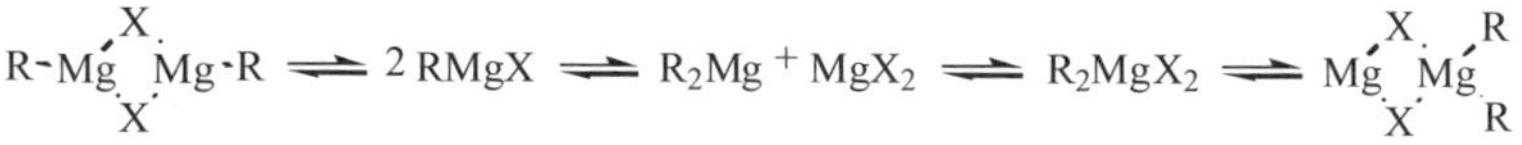

**Iron-Catalyzed Cross-Coupling Reactions**

B. Scheiper, M. Bonnekessel, H. Krause, A. Fürstner *Journal of Organic Chemistry* **2004**, _69_, 3943

## Examples:

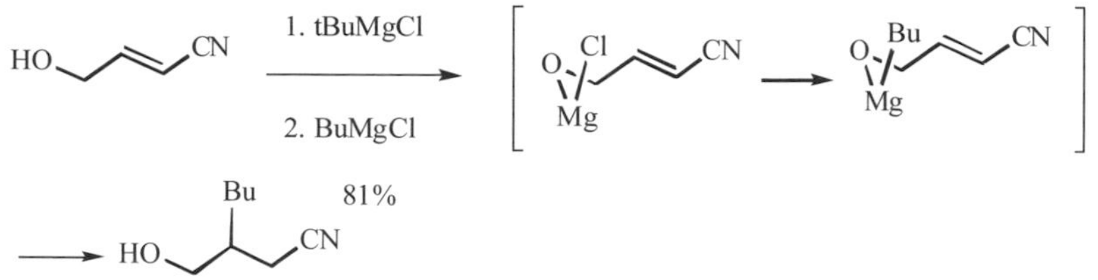

F. F. Fleming, Q. Wang, Z. Zhang, O. W. Steward, *Journal of Organic Chemistry* **2002**, <u>67</u>, 5953

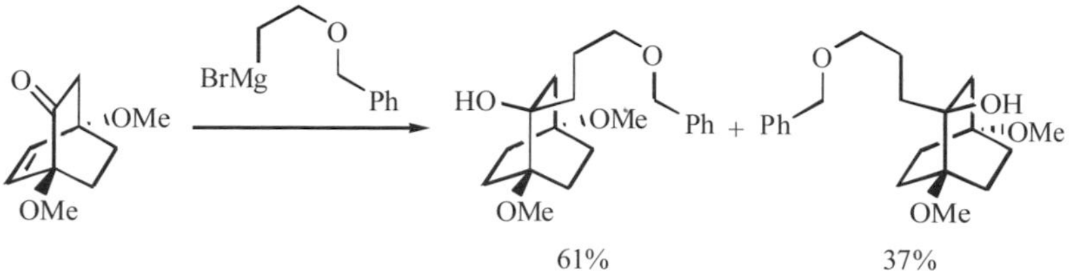

61% 37%

T. Uyehara, T. Marayama, K. Sakai, M. Ueno, T. Sato, *Tetrahedron Letters* **1996**, <u>37</u>, 7295

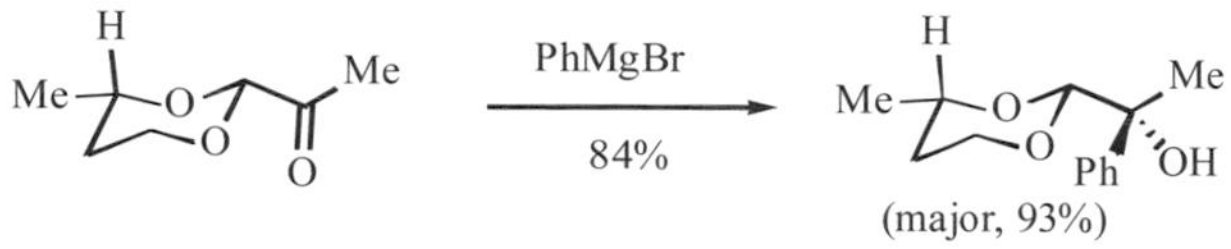

84%

(major, 93%)

W. F. Bailey, D. P. Reed, D. R. Clark, G. N. Kapur, *Organic Letters* **2001**, <u>3</u>, 1865

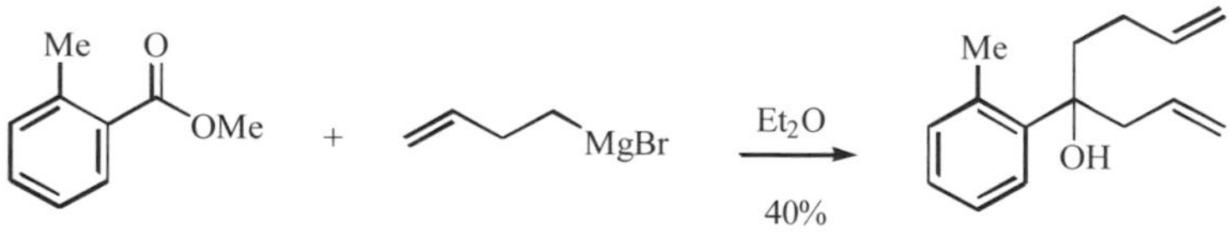

40%

P. A. Wender, T. M. Dore, M. A. Delong, *Tetrahedron Letters* **1996**, <u>37</u>, 7687

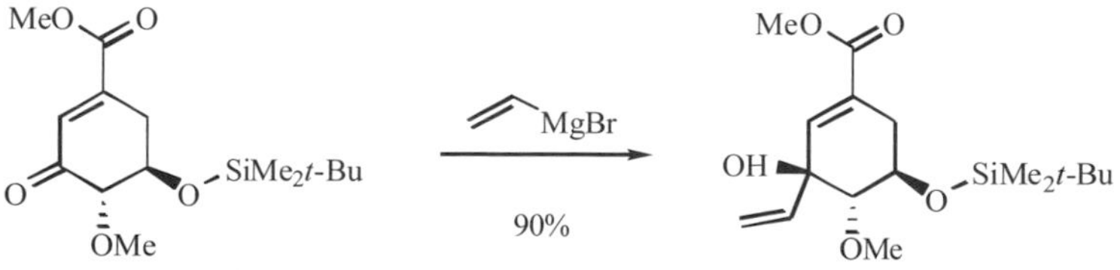

90%

S. Hanessian, J. Pan, A. Carnell, H. Bouchard, L. Lesage, *Journal of Organic Chemistry* **1997**, <u>62</u>, 465

There are a number of reactions employing **Grignard reagents** that have their own names:

## Bénary Reaction

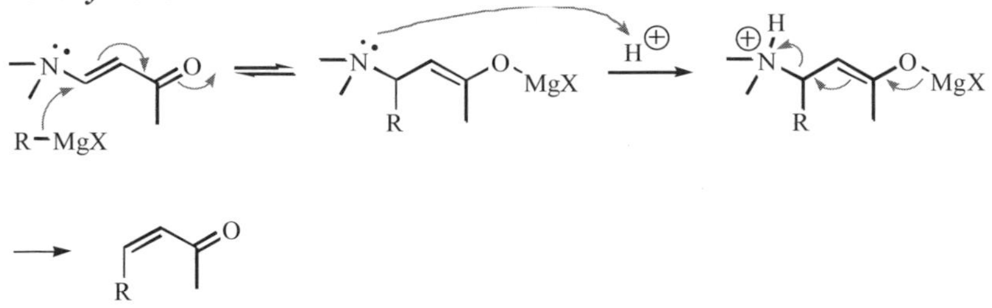

## Bodroux Reaction

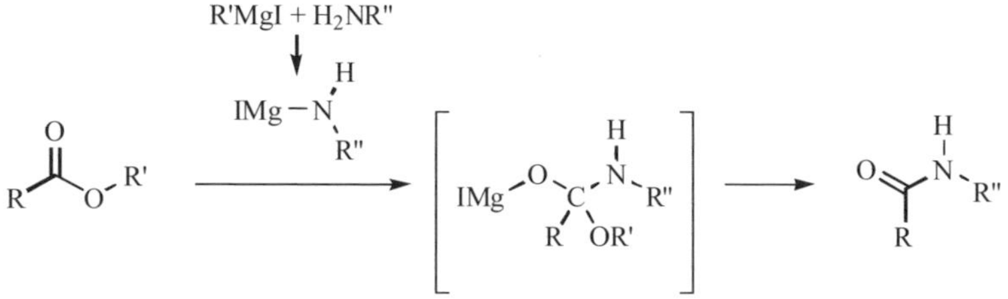

## Bouveault Reaction

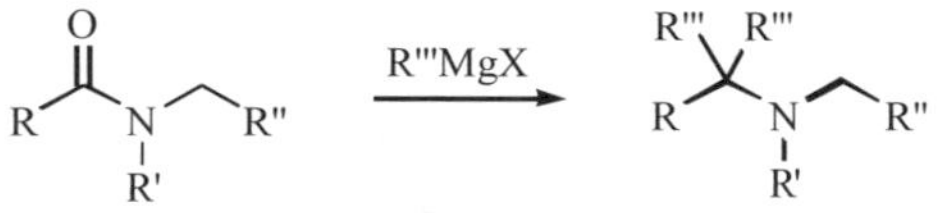

## Dzhemilev Reaction

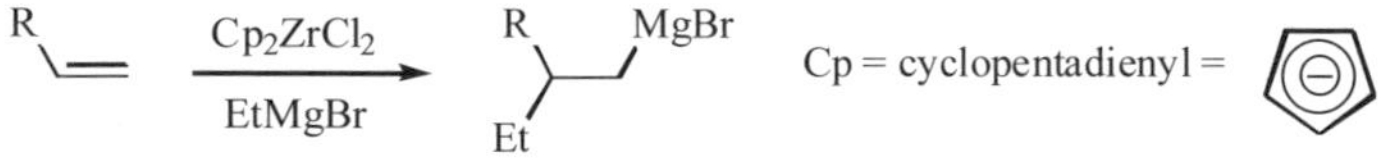

Cp = cyclopentadienyl =

**Proposed Mechanism:**

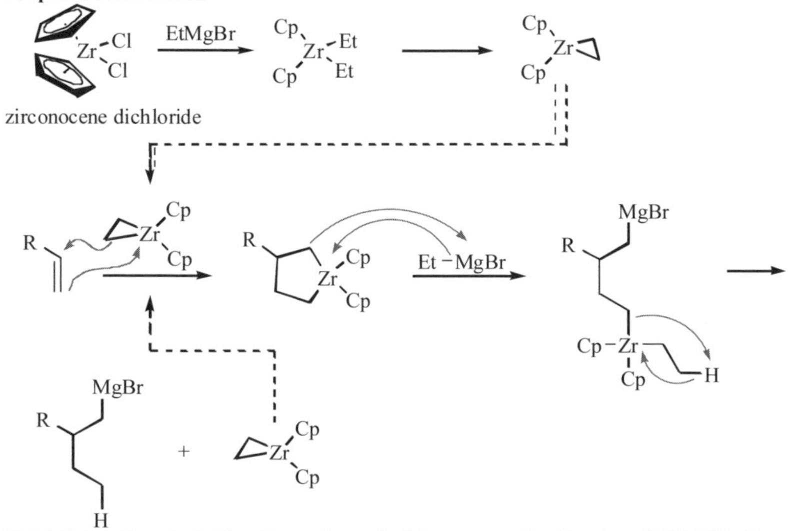

zirconocene dichloride

U. M. Dzhemilev, O. S. Vostrikova, *Journal of Organometallic Chemistry* **1985**, <u>285</u>, 43

**Example:**
Titanium Analog

1. Ti($i$-PrO)$_4$, EtMgBr
2. EtO-CHO

41%

G.-D. Tebben, K. Rauch, C. Stratmann, C. M. Williams, A. deMeijere, *Organic Letters* **2003**, <u>5</u>, 483

## *Kulinkovich Reaction*

**Examples:**

EtMgBr
Ti($i$-PrO)$_4$

70-80%

A. Esposito, M. Taddei, *Journal of Organic Chemistry* **2000**, <u>65</u>, 9245

EtMgBr
Cl-Ti(O$i$-Pr)$_3$

62%

S. Y. Cha, J. K. Cha, *Organic Letters* **2000**, <u>2</u>, 1337

## Other Examples of novel *Grignard* use:

FeCl$_3$

72%

M. Nakkamura, K. Matsuo, T. Inoue, E. Nakamura, *Organic Letters* **2003**, <u>5</u>, 1373

$t$-BuMgCl

64%

F. F. Fleming, Z. Zhang, Q. Wang, O. W. Steward, *Organic Letters* **2002**, <u>4</u>, 2493

# Grob Fragmentation

## The Reaction:

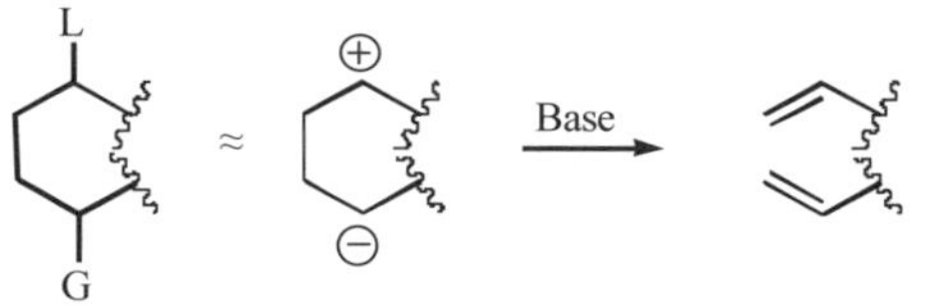

G = OH, Br, BH₂, BR₂
L = Cl, Br, OTs

$G = OH, Br, BH_2, BR_2$
$L = Cl, Br, OTs$

## Proposed Mechanism:

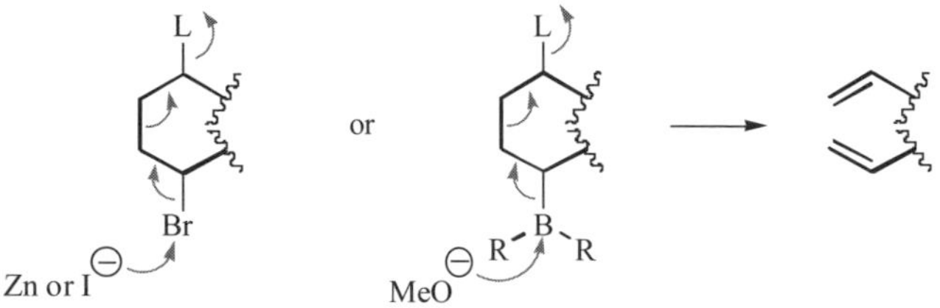

or

Zn or I⁻

MeO⁻    R–B–R

## Notes:

Proper stereoelectronic effects are required, but can lead to a variety of interesting structural products. For a somewhat dated, but readable account, see: P. Deslongchamps, *Stereoelectronic Effects in Organic Chemistry*, Pergamon Press, Oxford, England, 1983

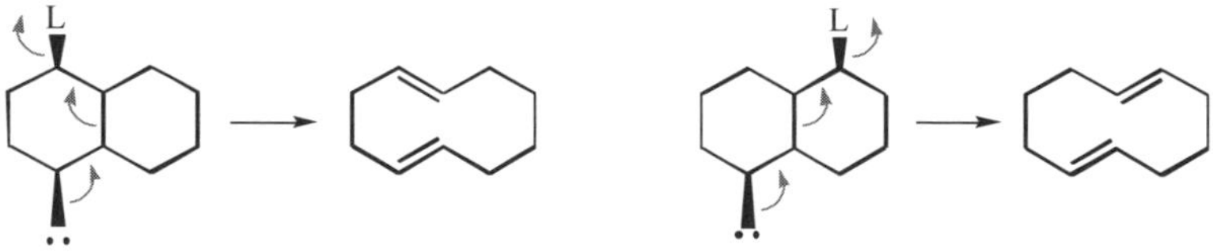

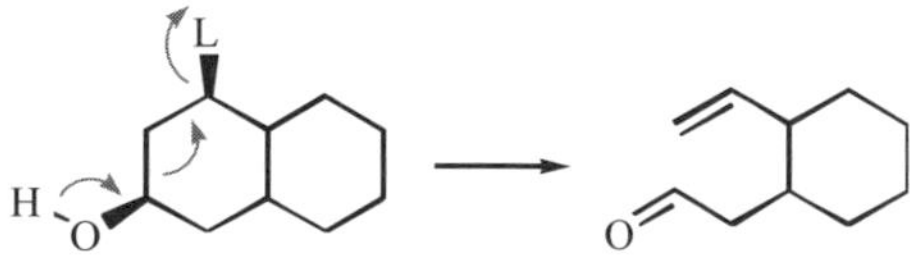

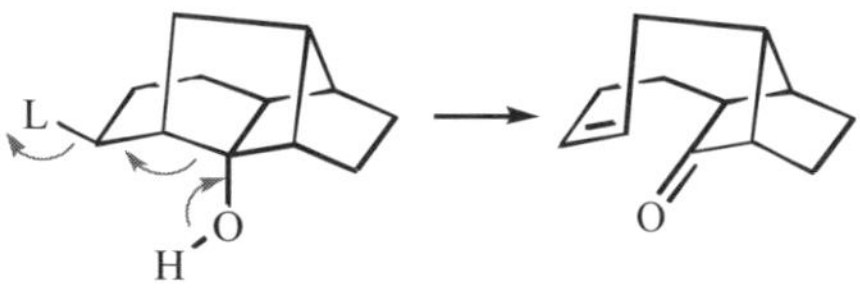

The stereoelectronic bond orientations are critical to the success of these reactions. This is easily seen in a generic *Marshall Boronate Fragmentation*.

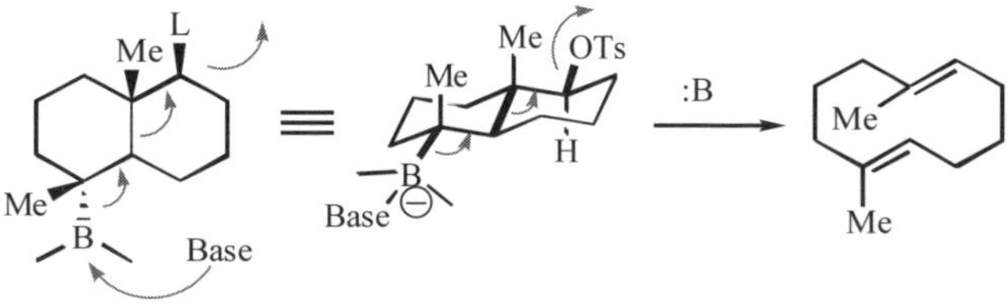

J. A. Marshall, G. L. Bundy, *Journal of the American Chemical Society* **1966**, 88, 4291

## Examples:

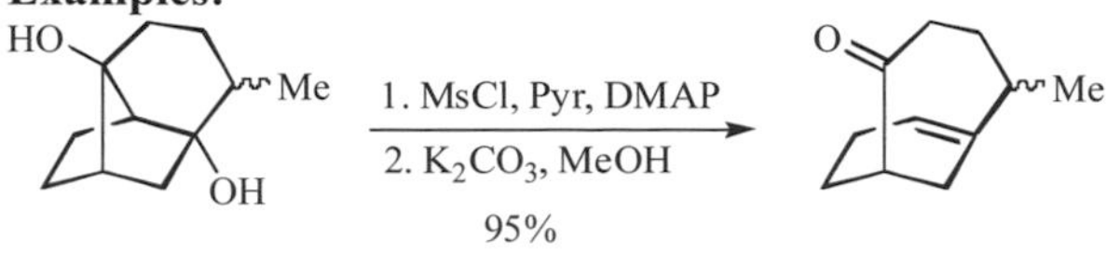

1. MsCl, Pyr, DMAP
2. K₂CO₃, MeOH
95%

J. T. Njardarson, J. L. Wood, *Organic Letters* **2001**, <u>3</u>, 2431

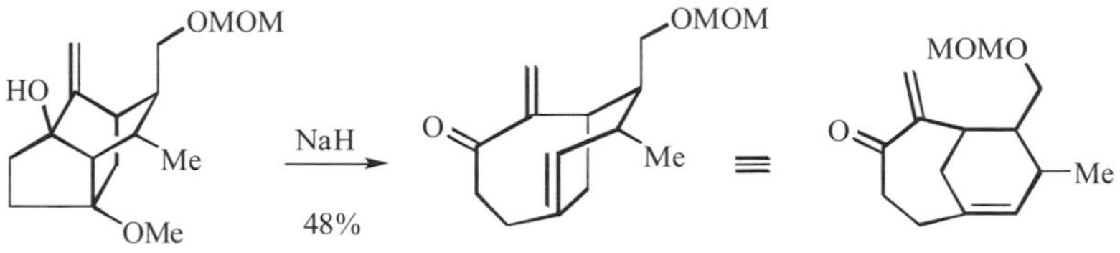

NaH
48%

T. Yoshimitsu, M. Yanagiya, J. Nagaoka, *Tetrahedron Letters* **1999**, <u>40</u>, 5215

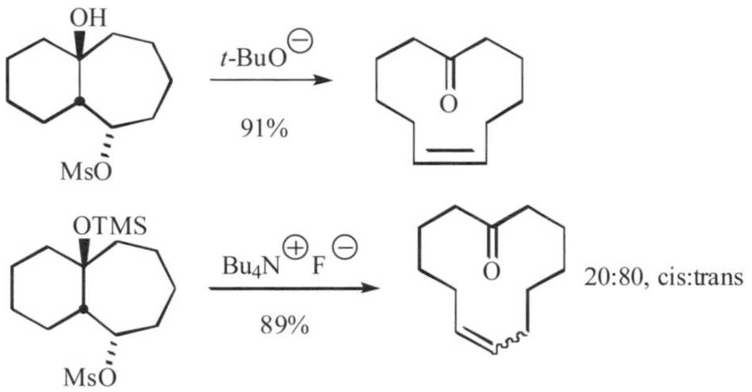

*t*-BuO⊖
91%

Bu₄N⊕ F⊖
89%

20:80, cis:trans

W. Zhang, P. Dowd, *Tetrahedron Letters* **1996**, <u>37</u>, 957

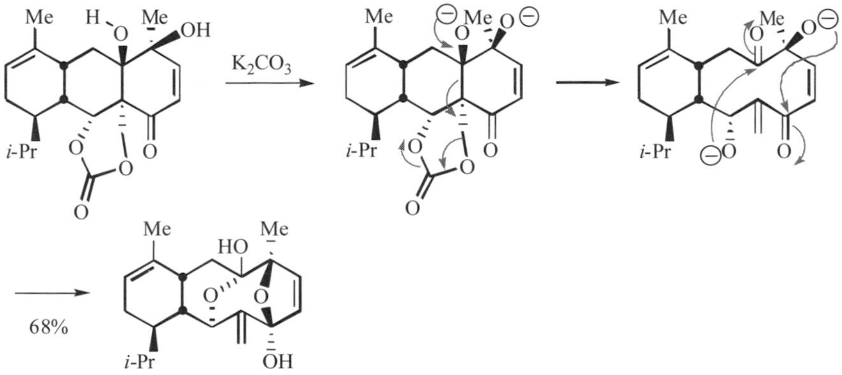

K₂CO₃

68%

J. D. Winkler, K. J. Quinn, C. H. MacKinnon, S. D. Hiscock, E. C. McLaughlin, *Organic Letters* **2003**, <u>5</u>, 1805

# Hajos-Weichert Reaction

## The Reaction:

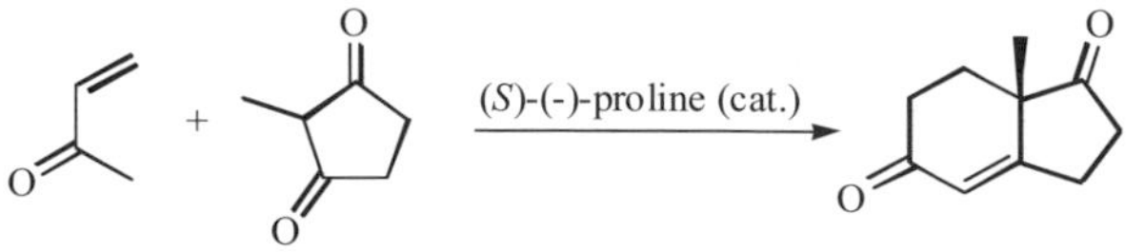

## Proposed Mechanism:

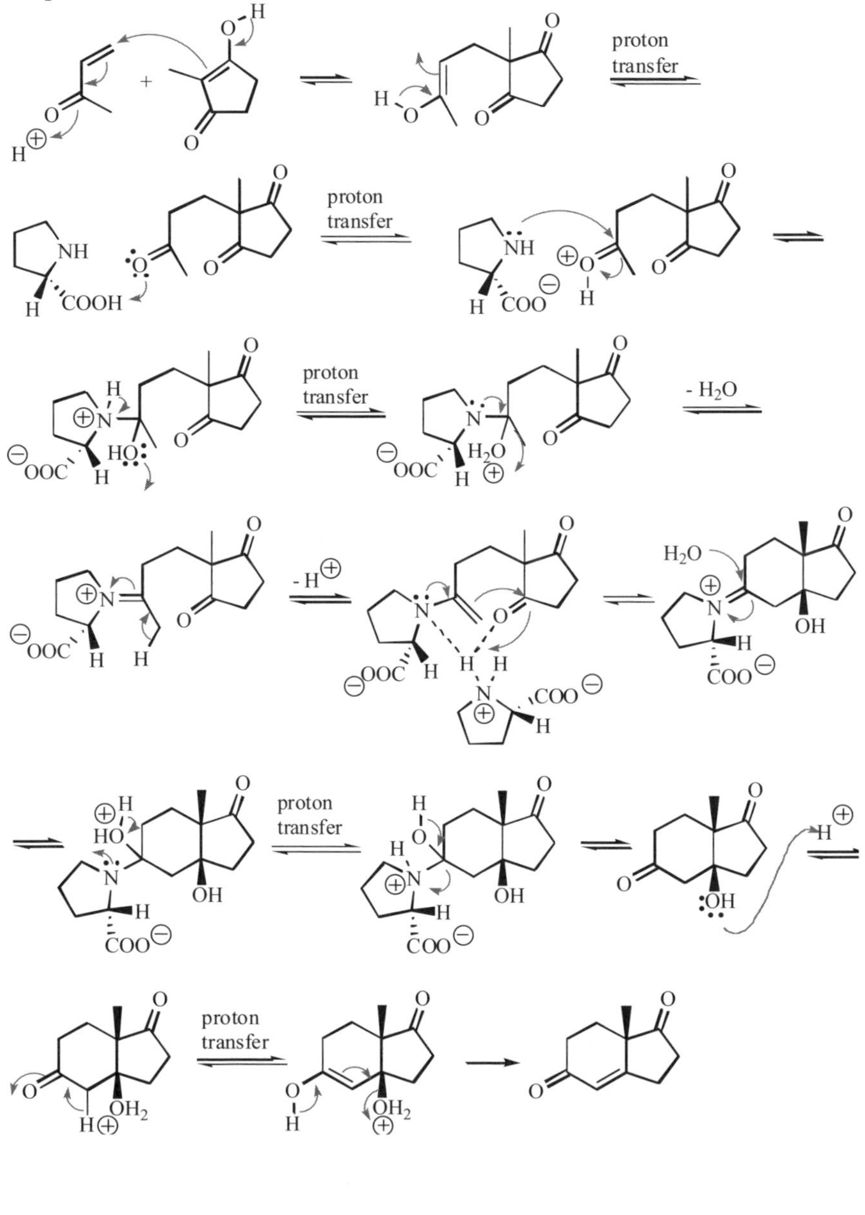

## Notes:

See: S. Wallbaum, J. Martens, *Encyclopedia of Reagents for Organic Synthesis*, John Wiley and Sons, Inc., L. A. Paquette, Ed., New York, 1995, <u>6</u>, 4301 for a discussion on proline-initiated ***aldol*** cyclizations, including chemical and optical yields.

## Examples:

The preparation of the ***Hajos-Parrish ketone:***

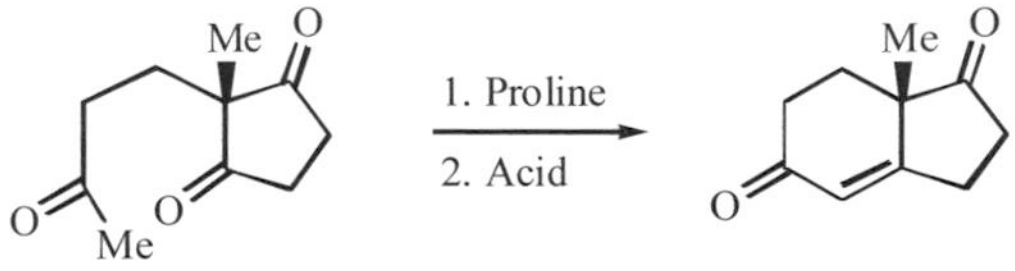

Z. G. Hajos, D. R. Parrish, *Journal of Organic Chemistry* **1974**, <u>39</u>, 1615

Phenylalanine can also be used as the source of chirality, sometimes with improved yields. See: H. Hagiwara, H. Uda, *Journal of Organic Chemistry* **1988**, <u>53</u>, 2308 and S. Takahashi, T. Oritani, K. Yamashita, *Tetrahedron* **1988**, <u>44</u>, 7081

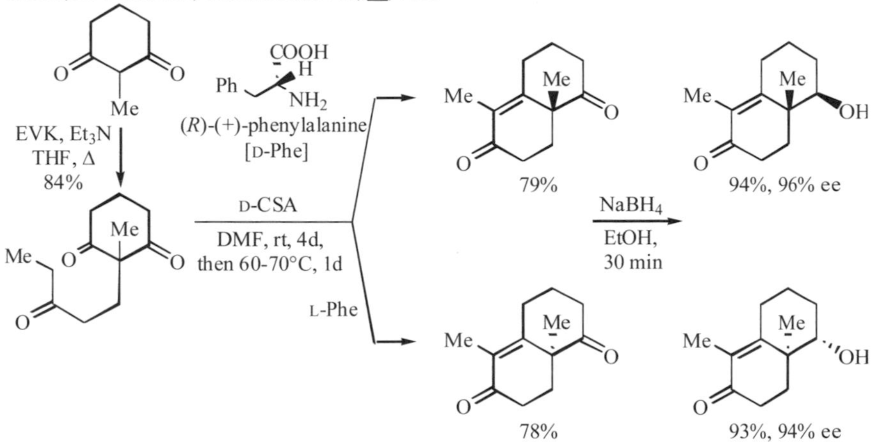

F. G. Favaloro, Jr., T. Honda, Y. Honda, G. W. Gribble, N. Suh, R. Risingsong, M. B. Sporn, *Journal of Medicinal Chemistry* **2002**, <u>45</u>, 4801

# Haller-Bauer Reaction

## The Reaction:

non-enolizable ketone,
typically = Ar

## Proposed Mechanism:

## Notes:

This is a reaction of non-enolizable ketones. Optical activity in the R- group is maintained.

See: G. Mehta, R. V. Venkateswaran, *Tetrahedron* **2000**, _56_, 1399. In this updated Review, the authors indicate the inclusion of alkoxide cleavages as *Haller-Bauer*. March describes these as "Hydro-de-Acylation" reactions. See also: M. B. Smith, J. March in *March's Advanced Organic Chemistry*, 5th ed., John Wiley and Sons, Inc., New York, 2001, p. 814; K. E. Hamlin, A. W. Weston, *Organic Reactions* **9**, 1.

L. A. Paquette, J. P. Gilday, *Organic Preparations and Procedures International* **1990**, _22_, 167

For an improved procedure see: W. Kaiser, *Synthesis* **1975**, 395

## Examples:

M = Na > M = K for reaction ee

L. A. Paquette, G. D. Maynard, C. S. Ra, M. Hoppe, *Journal of Organic Chemistry* **1989**, _54_, 1408

LiNH$_2$
NaNH$_2$
KO*t*-Bu          66                                29                          89

J. P. Gilday, J. C. Gallucci, L. A. Paquette, *Journal of Organic Chemistry* **1989**, _54_, 1399

K. Ishihara, T. Yano, *Organic Letters* **2004**, *6*, 1983

O. Arjona, R. Medel, J. Plumet, *Tetrahedron Letters* **2001**, *42*, 1287

G. Mehta, K. S. Reddy, A. C. Kunwar, *Tetrahedron Letters* **1996**, *37*, 2289

EWG = *m*-nitro     87%
      *p*-nitro     95%

N. Zhang, J. Vozzolo, *Journal of Organic Chemistry* **2002**, *67*, 1703

C. Mehta, D. S. Reddy, *Journal of the Chemical Society, Perkin Transactions 1*, **2001**, 1153

# Hass-Bender Reaction

## The Reaction:

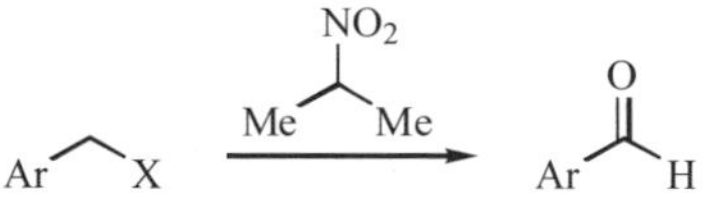

## Proposed Mechanism:

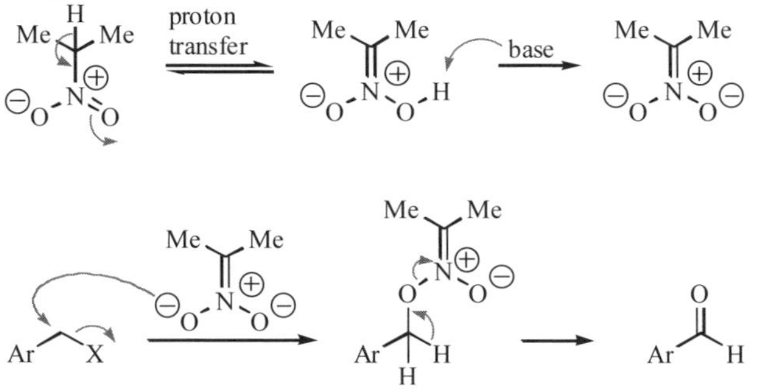

## Notes:

See related reactions; *__Aldehyde Syntheses__*

## Examples:

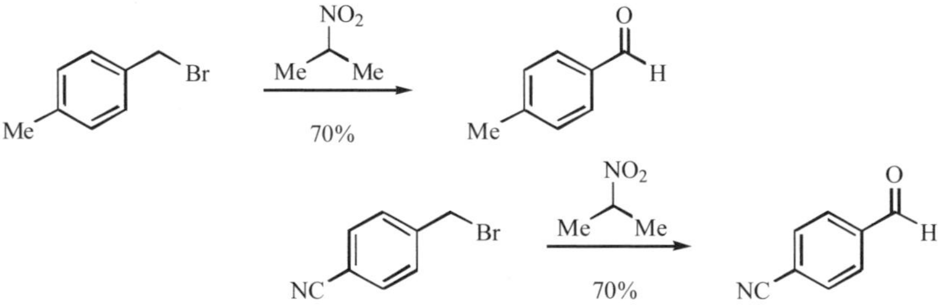

The authors state that non-aromatic ketones have been made.
H. B. Hass, M. L. Bender *Journal of the American Chemical Society* **1949**, <u>71</u>, 1767

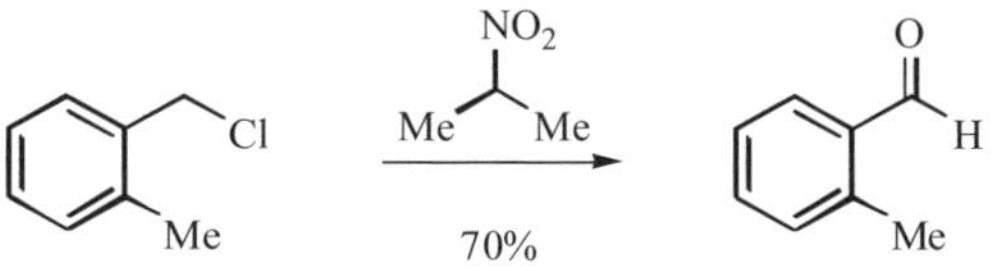

H. B. Hass M. L. Bender, *Organic Syntheses, Coll. Vol 4*, **1963**, 932

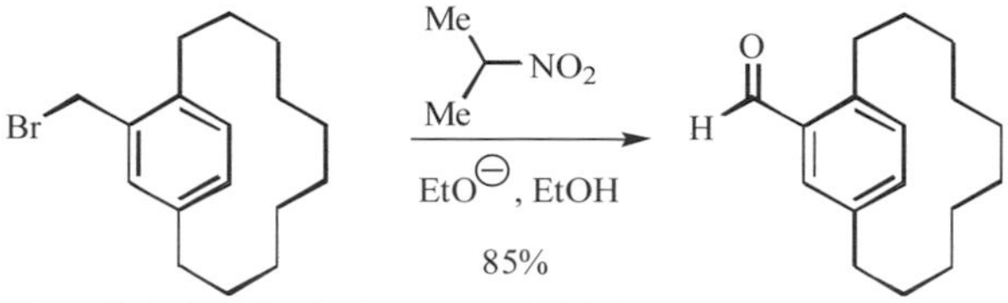

85%

Note: Only 20% by the **Sommelet Oxidation**.
A. T. Blomquist, R. E. Stahl, J. Meinwald, B. H. Smith, *Journal of Organic Chemistry* **1961**, 26, 1687

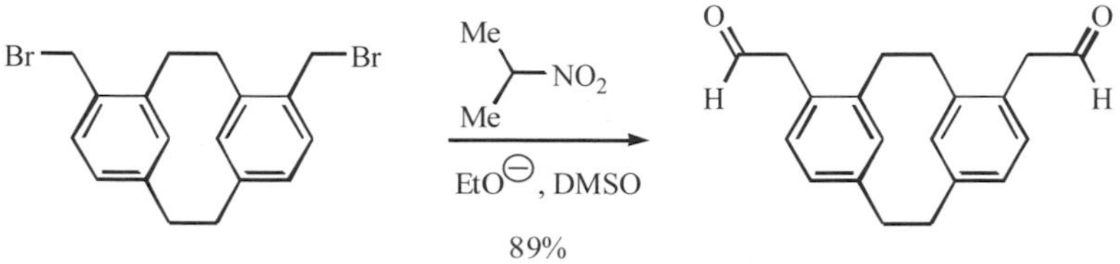

89%

Note: Only 57% by the **Sommelet Oxidation**.
S. Akabori, T. Sato, K. Hata, *Journal of Organic Chemistry* **1968**, 33, 3277

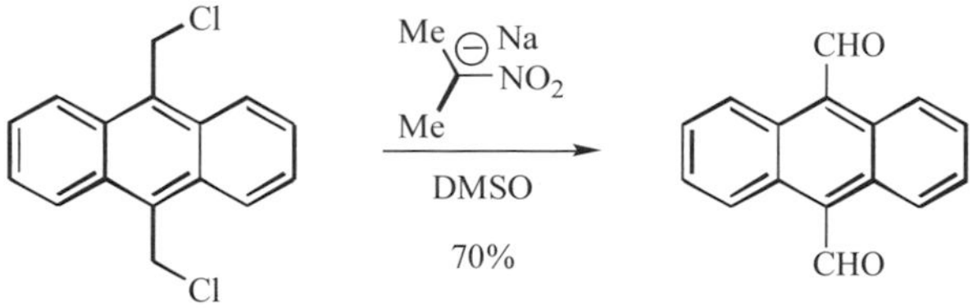

70%

Note: Without DMSO, no product is observed.
B. H. Klandermann, *Journal of Organic Chemistry* **1966**, 31, 2618

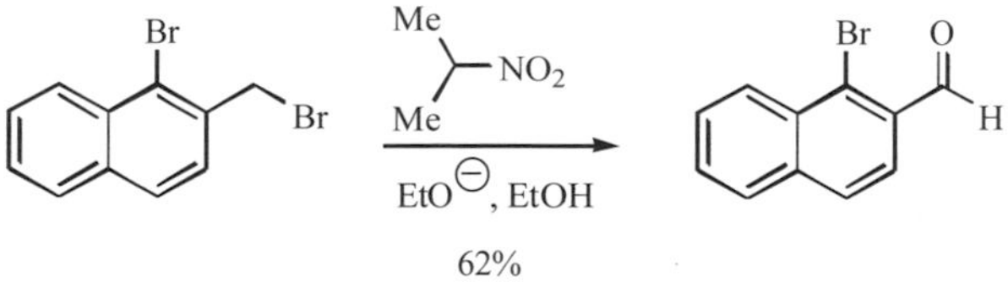

62%

J. Clayden, C. McCarthy, N. Westlund, C. S. Frampto *Journal of the Chemical Society, Perkin Transactions I* **2000**, 1363

# Heck Reaction

## The Reaction:

$$R-X \quad + \quad \overset{H}{\underset{R''}{\diagdown}}C=C\overset{R'}{\underset{R'''}{\diagup}} \quad \xrightarrow[\text{base}]{\text{Pd(0)}} \quad \overset{R}{\underset{R''}{\diagdown}}C=C\overset{R'''}{\underset{R'}{\diagup}}$$

## Proposed Mechanism:

## Notes:

T. Laue, A. Plagens, *Named Organic Reactions,* John Wiley and Sons, Inc., New York, 1998, pp. 144-147; M. B. Smith, J. March in *March's Advanced Organic Chemistry*, 5$^{th}$ ed., John Wiley and Sons, Inc., New York, 2001, p. 930; J. T. Link, *Organic Reactions* **60**, 2.


For a review on applications to Natural Product synthesis, see:
A. B. Dounay, L. E. Overman, *Chemical Reviews* **2003**, _103_, 2945

## The *Overman Spirocyclization*:

M. M. Abelman, L. E. Overman *Journal of the American Chemical Society* **1988**, _110_, 2328

## Examples:

Silver ion is sometimes added:

$$\underset{R}{\overset{Ph_3P}{\diagdown}}Pd\underset{PPh_3}{\overset{X}{\diagup}} \xrightarrow{Ag^{\oplus}} \left[\underset{R}{\overset{Ph_3P}{\diagdown}}Pd - PPh_3\right]^{\oplus} + AgX$$

The charged species is carried through the reaction until the last step when $H^{\oplus}$ is eliminated.

|  | 1 | : | 1 |
| with AgNO3 added: | 26 | : | 1 |

M. M. Abelman, T. Oh, L. E. Overman *Journal of Organic Chemistry* **1987**, <u>52</u>, 4130

64%

In this case, the silver suppresses desilylation. Without silver included, the product was obtained in 12% yield and the major product was the above right.

K. Karabelas, A. Hallber *Journal of Organic Chemistry* **1988**, <u>53</u>, 4909

C. Y. Hong, L. E. Overman, *Tetrahedron Letters* **1994**, <u>35</u>, 3453

F. E. Ziegler, U. R. Chakroborty, R. B. Weisenfield, *Tetrahedron* **1981**, <u>37</u>, 4035

# Hell-Volhard-Zelinski Reaction

## The Reaction:

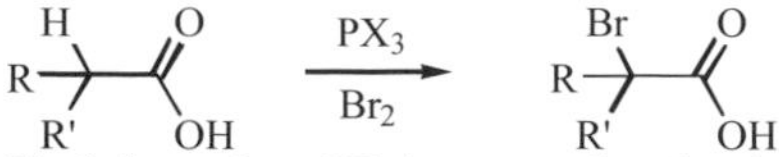

The halogen from $PX_3$ is not transferred to the alpha position.

## Proposed Mechanism:

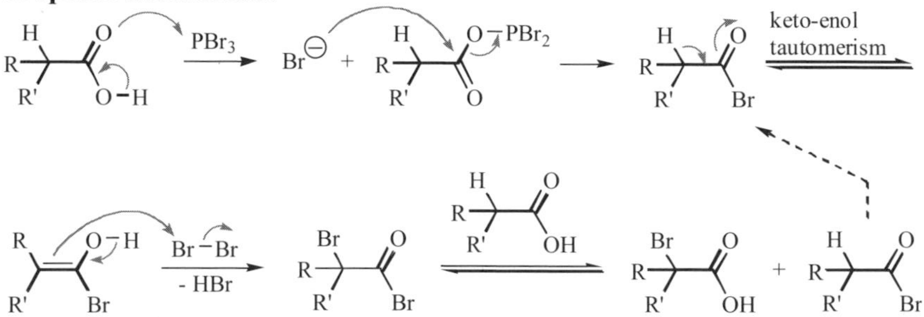

## Notes:

M. B. Smith, J. March in *March's Advanced Organic Chemistry*, 5$^{th}$ ed., John Wiley and Sons, Inc., New York, 2001, p. 777; T. Laue, A. Plagens, *Named Organic Reactions*, John Wiley and Sons, Inc., New York, 1998, pp. 147-149. These authors note that α-F or α-I cannot be obtained by this method,

Some describe:

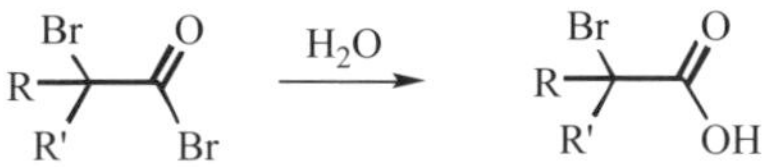

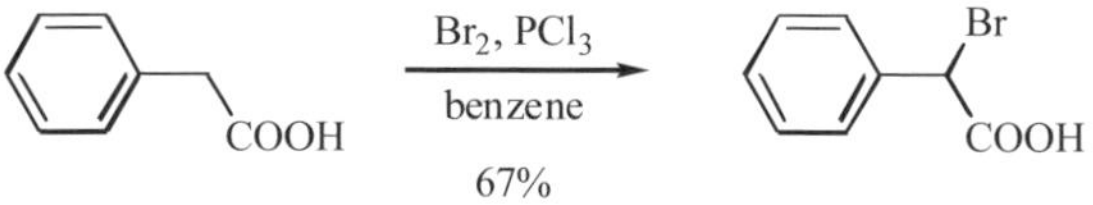

There is no water in the experimental.

L. A. Carpino, L. V. McAdams, III, *Organic Syntheses*, <u>CV6</u>, 403

**Examples:**

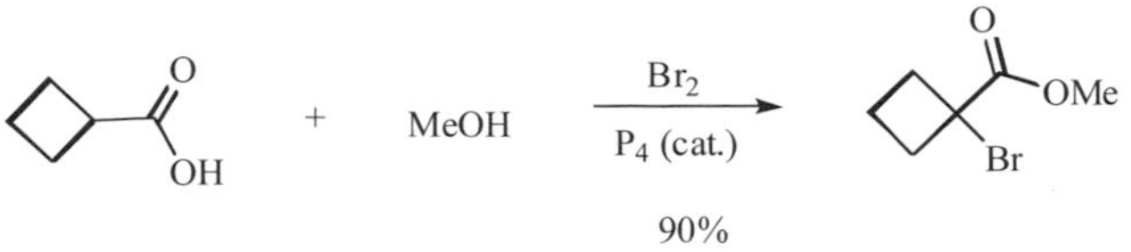

90%

K. Estieu, J. Ollivier, J. Salauen, *Tetrahedron Letters* **1996**, <u>37</u>, 623

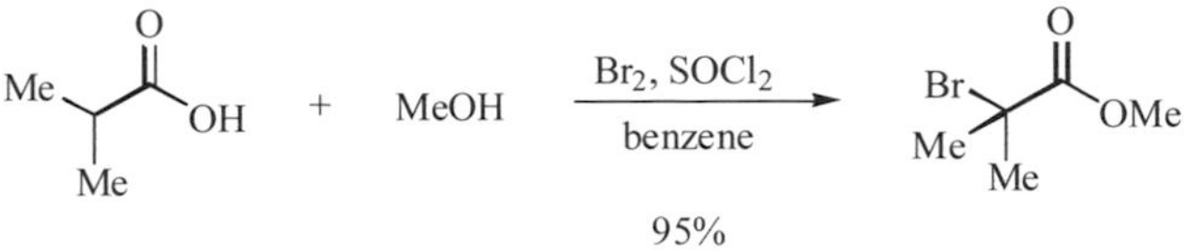

95%

Y. Ogata. T. Sugimaot, *Journal of Organic Chemistry* **1978**, <u>43</u>, 3684

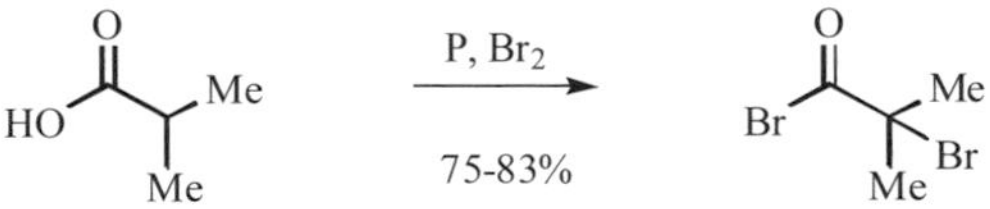

75-83%

C. W. Smith, D. G. Norton, *Organic Syntheses* <u>CV4</u>, 348

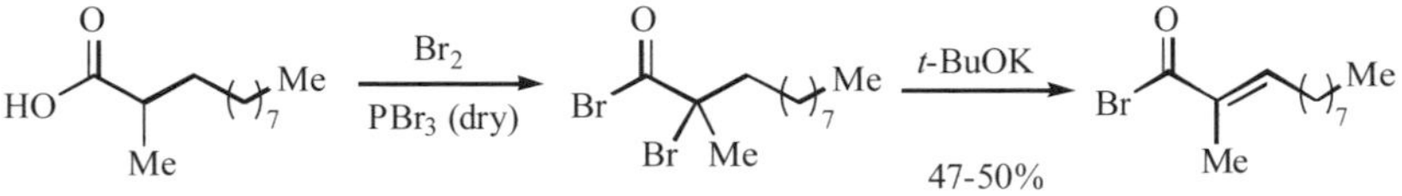

47-50%

C. F. Allen, M. J. Kalm, *Organic Syntheses* <u>CV4</u>, 398

# Henry Reaction / Kamlet Reaction

## The Reaction:

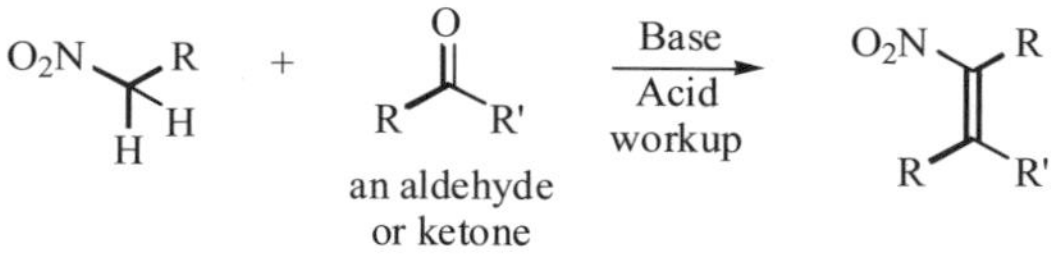

an aldehyde
or ketone

## Proposed Mechanism:

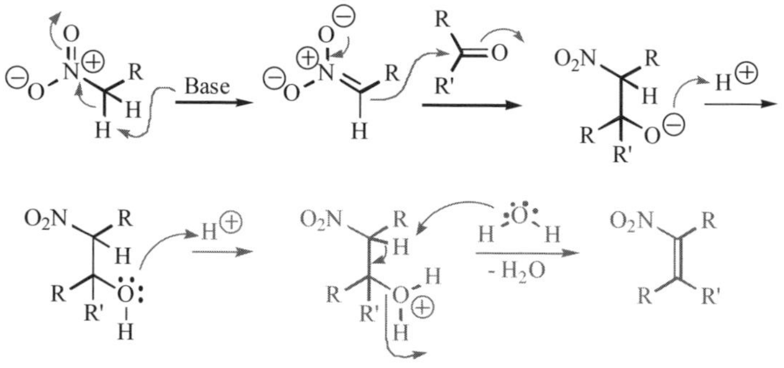

## Notes:

The reaction is mechanistically similar to the *Aldol reaction*.

See: V. K. Ahluwalia, R.K. Parashar, *Organic Reaction Mechanisms*, Alpha Science International Ltd., Pangbourne, U.K., 2002, p. 329

## Examples:

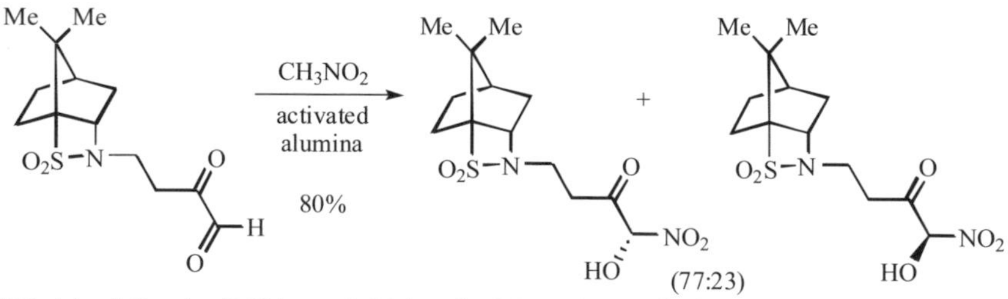

$CH_3NO_2$
activated alumina

80%

(77:23)

I.Kudyba, J. Raczko, Z. Urbanczyk-Lipkowska J. Jurczak, *Tetrahedron* **2004**, <u>60</u> 4807

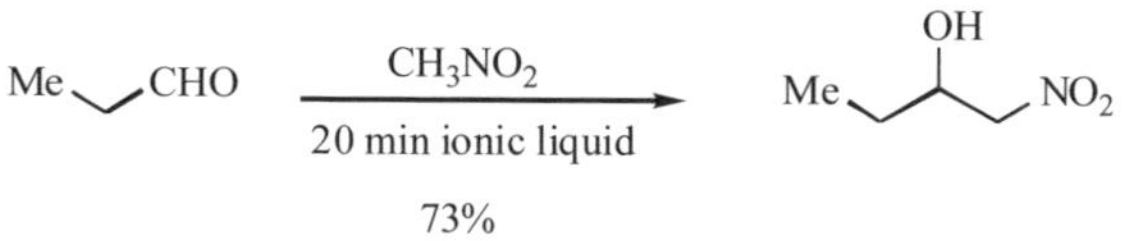

$Me$ $CHO$   $\xrightarrow[\text{20 min ionic liquid}]{CH_3NO_2}$

73%

T. Jiang, H. Gao, B. Han, G. Zhao, Y. Chang, W. Wu, L. Gao, G. Yang, *Tetrahedron Letters* **2004**, <u>45</u>, 2699

Y.-W. Zhong, P. Tian, G.-Q. Lin, *Tetrahedron: Asymmetry* **2004**, <u>15</u>, 771

## *Aza-Henry Reaction*:

T. Okino, S. Nakamura, T. Furukawa, Y. Takemoto, *Organic Letters* **2004**, <u>6</u>, 625

R. G. Soengas, J. C. Estevez, R. J. Estevez, *Organic Letters* **2003**, <u>5</u>, 4457

# Heterocyclic Syntheses

## Pyridine Syntheses

### *Chichibabin Pyridine Synthesis*

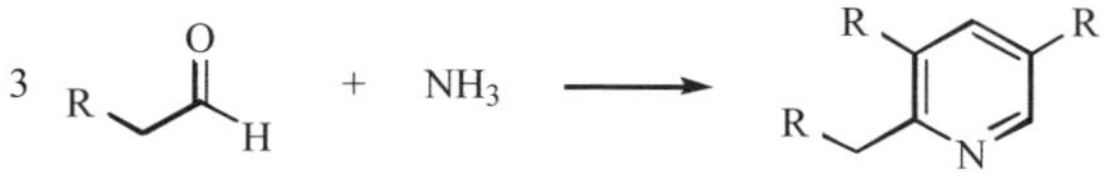

$$3 \quad R\overset{O}{\diagup}\diagdown_H \quad + \quad NH_3 \quad \longrightarrow$$

**Proposed Mechanism:**

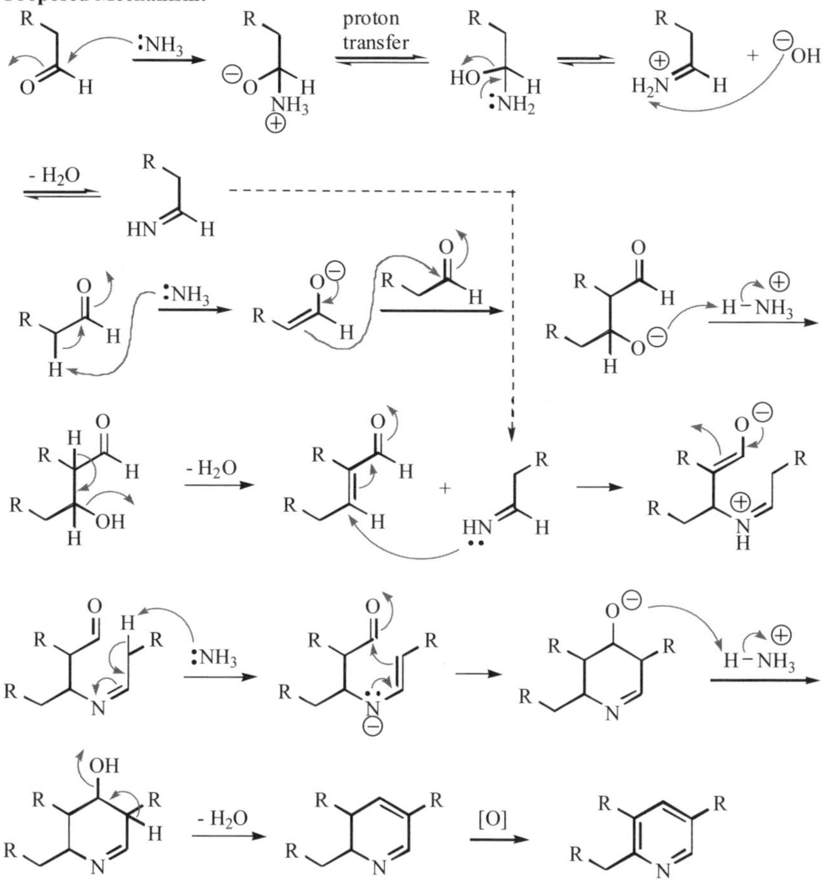

### *Guareschi-Thorpe Condensation*

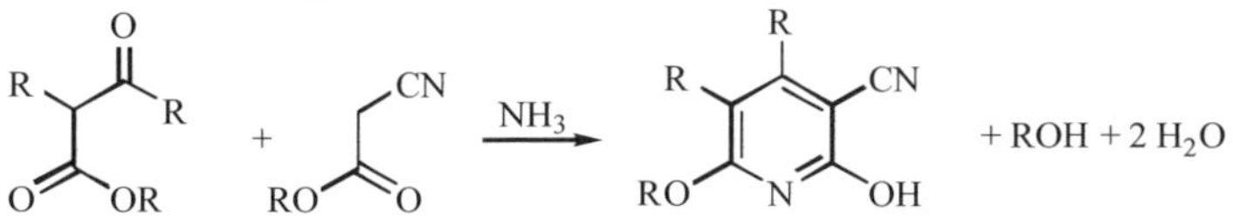

$$+\ ROH + 2\ H_2O$$

## *Hantzsch (Dihydro) Pyridine Synthesis*

**Proposed Mechanism:**

## Krohnke Pyridine Synthesis

**Proposed Mechanism:**

$$NH_4OAc \rightleftharpoons HOAc + NH_3$$

# Furan, Pyrrole, Thiophene Syntheses

## Furan Syntheses

### Feist-Benary Furan Synthesis

## Paal-Knorr Furan Synthesis

or other Lewis Acids eg $P_2O_5$

## Garst-Spencer Furan Annulation

## Pyrrole Syntheses

## Barton-Zard Pyrrole Synthesis

## Hantzsch Pyrrole Synthesis

## Knorr Pyrrole Synthesis

$$+ \; 2 \, H_2O$$

## Paal-Knorr Pyrrole Synthesis

## Piloty-Robinson Synthesis

$$+ \; NH_3$$

# Thiophene Syntheses

## Fiesselmann Thiophene Synthesis

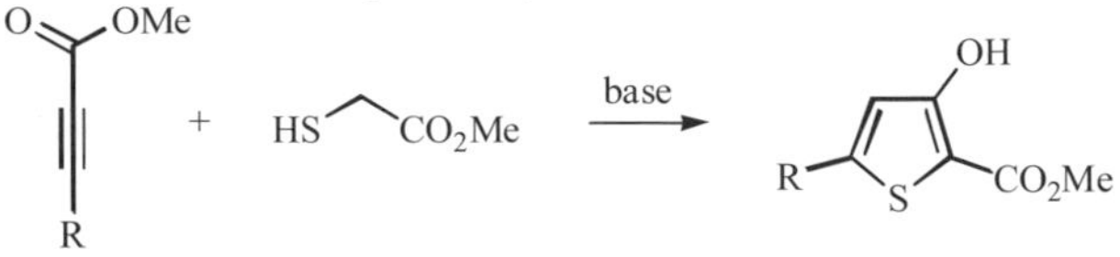

**Proposed Mechanism:**

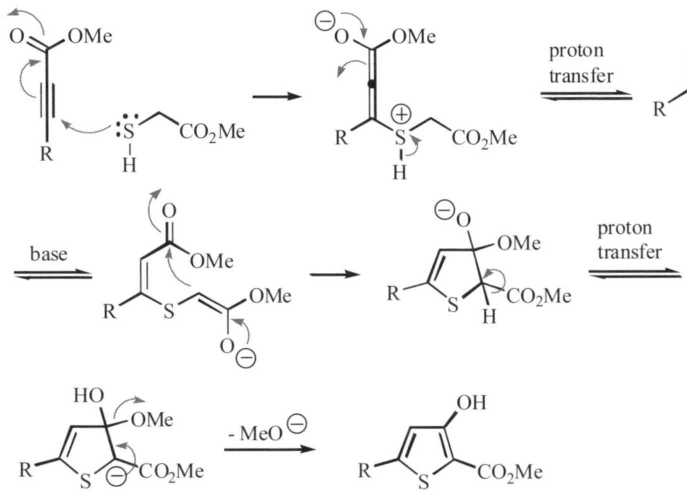

## Gewald Aminothiophene Synthesis

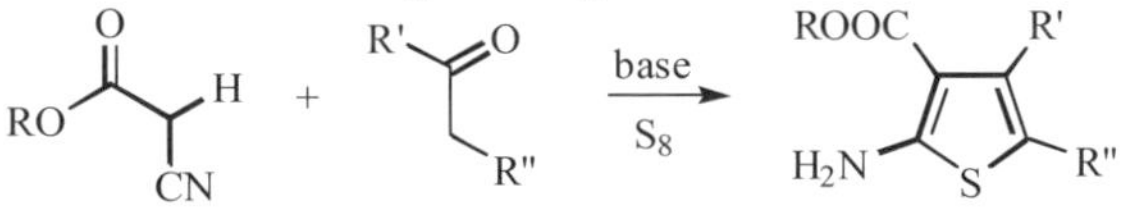

**Proposed Mechanism:**

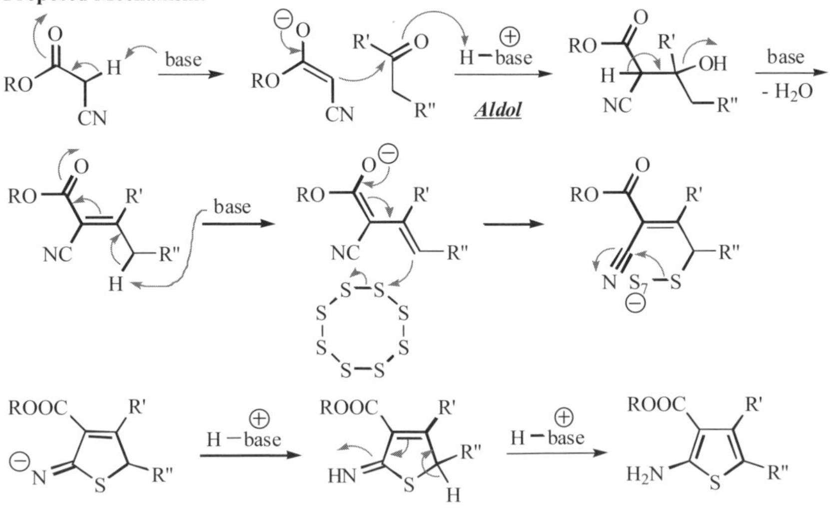

## *Hinsberg Synthesis of Thiophene Derivatives*

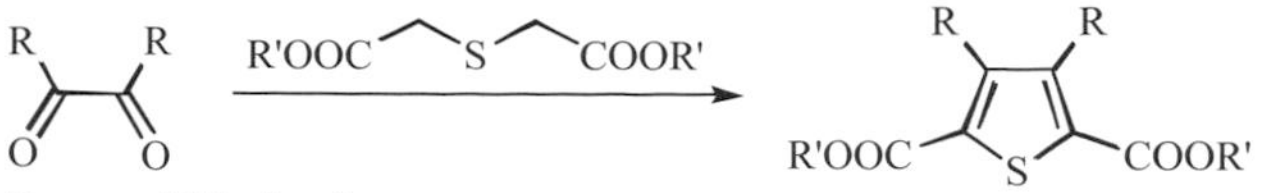

**Proposed Mechanism:**

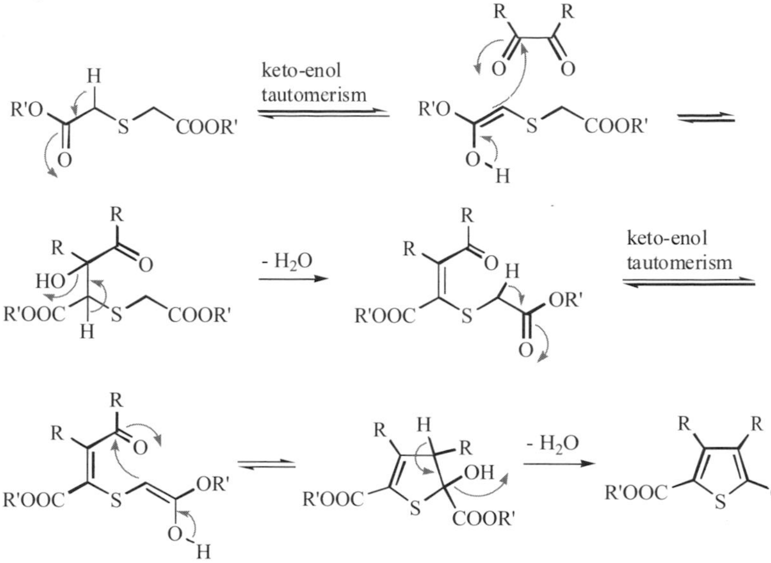

## *Vollhard-Erdmann Cyclization*

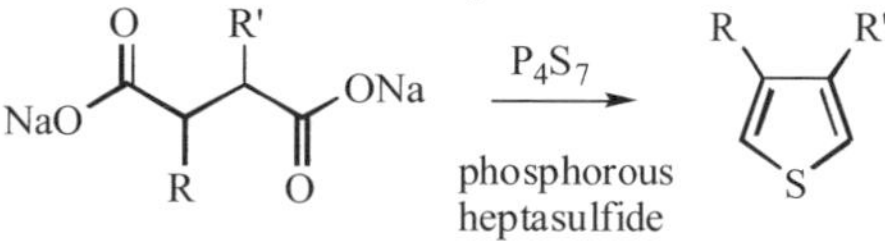

phosphorous
heptasulfide

# Indole Syntheses

## *Baeyer-Drewson Indigo Synthesis*

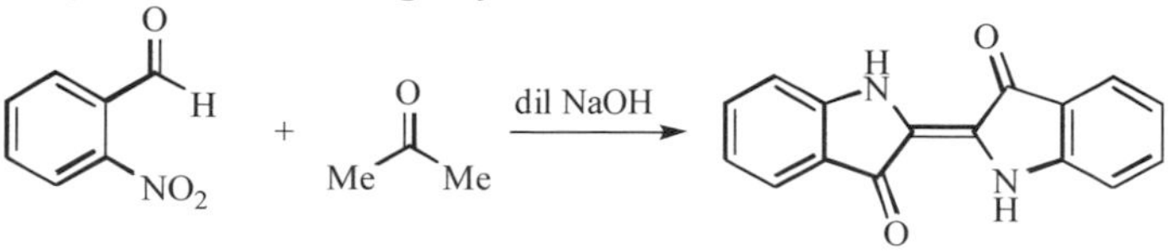

## <u>*Bartoli Indole Synthesis*</u>

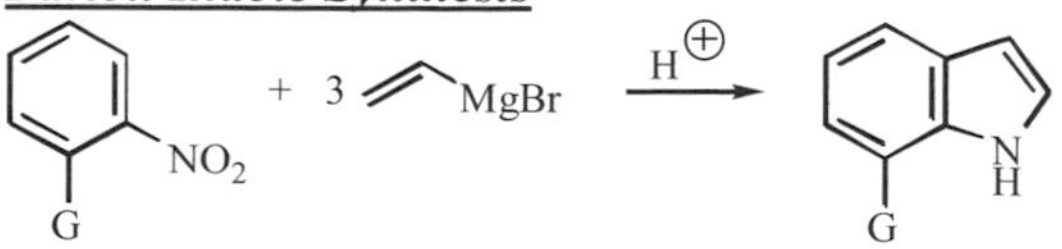

## ***Bischler-Möhlau Indole Synthesis***

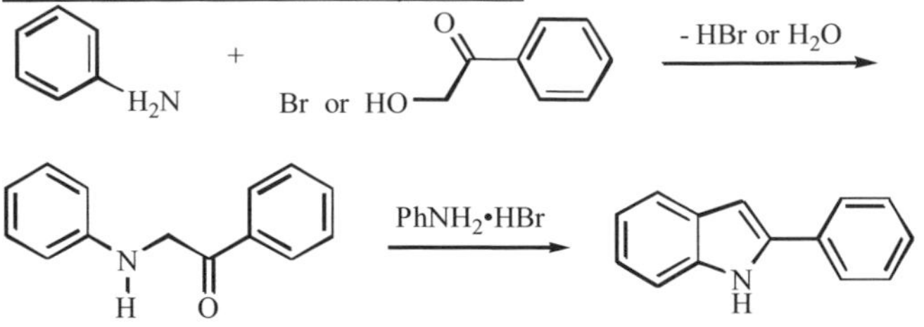
H2N
Br or HO
O
- HBr or H2O
N
H
O
PhNH2•HBr
N
H

## ***Fischer Indole Synthesis***

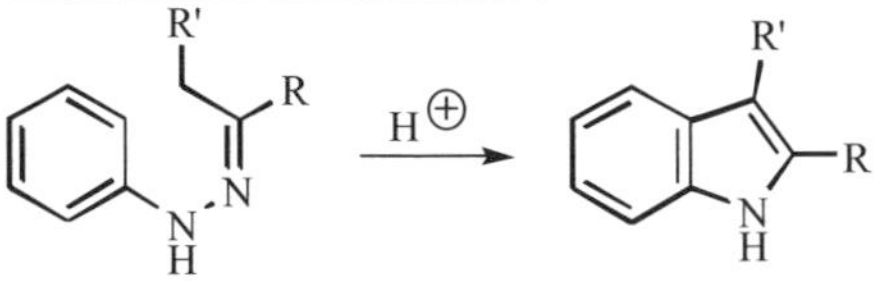
R'
R
N
N
H
H⊕
R'
R
N
H

## ***Furstner Indole Synthesis***

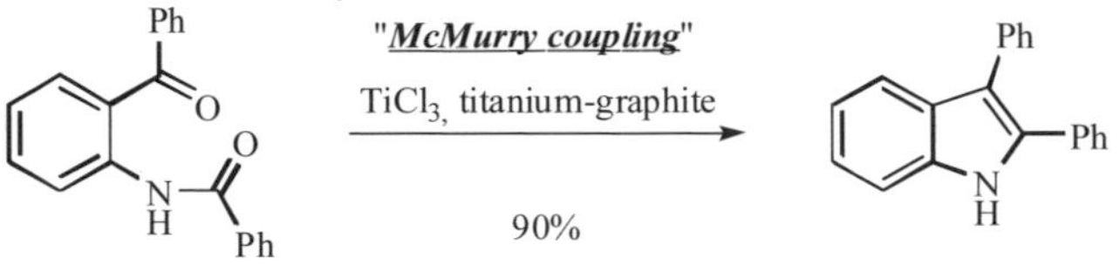
Ph
O
O
N
H
Ph
"McMurry coupling"
TiCl3, titanium-graphite
90%
Ph
Ph
N
H

## ***Gassman Indole Synthesis***

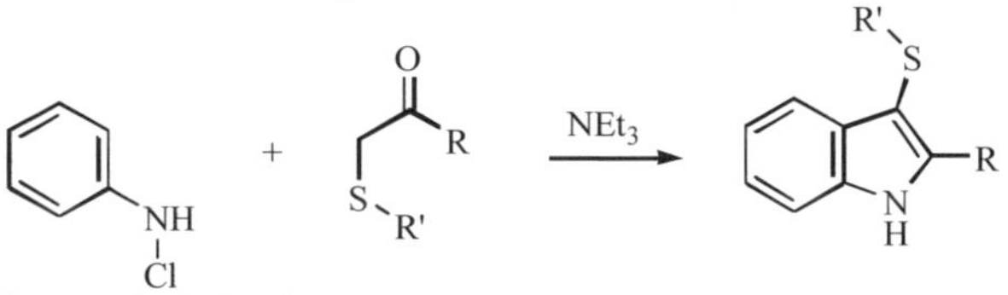
NH
Cl
+
O
R
S
R'
NEt3
R'
S
R
N
H

**Proposed Mechanism:**

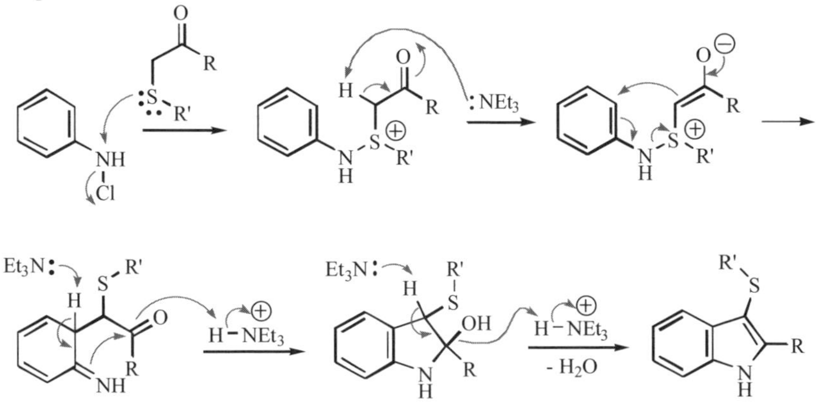
O
R
S
R'
NH
Cl
H
O
R
N
H
S
R'
: NEt3
O
R
N
H
S
R'
Et3N:
S
R'
H
O
R
NH
H−NEt3
Et3N:
R'
H
S
OH
N
H
R
H−NEt3
- H2O
R'
S
R
N
H

## ***Hegedus Indole Synthesis***

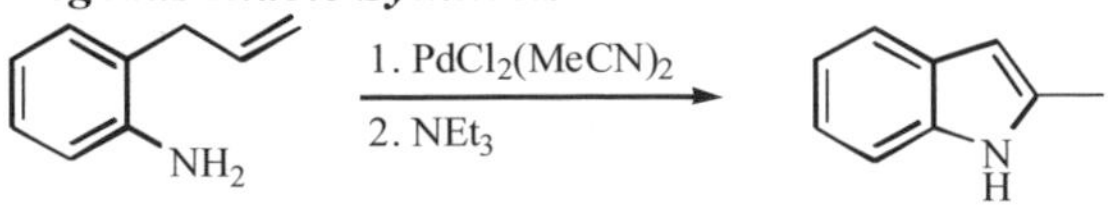
NH2
1. PdCl2(MeCN)2
2. NEt3
N
H

**Proposed Mechanism:**

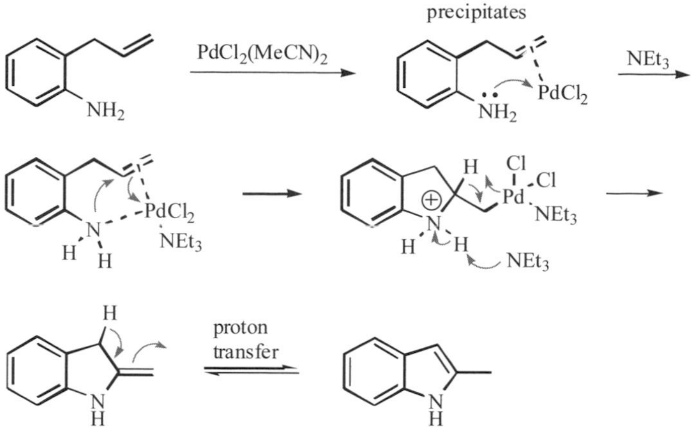
precipitates
PdCl2(MeCN)2
NEt3
PdCl2
NH2
NEt3
proton
transfer
N
H

## *Larock Indole Synthesis*

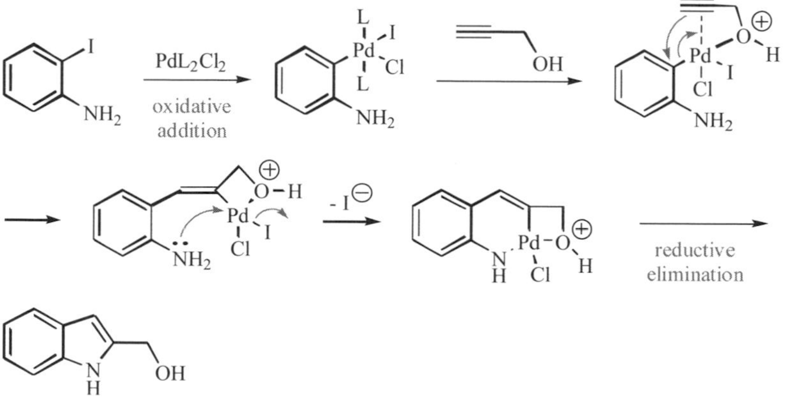
PdL2Cl2
oxidative
addition
OH
NH2
-I
reductive
elimination

## *Leimgruber-Batcho Indole Synthesis*

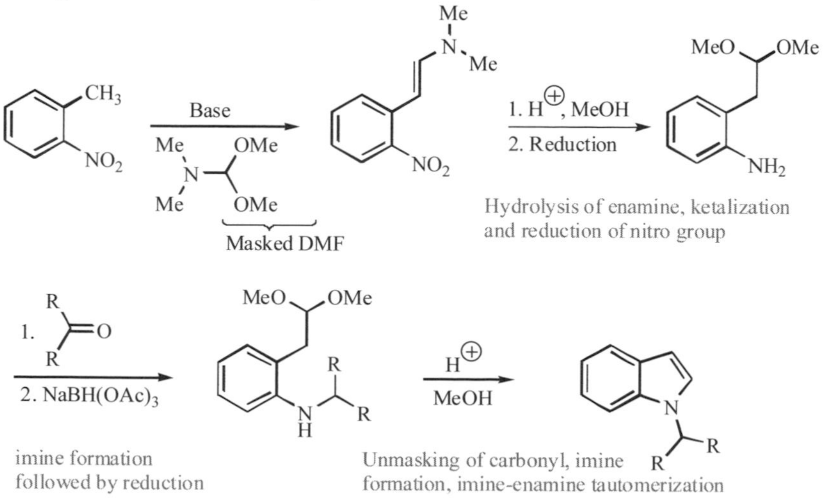
Base
Masked DMF
1. H+, MeOH
2. Reduction
Hydrolysis of enamine, ketalization
and reduction of nitro group
2. NaBH(OAc)3
imine formation
followed by reduction
MeOH
Unmasking of carbonyl, imine
formation, imine-enamine tautomerization

## Madelung Indole Synthesis

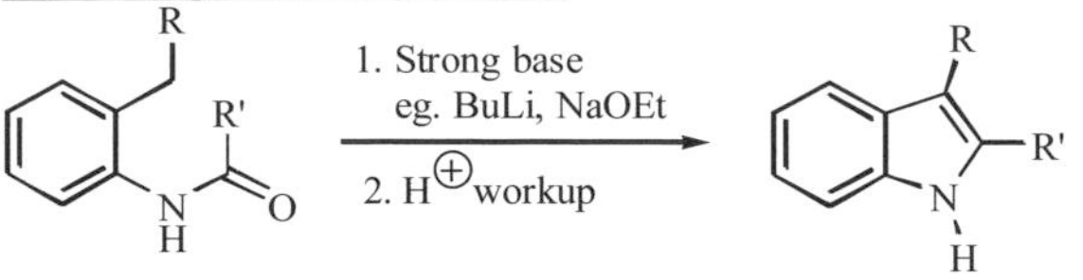

R
R'
N
H
O
1. Strong base
eg. BuLi, NaOEt
2. H⊕ workup
R
R'
N
H

## Mori-Ban Indole Synthesis

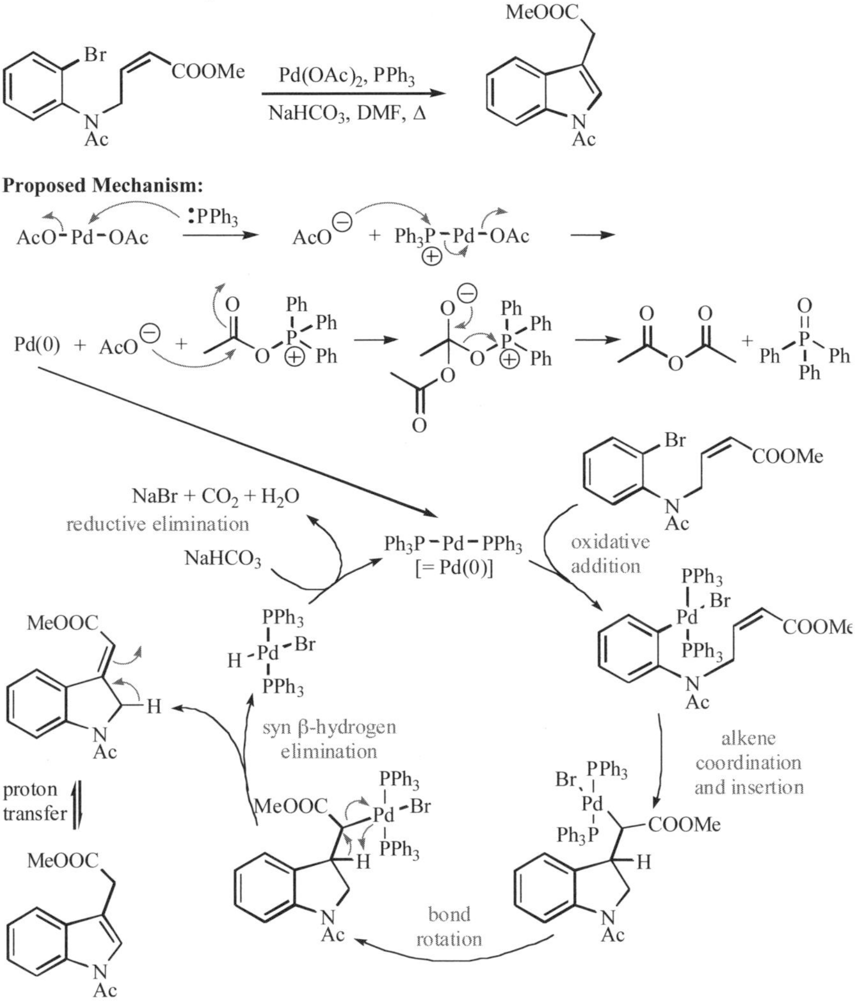

Br
COOMe
N
Ac
Pd(OAc)2, PPh3
NaHCO3, DMF, Δ
MeOOC
N
Ac
Proposed Mechanism:
AcO-Pd-OAc
:PPh3
AcO⊖ + Ph3P-Pd-OAc
Pd(0) + AcO⊖ +
O
Ph
Ph
Ph
NaBr + CO2 + H2O
reductive elimination
NaHCO3
Ph3P-Pd-PPh3
[= Pd(0)]
oxidative addition
PPh3
Br
Pd
PPh3
COOMe
alkene coordination and insertion
MeOOC
H
proton transfer
syn β-hydrogen elimination
H-Pd-Br
bond rotation
Ph3P
COOMe
Ac

## Nenitzescu Indole Synthesis

O
O
+
COOR
HN
R
HO
COOR
N
R

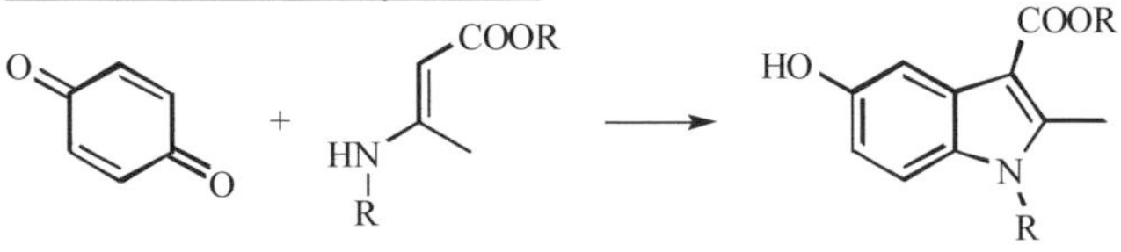

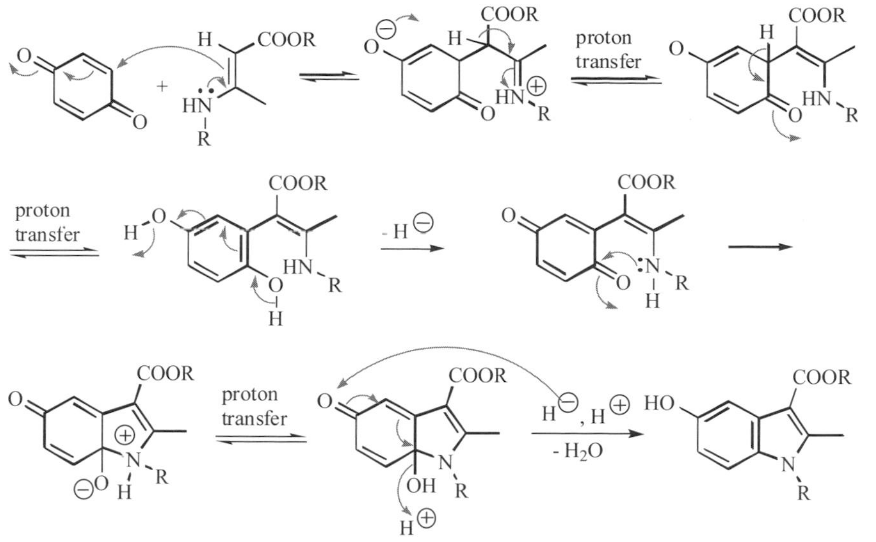


### ***Reissert Indole Synthesis***

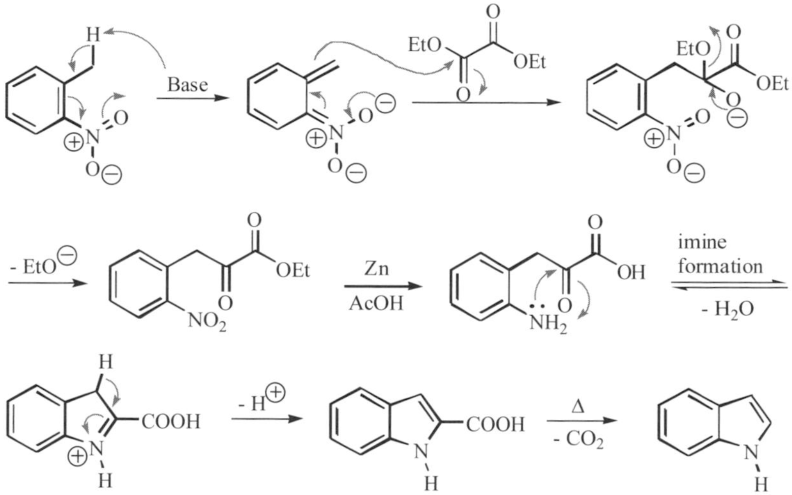


### ***Smith Indole Synthesis***

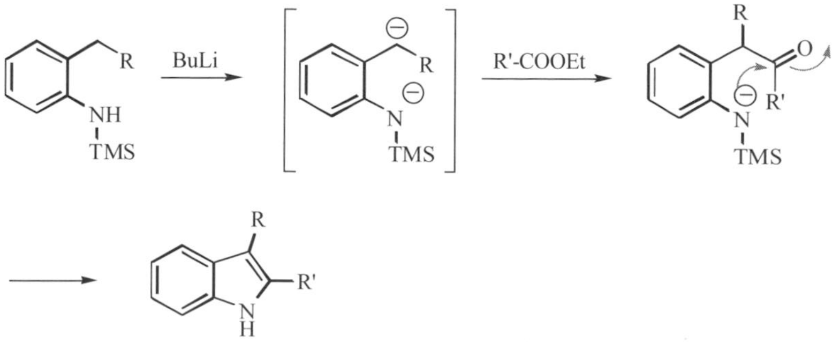

## *Wittig Indole Synthesis*

## Quinoline-Isoquinoline Syntheses

### *Bischler-Napieralski Reaction*

### *Camps Quinoline Synthesis*

The identity of R influences the product distribution.

### *Combes Quinoline Synthesis*

### *Conrad-Limpach Reaction*

### *Doebner Reaction*

### *Doebner-von Miller Reaction*

## *Friedlander Quinoline Synthesis*

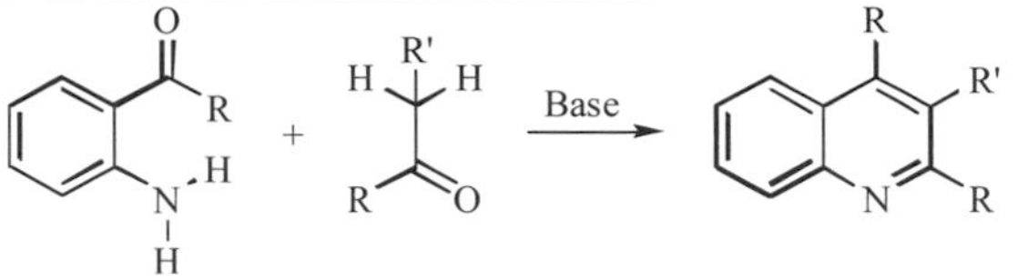

## *Gabriel-Colman Rearrangement*

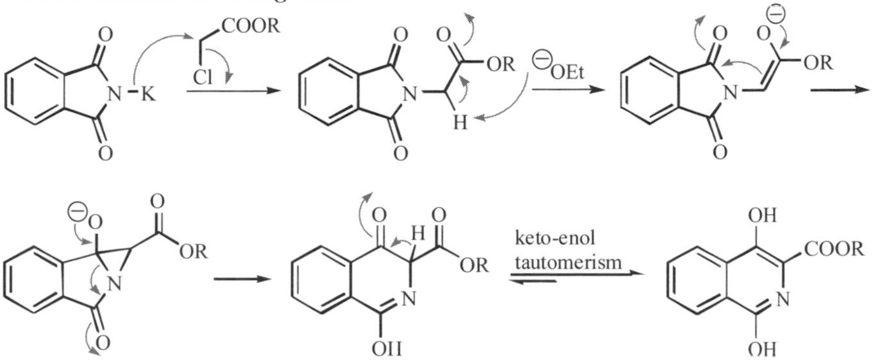

keto-enol
tautomerism

## *Gould-Jacobs Reaction*

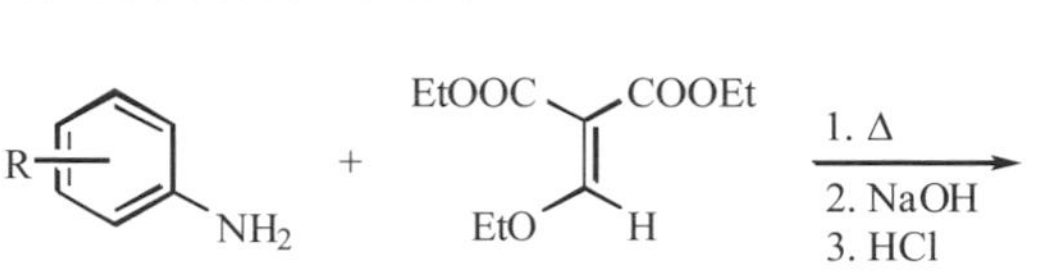

1. Δ
2. NaOH
3. HCl

## *Knorr Quinoline Synthesis*

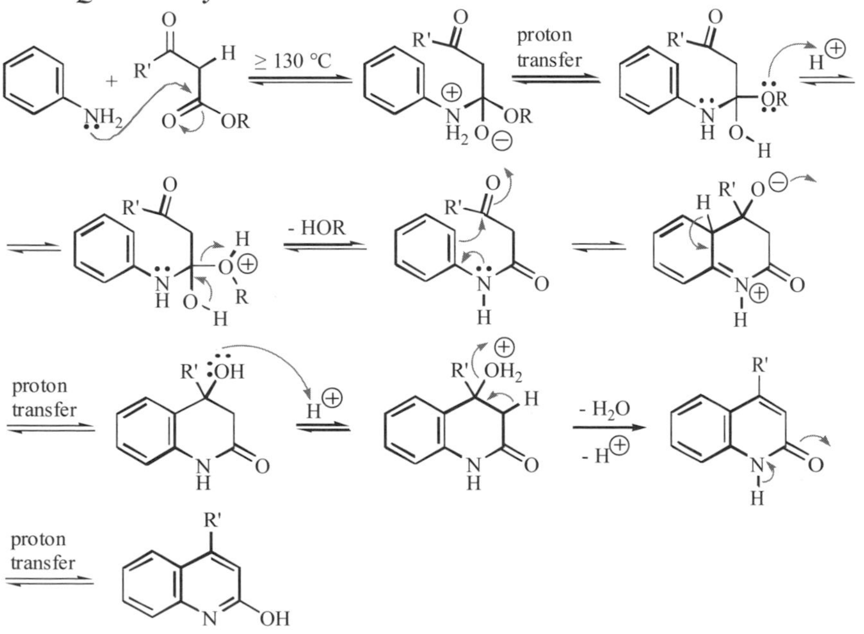

≥ 130 °C

proton
transfer

- HOR

proton
transfer

- H₂O
- H⁺

proton
transfer

## Niementowski Quinoline Synthesis

## Pictet-Gams Isoquinoline Synthesis

## Pictet-Spengler Isoquinoline Synthesis

## Pomeranz-Fritsch Reaction

## Riehm Quinoline Synthesis

## Schlittler-Müller Reaction (Modification of the Pomeranz-Fritsch Reaction)

## Skraup Reaction

# Cinnoline Syntheses

## *Borsche Cinnoline Synthesis*

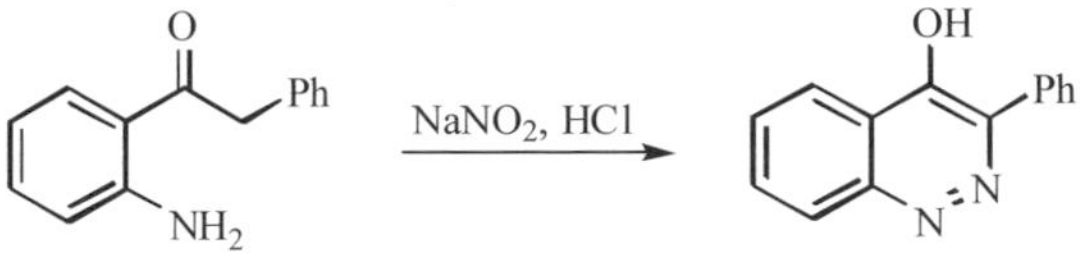

**Proposed Mechanism:**

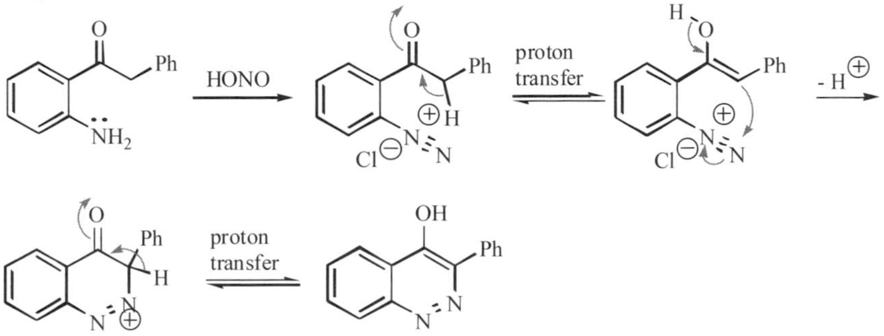

## *von Richter (Cinnoline) Synthesis*

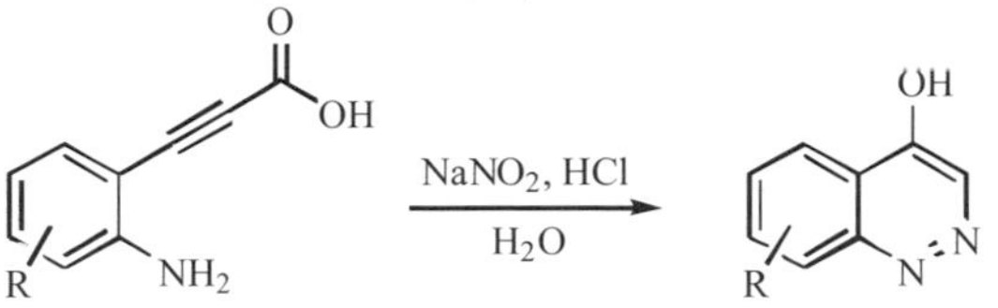

**Proposed Mechanism:**

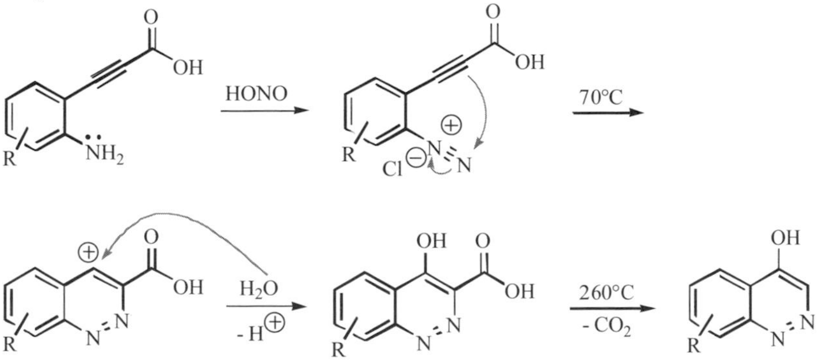

## *Widman-Stoermer (Cinnoline) Synthesis*

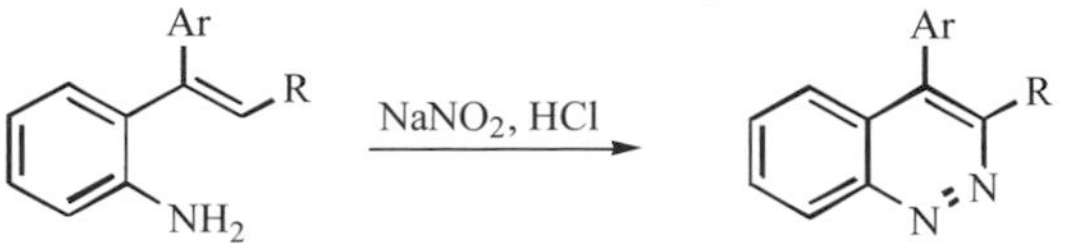

**Proposed Mechanism:**

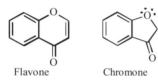

# Flavone-Chromone Syntheses

## *General Structures*

Flavone          Chromone

## *Algar-Flynn-Oyamada Reaction*

$$\xrightarrow[\text{HO}^{\ominus}]{\text{H}_2\text{O}_2}$$

*ortho*-hydroxyphenyl styryl ketones              flavonols

## *Allan-Robinson Reaction*

Base

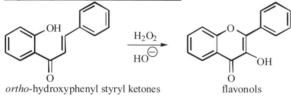

*Kostanecki Acylation* product, a coumarin

*Allan-Robinson Reaction* product, a chromone

## *Auwers Synthesis*

## *Pechmann Condensation*

## *Simonis Chromone Cyclization*

## *Wessely-Moser Rearrangement*

## *Perkin Rearrangement (Coumarin-Benzofuran Ring Contraction)*

# Miscellaneous

### Hoch-Campbell Aziridine Synthesis

### Brackeen Imidazole Synthesis

### Bredereck Imidazole Synthesis

### Knorr Pyrazole Synthesis

### Pechmann Pyrazole Synthesis

### Einhorn-Brunner Reaction

### Pellizzari Reaction

## Bernthsen Acridine Synthesis

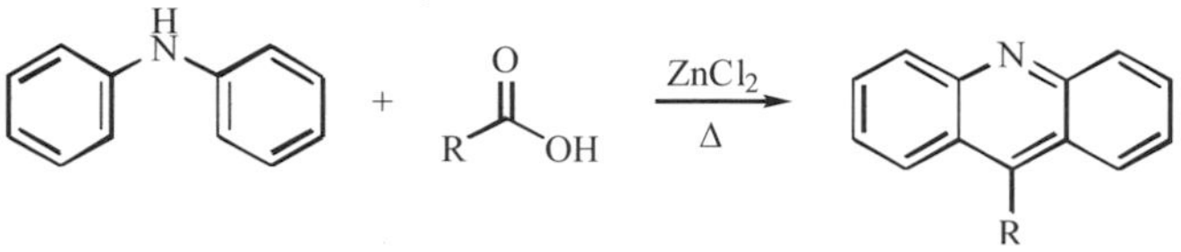

## Bischler Quinoxaline Synthesis

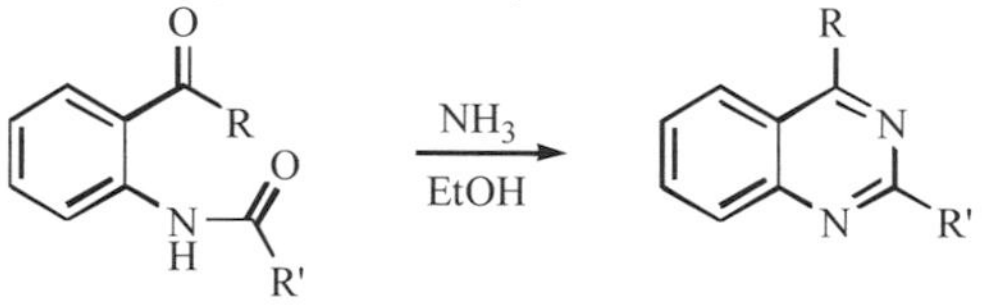

## Gutknecht Pyrazine Synthesis

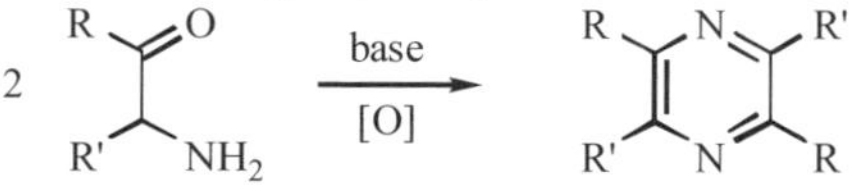

## Pinner Triazine Synthesis

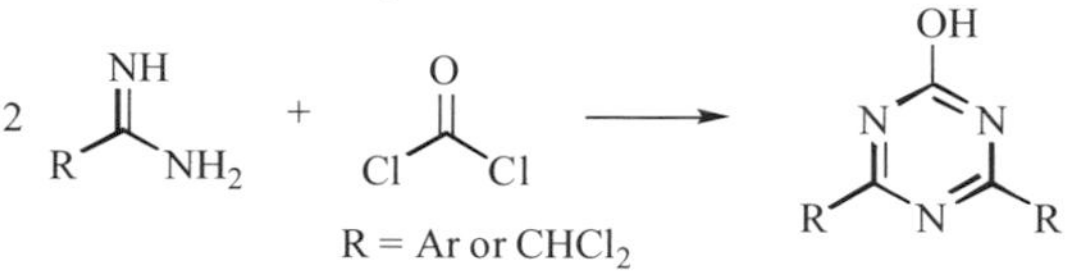

R = Ar or CHCl$_2$

## Dornow-Wiehler Oxazole Synthesis

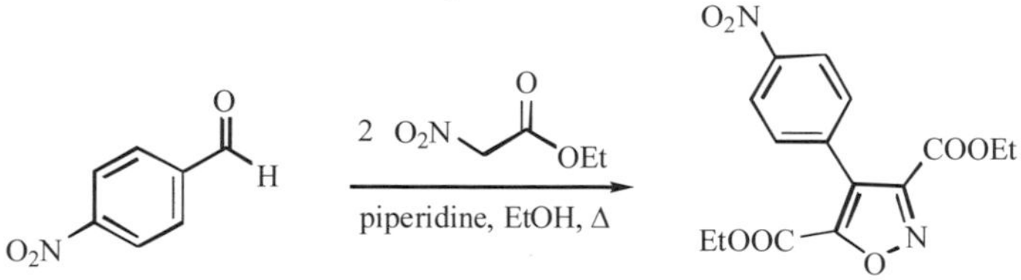

2 O$_2$N $\diagup$ OEt

piperidine, EtOH, Δ

## Fischer Oxazole Synthesis

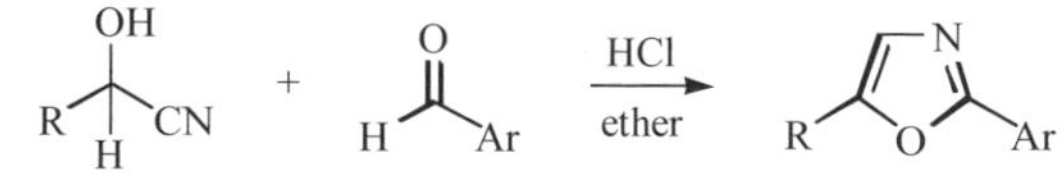

a cyanohydrin

## Hiyama Cross-Coupling Reaction

**The Reaction:**

$$R-X \quad + \quad F_3Si-R' \quad \xrightarrow[\text{base or } F^{\ominus}]{\text{Pd(0)}} \quad R-R'$$

**Proposed Mechanism:**

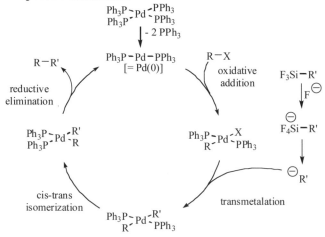

**Notes:**

## *Denmark-Mori Modification*

See: S. E. Denmark, R. F. Sweis, *Accounts of Chemical Research* **2002**, <u>35</u>, 835

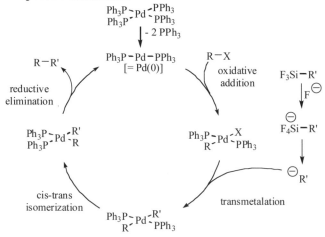

In this account they summarize related coupling reactions:

| | | |
|---|---|---|
| *Stille*-*Migata*-*Kosugi* | $M = SnR_3$ | |
| *Suzuki*-*Miyaura* | $M = BR_2$ or $B(OR)_2$ | |
| *Hiyama* | $M = SiR_{(3-n)}F_{(n)}$ | |
| *Tamao-Ito* | $M = Si(OR)_3$ | |

# Examples:

L. G. Quan, J. K. Cha, *Journal of the American Chemical Society* **2002**, <u>124</u>, 12424

71% yield, in a one-pot process

K. Itami, T. Nakami, J. Yoshida, *Journal of the American Chemical Society* **2001**, <u>123</u>, 5600

93%

S. Riggleman, P. DeShong, *Journal of Organic Chemistry* **2003**, <u>68</u>, 8106

# Hofmann Elimination (Exhaustive Methylation, Degradation)

## The Reaction:

$$
\begin{array}{c}
\text{H} \quad\quad \text{X} \\
\diagdown\!\!\!\diagup\!\!\!\diagdown \\
\end{array}
\quad
\begin{array}{l}
\text{1. NR}_3 \\
\text{2. Ag}_2\text{O , H}_2\text{O} \\
\hline
\text{3. } \Delta
\end{array}
\quad\longrightarrow\quad
\diagup\!\!=\!\!\diagdown
$$

generally gives the less
substituted alkene

## Proposed Mechanism:

Notes:

T. Laue, A. Plagens, *Named Organic Reactions*, John Wiley and Sons, Inc., New York, 1998, pp. 149-153: M. B. Smith, J. March in *March's Advanced Organic Chemistry*, 5$^{th}$ ed., John Wiley and Sons, Inc., New York, 2001, pp 1331-1333; A. C. Cope, E. R. Turnbull, *Organic Reactions* **11**, 5; S. H. Pine, *Organic Reactions* **18**, 4.

With amines, the first step is generally exhaustive methylation.

This reaction is conducted in a solvent that won't separate the ion pairs. This necessitates an eclipsed / syn orientation for elimination to take place. If there is a choice of eclipsed states, elimination will take place during the most stable, which generally gives the less substituted alkene.

The starting material can also be achieved with an amine that is exhaustively alkylated with alkyliodide. The alkyl group must not have β hydrogens, methyl and phenyl are common.

The reaction can be carried out with leaving groups other than amines.

**Hofmann Rule:** *Elimination from quaternary ammonium and tertiary sulfonium salts generally provides the lesser-substituted alkene as major product.*

## Emdé Degradation / Reduction

## Examples:

R-NH2
R'-CHO,
NaBH(OAc)3, HOAc
MeI
Base

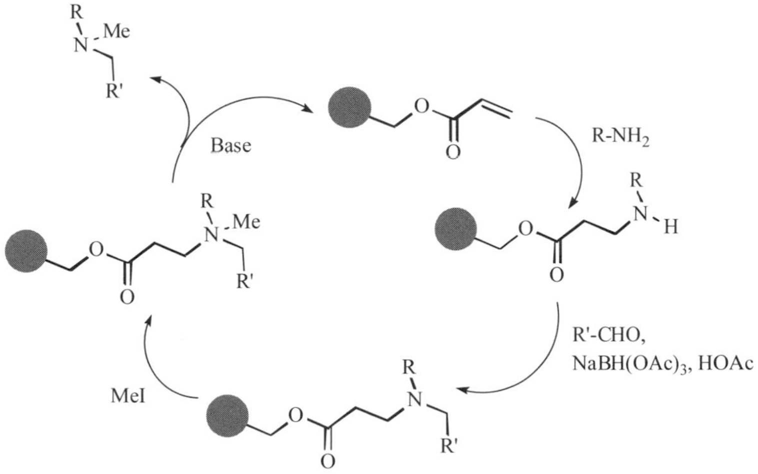

C. Alhambra, J. Castro, J. L. Chiara, E. Fernandez, A. Fernandez-Mayoralas, J. M. Fiandor, S. Garcıa-Ochoab, M. D. Martın-Ortegaa,*Tetrahedron Letters 2001,* 42, 6675

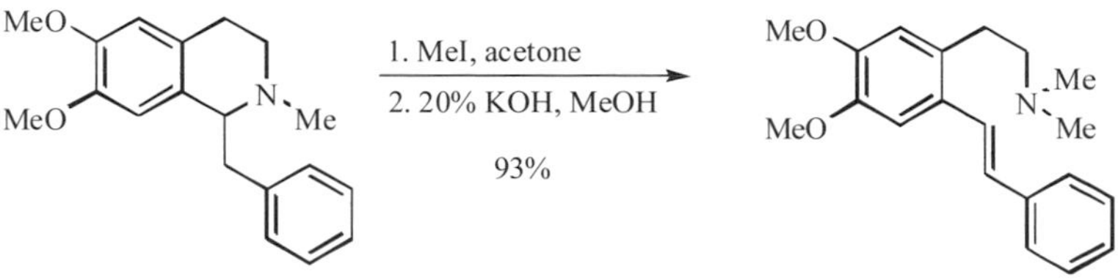

1. MeI, acetone
2. 20% KOH, MeOH
93%

S. V. Kini, M. M. V. Ramana, *Tetrahedron Letters* **2004**, 45, 4171

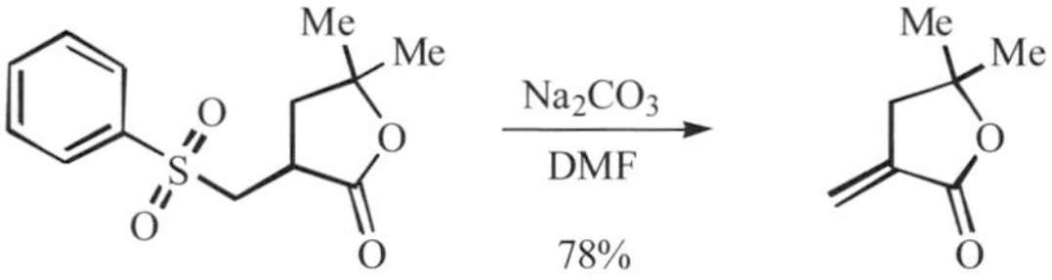

Na2CO3
DMF
78%

T. Shono, Y. Matsumara, S. Kashimura, K. Hatanaka, *Journal of the American Chemical Society* **1979**, 101, 4752

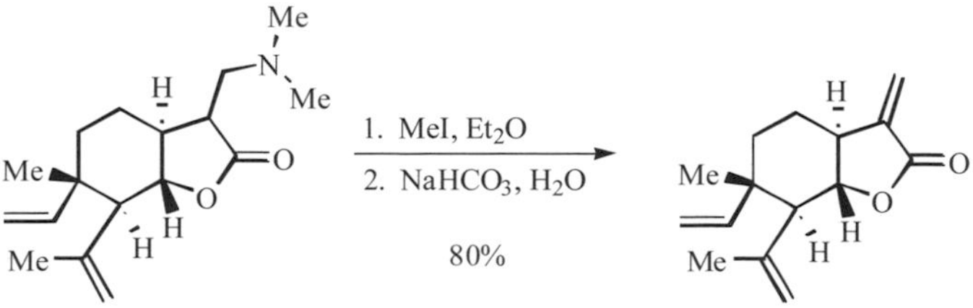

1. MeI, Et2O
2. NaHCO3, H2O
80%

T. C. Jain, C. M Banks, J. E. McCloskey, *Tetrahedron* **1976**, 32, 765

# Hofmann Isonitrile Synthesis (Carbylamine Reaction)

## The Reaction:

$$R-NH_2 \ + \ CHCl_3 \ + \ 3\,NaOH \ \longrightarrow \ R-N\equiv C\colon \ + \ 3\,NaCl \ + \ 3\,H_2O$$

## Proposed Mechanism:

$$CHCl_3 \ + \ NaOH \ \longrightarrow \ \colon CCl_2 \ + \ NaCl \ + \ H_2O$$

## Notes:

This is a carbene reaction.

**See:** V. K. Ahluwalia, R. K. Parashar, *Organic Reaction Mechanisms*, Alpha Science International Ltd., Pangbourne, U.K., 2002, p. 333

## Examples:

PTC = phase transfer conditions

Under different reaction conditions:

J. Zakrzewski, J. Jezierska, J. Hupko, *Organic Letters* **2004**, <u>6</u>, 695

$$R-NH_2 \xrightarrow[\underset{Et_3NBnCl^{\ominus}, \ 50\% \ NaOH}{\oplus}]{CHCl_3 \ (alcohol \ free), \ CH_2Cl_2} R-N\equiv C$$

R = *n*-Bu   60%
R = Bn    55%

W. P. Weber, G. W. Gokel, *Tetrahedron Letters* **1972**, <u>17</u>, 1637

# Hofmann Rearrangement

## The Reaction:

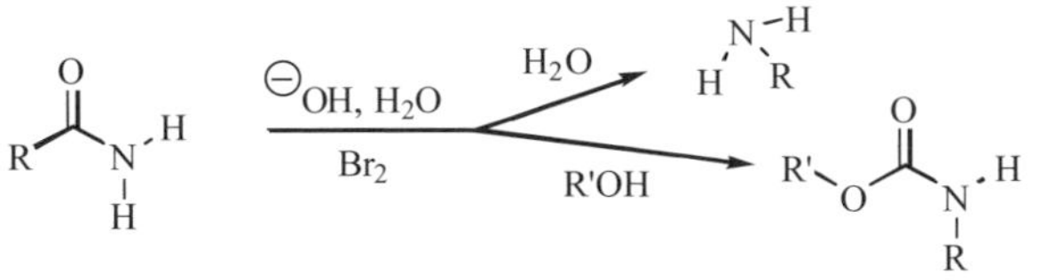

## Proposed Mechanism:

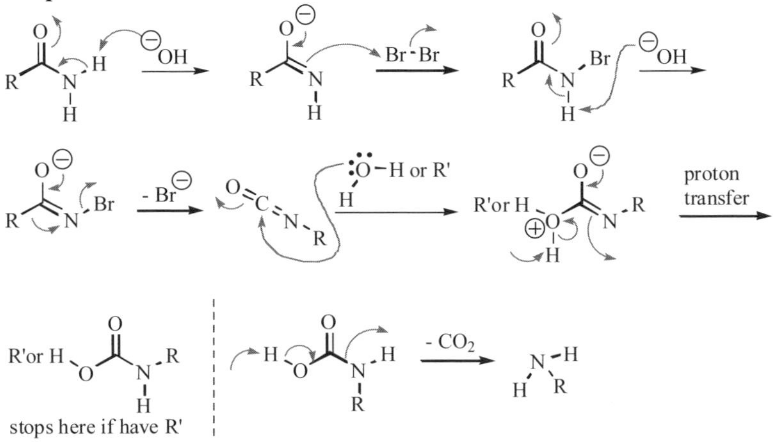

R'or H on left structure:
**stops here if have R'**

$- CO_2$

## Notes:
M. B. Smith, J. March in *March's Advanced Organic Chemistry*, 5th ed., John Wiley and Sons, Inc., New York, 2001, pp 1380, 1384; T. Laue, A. Plagens, *Named Organic Reactions*, John Wiley and Sons, Inc., New York, 1998, pp, 153-154; E. L. Wallis, J. F. Lane, *Organic Reactions* **3**, 7.

The workup can also be with $RNH_2$ to give urea derivatives.

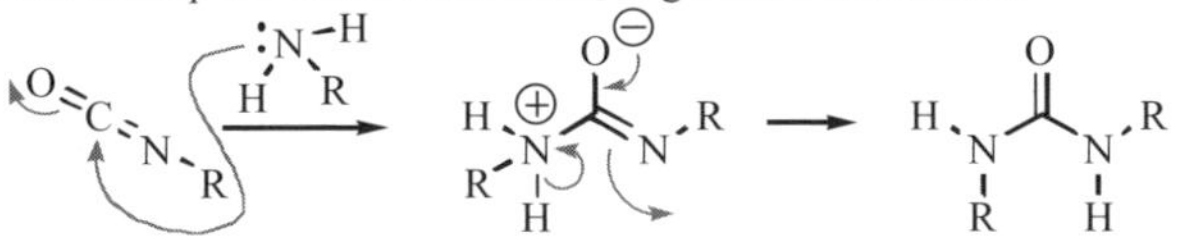

It has been found that $PhI(OAc)_2$ [iodosobenzene diacetate] and $PhI(CF_3CO_2)_2$ [bis(trifluoroacetoxy)iodosobenzene] are much better reagents for the reaction. See: L.-h. Zhang, G. S. Kauffman, J. A. Pesti, J. Yin, *Journal of Organic Chemistry* **1997**, *62*, 6918

The mechanism (abbreviated below) for the $PhI(CF_3CO_2)_2$ reaction has been discussed: R. H. Boutin G. M. Loudon, *Journal of Organic Chemistry* **1984**, *49*, 4211

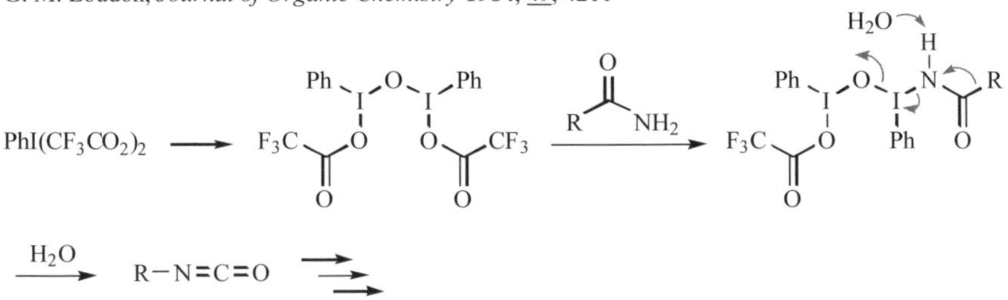

## Examples:

73%

98% ee

M.Yokoyaa, M. Kashiwagi, M. Iwasaki, K. Fuhshuku,H.Ohtab, T. Sugaia,
*Tetrahedron: Asymmetry* **2004**, <u>15</u>, 2817

1. PhI(OAc)$_2$
2. KOH, *i*-PrOH

78%

J. W. Hilborn, Z.-H. Lu, A. R. Jurgens, Q. K. Fang, P. Byers, S. A. Wald, C. H. Senanayaka,
*Tetrahedron Letters* **2001**, <u>42</u>, 8919

PhI(CF$_3$CO$_2$)$_2$

H$_2$
Pd/C

50% overall

K. G. Poullennec, D. Romo, *Journal of the American Chemical Society* **2003**, <u>125</u>, 6344

Et$_2$O

40%

D. E. DeMong, R. M. Williams, *Journal of the American Chemical Society* **2003**, <u>125</u>, 8561

AgOAc, NBS
DMF

77%

T. Hakogi, M. Taichi, S. Katsumura, *Organic Letters* **2003**, <u>5</u>, 2801

# Hofmann-Loffler-Freytag Reaction

## The Reaction:

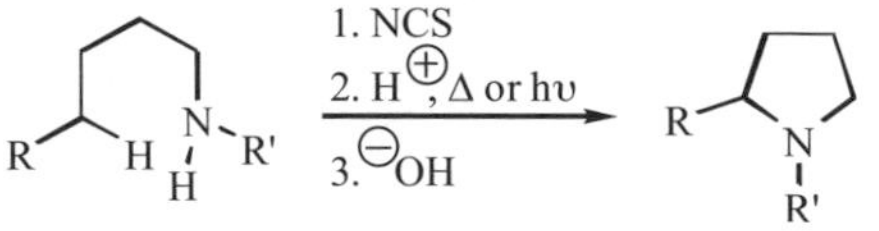

## Proposed Mechanism:

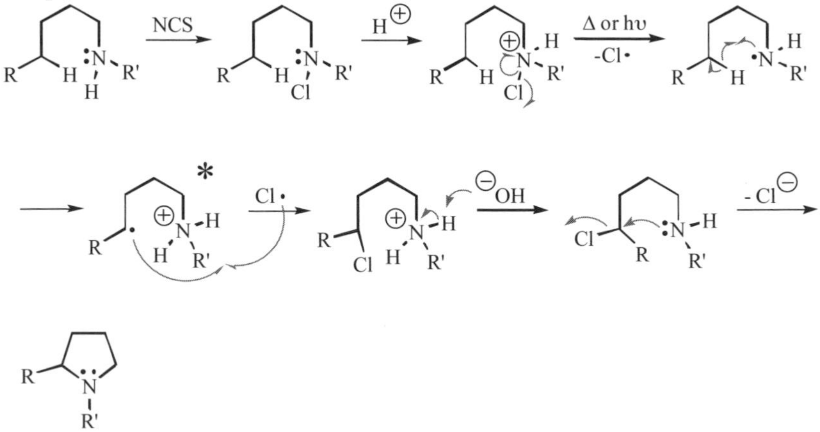

## Notes:

See Reagents: **NCS**

\* This step can serve as the propagation for a series of reactions:

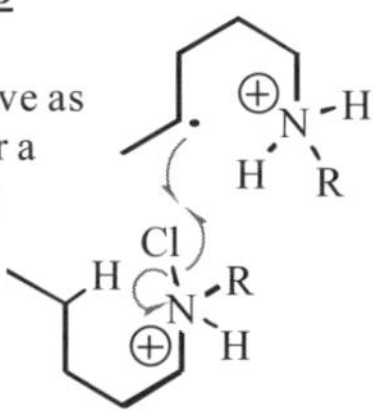

## Examples:

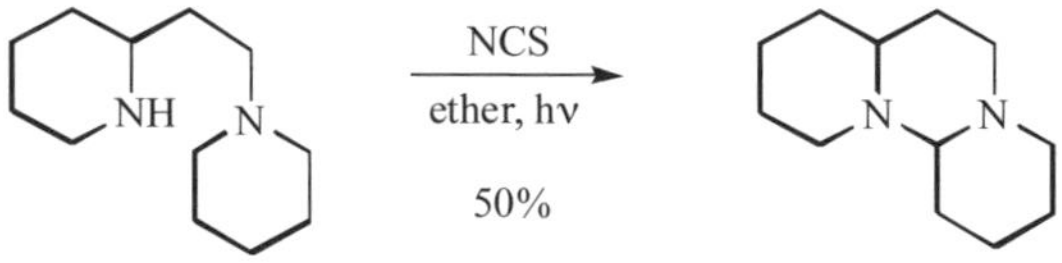

50%

K. Kimura, Y. Ban, *Synthesis* **1978**, 201

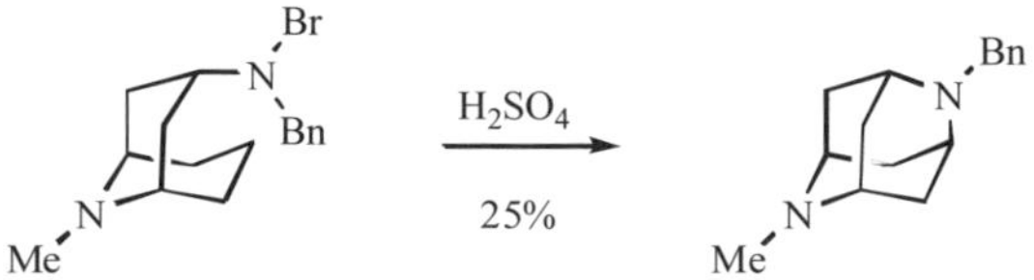

25%

R.-M. Dupeyre, A. Rassat, *Tetrahedron Letters* **1973**, 14, 2699

E. J. Corey, W.R. Hertler, *Journal of the American Chemical Society* **1960**, <u>82</u>, 1657

R. Furstoss, P. Teissier, B. Waegell, *Tetrahedron Letters* **1970**, <u>11</u>, 1263

S. L. Titouani, J.-P. Lavergne, P. Viallefont, *Tetrahedron* **1980**, <u>36</u>, 2961

# Hofmann-Martius Rearrangement

## The Reaction:

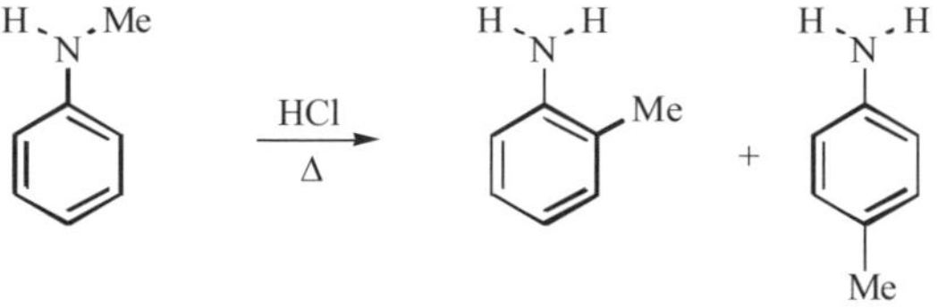

## Proposed Mechanism:

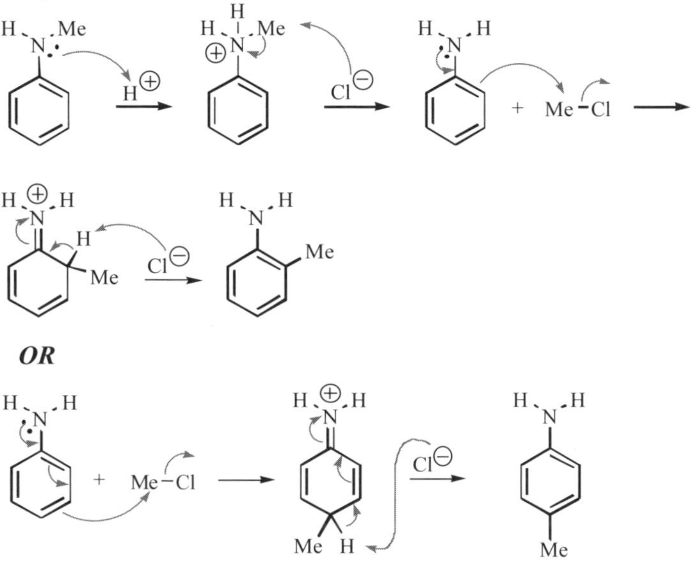

### OR

## Notes:

M. B. Smith, J. March in *March's Advanced Organic Chemistry*, 5th ed., John Wiley and Sons, Inc., New York, 2001, p. 729.

The ***Reilly-Hickinbottom Rearrangement*** uses CoCl$_2$, CdCl$_2$ or ZnCl$_2$ and the amine rather than protic acid and the amine salt.

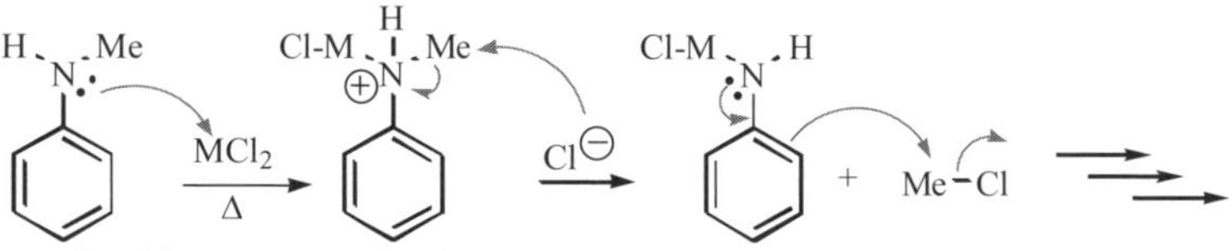

M = Co, Cd, Zn;  $\Delta$ = 200-350 °C

# Orton Rearrangement

## The Reaction:

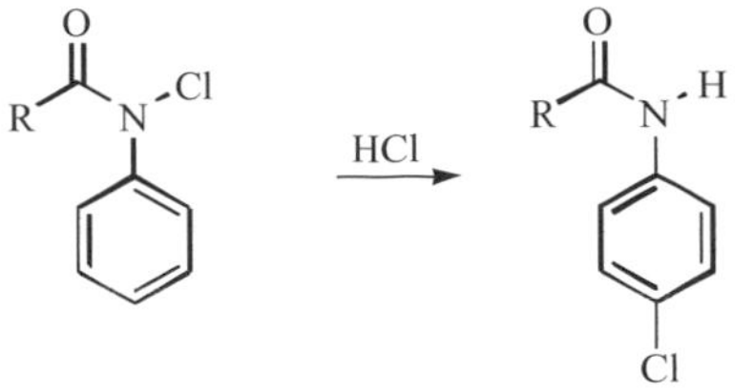

## Proposed Mechanism:

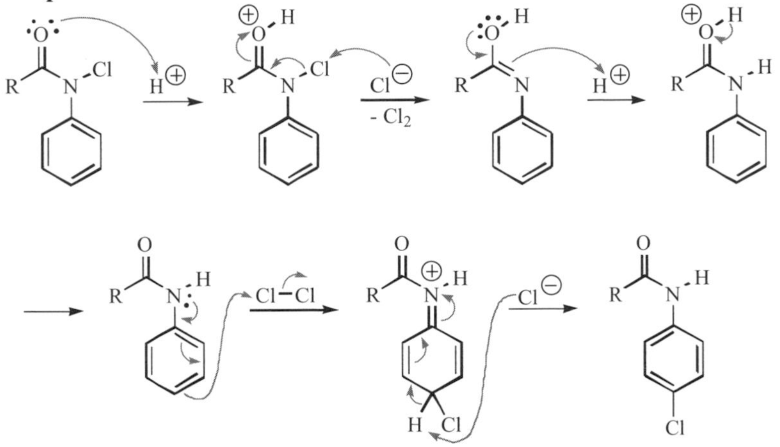

R. S. Neale, N. L. Marcus, R.G. Schepers, *Journal of the American Chemical Society* **1966**, <u>88</u>, 305
The authors describe a radical reaction for

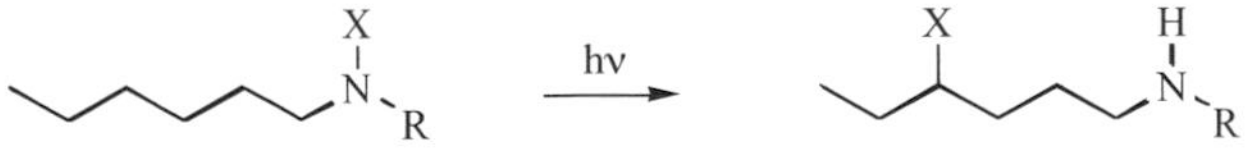

See: R. S. Neale, N. L. Marcus, *Journal of Organic Chemistry* **1969**, <u>34</u>, 1808 for related reactions
of *N*-halo-cyano compounds. A radical process is discussed.

# Hooker Reaction

## The Reaction:

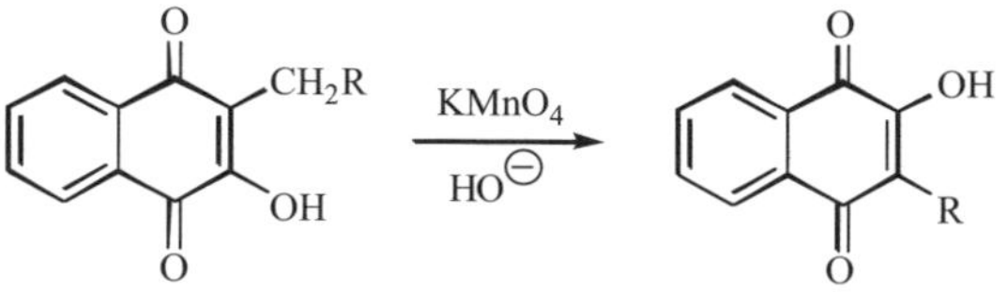

## Proposed Mechanism:

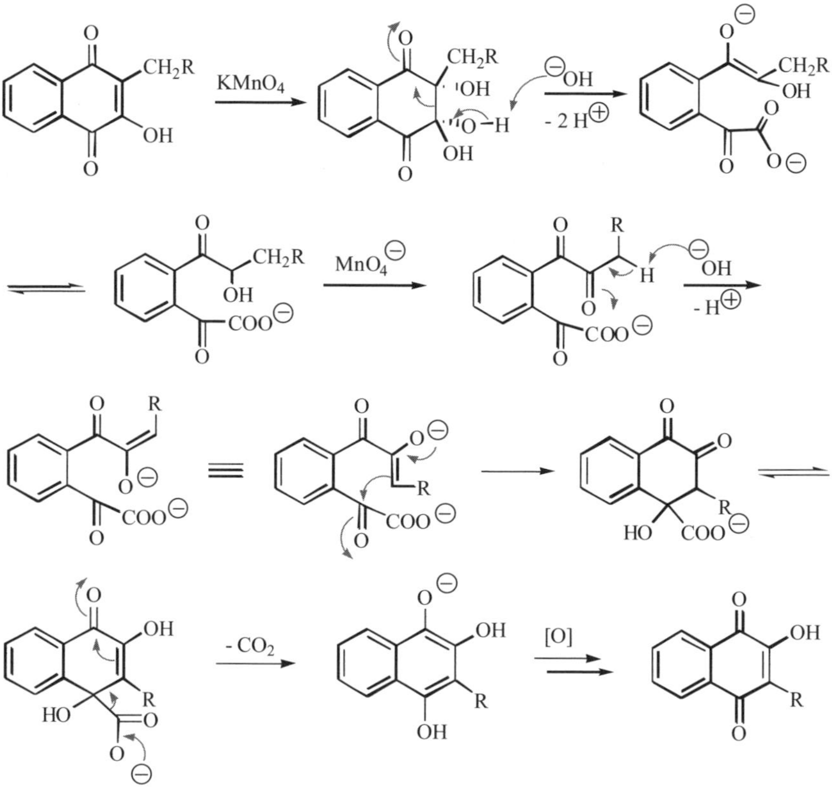

## Examples:

K. H. Lee, H. W. Moore, *Tetrahedron Letters* **1993**, 34, 235

K. H. Lee, H. W. Moore, *Tetrahedron Letters* **1993**, 34, 235

L. F. Fieser, M. Fieser, *Journal of the American Chemical Society* **1948**, 70, 3215

# Horner-Wadsworth-Emmons-Wittig Reaction

## The Reaction:

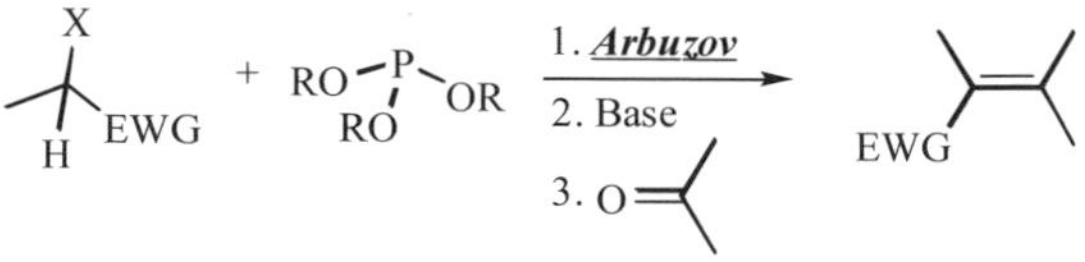

## Proposed Mechanism:

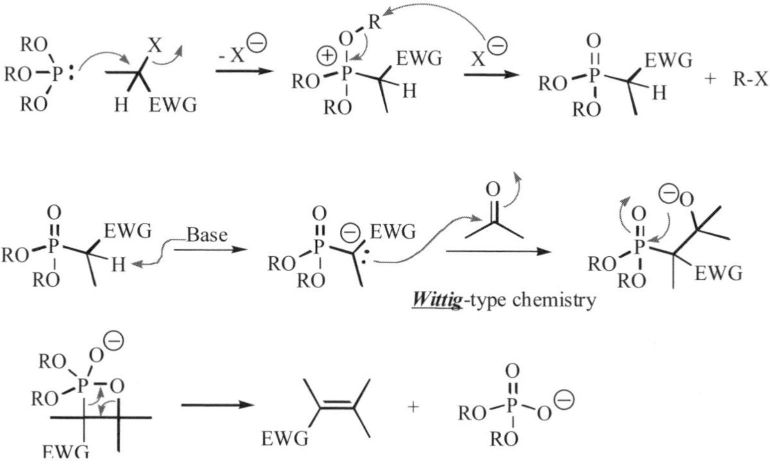

*Wittig*-type chemistry

## Notes:

M. B. Smith, J. March in *March's Advanced Organic Chemistry*, 5th ed., John Wiley and Sons, Inc., New York, 2001, p. 1233.

See also the **Masamune-Roush** conditions for this reaction. Generally useful for base-sensitive aldehydes.

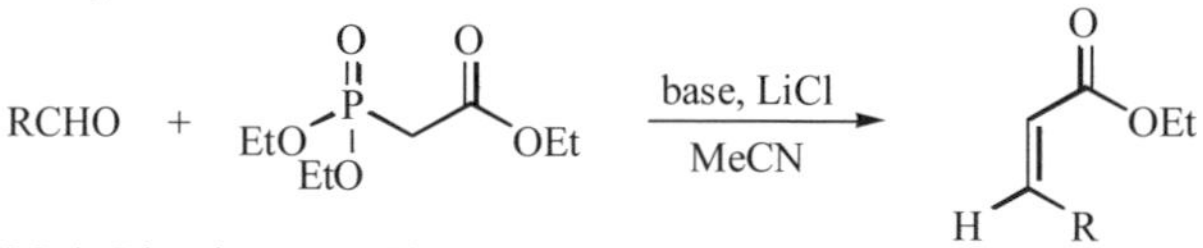

M. A. Blanchette, W. Choy, J. T. Davis, A. P. Essenfeld, S. Masamune, W. R. Roush, T. Sakai, *Tetrahedron Letters* **1984**, 25, 2183

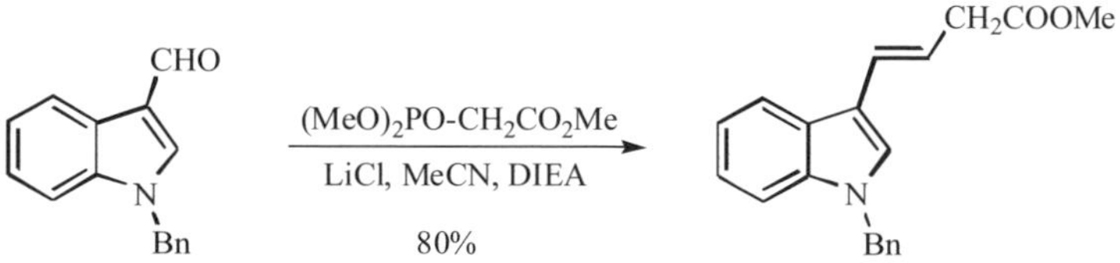

N. P. Pavri, M. L. Trudell, *Journal of Organic Chemistry* **1997**, 62, 2649

## Examples:

A. Srikrishna, D. Vijaykumar, T. J. Reddy, *Tetrahedron* **1997**, <u>53</u>, 1439

H. Kiyota, D. J. Dixon, C. K. Luscombe, S. Hettstedt, S. V. Ley, *Organic Letters* **2002**, <u>4</u>, 322

A. Samarat, V. Fargeas, J. Villieras, J. Lebreton, H. Amri, *Tetrahedron Letters* **2001**, <u>42</u>, 1273

I. Vemura, H. Miyagawa, T. Veno, *Tetrahedron* **2002**, <u>58</u>, 2351

J. Knol, B. L. Feringa, *Tetrahedron Letters* **1997**, <u>38</u>, 2527

# Houben-Hoesch Reaction / Hoesch Reaction

## The Reaction:

Lewis Acids include $ZnCl_2$, $AlCl_3$, $BCl_3$, etc.

## Proposed Mechanism:

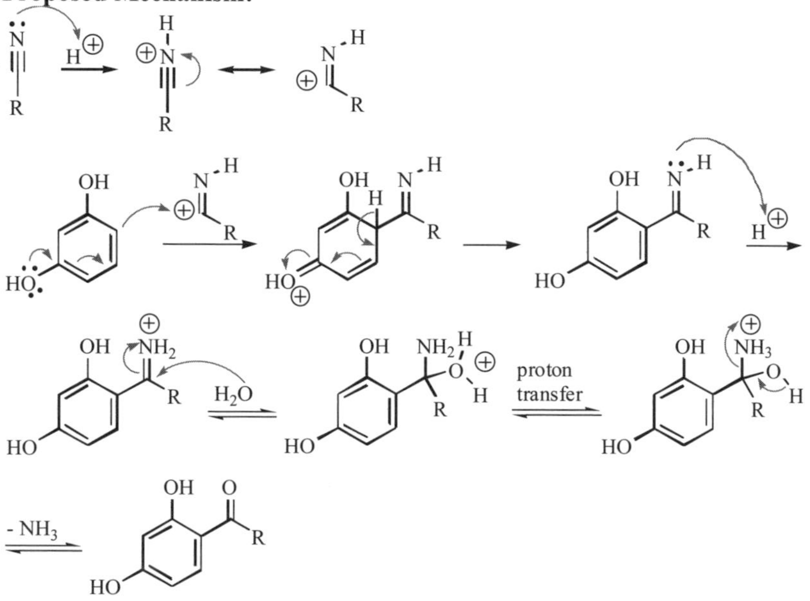

## Notes:

M. B. Smith, J. March in *March's Advanced Organic Chemistry*, 5th ed., John Wiley and Sons, Inc., New York, 2001, p.722; P. E. Sperri, A. S. DuBois, *Organic Reactions* **5**, 9.

The *Hoesch Reaction (Acylation)*:

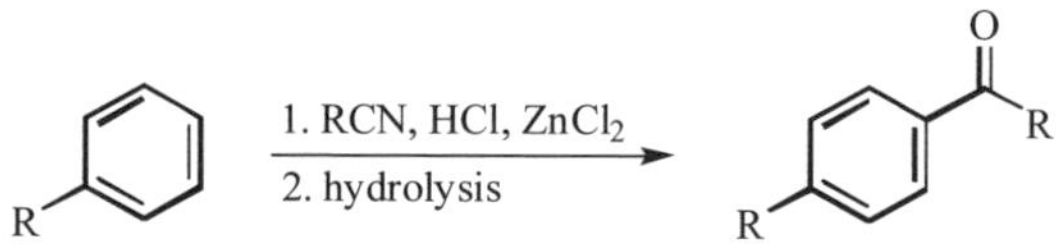

$R = OH, OR, NR_2$

The *Houben-Hoesch Reaction* refers specifically to phenols as substrates, where the reaction is generally most useful.

## Houben-Fischer Synthesis

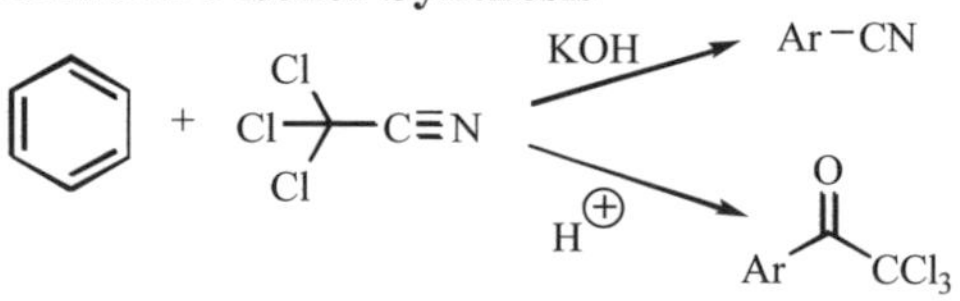

## Proposed Mechanism:

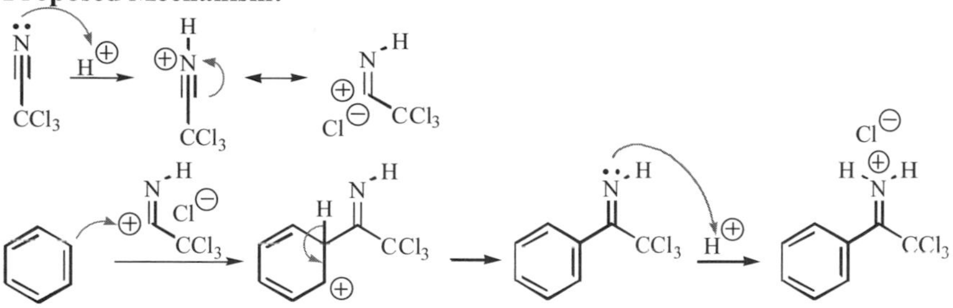

### Basic Hydrolysis:

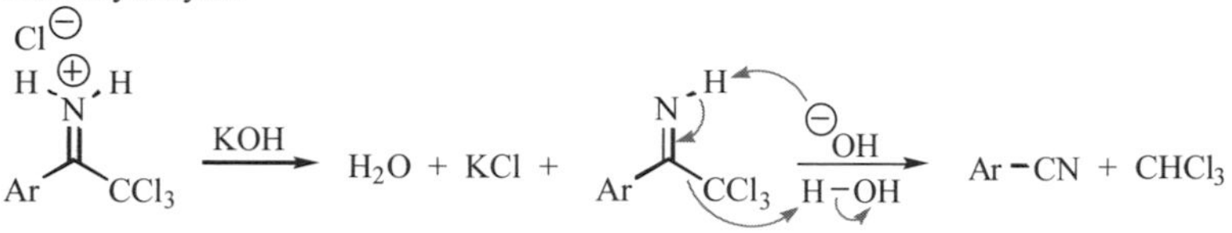

$$Ar-CN + CHCl_3$$

### Acidic Hydrolysis:

## Examples:

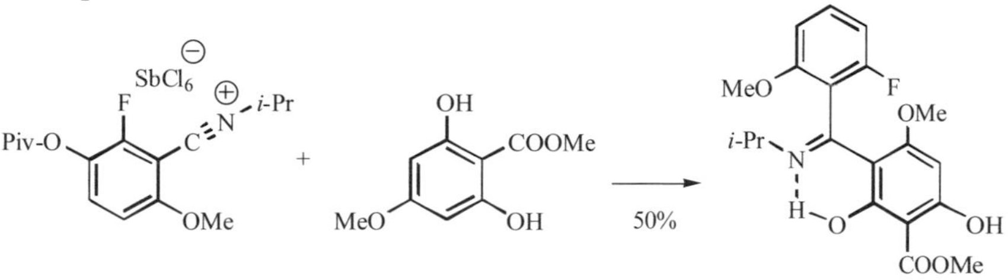

50%

D. W. Udwary, L. K. Casillas, C. A. Townsend *Journal of the American Chemical Society* **2002**, <u>124</u>, 5294

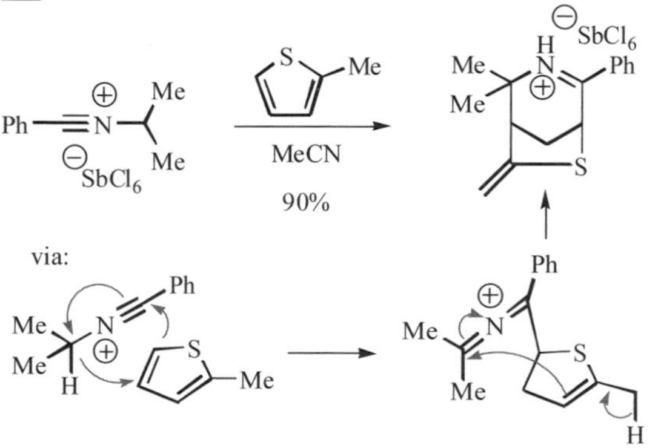

90%

via:

J. Henninger, K. Polborn, H. Mayr, *Journal of Organic Chemistry* **2000**, <u>65</u>, 3569

# Hunsdiecker Reaction (Borodine Reaction, Borodine-Hunsdiecker Reaction)
A. Borodin, composer, is best noted for the work "*Prince Igor*".

## The Reaction:

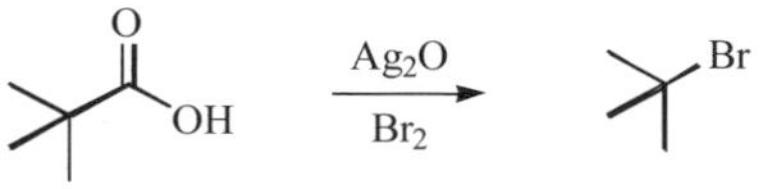

$$Ag_2O \quad / \quad Br_2$$

## Proposed Mechanism:

$$Ag_2O + H_2O \rightleftharpoons AgOH + HO^{\ominus} + Ag^{\oplus}$$

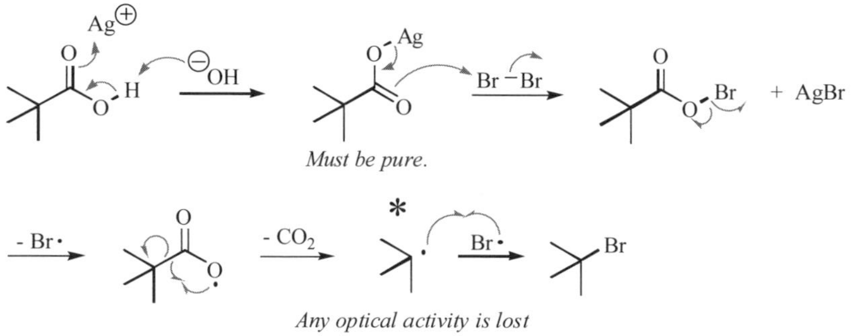

*Must be pure.*

*Any optical activity is lost*

\* This step can also be the propagation step:

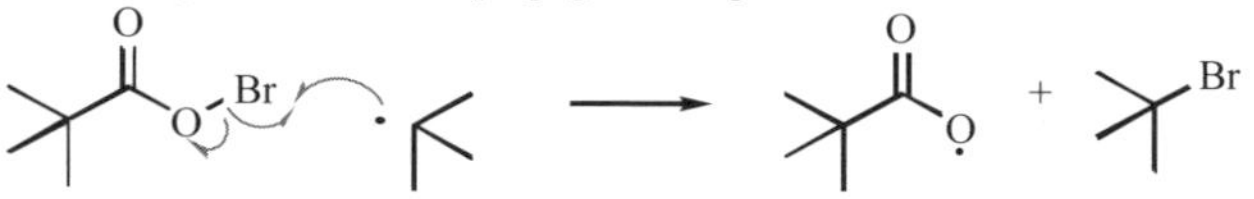

## Notes:
M. B. Smith, J. March in *March's Advanced Organic Chemistry*, 5th ed., John Wiley and Sons, Inc., New York, 2001, pp 942-943; T. Laue, A. Plagens, *Named Organic Reactions*, John Wiley and Sons, Inc., New York, 1998, pp. 155-157; C. V. Wilson, *Organic Reactions* **9**, 5.

The *Christol-Firth Modification* or *Christol-Firth-Hunsdiecker Reaction* uses HgO in place of Ag$_2$O. In this modification it is not necessary to isolate an intermediate.

The *Simonini Reaction* differs by the ratio of I$_2$. The products are significantly different, this reaction produces esters:

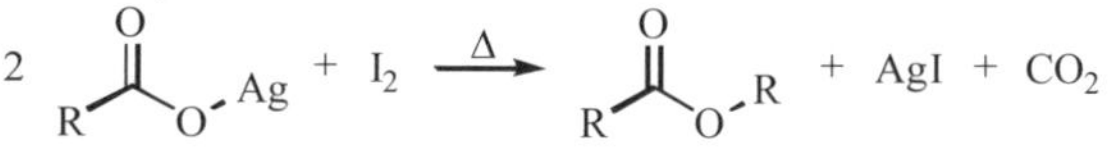

$$2 \quad R\text{-}C(O)\text{-}O\text{-}Ag + I_2 \xrightarrow{\Delta} R\text{-}C(O)\text{-}O\text{-}R + AgI + CO_2$$

## Examples:

$$\text{Tr} = \text{trityl} \quad \text{---CPh}_3$$

Et$_3$N •3HF

NBS

62%

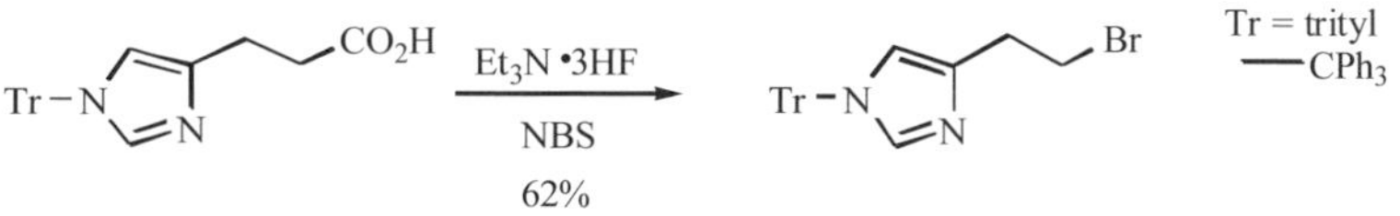

B. Dolensky, K. L. Kirk, *Journal of Organic Chemistry* **2002**, <u>67</u>, 3468

94%

D. Naskar, S. Chowdhury, S. Roy, *Tetrahedron Letters* **1998**, _39_, 699

85%

A. R. Al Dulayymi, J. R. Al Dulayymi, M. S. Baird, M. E. Gerrard, G. Koza, S. D. Harkins, E. Roberts, *Tetrahedron* **1996**, _52_, 3409

69%

M. S. Baird, H. L. Fitton, W. Clegg, A. McCamley, *Journal of the Chemical Society: Perkin Transactions I* **1993**, 321

71%

K. B. Wilberg, M. G. Matturo, P. J. Okarma, M. E. Jason, *Journal of the American Chemical Society* **1984**, _106_, 2194

90%

J. P. Das, S. Roy, *Journal of Organic Chemistry* **2002**, _67_, 7861

40%

G. M. Lampman, J. C. Aumiller, *Organic Syntheses* **1971**, 206

65%

L. DeLuca, G. Giacomelli, G. Porcu, M. Taddei, *Organic Letters* **2001**, _3_, 855

# Ireland-Claisen Rearrangement

## The Reaction:

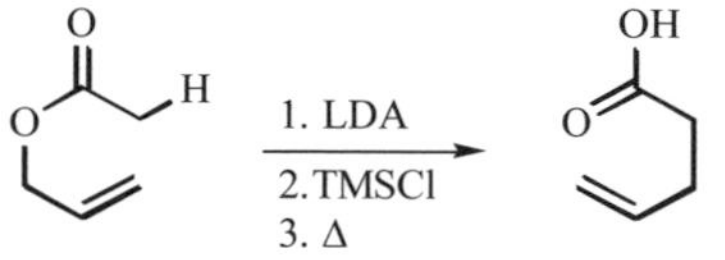

## Proposed Mechanism:

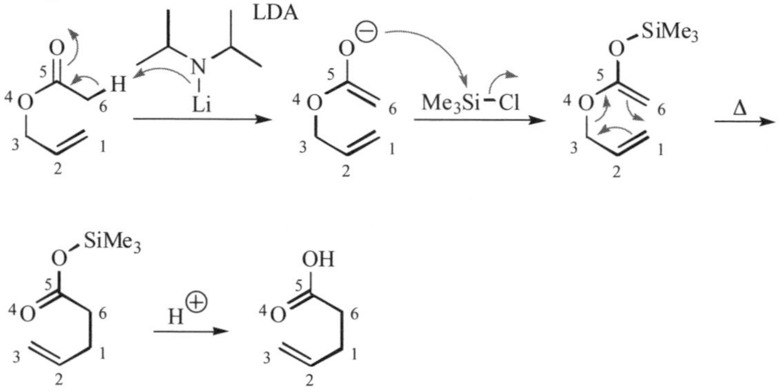

## Notes:

M. B. Smith, J. March in *March's Advanced Organic Chemistry*, 5th ed., John Wiley and Sons, Inc., New York, 2001, p. 1452.

## Examples:

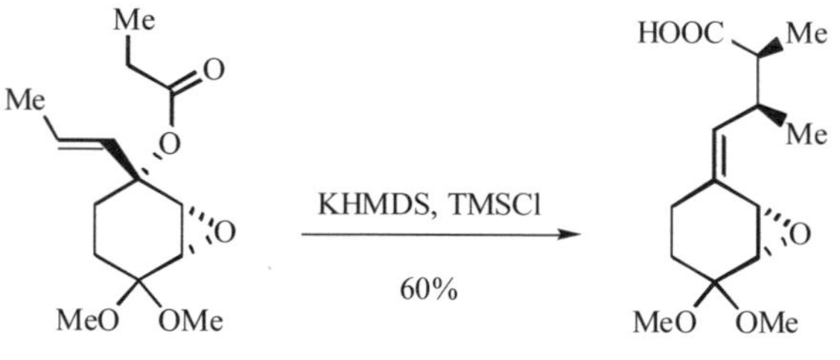

KHMDS, TMSCl

60%

S.-p. Hong, M. C. McIntosh, *Organic Letters* **2002**, *4*, 19

S. Höck, F. Koch, H.-J. Borschberg *Tetrahedron: Asymmetry* **2004**, <u>15</u>, 1801

J. C. Gilbert, J. Yin , F. H. Fakhreddine, M. L. Karpinski *Tetrahedron* **2004**, <u>60</u>, 51

P. Magnus, N. Westwood, *Tetrahedron Letters* **1999**, <u>40</u>, 4659

P. A. Jacobi, Y. Li, *Organic Letters* **2003**, <u>5</u>, 701

# Ivanov Reaction

## The Reaction:

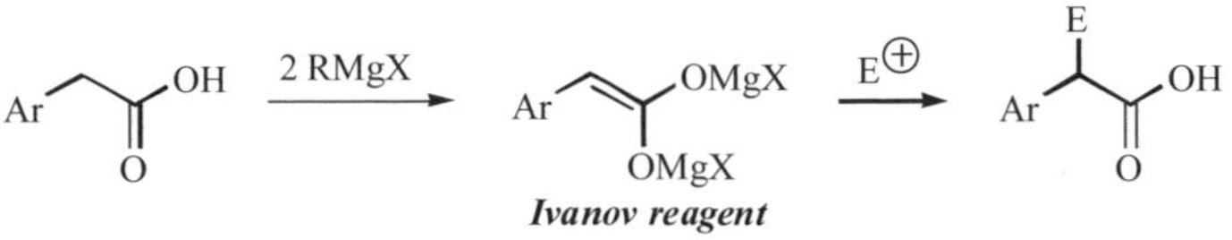

*Ivanov reagent*

## Proposed Mechanism:

*Ivanov reagent*

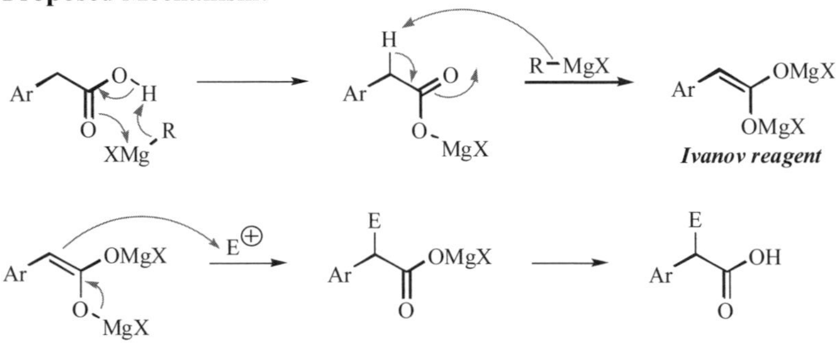

## Notes:

The stereochemistry of the **Ivanov Reactions**:
H. E. Zimmerman, M. D. Traxler; *Journal of the American Chemical Society* **1957**, <u>79</u>, 1920

The reaction is particularly useful with the electrophile being an aldehyde or ketone:

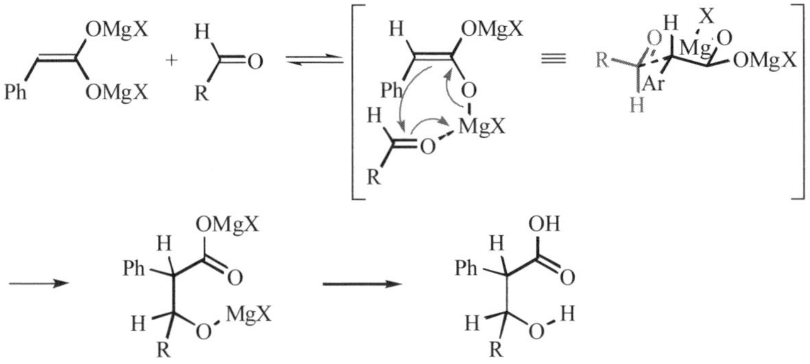

| R | Rel Rate |
|---|---|
| *i*-Pr | 6.3 |
| *t*-Bu | 1 |

J. Toullec, M. Mladenova, C. F. Gaudemar-Bardone, C. B. Blagoev, *Tetrahedron Letters* **1983**, <u>24</u>, 589

## Examples:

53%

F. F. Blicke, H. Zinnes, *Journal of the American Chemical Society* **1955**, <u>77</u>, 6247

M. Mladenova, B. Blagoev, M. Gaudemar, F. Gaudemar-Bardone, J. Y. Lallemand *Tetrahedron* **1981**, <u>37</u>, 2157

Y. A. Zhdanov, G. V. Bogdanova, O. Y. Riabuchina *Carbohydrate Research* **1973**, <u>29</u>, 274

# Jacobsen-Katsuki Epoxidation

## The Reaction:

E. N. Jacobsen, L. Deng, Y. Furukawa, L. E. Martinez *Tetrahedron* **1994**, <u>50</u>, 4323

## Proposed Mechanism:

The mechanistic details are still being studied. Two interpretations most thought about include:

A recent study provides another interpretation as a possibility:

Product

Y. G. Abashkin, S. K. Burt, *Organic Letters* **2004**, <u>6</u>, 59

See: D. L. Hughes, G. B. Smith, J. Liu, G. C. Dezeny, C. H. Senanayake, R. D. Larsen, T. R. Verhoeven, P. J. Reider, *Journal of Organic Chemistry* **1997**, <u>62</u>, 2222, for details of the mechanistic interpretation.

## Notes:

M. B. Smith, J. March in *March's Advanced Organic Chemistry*, 5th ed., John Wiley and Sons, Inc., New York, 2001, p. 1053.

Selectivity depends on substrate structure:

| Ar | | |
|---|---|---|
| Ar = *p*-MeO-phenyl | ~ 11 : | 1 |
| Ar = *p*-NO$_2$-phenyl | ~ 1 : | 4 |

E. N. Jacobsen, L. Deng, Y. Furukawa, L. E. Martinez, *Tetrahedron* **1994**, <u>50</u>, 4323

The oxygen has its origins in the NaOCl:

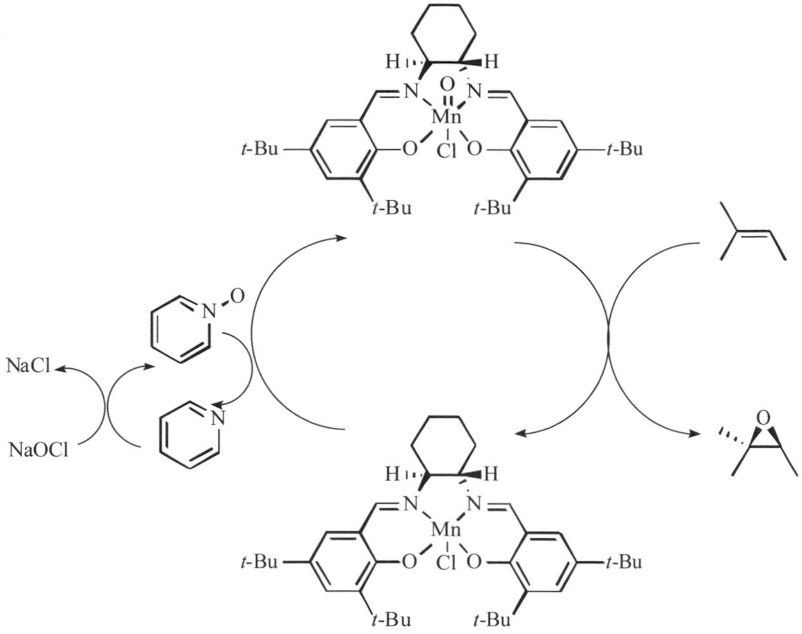

**Examples:**

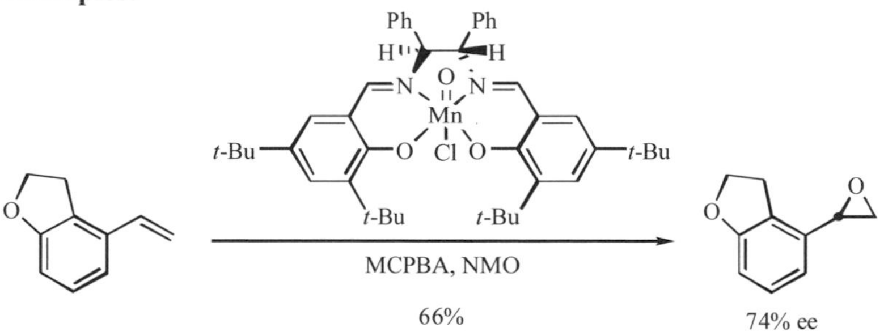

66%

74% ee

J. S. Prasad, T. Vu, M. J. Totleben, G. A. Crispino, D. J. Kacsur, S. Swaminathan, J. E. Thornton, A. Fritz, A. K. Singh, *Organic Process Research & Development* **2003**, *7*, 821

# Japp-Klingemann Reaction

## The Reaction:

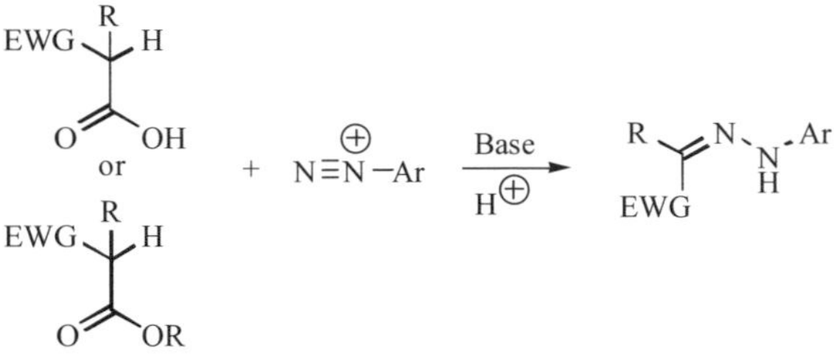

## Proposed Mechanism:

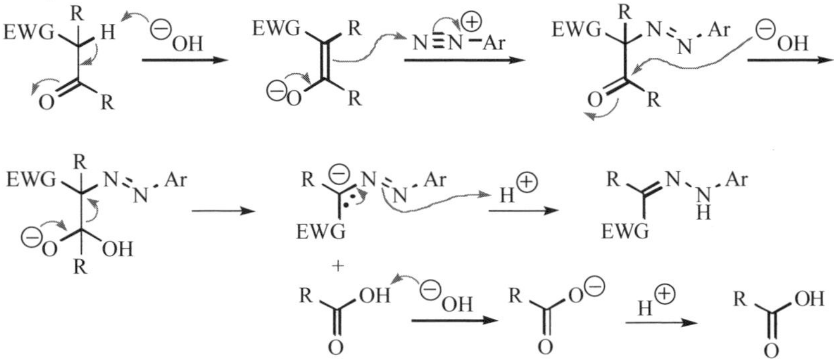

The carboxylic acid byproduct would exist as a carboxylate until acid is added.

## Notes:

M. B. Smith, J. March in *March's Advanced Organic Chemistry*, 5<sup>th</sup> ed., John Wiley and Sons, Inc., New York, 2001, p. 779; T. Laue, A. Plagens, *Named Organic Reactions*, John Wiley and Sons, Inc., New York, 1998, pp. 161-163; R. R. Philips, *Organic Reactions* **10**, 2.

## Examples:

A variation where the intermediate is reduced:

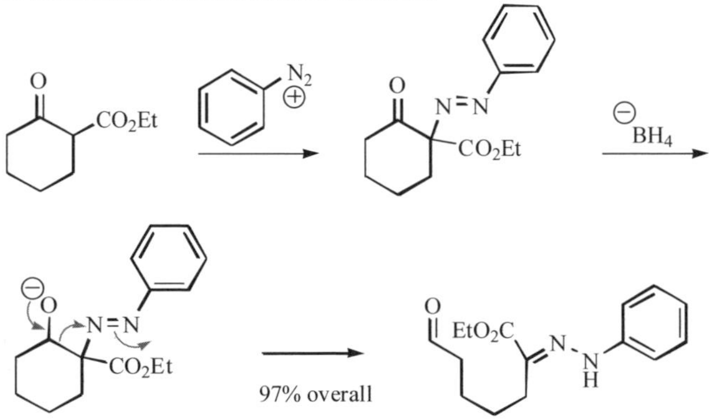

97% overall

A. P. Kosikowski, W. C. Floyd, *Tetrahedron Letters* **1978**, <u>19</u>, 19

Y. Bessard, *Organic Process Research & Development* **1993**, 2, 214

H. C. Yao, P. Resnick, *Journal of the American Chemical Society* **1962**, 84, 3514

D. Shapiro, R. A. Abramovitch, *Journal of the American Chemical Society* **1955**, 77, 6690

# Johnson Polyene Cyclization

## The Reaction:

The reactions encompass the attempts to mimic the suggested biosynthetic cyclization for multi-ring terpenes and steroids, such as:

squalene 2,3-epoxide

lanosterol → cholesterol

W. S. Johnson, *Accounts of Chemical Research* **1968**, <u>1</u>, 1

## Proposed Mechanism:

These are generally cationic processes where correct alignment of alkenes bonds will allow for ring formation:

## Notes:

M. B. Smith, J. March in *March's Advanced Organic Chemistry*, 5th ed., John Wiley and Sons, Inc., New York, 2001, p. 1019.

Success in this reaction requires control of the *E/Z* configuration of the alkene bonds and often a good terminator step.

## Examples:

BF₃ Et₂O

79%

M. Franck-Neumann, P. Geoffroy, D. Hanss *Tetrahedon Letters* **2002**, <u>43</u>, 2277

P. V. Fish, A. R. Sudhakar, W. S. Johnson, *Tetrahedron Letters* **1993**, <u>34</u>, 7849

A. B. Smith, III, T. Kinsho *Tetrahedron Letters* **1996**, <u>37</u>, 6461

S. R. Harring and T. Livinghouse, *Journal of Organic Chemistry* **1997**, <u>62</u>, 6388

E. J. Corey, H. B. Wood, Jr., *Journal of the American Chemical Society* **1996**, <u>118</u>, 11982

G. Demailly, G. Solladie, *Tetrahedron Letters* **1980**, <u>21</u>, 3355

# Johnson-Claisen Rearrangement (Johnson Orthoester Rearrangement)

## The Reaction:

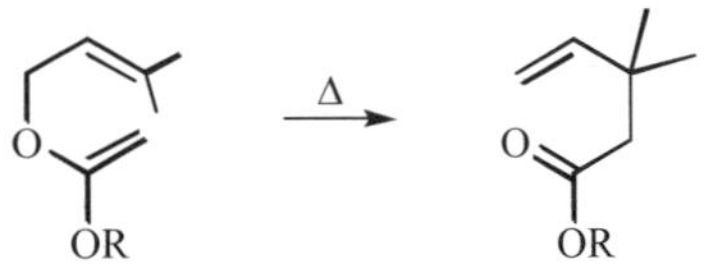

## Proposed Mechanism:

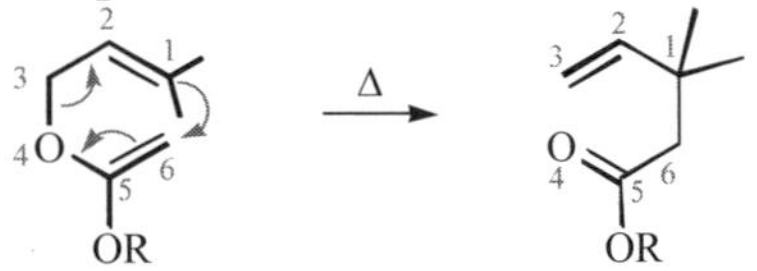

## Notes:

Preparation of starting material:

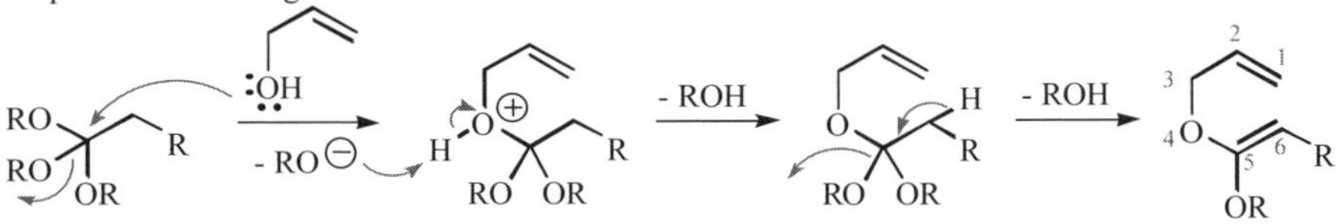

## Examples:

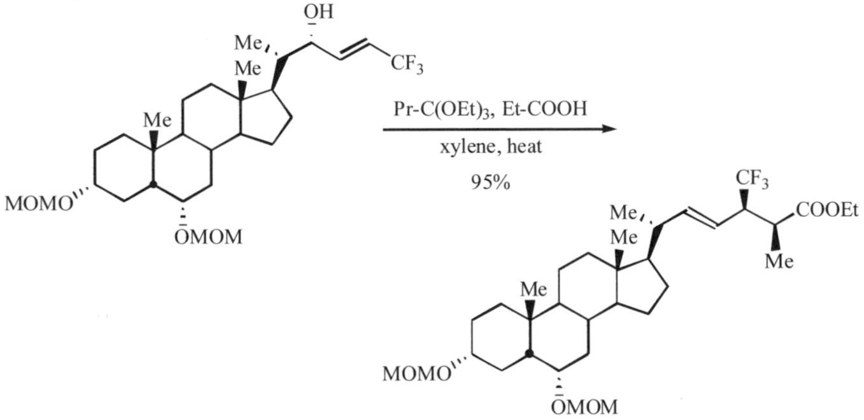

B. Jiang, Y. Liu, W.-s. Zhou, *Journal of Organic Chemistry* **2000**, <u>65</u>, 6231

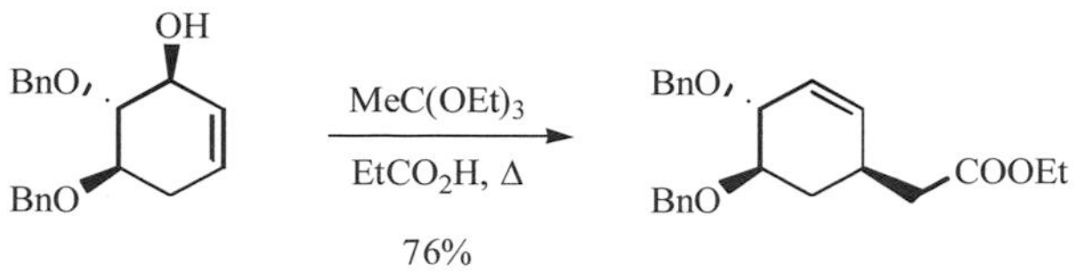

76%

H. Fuwa, Y. Okamura, H. Natsugari, *Tetrahedron* **2004**, <u>60</u>, 5341

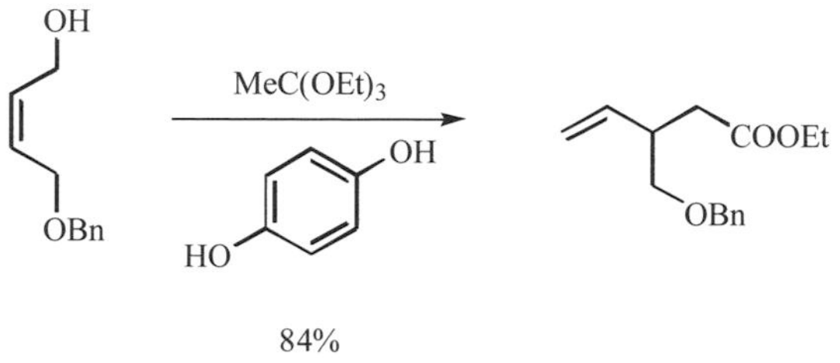

84%

E. Marotta, P. Righi, G. Rosini, *Organic Letters* **2000**, <u>2</u>, 4145

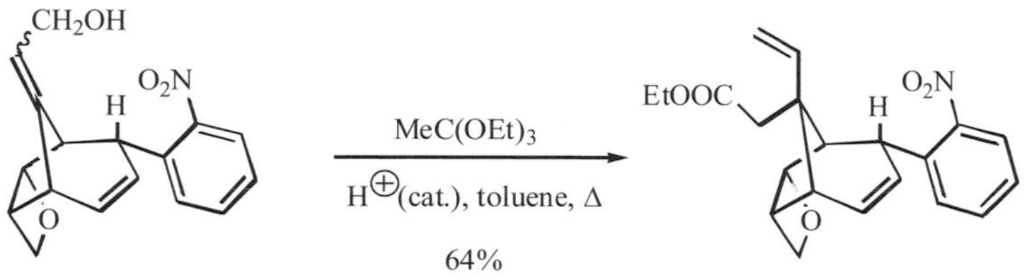

64%

F. W. Ng, H. Lin, S. J. Danishefsky, *Journal of the American Chemical Society* **2002**, <u>124</u>, 9812

# Jones Oxidation

## The Reaction:

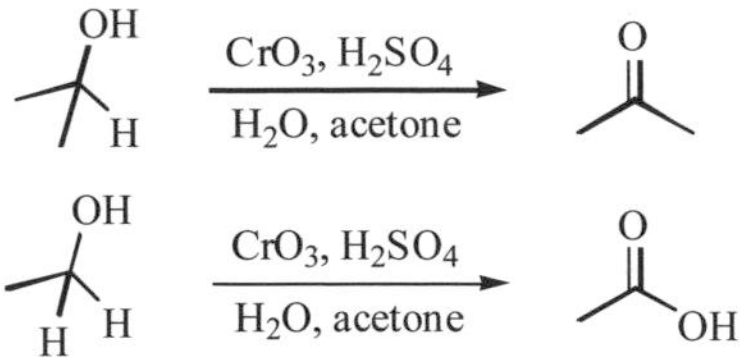

## Proposed Mechanism:

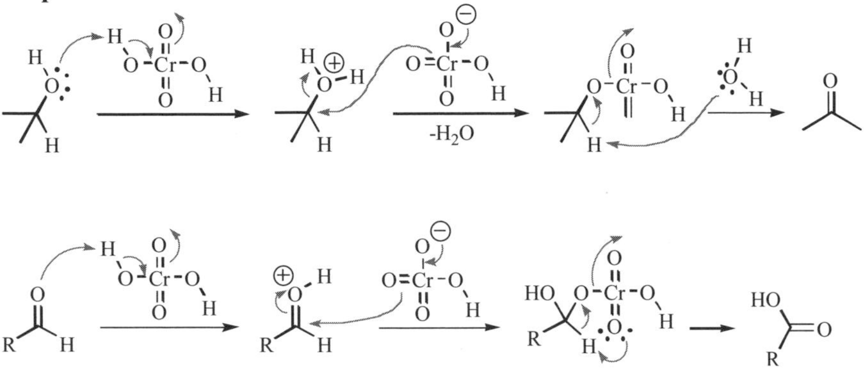

## Notes:

### *Chromium Reagent Formation*:

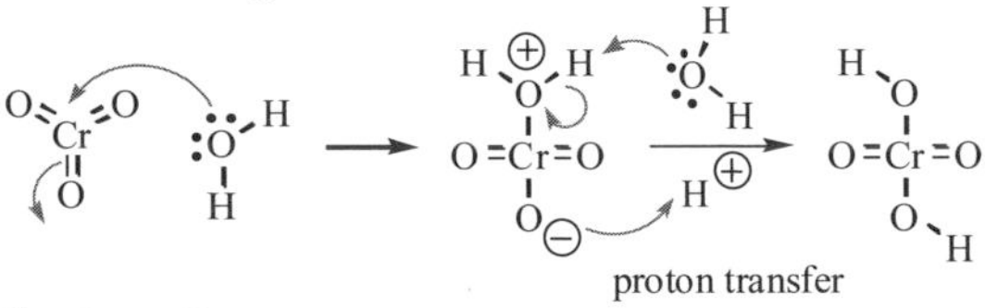

proton transfer

See *Jones Reagent*.

This is a useful and simple reaction to carry out. The oxidizing solution can be slowly titrated into the reaction flask until the color of the oxidant persists.

Primary alcohols are readily converted to acids, secondary alcohols to ketones. 1,2-Diols can suffer fragmentation. Allylic alcohols are readily oxidized.

## Collins Oxidation

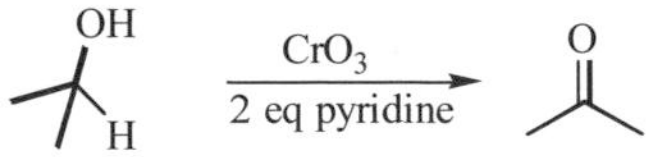

## Sarrett Oxidation

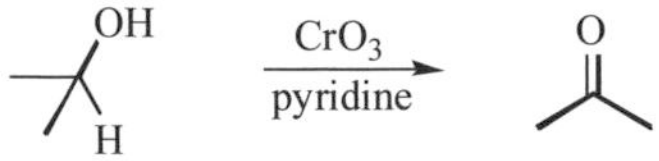

Note: Must add $CrO_3$ to pyridine solution.

## Examples:

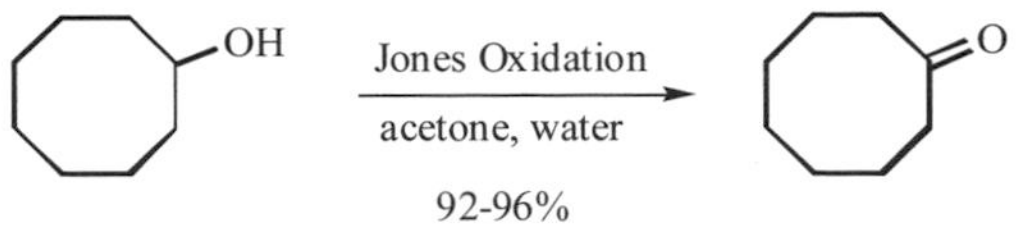

Jones Oxidation

acetone, water

92-96%

E. J. Eisenbraun, *Organic Syntheses,* CV5, 597

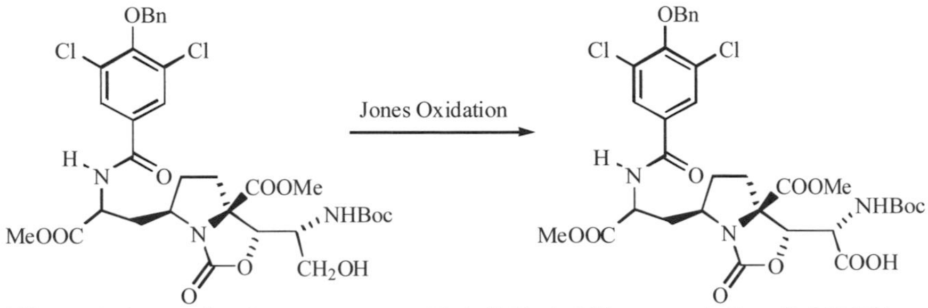

Jones Oxidation

This was the last step in a four-step sequence. No individual yields presented. Overall yield 54%

D. Ma and J. Yang, *Journal of the American Chemical Society* **2001**, 123, 9706

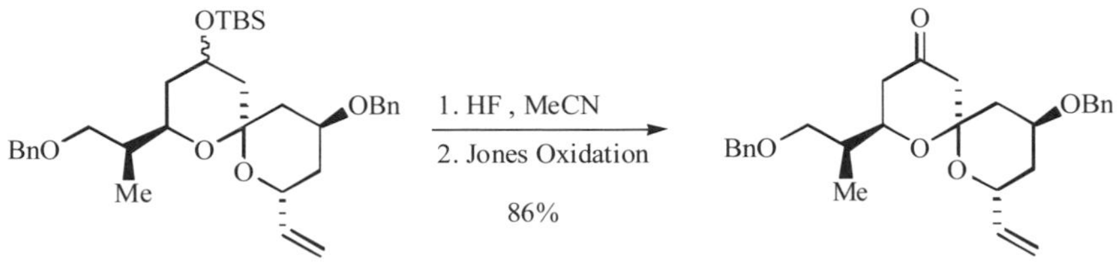

1. HF , MeCN
2. Jones Oxidation

86%

M. T. Crimmins, C. A. Carroll, B. W. King, *Organic Letters* **2000**, 2, 597

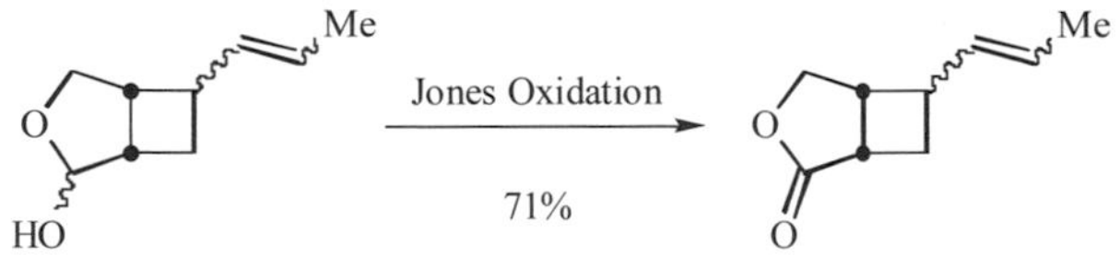

Jones Oxidation

71%

J. Panda, S. Ghosh, S. Ghosh *Journal of the Chemical Society, Perkin Transactions 1* **2001**, 3013

# Julia-Bruylants Cyclopropyl Carbinol Rearrangement

## The Reaction:

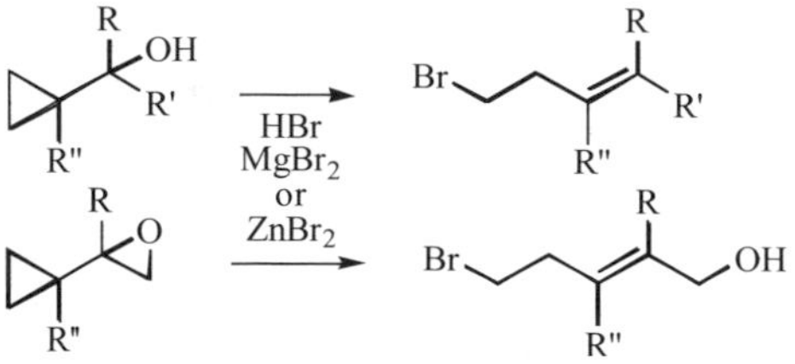

## Proposed Mechanism:

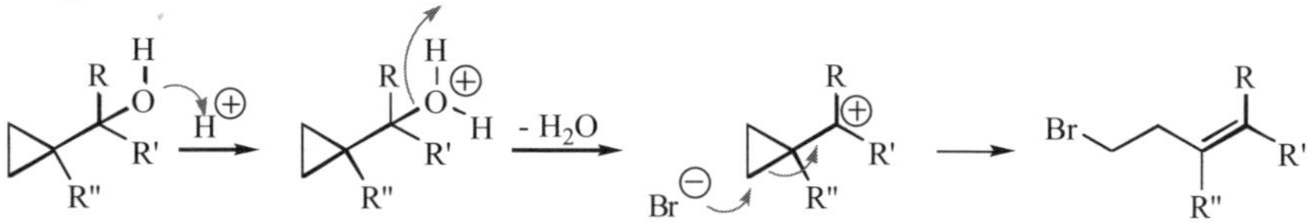

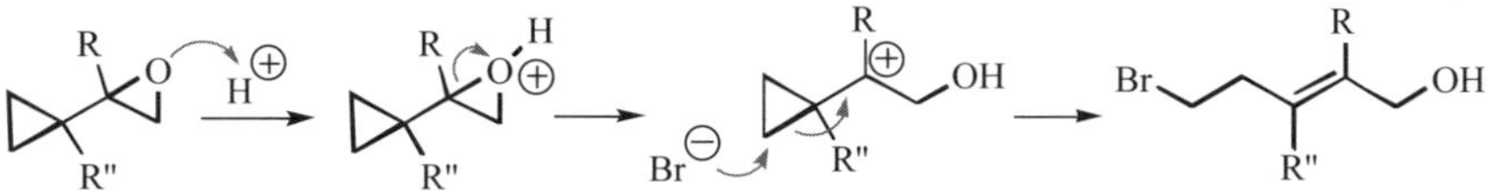

## Examples:

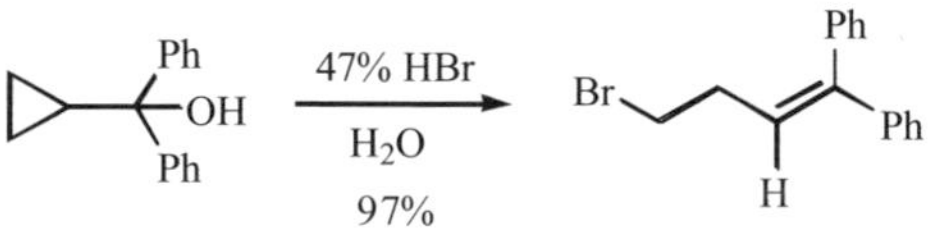

A. Van der Bent, A. G. S. Blommaert, C. T. M. Melman, A. P. IJzerman, I. Van Wijngaarden, W. Soudijn, *Journal of Medicinal Chemistry* **1992**, <u>35</u>, 1042

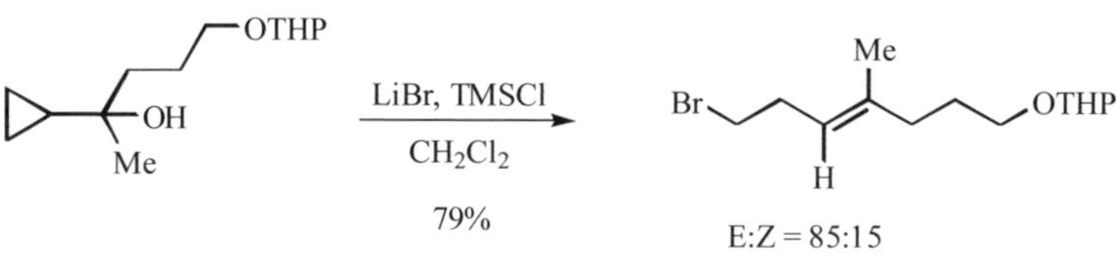

Z. Liu, W. Z. Li, Y. Li, *Tetrahedron: Asymmetry* **2001**, <u>12</u>, 95

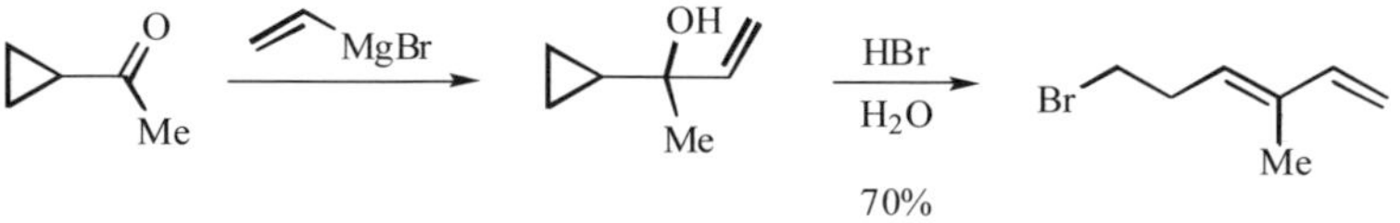

P. A. Wender, J. M. Nuss, D. B. Smith, A. Suarez-Sobrina, J. Vagberg, D. Decosta, J. Bordner, *Journal of Organic Chemistry* **1997**, <u>62</u>, 4908

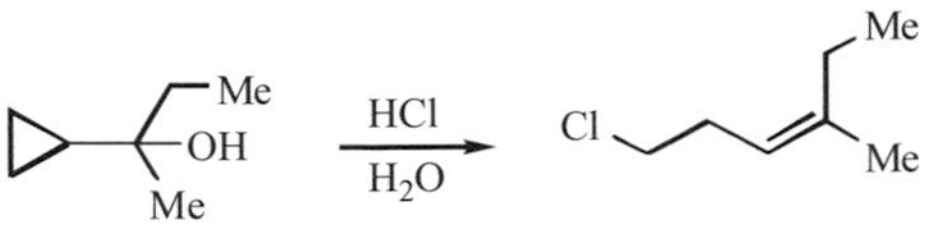

90%

C. A. Henrick, F. Schaub, J. B. Siddall, *Journal of the American Chemical Society* **1972**, <u>94</u>, 5374

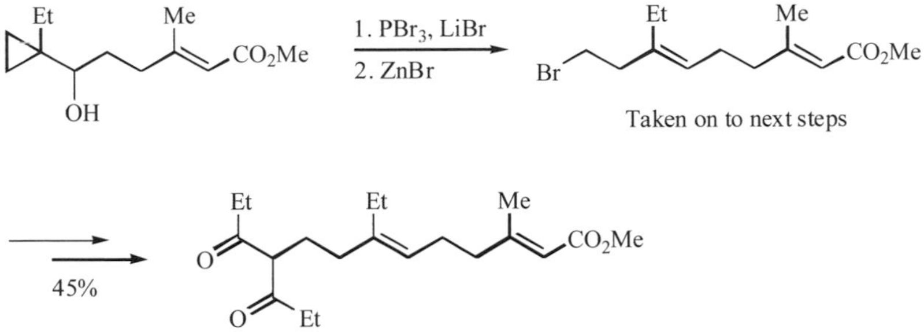

Taken on to next steps

45%

W. S. Johnson, T.-T. Li, D. J. Faulkner, S. F. Campbell, *Journal of the American Chemical Society* **1968**, <u>90</u>, 6225

# Julia Olefination (Julia-Lythgoe Olefination)

## The Reaction:

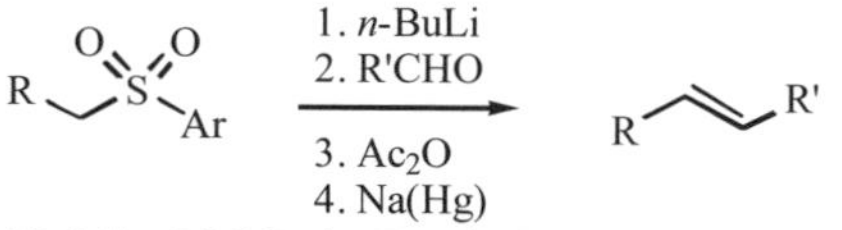

1. *n*-BuLi
2. R'CHO
3. Ac$_2$O
4. Na(Hg)

M. Julia, J. M. Paris, *Tetrahedron Letters* **1973**, <u>14</u>, 4833

## Proposed Mechanism:

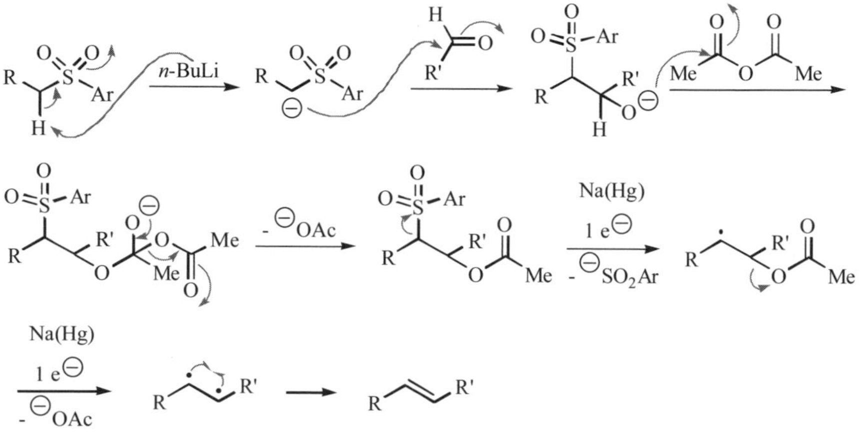

## Julia Coupling

Utilizes the first part of the reaction sequence:

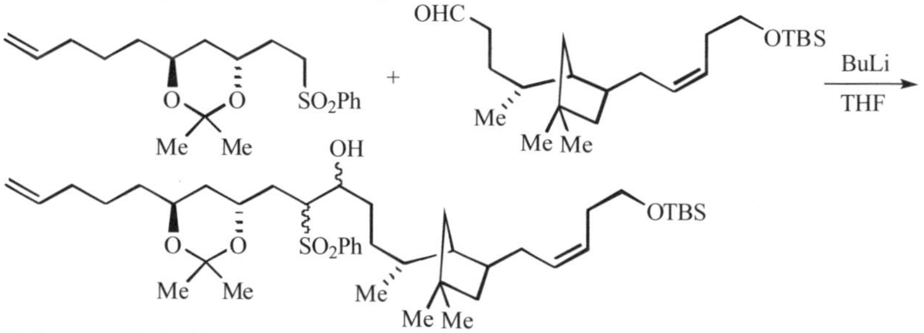

BuLi
THF

K. Suenaga, K. Araki, T. Sengoku, D. Uemura, *Organic Letters* **2001**, <u>3</u>, 527

## Examples:

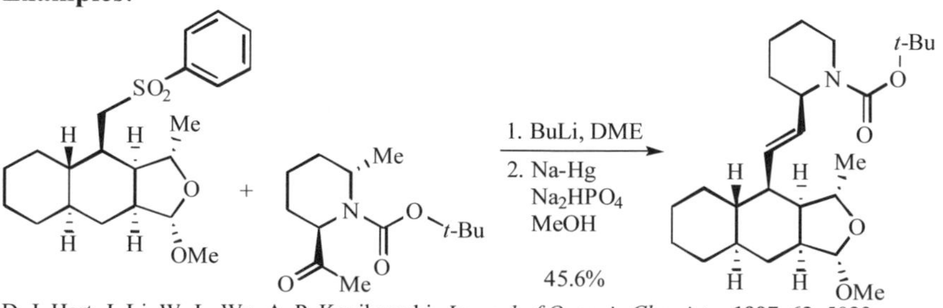

1. BuLi, DME
2. Na-Hg
   Na$_2$HPO$_4$
   MeOH

45.6%

D. J. Hart, J. Li, W.-L. Wu, A. P. Kozikowski, *Journal of Organic Chemistry* **1997**, <u>62</u>, 5023

M. Z. Hoemann, K. A. Agriosw, J. Aube, *Tetrahedron Letters* **1996**, <u>37</u>, 953

(2:1)

I. E. Marko, F. Murphy, S. Donlan, *Tetrahedron Letters* **1996**, <u>37</u>, 2089

1-methyl-1-methoxyethyl ether = MIP

G. Zanoni, A. Porta, F. Castronovo, G. Vidari, *Journal of Organic Chemistry* **2003**, <u>68</u>, 6005

# Kahne Glycosylation

## The Reaction:

1. Tf$_2$O, CH$_2$Cl$_2$, or toluene
2. R'OH
3. Base

Base =

D. Kahne, S. Walker, Y. Cheng, D. Van Engen *Journal of the American Chemical Society* **1989**, 111, 6881

## Proposed Mechanism:

## Examples:

1. Tf$_2$O, toluene

2.

Me ⎯ Me

Me OH

3.

*t*-Bu ⎯ N ⎯ *t*-Bu

(70%, α:β = 2:1)

D. Kahne, S. Walker, Y. Cheng, D. Van Engen *Journal of the American Chemical Society* **1989**, 111, 6881

1. Tf$_2$O, toluene

2.

Me ⎯ Me

Me OH

3.

*t*-Bu ⎯ N ⎯ *t*-Bu

*(part of a large sequence, no yield for this step)*

70%, (α:β = 2:1)

L. Yan, C. M. Taylor, R. Goodnow, Jr., D. Kahne, *Journal of the American Chemical Society* **1994**, 116, 6953

1. Tf$_2$O

2.   MeO
     N$_3$ ⎯ OMe
     OH

Me

3.

*t*-Bu ⎯ N ⎯ *t*-Bu

95%

D. J. Silva, H. Wang, N. M. Allanson, R. K. Jain, M. J. Sofia, *Journal of Organic Chemistry* **1999**, 64, 5926

# Keck Macrolactonization

**The Reaction:**

**Proposed Mechanism:**

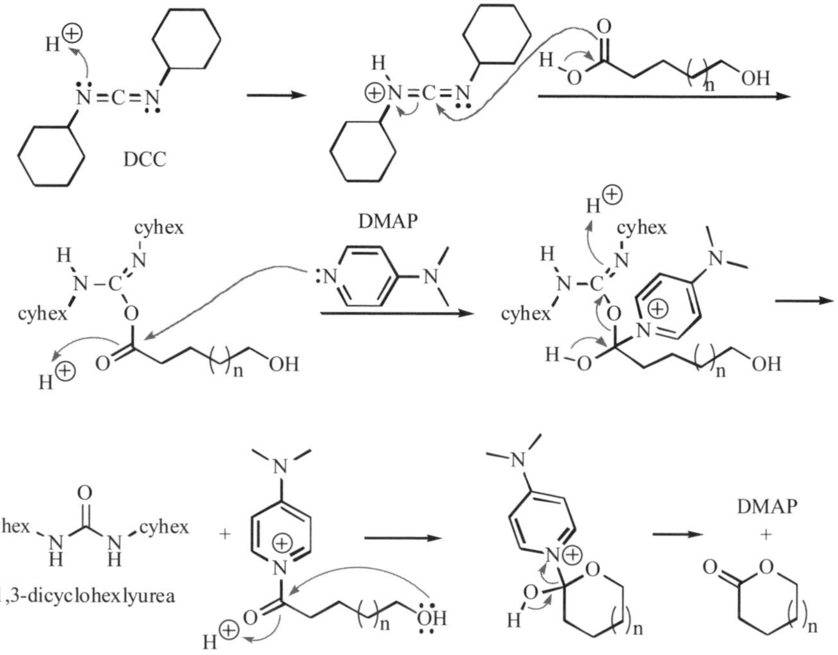

## Notes:

See ***Macrolactonization*** for related methods.

This reaction is an extension of the ***Steglich esterification reaction***. See: E. P. Boden, G. E. Keck, *Journal of Organic Chemistry* **1985**, _50_, 2394.

For another approach to macrolactonization, see:

# Mukaiyama Esterification

**The Reaction:**

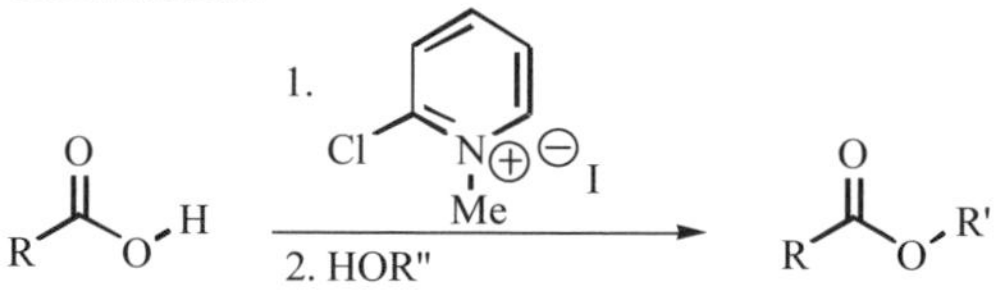

**Proposed Mechanism:**

**Notes:**
See: ***Mukaiyama's Reagent*** for examples.

## Examples (of the Keck Macrolactonization):

F. Fujiwara, D. Awakura, M. Tsunashima, A. Nakamura, T. Honma, A. Murai, *Journal of Organic Chemistry* **1999**, <u>64</u>, 2616

| n | % |
|---|---|
| 8 | 52 |
| 12 | 96 |

G. E. Keck, C. Sanchez, C. A. Wagner, *Tetrahedron Letters* **2000**, <u>41</u>, 8673

# Kennedy Oxidative Cyclization

## The Reaction:

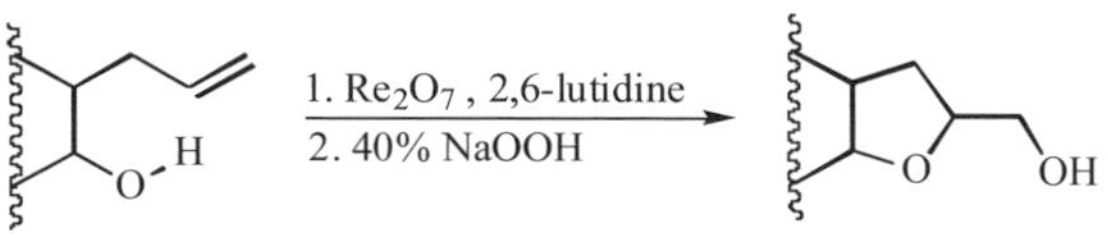

1. Re$_2$O$_7$ , 2,6-lutidine
2. 40% NaOOH

## Proposed Mechanism:

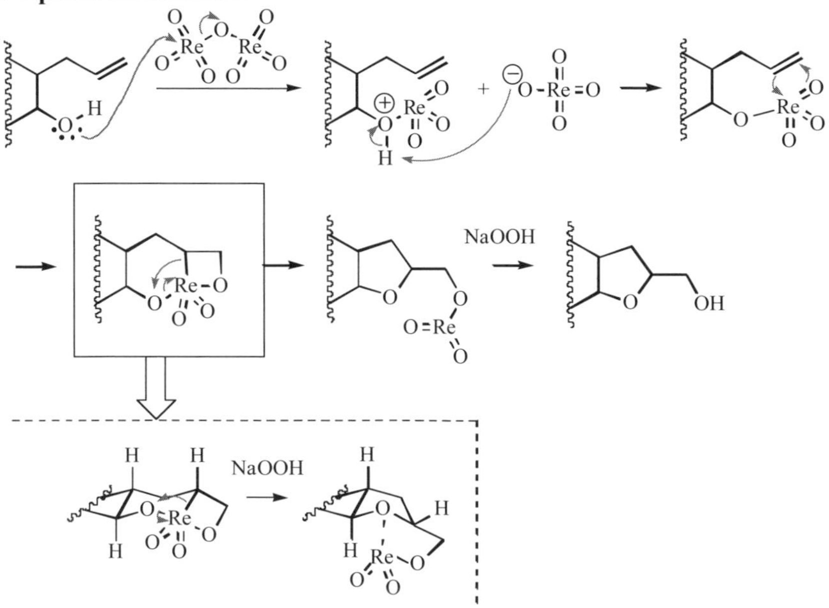

NaOOH

## Notes:

For comments on mechanism, see: S. Tang, R. M. Kennedy, *Tetrahedron Letters* **1992**, <u>33</u>, 5299, 5303 ; S. C. Sinha, E. Keinan, S. C. Sinha, *Journal of the American Chemical Society* **1998**, <u>120</u>, 9076

## Examples:

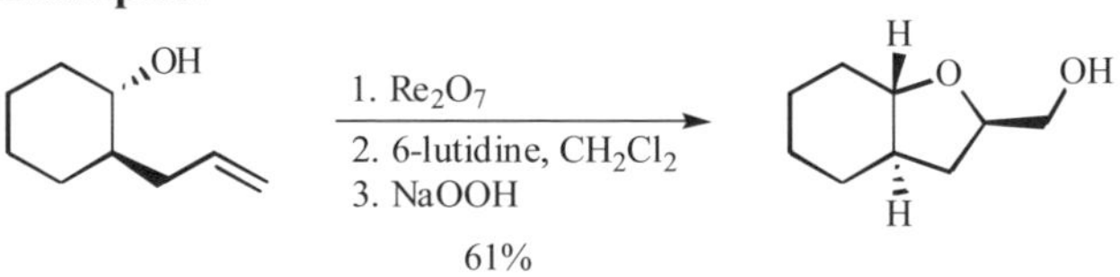

1. Re$_2$O$_7$
2. 6-lutidine, CH$_2$Cl$_2$
3. NaOOH

61%

S. Tang, R. M. Kennedy, *Tetrahedron Letters* **1992**, <u>33</u>, 5299

73.9%  13.1%

S. Tang, R. M. Kennedy, *Tetrahedron Letters* **1992**, <u>33</u>, 5299.

S. C. Sinha, E. Keinan, S. C. Sinha, *Journal of the American Chemical Society* **1998**, <u>120</u>, 9076

S. C. Sinha, A. Sinha, S. C. Sinha, E. Keinan, *Journal of the American Chemical Society* **1997**, <u>119</u>, 12014

There are other oxidative cyclizations related to the Kennedy protocol:

For example:

A. R. L. Cecil, Y. Hu, M. J. Vinent, R. Duncan, R. C. D. Brown, *Journal of Organic Chemistry* **2004**, <u>69</u>, 3368

# Knoevenagel Condensation (Reaction)

## The Reaction:

no α protons        a methylene with two
                    electron withdrawing groups

## Proposed Mechanism:

## Notes:

See: R. Bruckner, *Advanced Organic Chemistry*, Harcourt/Academic Press, San Diego, 2002, pp 419-422, for an interesting discussion of the active nucleophile in this reaction.

Another interpretation involves the formation of an intermediate iminium species as the electrophile: T. Laue, A. Plagens, *Named Organic Reactions*, John Wiley and Sons, Inc., New York, 1998, 164-167.

Iminium ion

M. B. Smith, J. March in March's *Advanced Organic Chemistry*, 5$^{th}$ ed., John Wiley and Sons, Inc., New York, 2001, pp 1225-1228; G, Jones, *Organic Reactions* **15**, 2.

When the amine catalyst is specifically pyridine, the reaction is known as the ***Doebner Modification of the Knoevenagel Reaction***:

**Examples:**

yields not reported

N. S. Reddy, M. R. Mallireddigari, S. Cosenza, K. Gumireddy, S. C. Bell, E. P. Reddy, M.V.R. Reddy, *Bioorganic & Medicinal Chemistry Letters* **2004**, *14*, 4093

90 - 95%

L. F. Tietze, U. Beifuss, *Organic Syntheses* CV 9, 310

Use of ionic liquids:
Solvent used several times, without additional base, to carry out the reaction.

96%

P. Formentin, H. Garcia, A. Leyva, *Journal of Molecular Catalysis A: Chemical* **2004**, *214*, 137

Tandem *Ugi-Knoevenagel Reaction*:

78%

S. Marcaccini, R. P. M. Cruz Pozo, S. Basurto, M. Garcı´a-Valverdeb, T. Torrobab, *Tetrahedron Letters* **2004**, *45*, 3999

75%

H. Hu, T. J. Harrison, P. D. Wilson, *Journal of Organic Chemistry* **2004**, *69*, 3782

42 - 46%

P. J. Jessup, C. B. Petty, J. Roos, L. E. Overman, *Organic Syntheses*, CV6, 95

# Knorr Pyrrole Synthesis

## The Reaction:

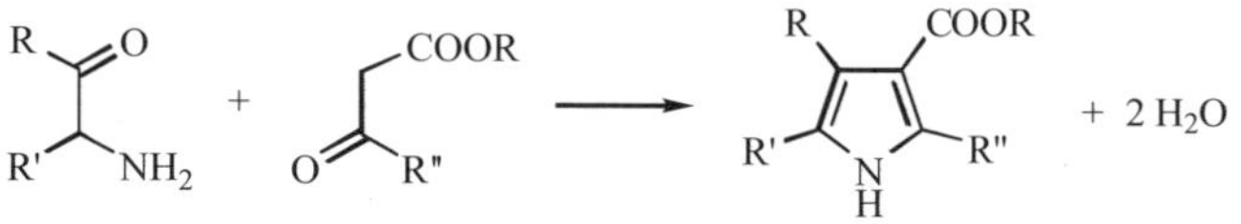

## Proposed Mechanism:

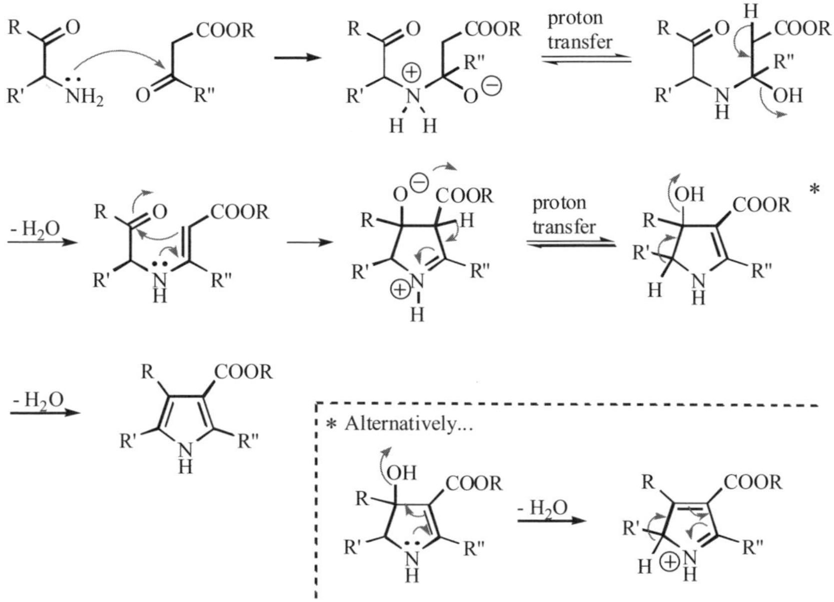

## Notes:

T. Laue, A. Plagens, *Named Organic Reactions*, John Wiley and Sons, Inc., New York, 1998, pp. 168-170.

## Examples:

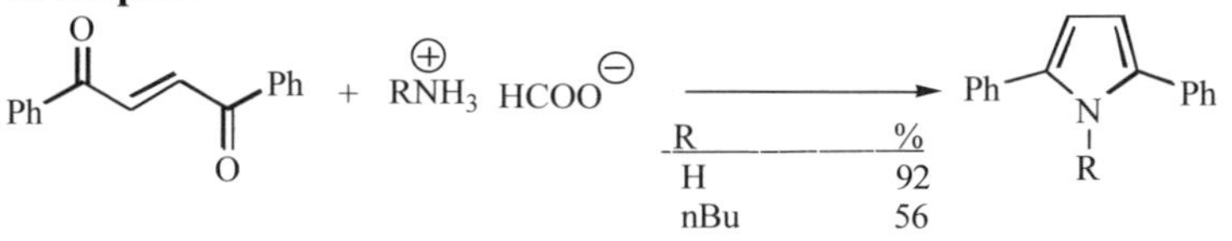

| R | % |
|---|---|
| H | 92 |
| nBu | 56 |

H. Surya P. Rao, S. Jothilingam, *Tetrahedron Letters* **2001**, 42, 6595

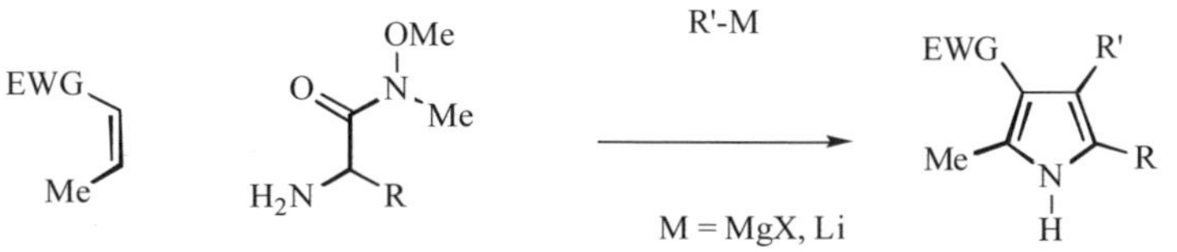

Weinreb Amide

A. Alberola, A. G. Ortega, M. L. Sidaba, C. Safiudo, *Tetrahedron* **1999**, 55, 6555

J. B. Hendrickson, R. Rees, *Journal of the American Chemistry Society* **1961**, 83, 1250

71% for both steps

J. M. Manley, M. J. Kalman, B. G. Conway, C. C. Ball, J. L. Havens, R. Vaidyanathan, *Journal of Organic Chemistry* **2003**, 68, 6447

63%

R. K. Bellingham, J. S. Carey, N. Hussain, D. O. Morgan, P. Oxley, L. C. Powling *Organic Process Research & Development* **2004**, 8, 279

1. HONO
2. Zn HOAc

55 - 60%

H. Fischer, *Organic Syntheses* CV3, 513

# Koch-Haaf Carbonylation (Reaction)

## The Reaction:

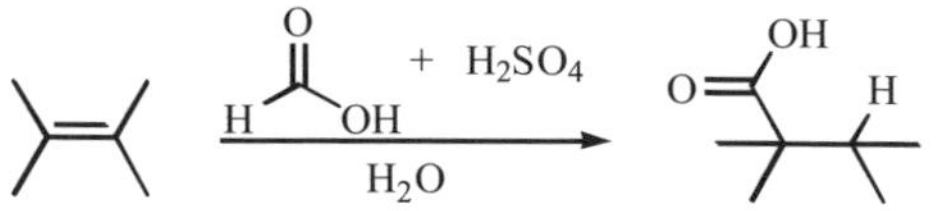

## Proposed Mechanism:

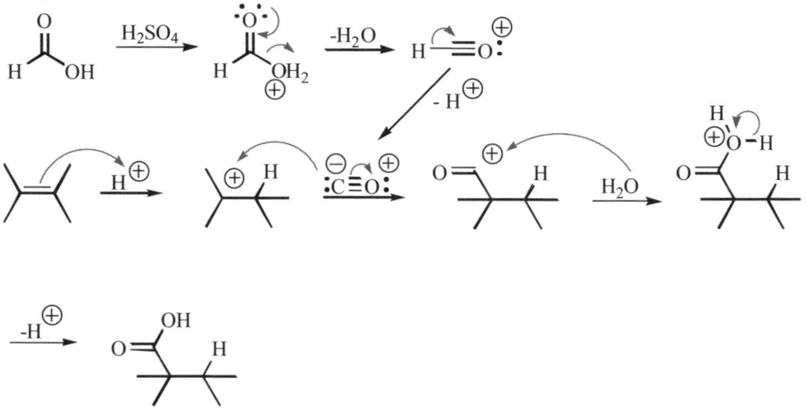

The carbocation can also come from an alcohol.

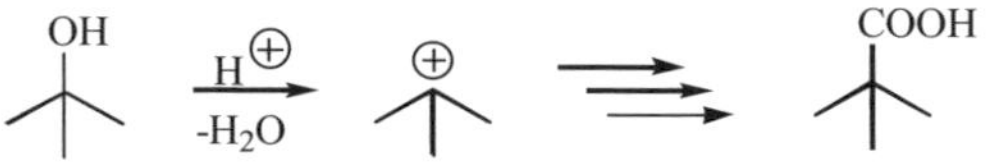

## Notes:

M. B. Smith, J. March in *March's Advanced Organic Chemistry*, 5th ed., John Wiley and Sons, Inc., New York, 2001, pp. 564, 1035.

The acid-catalyzed hydrocarboxylation of an alkene is known as the **Koch Reaction**. When the source of both the CO and the $H_2O$ is formic acid, the process is called the **Koch-Haaf Carbonylation**.

## Examples:

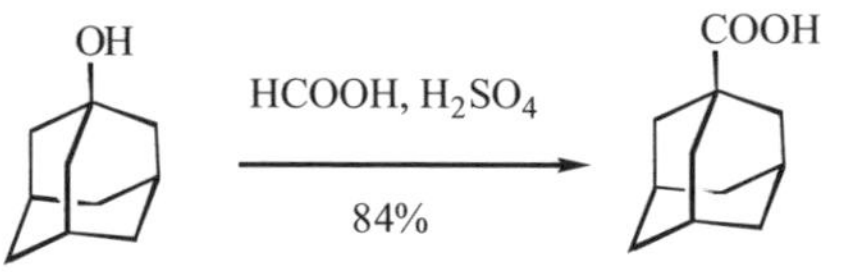

H. Langhals, I. Vergelsberg, C. Riichardt, *Tetrahedron Letters* **1981**, <u>22</u>, 2365

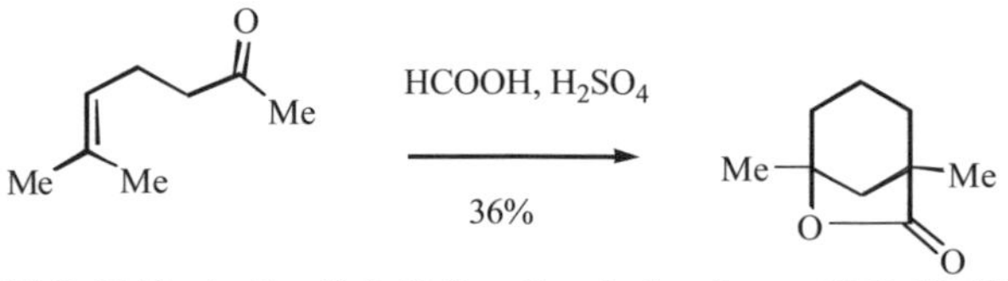

M. E. N. Nambudiry, G. S. K. Rao, *Tetrahedron Letters* **1972**, <u>13</u>, 4707

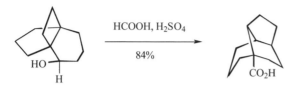

D. J. Raber, R. C. Fort, E. Wiskoit, C. W. Woodworth, P. v. R. Schleyer, J. Weber, H. Steer, *Tetrahedron* **1971**, 27, 3

K. Kakiuchi, M. Ue, I. Wakaki, Y. Tobe, Y. Odaira, M. Yasuda, K. Shimal, *Journal of Organic Chemistry* **1986**, 51, 282

# Köenigs-Knorr Synthesis / Reaction / Method / Glycosidation

## The Reaction:

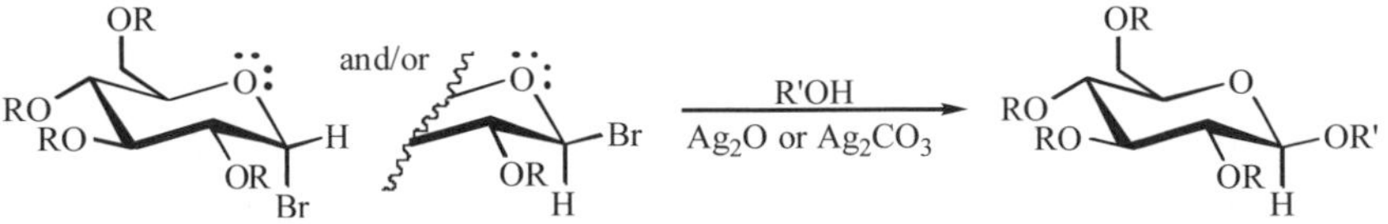

## Proposed Mechanism:

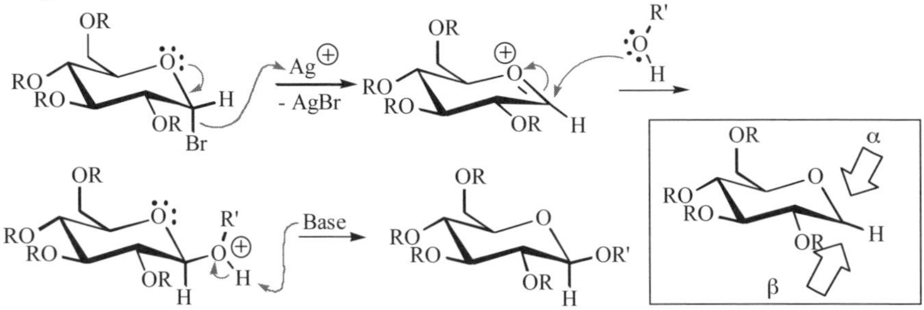

α– attack preferred.

## Notes:

An adjacent acetoxy group can participate in the reaction:

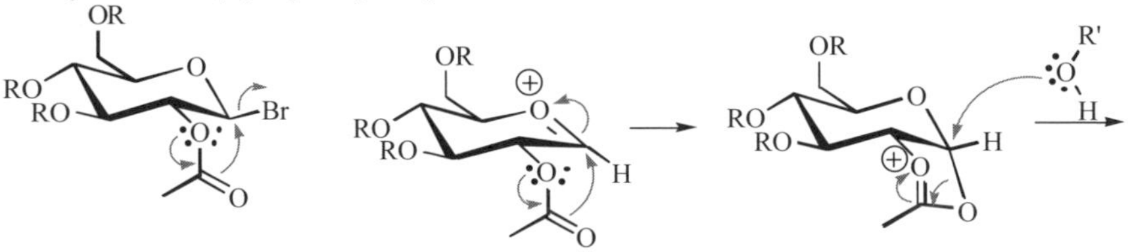

In the early works, Hg salts were often used in place of the Ag salts to initiate the reaction. This has been called the *Zemplen-Helferich modification*.

## Examples:

An improvement over the traditional method to avoid use of Hg salts.

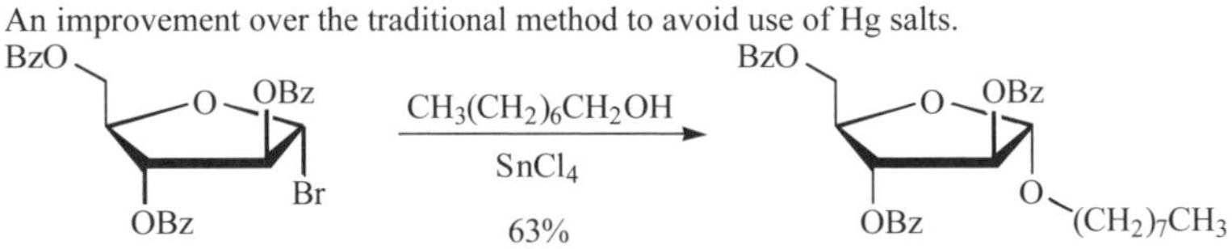

$CH_3(CH_2)_6CH_2OH$

$SnCl_4$

63%

The procedure also replaces anomeric acetate groups.

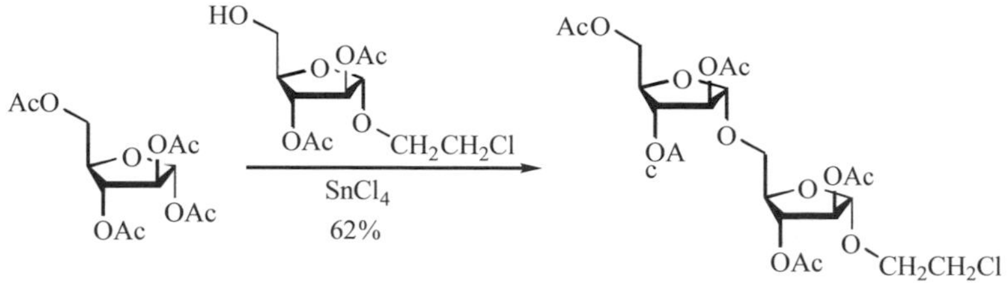

A. K. Pathak, Y. A. El-Kattan, N. Bansal, J. A. Maddry, R. C. Reynolds, *Tetrahedron Letters* **1998**, 39, 1497

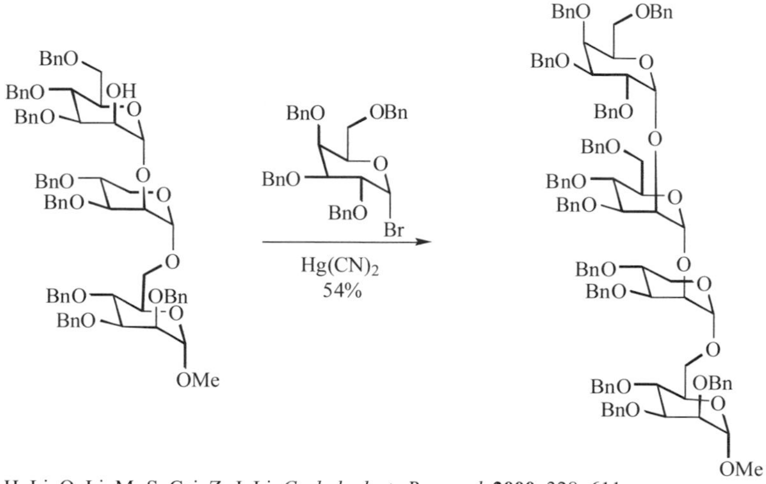

H. Li, Q. Li, M.-S. Cai, Z.-J. Li, *Carbohydrate Research* **2000**, 328, 611

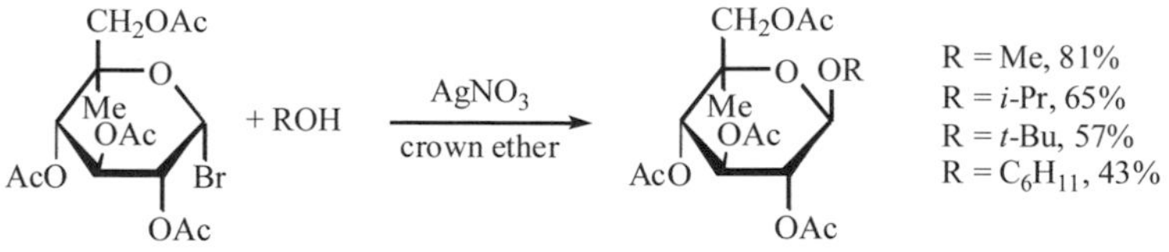

R = Me, 81%
R = *i*-Pr, 65%
R = *t*-Bu, 57%
R = C₆H₁₁, 43%

A. Knochel, G. Rudopph, J. Thiem, *Tetrahedron Letters* **1974**, 15, 55

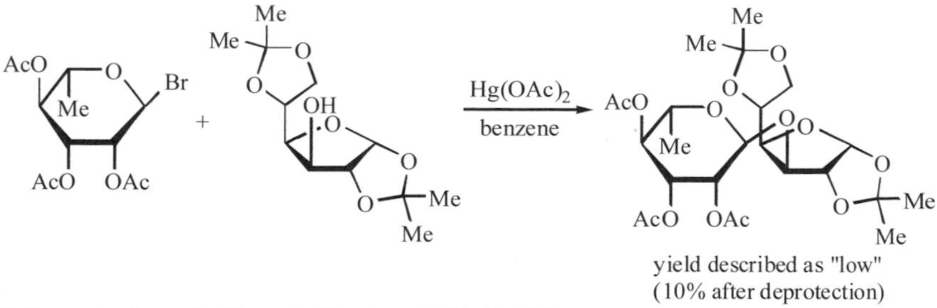

yield described as "low"
(10% after deprotection)

F. Imperato, *Journal of Organic Chemistry* **1976**, 41, 3478

# Kolbé Electrolysis

## The Reaction:

$$2 \quad \underset{R}{\overset{O}{\|}}\!\!-O\!-H \quad \xrightarrow[\text{2. electrolysis}]{\text{1. Base}} \quad R\!-\!R \ + CO_2$$

## Proposed Mechanism:

$$2 \quad \underset{R}{\overset{O}{\|}}\!\!-O\!-H \quad \xrightarrow{\text{Base}} \quad 2 \quad \underset{R}{\overset{O}{\|}}\!\!-O^{\ominus} \ M^{\oplus} \quad \xrightarrow[\text{anode}]{-e^{\ominus}} \quad 2 \quad \underset{R}{\overset{O}{\|}}\!\!-O^{\cdot} \quad \xrightarrow{-CO_2}$$

$$\text{electrolysis}$$

$$R\!\cdot \quad \cdot R \quad \longrightarrow \quad R\!-\!R$$

## Notes:

M. B. Smith, J. March in *March's Advanced Organic Chemistry*, 5th ed., John Wiley and Sons, Inc., New York, 2001, p. 942; T. Laue, A. Plagens, *Named Organic Reactions*, John Wiley and Sons, Inc., New York, 1998, pp. 170-171.

1. The **Brown-Walker Electrolysis** is useful for dimerizing half-acid-esters:

$$2 \quad \underset{RO}{\overset{O\ \ \ O}{\|\ \ \ \|}}\!\!\underset{n}{()}\!\!-OH \quad \longrightarrow \quad \underset{RO}{\overset{O}{\|}}\!\!\underset{n}{()}\!\!\underset{m}{()}\!\!\underset{O}{\overset{}{\|}}\!\!-OR$$

2. Mixed coupling can be accomplished by relying on statistics and using the less-costly acid in excess. Prediction of yield of R-R' for the reaction:

$$R\text{-COOH} + R'\text{-COOH} \longrightarrow R\text{-R} + R'\text{-R'} + R\text{-R'}$$

YIELD = $100N / (1+N)$, assuming a 1:N ratio of the two acids.

H.J. Schafer, in *Comprehensive Organic Synthesis*, B. M. Trost, editor-in-chief, Pergamon Press, Oxford, 1991, **3**, Chapter 2.8

See: M.A. Iglesias-Arteaga, E. Juaristi, F. J. Gonzalez, *Tetrahedron* **2004**, <u>60</u>, 3605 for a detailed discussion of the electrochemistry.

## Examples:

K. Schierle, J. Hopke, M.-L. Niedt, W. Boland, E. Steckhan, *Tetrahedron Letters* **1996**, <u>37</u>, 8715

$$C_{12}H_{25}\text{-CO}_2H \ + \quad \xrightarrow[\text{Et}_3N]{\text{electrolysis}}$$

40%

A. Forster, J. Fitremann, P. Renaud, *Tetrahedron Letters* **1998**, <u>39</u>, 7097

J. Hiebl, H. Kollmann, F. Rovenszky, K. Winkler *Bioorganic & Medicinal Chemistry Letters* **1997**, 7, 2963

L. Becking, H. J. Schafer, *Tetrahedron Letters* **1988**, 29, 2797

# Kolbé-Schmitt Reaction (Kolbé Synthesis, Schmitt Synthesis)

## The Reaction:

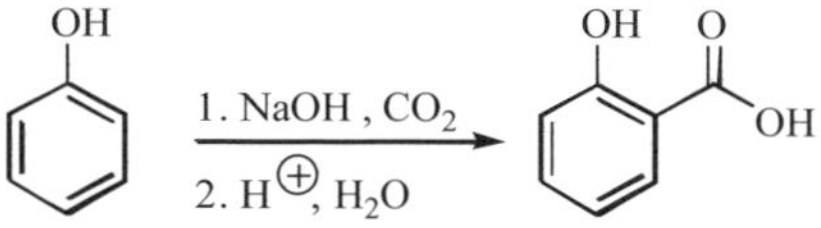

$$\text{OH} \xrightarrow[\text{2. H}^{\oplus}, \text{H}_2\text{O}]{\text{1. NaOH , CO}_2} \text{2-hydroxybenzoic acid}$$

## Proposed Mechanism:

keto-enol tautomerism

and aromatization

$$\xrightarrow{\text{H}^{\oplus}}$$

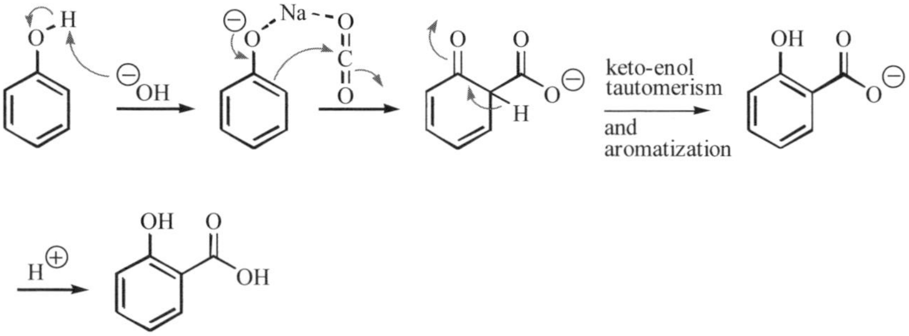

## Notes:

Recent work shows that a phenoxide-$CO_2$ complex is formed competitively with direct carboxylation. This does not go on to form products, but rather decomposes to phenoxide and carbon dioxide. (Y. Kosugi, Y. Imaoka, F. Gotoh, M. A. Rahim, Y. Matsui, K. Sakanishi, *Organic and Biomolecular Chemistry* **2003**, <u>1</u>, 817)

### *Lederer-Manasse Reaction*

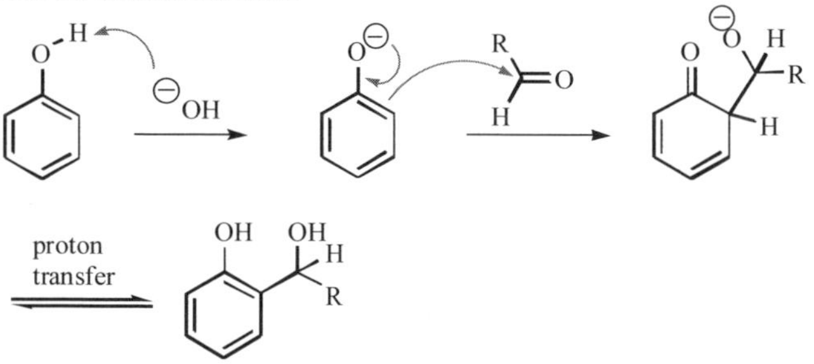

proton transfer

## Examples:

A. C. Regan, J. Staunton, *Journal of the Chemical Society, Chemical Communications* **1987**, 520

A. Fuerstner, N. Kindler, *Tetrahedron Letters* **1996**, 37, 7005

M. Hauptschein, E. A. Nodiff, A. J. Saggiomo, *Journal of the American Chemical Society* **1954**, 76, 1051

W. H Meek, C. H. Fuchsman, *Journal of Chemical and Engineering Data* **1969**, 14, 388

# Kornblum Aldehyde Synthesis

## The Reaction:

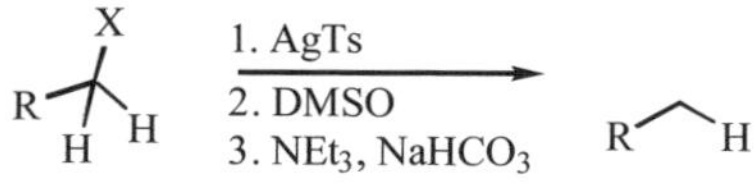

$$\underset{\substack{H\phantom{xx}H}}{\overset{X}{R\diagdown\!\!\!\!\diagup\!\!\!\!\diagdown H}} \quad \begin{array}{l} \text{1. AgTs} \\ \overline{\text{2. DMSO}} \\ \text{3. NEt}_3, \text{NaHCO}_3 \end{array} \quad R\diagdown\!\!\!\!\diagup H$$

X = I, Br, OTs

## Proposed Mechanism:

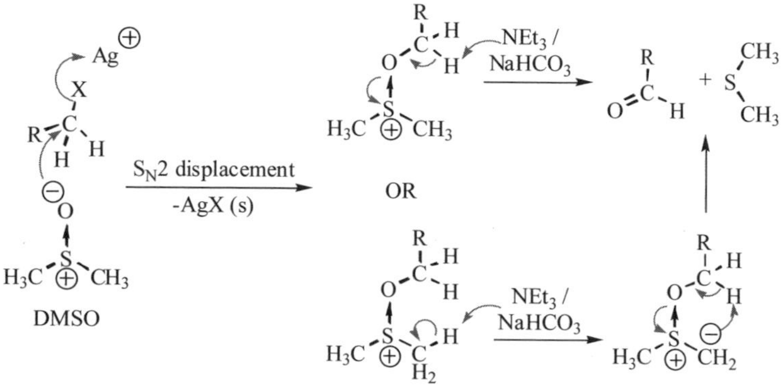

## Notes:
T. T. Tidwell, *Organic Reactions* **39**, 3

The ***Hass or Hass-Bender Reaction*** will specifically convert a benzylic halide to an aldehyde:

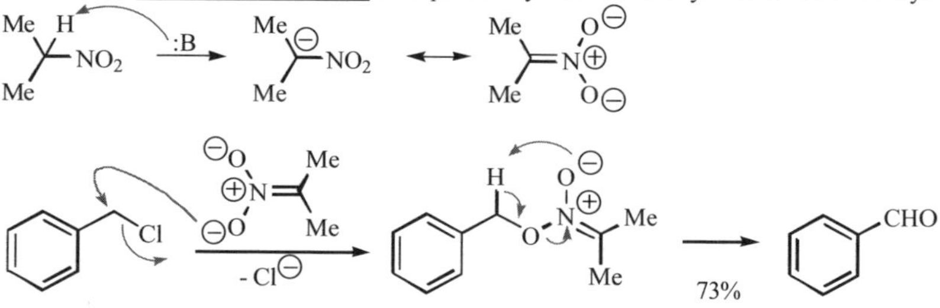

H. B. Hass, M. L. Bender, *Journal of the American Chemical Society* **1949**, <u>71</u>, 1767

## Examples:

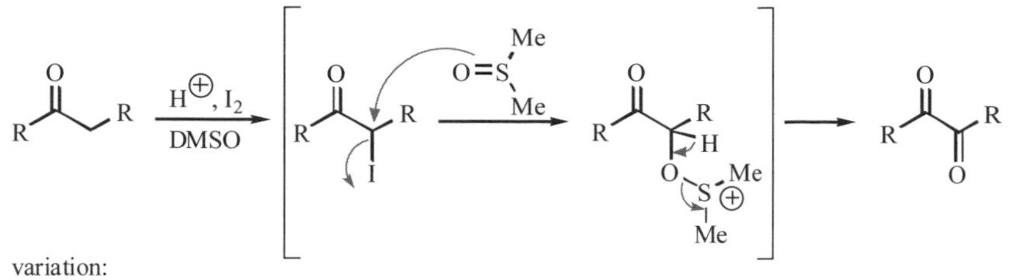

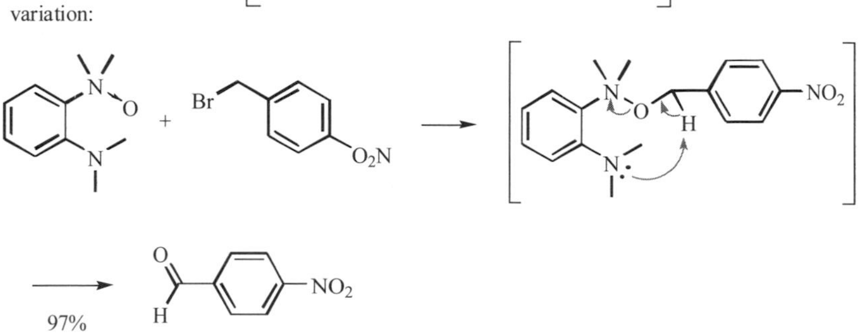

variation:

97%

S. Chandrasekhar, M. Sridhar, *Tetrahedron Letters* **2000**, <u>41</u>, 5423

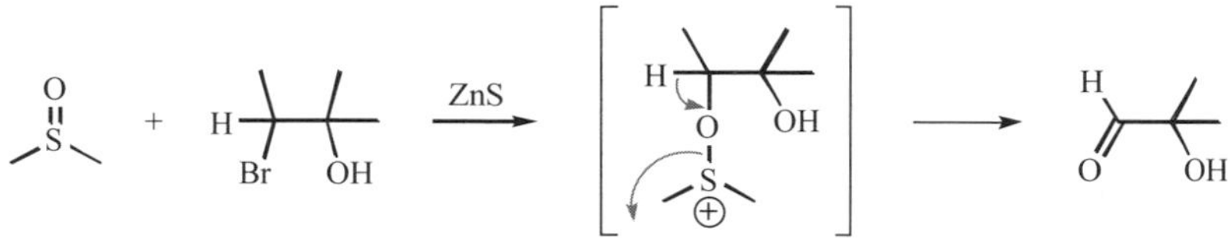

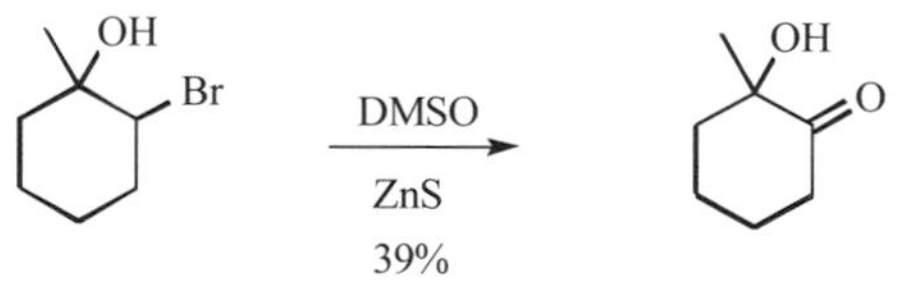

39%

B. K. Bettadaiah, K. N. Burudutt, P. Srinivas, *Journal of Organic Chemistry* **2003**, <u>68</u>, 2460

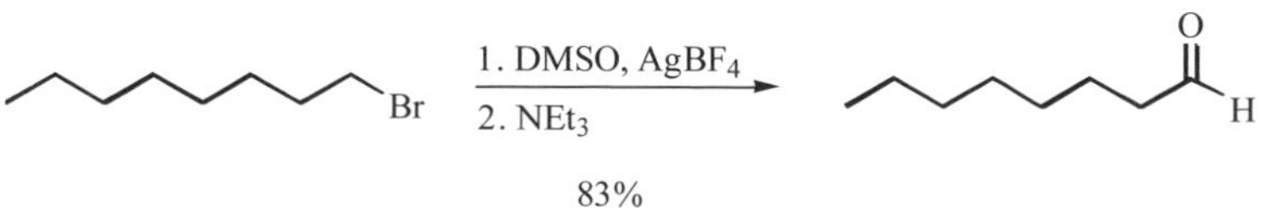

1. DMSO, AgBF$_4$
2. NEt$_3$

83%

B. Ganem, R. K. Boeckman, *Tetrahedron Letters* **1974**, 917

# Kostanecki Acylation

## The Reaction:

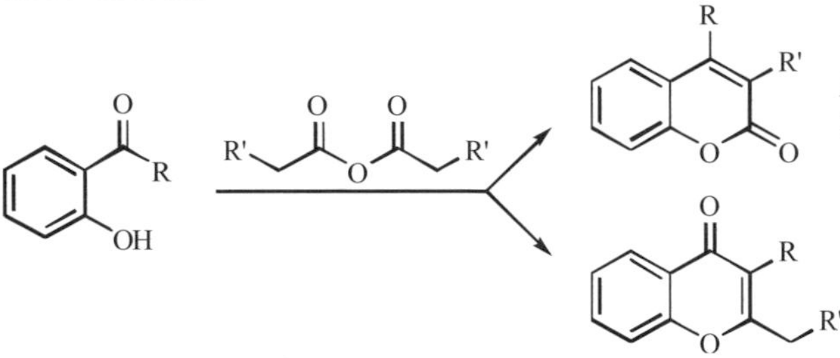

*Kostanecki Acylation* product, a coumarin

*Allan-Robinson Reaction* product, a chromone

## Proposed Mechanism:

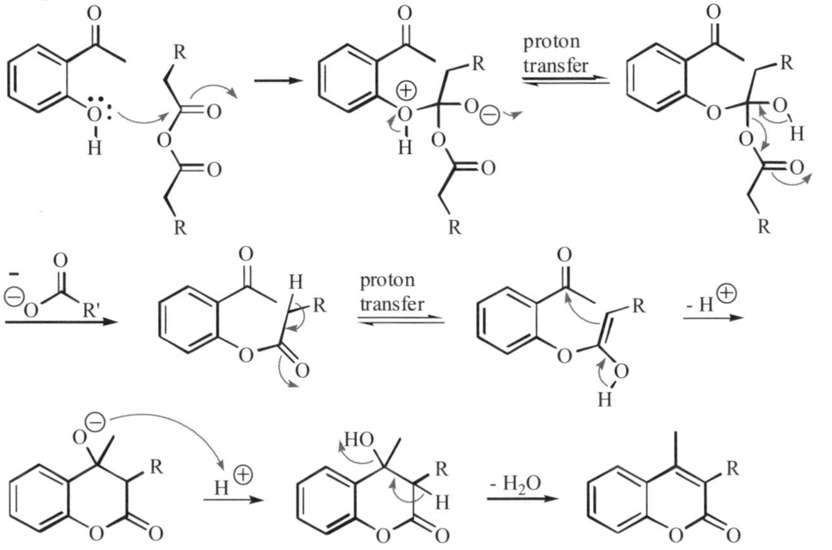

proton transfer

proton transfer

*Kostanecki Acylation* Product = Coumarin

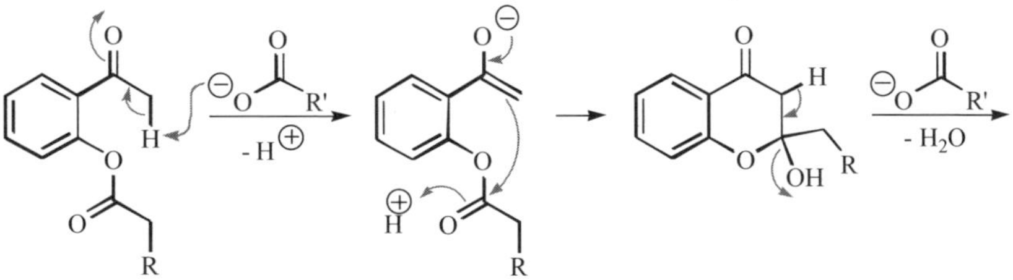

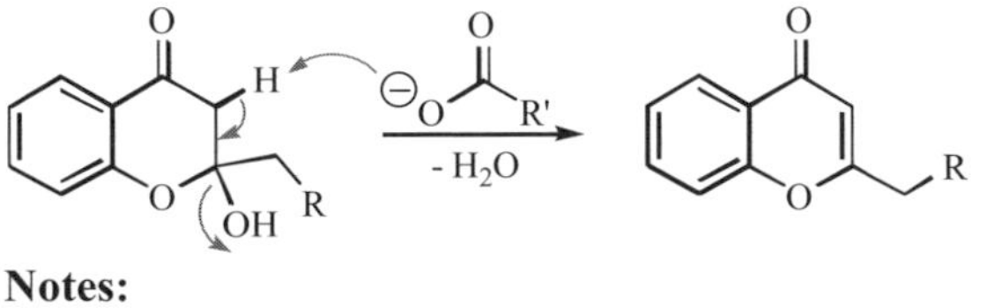

*Allan-Robinson Reaction* Product = Chromone

## Notes:

There is competition between products from *Allan - Robinson* and *Kostanecki Acylation*. A search of literature shows that both product types are often classified as the latter. This is exemplified with examples, all of which are searchable under *Kostanecki Acylation*.

## Examples:

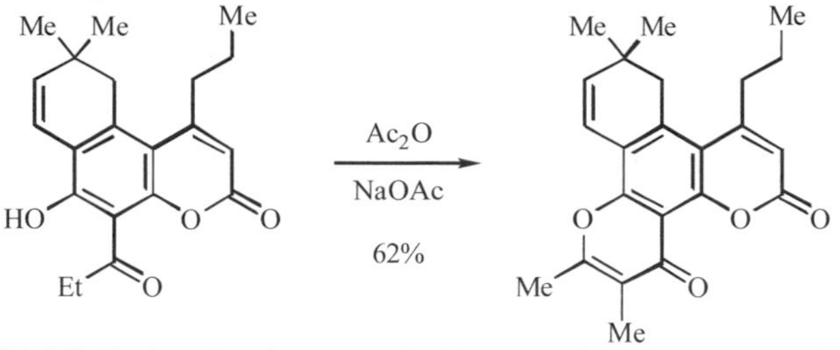

M. I. T. Flavin, J. D. Rizzo, A. Khilevich, A. Kucherenko, A. K. Sheinkman, V. Vilaychack, L. Lin, W. Chen, E. M. Greenwood, T. Pengsuparp, J. M. Pezzuto, S. H. Hughes, T. M. Flavin, M. Cibulski, W. A. Boulanger, R. L. Shone, Z.- Q. Xu, *Journal of Medicinal Chemistry* **1996**, <u>39</u>, 1303

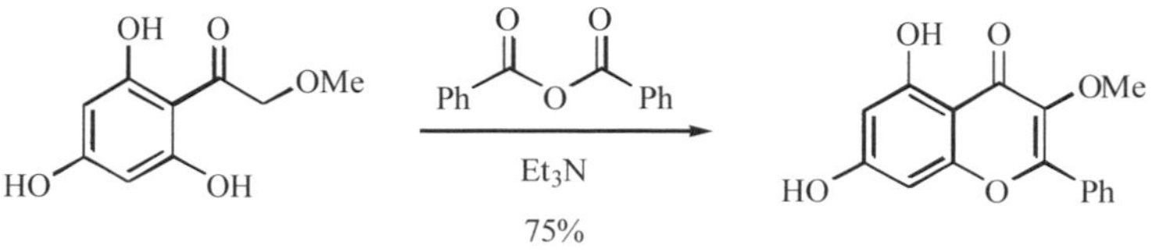

J. H. Looker, J. H. McMechan, J. W. Mader, *Journal of Organic Chemistry*, **1978**, <u>43</u>, 2344

Using the Scheme shown below, a series of substituted chromones was prepared. Yields ranged from 13 - 67%

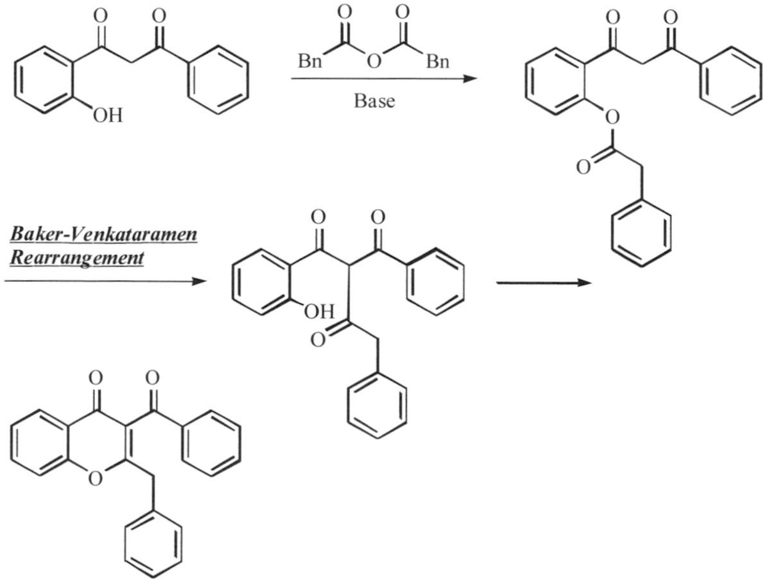

V. Rossollin, V. Lokshin, A. Samat, R. Guglielmetti, *Tetrahedron* **2003**, <u>59</u>, 7725

# Krapcho Dealkoxycarbonylation

## The Reaction:

$$MX = ( \text{Li or Na} )MCl, \ LiI, \ (\text{Na or K}) \ MCN$$

## Proposed Mechanism:

As evidence for an intermediate anion, there have been applications where an electrophile is captured.

## Notes:

In solvents such as DMSO and DMF, the Cl⁻ is not solvated; thus, it's nucleophilicity is enhanced for the required $S_N2$ displacement.

This reaction is greatly accelerated under microwave conditions.

R = H  8 min, 96%

R = Bu  20 min, 89%

L. Perreux, A. Loupy, *Tetrahedron* **2001**, <u>57</u>, 9199.

## Examples:

NaCl, DMSO, H₂O
Δ

*(no yield given. The reaction product taken directly to the next step.)*

P. J. Garratt, J. R. Porter, *Journal of Organic Chemistry* **1986**, <u>51</u>, 5450

NaCl, H₂O → DMSO, 92%

NaCl, H₂O → DMSO, 77%

(1.7 : 1 mixture)

G. H. Posner, E. M. Shulman-Roskes, *Tetrahedron* **1992**, <u>48</u>, 4677

KCN, DMSO, 93%

P. A. Evans, L. J. Kennedy, *Journal of the American Chemical Society* **2001**, <u>123</u>, 1234

DMSO, H₂O, Δ, 99%

D. A. Evans, K. A. Scheidt, C. W. Downey, *Organic Letters* **2001**, <u>3</u>, 3009

DMSO, H₂O, 88%

A. Toro, P. Nowak, P. Deslongchamps, *Journal of the American Chemical Society* **2000**, <u>122</u>, 4526

H₂O  DMF, mic rowave, 92%

Via direct attack of the carbonyl by water, followed by decarboxylation of the resulting half acid.

D. P. Curran, Q. Zhang, *Advanced Synthesis and Catalysis* **2003**, <u>345</u>, 329

# Kröhnke Pyridine Synthesis

## The Reaction:

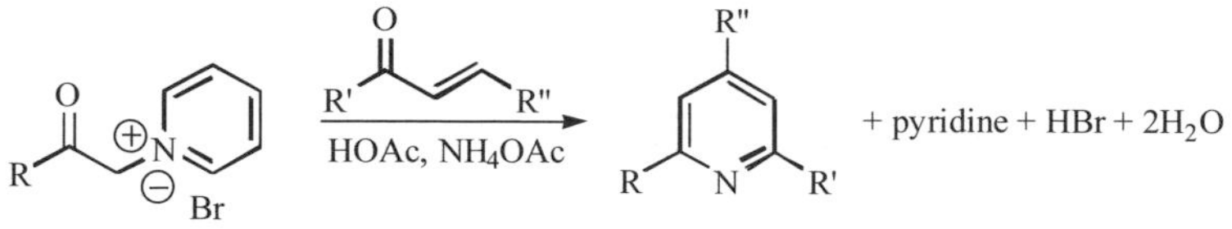

+ pyridine + HBr + 2H$_2$O

## Proposed Mechanism:

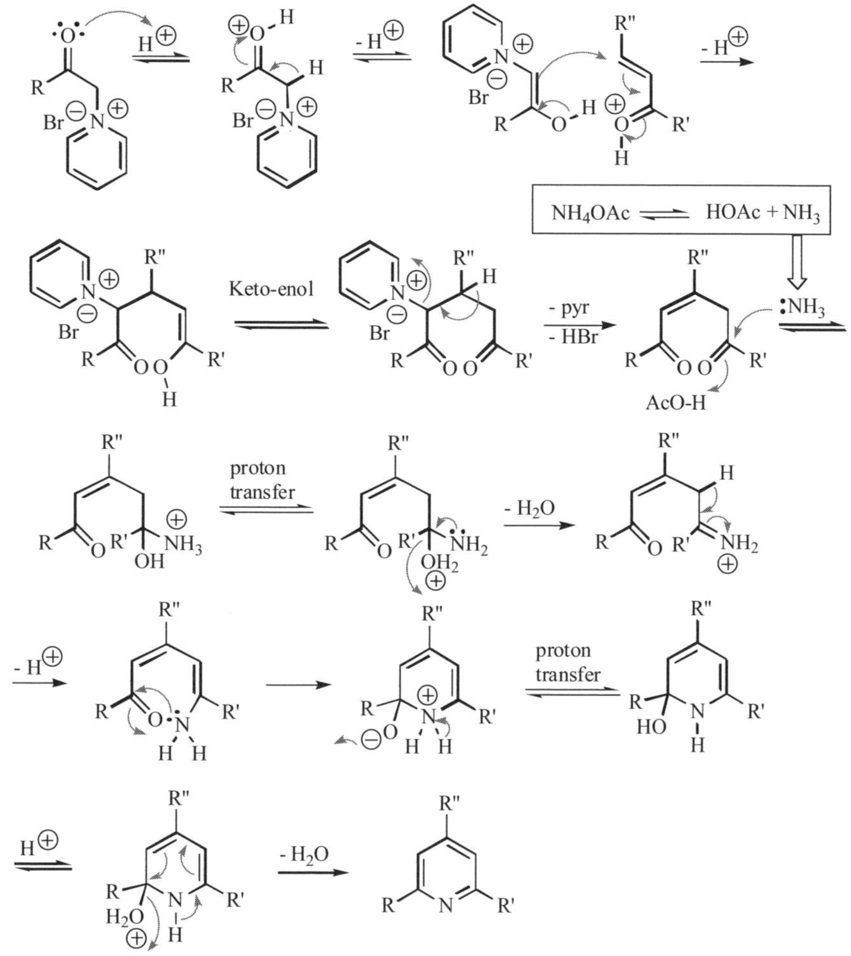

NH$_4$OAc $\rightleftharpoons$ HOAc + NH$_3$

Keto-enol

- pyr
- HBr

:NH$_3$

AcO-H

proton
transfer

- H$_2$O

- H$^\oplus$

proton
transfer

H$^\oplus$

- H$_2$O

## Examples:

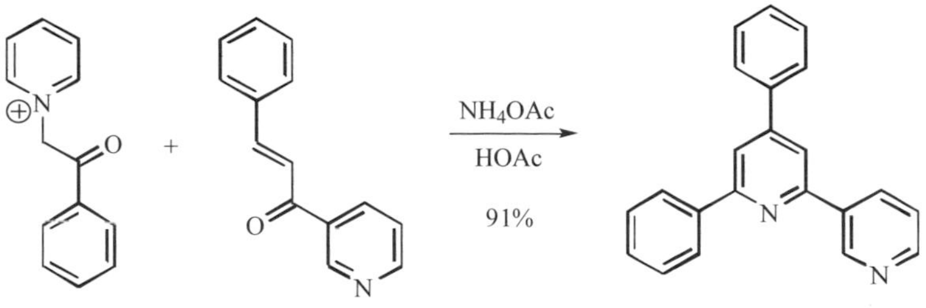

NH₄OAc / HOAc

91%

F. Krohnke, *Synthesis* **1976**, 1 (Review)

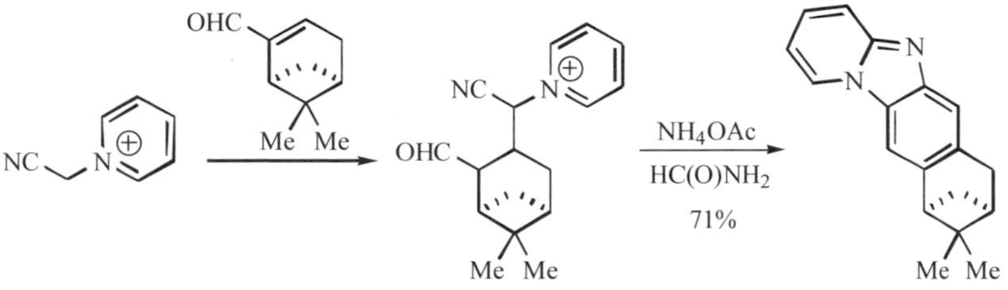

NH₄OAc / HC(O)NH₂

71%

N. C. Fletcher, D. Abeln, A. von Zelewsky *Journal of Organic Chemistry* **1997**, 62, 8577

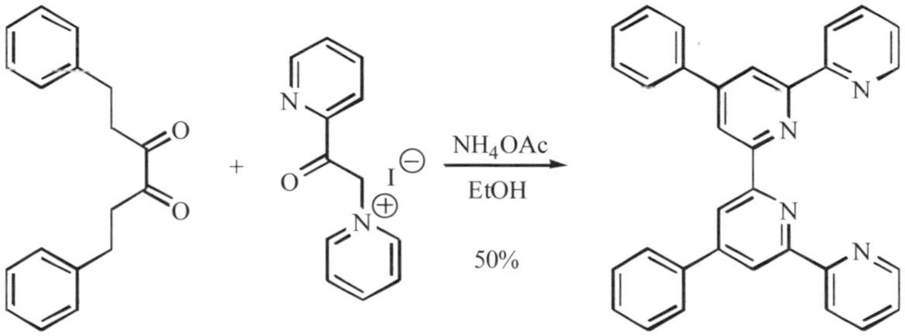

NH₄OAc / EtOH

50%

E. C. Constable, M. J. Hannon, D. R. Smith, *Tetrahedron Letters* **1994**, 35, 6657

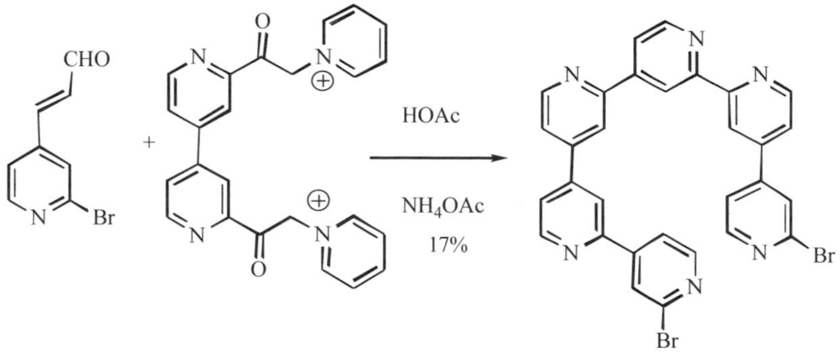

HOAc

NH₄OAc

17%

T. R. Kelly, Y.-J. Lee, R. J. Mears, *Journal of Organic Chemistry.* **1997**, 62, 2774

# Kucherov Reaction

## The Reaction:

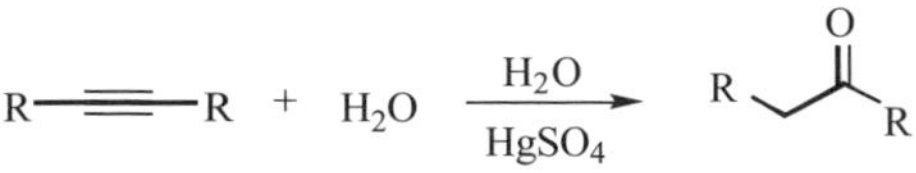

## Proposed Mechanism:

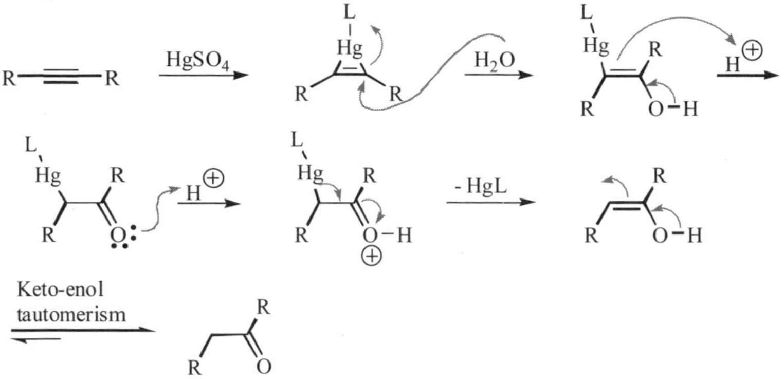

HgSO4
H2O
-HgL
Keto-enol
tautomerism

## Notes:

See also: ***Oxy-and Solvomercuration***

A discussion about the protonation of acetylene: V. Lucchini and G. Modena, *Journal of the American Chemical Society* **1990**, 112, 6291

Other alkyne hydration reactions:

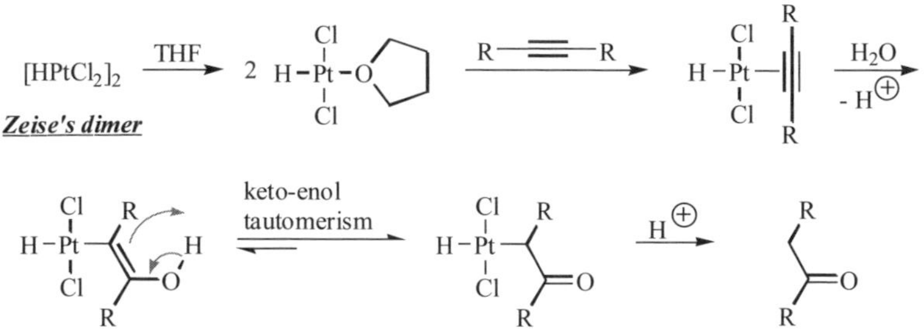

[HPtCl2]2
THF
Zeise's dimer
H2O
keto-enol
tautomerism

By this method:

[HPtCl2]2
THF
93%
50%
2 : 1

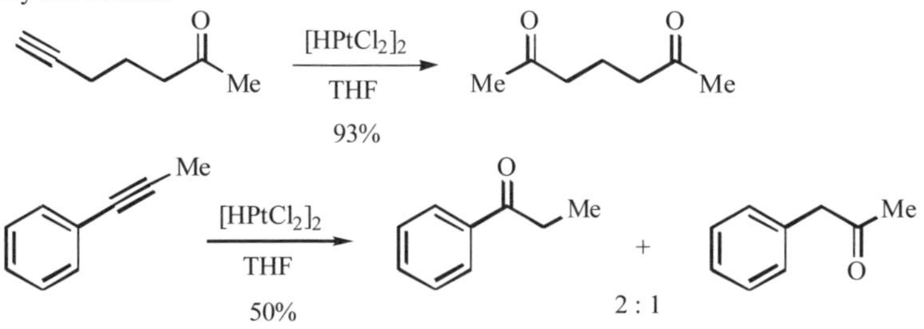

W. Hiscox, P. W. Jennings, *Organometallics* **1990** 9, 1997

## Examples:

Anti-Markovnikov hydration of a terminal alkyne:

$$\text{Me}\diagdown\diagdown\diagup\equiv\text{—H} \xrightarrow[\textit{i-}\text{PrOH, under Ar}]{\text{RuCpCl(dppe), H}_2\text{O}} \text{Me}\diagdown\diagup\diagdown\diagup\diagdown\text{CHO}$$

>99%

T. Suzuki, M. Tokunaga, Y. Wakatsuki, *Organic Letters* **2001**, <u>3</u>, 735

$$\xrightarrow[\text{2. H}_2\text{O}]{\text{1. PhHgOH, CHCl}_3}$$

49%

V. Janout, S. L. Regen, *Journal of Organic Chemistry* **1982**, <u>47</u>, 3331

## Hydration of a nitrile:

$$\text{R—}\equiv\text{N} \xrightarrow[\text{H}_2\text{O}]{\text{Hg(OAc)}_2} \text{R—}\overset{\oplus}{\equiv}\text{N—HgOAc} \xrightarrow[\text{H}_2\text{O}]{\overset{\ominus}{\text{OAc}}}$$

$$\left[ \underset{\text{R}}{\overset{\text{AcO}}{>}}{=}\underset{\text{HgOAc}}{\text{N}} \quad \text{or} \quad \underset{\text{R}}{\overset{\text{HO}}{>}}{=}\underset{\text{HgOAc}}{\text{N}} \right] \xrightarrow{\text{H}_2\text{O}} \underset{\text{R}}{\overset{\text{HO}}{>}}{=}\underset{\text{H}}{\text{N}}$$

$$\xrightarrow[\text{tautomerism}]{\text{keto-enol}} \underset{\text{R}}{\overset{\text{O}}{\diagup\diagdown}}\text{NH}_2$$

By this method:

$$\text{(C}_6\text{H}_5\text{)CN} \longrightarrow \text{(C}_6\text{H}_5\text{)C(=O)NH}_2$$

73%

$$\text{Ph}\diagdown\triangle\diagup\text{CN} \longrightarrow \text{Ph}\diagdown\triangle\diagup\text{C(=O)NH}_2$$

92%

K. Maeyama, N. Iwasawa, *Journal of the American Chemical Society* **1998**, <u>120</u>, 1928

# Kumada Coupling Reaction

## The Reaction:

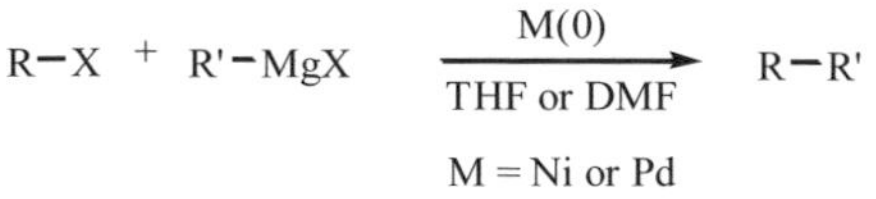

$$R-X \ + \ R'-MgX \ \xrightarrow[\text{THF or DMF}]{M(0)} \ R-R'$$

$$M = Ni \text{ or } Pd$$

## Proposed Mechanism:

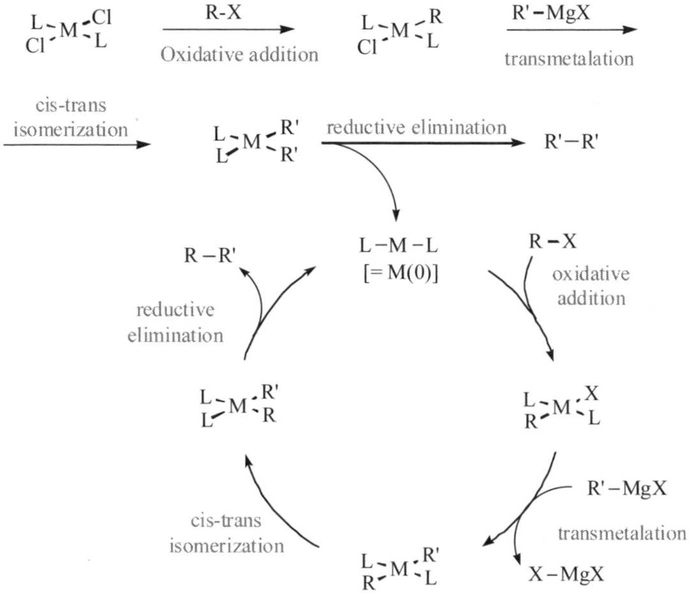

## Notes:

For R-X: When M = Ni, Cl > Br > I; when M = Pd, I > Br > Cl

Reaction is limited by any functional group that will normally react with a **_Grignard reagent_**.

## Examples:

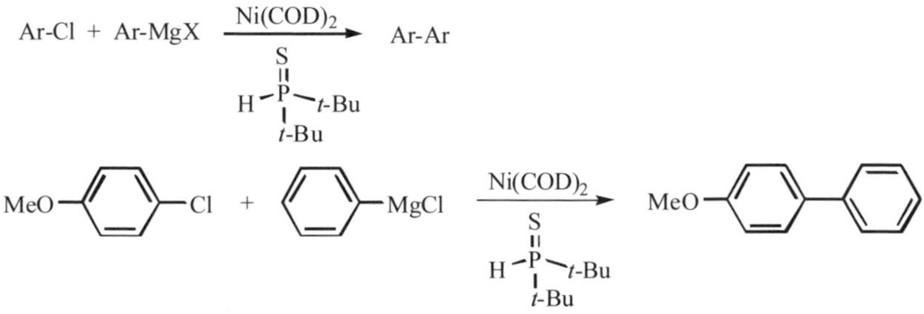

G. Y. Li, W. J. Marshall, *Organometallics* **2002**, <u>21</u>, 590

An important influence of butadiene on the reaction has been noted:

$$n\text{-}C_{10}H_{21}Br + n\text{-}BuMgBr \xrightarrow[\text{Me}]{\text{NiCl}_2} n\text{-}C_{14}H_{30}$$

PdCl$_2$ catalysis provides only 38% yield.

100%

+  $n$-BuMgCl $\xrightarrow[\text{100%}]{\text{NiCl}_2}$

**Mechanistically:**

J. Terao, H. Watanabe, A. Ikumi, H. Kuniyasu, N. Kambe, *Journal of the American Chemical Society* **2002**, <u>124</u>, 4222

## Examples:

P. Walla, C. O. Kappe, *Chemical Commmunications* **2004**, 564

K. Tamoa, S. Kodama, I. Nakajima, M. Kumada, A. Minato, K. Suzuki, *Tetrahedron* **1982**, <u>38</u>, 3347

# Lemieux-Johnson Oxidation

## The Reaction:

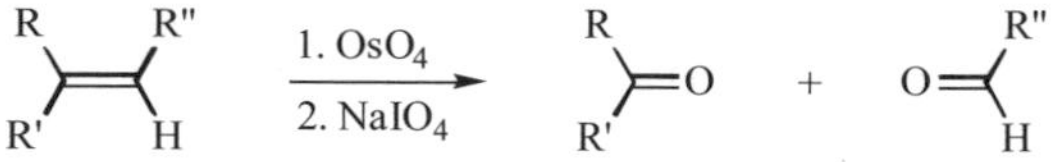

## Proposed Mechanism:

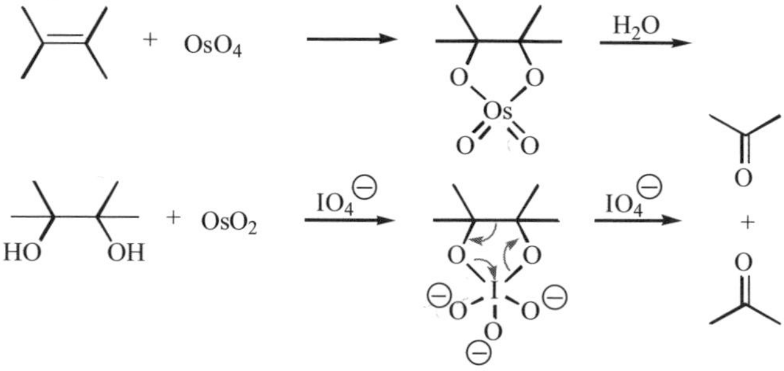

The NaIO$_4$ serves two purposes: (1) cleavage of the diol generated from the OsO$_4$ and, (2) reoxidation of the reduced osmium.

## Notes:

See also the **Lemieux-von Rudloff Reagent**, where catalytic KMnO$_4$ is used in place of OsO$_4$:

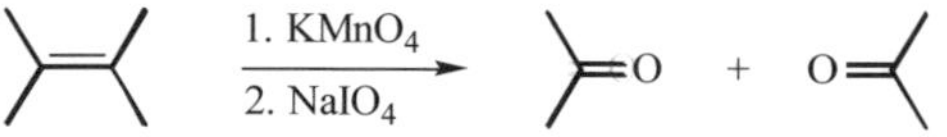

The **Malaprade Periodic Acid Oxidation Reaction** oxidizes 1,2 diols or 2-amino alcohols with periodic acid:

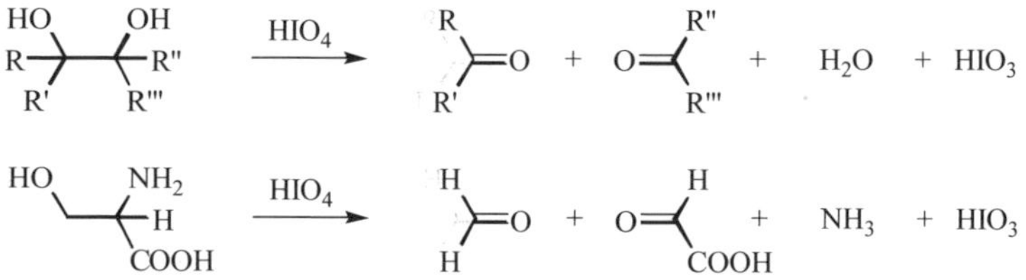

B. H. Nicolrt, L. A. Shinn, *Journal of the American Chemical Society* **1939**, _61_, 1615

Recent example of **Malaprade Reaction**:

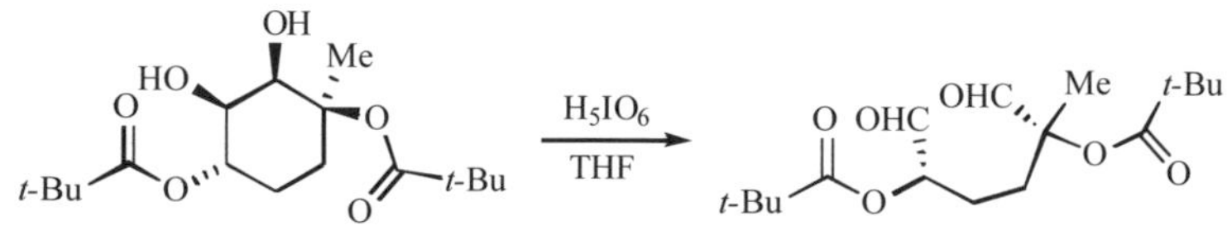

L. Vares, T. Rein, *Journal of Organic Chemistry* **2002**, _67_, 7226

**Criegee Glycol Oxidation**

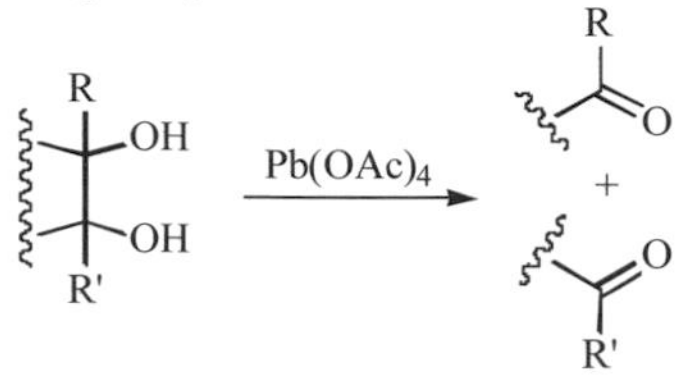

See: **Criegee Reagent**

## Examples:

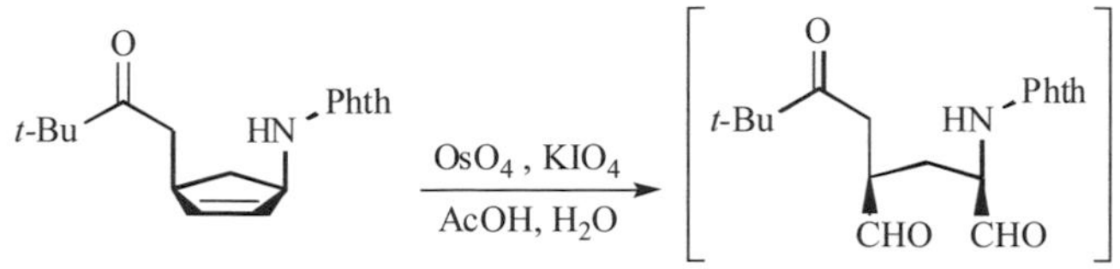

Not isolated; carried on to the next step.

F. A. Luzzio, A. V. Mayrov, W. D. Figg, *Tetrahedron Letters* **2000**, 41, 2275

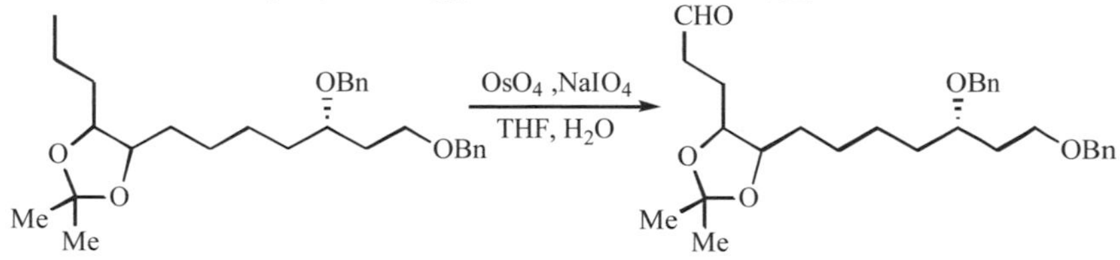

Taken directly to the next step.

S. Takahashi, A. Kubota, T. Nakata, *Organic Letters* **2003**, 5, 1353

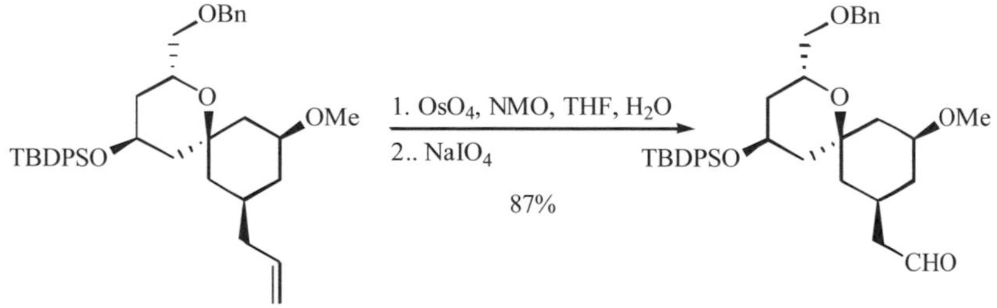

1. OsO$_4$, NMO, THF, H$_2$O
2. NaIO$_4$

87%

D. Zuev, L. A. Paquette, *Organic Letters* **2000**, 2, 679

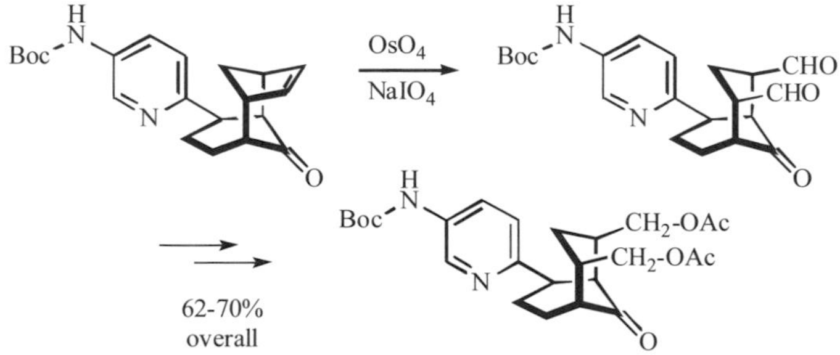

62-70%
overall

M. J. Sung, H. I. Lee, Y. Chong, J. K. Cha, *Organic Letters* **1999**, 1, 2017

# Leuckart Reaction / Reductive Amination

## The Reaction:

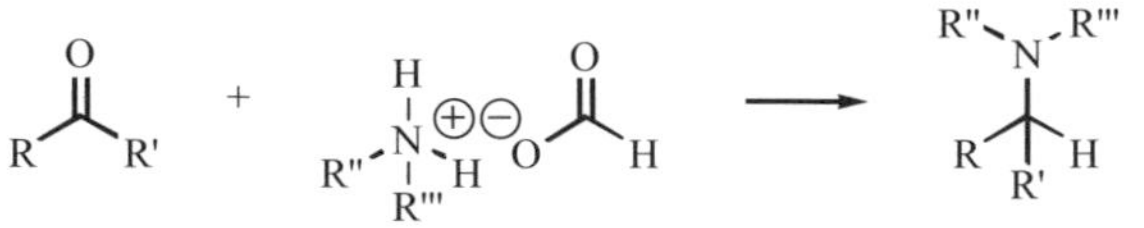

## Proposed Mechanism:

A common mechanistic interpretation involves hydride transfer in the reduction of the immonium ion intermediate:

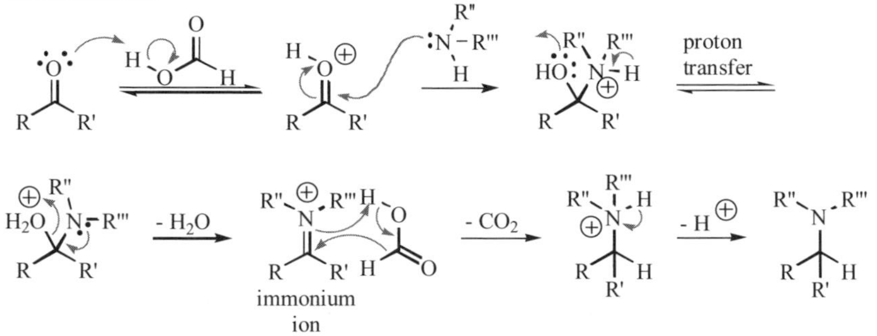

immonium
ion

Another interpretation:
See: P. I. Awachie, V. C. Agwada, *Tetrahedron* **1990**, <u>46</u>, 1899

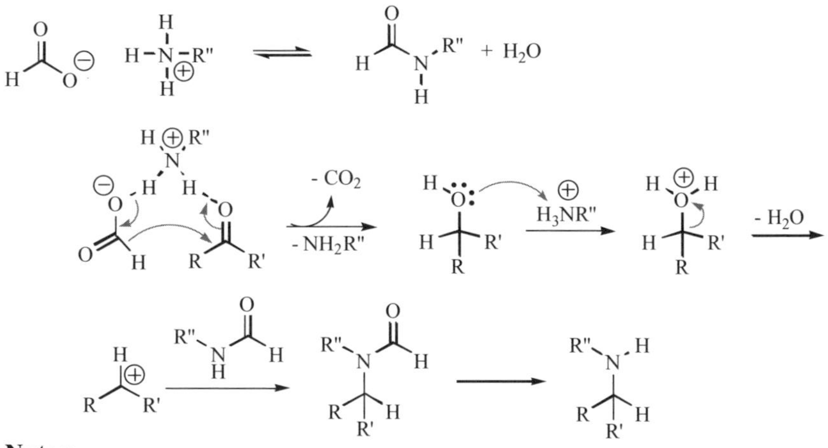

## Notes:

This reaction refers to the case when a ketone or aldehyde is reductively aminated, using ammonium formate or another amine salt of formic acid.

When conducted in excess formic acid, so that the formic acid is the hydride source, the reaction is called the ***Wallach Reaction***.

See also the ***Eschweiler-Clarke Methylation*** and the ***Borche Reduction***.

## Examples:

73%

B. M. Adger, U. C. Dyer, I. C. Lennon, P. D. Tiffin, S. E. Ward, *Tetrahedron Letters* **1997**, <u>38</u>, 2153

M. Kitamura, D. Lee, S. Hayashi, S. Tanaka, M. Yoshimura, *Journal of Organic Chemistry* **2002**, <u>67</u>, 8685

M. Allegretti, R. Anacardio, M. C. Cesta, R. Curti, M. Mantovanini, G. Nano, A. Topai, G. Zampella, *Organic Process Research & Development* **2003**, <u>7</u>, 209

# Lieben Haloform (Iodoform) Reaction

**The Reaction:**

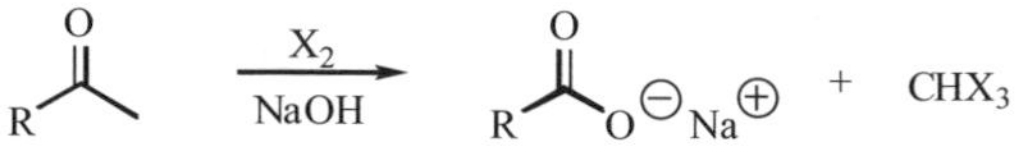

**Proposed Mechanism:**

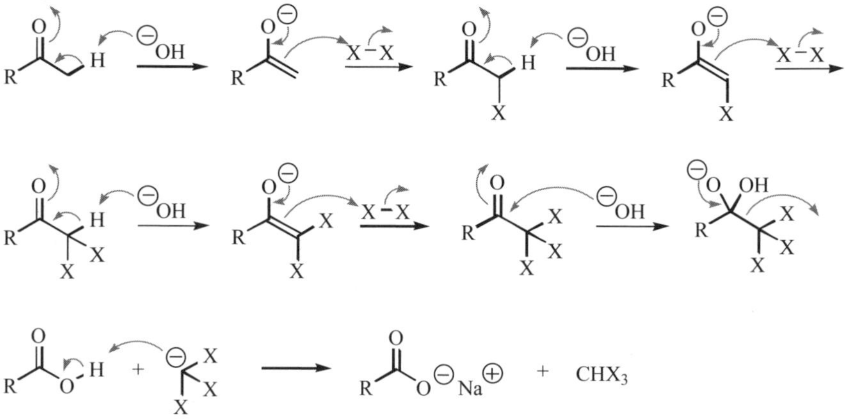

**Notes:**

$Cl_2$ and $Br_2$ can also be used besides $I_2$.

This reaction has long been considered a chemical test.

Since the reaction conditions are oxidative, alcohols such as

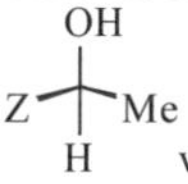

will oxidize to the corresponding carbonyl compound and give a positive iodoform test.

## Examples:

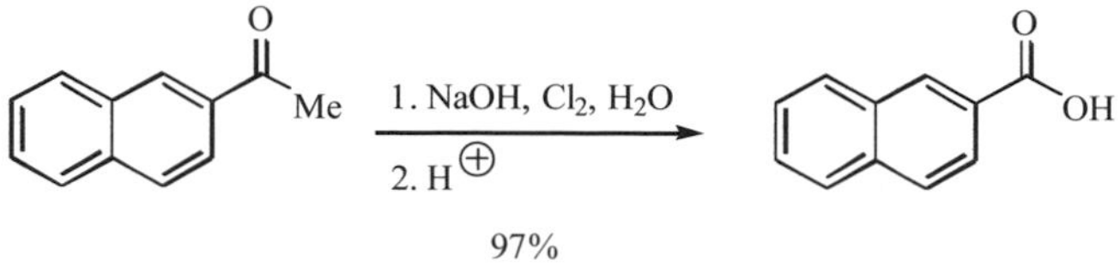

97%

M. S. Newman, H. L. Holme, *Organic Syntheses*, <u>CV2</u>, 428

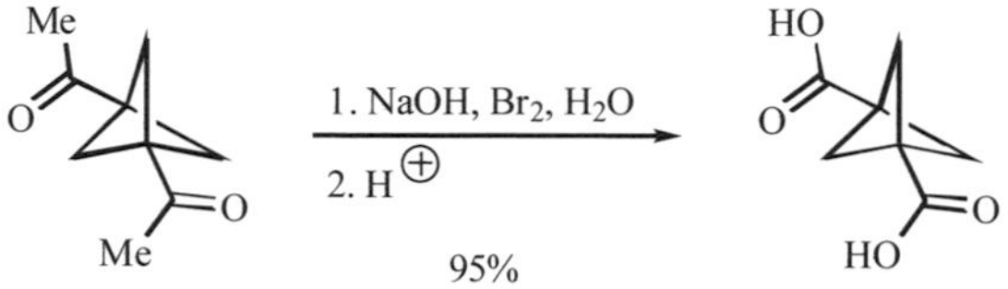

95%

M. D. Levin, P. Kosznski, J. Michl, *Organic Syntheses*, <u>77</u>, 249

# Liebeskind–Srogl Coupling

## The Reaction:

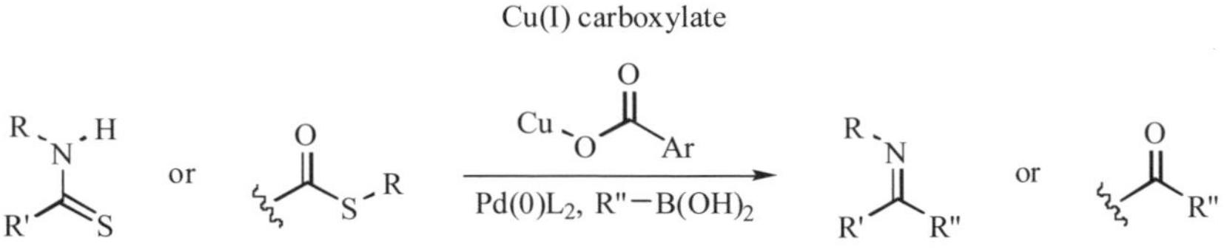

Cu(I) carboxylate

Cu(I) carboxylates include:

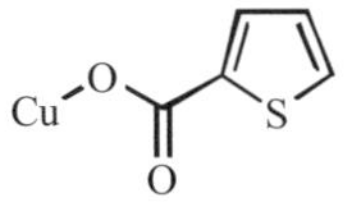

CuTC = Cu(I)thiophene-2-carboxylate 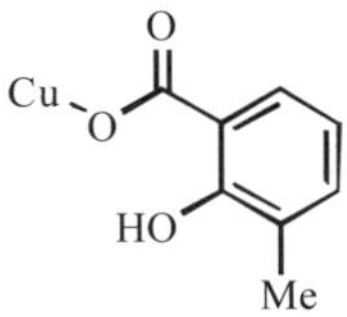    CuMeSal = copper(I) 3-methylsalicylate

## Proposed Mechanism:

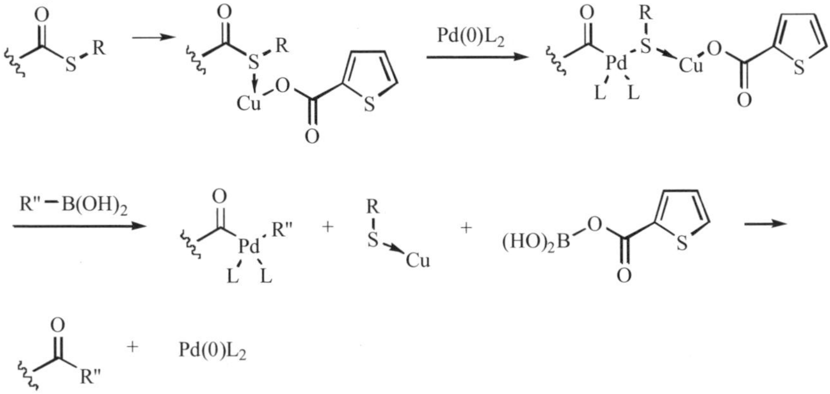

## Notes:

The reaction proceeds under nonbasic conditions.

## Examples:

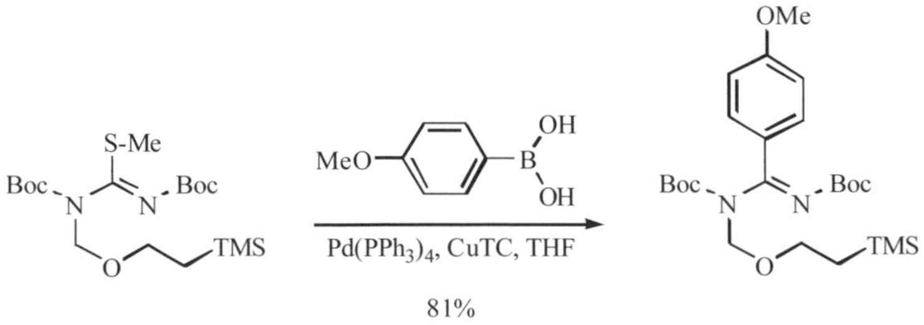

81%

C. L. Kusturin, L. S. Liebeskind, W. L. Neumann, *Organic Letters* **2002**, <u>4</u>, 983

A. Lengar, C. O. Kappe, *Organic Letters* **2004**, <u>6</u>, 771

C. Kusturin, L. S. Liebeskind, H. Rahman, K. Sample, B. Schweitzer, J. Srogl, W. L. Neumann, *Organic Letters* **2003**, <u>5</u>, 4349

***Thioimidate-Kumuda Cross-Coupling*** variation:

D. M. Mans, W. H. Pearson, *Journal of Organic Chemistry* **2004**, <u>69</u>, 6419

# Lindlar Reduction

## The Reaction:

R—≡—R $\xrightarrow[\text{CaCO}_3, \text{PbAc}_2 \text{ or quinoline}]{\text{1 equiv. H}_2 \text{ Pd} /}$

$$\begin{array}{cc} R & R \\ \diagdown & \diagup \\ & \\ \diagup & \diagdown \\ H & H \end{array}$$

= quinoline

## Proposed Mechanism:

$$R-\overset{O}{\underset{H}{C}}=\overset{O}{\underset{H}{C}}-R \longrightarrow \begin{array}{cc} R & R \\ \diagdown & \diagup \\ & \\ \diagup & \diagdown \\ H & H \end{array}$$

## Notes:

Generally an alkene will reduce to an alkane, however, the "poisoned" catalyst suppresses the ability for this to readily occur.

See **_Lindlar catalyst_**

## Examples:

$\xrightarrow[\text{Lindlar}]{\text{H}_2, \text{MeOH}}$ 74%

T. Lindel, M. Hochgurtel, *Journal of Organic Chemistry* **2000**, <u>65</u>, 2806

$\xrightarrow[\substack{\textit{Lindlar catalyst} \\ \text{quinoline, MeOH} \\ 32\%}]{\text{H}_2}$

Mixture of *cis,cis, cis,* and *cis, cis, trans*

A. Tai, F. Matsumura, H. C. Coppel, *Journal of Organic Chemistry* **1969**, <u>34</u>, 2180

Quinoline was added to the reduction mixture.
A. Furstner, T. Dierkes, *Organic Letters* **2000**, <u>2</u>, 2463

T. Itoh, N. Yamazaki, C. Kibayashi, *Organic Letters* **2002**, <u>4</u>, 2469

K. Suenaga, K. Araki, T. Sengoku, D. Uemura, *Organic Letters* **2001**, <u>3</u>, 527

# Lössen Rearrangement

## The Reaction:

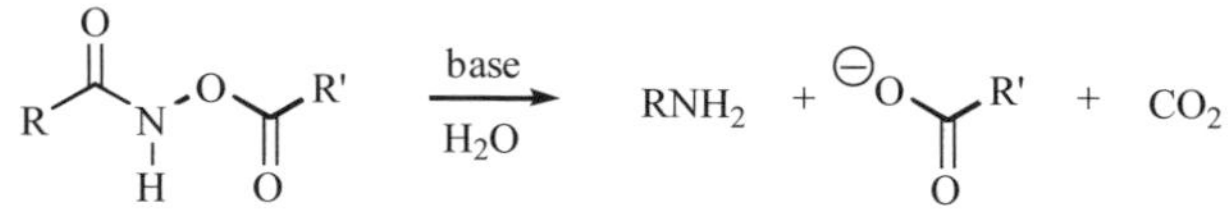

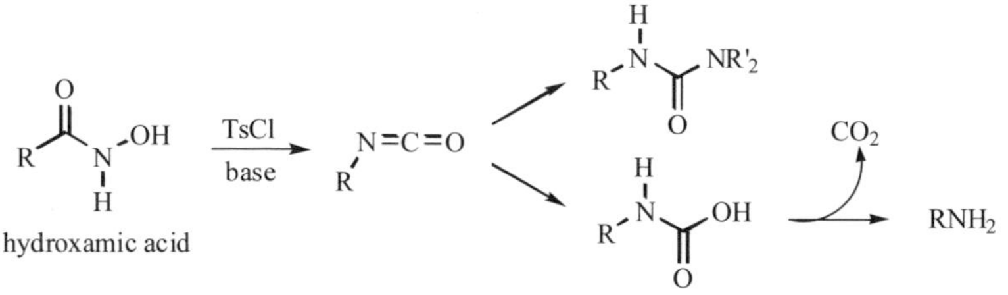

hydroxamic acid

## Proposed Mechanism:

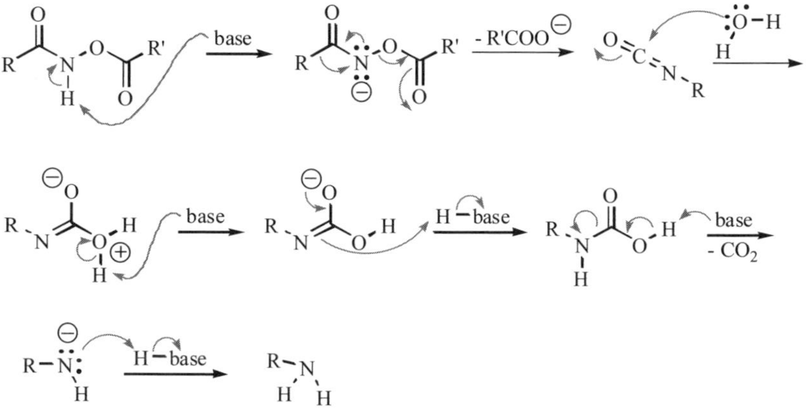

## Notes:

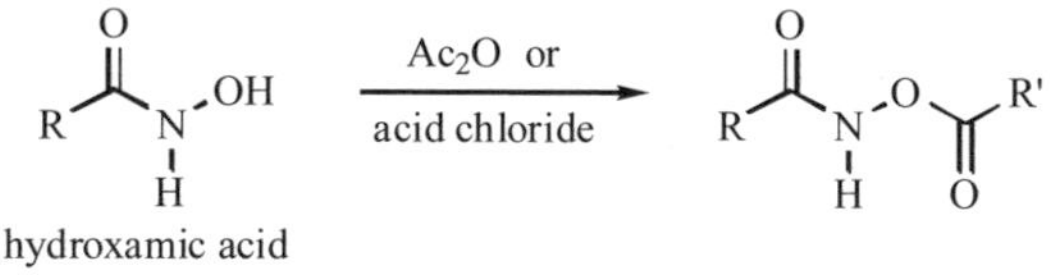

hydroxamic acid

## Examples:

9.3g                    7.0g

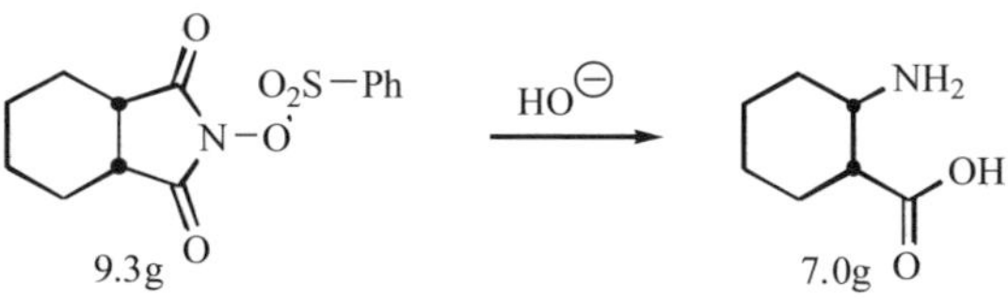

L. Baue, S. V. Miarha, *Journal of Organic Chemistry* **1959**; <u>24</u>, 1293

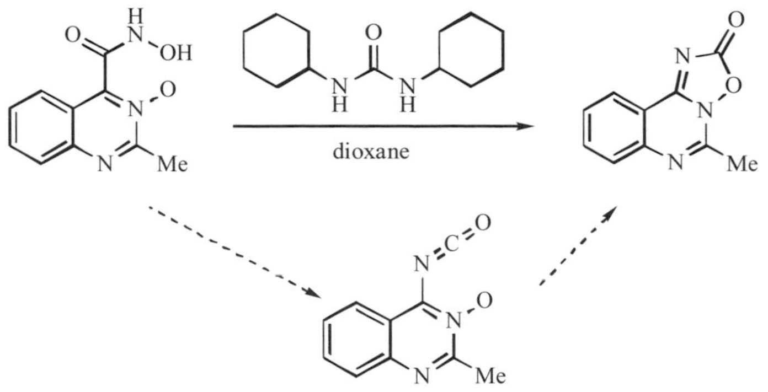

J.Bergman, J.-O. Lindstriim, *Tetrahedron Letters* **1976**, 17, 3615

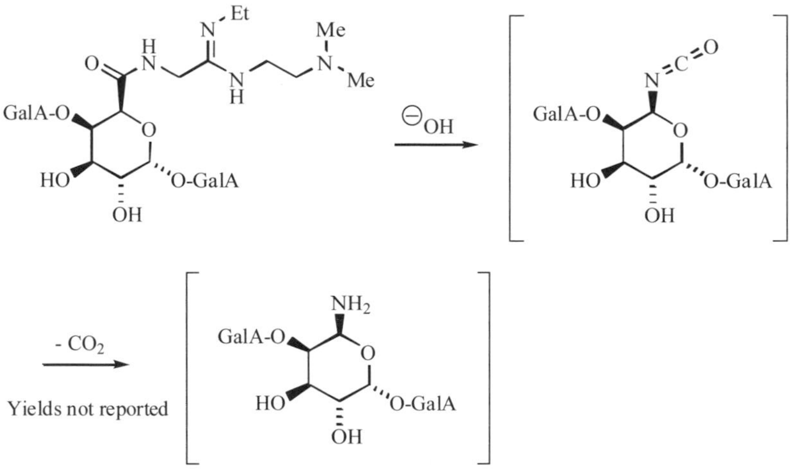

- CO₂

Yields not reported

P. W. Needs, N. M. Rigby, S. G. Ring, A. J. MacDougall, *Carbohydrate Research*, **2001**, 333, 47

# Luche Reduction

## The Reaction:

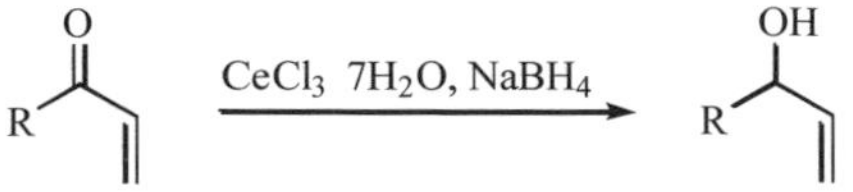

O
‖
R — (C) — ‖        CeCl₃ 7H₂O, NaBH₄ →        OH
                                                    |
                                              R — (CH) — ‖

## Proposed Mechanism:

A. L. Gemal, J. -L. Luche, *Journal of the American Chemical Society* **1981**, <u>103</u>, 5454

$$CeCl_3 + NaBH_4 \longrightarrow HCeCl_2$$

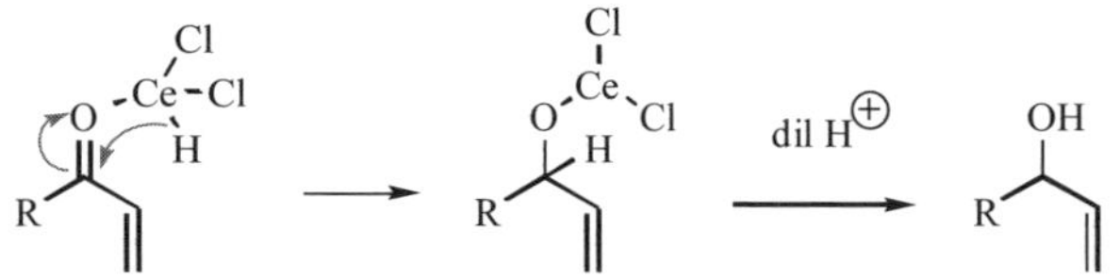

The cerium reagent coordinates to the carbonyl, making only a 1,2 addition possible.

## Notes:

The major use of the **Luche conditions** is found in applications where conjugate addition needs to be suppressed. However, there are a number of reports where stereoselectivity has been modified with this reducing protocol.

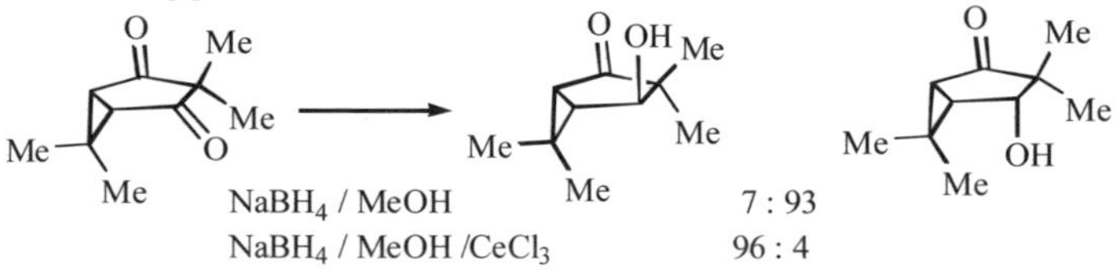

NaBH₄ / MeOH
NaBH₄ / MeOH /CeCl₃

7 : 93
96 : 4

Reported in: L. A. Paquette, *Encyclopedia of Reagents for Organic Synthesis*, John Wiley and Sons, Inc., L. A. Paquette, Ed., New York, 1995, **2**, 1031

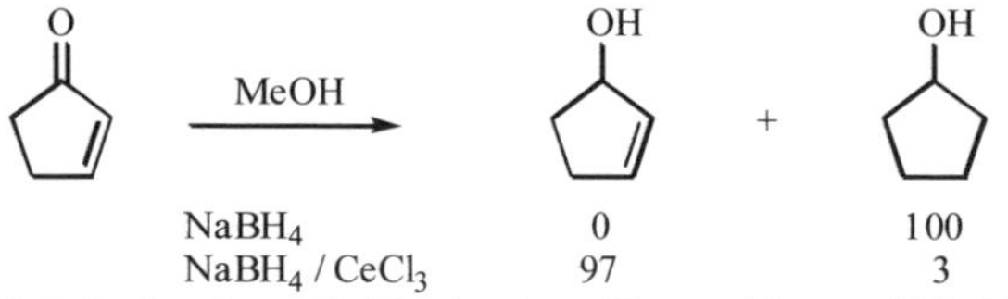

|              | OH      | OH      |
|--------------|---------|---------|
|              | (+)     |         |
| NaBH₄        | 0       | 100     |
| NaBH₄ / CeCl₃| 97      | 3       |

J. L. Luche, *Journal of the American Chemical Society* **1978**, <u>100</u>, 2226

## Examples:

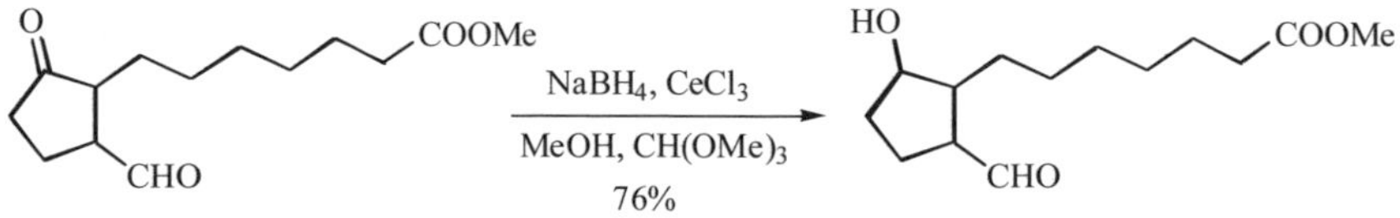

NaBH₄, CeCl₃
——————————
MeOH, CH(OMe)₃
76%

A. L. Gemal, J. L. Luche, *Journal of Organic Chemistry* **1979**, <u>44</u>, 4187

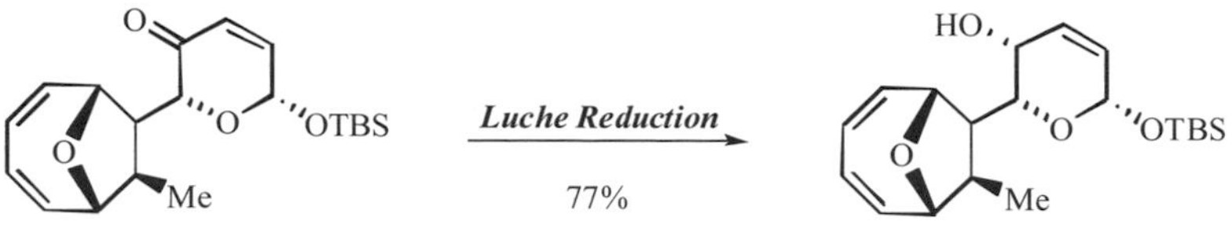

K. Takao, G. Watanabe, H. Yasui, K. Tadano, *Organic Letters* **2002**, 4, 2941

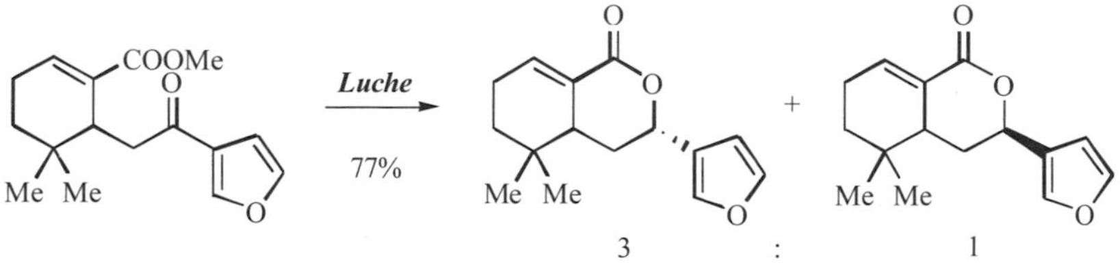

3  :  1

K. Agapiou, M. J. Krische, *Organic Letters* **2003**, 5, 1737

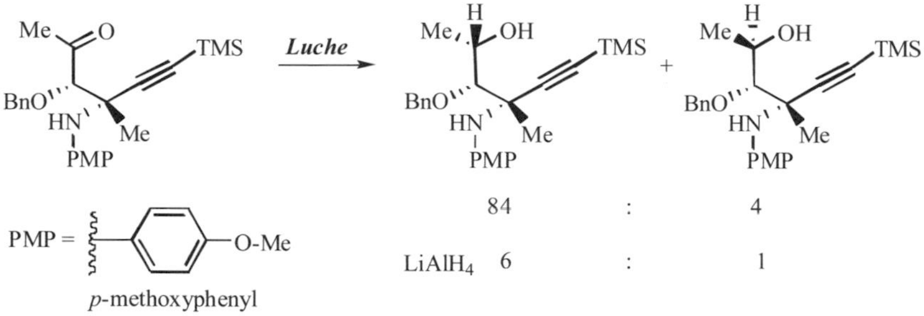

84  :  4

LiAlH₄  6  :  1

PMP = O-Me

*p*-methoxyphenyl

W. W. Cutchins, F. E. McDonald, *Organic Letters* **2002**, 4, 749

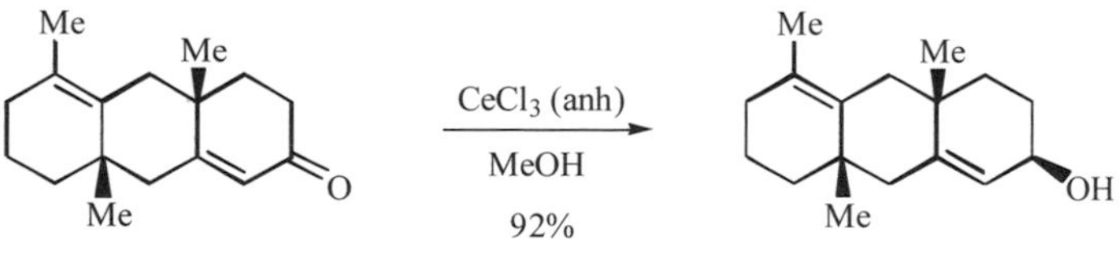

L. A. Paquette, D. T. Belmont, Y. -L. Hsu, *Journal of Organic Chemistry* **1985**, 50, 4667

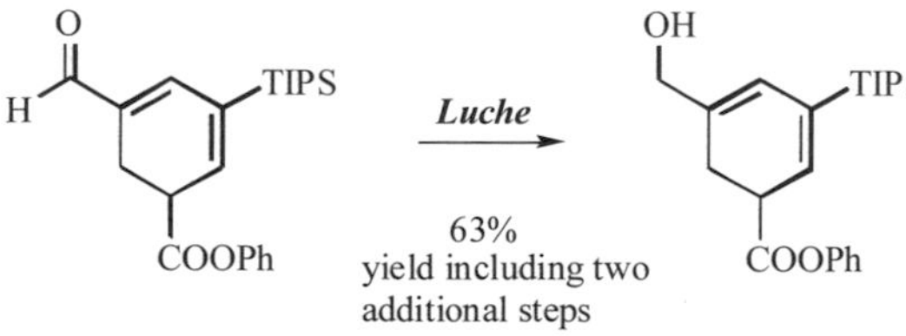

63%
yield including two
additional steps

D. L. Comins, A. L. Williams, *Organic Letters* **2001**, 3, 3217

# Macrolactonization Methods

See: R. H. Boeckman, Jr., S. W. Goldstein, *in The Total Synthesis of Natural Products,* edited by J. ApSimon, John Wiley and Sons, Inc., New York, 1988, Chapter 1, for a useful review of macrocyclic lactone syntheses.

A key feature of most macrolactonization protocols is enhancement of the acyl group.

## The Reaction:

### *Corey-Nicoloau Macrocyclization*

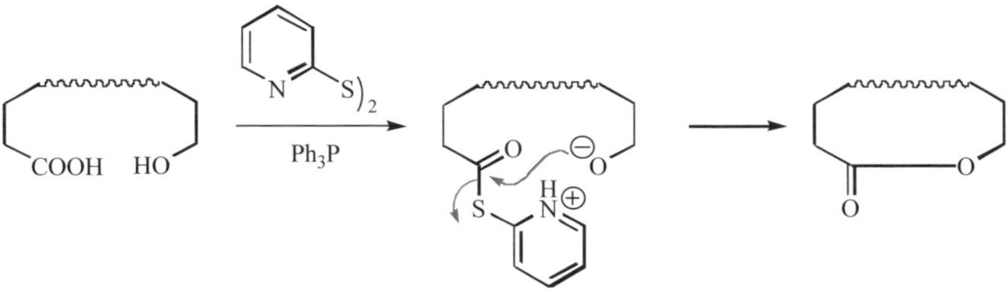

### *Keck Macrolactonization*

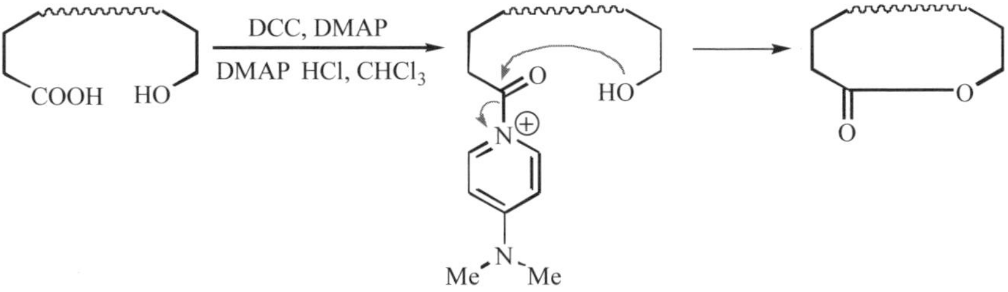

### *Masamune Macrolactonization*

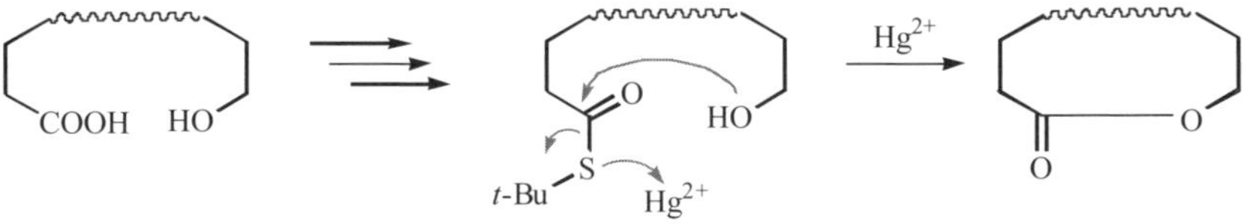

### *Mitsunobu Macrolactonization*
   See: *Mitsunobu Lactonization*

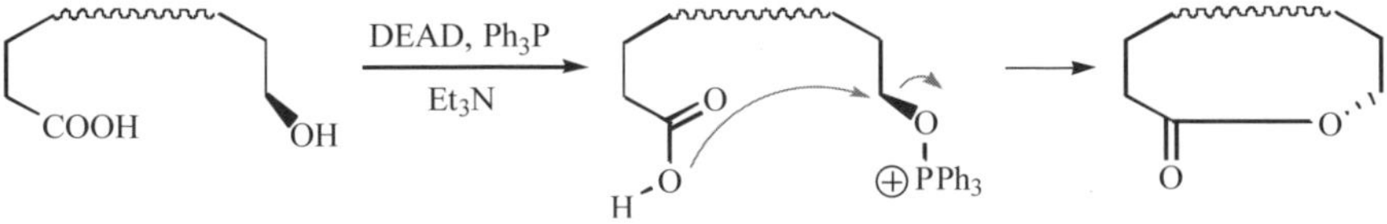

### Mukaiyama's Macrolactonization
#### See *Mukaiyama's Reagent*

### Yamaguchi's Macrolactonization
#### See: *Yamaguchi's Esterification*

**2,4,6-trichlorobenzoyl chloride**
(*Yamaguchi reagent*)

### Yamamoto's Macrolactonization
#### See: *Yamamoto's Reagent*, *Yamamoto's Esterification*

# Madelung Indole Synthesis

**The Reaction:**

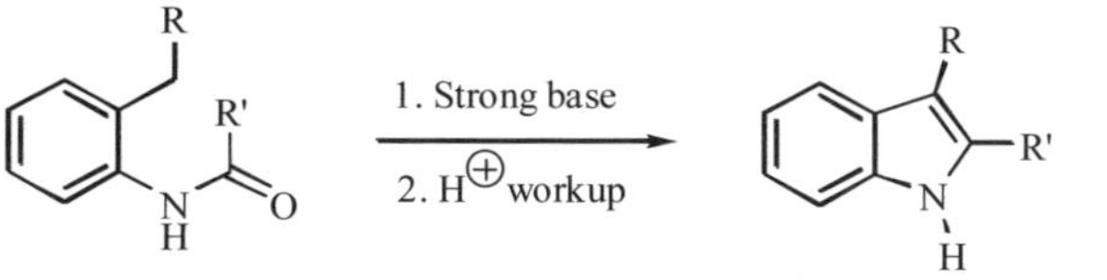

**Proposed Mechanism:**

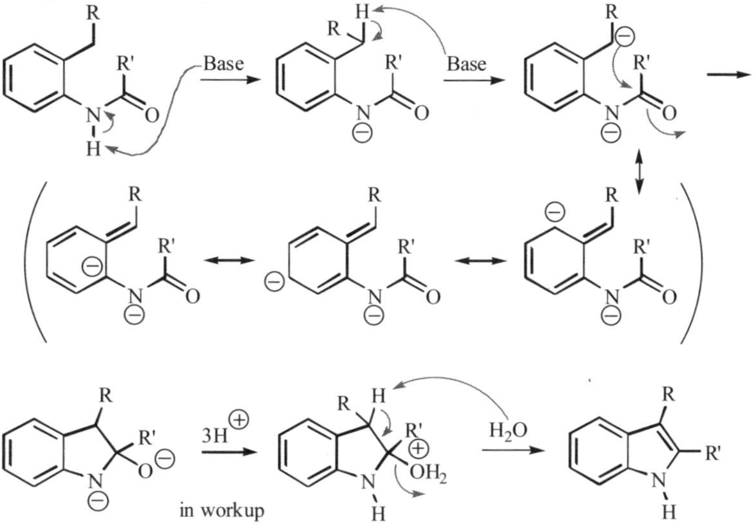

in workup

**Notes:**

See other Indole Syntheses, in ***Heterocyclic Syntheses***

**Examples:**

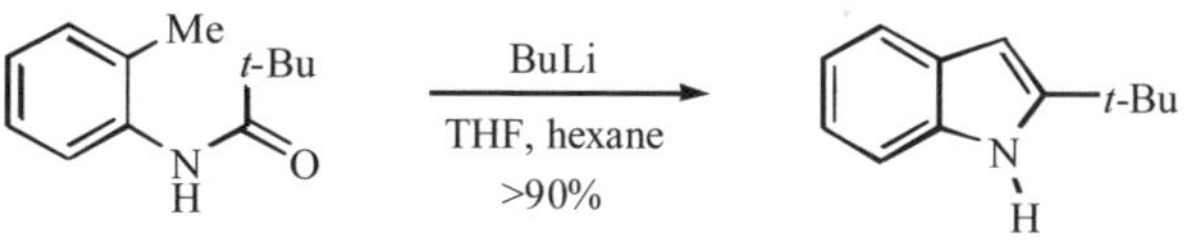

W. Fuhrer, H. W. Gschwend, *Journal of Organic Chemistry* **1979**, <u>44</u>, 1133

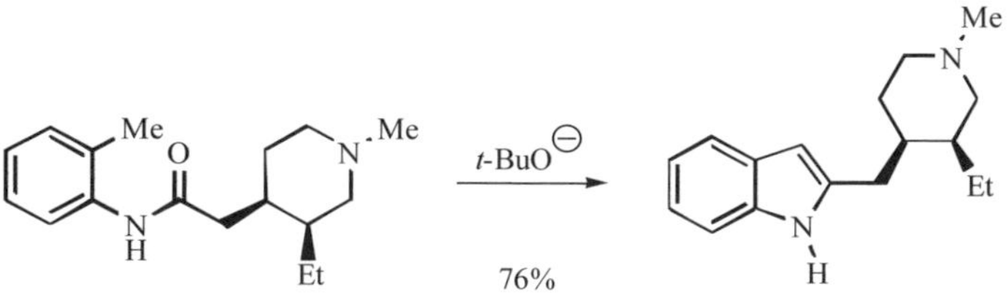

76%

A. Wu, V. Snieckus, *Tetrahedron Letters* **1975**, <u>16</u>, 2057

F. T. Tyson, *Organic Syntheses* **1943**, <u>23</u>, 42

W. J. Houlihan, V. A. Parrino, Y. Uike, *Journal of Organic Chemistry* **1981**, <u>46</u>, 4511

# Malonic Ester Synthesis

## The Reaction:

R often = ethyl = malonic ester

## Proposed Mechanism:

The methylene protons, being influenced by two carbonly groups, are the most acidic.

This part of the sequence is an enolate alkylation.

A second alkylation may occur (with a different electrophile if desired.)

Under basic conditions, the both esters are saponified to acids. (Arrows are shown for one.)

Carboxylic acids exist as carboxylate ions in base.

The carboxylates are protonated in acid and with heat, α-keto acids can decarboxylate, liberateing $CO_2$

The resulting enol undergoes keto-enol tautomerism to give the final acid product.

## Notes:

M. B. Smith, J. March in *March's Advanced Organic Chemistry*, 5th ed., John Wiley and Sons, Inc., New York, 2001, p. 549; T. Laue, A. Plagens, *Named Organic Reactions*, John Wiley and Sons, Inc., New York, 1998, pp. 178-181

Malonitrile provides similar chemistry

B. Koenig, W. Pitsch, I. Dix, P. G. Jones, *Synthesis* **1996**, 446

## Examples:

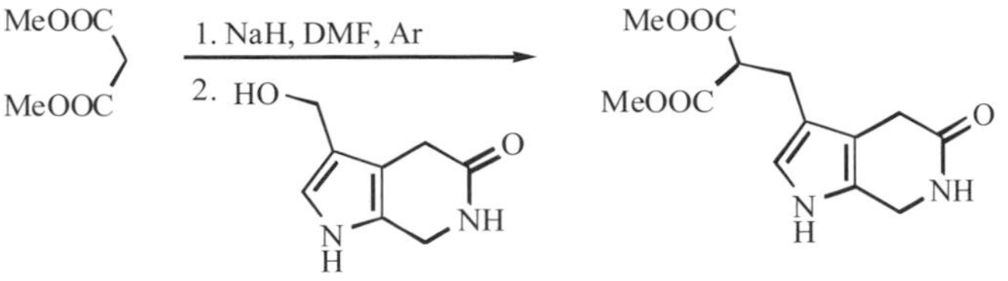

C. Y. DeLeon, B. Ganem, *Journal of Organic Chemistry* **1996**, <u>61</u>, 8730

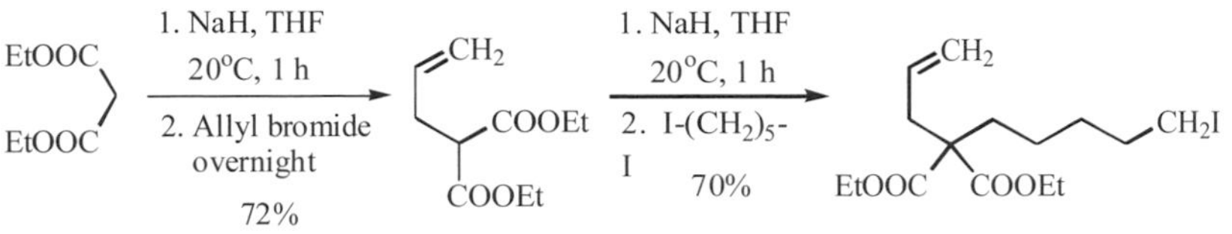

T. Yamazaki, A. Kasatkin, Y. Kawanaka, F. Sata, *Journal of Organic Chemistry* **1996**, <u>61</u>, 2266

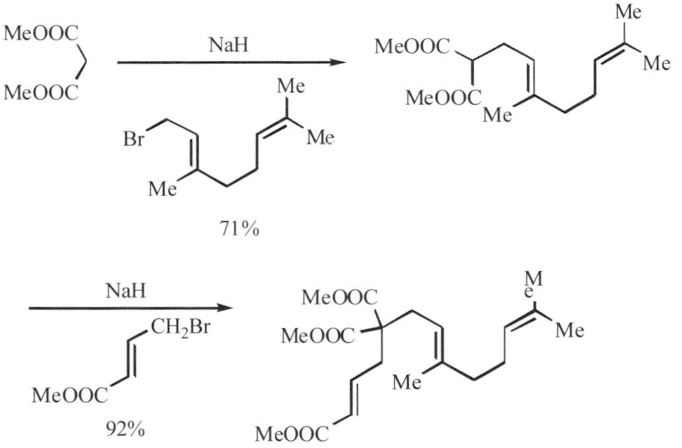

71%

92%

T. Takasu, S. Maiti, A. Katsumata, M. Mihara, *Tetrahedron Letters* **2001**, <u>42</u>, 2157

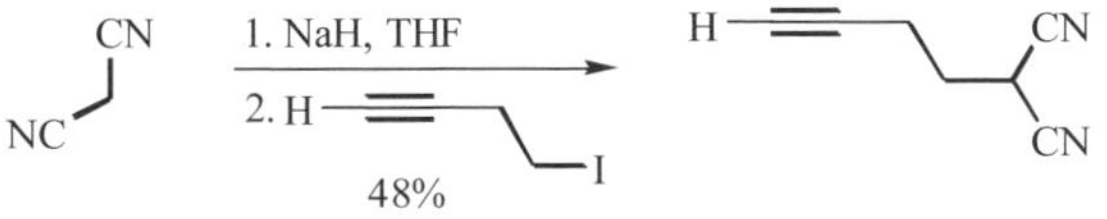

48%

E. C. Taylor, J. E. Macor, L. G. French, *Journal of Organic Chemistry* **1991**, <u>56</u>, 1807

# Mannich Reaction

## The Reaction:

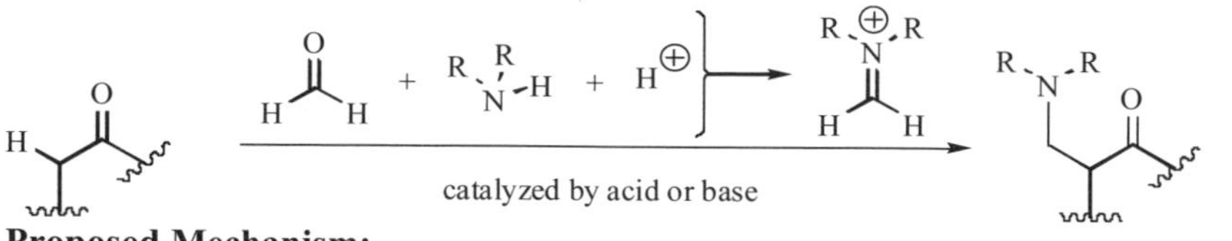

catalyzed by acid or base

## Proposed Mechanism:

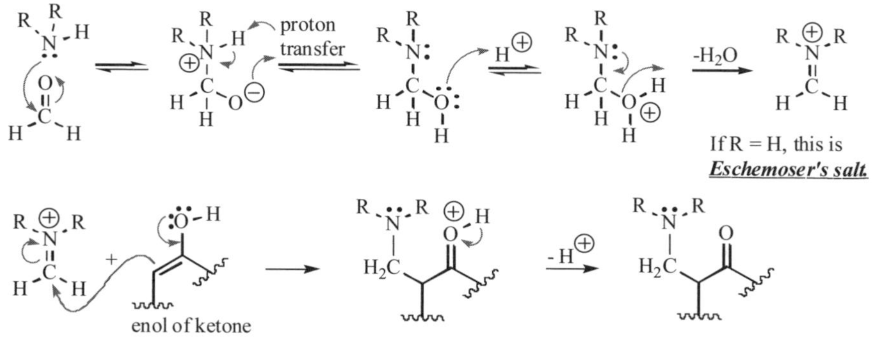

If R = H, this is
*Eschemoser's salt.*

enol of ketone

## Notes:

M. B. Smith, J. March in *March's Advanced Organic Chemistry*, 5th ed., John Wiley and Sons, Inc., New York, 2001, p. 1189; T. Laue, A. Plagens, *Named Organic Reactions*, John Wiley and Sons, Inc., New York, 1998, pp. 182-184; F. F. Blicke, *Organic Reactions* 1, 10; J. H. Brewster, E. L. Eliel, *Organic Reactions* 7, 3.

The imminium ion can be trapped by other nucleophiles:

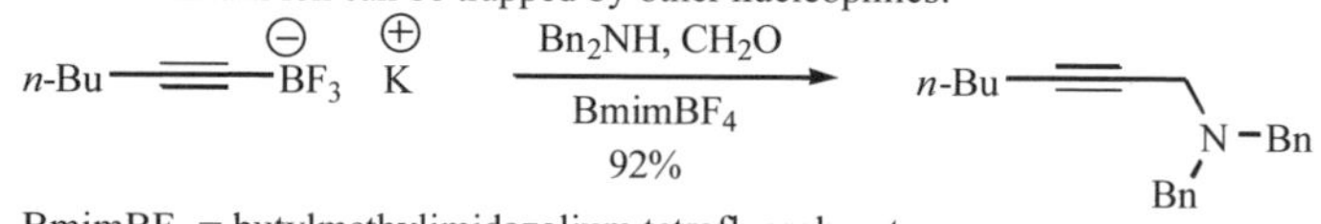

BmimBF$_4$ = butylmethylimidazolium tetrafluoroborate
G. W. Kabalka, B. Venkataiah, G. Dong *Tetrahedron Letters* 2004, 45, 729

## Mannich-Eschenmoser Methylenation

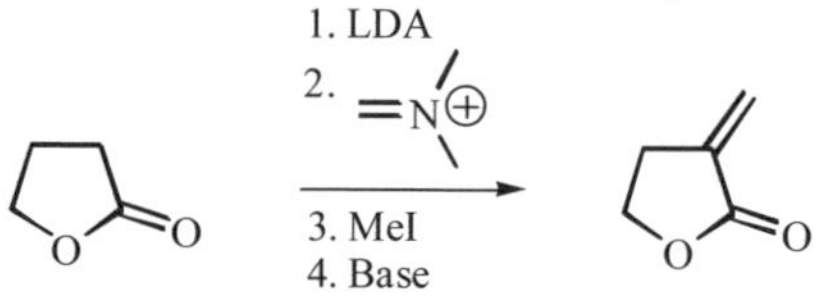

## Examples:

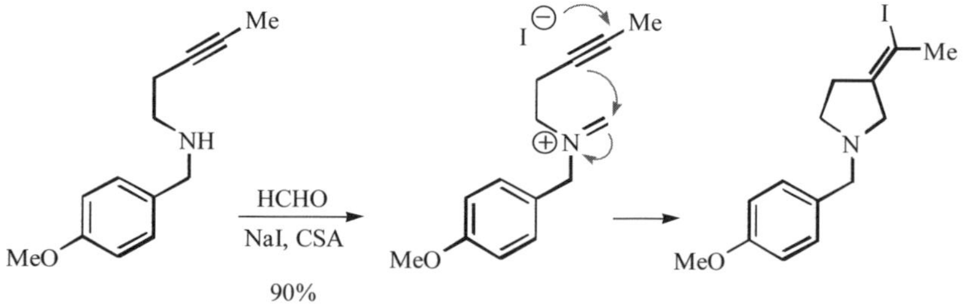

H. Arnold, L. E. Overman, M. J. Sharp, M. C. Witschel *Organic Syntheses* 1992, 70, 111

S. Liras, C. L. Lynch, A. M. Fryer, B. T. Vu, S. F. Martin, *Journal of the American Chemical Society* **2001**, 123, 5918

S. Matsunaga, N. Kumagai, S. Harada, M. Shibasaki, *Journal of the American Chemical Society* **2003**, 125, 4712

B. List, *Journal of the American Chemical Society* **2000**, 122, 9336

# Marschalk Reaction

**The Reaction:**

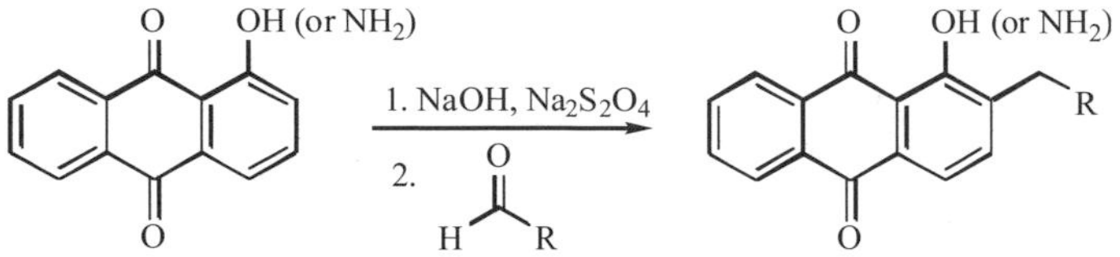

**Proposed Mechanism:**

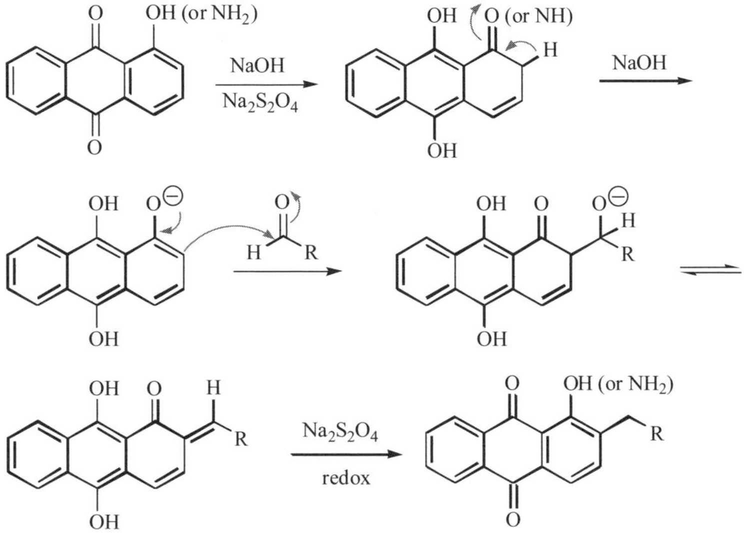

**Examples:**

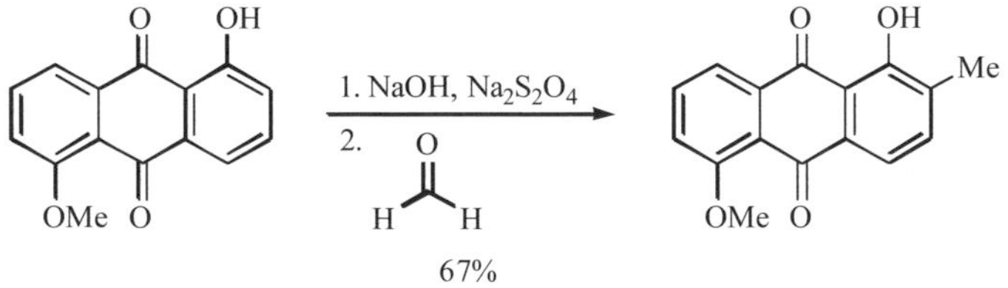

67%

K. Krohn, W. Baltus, *Tetrahedron* **1988**, <u>44</u>, 49

K. Muller, R. Altmann, H. Prinz, *European Journal of Medicinal Chemistry* **2002**, <u>37</u>, 83

K. Krohn, E. Broser, *Journal of Organic Chemistry* **1984**, <u>49</u>, 3766

A. B. Argade, A. R. Mehendale, N. R. Ayyangar, *Tetrahedron Letters* **1986**, <u>27</u>, 3529

28% trans; 21% cis

K. Krohn, W. Priyono, *Tetrahedron* **1984**, <u>40</u>, 4609

# McFadyen-Stevens Aldehyde Synthesis (McFadyen-Stevens Reduction)

## The Reaction:

R = Ar or alkyl with no α-protons

## Proposed Mechanism:

Alternate carbene mechanism:

## Notes:

M. B. Smith, J. March in *March's Advanced Organic Chemistry*, 5th ed., John Wiley and Sons, Inc., New York, 2001, p. 534.

Starting material preparation:

## Examples:

Na₂CO₃, ethylene glycol

Δ, powdered glass

62%

M. S. Newman, E. G. Caflisch, Jr., *Journal of the American Chemical Society* **1958**, <u>80</u>, 862

Ⓞ-CH₂-O-C-C=C-C-C-N-N-S-⟨phenyl⟩   **_McFayden-Stevens_** ⟶

yield not reported

Ⓞ-CH₂-O-C-C=C-C-CHO

F. Gavina, A. M. Costero, A. M. Gonzalez, *Journal of Organic Chemistry* **1990**, 55, 2060

R-C-N-N-S-⟨aryl⟩   $\xrightarrow[\Delta]{K_2CO_3, \text{ MeOH}}$   R-CHO

For most systems
yields about 50%

C. C. Dudman, P. Grice, C. B. Reese, *Tetrahedron Letters* **1980**, 21, 4645

Bu-C-N-N-S-⟨aryl⟩-Me   $\xrightarrow[\text{ethylene glycol, water (1:1)}]{HO\text{-}CH_2\text{-}CH_2\text{-}O^{\ominus} \ Na^{\oplus}}$   Bu-CHO

10%

H. Babad, W. Herbert, A. W. Stiles, *Tetrahedron Letters* **1966**, 7, 2927

# McLafferty Rearrangement

## The Reaction:

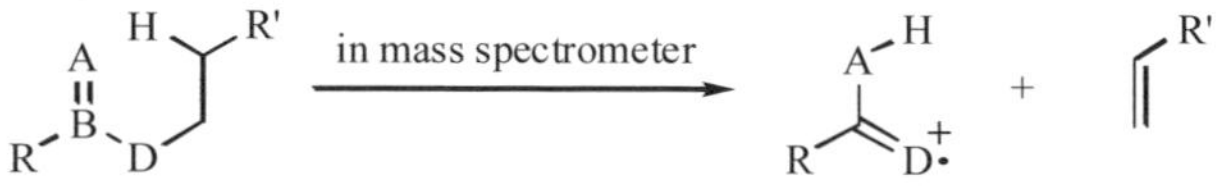

A, B, X = carbon or heteroatom

## Proposed Mechanism:

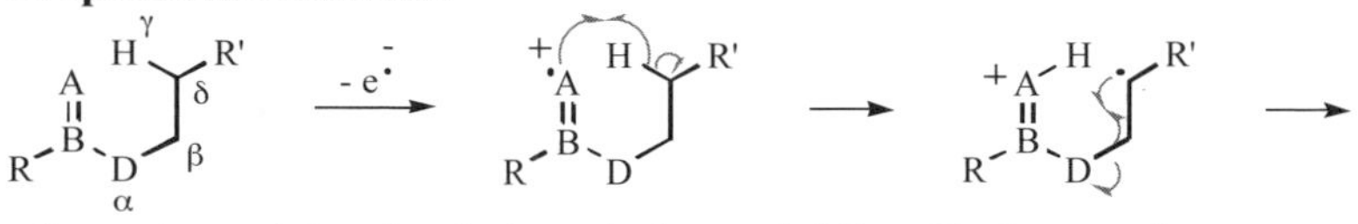

After an electron is lost, the γ hydrogen is abstracted, followed by fragmentation.

D. G. I. Kingston, J. T. Bursey, M. M. Bursey, *Chemical Reviews* **1974**, 74, 215

## Notes:

The reaction is most commonly associated with the mass spectral fragmentations of carbonyl derivatives.

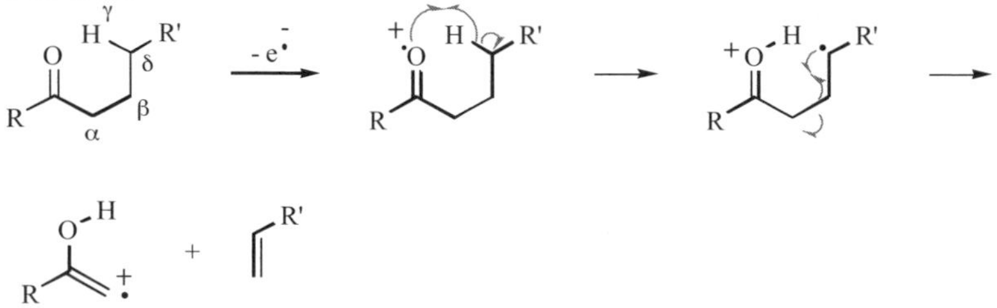

A secondary hydrogen atom will migrate about ten times better than a primary hydrogen atom.

The reaction is often compared to a similar photochemical reaction:

## Norrish Type II Cleavage

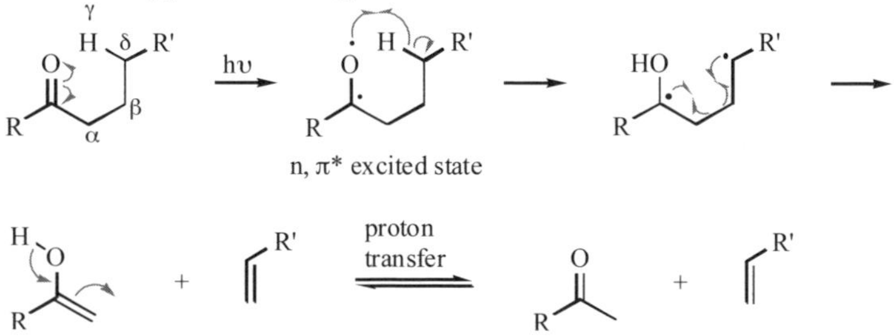

n, π* excited state

There is also the **Norrish Type I Cleavage**:

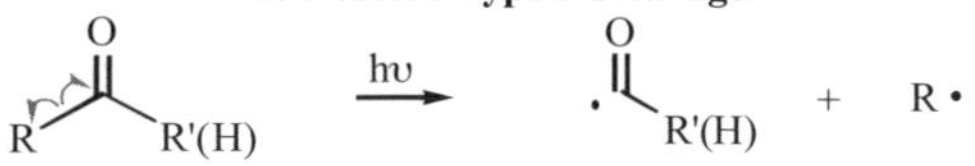

Rather than fragmentation of **Norrish Type II** biradicals, cyclization can occur. This is called the **Yang Cyclization**:

n, π* excited state

## Examples:

A double **McLafferty Rearrangement**:

G. Eadon, *Journal of the American Chemical Society* **1972**, _94_, 8938

The utility of the **McLafferty rearrangement** in assigning structures is easily documented: For each example, the important atoms are shown **in bold**.

J. E. Baldwin, S. P. Romeril, V. Lee, T. D. W. Claridge, *Organic Letters* **2001**, _3_, 1145

N. Carballeira, V. Pagan, *Journal of Natural Products* **2001**, _64_, 620

W. Engel, *Journal of Agricultural and Food Chemistry* **2002**, _50_, 1686

# McMurry (Olefination) Reaction

## The Reaction:

$$2 \quad \overset{O}{\underset{}{\bigwedge}} \quad \xrightarrow[\substack{\text{K, Zn-Cu, LiAlH}_4 \\ \text{or a similar reducing} \\ \text{reagent}}]{\text{TiCl}_3} \quad \bigvee$$

## Proposed Mechanism:

Ti(0)                         Ti(I)   Ti(I)                     Ti(I)   Ti(I)

$\xrightarrow{2X}$                                             $\longrightarrow$                       $\xrightarrow{\substack{\text{homolytic} \\ \text{cleavage}}}$

Ti(II)  Ti(I)                  Ti(II)  Ti(II)  $\equiv$  2 Ti(II)O

$\longrightarrow$                                             $+ \quad \bigvee$

## Notes:

M. B. Smith, J. March in *March's Advanced Organic Chemistry*, 5th ed., John Wiley and Sons, Inc., New York, 2001, p. 1561; T. Laue, A. Plagens, *Named Organic Reactions*, John Wiley and Sons, Inc., New York, 1998, pp. 184-187

See: ***McMurry Reagent***

The reaction of a ketone with an amide provides a unique entry into the indole skeleton (***Furstner Indole Synthesis***).

$$\xrightarrow[\text{TMSCl}]{\text{TiCl}_3 \text{ , Zn}}$$

Example:

63%

A. Furstner, B. Bogdanovic, *Angewandte Chemie International Edition in English* **1996**, *35*, 2442; A. Furstner, D. N. Jumbam, *Tetrahedron* **1992**, *48*, 5991

## Examples:

$$\xrightarrow[\text{THF}]{\text{TiCl}_4, \text{ Zn}}$$

85%

E. Lee, C. H. Yoon, *Tetrahedron Letters* **1996**, *37*, 5929

A. S. Kende, S. Johnson, P. San Filippo, J. C. Hodges, L. N. Jungheim, *Journal of the American Chemical Society* **1986**, <u>108</u>, 3513

A. Furstner, O. R. Thiel, N. Kindlar, B. Bartkowska, *Journal of Organic Chemistry* **2000**, <u>65</u>, 7990

F. B. Mallory, K. E. Butler, A. Bérubé, E. D. Luzik, Jr., C. W. Mallory, E. J. Brondyke, R. Hiremath, P. Ngo, P. Carroll, *Tetrahedron* **2001**, <u>57</u>, 3715

T. Eguchi, K. Ibaragi, K. Kakinuma, *Journal of Organic Chemistry* **1998**, <u>63</u>, 2689

# Meerwein-Ponndorf-Verley Reduction

## The Reaction:

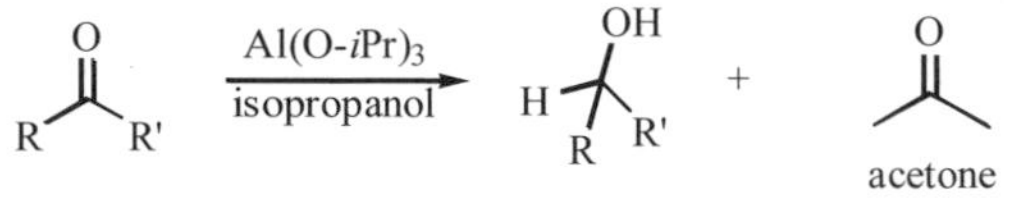

This is the reverse of the ***Oppenauer Oxidation***.

## Proposed Mechanism:

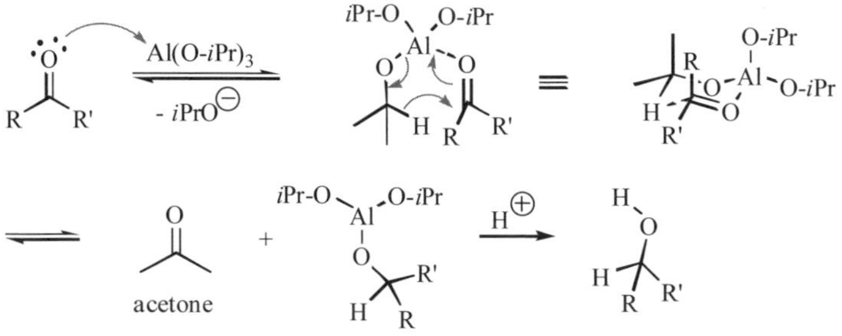

## Notes:

M. B. Smith, J. March in *March's Advanced Organic Chemistry*, 5th ed., John Wiley and Sons, Inc., New York, 2001, p. 1199; T. Laue, A. Plagens, *Named Organic Reactions*, John Wiley and Sons, Inc., New York, 1998, pp. 187-188; A. Wilds, *Organic Reactions* **2**, 5.

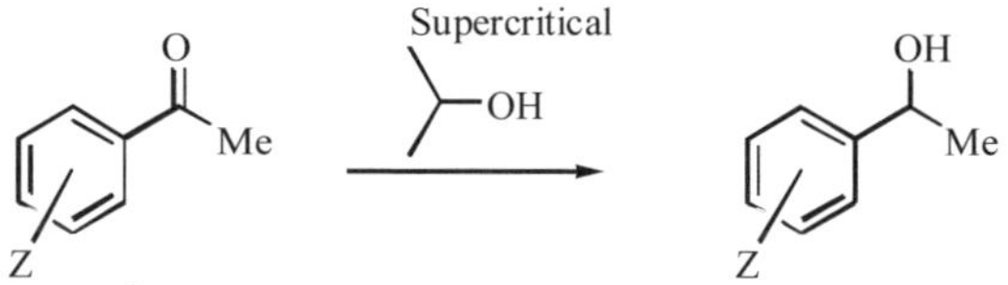

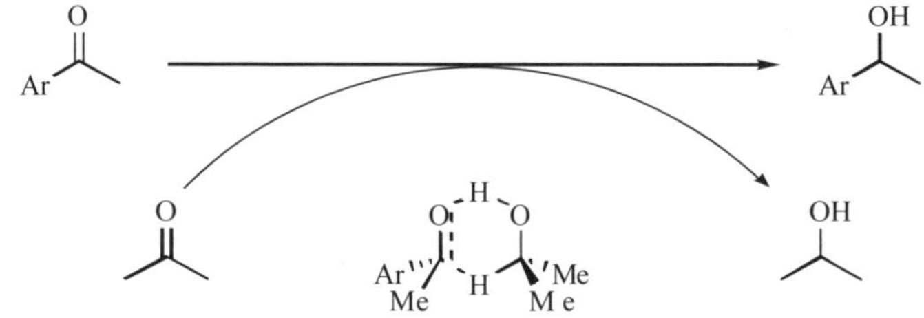

A Hamett analysis provides $\rho$ = 0.33; suggestive of a not-strongly ionized transition state.
T. Kamitanaka, T. Matsuda, T. Haradon, *Tetrahedron Letters* **2003**, 44, 4551

**Examples:**

Al(O-*i*-Pr)$_3$
toluene

79%

Note also the oxidation

T. K. M. Shing, C. M. Lee, H. Y. Lo, *Tetrahedron Letters* **2001**, <u>42</u>, 8361

Simple alcohol epimerization via a redox system:

1. [structure], NaH, THF

2. Ni(OAc)$_2$, NaH, THF
3. Add substrate

96%

96 : 4
By "classical" M-P-V:   84 : 16

R. Vanderesse, G. Feghouli, Y. Fort, P. Caubere, *Journal of Organic Chemistry* **1990**, <u>55</u>, 5916

A catalytic reaction:

AlMe$_3$ (10 mol%)
4 *i*-PrOH, toluene

82%

AlMe$_3$ (10 mol%)
4 *i*-PrOH, toluene

91%

E. J. Campbell, H. Zhou, S. T. Nguyen, *Organic Letters* **2001**, <u>3</u>, 2391

# Meinwald Rearrangement

## The Reaction:

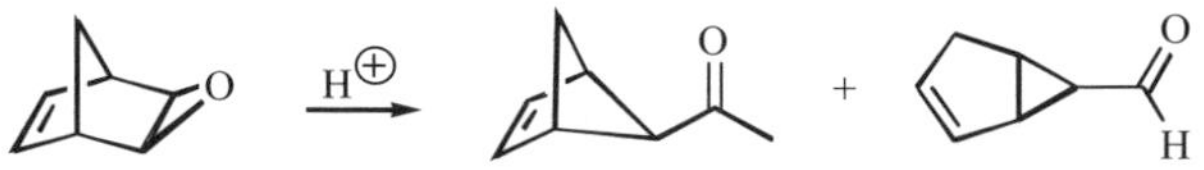

## Proposed Mechanism:

path A
and/or
path B

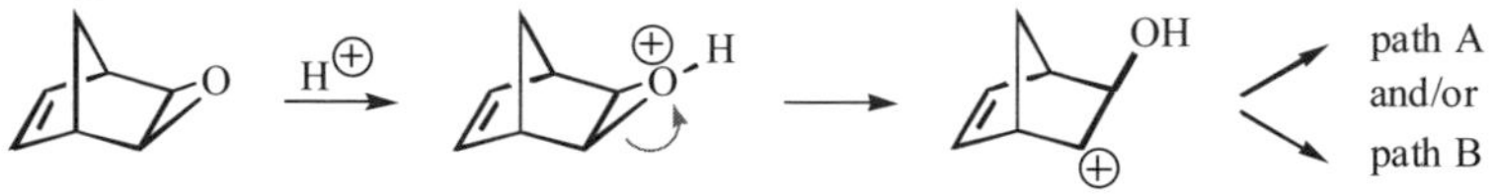

path A

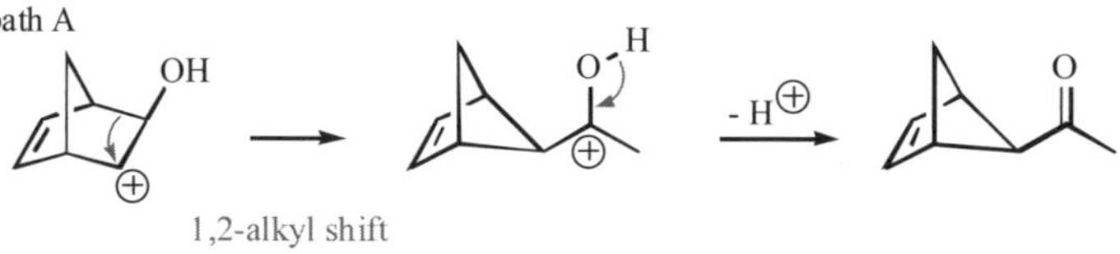

1,2-alkyl shift

path B

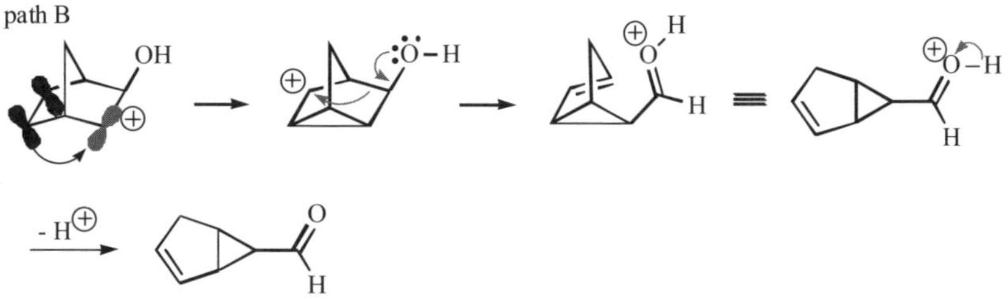

## Notes:

M. B. Smith, J. March in *March's Advanced Organic Chemistry*, 5th ed., John Wiley and Sons, Inc., New York, 2001, p. 1398.

A more generic reaction generally associated with this Name Reaction is:

Lewis
ac id

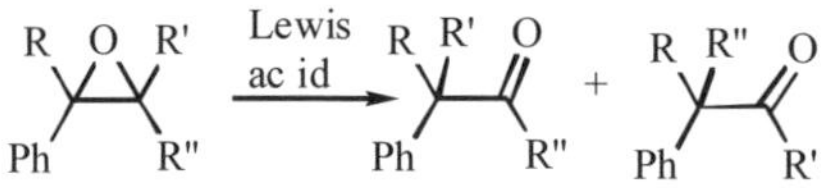

## Examples:

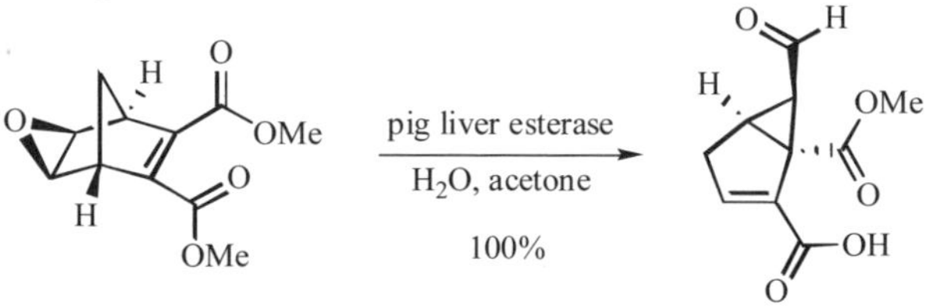

pig liver esterase
$H_2O$, acetone
100%

S. Niwayama, S. Kobayashi, M. Ohno, *Journal of the American Chemical Society* **1994**, 116, 3290

Y. Kita, A. Furukawa, J. Futamura, K. Higuchi, K. Ueda, H. Fujioka, *Tetrahedron* **2001**, <u>57</u>, 815

A.D. Baxter, S. M. Roberts, M. Roberts, F. Scheinmann, B. J. Wakefield, R. F. Newton, *Journal of the Chemical Society, Chemical Communications* **1983**, 932

Use of IrCl₃ catalyst:

I. Karame, M. L. Tommasino, M. Lemaire, *Tetrahedron Letters* **2003**, <u>44,</u> 7687

# Meisenheimer Rearrangement

## The Reaction:

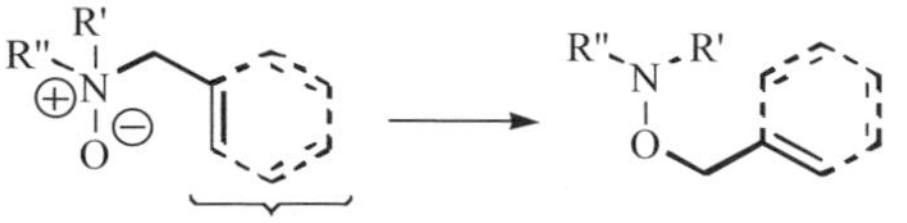

Usually allyl or benzyl.

## Proposed Mechanism:

Ionic and radical mechanisms have been proposed:

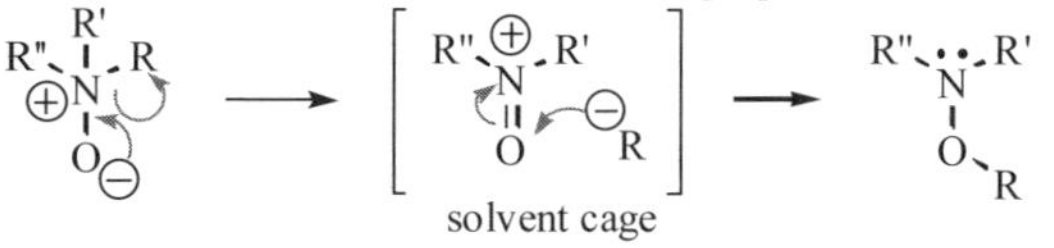

solvent cage

or

solvent cage

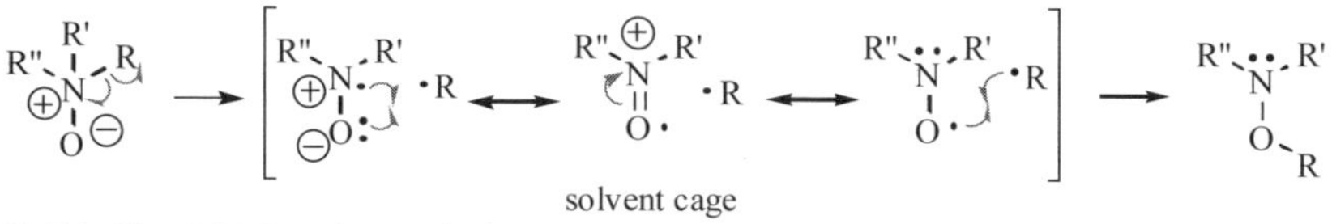

U. Schöllkopf, M. Patsch, H. Schäfer, *Tetrahedron Letters* **1964**, *5*, 2515;
N. Castagnoli, Jr., J. C. Craig, A. P. Melikian, S. K. Roy, *Tetrahedron* **1970**, *26*, 4319

## Notes:

M. B. Smith, J. March in *March's Advanced Organic Chemistry*, 5th ed., John Wiley and Sons, Inc.,
New York, 2001, p. 1420

A 2,3-sigmatropic rearrangement is possible:

## Examples:

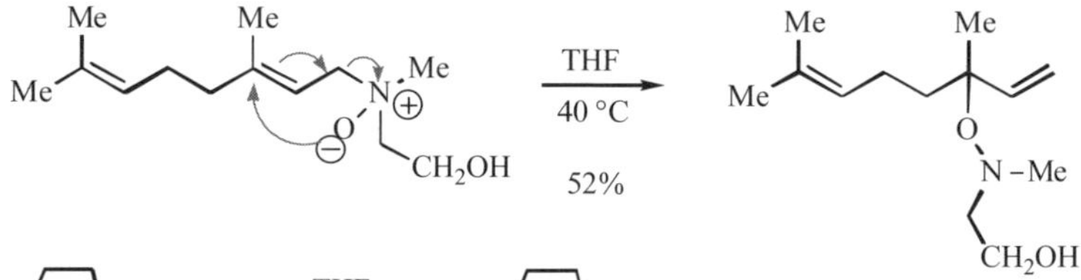

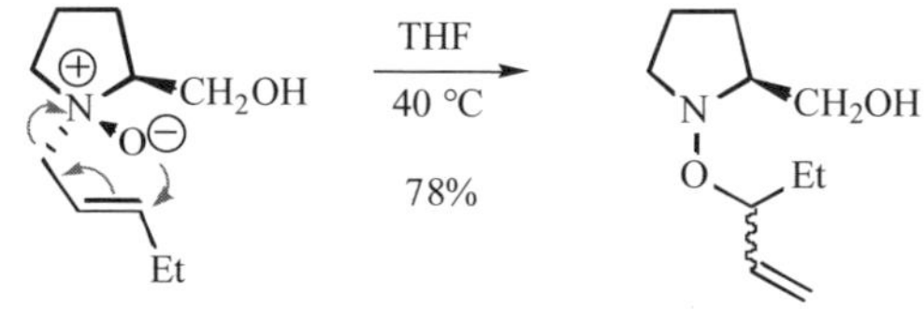

J. E. H. Buston, I. Coldham, K. R. Mulholland, *Tetrahedron: Asymmetry* **1998**, *9*, 1995

MeO

MeO

N–Me

mesitylene

*n*-PrCN

51%

MeO

MeO

N–Me

T. S. Bailey, J. B. Bremner, D. C. Hockless, B. W. Skelton, A. H. White, *Tetrahedron Letters* **1994**, 35, 2409

HO COOMe

N–Boc

N

Me

Cl

1. MCPBA, CH$_2$Cl$_2$

2. HCl, EtOAc

61%

HO COOMe

N–Boc

O

N

Me

Cl

C. Didier, D. J. Critcher, N. D. Walshe, Y. Kojima, Y. Yamauchi, A. G. M. Barrett, *Journal of Organic Chemistry* **2004**, 69, 7875

N

Me

AcNMe$_2$

62%

O

O–N

Me

S. Saba, P. W. Domkowski, F. Firooznia, *Synthesis* **1990**, 921

Me Me

Me N

Et

1. MCPBA, CH$_2$Cl$_2$

2. THF, 60 °C

80%

Me Me

Me N

O

Et

J. E. H. Buston, I. Coldham, K. R. Mulholland, *Journal of the Chemical Society, Perkin Transactions* **1999**, 2327

Ph CH$_2$OH

Me N–O

Ph

Acetone

rt

Quantitative

Ph CH$_2$OH

Me N O

Ph

J. Blanchet, M. Bonin, L. Micouin, H.-P. Husson, *Tetrahedron Letters* **2000**, 41, 8279

# Meta Photocycloaddition Reaction of Arenes

## The Reaction:

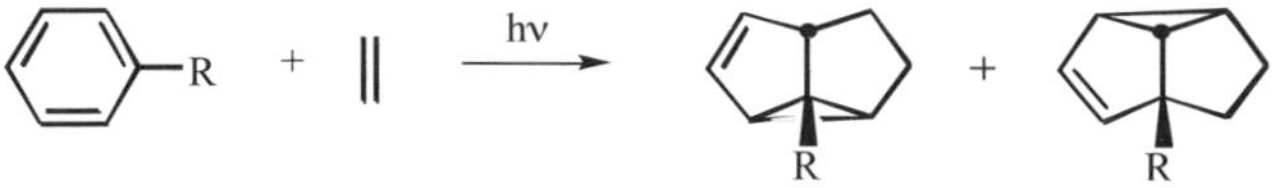

## Proposed Mechanism:

A hypothetical sequence to
show connectivity:

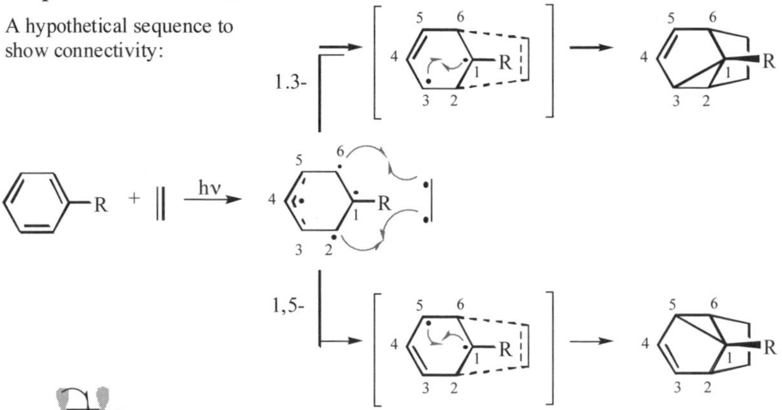

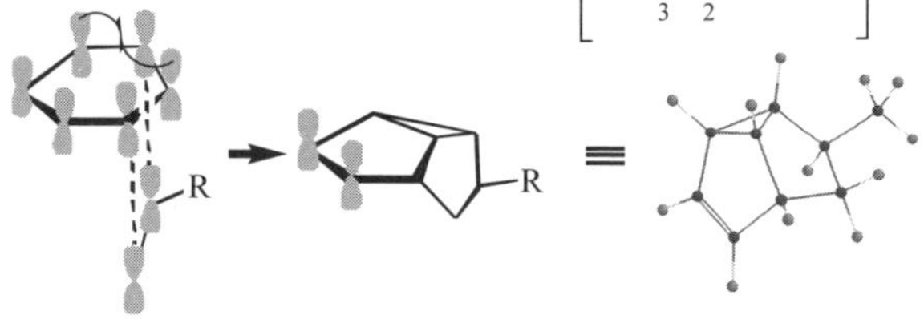

## Notes:

See: http://www.stanford.edu/group/pawender/html/photo.html for a useful overview of the
important work of the Wender group and its use of this reaction.

In this drawing it is apparant that the ring has excellent potential for synthesis.
By proper design of cyclopropane cleavage one can arrive at a number of important ring systems:

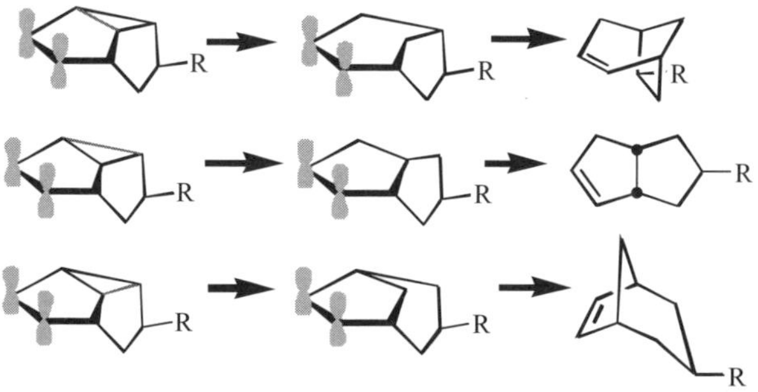

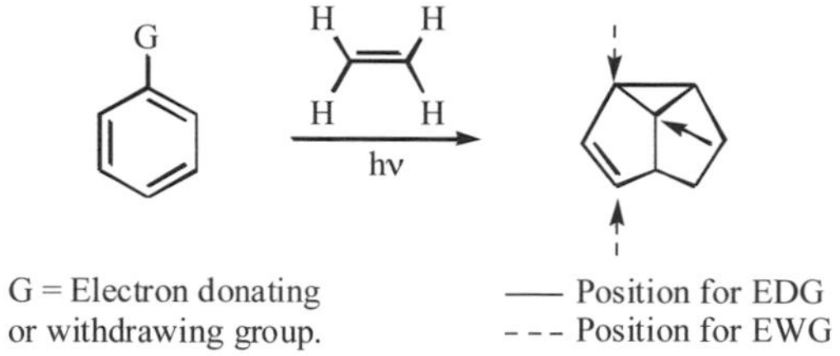

G
H
H
H
H
hν
G = Electron donating
or withdrawing group.
— Position for EDG
- - - Position for EWG

## Examples:

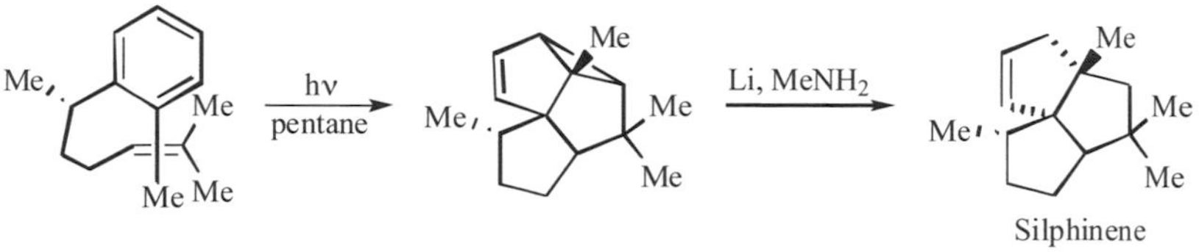

Me
Me
Me
Me
hν
pentane
Me
Me
Me
Me
Li, MeNH2
Me
Me
Me
Me
Silphinene

P. A. Wender, R. J. Ternansky, *Tetrahedron Letters* **1985**, 26, 2625

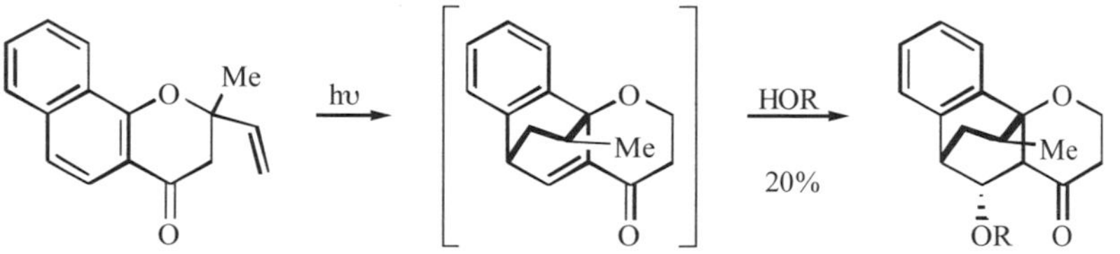

O
Me
O
hν
O
Me
O
HOR
20%
O
Me
OR
O

G. P. Kalena, P. Pradhan, V. S. Puranik, A. Banerji, *Tetrahedron Letters* **2003**, 44, 2011

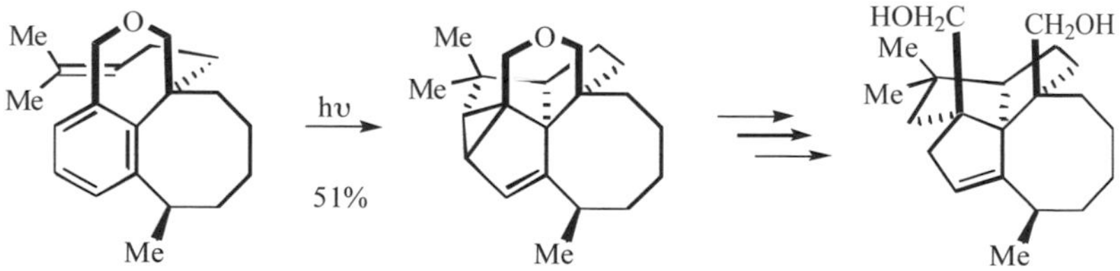

Me
O
Me
Me
hν
51%
Me
O
Me
Me
HOH2C
CH2OH
Me
Me
Me

P. A. Wender, T. W. von Geldern, B. H. Levine, *Journal of the American Chemical Society* **1988**, 110, 4858

# Meyer-Schuster propargyl alcohol rearrangement

## The Reaction:

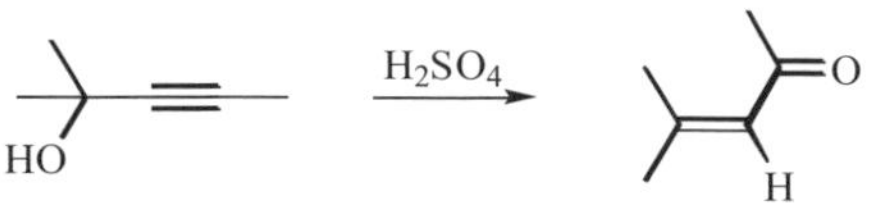

## Proposed Mechanism:

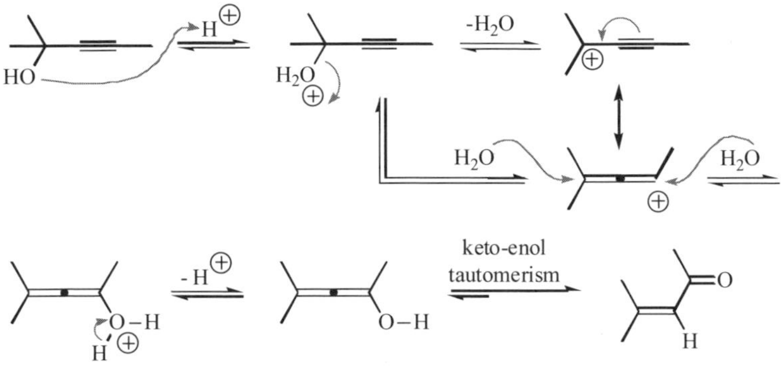

M. Edens, D. Boerner, C. R. Chase, D. Nass, M. D. Schiavelli, *Journal of Organic Chemistry* **1977**, <u>42</u>, 3403

## Notes:

The ***Meyer-Schuster Rearrangement*** is similar to the ***Rupe Rearrangement***.

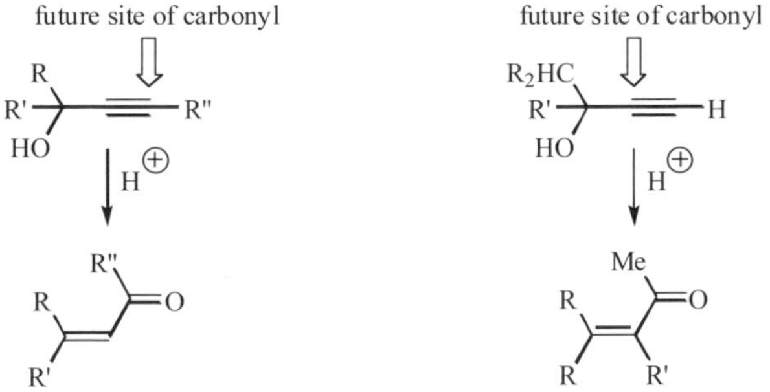

When R" = H, an aldehyde is the product.            No aldehyde product
R' may be H

## Examples:

D. Crich, S. Natarajan, J. Z. Crich, *Tetrahedron* **1997**, <u>53</u>, 7139

HCl, EtOH

acetone

No yield given          **Trt = trityl = Ph₃C-**

H. Stark, B. Sadek, M. Krause, A. Huls, X. Ligneau, C. R. Ganellin, J.-M. Arrang, J.-C. Schwartz, W. Schunak, *Journal of Medicinal Chemistry* **2000**, <u>43</u>, 3987

PPSE *

54%

* polyphosphoric acid trimethylsilylester

M.Yoshimatsu, M. Naito, M. Kawahigashi, H. Shimizu, T. Kataoka, *Journal of Organic Chemistry* **1995**, <u>60</u>, 4798

H₂SO₄

50%

G. R. Brown, D. M. Hollinshead, E. S. E. Stokes, D. S. Clarke, M. Eakin, A. J. Foubister, S. C. Glossop, D. Griffiths, M. C. Johnson, F. McTaggart, D. J. Mirrlees, G. J. Smith, R. Wood, *Journal of Medicinal Chemistry* **1999**, <u>42</u>, 1306

# Michael Reaction

## The Reaction:

1,4 additions to enones

## Proposed Mechanism:

See: T. Poon, B. P. Mundy, T. W. Shattuck, *Journal of Chemical Education,* **2002**, <u>79</u>, 264

## Notes:

M. B. Smith, J. March in *March's Advanced Organic Chemistry*, 5$^{th}$ ed., John Wiley and Sons, Inc., New York, 2001, pp. 1022-1024; T. Laue, A. Plagens, *Named Organic Reactions*, John Wiley and Sons, Inc., New York, 1998, pp. 189-191; E. D. Bergmann, D. Ginsburg, R. Pappo, *Organic Reactions* **10**, 3; T. Mukaiyama, S. Kobayashi, *Organic Reactions* **46**, 1; R. D. Little, M. R. Masjedizadeh, O. Wallwuist, J. I. McLoughlin, *Organic Reactions* **47**, 2.

In a strict sense the name of this reaction should only be applied to C-based 1,4-nucleophilic addition reactions. There have been interpretations to include other hetero-based nucleophiles.

For these reactions one can always consider the potential competition of 1,2- vs. 1,4- addition:

## Examples:

(92% ee)

K. Majima, R. Takita, A. Okada, T. Ohshima, M. Shibasaki, *Journal of the American Chemical Society* **2003**, <u>125</u>, 15837

An *Aza-Michael Addition*:

Ionic liquids: ***BmimPF$_6$***, BmimBr, BmimBF$_4$
Catalysts: Et$_3$N, pyridine

L.-W. Xu, L. Li, C.-G. Xia, S.-L. Zhou, J.-W. Li, *Tetrahedron Letters* **2004**, <u>45</u>, 1219

NaH
DMF
55%

D. M. Gordon, S. J. Danishefsky, G. K. Schulte, *Journal of Organic Chemistry* **1992**, <u>57</u>, 7052

COOEt

NaH
EtOH
76%

H₂C
Me

COOEt

Me

S. Nara, T. Toshima, A. Ichihara, *Tetrahedron* **1997**, <u>53</u>, 9509

LiClO₄ accelerated reactions under solvent-free conditions:

NH  +  H₂C$\diagup$CO₂Me

LiClO₄
86%

N$\diagdown$CO₂Me

NH  +  H₂C$\diagup$CN

LiClO₄
82%

N$\diagdown$CN

N. Azizi, M. R. Saidi, *Tetrahedron* **2004**, 60, 383

O
H
+
N-H
Mbs

O
Me
Me

Bn⁺-NEt₃Cl⁻
K₂CO₃, CHCl₃
84%

O
H
O⊖
Me
N
Mbs
Me

O
Me
N
Mbs
Me

MeO—⟨ ⟩—SO₂—
Mbs

K. Makino, O. Hara, Y. Takiguchi, T. Katano, Y. Asakawa, K. Hatano, Y. Hamada, *Tetrahedron Letters* **2003**, <u>44</u>, 8925

# Mislow-Evans Rearrangement (Evans (-Mislow) Rearrangement)

## The Reaction:

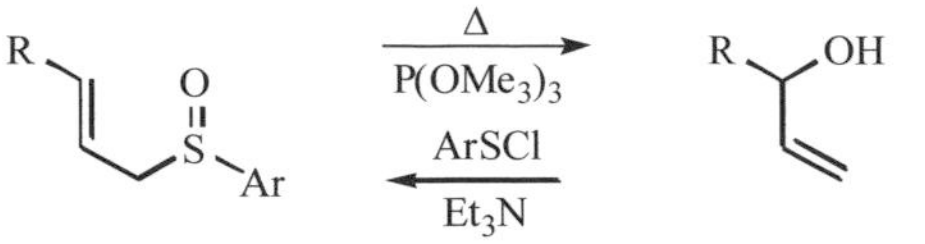

## Proposed Mechanism:

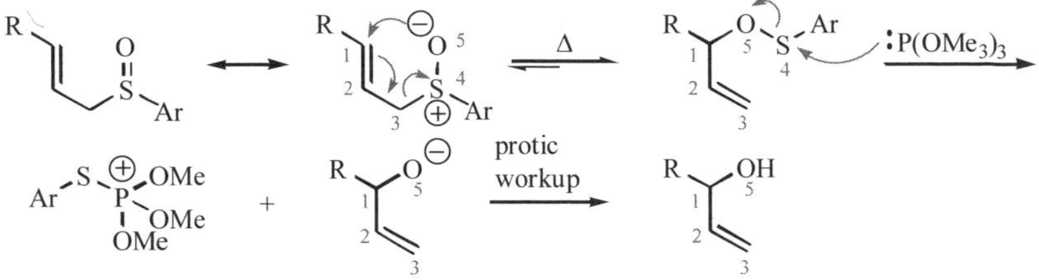

## Notes:
M. B. Smith, J. March in *March's Advanced Organic Chemistry*, 5$^{th}$ ed., John Wiley and Sons, Inc., New York, 2001, p. 1455.

The reverse reaction is accomplished by a reaction with ArSCl and thermal rearrangement.

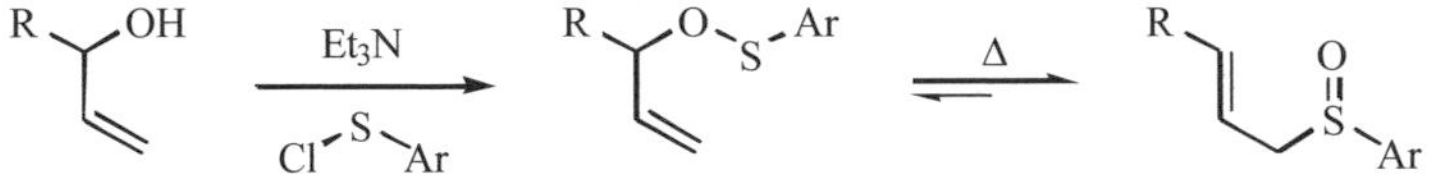

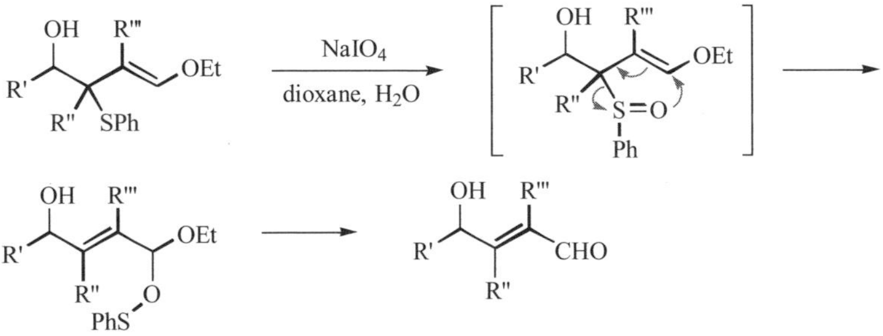

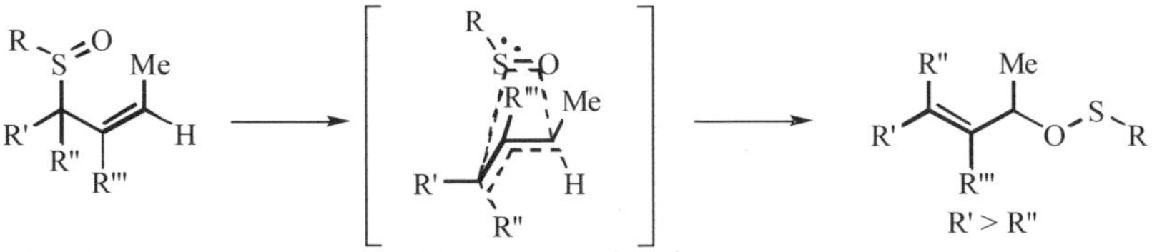

The selectivity for *E*-isomers should increase as $R^1 > R^2$ increases.

T. Sato, J. Otera, H. Nozaki, *Journal of Organic Chemistry* **1989**, <u>54</u>, 2779

## Examples:

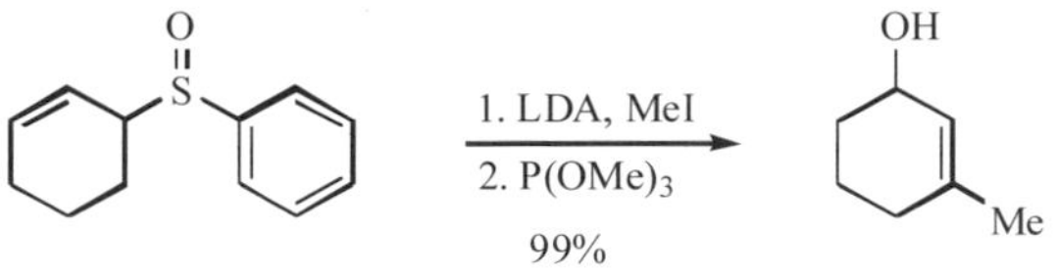

1. LDA, MeI
2. P(OMe)₃

99%

D. A. Evans, G. C. Andrews, T. T. Fujimi, D. Wells, *Tetrahedron Letters* **1973**, <u>14</u>, 1385

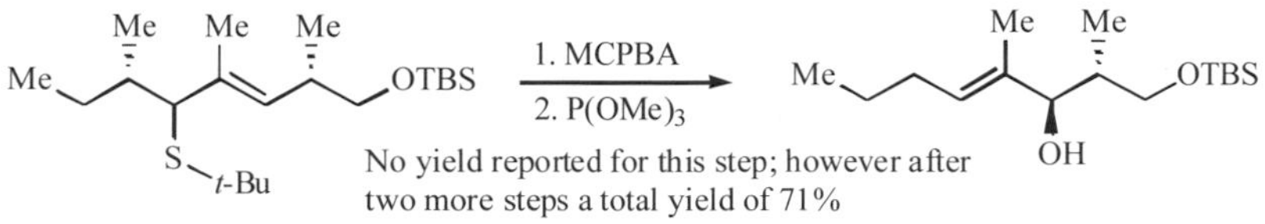

1. MCPBA
2. P(OMe)₃

No yield reported for this step; however after two more steps a total yield of 71%

A. K. Mapp, C. H Heathcock, *Journal of Organic Chemistry* **1999**, <u>64</u>, 23

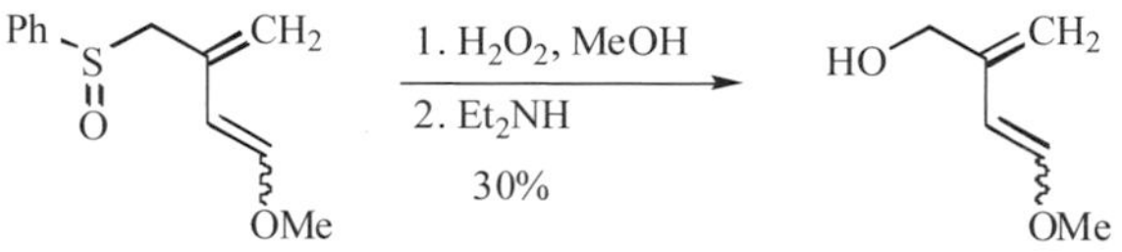

1. H₂O₂, MeOH
2. Et₂NH

30%

T. Mandai, K. Osaka, M. Kawagishi, M. Kawada, J. Otera, *Journal of Organic Chemistry* **1984**, <u>49</u>, 3595

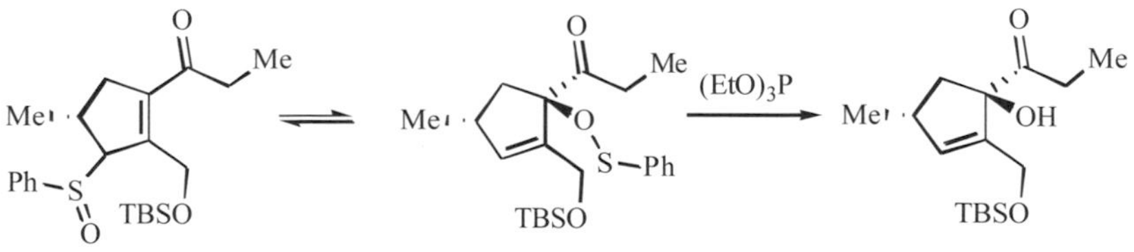

(EtO)₃P

Reported in a Review: A. B. Smith, III, C. M. Adams, *Accounts of Chemical Research* **2004**, <u>37</u>, 365

# Mitsunobu Lactonization

## The Reaction:

OH
COOH
DEAD, Ph3P
THF
n
O
O

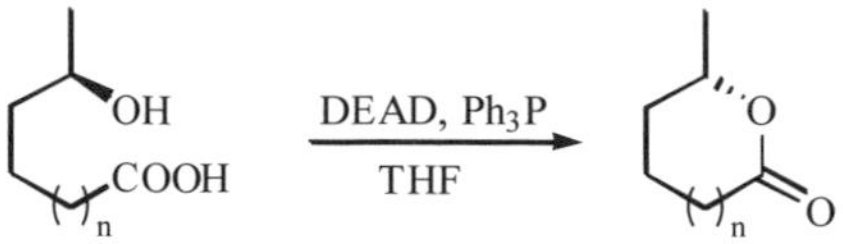

## Proposed Mechanism:

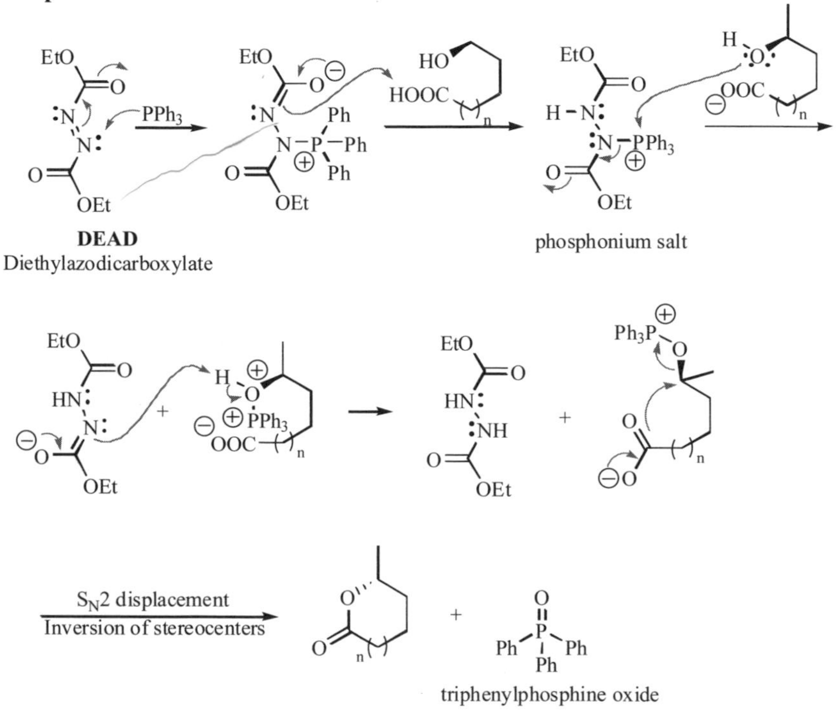

EtO
O
N
N
PPh3
OEt
DEAD
Diethylazodicarboxylate
Ph
N−P−Ph
Ph
HO
HOOC
H−N
N−PPh3
phosphonium salt
H
OOC
n
HN
NH
Ph3P
SN2 displacement
Inversion of stereocenters
Ph
triphenylphosphine oxide

## Notes:

See: ***Mitsunobu Reaction*** and ***Macrolactonization Methods***.

This lactonization process inverts the stereochemistry of the alcohol portion of the lactone.

## Examples:

S. Takahashi, A. Kubota, T. Nakata, *Organic Letters* **2003**, 5, 1353

A. B. Smith, III, G. A. Sulikowski, K. Fujimoto, *Journal of the American Chemical Society* **1989**, 111, 8039

M. T. Crimmins, M. G. Stanton, S. P. Allwein, *Journal of the American Chemical Society* **2002**, 124, 5959

# Mitsunobu Reaction

## The Reaction:

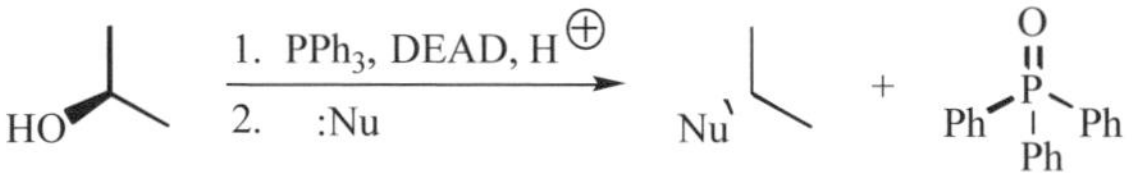

## Proposed Mechanism:

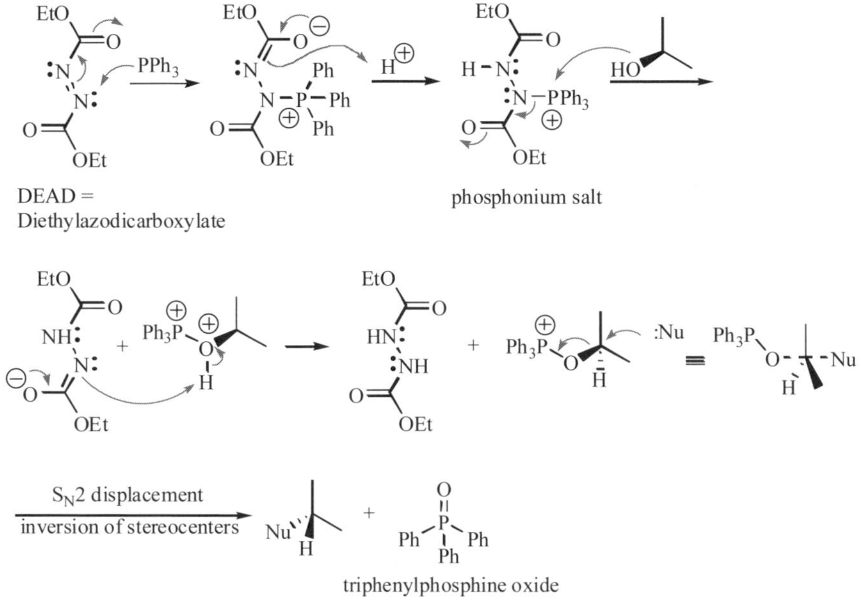

DEAD =
Diethylazodicarboxylate

phosphonium salt

$S_N2$ displacement
inversion of stereocenters

triphenylphosphine oxide

## Notes:

M. B. Smith, J. March in *March's Advanced Organic Chemistry*, 5th ed., John Wiley and Sons, Inc., New York, 2001, p. 486; T. Laue, A. Plagens, *Named Organic Reactions*, John Wiley and Sons, Inc., New York, 1998, pp. 192-194; D. L. Hughes, *Organic Reactions* **42**, 2

A number of nucleophilic
displacements from the **Mitsunobu**
intermediate are possible:

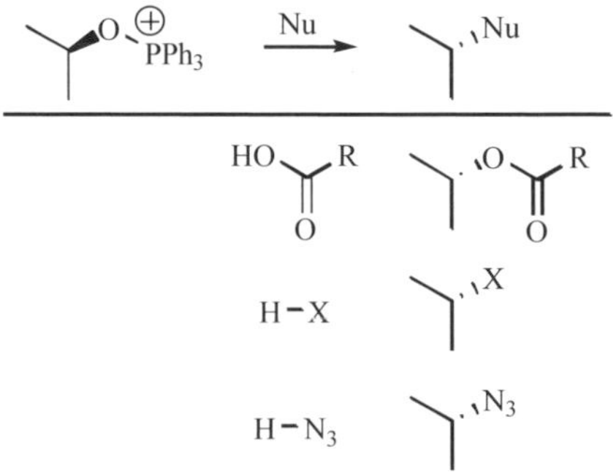

When a carboxylic acid is used as the nucleophile, simple ester hydrolysis releases the alcohol, providing an inversion of alcohol stereochemistry.

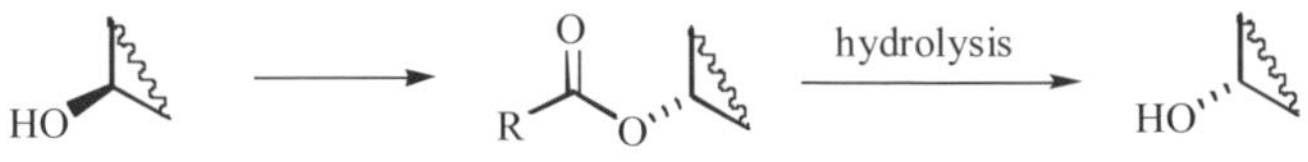

## Examples:

1. DEAD, Ph₃P, ArCO₂H

2. NaOH

99% overall

M. T. Crimmins, J. M. Pace, P. G. Nantermet, A. S. Kim-Meade, J. B. Thomas, S. H. Watterson, A. S. Wagman, *Journal of the American Chemical Society* **1999**, <u>121</u>, 10249

Ph₃P, DIAD

*p*-NO₂-ArCO₂H

Taken directly to next step

K. Kadota, M. Takeuchi, T. Taniguchi, K. Ogasawdra, *Organic Letters* **2001**, <u>3</u>, 1769

DIAD, PPh₃, THF

91%

N. Defacqz, V. Tran-Trieu, A. Cordi, J. Marchand-Brynaert, *Tetrahedron Letters* **2003**, <u>44</u>, 9111

*Mitsunobu Etherification*:

DIAD, PPh₃, THF

52%

S. J. Gregson, P. W. Howard, D. R. Gullick, A. Hamaguchi, K. E. Corcoran, N. A. Brooks, J. A. Hartley, T. C. Jenkins, S. Patel, M. J. Guille, D. E. Thurston, *Journal of Medicinal Chemistry* **2004**, <u>47</u>, 1161

A general protocol:

1. Mitsunobu

2. Deprotect

N. Brosse, A. Grandeury, B. Jamart-Gregoire, *Tetrahedron Letters* **2002**, <u>43</u>, 2009

# Miyaura Boration Reaction

## The Reaction:

$$Ar-I \quad + \quad \text{(bis(pinacolato)diboron)} \quad \xrightarrow[\text{base}]{Pd(0)} \quad Ar-B\text{(pinacolboronate)}$$

## Proposed Mechanism:

reductive
elimination

$Ph_3P-Pd-PPh_3$
$[= Pd(0)]$

$Ar-I$

oxidative
addition

$Ph_3P\text{-}Pd\text{-}I$
$Ar\text{-}PHPh_3$

base

$^{\ominus}OAc$

base

$Ph_3P\text{-}Pd\text{-}OAc$
$Ar\text{-}PPh_3$

$Ph_3P\text{-}Pd\text{-}B\text{-}O$
$Ph_3P\text{-}Ar$

cis-trans
isomerization

$Ph_3P\text{-}Pd\text{-}B\text{-}O$
$Ar\text{-}PPh_3$

transmetalation

$I-B\text{(pinacolboronate)}$

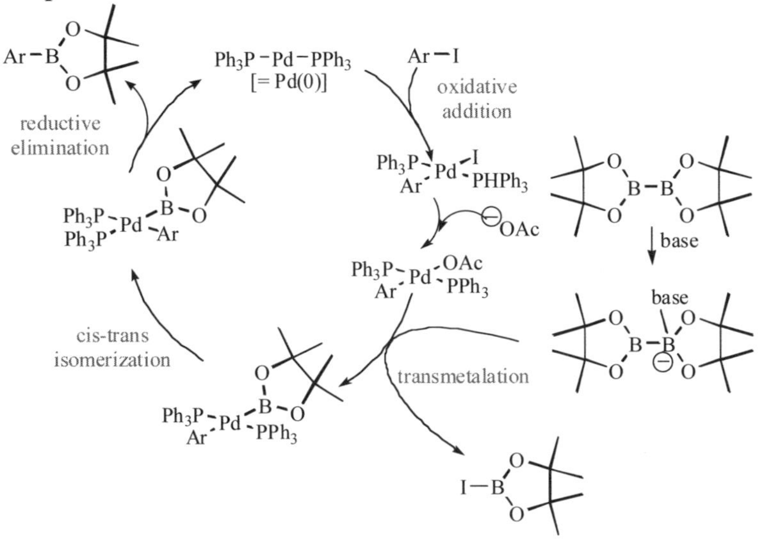

## Examples:

T. Ishiyama, Y. Itoh, T. Kitano, N. Miyaura, *Tetrahedron Letters* **1997**, <u>38</u>, 3447

T. Ishiyama, M. Murata, N. Miyaura, *Journal of Organic Chemistry* **1995**, <u>60</u>, 7508

# Morin Rearrangement

## The Reaction:

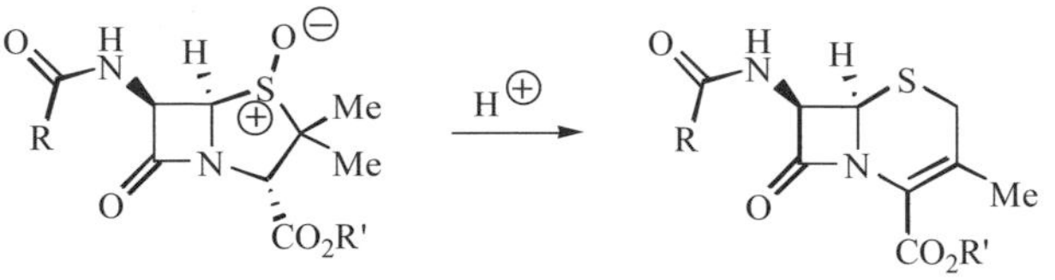

## Proposed Mechanism:

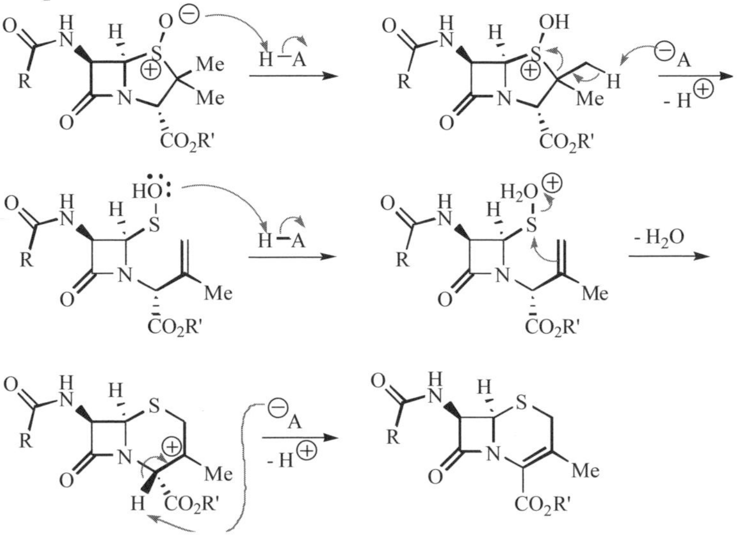

## Notes:

R. B. Morin, B. G. Jackson, R. A. Mueller, E. R. Lavagnino, W. B. Scanlon, S. L. Andrews, *Journal of the American Chemical Society* **1963**, <u>85</u>, 1896

An intermediate of this rearrangement has been trapped:
J. D. Freed, D. J. Hart, N. A. Magomedov, *Journal of Organic Chemistry* **2001**, <u>66</u>, 839

## Examples:

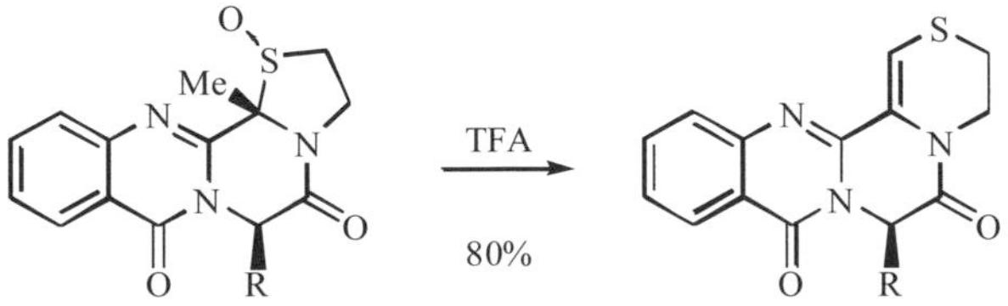

D. J. Hart, N. Magomedov, *Journal of Organic Chemistry* **1999**, <u>64</u>, 2990

T. Fekner, J. E. Baldwin, R. M. Adlington, D. J. Schofield, *Journal of the Chemical Society, Chemical Communications* **1996**, 1989

A. Nudelman, R. J. McCaully, *Journal of Organic Chemistry* **1977**, <u>42</u>, 2887

J. P. Clayton, J. H. C. Nayler, M. J. Pearson, R. Southgate, *Journal of the Chemical Society, Perkin Transactions 1* **1974**, 22

V. Farina, J. Kant, *Tetrahedron Letters* **1992**, <u>33</u>, 3559

# Mukaiyama Reaction

## The Reaction:

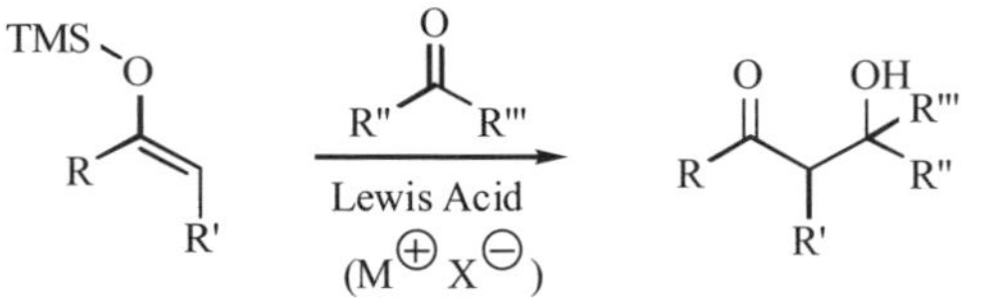

## Proposed Mechanism:

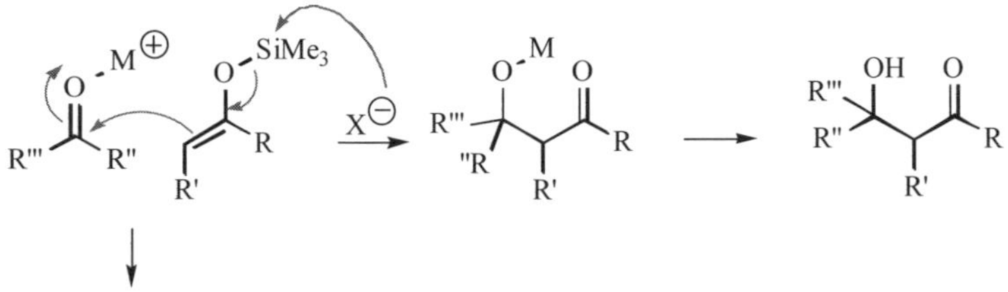

alternate view to address stereochemistry, R''' = H

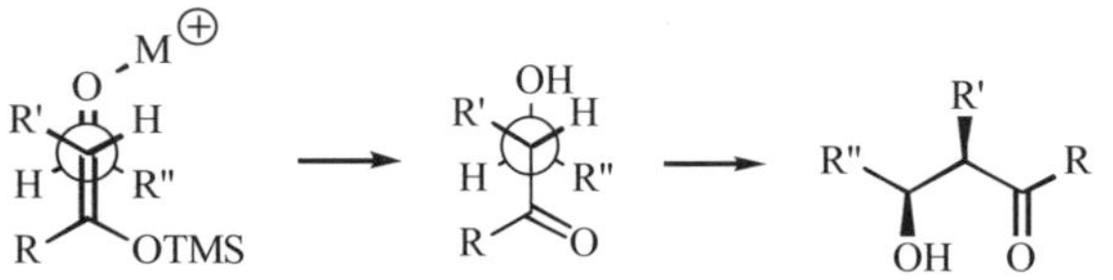

## Notes:

M. B. Smith, J. March in *March's Advanced Organic Chemistry*, 5$^{th}$ ed., John Wiley and Sons, Inc., New York, 2001, p.1223.

A variant with enones: the ***Mukaiyama Michael Reaction***:

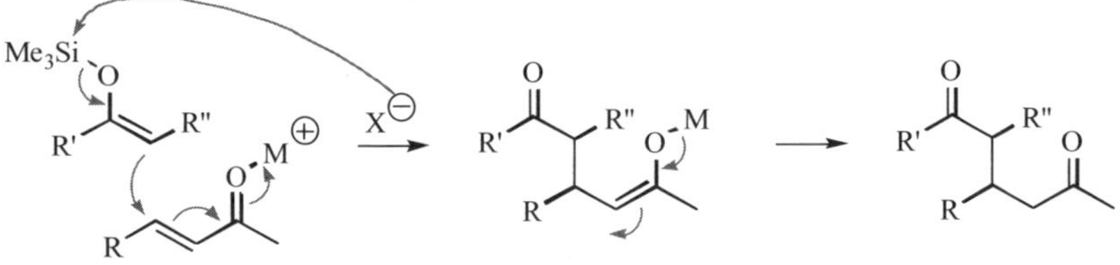

## Examples:

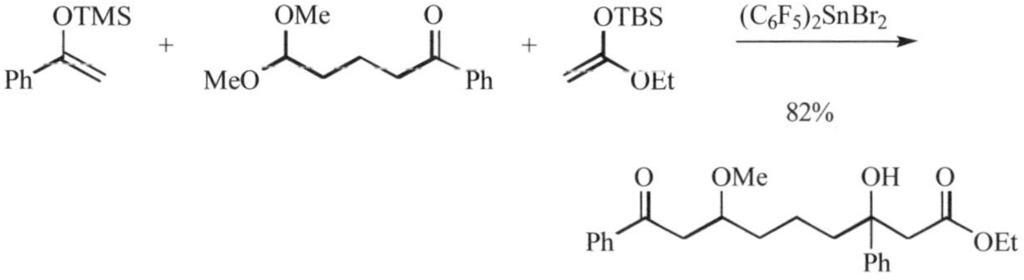

82%

J.-x. Chen, J. Otera, *Angewandte Chemie International Edition in English* **1998**, *37*, 91

TBDMS–O / t-BuO –C(Me)= + (2-carbethoxycyclopentenone) CO$_2$Et

1. TiCl$_2$(O$i$-Pr)$_2$,
2. HCl

50%

A. Ishii, J. Kojima, K. Mikami, *Organic Letters* **1999**, 1, 2013

2 (α-methylstyrene OTMS)

InCl$_3$ (cat.), trace H$_2$O
solvent free

72%

Control of water with the InCl$_3$ is critical to the reaction.

S. Chancharunee, P. Perlmutter, M. Statton, *Tetrahedron Letters* **2002**, 44, 5683

OBn O / Me, H + Me–C(OTMS)=C(Br)–OMe

MgBr$_2$
Et$_2$O

65%

OBn OH O / Me Me Br–OMe  +  OBn OH O / Me Me Br–OMe

> 20    :    1

Y. Guindon, K. Houde, M. Prevost, B. Cardinal-David, B. Daoust, M. Benchegroun, B. Guerin, *Journal of the American Chemical Society* **2001**, 123, 8496

CO$_2$Me ... OTMS phthalimide  +  NC–CH$_2$CH$_2$–CO$_2$Me

TiCl$_4$
CH$_2$Cl$_2$

87%

MeO$_2$C ... CO$_2$Me / OH / CN / phthalimide

A. R. Chaperon, T. M. Engeloch, R. Neier, *Angewandte Chemie International Edition in English* **1998**, 37, 358

Si–O(furan)

Me–CH=CH–CHO  /  LA

Me–CH=CH–CHO  /  amine catalyst

Developed and tested a new approach to a common natural products subunit

S. P. Brown, N. C. Goodwin, D. W. C. MacMillan, *Journal of the American Chemical Society* **2003**, 125, 1192

# Mundy *N*-Acyllactam Rearrangement

## The Reaction:

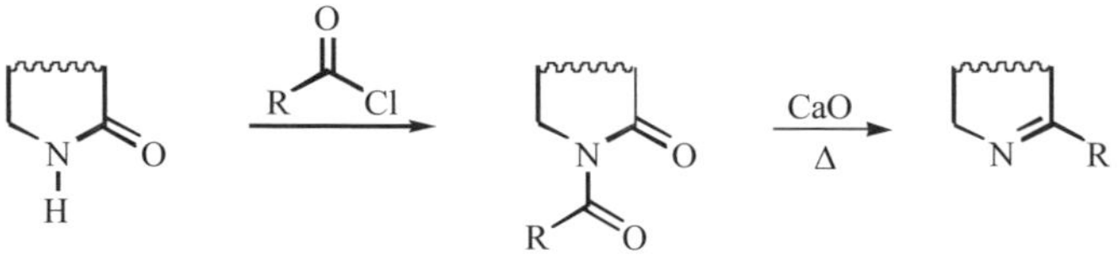

Reaction scheme: pyrrolidinone with N-H reacts with R-C(=O)-Cl to give N-acyl lactam, then CaO / Δ gives the pyrroline product.

## Proposed Mechanism:

Not established, however the path of $^{14}C$ label as follows:

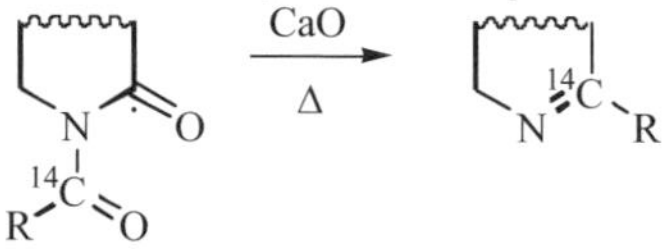

Scheme showing $^{14}C$ labeled N-acyl lactam reacting with CaO, Δ to give the $^{14}C$ labeled pyrroline.

## Notes:

This sequence can provide easy access to substituted pyrrolines. For example, the simple synthesis of 2-phenylpyrroline:

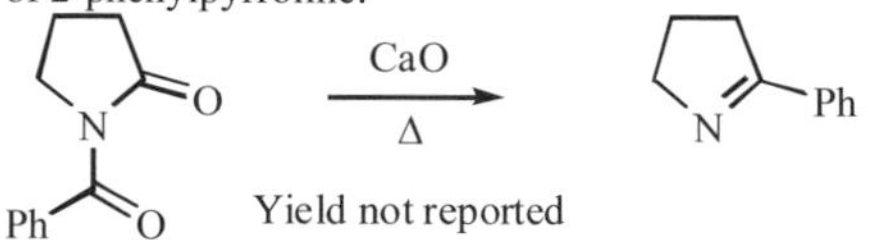

Scheme: N-benzoyl pyrrolidinone with CaO, Δ gives 2-phenylpyrroline. Yield not reported

B. P. Mundy, B. R. Larsen, L. F. McKenzie, G. Braden, *Journal of Organic Chemistry* **1972**, *37*, 1635

is also accomplished by:

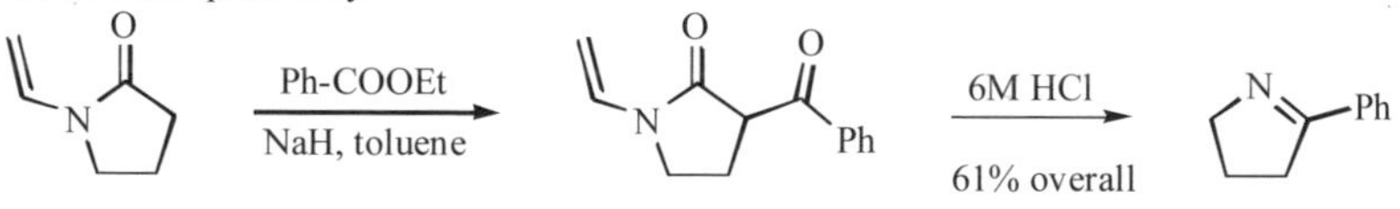

Scheme: N-vinyl pyrrolidinone with Ph-COOEt, NaH, toluene gives the benzoyl substituted intermediate, then 6M HCl gives 2-phenylpyrroline. 61% overall

K. L. Sorgi, C. A. Maryanoff, D. F. McComsey, B. E. Maryanoff, *Organic Syntheses*, Vol. 75, 215

## Examples:

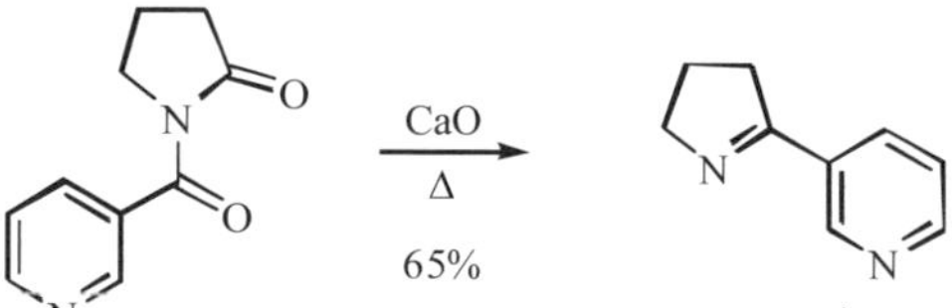

Scheme: N-(nicotinoyl) pyrrolidinone with CaO, Δ (65%) gives myosmine

*myosmine*

B. P. Mundy, B. R. Larsen, L. F. McKenzie, G. Braden, *Journal of Organic Chemistry* **1972**, *37*, 1635

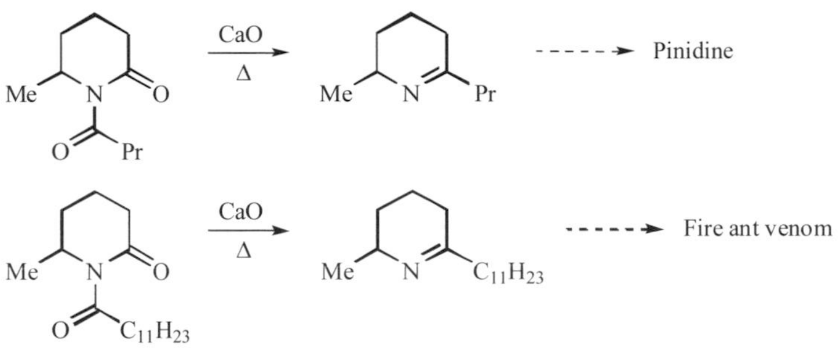

CaO
Δ
Me
N
O
O
Pr
Me
N
Pr
Pinidine
CaO
Δ
Me
N
O
O
$C_{11}H_{23}$
Me
N
$C_{11}H_{23}$
Fire ant venom


R. K. Hill, T. Yuri, *Tetrahedron* **1977**, 33, 1569

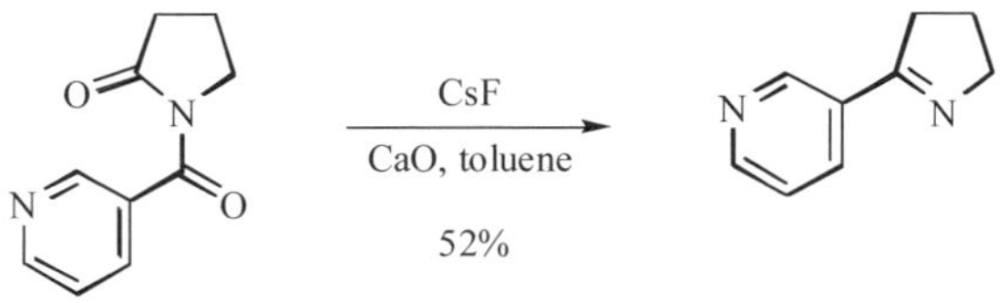

O
N
N
O
CsF
CaO, toluene
52%
N
N


D. Villemin, M. Hachemi, *Reaction Kinetics and Catalysis Letters* **2001**, 72, 3

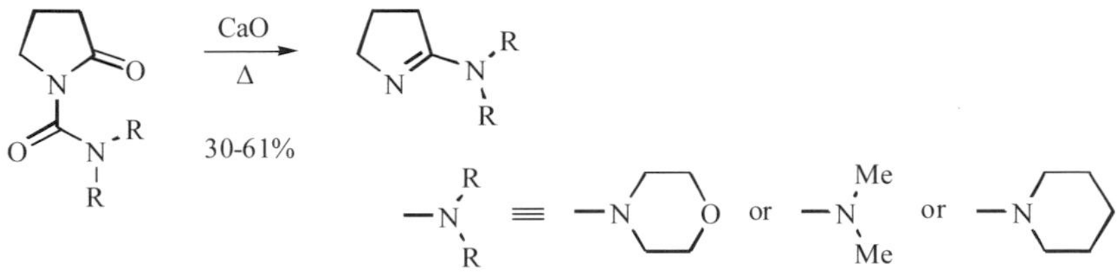

N
O
O
N
R
R
CaO
Δ
30-61%
N
N
R
R
or
or
Me
Me


S. Ravi, M. Easwatamourthy, *Oriental Journal of Chemistry* **2001**, 17, 349 (AN 2001:775237)

# Myers-Saito Cyclization / Schmittel Cyclization

## The Reaction:

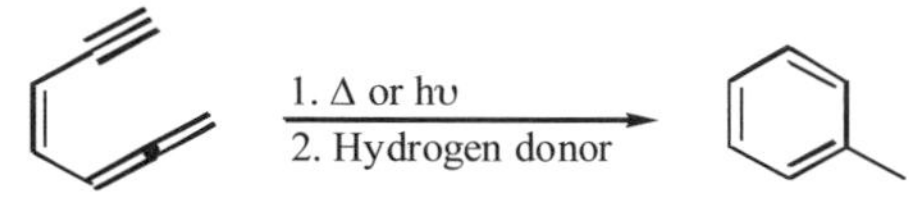

1. Δ or hυ
2. Hydrogen donor

allenyl enyne

## Proposed Mechanism:

H
H
H
H
H
+
+

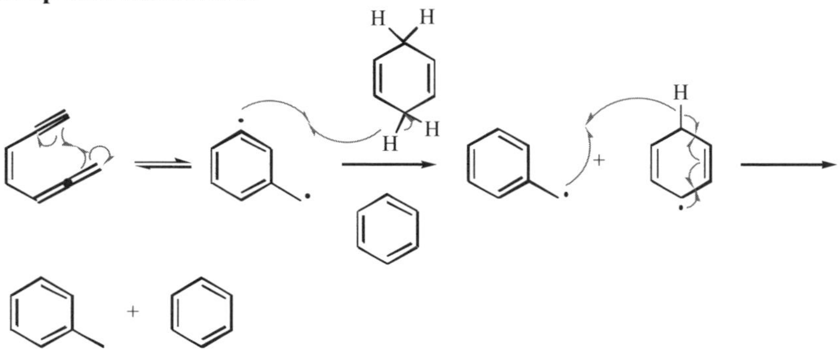

## Notes:

See related chemistry of ***Bergman Cyclization***:

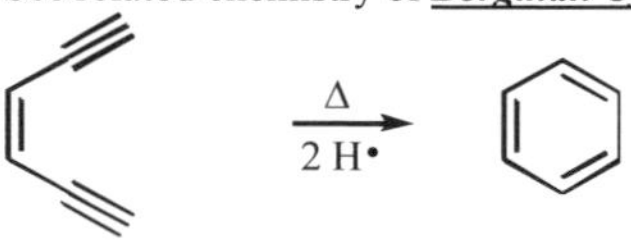

Δ
2 H•

## Examples:

O
Me
HS CO2Me
CD3CO2D, CD3OD
85%
O
Me
S
MeOOC
intermediate
D D
Me
O
S
D
MeOOC

P. A. Wender, M. J. Tebbe, *Tetrahedron* **1994**, 50, 1419

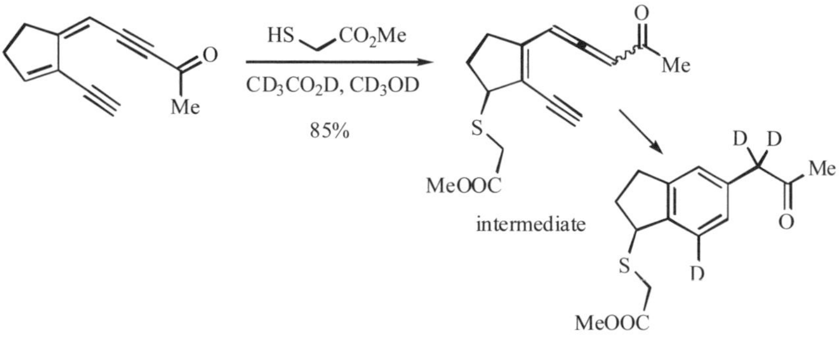

K. K. Wang, Z. Wang, A. Tarli, P. Gannett, *Journal of the American Chemical Society* **1996**, <u>118</u>, 10783

B. Liu, K. K. Wang, J. L. Petersen, *Journal of Organic Chemistry* **1996**, <u>61</u>, 8503

M. Schmittel, J.-P. Steffen, D. Auer, M. Maywald, *Tetrahedron Letters* **1997**, <u>38</u>, 611

# Nametkin Rearrangement

## The Reaction:

## Proposed Mechanism:

## Notes:

M. B. Smith, J. March in *March's Advanced Organic Chemistry*, 5$^{th}$ ed., John Wiley and Sons, Inc., New York, 2001, p. 1394.

A specific example of a ***Wagner - Meerwein rearrangement***
An interesting experiment to analyze for trapping carbocation:

A. G. Martinez, E. T. Vilar, A. G. Fraile, A. H. Fernandez, S. De La Moya Cerero,
F. M. Jimenez, *Tetrahedron* **1998**, 54, 4607

An examination of torsional factors in the rearrangement:

P. C. Moews, J. R. Knox, W. R. Vaughan, *Journal of the American Chemical Society* **1978**, <u>100</u>, 260

## Examples:

S. M. Starling, S. C. Vonwiller, J. N. H. Reek, *Journal of Organic Chemistry* **1998**, <u>63</u>, 2262

A. G. Martinez, E. Teso Vilar, A. G. Fraile, S. de la Cerero, M. E. R. Herrero, P. M. Ruiz, L. R. Subramanian, A. G. Gancedo, *Journal of Medicinal Chemistry* **1995**, <u>38</u>, 4474

# Nazarov Cyclization

## The Reaction:

## Proposed Mechanism:
### "Curly Arrow Formalism"

### "Electrocyclization Formalism"

## Notes:

M. B. Smith, J. March in *March's Advanced Organic Chemistry*, 5th ed., John Wiley and Sons, Inc., New York, 2001, p. 1021; T. Laue, A. Plagens, *Named Organic Reactions,* John Wiley and Sons, Inc., New York, 1998, pp. 195-196; K. L. Habermas, S. C. Denmark, T. K. Jones, *Organic Reactions* **45**, 1.

## Examples:

L. A. Paquette, H.-J. Kang, *Journal of the American Chemical Society* **1991**, <u>113</u>, 2610

J. A. Bender, A. M. Arif, F. G. West, *Journal of the American Chemical Society* **1999**, <u>32</u>, 7443

M. Meisch, L. Miesch-Gross, M. Franck-Neumann, *Tetrahedron* **1997**, <u>53</u>, 2103

K.-F. Cheng, M.-K. Cheung, *Journal of the Chemical Society, Perkin Transaction 1* **1996**, <u>11</u>, 1213

T. Minami, M. Makayama, K. Fujimoto, S. Matsuo, *Journal of the Chemical Society, Chemical Communications* **1992**, <u>2</u>, 190

J. Motoyoshiya, T. Yazaki. S. Hayashi, *Journal of Organic Chemistry* **1991**, <u>56</u>, 735. This work studied rearrangements during the cyclization process.

W. He, X. Sun, A. J. Frontier, *Journal of the American Chemical Society* **2003**, <u>125</u>, 14278

G. Liang, S. N. Gradl, D. Trauner, *Organic Letters* **2003**, <u>5</u>, 4931

# Neber Rearrangement

## The Reaction:

KOEt
$H_2O$
+ TsOH

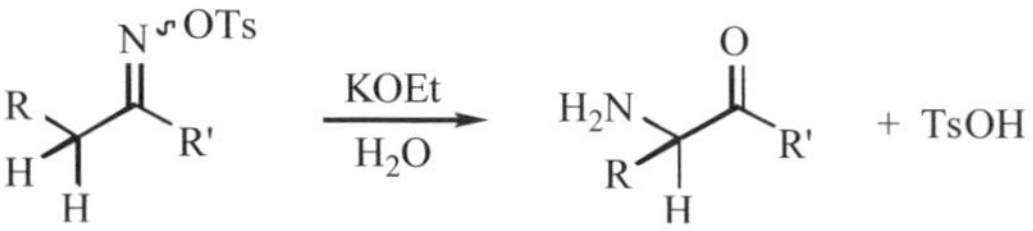

## Proposed Mechanism:

(a)

*Syn* or *anti* orientation gives same product
- OTs
$H_2O$
proton transfer
base

*Has been isolated*

protic workup

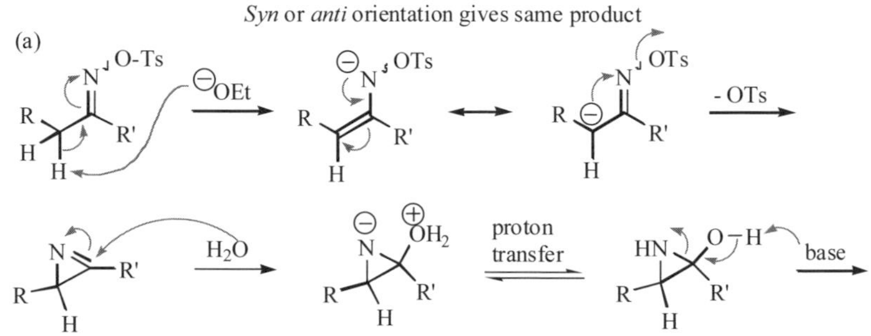

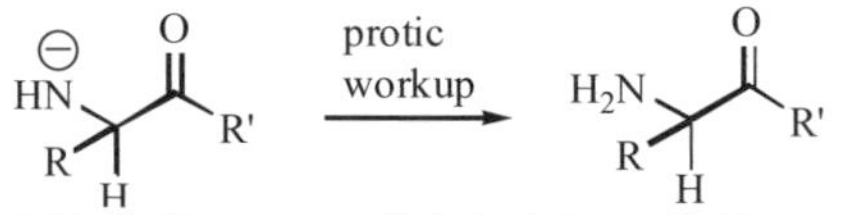

M. M. H. Verstappen, G. J. A. Ariaans, B. Zwanenburg, *Journal of the American Chemical Society* **1996**, 118, 8491

(b)

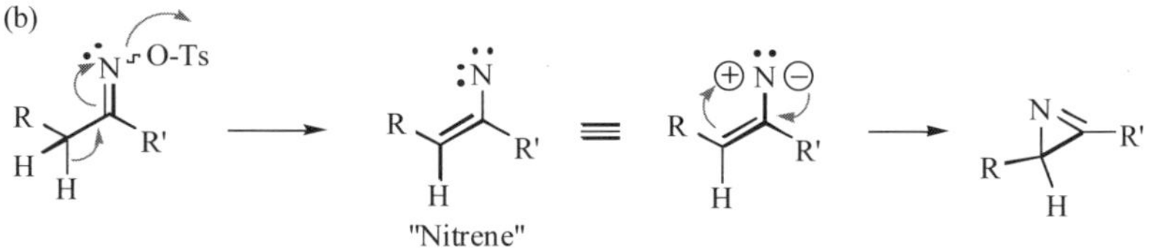

"Nitrene"

## Notes:

M. B. Smith, J. March in *March's Advanced Organic Chemistry*, 5th ed., John Wiley and Sons, Inc., New York, 2001, pp. 288, 1410; T. Laue, A. Plagens, *Named Organic Reactions*, John Wiley and Sons, Inc., New York, 1998, p. 197

The nitrene mechanism is not consistent with the outcome of a chiral-based study:
T. Ooi, M. Takahashi, K. Doda, K. Maruoka, *Journal of the American Chemical Society* **2002**, 124, 7640.

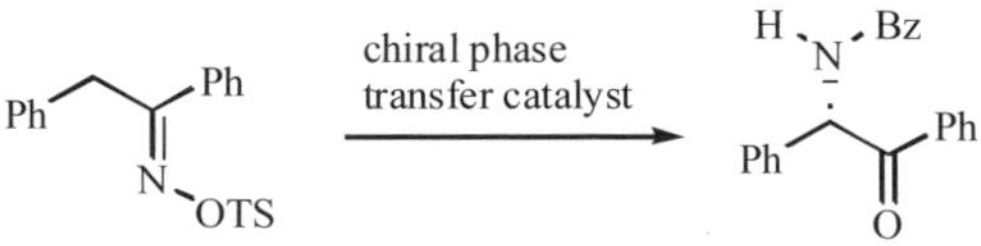

chiral phase transfer catalyst

Unlike the similar ***Beckmann rearrangement***, the stereochemistry of the oxime is not critical to the outcome of the reaction.

## Examples:

N. K. Garg, D. D. Caspi, B. M. Stoltz, *Journal of the American Chemical* **2004**, *126*, 9552

T. Ooi, M. Takahashi, K. Doda, K. Maruoka *Journal of the American Chemical* **2002**, *124*, 7640

J. Y. L. Chung, G.-J. Ho, M. Chartrain, D. Zhao, J. Leazer, R. Farr, M. Robbins, K. Emerson, D. J. Mathre, J. M. McNamaro, D. L. Hughes, E. J. J. Grabowski, P. J. Reider, *Tetrahedron Lett*ers **1999**, *40*, 6739

I. Lopez, A. Diez, M. Rubiralta, *Tetrahedron* **1996**, *52*,. 8581-

# Nef Reaction

## The Reaction:

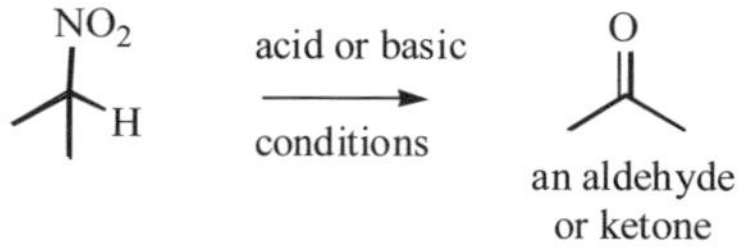

NO$_2$ → acid or basic conditions → an aldehyde or ketone

## Proposed Mechanism:

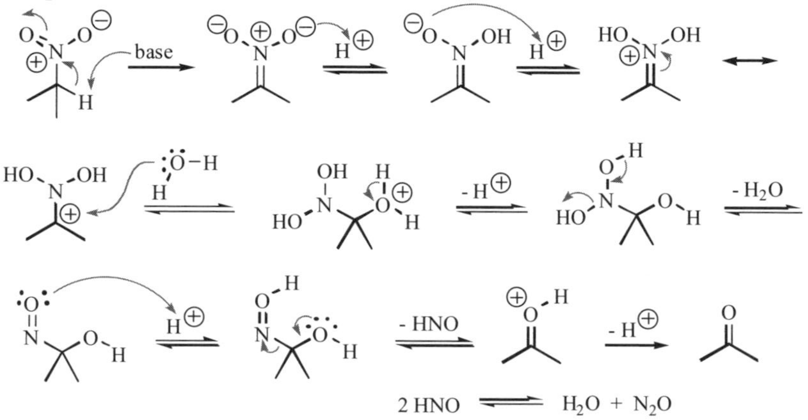

$2 \text{ HNO} \rightleftharpoons H_2O + N_2O$

## Notes:

TiCl$_3$ can be very effective for this reaction. This is called the **McMurry Modification of the Nef Reaction**.

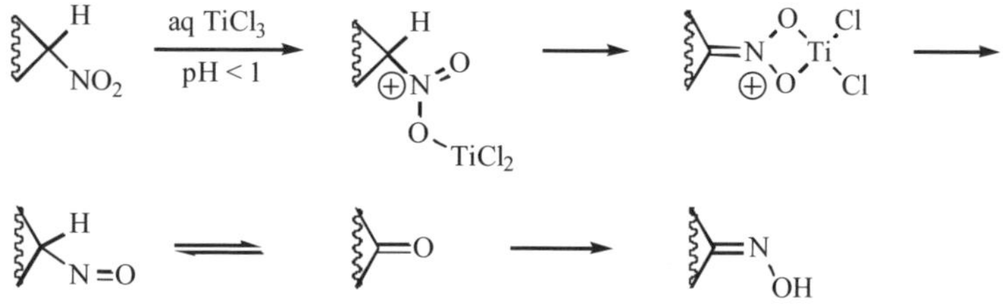

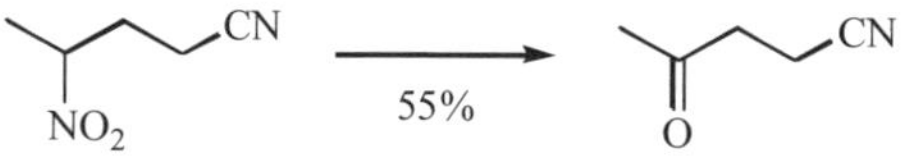

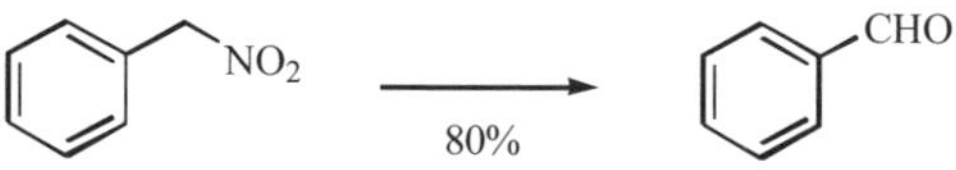

aq TiCl$_3$, pH < 1

55%

80%

J. E. McMurry, J. Melton, *Journal of Organic Chemistry* **1973**, <u>38</u>, 4367

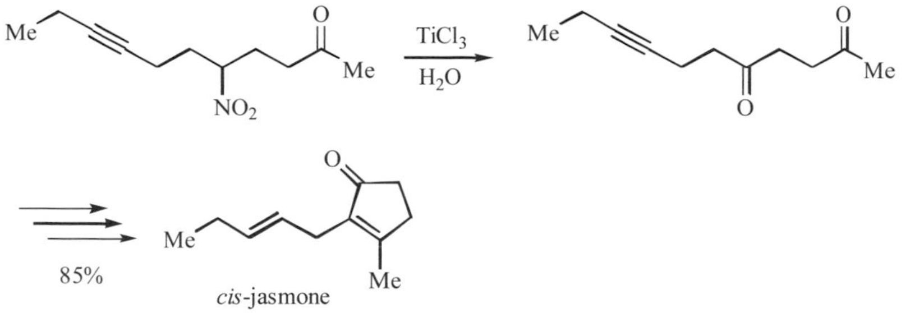

Basic permanganate can accomplish this conversion:

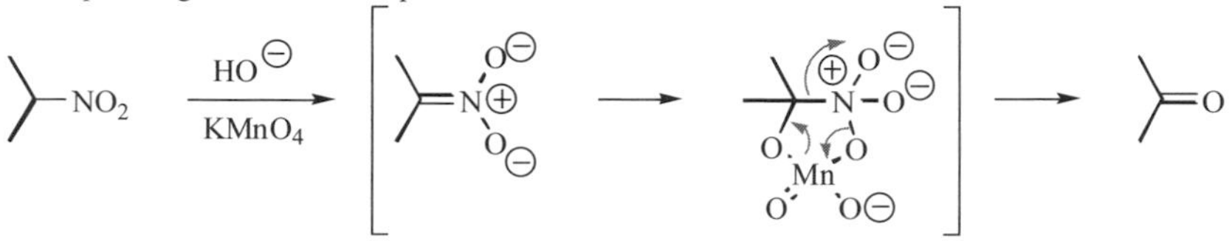

## Examples:

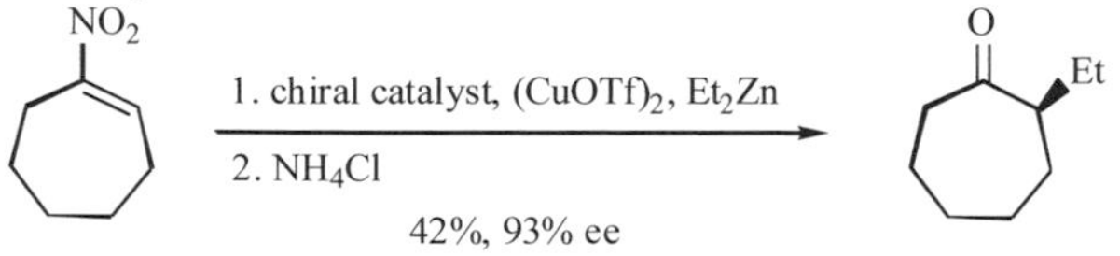

42%, 93% ee

C. A. Luchaco-Cullis, A. H. Hoveyda, *Journal of the American Chemical Society* **2002**, <u>124</u>, 8192

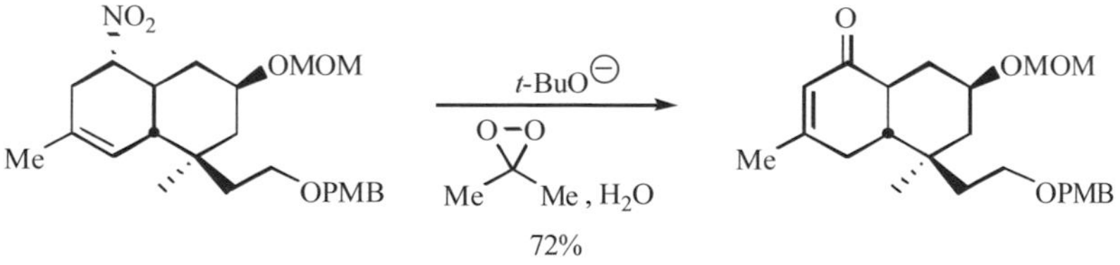

72%

R. Williams, T. A. Brugel, *Organic Letters* **2000**, <u>2</u>, 1023

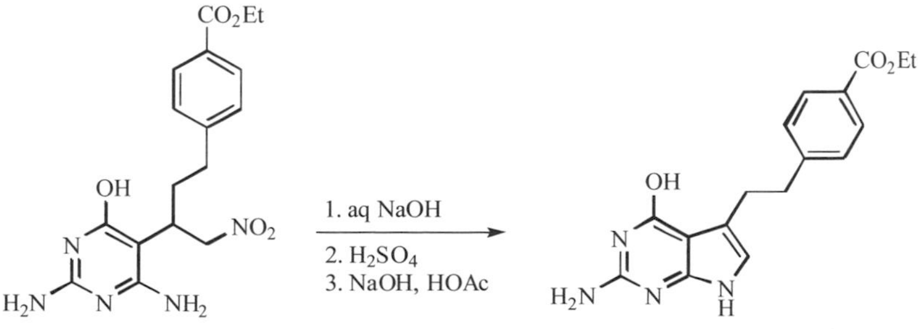

1. aq NaOH
2. $H_2SO_4$
3. NaOH, HOAc

E. C. Taylor, B. Liu, *Journal of Organic Chemistry* **2003**, <u>68</u>, 9938

# Negishi Coupling

## The Reaction:

$$R-ZnX \xrightarrow[\text{R'-X}]{\text{Ni or Pd}} R-R'$$

R = alkyl, alkenyl, aryl, allylic, benzylic
R' = Alkenyl, alkynyl, aryl, allylic, benzylic

E. Negishi, M. Kotora, C. Xu, *Journal of Organic Chemistry* **1997**, <u>62</u>, 8957

## Proposed Mechanism:

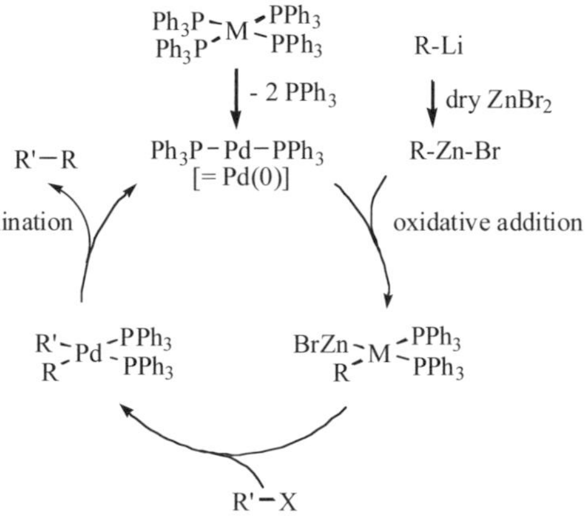

## Notes:

For unactivated halides:

$$R-Br \xrightarrow[]{\text{Zn, cat. } I_2} [R-ZnBr] \xrightarrow[\text{cat. Ni}]{Ar-X} R-Ar$$

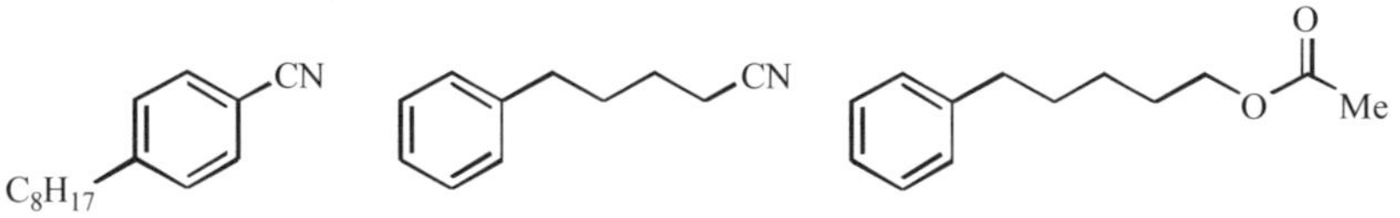

91%                                      98%                                      91%

S. Huo, *Organic Letters* **2003**, <u>5</u>, 423

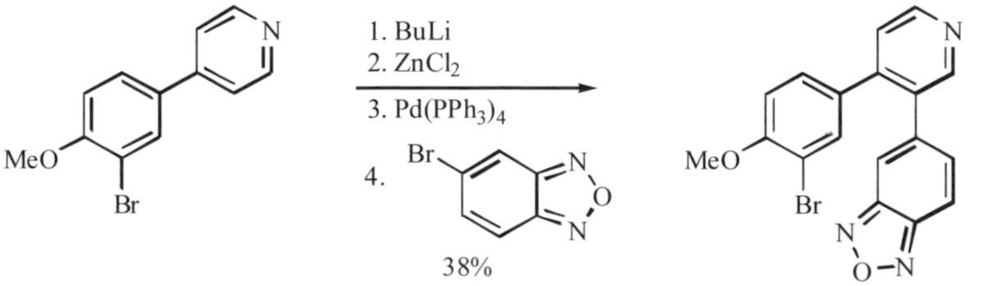

1. BuLi
2. ZnCl$_2$
3. Pd(PPh$_3$)$_4$
4.

38%

P. W. Manley, M. Acemoglu, W. Marterer, W. Pachinger, *Organic Processes and Research Development* **2003**, <u>7</u>, 436

## Examples:

X. Zeng, Q. Hu, M. Qian, E.-i.Negishi, *Journal of the American Chemical Society* **2003**, <u>125</u>, 13636

G. D. McAllister, R. J. K. Taylor, *Tetrahedron Letters* **2004**, <u>45</u>, 2551

J. M. Herbert, *Tetrahedron Letters* **2004**, <u>45</u>, 817

M. Yus, J. Gomis, *Tetrahedron Letters*, **2001**, <u>42</u>, 5721

J. A. Panek, T. Hu, *Journal of Organic Chemistry* **1997**, <u>62</u>, 4912

C. Dai, G. C. Fu, *Journal of the American Chemical Society* **2001**, <u>123</u>, 2719

# Nenitzescu Acylation

## The Reaction:

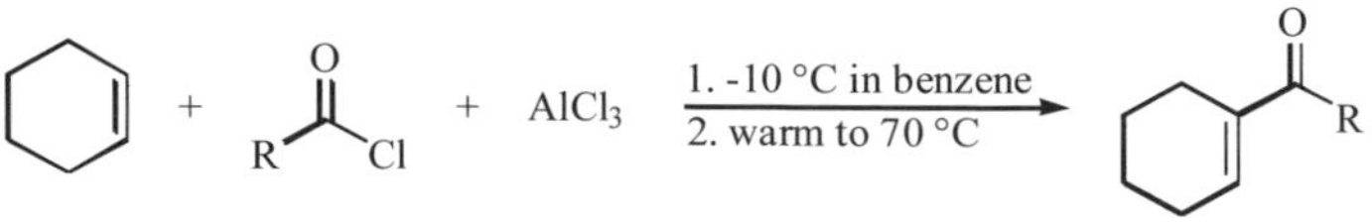

## Proposed Mechanism:

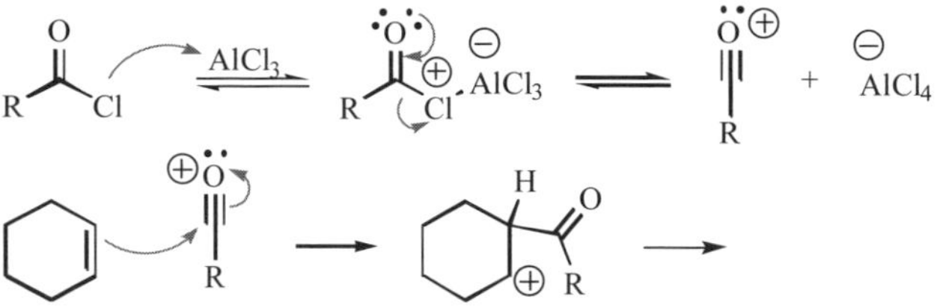

## Notes:

Note similarities to the ***Friedel-Crafts Acylation***.

## Examples:

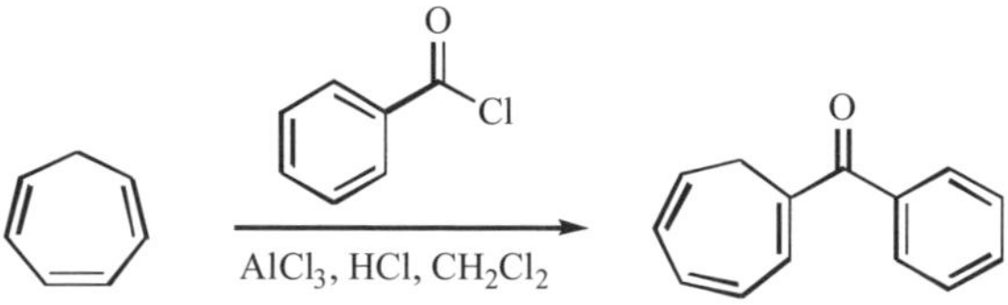

J. A. Blair, C. J. Tate, *Chemical Communications* **1969**, 1506

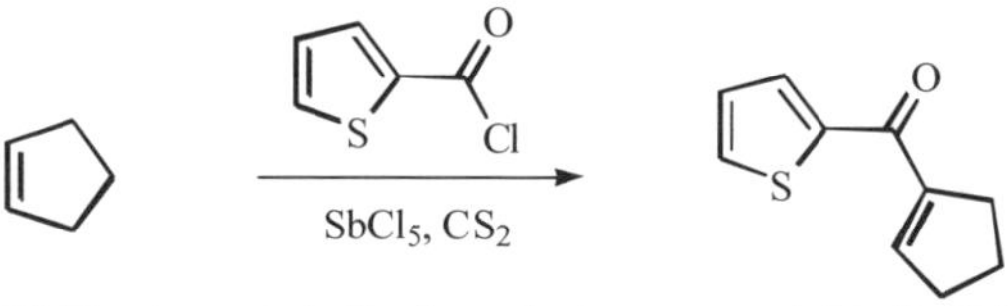

Yield not reported. Crude product was converted to the 2,4-D derivative.

L. H. Klemm, T. Largman, *Journal of the American Chemical Society* **1952**, 74, 4458

M. L. Patil, G. K. Jnaneshwara, D. P. Sabde, M. K. Dongare, A. Sudalai, F. H. Deshpande, *Tetrahedron Letters*, **1997**, <u>38</u>, 2137

D. Villemin, B. Labiad, *Synthetic Communications* **1992**, <u>22</u>, 3181 (AN 1993:168726)

K. E. Harding, K. S. Clement, *Journal of Organic Chemistry* **1984**, <u>49</u>, 3870

T. Shono, I. Sishiguchi, M. Sasaki, H. Ikeda, M. Kurita, *Journal of Organic Chemistry* **1983**, <u>48</u>, 2503

# Nenitzescu Indole Synthesis

## The Reaction:

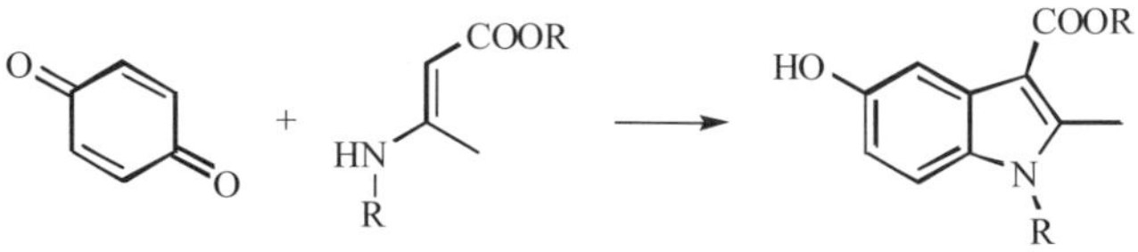

## Proposed Mechanism:

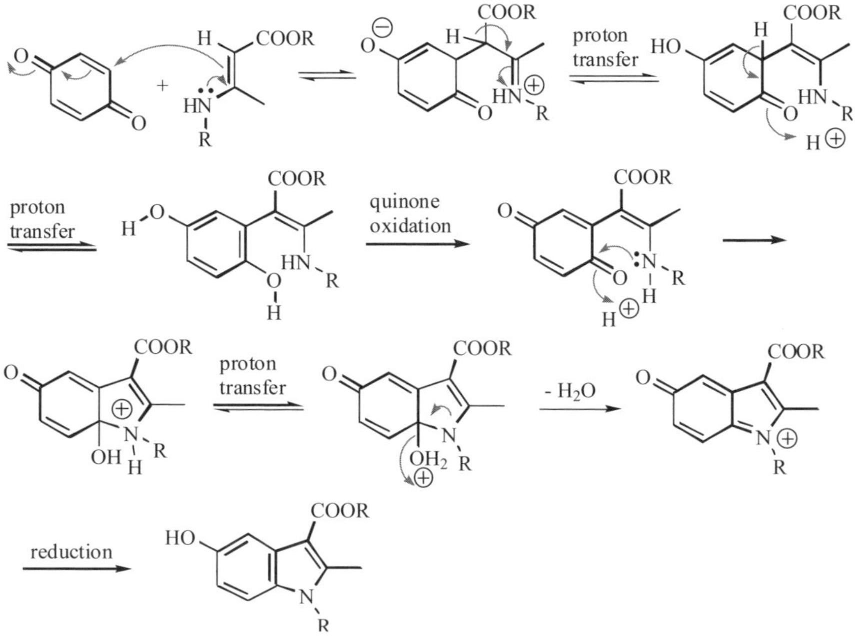

## Notes:

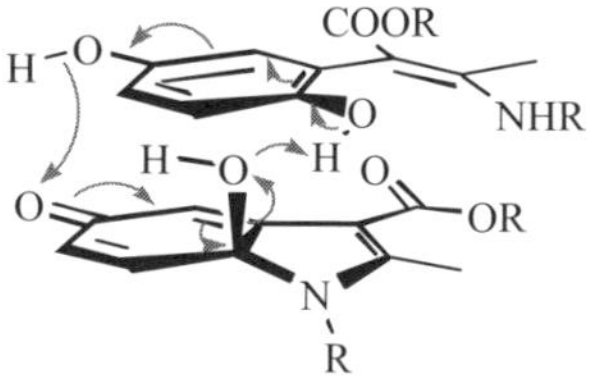

The hydride transfer may be a bimolecular oxidation / reduction process.

## Examples:

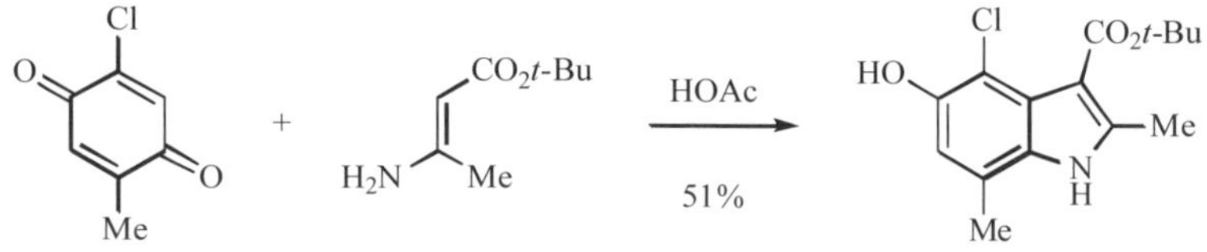

51%

J. F. Poletto, M. J. Weiss, *Journal of Organic Chemistry* **1970**, <u>35</u>, 1190

A large assortment of substitution patterns prepared:

HOAc

large series prepared

T. M. Boehme, C. E. Augelli-Szafran, H. Hallak, T. Pugsley, K Serpa, R. D. Schwarz, *Journal of Medicinal Chemistry* **2002**, 45, 3094

A solid-phase synthesis:

trimethylorthoformate

cleaved from resin with TFA

D. M. Ketcha, L. J. Wilson, D. E. Portlock, *Tetrahedron Letters* **2000**, 41, 6253

acetone

low yield

1 : 1 ratio attachment

R. Littell, G. R. Allen, Jr., *Journal of Organic Chemistry* **1968**, 33, 2064

A modification used to prepare carbazoles:

EtOH

Δ

55%

R. Littell, G. O.Morton, G. R. Allen, Jr., *Journal of the American Chemical Society* **1970**, 92, 3740

# Nicholas Reaction

## The Reaction:

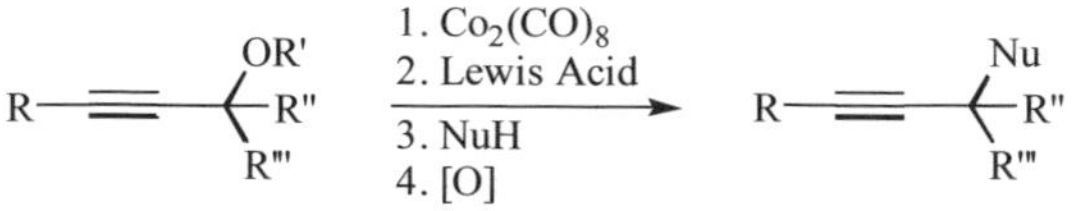

## Proposed Mechanism:

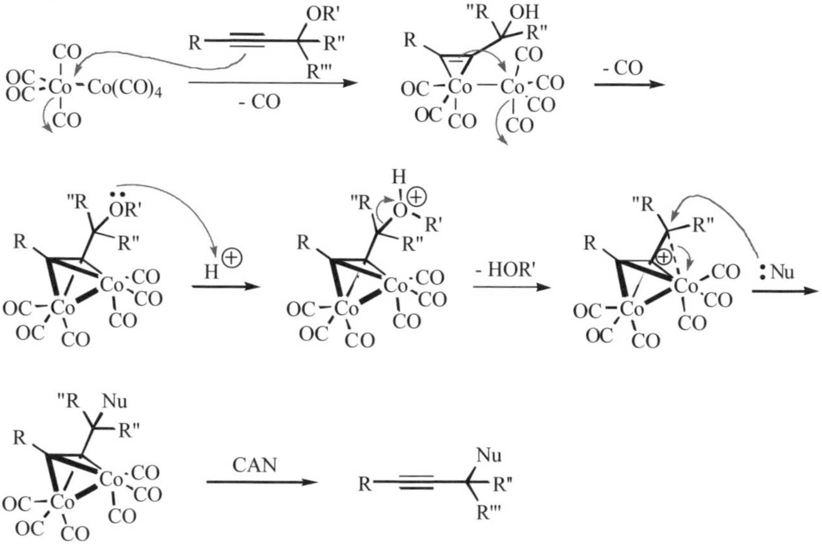

## Notes:

A Ru-catalyzed variation:

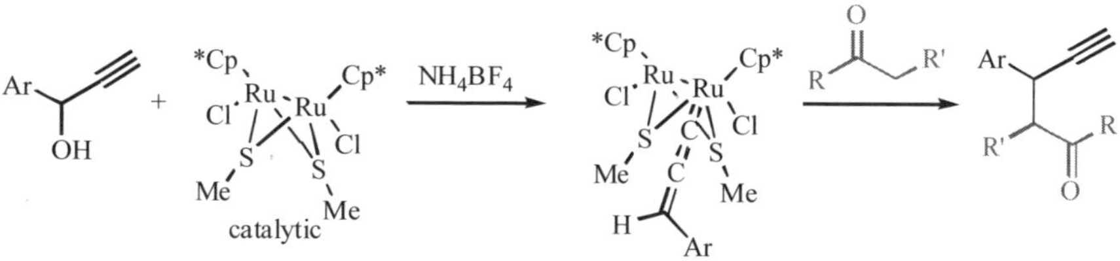

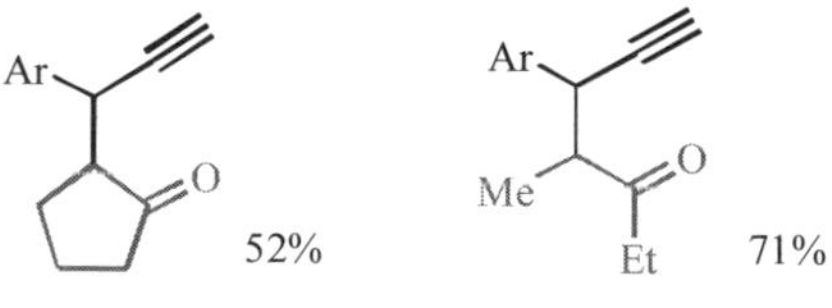

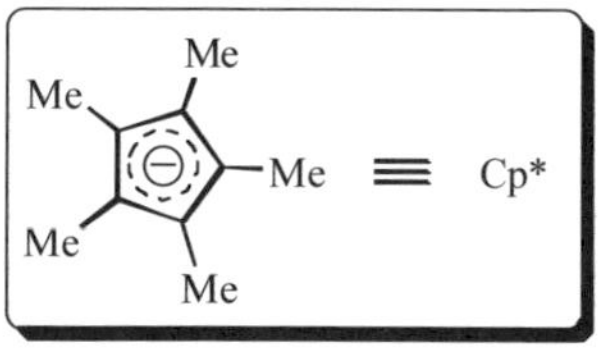

52%   71%

Y. Nishibayashi, I. Wakiji, Y. Ishii, S. Uemura, M. Hidai, *Journal of the American Chemical Society* **2001**, 123, 3393

## Examples:

J. A. Cassel, S. Leue, N. I. Gachkova, N. C. Kann, *Journal of Organic Chemistry* **2002**, <u>67</u>, 9460

F. R. P. Crisostomo, T. Martın, V. S. Martın, *Organic Letters* **2004**, <u>6</u>, 565

C. Mukai, H. Yamashita, M. Sassa, M. Hanaoka, *Tetrahedron* **2002**, <u>58</u>, 2755

# Niementowski Quinazoline Synthesis

## The Reaction:

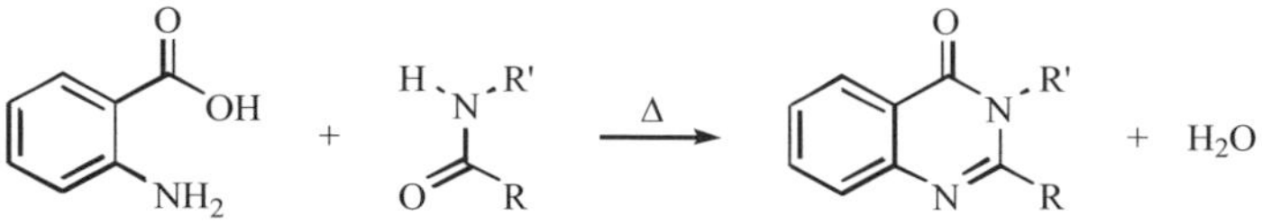

## Proposed Mechanism:

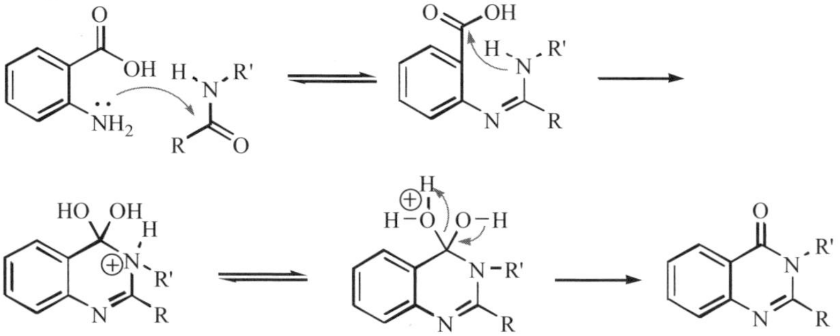

## Notes:

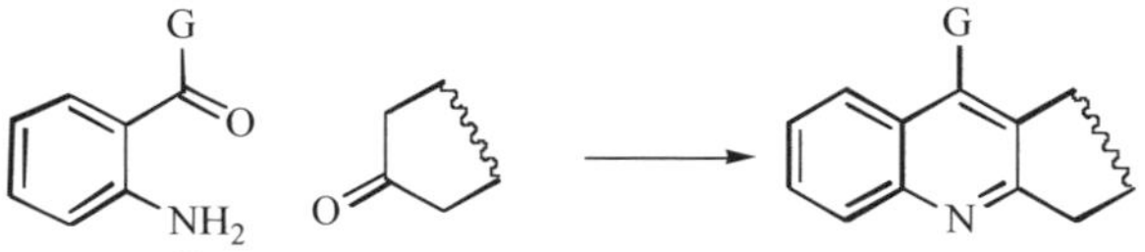

| G | Name Reaction |
|---|---|
| H, R | Friedlander |
| C(O)O | Pfitzinger |
| OH, OR | Niementowski |

R .J. Chong, M. A. Siddiqui, V. Snieckus, *Tetrahedron Letters* **1986**, <u>27</u>, 5323

## Examples:

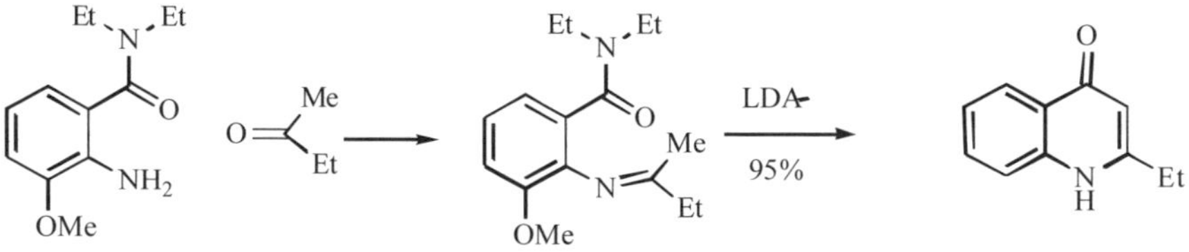

R .J. Chong, M. A. Siddiqui, V. Snieckus, *Tetrahedron Letters* **1986**, <u>27</u>, 5323

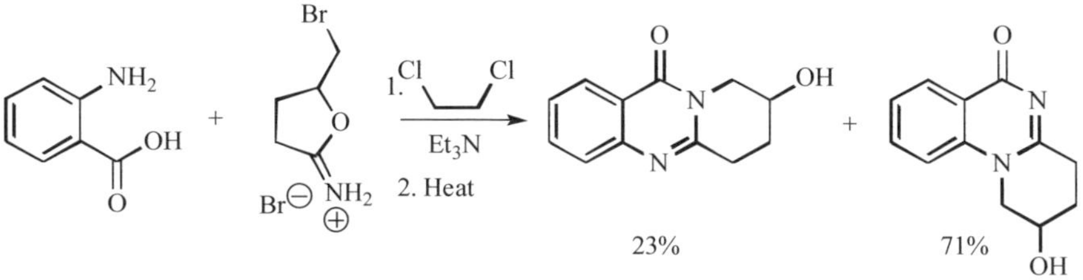

G. Fantin, M. Fogagnolo, A. Medici, P. Pedrini, *Journal of Organic Chemistry* **1993**, <u>58</u>, 741

L. Domon, C. Le Coeur, A. Grelard, V. Thiery, T. Besson, *Tetrahedron Letters* **2001**, <u>42</u>, 6671

J. F. Meyer, E. C. Wagner, *Journal of Organic Chemistry* **1943**, <u>8</u>, 239

The major reactant, formed in situ, was

T. Kametani, C. V. Loc, T. Higa, M. Koizumi, M. Ihara, K. Fukumoto, *Journal of the American Chemical Society* **1977**, <u>99</u>, 2306

F. Alexandre, A. Berecibar, R. Wrigglesworth, T. Bessonb, *Tetrahedron* **2003,** <u>59</u>, 1413

# Noyori Annulation

## The Reaction:

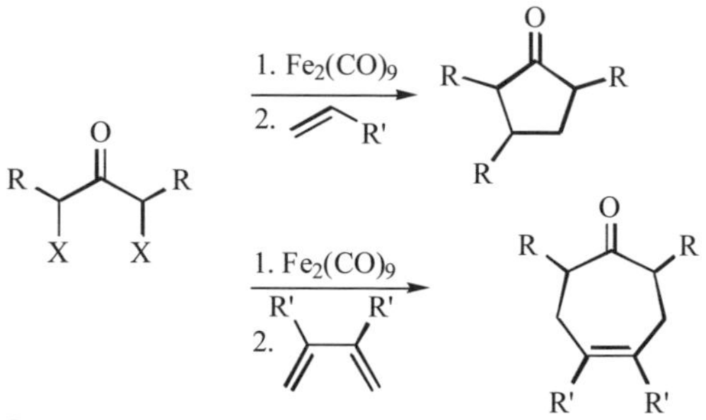

## Proposed Mechanism:

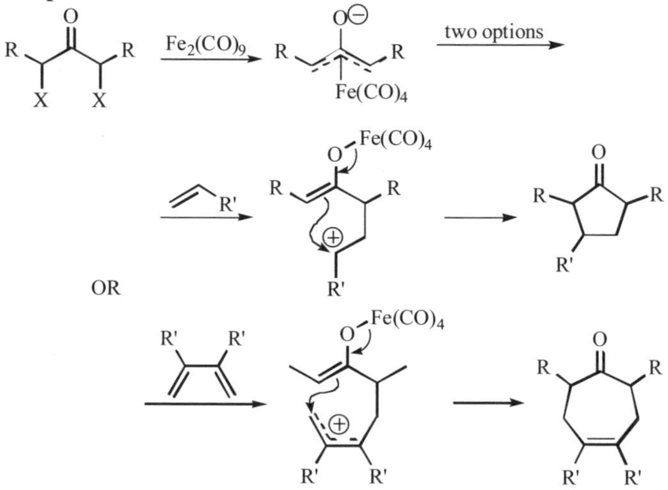

## Notes:

In the original work, reaction with enamines:

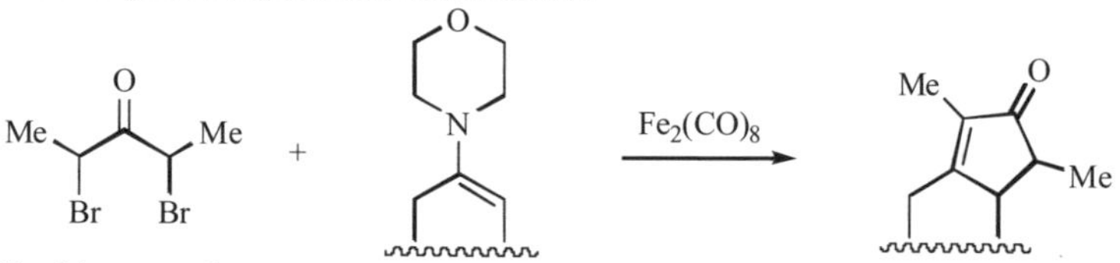

By this approach:

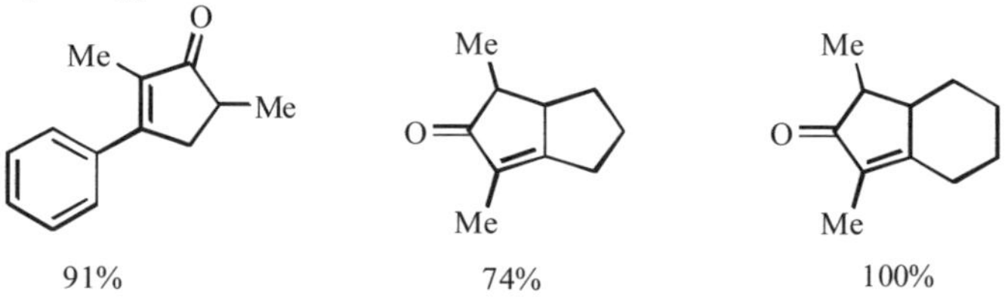

91%                                 74%                              100%

Y. Hayakawa, K. Yokoyama, R. Noyori, *Journal of the American Chemical Society* **1978**, <u>100</u>, 1799

# Examples:

L. S. Hegedus, M. S. Holden, *Journal of Organic Chemistry* **1985**, <u>50</u>, 3920

T. Ishizu, M. Mori, K. Kanematsu, *Journal of Organic Chemistry* **1981**, <u>46</u>, 526

Y. Hayakawa, K. Yokiyama, R. Noyori, *Journal of the American Chemical Society* **1978**, <u>100</u>, 1791

R. J. P. Corriu, J. J. E. Moreau, M. Pataud-Sat, *Journal of Organic Chemistry* **1990**, <u>55</u>, 2878

# Nozaki-Hiyama-Kishi Reaction

## The Reaction:

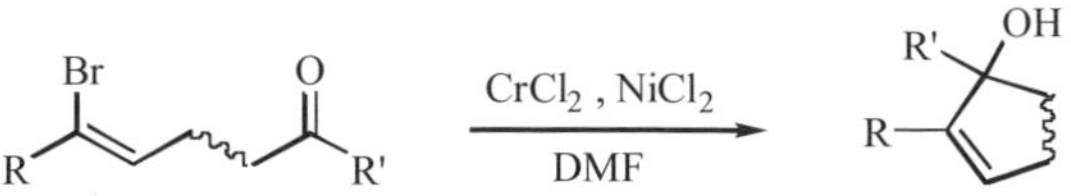

## Proposed Mechanism:

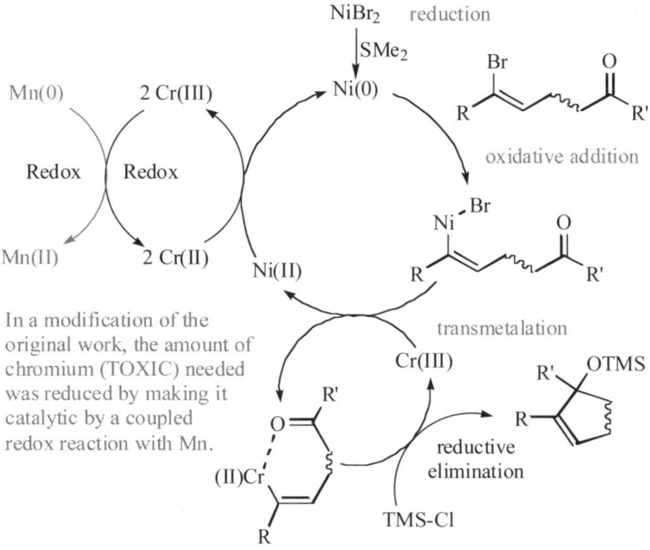

In a modification of the original work, the amount of chromium (TOXIC) needed was reduced by making it catalytic by a coupled redox reaction with Mn.

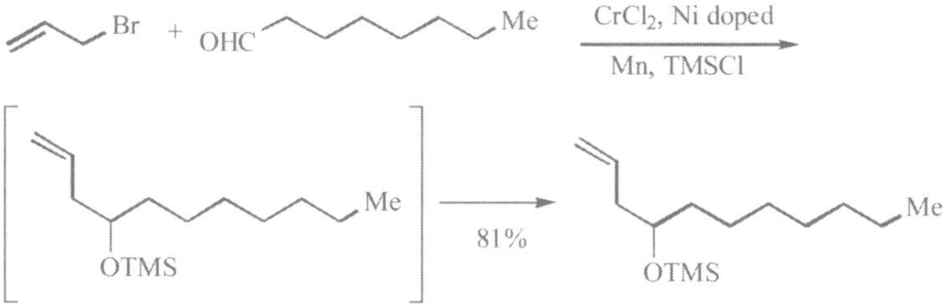

A. Furstner, N. Shi, *Journal of the American Chemical Society* **1996**, 118, 12349

## Notes:

P. Cintas, *Synthesis* **1992**, 248

Disubstituted alkenyl halides maintain streochemistry in the product:

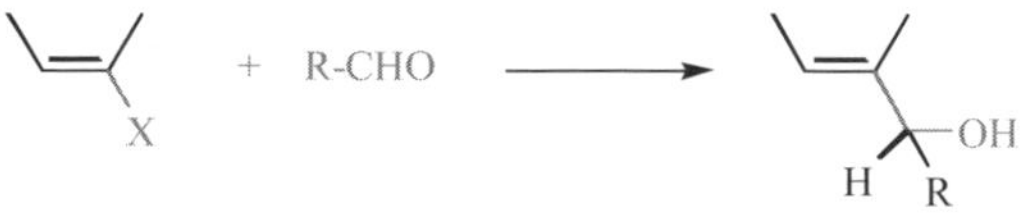

K. Takai, K. Kimura, T. Kuroda, T. Hiyama, H. Nozaki, *Tetrahedron Letters* **1983**, 24, 5281

The occasional observation of ketone products can be rationalized by a Cr-mediated ***Oppenauer oxidation***:

H. S. Schrekker, M. W. G. de Bolster, R. V. A. Orru, L. A. Wessjohann, *Journal of Organic Chemistry* **2002**, <u>67</u>, 1977

## Examples:

X.-T. Chen, S. K. Bhattacharya, B. Zhou, C. E. Gutteridge, T. R. R. Pettus, S. J. Danishefsky, *Journal of the American Chemical Society* **1999**, <u>121</u>, 6563

B. M. Trost, A. B. Pinkerton, *Journal of Organic Chemistry* **2001**, <u>66</u>, 7714

X.-Q. Tang, J. Montgomery, *Journal of the American Chemical Society* **2000**, <u>122</u>, 6950

R. A. Pilli, M. M. Victor, *Tetrahedron Letters* **2002**, <u>43</u>, 2815

# Oppenauer Oxidation

## The Reaction:

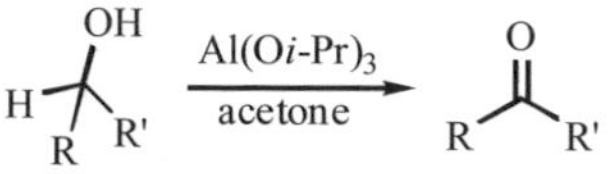

This is the reverse of the ***Meerwein-Ponndorf-Verley Reduction***.

## Proposed Mechanism:

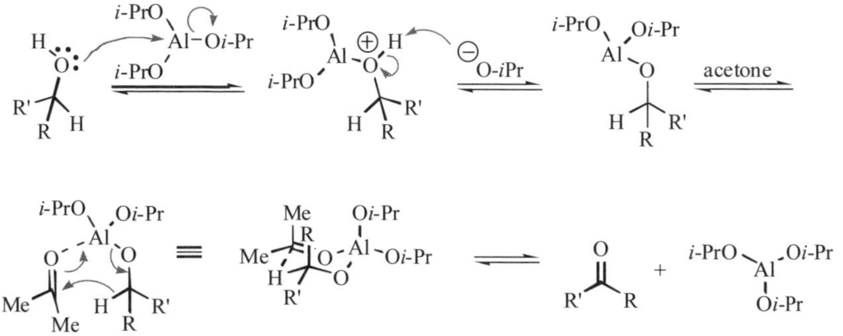

## Notes:

M. B. Smith, J. March in *March's Advanced Organic Chemistry*, 5th ed., John Wiley and Sons, Inc., New York, 2001, p. 1516; C. Djerassi, *Organic Reactions* **6**, 5

An important feature of this reaction is the lack of over oxidation.

### Examples:

A unique in-situ oxidation:

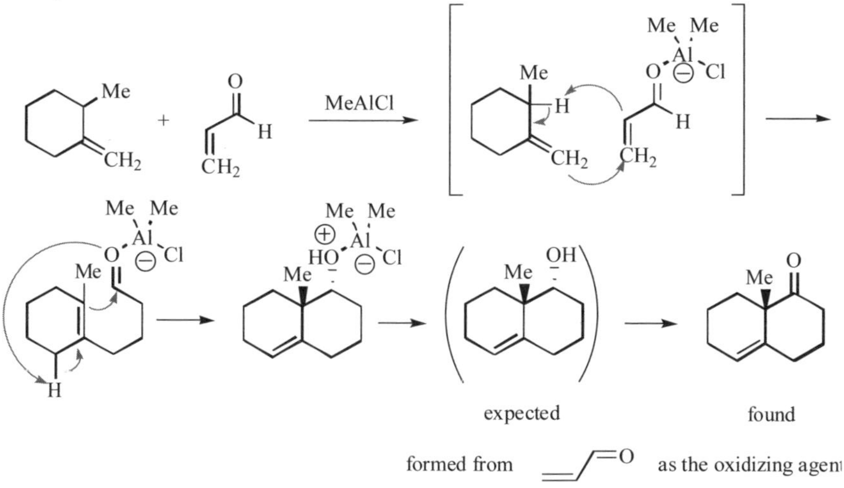

expected                          found

formed from _____ /=O  as the oxidizing agent

B. B. Snider, B. E. Goldman, *Tetrahedron* **1986**, _42_, 2951

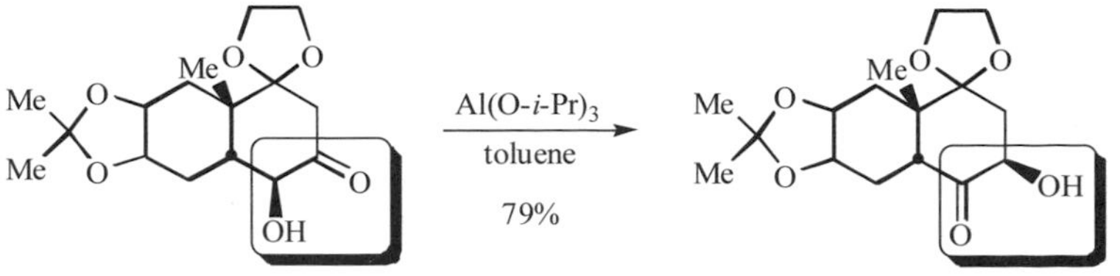

Al(O-*i*-Pr)$_3$
toluene
79%

Note also the oxidation

T. K. M. Shing, C. M. Lee, H. Y. Lo, *Tetrahedron Letters* **2001**, <u>42</u>, 8361

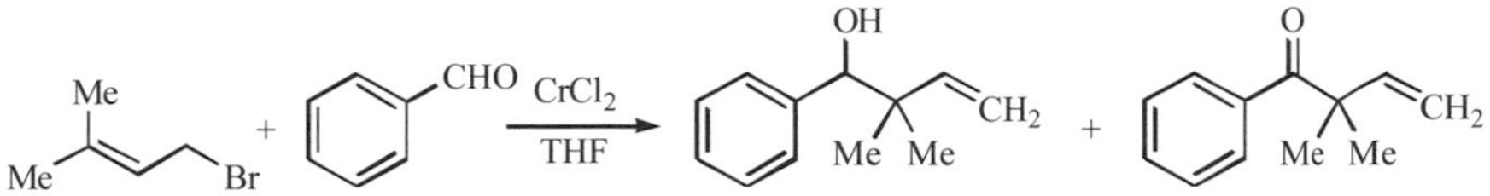

78%, 1:1

**Hiyama-Nozaki** product          **Oppenauer** product

The **Oppenauer** product formed from an interaction of the expected product reacting with benzaldehyde in a Cr-catalyzed oxidation

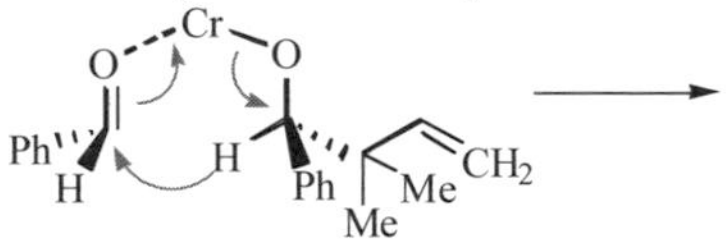

H. S. Schrekker, M. W. G. de Bolster, R. V. A. Orru, L. A. Wessjohann, *Journal of Organic Chemistry* **2002**, <u>67</u>, 1975

New oxidation system:

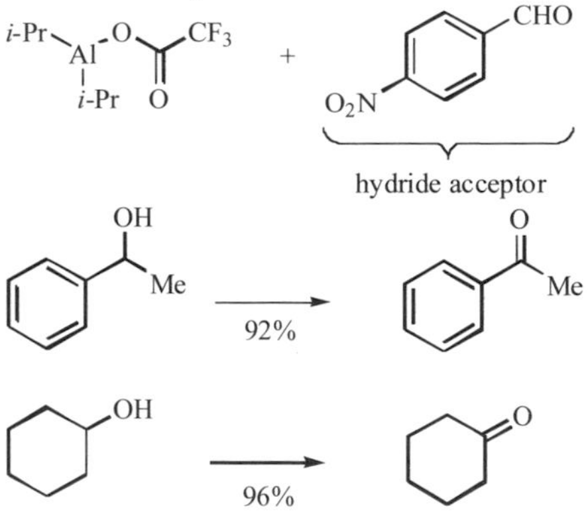

hydride acceptor

92%

96%

K. G. Akamanchi, B. A. Chaudhari, *Tetrahedron Letters* **1997**, <u>38</u>, 6925

# Overman Rearrangement

## The Reaction:

$$\xrightarrow[\text{2. }\Delta]{\text{1. NaH, } N\equiv\!\!\!-CCl_3}$$

## Proposed Mechanism:

serves as base; thus, only catalytic amounts of NaH are needed.

Catalytic amounts of Hg(II) or Pd(II) work in some cases.

## Notes:

Only catalytic amounts of NaH are needed ( see above).

## Examples:

1. Cl$_3$C-CN, NaH
2. Ph-Cl, K$_2$CO$_3$, $\Delta$

95%

M. Reilly, D. R. Anthony, C. Gallagher, *Tetrahedron Letters* **2003**, 44, 2927

H. Ovaa, J. D. C. Codée, B. Lastdrager, H. S. Overkleeft, G. A. van der Marel, J. H. van Boom, *Tetrahedron Letters* **1999**, <u>40</u>, 5063

T. J. Donohoe, K. Blades, M. Helliwell, P. R. Moore, J. J. G. Winter, *Journal of Organic Chemistry* **1999**, <u>64</u>, 2980

T. Nishikawa, M. Asai, N. Ohyabu, N. Yamamoto, Y. Fukuda, M. Isobe, *Tetrahedron* **2001**, <u>57</u>, 3875

# Oxy-Cope Rearrangement

## The Reaction:

keto-enol
tautomerism

## Proposed Mechanism:

keto-enol
tautomerism

## Notes:

Useful Review: L. A. Paquette, *Tetrahedron* **1997**, 53, 13971

FMO theory predicts, and it is observed, that the reaction is greatly accelerated if the alcohol proton
is removed (anion-accelerated). Many reactions of this type can be carried out at low temperature.

### Anion-accelerated Oxy-Cope

From FMO theory we imagine the *LUMO* portion of the reacting pair will be of the same energy in
both reactions; however, the *HOMO* is raised for the anion.

## Examples:

1. NaH, THF

2. MeOH

84%

L. A. Paquette, Z. Gao, Z. Ni., G. F. Smith, *Tetrahedron Letters* **1997**, 38, 1271

KH

THF

82%

S. F. Martin, J.-M. Assercq, R. E. Austin, A. P. Dantanarayana, J. R. Fishpaugh, C. Gluchowski, D.
E. Guinn, M. Hartmann, T. Tanaka, R. Wagner, J. B. White, *Tetrahedron* **1995**, 51, 3455

A.V. R. L. Sudha, M. Nagarajan, *Journal of the Chemical Society, Chemical Communications*, **1996**, 1359

J.-F. Devaux, I. Hanna, J.-Y. Lallemand, *Journal of Organic Chemistry* **1997**, 62, 5062

L. A. Paquette, D. R. Sauer, S. D. Edmondson, D. Friedrich, *Tetrahedron* **1994**, 50, 4071

2:1 *cis / trans*

B. H. White, M. L. Snapper, *Journal of the American Chemical Society* **2003**, 125, 14901

anion-accelerated oxy-Cope,    Diels-Alder,     anion-accelerated oxy-Cope

Michael

C. M. Tice, C. H. Heathcock, *Journal of Organic Chemistry* **1981**, 46, 9

# Oxy- and Solvomercuration

## The Reaction:

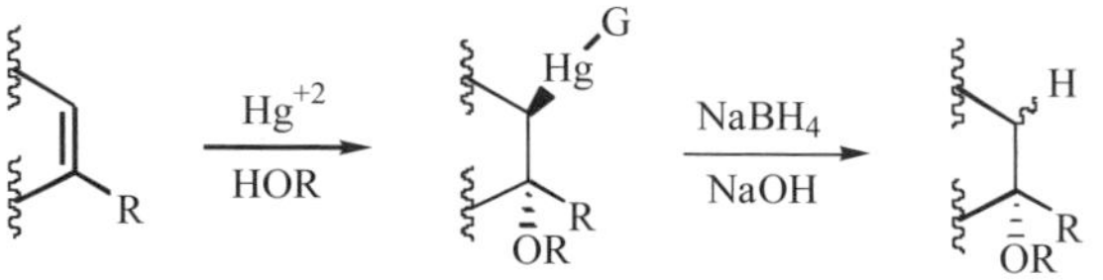

## Proposed Mechanism:

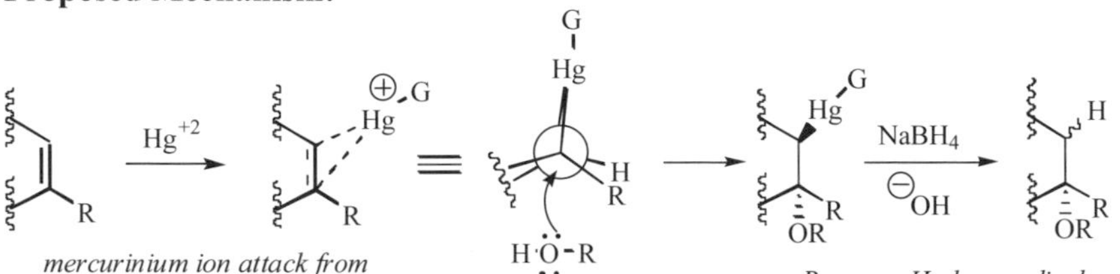

*mercurinium ion attack from
the less-hindered face.*

*Water or alcohol adds to the
more substituted carbon.*

*Removes Hg by a radical
mechanism and is
replaced by H radical.*

## Notes:

This is a useeful extension of the much older ***Hofmann-Sand reaction***, where an alkene reacts with
a mercuric salt:

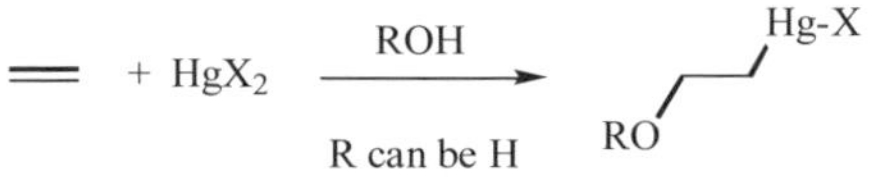

R can be H

Cyclopropane rings can undergo the oxymercuration/ demercuration reaction:

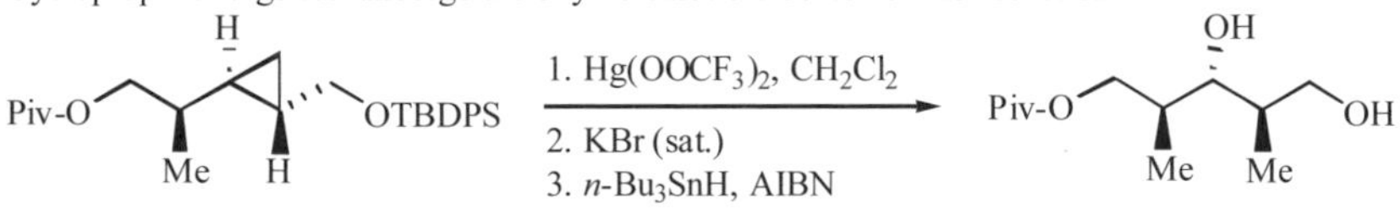

1. Hg(OOCF$_3$)$_2$, CH$_2$Cl$_2$

2. KBr (sat.)

3. $n$-Bu$_3$SnH, AIBN

92%

J. Cossy, N. Blanchard, C. Meyer, *Organic Letters* **2001**, <u>3</u>, 2567

An intramolecular reaction sometimes known as the ***Speckamp procedure***

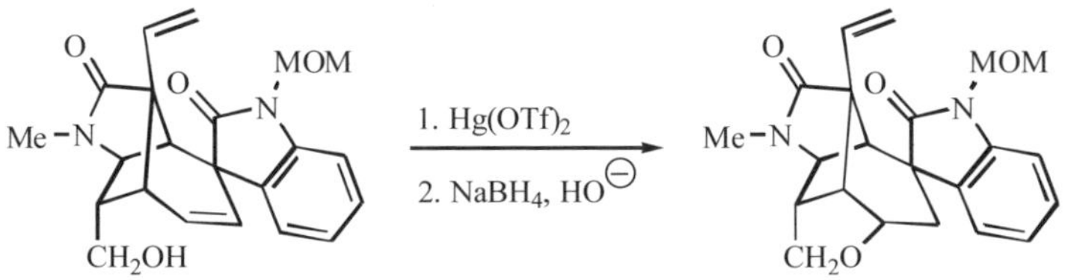

1. Hg(OTf)$_2$

2. NaBH$_4$, HO$^{\ominus}$

T. Fukuyama, G. Liu, *Journal of the American Chemical Society* **1996**, <u>118</u>, 7426

## Examples:

S. D. Dreher, J. L. Leighton, *Journal of the American Chemical Society* **2001**, <u>123</u>, 341

Chloral has a high propensity for forming hemiacetals

L. E. Overman, C. B. Campbell, *Journal of Organic Chemistry* **1974**, <u>39</u>, 1474

L. E. Overman, R. M. Burk, *Tetrahedron Letters* **1984**, <u>25</u>, 5739

N. Takaishi, Y. Fujikura, Y. Inamoto, *Journal of Organic Chemistry* **1975**, <u>40</u>, 3767

This early start of the investigations into this reaction provides useful insights into looking at reactivity.

H. C. Brown, P. Geoghegan, Jr., *Journal of the American Chemical Society* **1967**, <u>89</u>, 1522

# Paal-Knorr Furan Synthesis
## The Reaction:

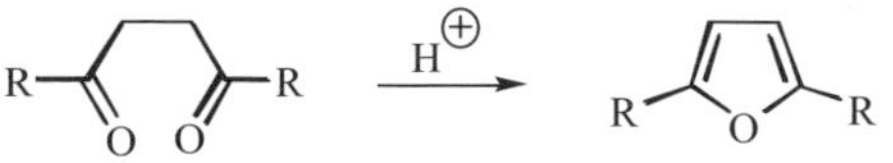

or other Lewis Acids

## Proposed Mechanism:

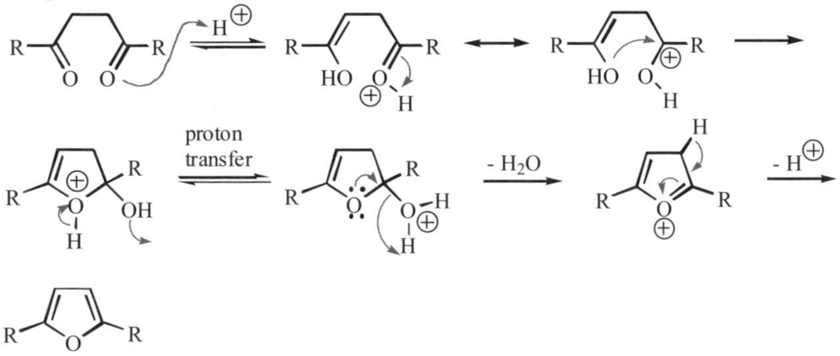

# Paal-Knorr Pyrrole Synthesis
## The Reaction:

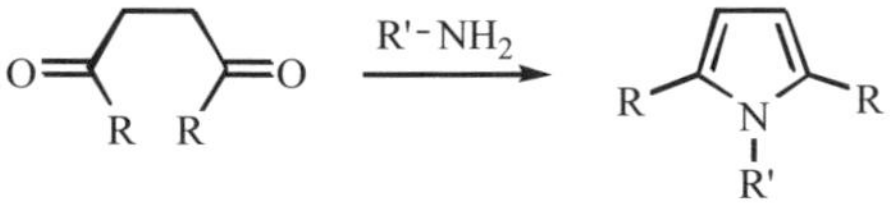

## Proposed Mechanism:

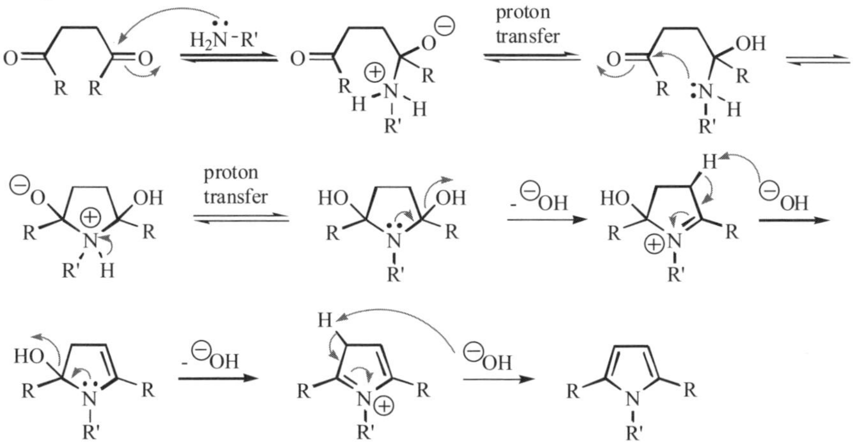

## Examples:

Me—C(=O)—CH2CH2—C(=O)—Me  +

*t*-BuNH2 $\xrightarrow[\text{benzene}]{\text{Ti(i-PrO)}_4}$ 22%

Me, N(*t*-Bu), Me pyrrole

PhCH2NH2 $\xrightarrow[\text{benzene}]{\text{Ti(i-PrO)}_4}$ 91%

Me, N(CH2Ph), Me pyrrole

Y. Dong, N. N. Pai, S. L. Ablaza, S.-X. Yu, S. Bolvig, D. A. Forsyth, P. W. LeQuesne, *Journal of Organic Chemistry* **1999**, <u>64</u>, 2657

$\xrightarrow[\text{hv, O}_2]{\text{Rose-Bengal}}$

***Kornblum-de la Mare Rearrangement***

N. Kornblum, H. E. de la Mare, *Journal of the American Chemical Society* **1951**, <u>73</u>, 880

***Lawesson's Reagent*** 62%

Ph, S, Ph

*n*-Bu-NH2 85%

Ph, N(Bu), Ph

C. E. Hewton, M. C. Kimber, D. K. Taylor, *Tetrahedron Letters* **2002**, <u>43</u>, 3199

Ph—CH=CH—CH2—C(=O)—Ph $\xrightarrow[\text{H}_2\text{SO}_4 \text{ (cat), microwave}]{\text{HCO}_2\text{H, 5\% Pd-C}}$ 95%

Ph, O, Ph

H. S. P. Rao, S. Jothilingam, *Journal of Organic Chemistry* **2003**, <u>68</u>, 5392

# Parham Cyclization

## The Reaction:

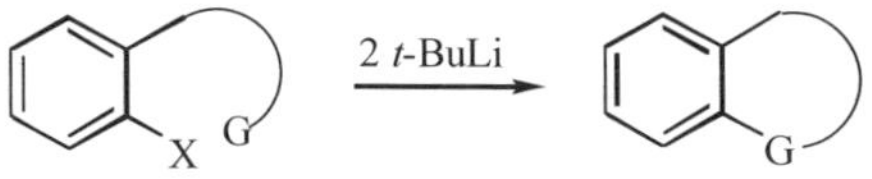

X = Br, I$_2$

G = electrophilic groups:
  -COOH, -CONR$_2$,
  -CH$_2$Br, -CH$_2$Cl,
  epoxide,

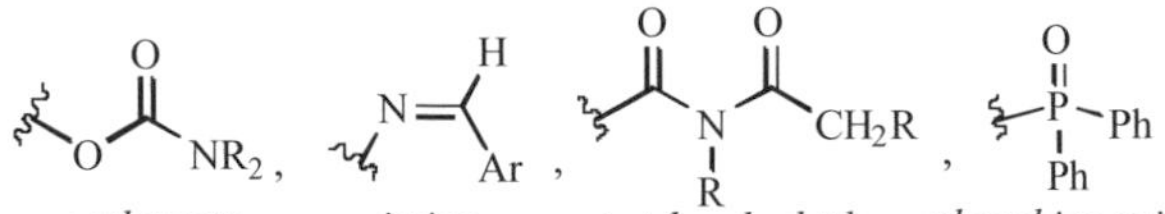

*carbamate*          *imine*          *imidoanhydride*     *phosphine oxide*

## Proposed Mechanism:

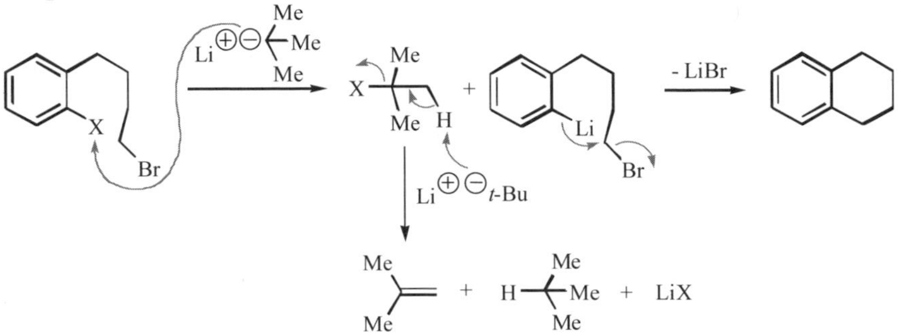

## Examples:

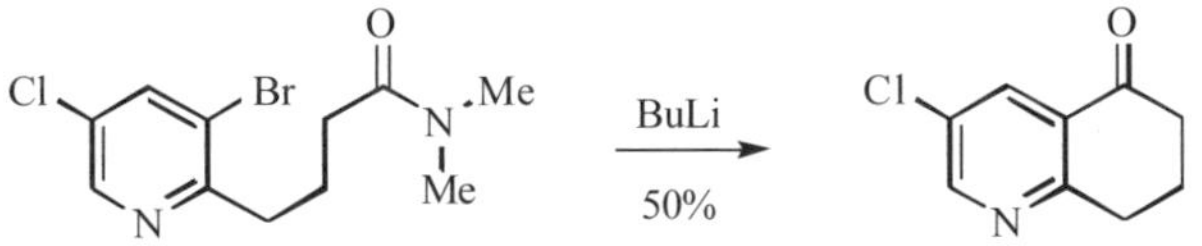

G. J. Quallich, D. E. Fox, R. C. Friedmann, C. W. Murtiashaw, *Journal of Organic Chemistry* **1992**, 57, 761

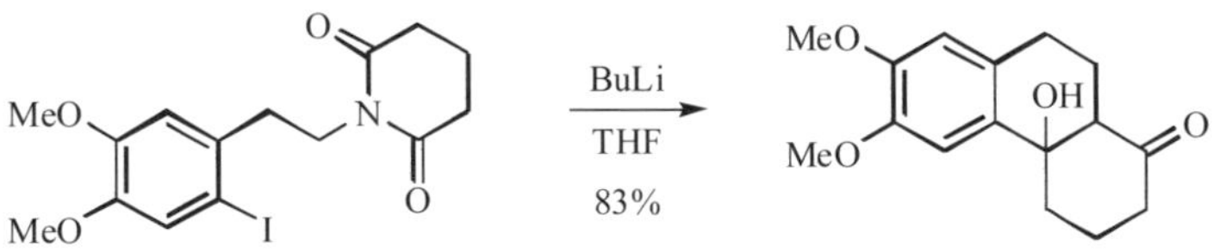

M. I. Collado, N. Sotomayor, M.-J. Villa, E. Lete, *Tetrahedron Letters* **1996**, 37, 6193

Using **_Weinreb amides_**:

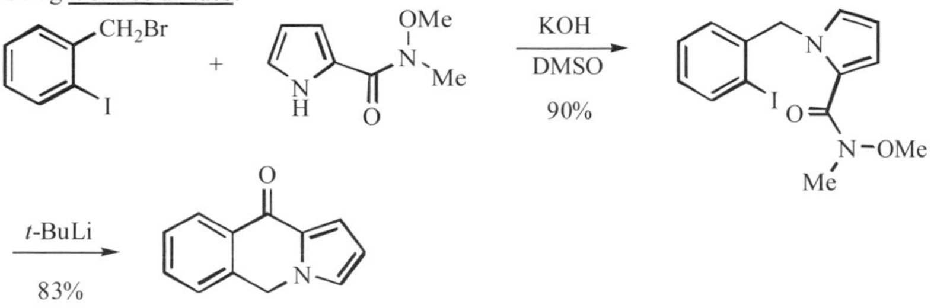

$t$-BuLi
——————→
83%

J. Ruiz, N. Sotomayer, E. Lete, *Organic Letters* **2003**, <u>5</u>, 1115

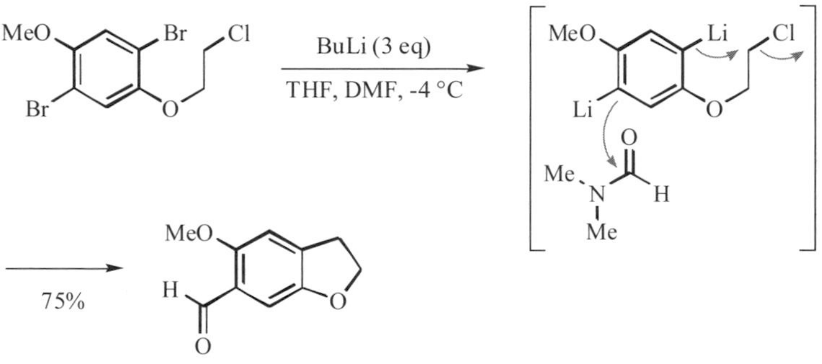

75%

M. Plotkin, S. Chen, P. G. Spoors, *Tetrahedron Letters* **2000**, <u>41</u>, 2269

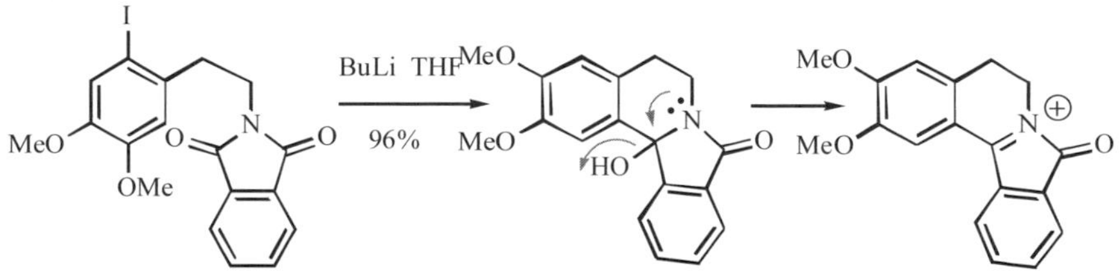

I. Osante, E. Lete, N. Sotomayor, *Tetrahedron Letters* **2004**, <u>45</u>, 1253

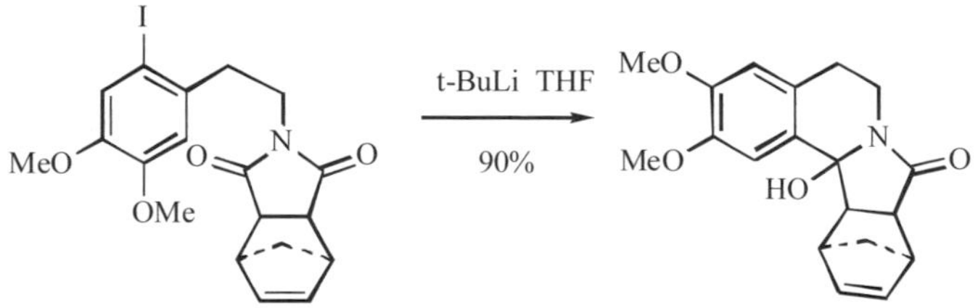

I. Gonza´lez-Temprano, I. Osante, E. Lete, N. Sotomayor, *Journal of Organic Chemistry* **2004**, <u>69</u>, 3875

# Parikh-Doering Oxidation

## The Reaction:

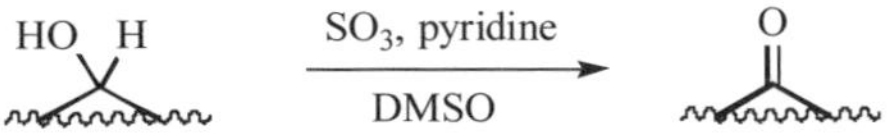

$$\text{HO} \quad \text{H} \xrightarrow[\text{DMSO}]{\text{SO}_3,\ \text{pyridine}} \text{O}$$

## Proposed Mechanism:

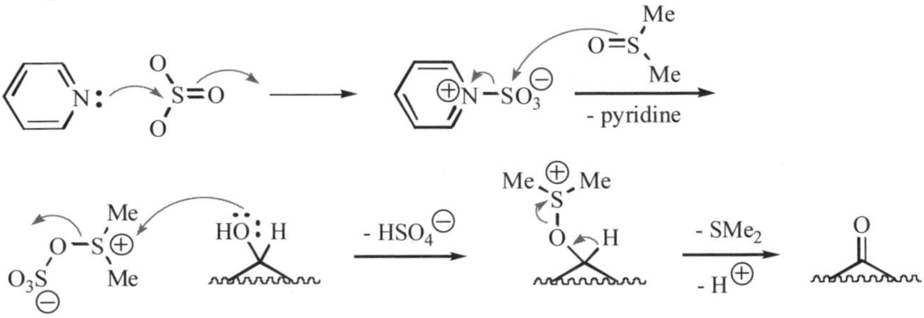

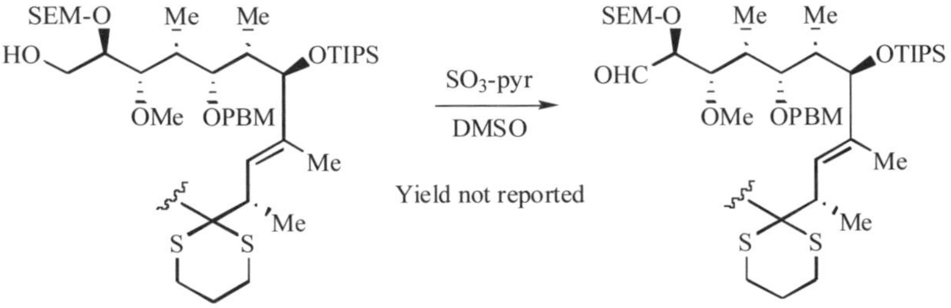

A. B. Smith, III, C. M. Adams, S. A. L. Barbosa, A. P. O. Degnan, *Journal of the American Chemical Society* **2003**, <u>125</u>, 350

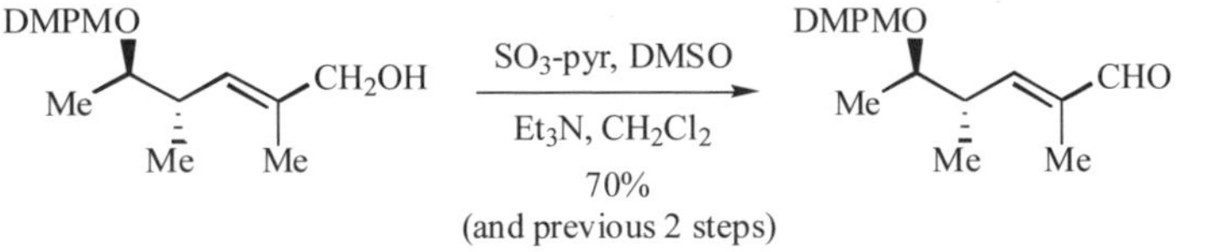

SO$_3$-pyr, DMSO

Et$_3$N, CH$_2$Cl$_2$

70%
(and previous 2 steps)

W. R. Roush, J. S. Newcom, *Organic Letters* **2002**, <u>4</u>, 4739

J. A. Panek, C. E. Masse, *Journal of Organic Chemistry* **1997,** 62, 8290

A. B. Smith, III, W. Zhu, S. Shirakami, C. Sfouggatakis, V. A. Doughty, C. S. Bennett, Y. Sakamoto, *Organic Letters* **2003,** 5, 761

M. J. Porter, N. J. White,.G. E. Howells, D. D. P. Laffan, *Tetrahedron Letters* **2004,** 45, 6541

# Passerini Reaction
## The Reaction:

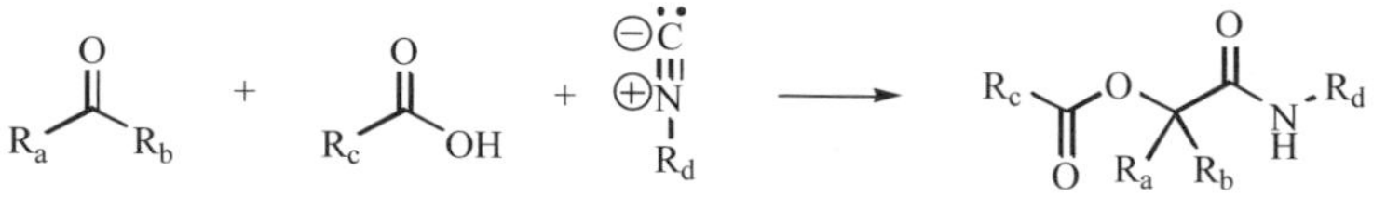

## Proposed Mechanism:

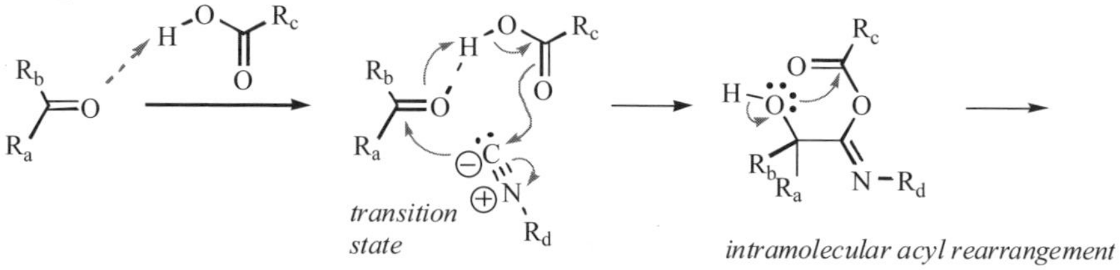

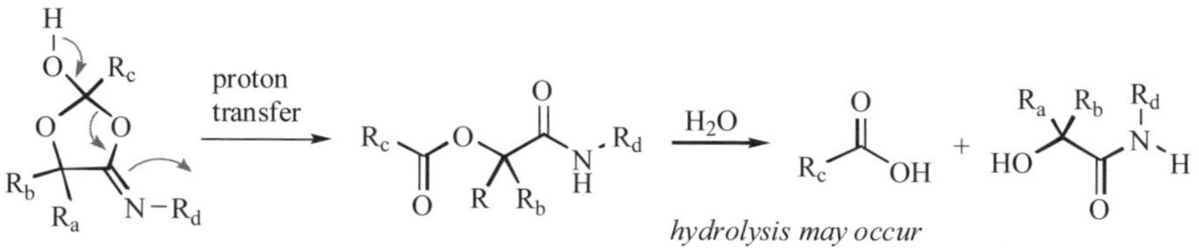

*transition state*

*intramolecular acyl rearrangement*

*hydrolysis may occur*

## Notes:
See: M. B. Smith, J. March in *March's Advanced Organic Chemistry*, 5th ed., John Wiley and Sons, Inc., New York, 2001, p. 1252

## Examples:

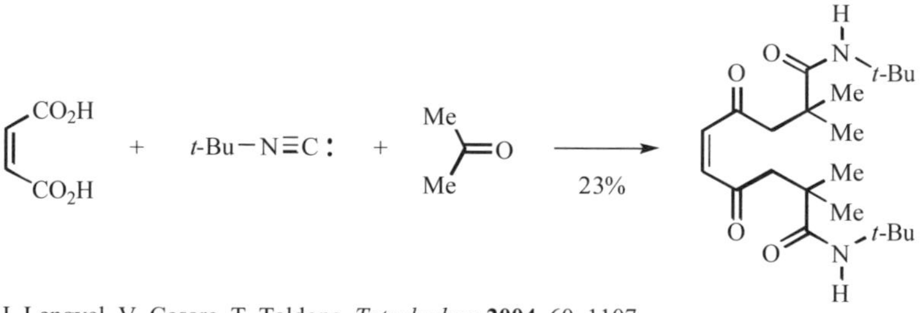

I. Lengyel, V. Cesare, T. Taldone, *Tetrahedron* **2004**, <u>60</u>, 1107

J. E. Sample, T. D. Owens, K. Nguyen, O. E. Levy, *Organic Letters* **2000**, <u>2</u>, 2769

| R | R' | Yield |
|---|-----|-------|
| Me- | Ph(CH$_2$)$_2$- | 32% |
| Me- | ▷ | 35% |

B. Henkel, B. Beck, B. Westner, B. Mejat, A. Domling, *Tetrahedron Letters* **2003**, <u>44</u>, 89

PMB =
*p*-methoxybenzyl

S. P. G. Costa, H. L. S. Maia, S. M. M. A. Pereira-Lima, *Organic & Biomolecular Chemistry* **2003**, <u>1</u>, 1475

# Paterno-Buchi Reaction
## The Reaction:

expect a mixture
of regioisomers

## Proposed Mechanism:

This is a stepwise reaction with a diradical intermediate.

## Notes:

T. Laue, A. Plagens, *Named Organic Reactions*, John Wiley and Sons, Inc., New York, 1998, pp. 209-210; M. B. Smith, J. March in *March's Advanced Organic Chemistry*, 5$^{th}$ ed., John Wiley and Sons, Inc., New York, 2001, pp 1249-1250

The reaction is not concerted, as evidenced by reactions of *cis* and *trans* alkenes giving the same product mixture.

## Examples:

M. C. de la Torre, I. Garcia, M. A. Sierra, *Journal of Organic Chemistry* **2003**, <u>68</u>, 6611

M. D'Auria, L. Emanuele, G. Poggi, R. Racioppi, G. Romaniello, *Tetrahedron* **2000**, <u>58</u>, 5045

S. A. Fleming, J. J. Gao, *Tetrahedron Letters* **1997**, <u>38</u>, 5407

S. Buhr, G. Axel, J. Lex, J. Mattay, J. Schroeer, *Tetrahedron Letters* **1996**, <u>37</u>, 1195

H. A. J. Car-less, J. Beanland, S. Mwesigye-Kibende, *Tetrahedron Letters* **1987**, <u>28</u>, 5933

# Pauson-Khand Cyclopentenone Annulation
## The Reaction:

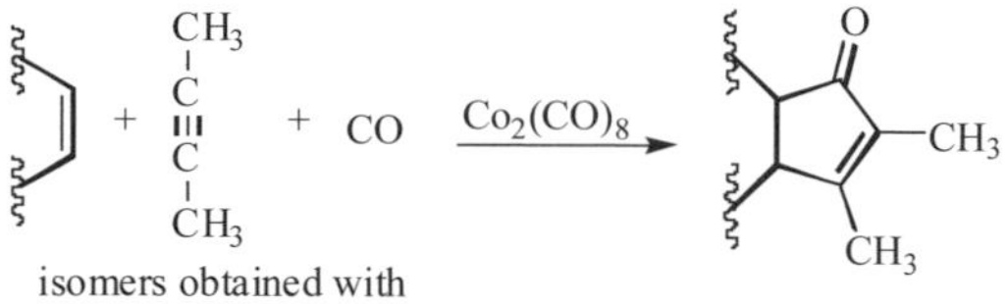

isomers obtained with
unsymmetrical alkyne

## Proposed Mechanism:

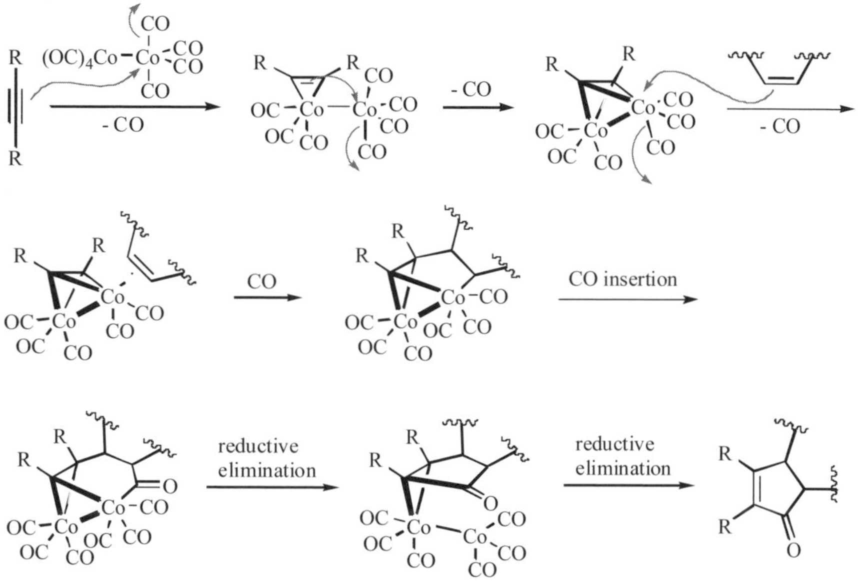

## Notes:

M. B. Smith, J. March in *March's Advanced Organic Chemistry*, 5th ed., John Wiley and Sons, Inc.,
New York, 2001, p. 1091; T. Laue, A. Plagens, *Named Organic Reactions*, John Wiley and Sons,
Inc., New York, 1998, pp. 210-213; N. E. Schore, *Organic Reactions* **40**, 1.

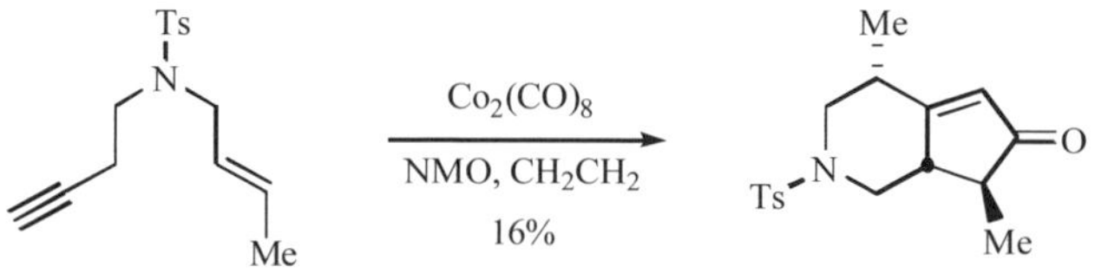

D. A. Ockey, M. A. Lewis, N. E. Schore, *Tetrahedron* **2003**, <u>59</u>, 5377

M. E. Krafft, J. A. Wright, L. V. R. Bonaga, *Tetrahedron Letters* **2003**, <u>44</u>, 3417

C. Mukai, M. Kobayashi, I. J. Kim, M. Hanacka, *Tetrahedron* **2002**, <u>58</u>, 5225

P. Magnus, M. J. Slater, L. M. Principe, *Journal of Organic Chemistry* **1989**, <u>54</u>, 5148

K. M. Brummond, D. Gao, *Organic Letters* **2003**, <u>5</u>, 3491;
K. M. Brummond, H. Chen, K. D. Fisher, A. D. Kerekes, B. Rickards, P. C. Sill, S. J. Geib, *Organic Letters,* **2002**, <u>4</u>, 1931

# Payne Rearrangement
## The Reaction:

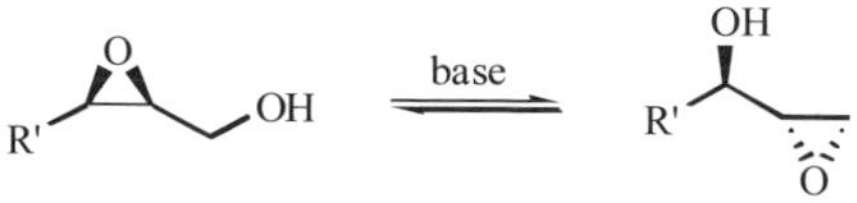

base = NaOH, KOH, NaOR, NaSR

## Proposed Mechanism:

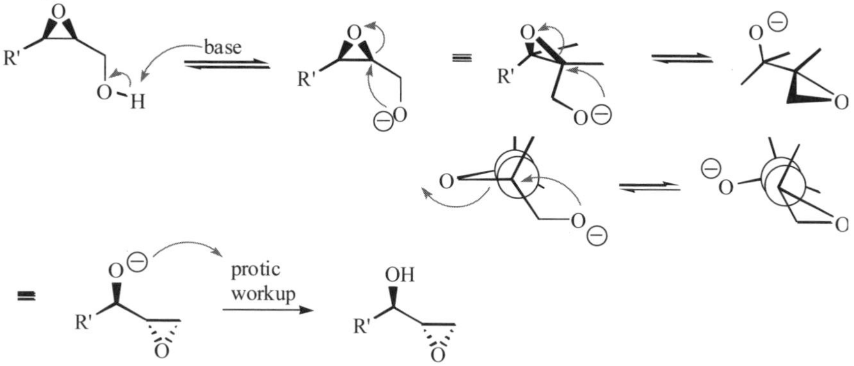

## Notes:
M. B. Smith, J. March in *March's Advanced Organic Chemistry*, 5th ed., John Wiley and Sons, Inc., New York, 2001, p. 481; R. M. Hanson, *Organic Reactions* **60**, 1

An *Aza-Payne reaction* is also possible:

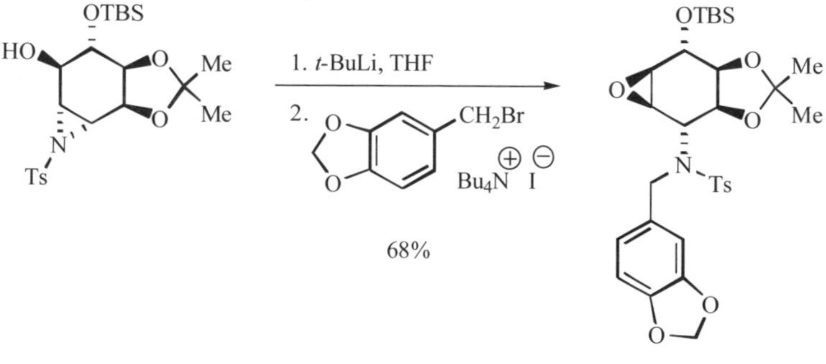

68%

U. Rinner, P. Siengalewicz, T. Hudlicky, *Organic Letters* **2002**, 4, 115

## Examples:

D. Herlem, F. Khuong-Huu, *Tetrahedron* **1997**, <u>53</u>, 673

V. Jaeger, D. Schroeter, B. Koppenhoefer, *Tetrahedron* **1991**, <u>47</u>, 2195

N. Cohen, B. L. Banner, R. J. Lopresti, F. Wong, M. Rosenberger, Y. Y. Liu, E. Thom, A. A. Liebman, *Journal of the American Chemical Society* **1983**, <u>105</u>, 3661

G. B. Payne, *Journal of Organic Chemistry* **1962**, <u>27</u>, 3819

B. M. Wang, Z. L. Song, C. A. Fan, Y. Q. Tu, Y. Shi, *Organic Letters* **2002**, <u>4</u>, 363

# Pechmann Condensation
## The Reaction:

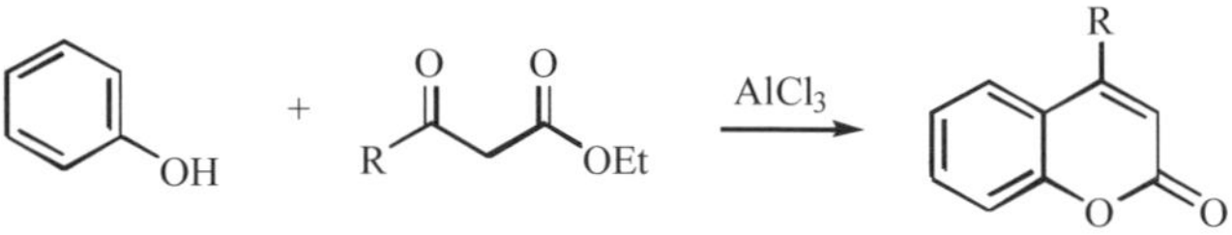

## Proposed Mechanism:

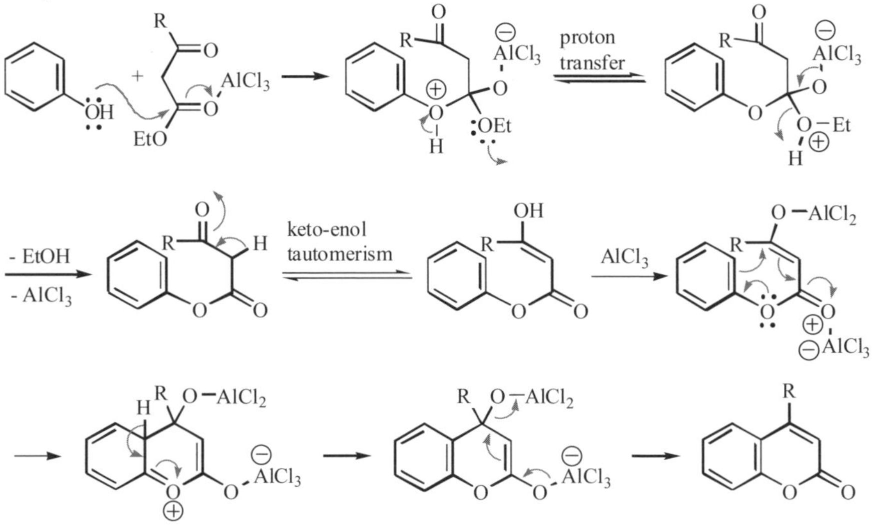

## Notes:

S. Sethna, R. Phadke, *Organic Reactions*, **7**, 1

This reaction provides a simple entry into the *coumarins*:

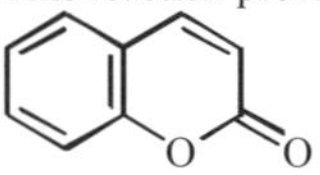

Many catalysts can be used, including sulfuric acid, aluminum chloride, phosphorous pentoxide, etc.

## Examples:

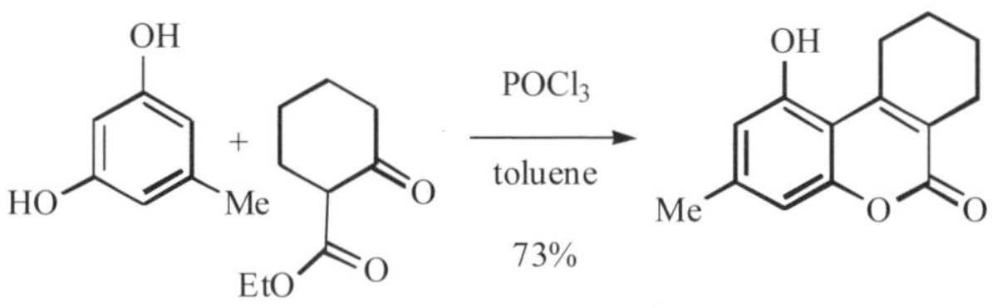

P. Selle`s, U. Mueller, *Organic Letters* **2004**, <u>6</u>, 277

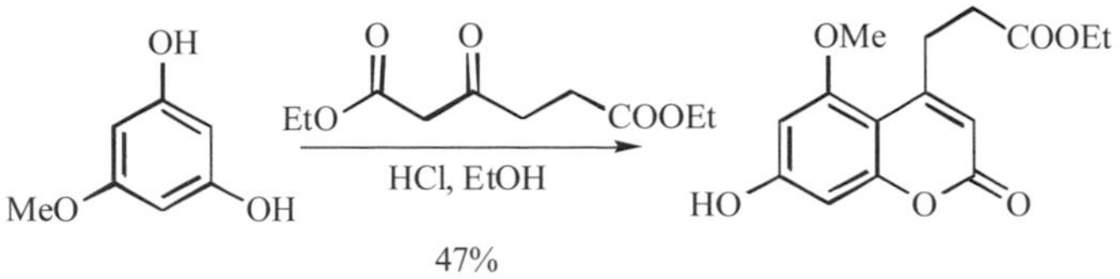

47%

B. M. Trost, F. D. Toste, *Journal of the American Chemical Society* **2003**, <u>125</u>, 3090

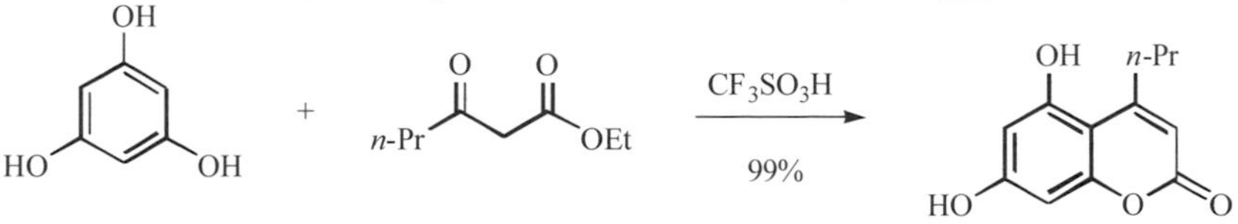

99%

B. Chenera, M. L. West, J. A. Finkelstein, G. B. Dreyer, *Journal of Organic Chemistry*, **1993**, <u>58</u>, 5605

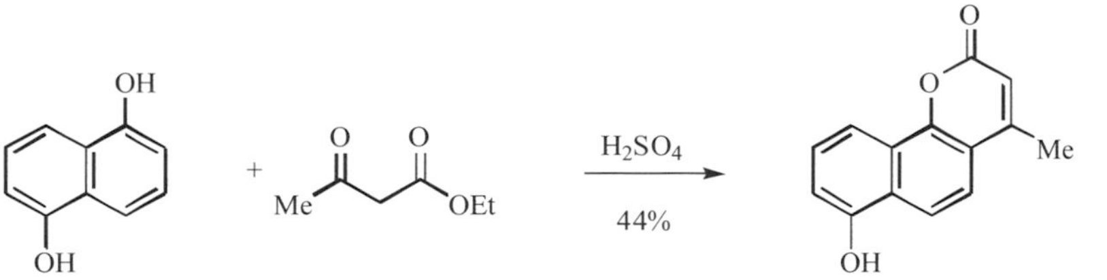

44%

W. Adam. X. Qian, C. R. Shah-Moeller, *Journal of Organic Chemistry* **1993**, <u>58</u>, 3769

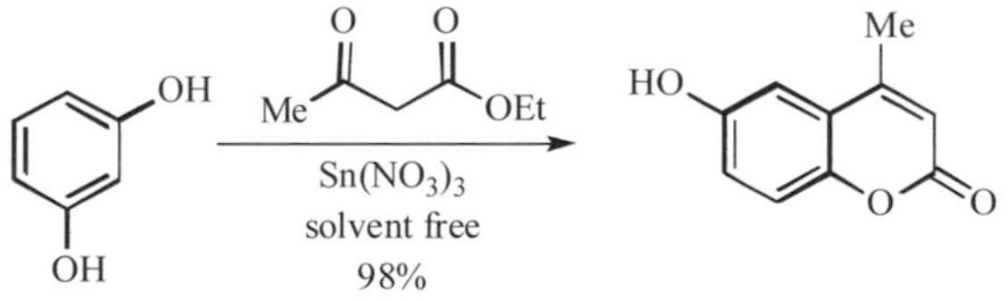

Sn(NO$_3$)$_3$
solvent free
98%

S. S. Bahekara, D, B. Shindeb,*Tetrahedron Letters* **2004, <u>45</u>,** 7999

The use of indium(III) can do the same reaction: D. S. Bose, A. P. Rudradas, M. H. Babu, *Tetrahedron Letters* **2002**, <u>43</u> , 9195.

# Perkin Reaction (Perkin cinnamic acid synthesis)
## The Reaction:

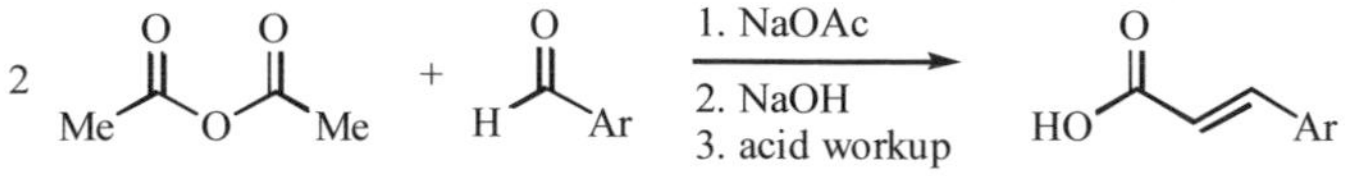

## Proposed Mechanism:

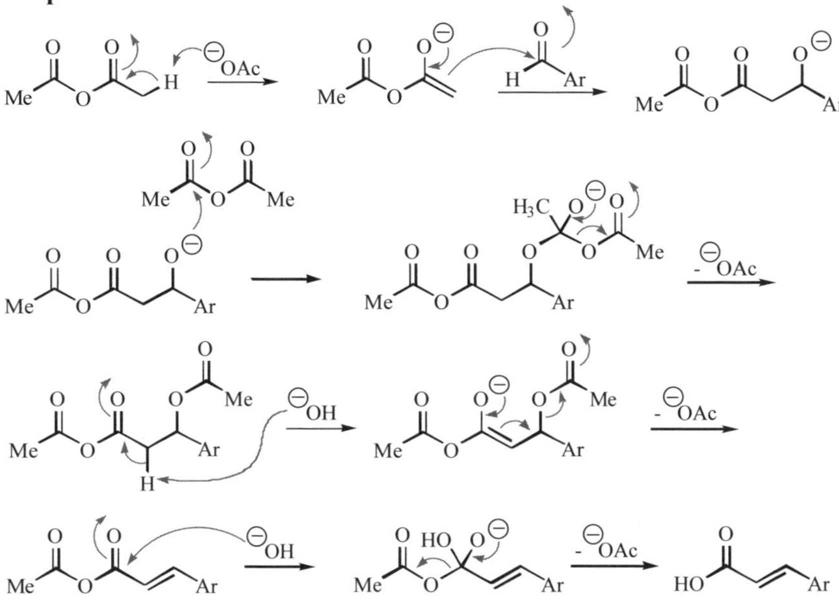

## Notes:
M. B. Smith, J. March in *March's Advanced Organic Chemistry*, 5th ed., John Wiley and Sons, Inc., New York, 2001, p.p. 1217, 1229; T. Laue, A. Plagens, *Named Organic Reactions*, John Wiley and Sons, Inc., New York, 1998, pp. 213-215; J. R. Johnson, *Organic Reactions* **1**, 8.

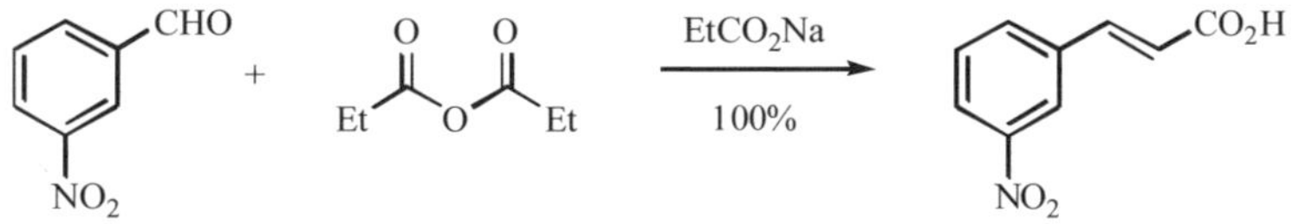

R. W. Maxwell, R. Adams, *Journal of the American Chemical Society* **1930**, <u>52</u>, 2967

G. Solladie, Y. Pasturel-Jacope, J. Maignan, *Tetrahedron* **2003**, <u>59</u>, 3315.

This method allows for the use of acids as a replacement for anhydrides.

C. I. Chiriac, F. Tanasa, M. Onciu, *Tetrahedron Letters* **2003**, <u>44</u>, 3579

R. F. Buckles, J. A. Cooper, *Journal of Organic Chemistry* **1965**, <u>30</u>, 1588

D. V. Sevenard, *Tetrahedron Letters* **2003**, <u>44</u>, 7119

# Petasis Reaction
**The Reaction:**

See: J. P. Tremblay-Morin, S. Raeppel, F. Gaudette, *Tetrahedron Letters* **2004**, _45_, 3471

This is an example of a **boronic-_Mannich reaction_**. The mechanism is not fully established; however it has been shown that an intermediate of the type:

is not involved since an **_Eschenmoser salt_** will not react.

A series was prepared using a reusable resin. Yields were good.
K. A. Thompson, D. G. Hall, *Chemical Communications* **2000**, 2379

N. A. Petasis, I. Akritopoulou, *Tetrahedron Letters* **1993**, <u>34</u>, 583

B. Jiang, C.-G. Yang, X.-H. Gu, *Tetrahedron Letters* **2001**, <u>42</u>, 2546

G. W. Kalbaka, B. Venkataiah, G. Dong, *Tetrahedron Letters* **2004**, <u>45</u>, 727

R. A. Batey, D. B. MacKay, V. Santhakumar, *Journal of the American Chemical Society* **1999**, <u>121</u>, 5075

# Peterson Olefination Reaction (Peterson Elimination, Silyl-Wittig Reaction)
## The Reaction:

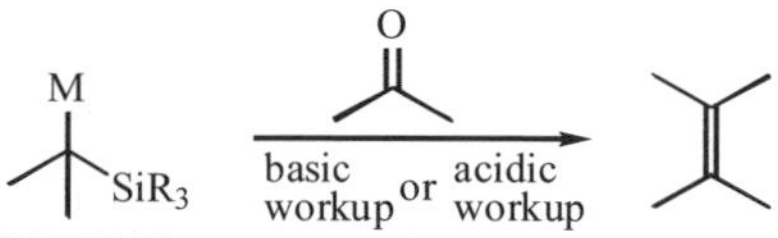

M = Lithium or Magnesium

## Proposed Mechanism:

Different alkenes can be obtained, depending on how the reaction is carried out. In both approaches the same intermediate is obtained:

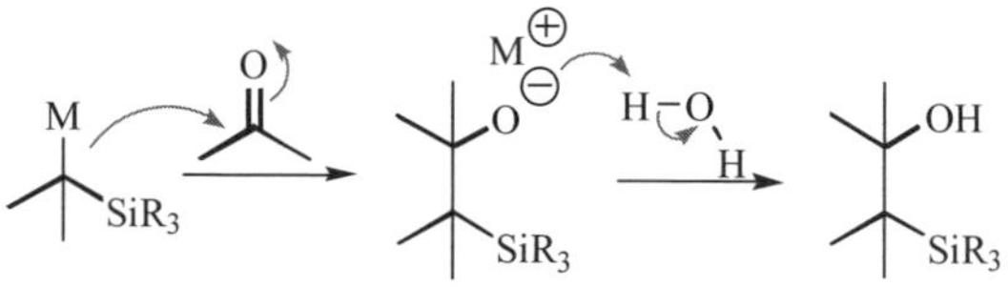

basic workup:

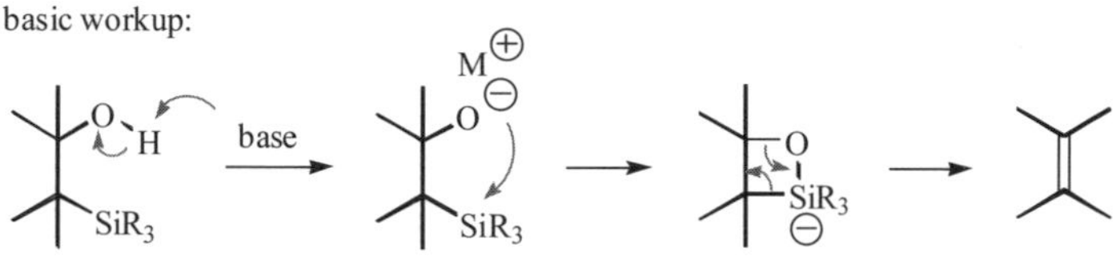

*cis* elimination

acidic workup:

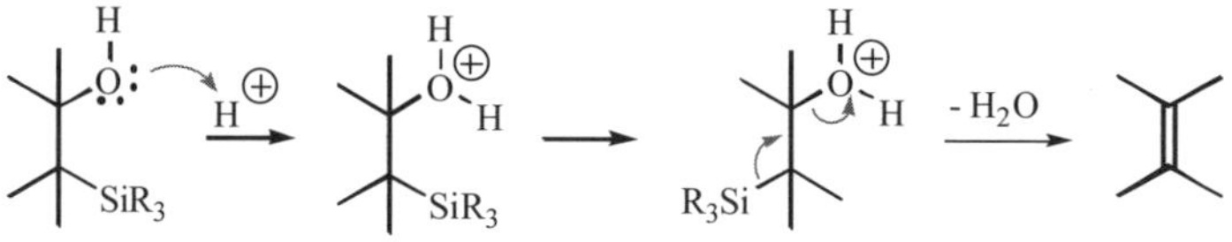

*anti* elimination

## Notes:

M. B. Smith, J. March in *March's Advanced Organic Chemistry*, 5$^{th}$ ed., John Wiley and Sons, Inc., New York, 2001, p. 1228; T. Laue, A. Plagens, *Named Organic Reactions*, John Wiley and Sons, Inc., New York, 1998, pp. 215-217; D. J. Ager, *Organic Reactions* **38**, 1.

For an altenative approach, see ***Tebbe Reagent***, ***Lombardo Reagent*** or the ***Wittig Reaction***. The steric requirements for the ***Peterson*** approach are less than for the ***Wittig*** approach.

A variation using fluoride-based elimination has been reported:

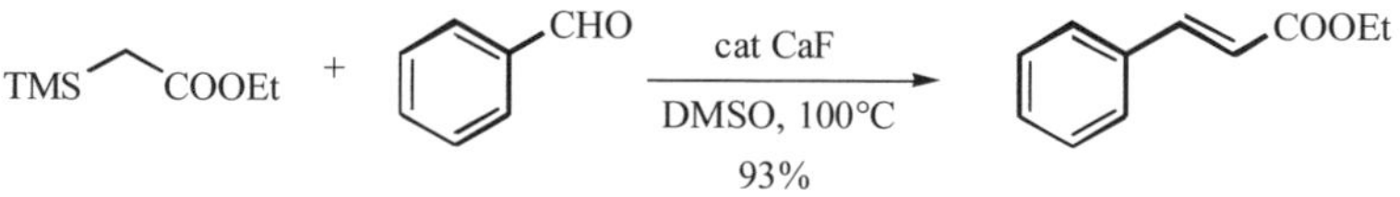

93%

M. Bellassoued, N. Ozanne, *Journal of Organic Chemistry* **1995**, 60, 6582

Unique control of stereochemistry:

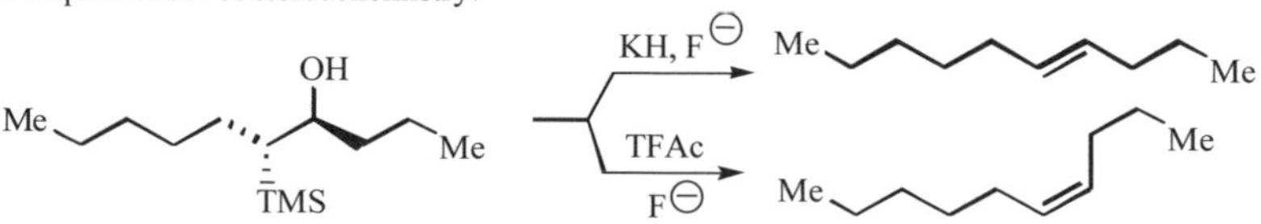

A. G. M. Barrett, J. A. Flygare, *Journal of Organic Chemistry* **1991**, 56, 638

## Examples:

lithium 1-(dimethylamino)naphthalide

I. Lengyel, V. Cesare, T. Taldone, *Tetrahedron* **2004**, <u>60</u>, 1107

J. E. Sample, T. D. Owens, K. Nguyen, O. E. Levy, *Organic Letters* **2000**, <u>2</u>, 2769

B. Henkel, B. Beck, B. Westner, B. Mejat, A. Domling, *Tetrahedron Letters* **2003**, <u>44</u>, 89

S. P. G. Costa, H. L. S. Maia, S. M. M. A. Pereira-Lima, *Organic & Biomolecular Chemistry* **2003**, <u>1</u>, 1475

# Pfau–Plattner Azulene Synthesis
## The Reaction:

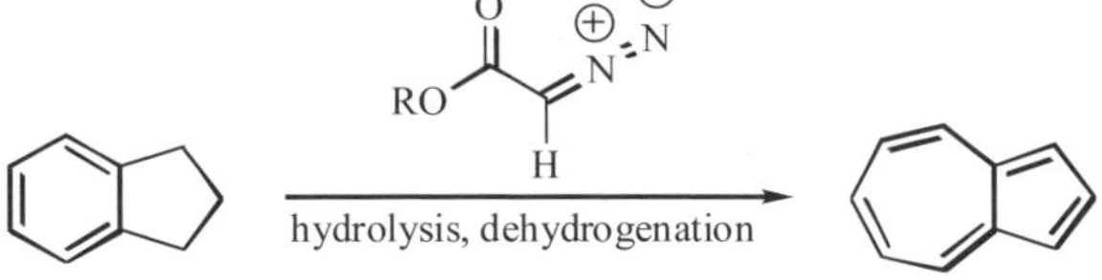

## Proposed Mechanism:

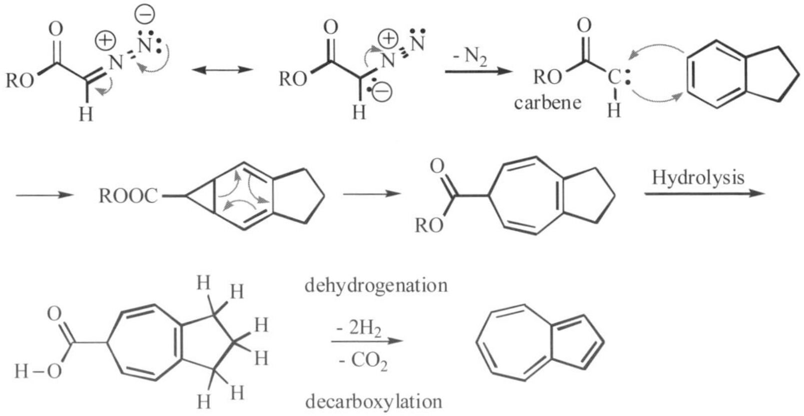

## Notes:
This chemistry is an application of the **Buchner Method of Ring Enlargement**:

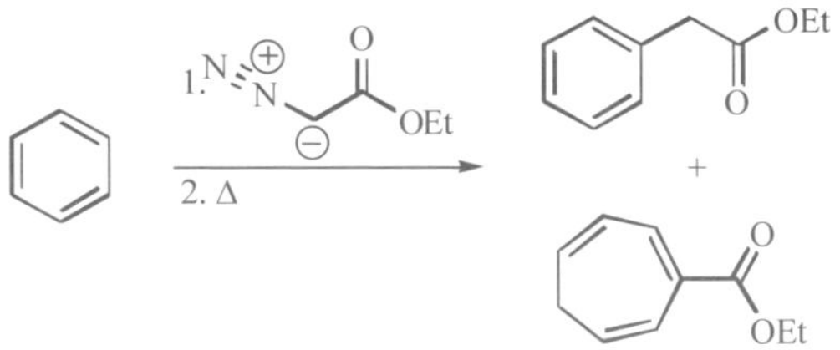

The reaction, as originally carried out, is of little contemporary use. Metal-mediated catalysis is more-often applied, where the reactive species are metal carbenoids.

**Buchner-Curtius-Schlotterbeck Reaction**

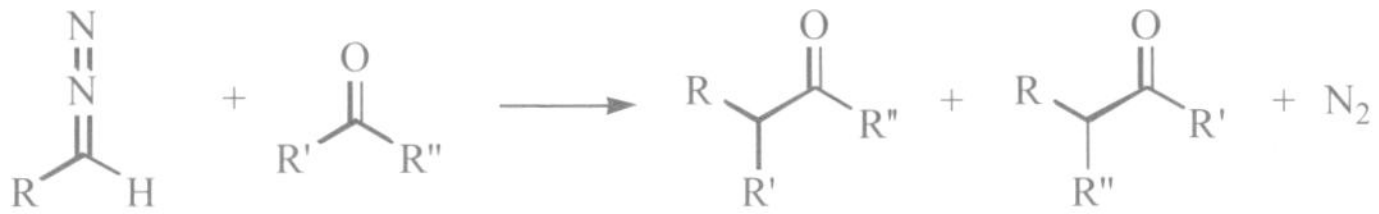

## Proposed Mechanism:

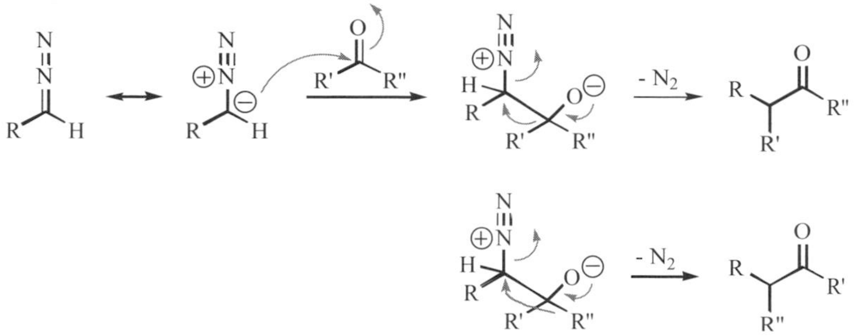

## Examples:

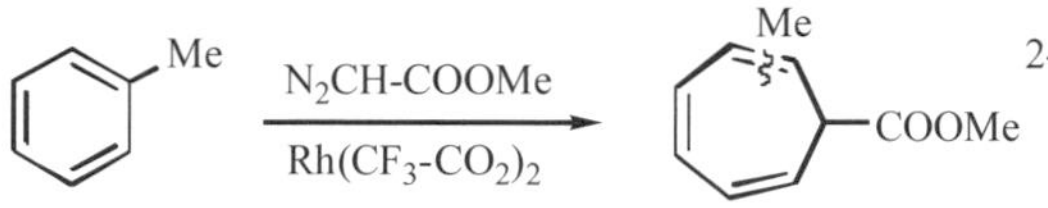

2-Me : 3-Me: 4-Me = 17 : 23 : 56

A. J. Anciaux, A. Demonceau, A. J. Hubert, A. F. Noels, N. Petiniot, P. Teyssie, *Journal of the Chemical Society, Chemical Communications* **1980**, 765

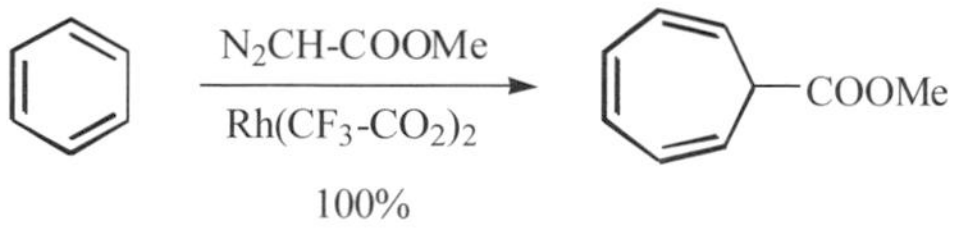

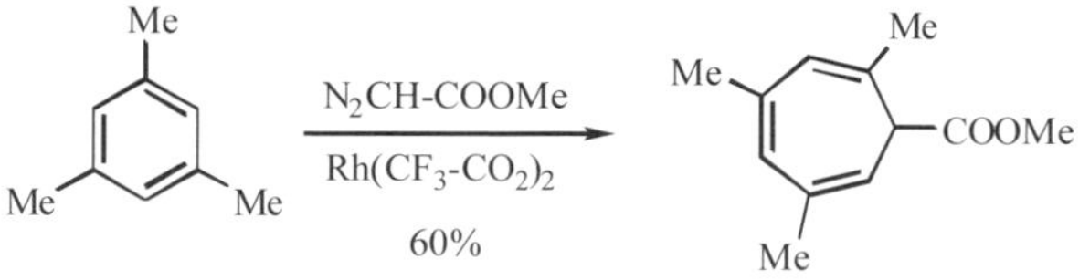

100%

60%

A. J. Anciaux, A. Demonceau, A. F. Noels, A. J. Hubert, R. Warin, P. Teyssi, *Journal of Organic Chemistry* **1981**, 46, 873

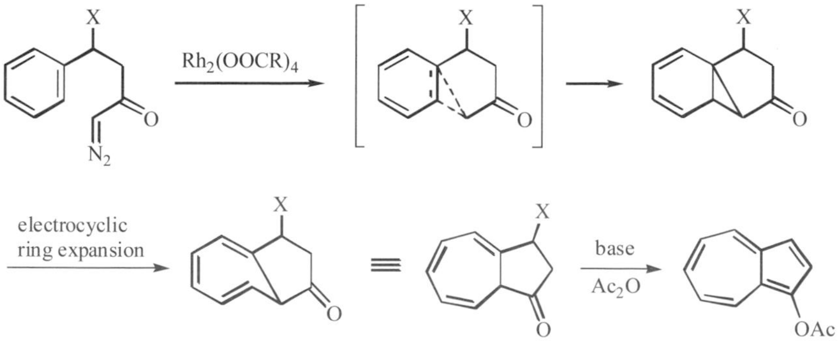

J. L. Kane, Jr., K. M. Shea, A. L. Crombie, R. L. Danheiser, *Organic Letters*, **2001**, 3, 1081

# Pfitzinger Reaction
## The Reaction:

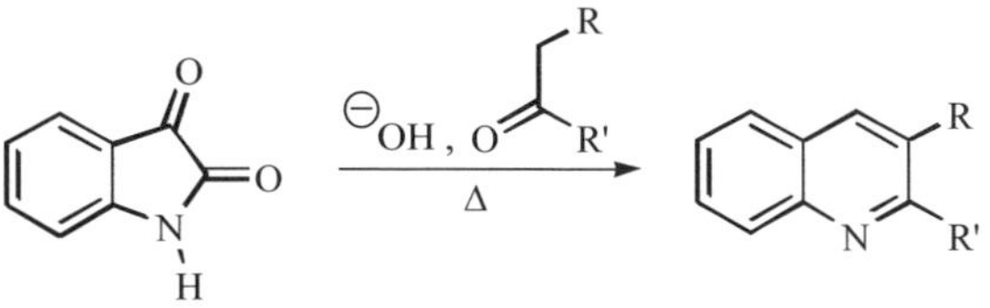

## Proposed Mechanism:

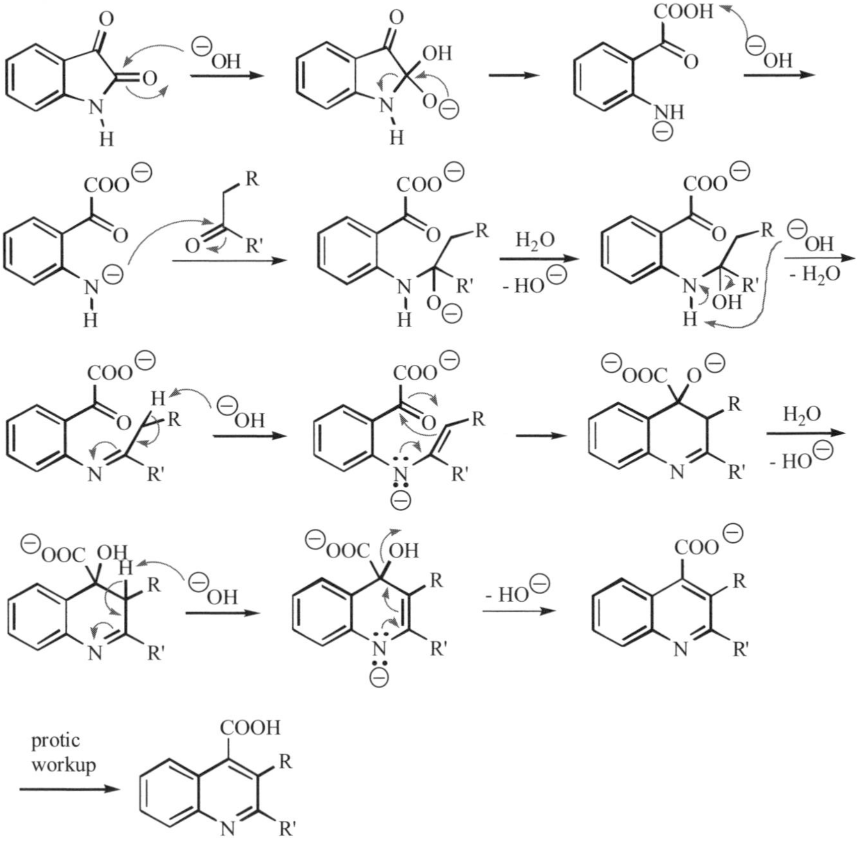

## Notes:
Decarboxylation can take place

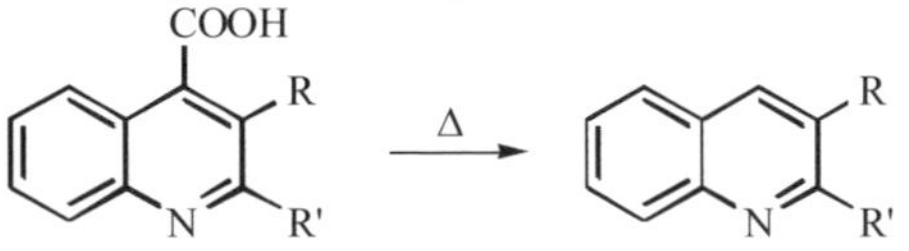

L. W. Deady, J. Â Desneves, A. J. Kaye, M. Thompson, G. J. Finlay, B. C. Baguley, W. A. Denny, *Bioorganic and Medicinal Chemistry* **1999**, <u>7</u>, 2801

J. A. Jongejan, R. P. Bexemer, J. A. Duine, *Tetrahedron Letters* **1988**, <u>29</u>, 3709

A. V. Ivachtchenko, A. V. Khvat, V. V. Kobak, V. M. Kysil, C. T. Williams, *Tetrahedron Letters* **2004**, <u>4</u>, 5473

L. W. Deady,  J. Desneves, A. J. Kaye, M,Thompson, G. J. Finlay, B. C. Baguley, W. A. Denny, *Bioorganic & Medicinal Chemistry* **1999**, <u>7</u>, 2801

This paper describes the preparation of a number of libraries constructed with this reaction.
A. V. Ivachtchenko, V. V. Kobak, A. P. Il'yin, A. S. Trifilenkov, A. A. Busel, *Journal of Combinatorial Chemistry* **2003**, <u>5</u>, 645

# Pfitzner-Moffatt Oxidation

## The Reaction:

$$\text{(structure: secondary alcohol)} \xrightarrow[\text{DMSO}]{\text{DCC, HX}} \text{(structure: ketone)}$$

## Proposed Mechanism:

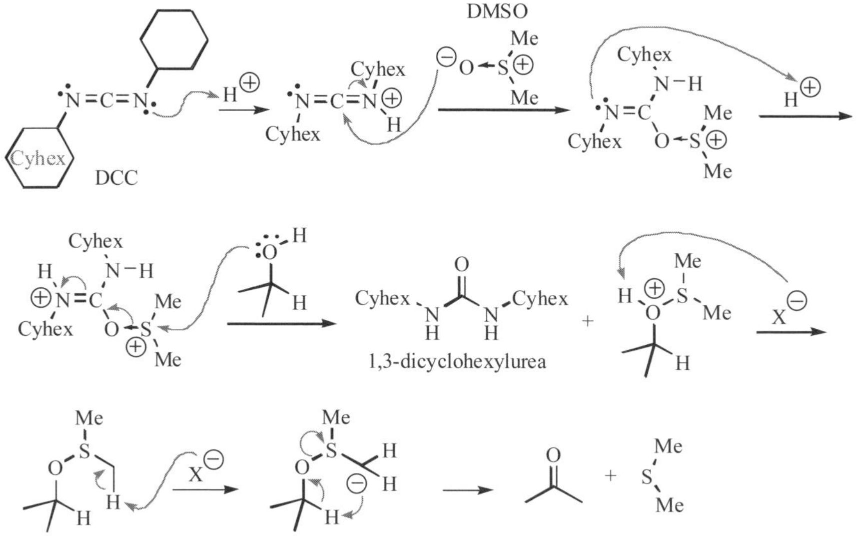

A sulfur ylide is formed with base which then abstracts the α proton, generating dimethylsulfide and the aldehyde or ketone.

## Notes:

T. T. Tidwell, *Organic Reactions* **39**, 3.

A number of oxidations belong to this general class.

Dimethyl sulfoxide oxidations: W. W. Epstein, F. W. Sweat, *Chemical Reviews* **1967**, <u>67</u>, 247

An "improved" reagent, based on the general idea associated with DMSO-based oxidations has been reported:

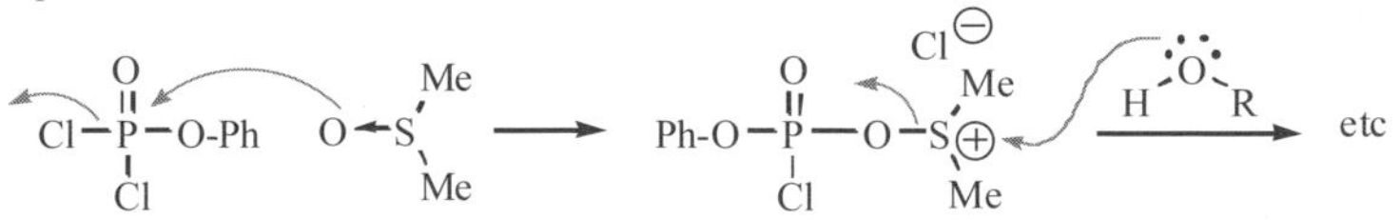

H.-J. Liu, J. M. Nyangulu, *Tetrahedron Letters* **1988**, <u>29</u>, 3167

## Examples:

R. Schworer, R. R. Schmidt, *Journal of the American Chemical Society* **2002**, <u>124</u>, 1632

R. Schworer, R. R. Schmidt, *Journal of the American Chemical Society* **2002**, <u>124</u>, 1632

B. Sauerbrei, J. Niggemann, S. Grisger, S. Lee, H. G. Floss, *Carbohydrate Research* **1996**, <u>280</u>, 223

F. Akahoshi, A. Ashimori, Y. Sakashita, M. Eda, T. Imada, M. Nakajima, N. Mitsutomi, S. Kuwahara, T. Ohtsuka, C. Fukaya, M. Miyazaki, N. Nakamur, *Journal of Medicinal Chemistry*, **2001**, <u>44</u>, 1297

# Pictet-Gams Isoquinoline Synthesis

## The Reaction:

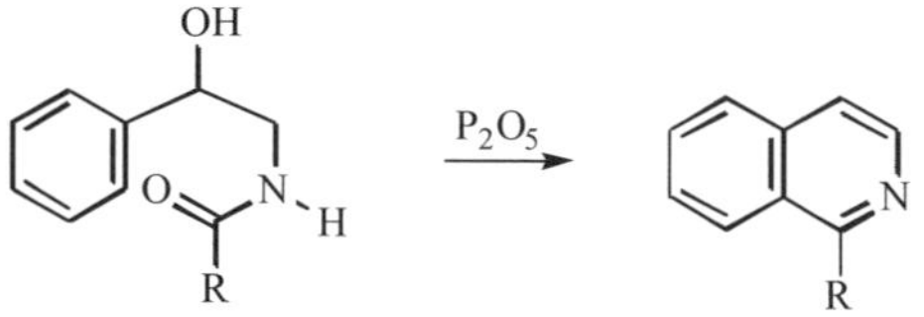

## Proposed Mechanism:

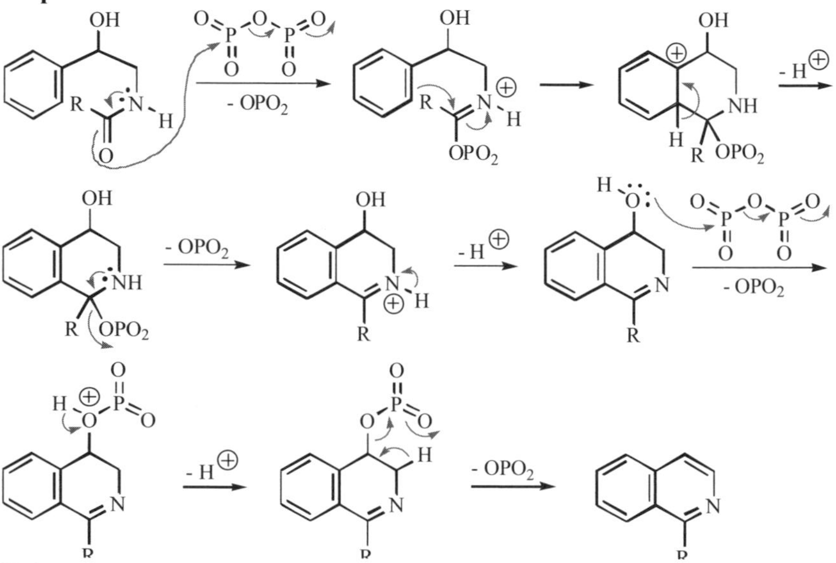

## Notes:

Other condensing agents can be used: Examples: POCl$_3$, Polyphosphoric acid

## Examples:

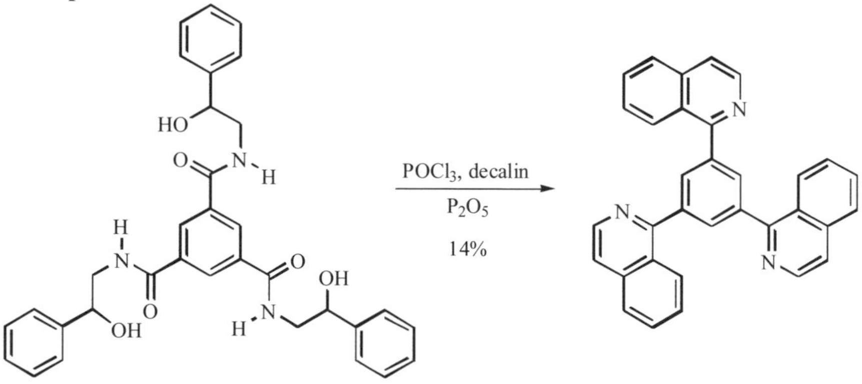

POCl$_3$, decalin

P$_2$O$_5$

14%

G. Dyker, M. Gabler, M. Nouroozian. P. Schulz, *Tetrahedron Letters* **1994**, <u>35</u>, 9697

J. R. Falck, S. Manna, C. Mioskowski, *Journal of Organic Chemistry* **1981**, <u>46</u>, 3742

A. O. Fitton. J. R. Frost, M. M. Zakaria, G. Andrew, *Journal of the Chemical Society, Chemical Communications* **1973**, 889

J. Wiejlard, E. F. Swanezy, E. Tashihan, *Journal of the American Chemical Society* **1949**, <u>71</u>, 1889

The CF$_3$ group appears to delay loss of the -OMe group under the reaction conditions.
L. Poszavacz, G. Simig, *Tetrahedron,* **2001**, <u>57</u>, 8573

# Pictet-Hubert Reaction / Morgan-Walls Reaction
## The Reaction:

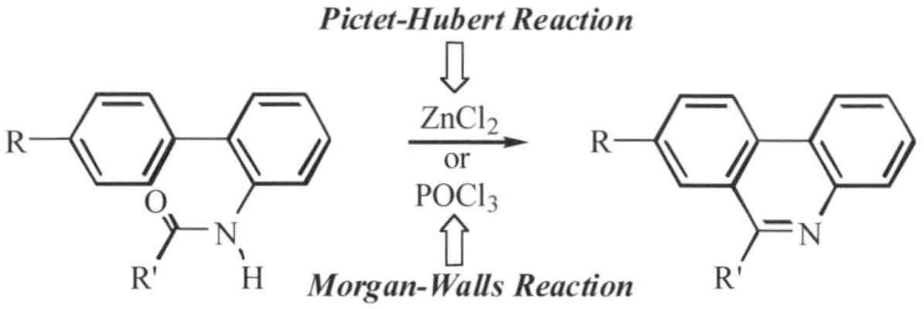

*Pictet-Hubert Reaction*

⇩

$$\text{R} \longrightarrow \xrightarrow[\text{POCl}_3]{\substack{\text{ZnCl}_2 \\ \text{or}}} \text{R} \longrightarrow$$

⇧ *Morgan-Walls Reaction*

## Proposed Mechanism:

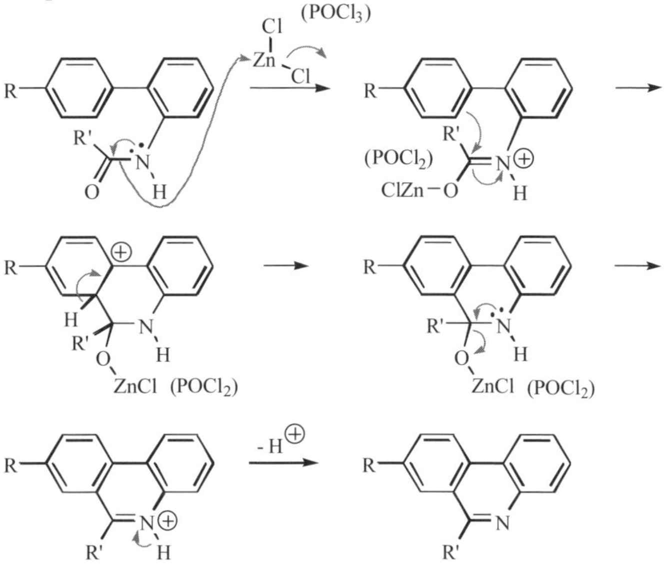

## Examples:

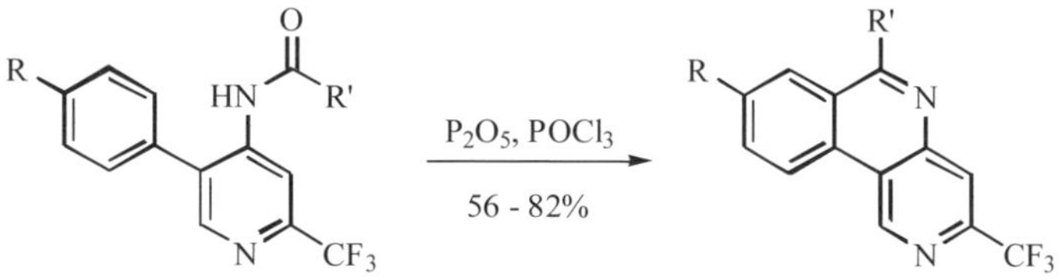

$$\xrightarrow[\text{56 - 82\%}]{\text{P}_2\text{O}_5,\ \text{POCl}_3}$$

A series of these products was prepared.

V. I. Tyvorskii, D. N. Bobrov, O. G. Kulinkovich, W. Aelterman, N. De Kimpe, *Tetrahedron* **2000**, <u>56</u>, 7313

H. Gilman, J. Eisch, *Journal of the American Chemical Society* **1957**, <u>79</u>, 4423

S. T. Mullins, N. K. Annan, P. R. Cook, G. Lowe, *Biochemistry* **1992**, <u>31</u>, 842

# Pictet-Spengler Isoquinoline Synthesis
## The Reaction:

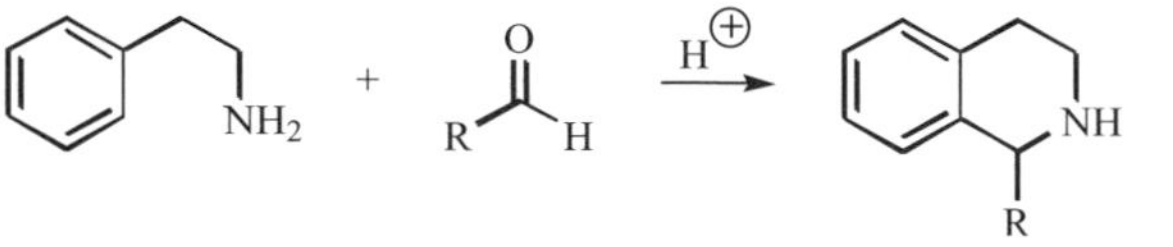

## Proposed Mechanism:

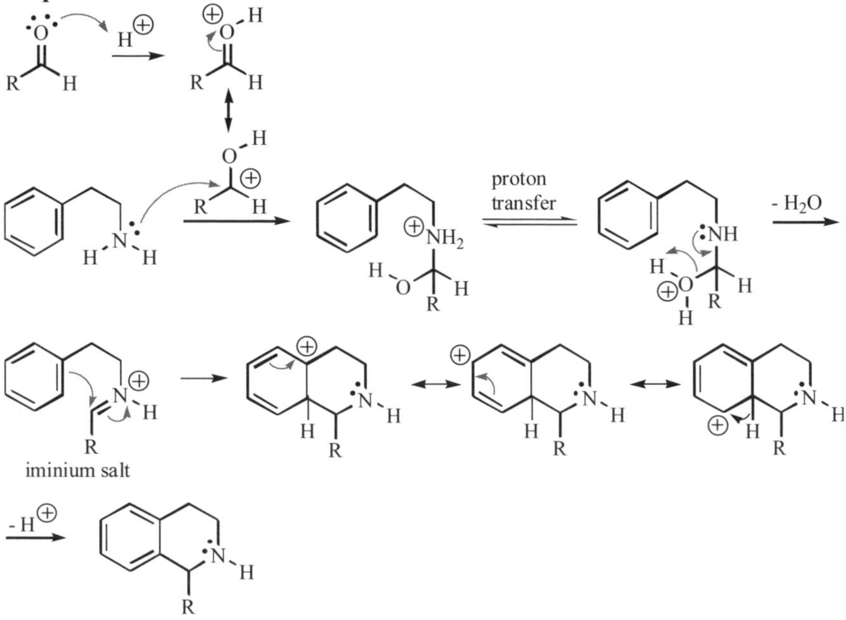

iminium salt

## Notes:
W. M. Whaley, T. Govindachar1, *Organic Reactions* **6**, 3.

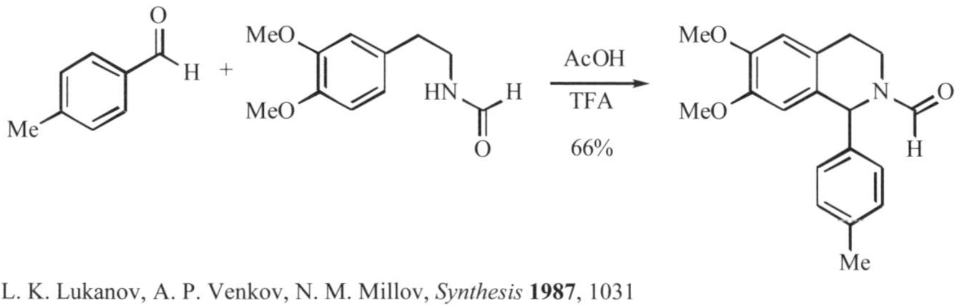

AcOH
TFA
66%

L. K. Lukanov, A. P. Venkov, N. M. Millov, *Synthesis* **1987**, 1031

N. Totomayor, E. Dominguez, E. Lete, *Tetrahedron* **1995**, <u>51</u>, 12159

A. Hegedus, Z. Hell, *Tetrahedron Letters* **2004**, <u>45</u>, 8553

B. E. Maryanoff, M. C. Rebarchak, *Synthesis* **1992**, 1245

F.-M. Kuo, M.-C. Tseng, Y.-H. Yen, Y.-H. Chu, *Tetrahedron* **2004**, <u>60</u>, 12075

# Piloty-Robinson (Pyrrole) Synthesis
## The Reaction:

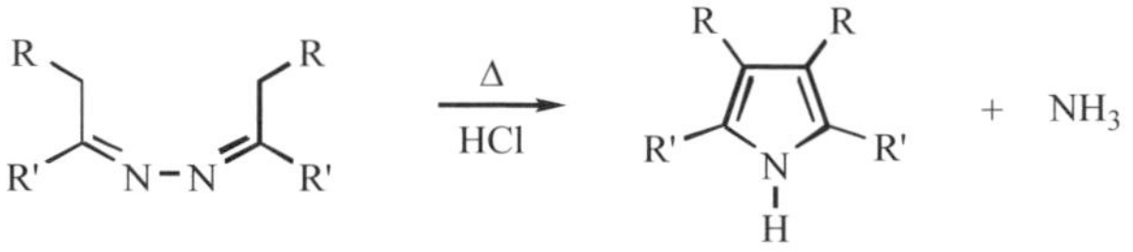

## Proposed Mechanism:

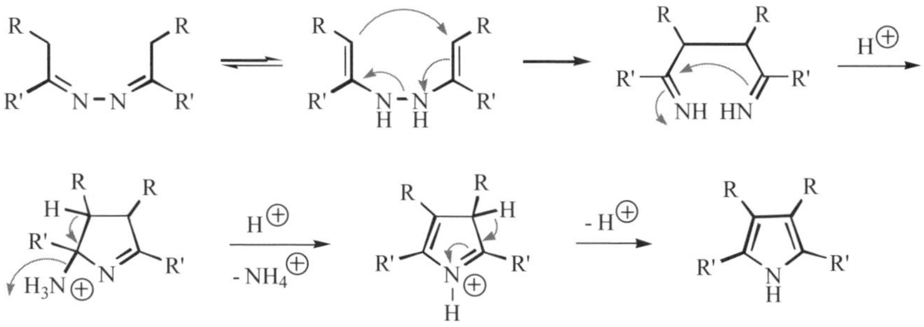

## Notes:

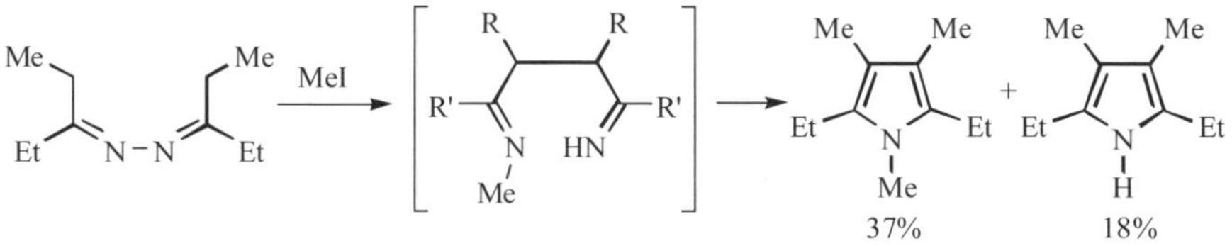

37%        18%

## Examples:

1. LDA, THF

2. H$_2$O

52%

via:

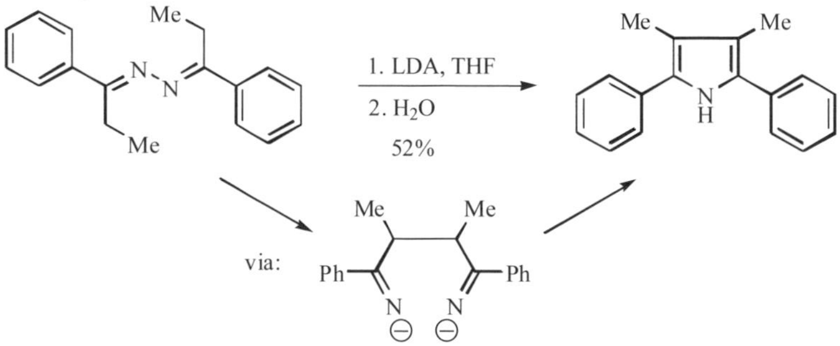

Z. Yoshida, T. Harada, Y. Tamura, *Tetrahedron Letters* **1976**, <u>17</u>, 3823

P. S. Portoghese, A. W. Lipkowski, A. E. Takemor, *Journal of Medicinal Chemistry* **1987**, <u>30</u>, 239

Y. Tamura, T. Harada, Z.-I. Yoshida, *Journal of Organic Chemistry* **1978**, <u>43</u>, 3370

J. L. Born, *Journal of Organic Chemistry* **1972**, <u>37</u>, 3952

# Pinacol Coupling Reaction

**The Reaction:**

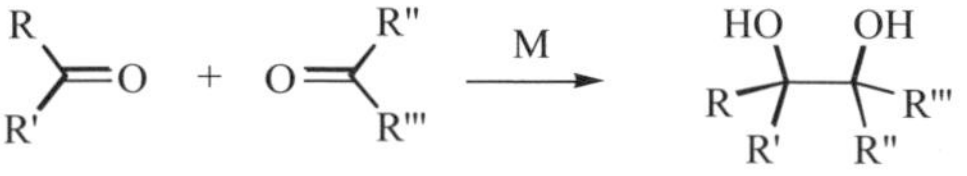

$$R_2C{=}O + O{=}CR''_2 \xrightarrow{M} \text{(pinacol product)}$$

**Proposed Mechanism:**

Most mechanistic interpretations have the metal providing an electron, followed by coupling of the ketyl moieties:

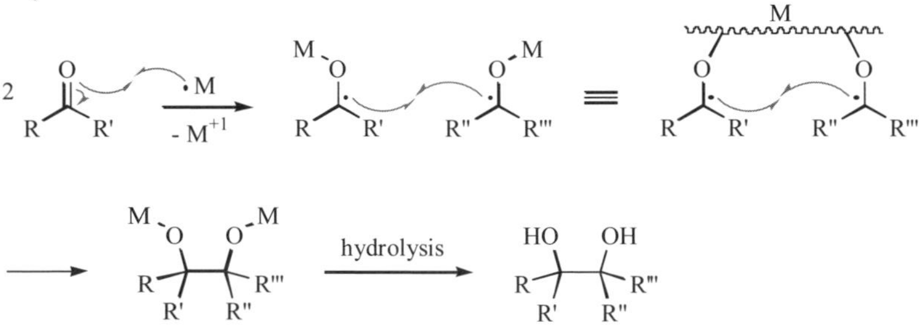

hydrolysis

**Notes:**

There are many questions still to be answered. What is the role of different metals in the reaction? What are the factors governing mixed coupling?

Typical reductive coupling reaction mixtures: Al / Hg, Mg / Hg, Mg / TiCl$_4$, SmI$_2$, Mg / MgI$_2$, CeI$_2$, Yb.

Influence of conditions on product composition:

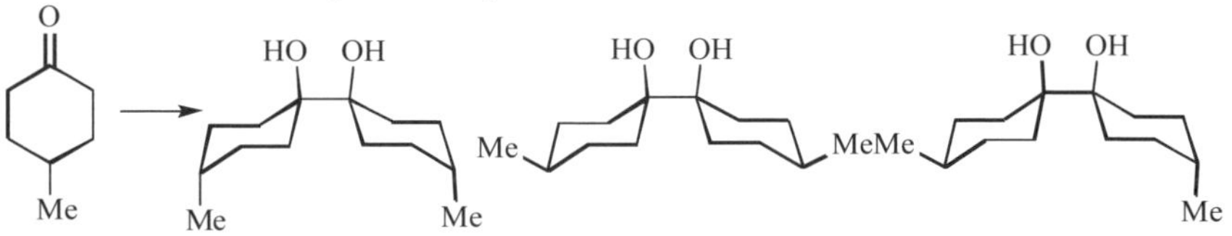

| | | | |
|---|---|---|---|
| [Hg-Al] | 43 | 13 | 44 |
| [Ti] | 4 | 63 | 33 |

B. P. Mundy, D. R. Bruss, Y. Kim, R. D. Larsen, R. J. Warnet, *Tetrahedron Letters* **1985**, 26, 3927

***Gomberg-Bachmann Pinacol Synthesis***

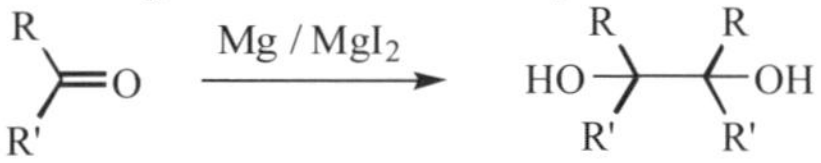

$$R_2C{=}O \xrightarrow{Mg / MgI_2} \text{(pinacol product)}$$

**Enantioselective Pinacol Coupling of Aldehydes**: A. Bensari, J. -L. Renaud, O. Riant, *Organic Letters* **2001**, 3, 3863

See: E. J. Corey, R. L. Danheiser, S. Chandrasekaran, *Journal of Organic Chemistry* **1976**, 41, 260; S. Talukdar, J.-M. Fang, *Journal of Organic Chemistry* **2001**, 66, 330; J. E. McMurry, W. Choy, *Journal of Organic Chemistry* **1978**, 43, 1800

The **Pinacol Coupling** of acetone gives 2,3-dimethyl-2,3-butanediol which is also known as pinacol and the source of the name of this reaction:

P. E. Eaton, P. G. Jobe, K. Nyi, *Journal of the American Chemical Society* **1980**, 102, 6636

E. J. Corey, R. L. Danheiser, S. Chandrasekaran, P. Siret, G. E. Keck, J. -L. Gras, *Journal of the American Chemical Society* **1978**, 100, 8031, 8034

A. Clerici, O. Porta, *Tetrahedron Letters* **1982**, 23, 3517

D. Ghiringhelli, *Tetrahedron Letters* **1983**, 24, 287

M. Yamashita, K. Okuyama, I. Kawasaki, S. Ohta, *Tetrahedron Letters* **1996**, 37, 7755

# Pinacol Rearrangement

## The Reaction:

## Proposed Mechanism:

## Notes:

T. Laue, A. Plagens, *Named Organic Reactions*, John Wiley and Sons, Inc., New York, 1998, pp. 217-218; M. B. Smith, J. March in *March's Advanced Organic Chemistry*, 5th ed., John Wiley and Sons, Inc., New York, 2001, pp 1396-1398

**Influence of temperature:**

|                |      |
|----------------|------|
| Conc. $H_2SO_4$ | 0°  |
| Conc. $H_2SO_4$ | 97° |

Major

Major

in warm acid

B. P. Mundy, R. Srinivasa, *Tetrahedron Letters* **1979**, 20, 2671

**Concentration Effects**

| Conc. $H_2SO_4$ | 0° | 0  | 5 | 95 |
|-----------------|----|----|---|----|
| 25%. $H_2SO_4$  | 0° | 30 | 1 | 69 |

B. P. Mundy, R. Srinivasa, R. D. Otzenberger, A. R. DeBernardis, *Tetrahedron Letters* **1979**, 20, 2673

A. Nath, J. Mal, R. V. Venkateswaran, *Journal of Organic Chemistry* **1996**, <u>61</u>, 4391

H. Nemoto, J. Miyuata, H. Hakamata, M. Nagamochi, K. Fukumoto, *Tetrahedron* **1995**, <u>51</u>, 5511

C. M. Marson, S. Harper, A.J. Walker, J. Pickering, J. Campbell, R. Wrigglesworth, S.J. Edge, *Tetrahedron* **1993**, <u>49</u>, 10339

B. P. Mundy, Y. Kim, R. J. Warnet, *Heterocycles* **1983**, <u>20</u>, 1727

| LA | RATIO | |
|---|---|---|
| ZnCl₂, CH₂Cl₂ | 43 | 57 |
| TiCl₄, ether | 100 | 0 |
| AlCl₃, toluene | 21 | 79 |

T. Bach, F. Eilers, *Journal of Organic Chemistry* **1999**, <u>64</u>, 8041

# Pinner Reaction
## The Reaction:

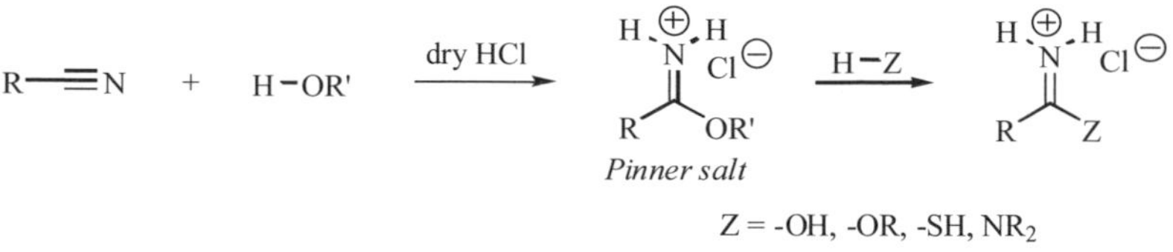

$$R-\!\!\equiv\!\!N \quad + \quad H-OR' \quad \xrightarrow{\text{dry HCl}}$$

Pinner salt

$$\xrightarrow{\text{H}-Z}$$

$$Z = \text{-OH, -OR, -SH, NR}_2$$

## Proposed Mechanism:

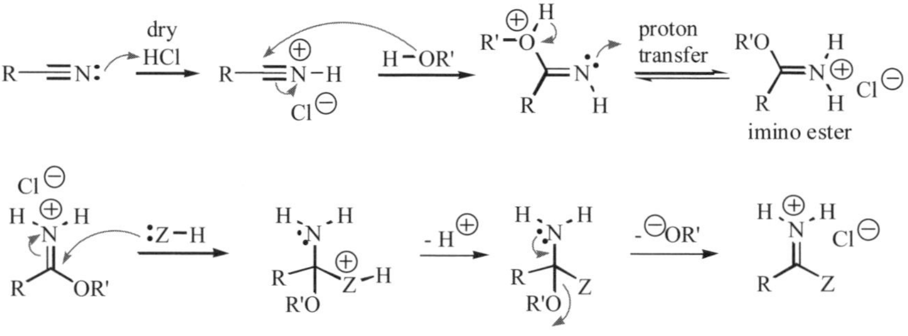

imino ester

## Notes:

M. B. Smith, J. March in *March's Advanced Organic Chemistry*, 5$^{th}$ ed., John Wiley and Sons, Inc., New York, 2001, p. 1183

Imino esters are also known as imidates or imino ethers

Normally the product is subjected to hydrolysis:

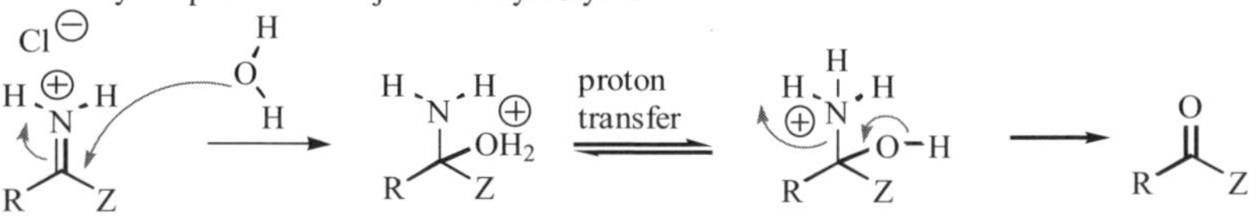

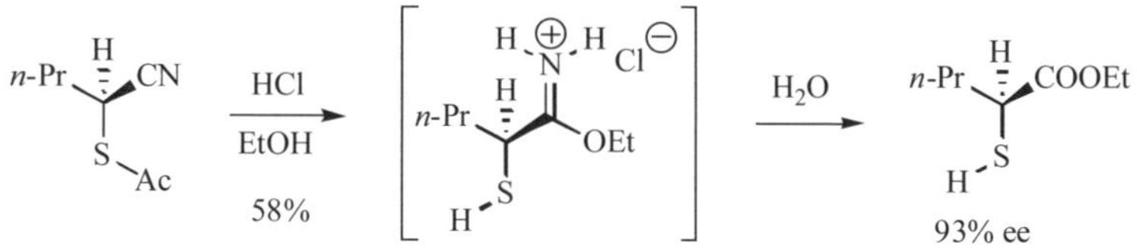

58%

93% ee

S. Gaupp, F. Effenberger, *Tetrahedron: Asymmetry* **1999**, _10_, 1777

L.-H. Zhang, J. C. Chung, T. D. Costello, I. Valvis, P. Ma, S. Kauffman, R. Ward, *Journal of Organic Chemistry* **1997**, 62, 2466

S. Caron, E. Vazquez, *Organic Process Research & Development* **2001**, 5, 587

**Intermolecular Pinner Reaction:**

Y. Hamada, O. Hara, A. Kawai, Y. Kohno, T. Shioiri, *Tetrahedron* **1991**, 47, 8635

S. K. Sharma, M. Tandon, J. W. Lown, *Journal of Organic Chemistry* **2001**, 66, 1030

# Pinnick Oxidation
## The Reaction:

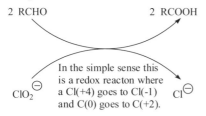

## Proposed Mechanism:

2 RCHO                                            2 RCOOH

$ClO_2^{\ominus}$  
In the simple sense this is a redox reacton where a Cl(+4) goes to Cl(-1) and C(0) goes to C(+2).  $Cl^{\ominus}$

## Notes:

Me      Me

Me      Me       serves as a scavanger for other oxidants.

Idea for reaction based on  
G. A. Krause, B. Roth, *Journal of Organic Chemistry* **1980**, <u>45</u>, 4825

## Examples:

Me

*Pinnick Oxidation*

90%

Me

OBn

$H_2C$   CHO

$H_2C$   $CO_2H$

OBn

Me   Me

Me

CHO

*Pinnick Oxidation*

90%

Me   Me

Me

$CO_2H$

B. S. Bal, W. E. Childers, Jr., H. W. Pinnick, *Tetrahedron* **1981**, <u>37</u>, 2091

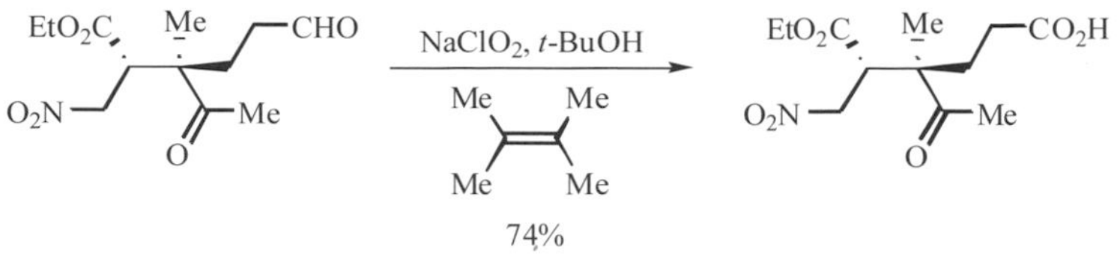

74%

J. Mulzer, D. Reither, *Organic Letters* **2000**, _2_, 3139

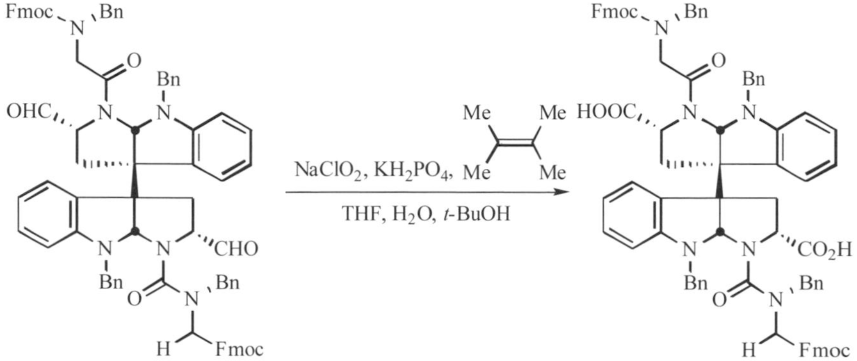

Yield not reported for this step. After two additional steps a total yield of 62%.
L. E. Overman, D. V. Paone, *Journal of the American Chemical Society* **2001**, <u>123</u>, 9465

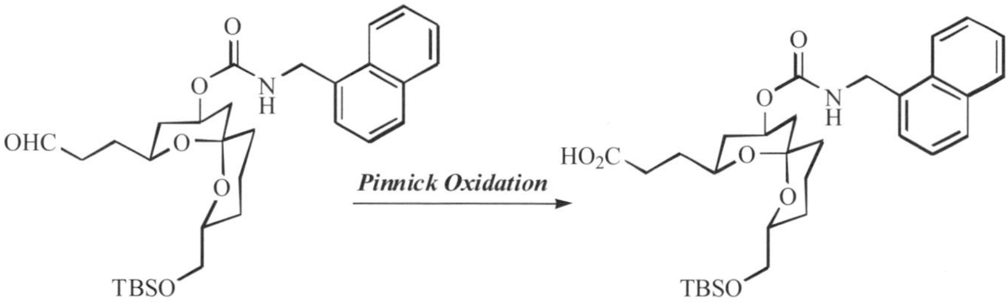

*Pinnick Oxidation*

B. A. Kulkarni, G. P. Roth, E. Lobkovsky, J. A. Porco, Jr., *Journal of Combinatorial Chemistry* **2002**, <u>4</u>, 56

# Polonovski Reaction
## The Reaction:

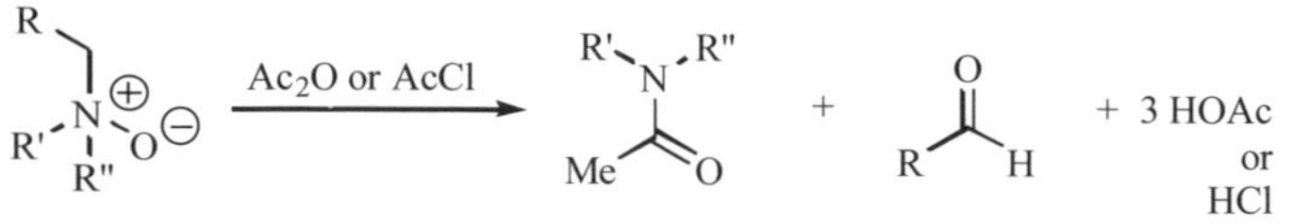

## Proposed Mechanism:

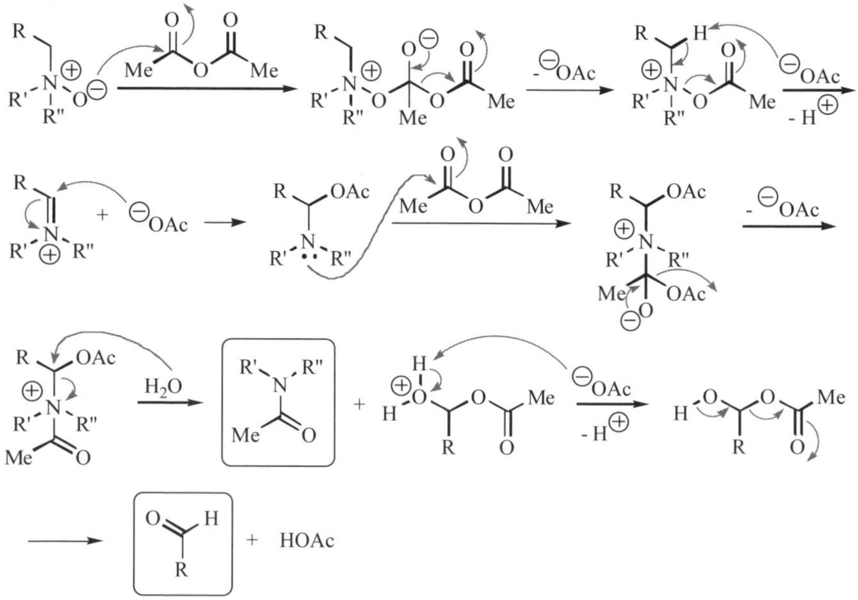

## Notes:
D. Grierson, *Organic Reactions* **39**, 2

## Examples:

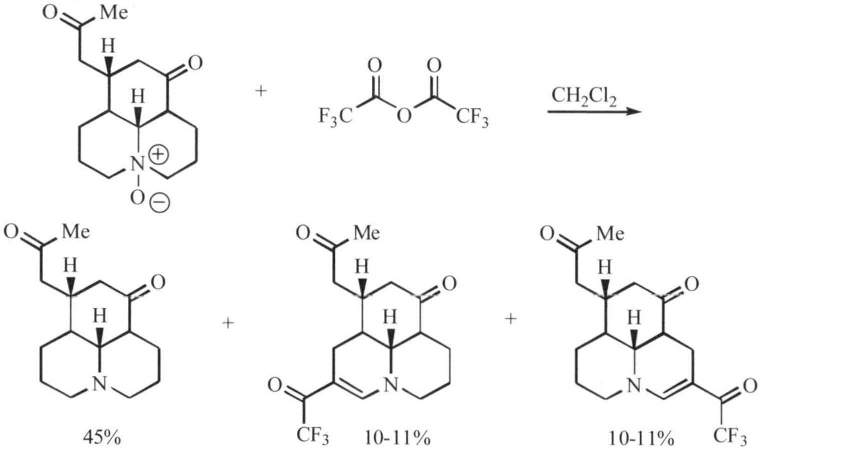

45%                                    CF₃   10-11%                 10-11%   CF₃

E. Wenkert, B. Chauncy, K. G. Dave, A. R. Jeffcoat, F. M Schell, H. P. Schenk, *Journal of the American Chemical Society* **1973**, 95, 8427

H. Morita, J. Kobayashi, *Journal of Organic Chemistry* **2002**, <u>67</u>, 5378

Y.-Y. Ku, D. Riley, H. Patel, C. X. Yang, J.-H. Liu, *Bioorganic and Medicinal Chemistry Letters*
**1997**, 7, 1203

"Non-classical Polonvski" used as a new procedure for demethylation.

K. McCamley, J. A. Ripper, R. D. Singer, P. J. Srammells, *Journal of Organic Chemistry*
**2002**, <u>68</u>, 9847

# Polonovski-Potier Reaction
## The Reaction:

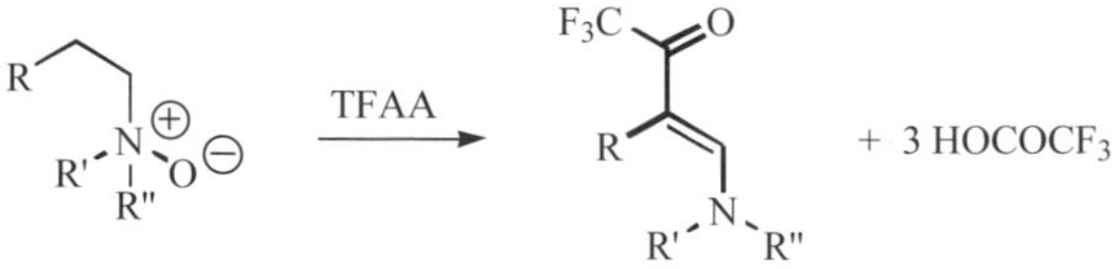

$$R \underset{R''}{\overset{R'}{\underset{|}{N^{\oplus}}} \cdot O^{\ominus}} \xrightarrow{\text{TFAA}} \underset{R' \cdot N \cdot R''}{R \overset{F_3C \diagdown \diagup O}{\diagdown}} + 3\ \text{HOCOCF}_3$$

## Proposed Mechanism:

TFAA = Trifluoroacetic anhydride

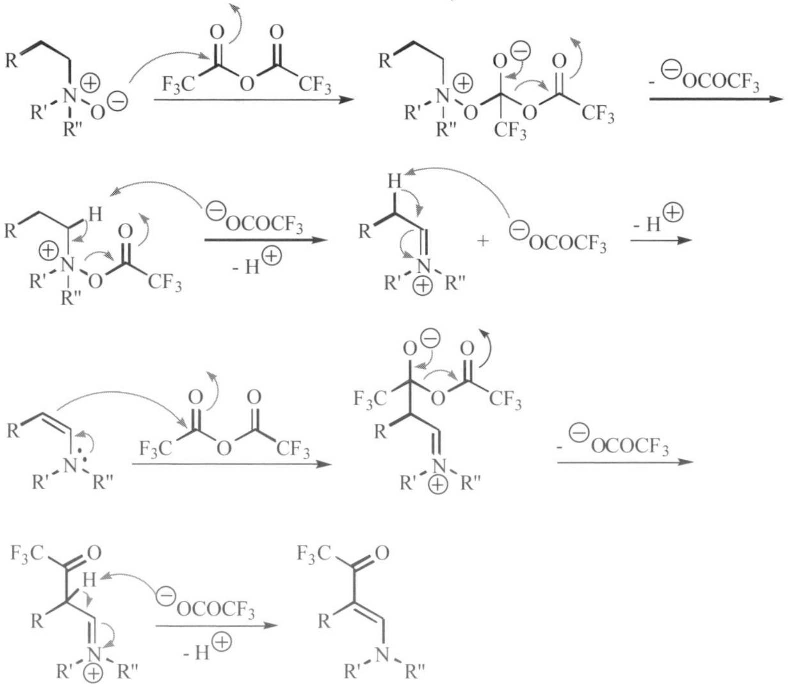

## Examples:

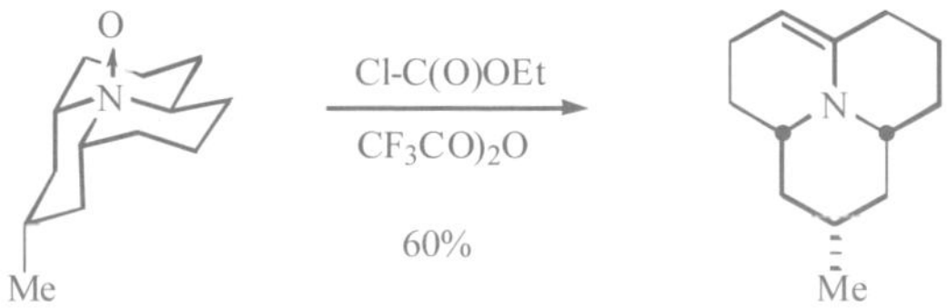

Cl-C(O)OEt

CF$_3$(CO)$_2$O

60%

B. Tursch, D. Daloze, J. C. Brakeman, C. Hootele, J. M. Pasteels, *Tetrahedron* **1975**, <u>31</u>, 1541.

R. Jokela, M. Halonen, M. Lounasmaa, *Heterocycles* **1994**, <u>38</u>, 189

E. Wenkert, B. Chauncy, K. G. Dave, A. R. Jeffcoat, F. M Schell, H. P. Schenk, *Journal of the American Chemical Society* **1973**, <u>95</u>, 8427

# Polycarbocycle Syntheses
## *Bogert-Cook Synthesis*
## The Reaction:

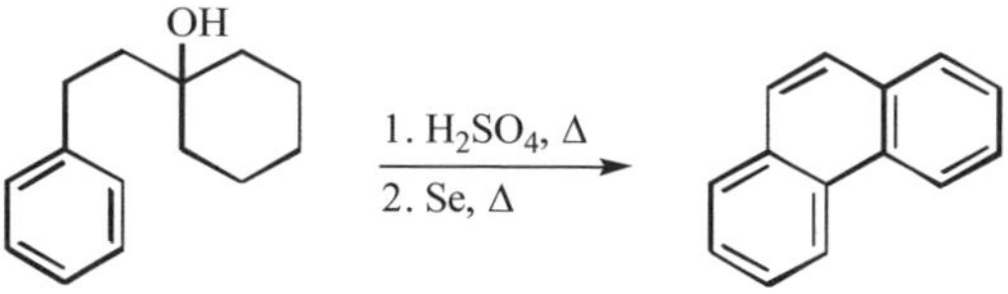

$$\xrightarrow[\text{2. Se, } \Delta]{\text{1. H}_2\text{SO}_4, \Delta}$$

## Proposed Mechanism:

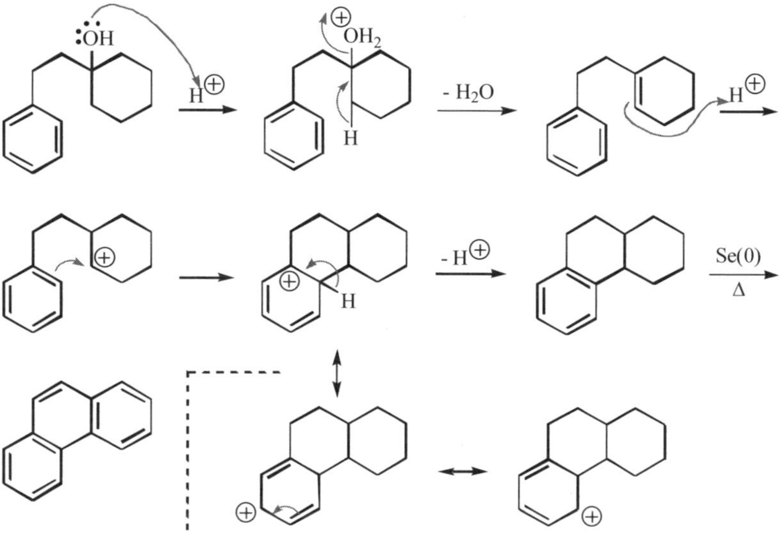

## Notes:
The Bardhan-Sengupta Phenanthrene Synthesis uses $P_2O_5$ in place of $H_2SO_4$.

When the alcohol containing portion is acyclic, the reaction is known as the ***Bogert Reaction***:

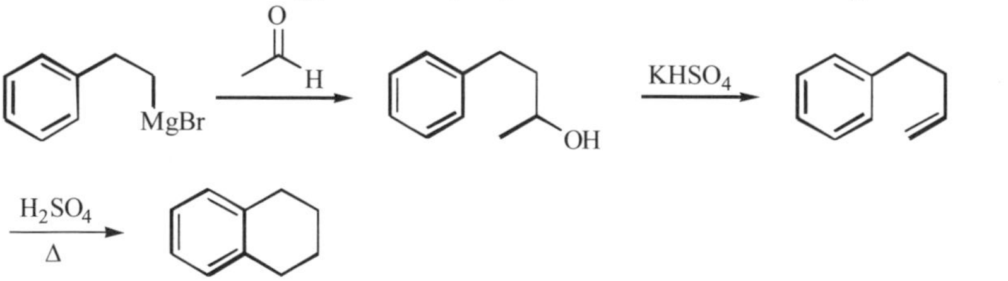

## *Darzens Synthesis of Tetralin Derivatives*

### The Reaction:

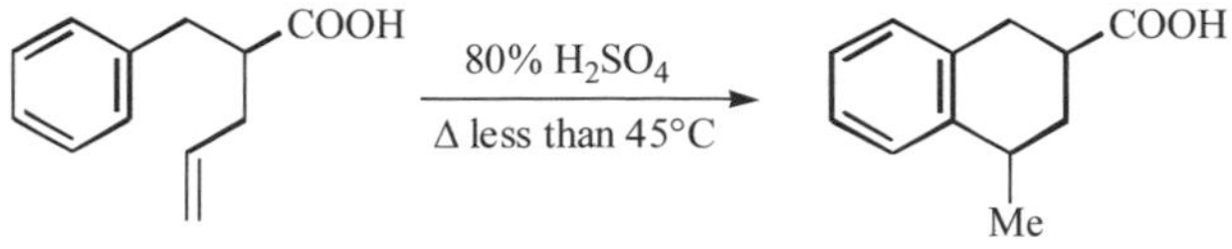

COOH
80% $H_2SO_4$
Δ less than 45°C
COOH
Me

### Proposed Mechanism:

COOH
H⊕
COOH
Me
COOH
H
Me
- H⊕
COOH
Me
COOH
Me
COOH
Me

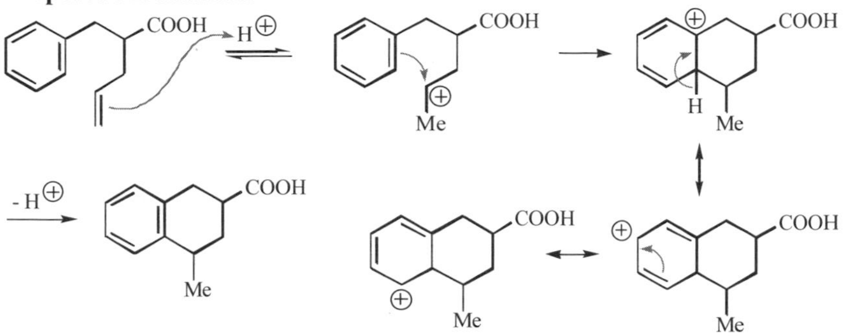

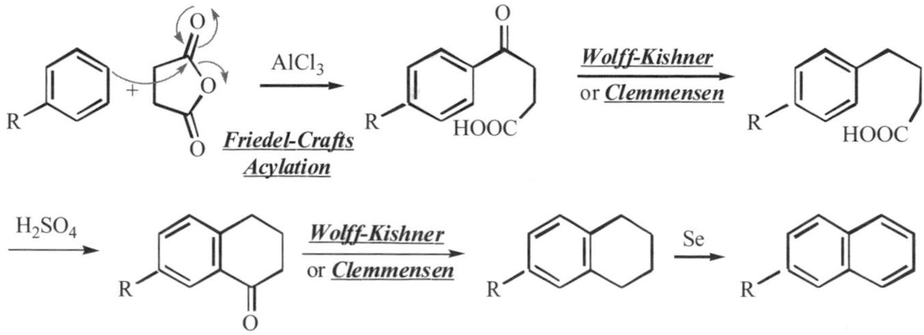

## *Haworth Reaction*

### The Reaction:

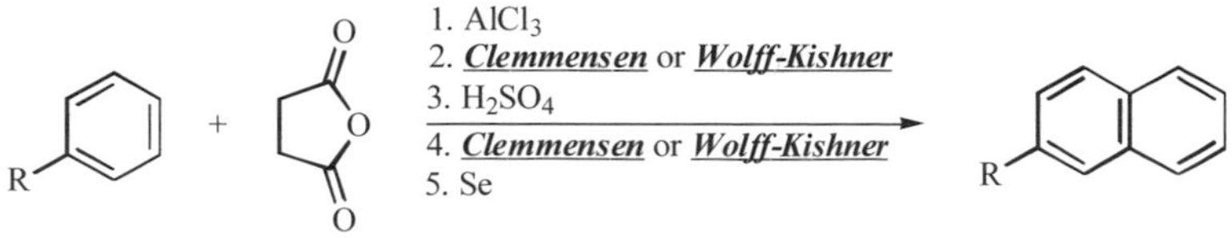

R
O
O
O
1. $AlCl_3$
2. Clemmensen or Wolff-Kishner
3. $H_2SO_4$
4. Clemmensen or Wolff-Kishner
5. Se
R

### Proposed Mechanism:

R
O
O
O
$AlCl_3$
Friedel-Crafts Acylation
O
R
HOOC
Wolff-Kishner
or Clemmensen
R
HOOC
$H_2SO_4$
R
O
Wolff-Kishner
or Clemmensen
R
Se
R

### Notes:

On naphthalene based starting materials, a phenanthrene is produced and the reaction is called the ***Hayworth Phenanthrene Synthesis***:

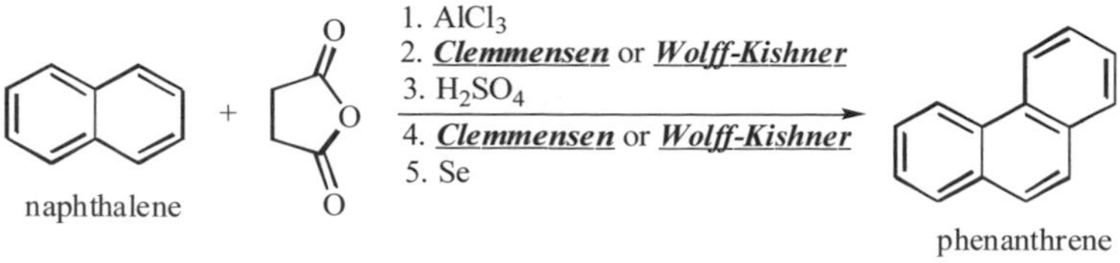

naphthalene
O
O
O
1. $AlCl_3$
2. Clemmensen or Wolff-Kishner
3. $H_2SO_4$
4. Clemmensen or Wolff-Kishner
5. Se
phenanthrene

# Pomeranz-Fritsch Reaction
## The Reaction:

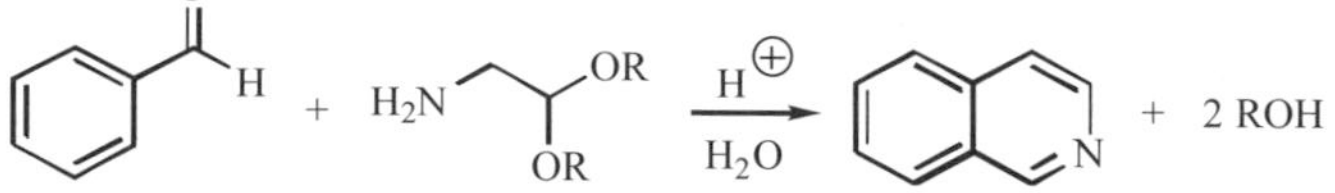

## Proposed Mechanism:

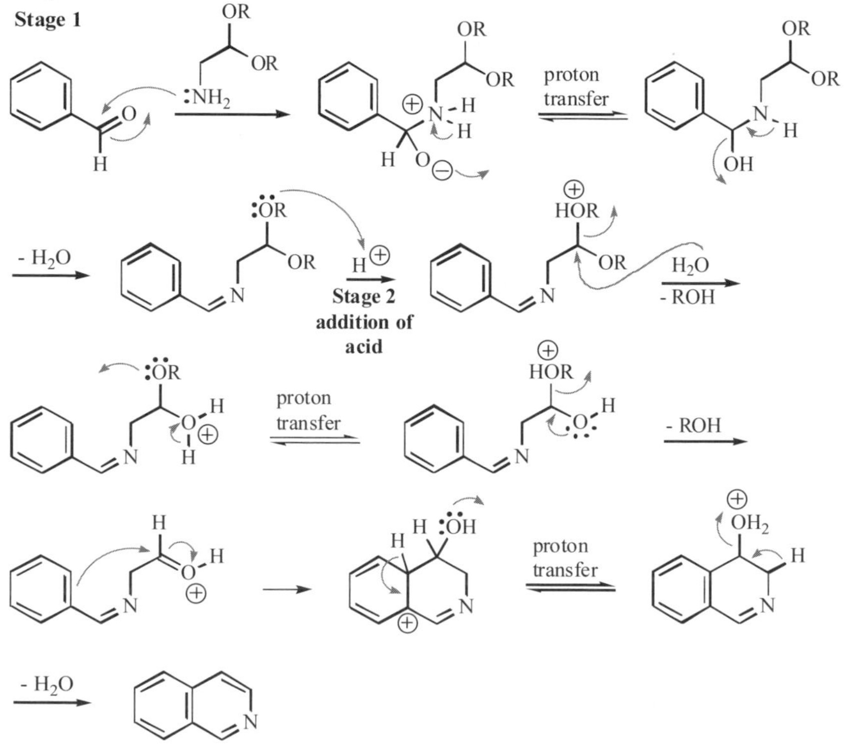

**Stage 1**

## Notes:
W. J. Gensler, *Organic Reactions* **6**, 4
## Schlittler-Muller Modification:

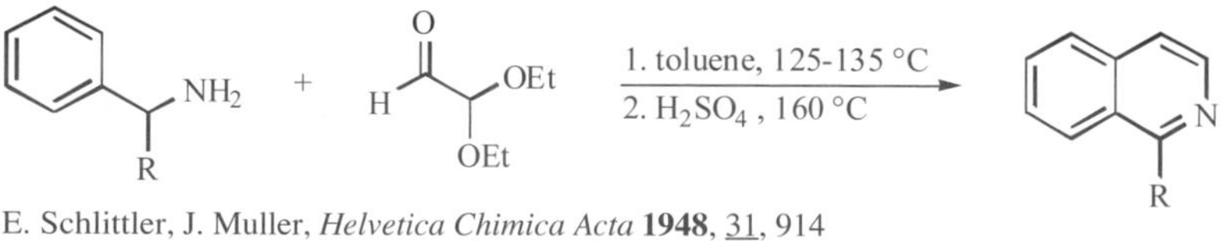

E. Schlittler, J. Muller, *Helvetica Chimica Acta* **1948**, <u>31</u>, 914

# Examples:

R. Hirkenkorn, *Tetrahedron Letters* **1991**, <u>32,</u> 1775

J. B. Hendrickson, C. Rodriquez, *Journal of Organic Chemistry* **1983**, <u>48</u>, 3344

M. J. Bevis, E. J. Fobes, B. C. Uff, *Tetrahedron* **1969**, <u>25</u>, 1585

M. J. Bevis, E. J. Forbes, N. N. Naik, B. C. Uff, *Tetrahedron* **1971**, <u>27</u>, 1253

D. L. Boger, C. E. Brotherton, M. D. Kelley, *Tetrahedron* **1981**, <u>37</u>, 3977

# Prévost Reaction (Glycolization)

**The Reaction:**

I2
2
O
R
O
Ag
OH
HO

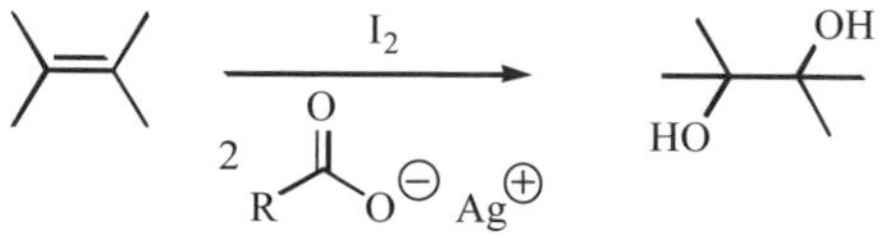

**Proposed Mechanism:**

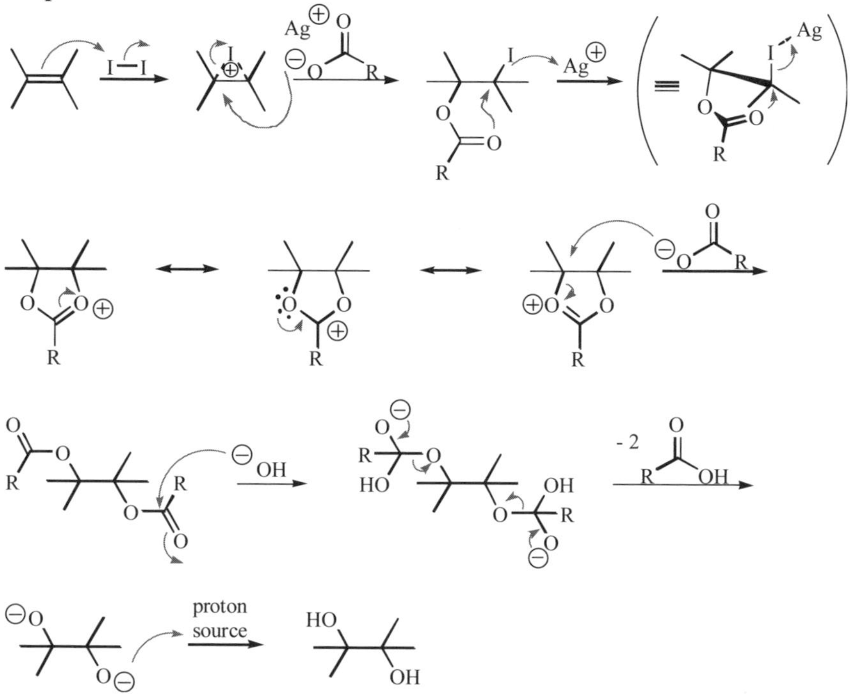

I—I
Ag
O
R
OH
- 2
HO
proton
source

**Notes:**

Me
H
RCOO
I
Trans-diaxial opening
of intermediate
Initial I2 attack will be
from the less-hindered face
R
O
HO
HO

V. K. Ahluwalia, R. K. Parashar, *Organic Reaction Mechanisms,* Alpha Science International Ltd., Pangbourne, U.K., 2002, pp. 189-191; M. B. Smith, J. March in *March's Advanced Organic Chemistry*, 5th ed., John Wiley and Sons, Inc., New York, 2001, pp 1049-1050

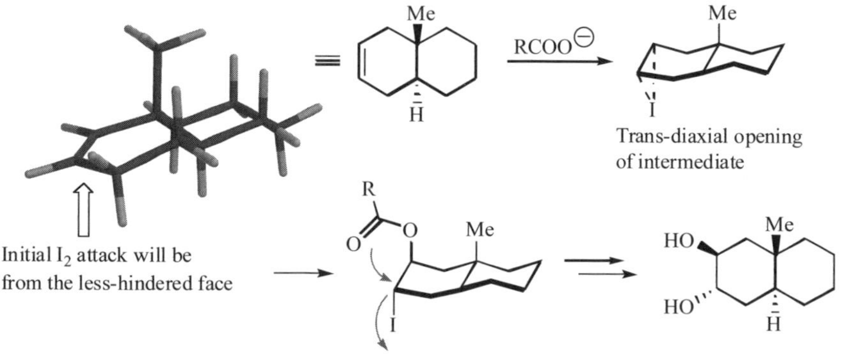

**Examples:**

B. Zajc, *Journal of Organic Chemistry* **1999**, <u>64</u>, 1902

H.-F.Chang, B. P. Cho, *Journal of Organic Chemistry* **1999**, <u>64</u>, 9051

# Prilezhaev (Prileschajew) Reaction / Epoxidation

## The Reaction:

## Proposed Mechanism:

## Notes:

M. B. Smith, J. March in *March's Advanced Organic Chemistry,* 5th ed., John Wiley and Sons, Inc., New York, 2001, p. 1052; T. Laue, A. Plagens, *Named Organic Reactions*, John Wiley and Sons, Inc., New York, 1998, pp. 218 - 220

The name for this "Named Reaction" is not used much.

The reaction had its origins with the use of peroxyorganic acids. The simplest, formic, peracetic, and perbenzoic acid, are not so much in contemporary use. The workhorse of peroxy acids is *m*-chloroperoxybenzoic acid (*m*CPBA).

See other named reactions and reagents:

*Oxone, DMD, Sharpless Epoxidation, Jacobsen-Katsuki Epoxidation, Corey-Chaykovsky reagent and Reaction, Shi (Asymmetric) Epoxidation.*

## Examples:

T. Nishikawa, D. Urabe, K. Yoshida, T. Iwabuchi, M. Asai, M. Isobe, *Organic Letters* **2002**, *4*, 2679

| | | |
|---|---|---|
| *m*CPBA | 15 | 85 |
| *DMD* | 91 | 9 |

W. Adam, A. Pastor, K. Peters, E. Peters, *Organic Letters* **2000**, *2*, 1019

F. M. Hauser, H. Yin, *Organic Letters* **2000**, <u>2</u>, 1045

|  | mCPBA | 3 | : | 2 |
|--|-------|---|---|---|
|  | *__Jacobsen__* | 6 | : | 1 |

Y. Chen, J. B. Evarts, Jr., E. Torres, P. L. Fuchs, *Organic Letters* **2002**, <u>4</u>, 3571

S. Aragones, F. Bravo, Y. Diaz, M. I. Matheu, S. Castillon, *Tetrahedron: Asymmetry* **2003**, <u>14</u>, 1847

W. Li, P. L. Fuchs, *Organic Letters* **2003**, <u>5</u>, 2849

# Prins Reaction
## The Reaction:

Products sometimes seen with the reaction

## Proposed Mechanism:

## Notes:

T. Laue, A. Plagens, *Named Organic Reactions*, John Wiley and Sons, Inc., New York, 1998, pp. 220-222; M. B. Smith, J. March in *March's Advanced Organic Chemistry*, 5th ed., John Wiley and Sons, Inc., New York, 2001, pp 1241-1242.; A. L. Henne, *Organic Reactions* **2**, 6; D. F. DeTar, *Organic Reactions* **9**, 7.

## Examples:

D. J. Kopecky, S. D. Rychnovsky, *Journal of the American Chemical Society* **2001** 123, 8420

M. J. Cloninger, L. E. Overman, *Journal of the American Chemical Society* **1999**, 121, 1092

A. P. Dobbs, S. Martinovic, *Tetrahedron Letters* **2002**, 43, 7055

57%

D. L. Aubele, C. A. Lee, P. E. Floreancig, *Organic Letters* **2003**, <u>5</u>, 4521

**Tandem Prins-<u>Pinacol</u> Strategy:**

L. E. Overman, L. D. Pennington, *Journal of Organic Chemistry* **2003**, <u>68</u>, 7143

D. J. Hart, C. E. Bennett, *Organic Letters* **2003**, <u>5</u>, 1499

S. R. Crosby, J. R. Harding, C. D. King, G. D. Parker, C. L. Willis, *Organic Letters* **2002**, <u>4</u>, 3407

# Pschorr Arylation (Reaction)
## The Reaction:

$$\xrightarrow[\text{Cu (0)}]{\text{NaNO}_2,\ \text{H}^{\oplus}}$$

$$G = \ \text{HC}=\text{CH},\ \text{H}_2\text{C}-\text{CH}_2,\ \overset{\overset{\text{O}}{\|}}{\text{C}},\ \text{NH},\ \text{CH}_2,\ \text{others}$$

## Proposed Mechanism:

$$\xrightarrow{\text{HONO}}$$

For a discussion of the reactions under acidic or basic conditions, see: D. F. Detar, S. V. Sagmanli, *Journal of the American Chemical Society* **1951**, <u>73</u>, 3240

*In acid solution:*

$$\xrightarrow{-\text{N}_2}$$

*In neutral or basic solution:*

$$\xrightarrow{\text{Cu(0)}}$$

## Notes:
Reactions are often difficult to control, with many side reactions observed.

## Examples:

1. NaNO₂, H₂SO₄
2. Cu-Sn, Na₂SO₄
65%

S. G. R. Guinot, J. D. Hepworth, M. Wainwright, *Journal of Chemical Research, Synopses* **1997**, 183 (AN 1977: 481656)

1. *i*-Pr(CH₂)₂-ONO, AcOH
2. hydroquinone
42%

W. Williams, X. Sun, D. Jabaratnam, *Journal of Organic Chemistry* **1997**, <u>62</u>, 4364

1. NaNO$_2$, HCl
2. Cu, NH$_4$OH

10%

NaNO$_2$, H$_2$SO$_4$

benzene

25%

D. F. Detarand, S. V. Sagmanli, *Journal of the American Chemical Society* **1950**, <u>72</u>, 965

HONO

70%

P. Tapolcsanyi, B. U. W. Maes, K. Monsieurs, G. L. F. Lemiere, Z. Riedl, G. Hajos, B. Van den Driessche, R. A. Dommisse, P. Matyus, *Tetrahedron* **2003**, <u>59</u>, 5919

CuI

G = O, S, SO, SO$_2$, CO, CH$_2$

Mixtures were often obtained. Report provides calculational data.
S. Karady, J. M. Cummins, J. J. Dannenberg, E. del Rio, P. G. Dormer, B. F. Marcune, R. A. Reamer, T. L. Sordo, *Organic Letters* **2003**, <u>5</u>, 1175

Cg(I)Z

Solvent-H

Z$^{\ominus}$

A. H. Lewin, T. Cohen, *Journal of Organic Chemistry* **1967**, <u>32</u>, 3844

# Pudovik Reaction

## The Reaction:

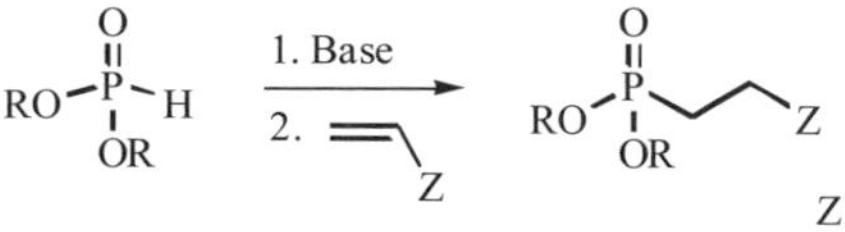

Z = electron acceptor

## Proposed Mechanism:

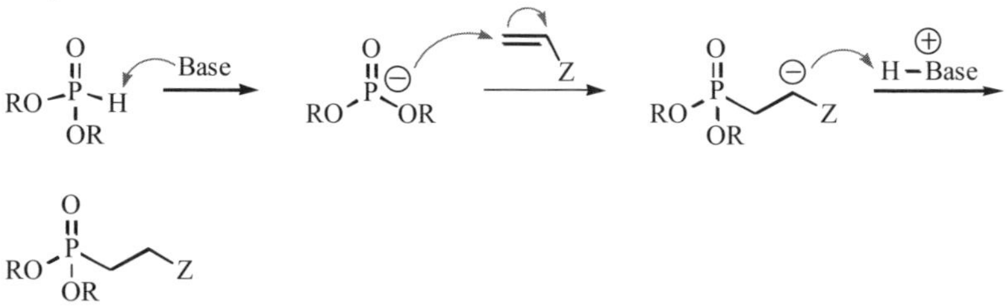

## Notes:

A versatile and general method for generating C-P bonds; however, other connections are possible.

## Examples:

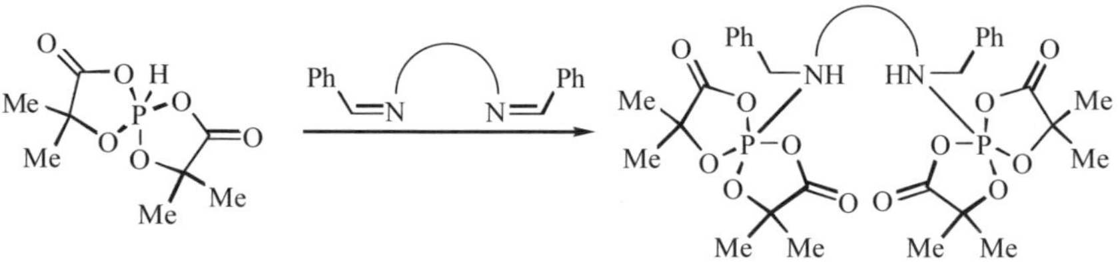

V. Pham, W. Zhang, V.Chen, V. Whitney, J. Yao, D. Froese, A. D. Friesen, J. M. Diakur, W. Haque, *Journal of Medicinal Chemistry* **2003**, 46, 3680

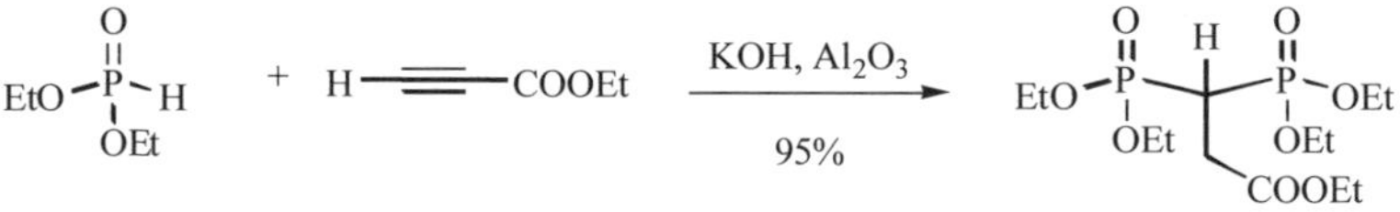

K. Vercruysse-Moreira, C. Déjugnat, G. Etemad-Moghada, *Tetrahedron* **2002**, 58, 5651

BuO–P(=O)(OBu)–H

1. Na, hexane
2. Cl⌒= (allyl chloride)
39 grams

→ BuO–P(=O)(OBu)–CH₂CH=CH₂
17 grams

D. Albouy, M. Lasptras, G. Etemad-Moghadam, M. Koenig, *Tetrahedron Letters* **1999**, <u>40</u>, 2311

EtO–P(=O)(OEt)–H

EtMgBr
→ EtO–P(=O)(OEt)–MgBr

Me–C(=O)–Me
97%
→ EtO–P(=O)(OEt)–C(Me)(Me)OH

G. M. Kosolapoff, *Journal of the American Chemical Society* **1951**, <u>73</u>, 4040

(pyridine-dioxine CHO compound)

*t*-BuO–P(=O)(O⁻)–O*t*-Bu  Na⁺

THF
82%

→ (pyridine-dioxine CH(OH)–P(=O)(O*t*-Bu)₂ compound)

O. Gawron, C. Grelecki, W. Reilly, J. Sands, *Journal of the American Chemical Society* **1953**, <u>75</u>, 3591

# Pummerer Rearrangement
## The Reaction:

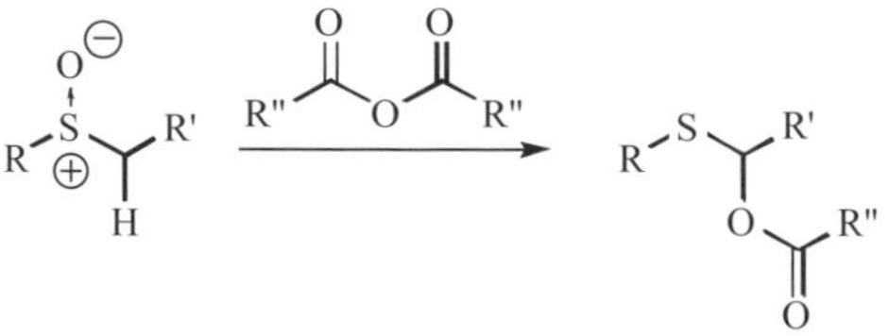

## Proposed Mechanism:

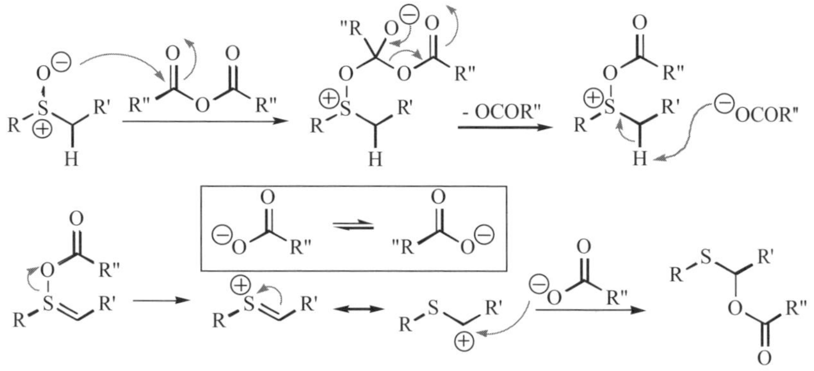

## Notes:
M. B. Smith, J. March in *March's Advanced Organic Chemistry*, 5th ed., John Wiley and Sons, Inc., New York, 2001, p 1566; O. DeLucchi, U. Miotti, G. Modena, *Organic Reactions* **40**, 3.

## Examples:

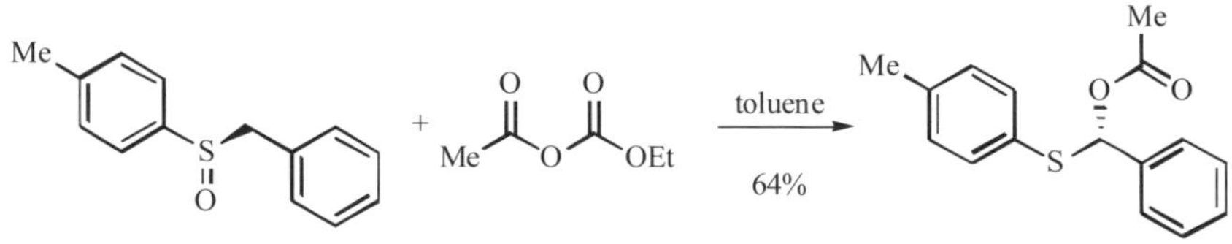

N. Shibata, M. Matsugi, N, Kawano, S. Fukui, C. Jujimori, K. Gotanda. K. Murata, Y. Kita, *Tetrahedron: Asymmetry* **1997**, 8, 303

Solid phase Pummerer cyclization:

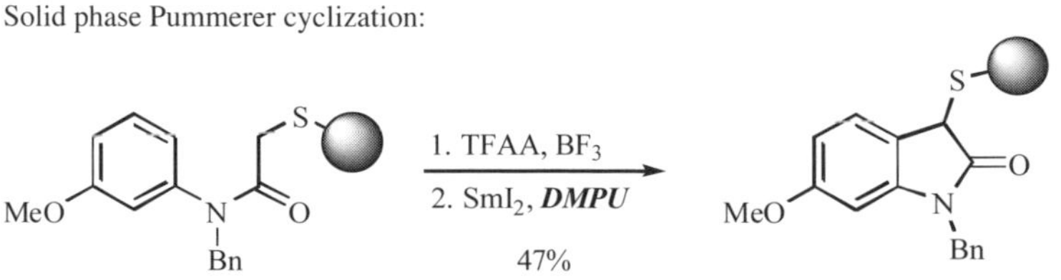

TFAA = Trifluoroaceticanhydride, *DMPU =N,N'-Dimethylpropyleneurea*
L. A. McAllister, S. Brand, R. de Gentile, D. J. Procter, *Chemical Communications* **2003**, 2380

T. K. Sarkar, N. Panda, S. Basak, *Journal of Organic Chemistry* **2003**, <u>68</u>, 6919

J. T. Kuethe, A. Padwa, *Journal of Organic Chemistry* **1997**, <u>62</u>, 774

Trapping the intermediate in a ***Mannich reaction***.

A. Padwa, M. D. Danca, K. I. Hardcastle, M. S. McClure, *Journal of Organic Chemistry* **2003**, <u>68</u>, 929

P. Magnus, T. Rainey, V. Lynch, *Tetrahedron Letters* **2003**, <u>44</u>, 2459

# Ramberg-Backlund Olefin Synthesis

## The Reaction:

$$X \overset{O\ O}{\underset{}{\overset{\backslash\!\!/}{S}}} H \quad \xrightarrow{\text{base}} \quad \text{alkene} \quad + \text{ SO}_2 \text{ (g)}$$

X = Cl, Br, I

## Proposed Mechanism:

$$X \overset{O\ O}{\underset{}{\overset{\backslash\!\!/}{S}}} \overset{}{H} \xleftarrow{\text{base}} \quad X \overset{O\ O}{\underset{}{\overset{\backslash\!\!/}{S}}} \ominus \quad \xrightarrow{-X \ominus} \quad \overset{O\ \ O}{\underset{}{\overset{}{S}}} \quad \xrightarrow{-SO_2} \quad \text{alkene}$$

these intermediates
have been isolated

## Notes:

M. B. Smith, J. March in *March's Advanced Organic Chemistry*, 5th ed., John Wiley and Sons, Inc., New York, 2001, p. 1342; T. Laue, A. Plagens, *Named Organic Reactions*, John Wiley and Sons, Inc., New York, 1998, pp. 223-224; L. A. Paquette, *Organic Reactions* **25**, 1; R. J. K. Taylor, G. Casey, *Organic Reactions* **62**, 2

Starting material preparation:

$$H \overset{\cdot\cdot}{\underset{}{S}} H \quad \xrightarrow{\text{oxidize}} \quad H \overset{O\ O}{\underset{}{\overset{\backslash\!/}{S}}} H \quad \xrightarrow[\text{NBS}]{\overset{\text{NCS}}{\text{or}}} \quad X \overset{O\ O}{\underset{}{\overset{\backslash\!/}{S}}} H$$

*Meyers Procedure*: Halogenates and eliminates in-situ

$$\xrightarrow[\text{CCl}_4,\ t\text{-BuOH}]{\text{Powdered KOH}}$$

C. Y. Meyers, A. M. Malte, W. S. Matthews, *Journal of the American Chemical Society* **1969**, <u>91</u>, 7510

*Vedejs Procedure*:

$$\xrightarrow[\text{DME}]{\text{NaH, CCl}_3\text{-CCl}_3}$$

75%

E. Vedejs, S. Singer, *Journal of Organic Chemistry* **1978**, <u>43</u>, 4884

There is a stereochemical preference for the "W" arrangement of proton and leaving group.

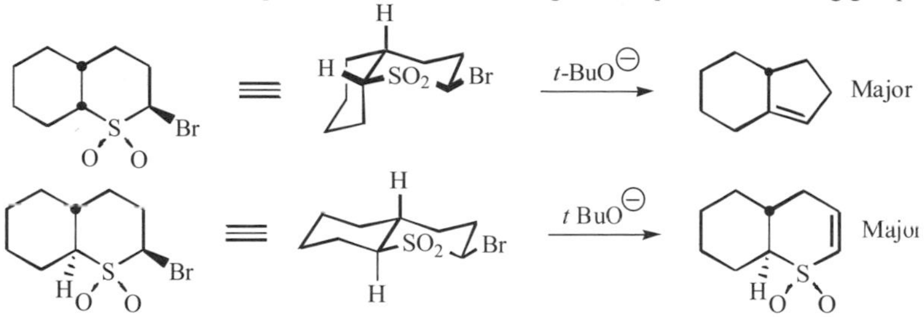

R. J. K. Taylor, G. Casy, *Organic Reactions* **2003**, <u>62</u>, Chapter 2, pp361-362

## Examples:

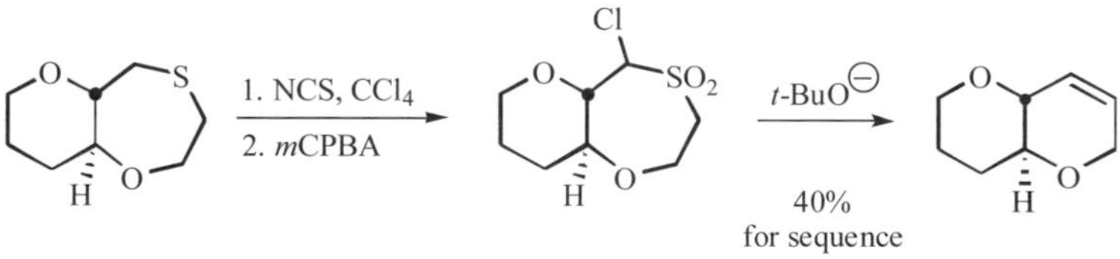

40%
for sequence

E. Alvarex, M. Delgado, M. T. Diaz, L. Harxing, R. Perez, J. D. Martin, *Tetrahedron Letters* **1996**, <u>37</u>, 2865

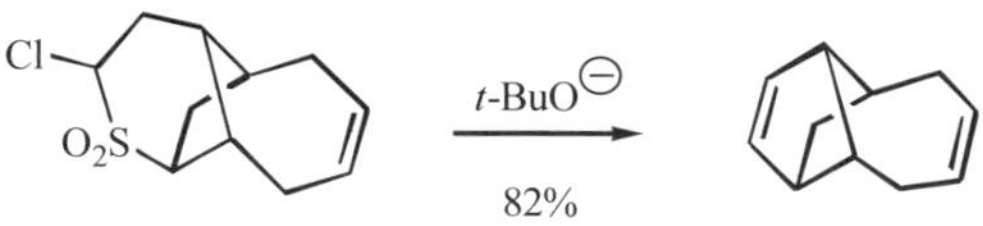

82%

J. Dressel, K. L. Chasey, L. A. Paquette, *Journal of the American Chemical Society* **1988**, <u>110</u>, 5479

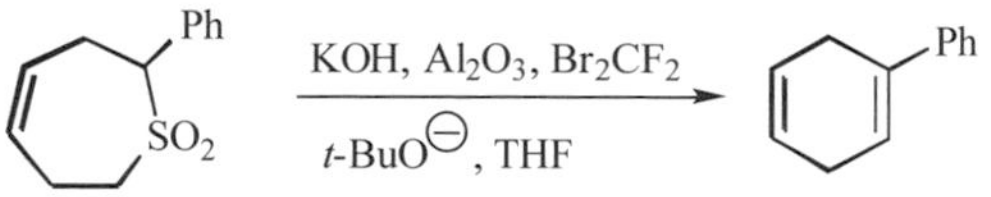

Q. Yao, *Organic Letters* **2002**, <u>4</u>, 427

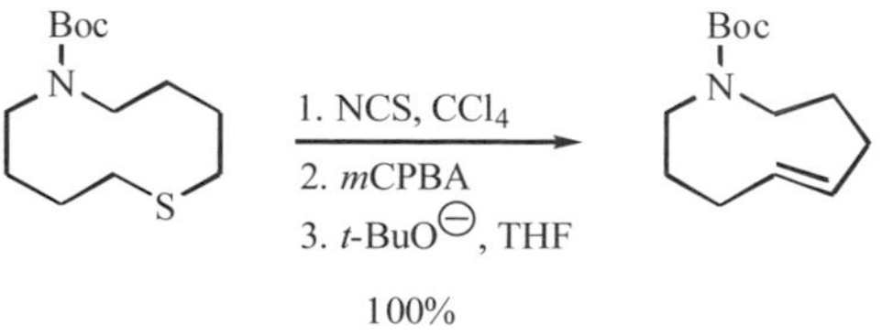

100%

D. I. MaGee, E. J. Beck, *Journal of Organic Chemistry* **2000**, <u>65</u>, 8367

# Reetz Alkylation

## The Reaction:

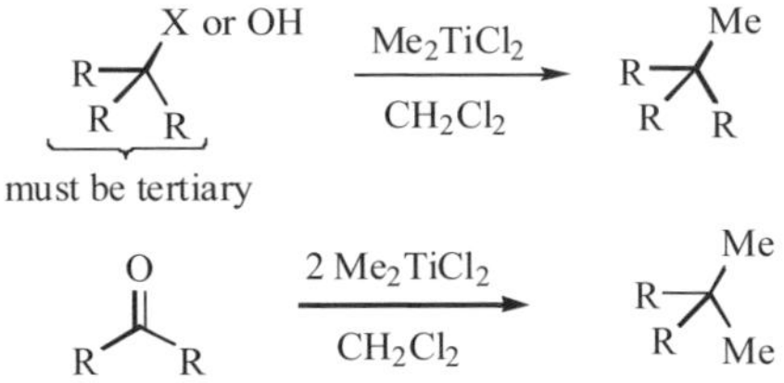

$$\underset{\underbrace{\text{R} \quad \text{R}}_{\text{must be tertiary}}}{\overset{\text{X or OH}}{R-}} \xrightarrow[\text{CH}_2\text{Cl}_2]{\text{Me}_2\text{TiCl}_2} \overset{\text{Me}}{\underset{\text{R} \quad \text{R}}{R-}}$$

$$\underset{\text{R} \quad \text{R}}{\overset{\text{O}}{\bigparallel}} \xrightarrow[\text{CH}_2\text{Cl}_2]{2\ \text{Me}_2\text{TiCl}_2} \overset{\text{Me}}{\underset{\text{R} \quad \text{Me}}{R-}}$$

## Proposed Mechanism:

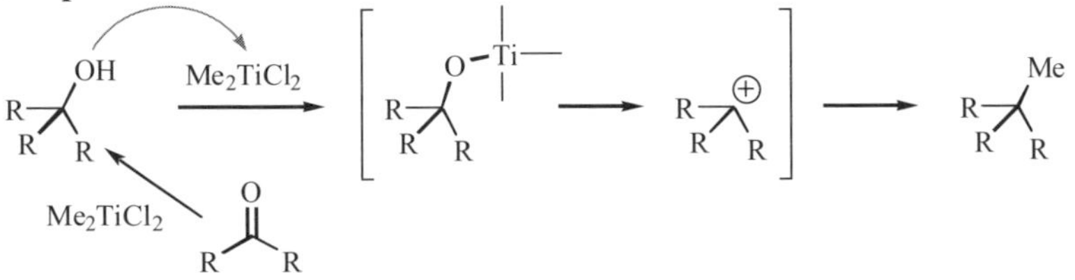

## Notes:

Original Publication: M. T. Reetz, J. Westermann, R. Steinbach, *Journal of the Chemical Society: Chemical Communications* **1981**, 237

Reagent Preparation:
$(\text{Me})_2\text{Zn} + \text{TiCl}_4 \longrightarrow \text{Me}_2\text{TiCl}_2$

Evidence for a carbocation rearrangement during the ***Reetz reaction***:

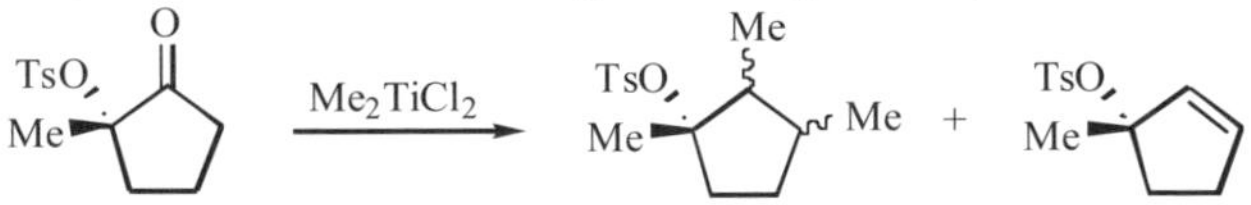

Optically active                                              each product was racemic

G. H. Posner, T. P. Kogan, *Journal of the Chemical Society: Chemical Communications* **1983**, 1481

## Examples:

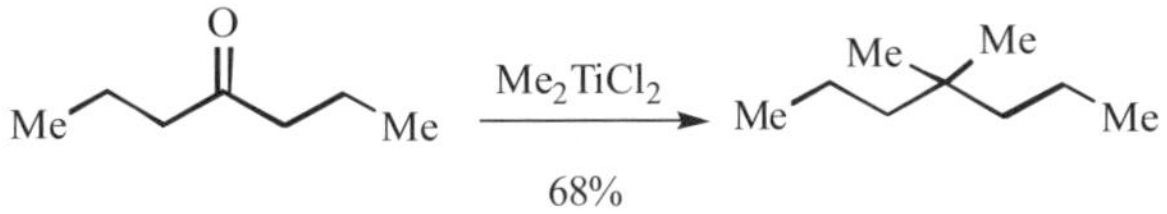

M. T. Reetz, R. Steinbach, B. Wenderoth, *Synthetic Communications* **1981**, <u>11</u>, 261 (AN 1981:406655)

Me₂TiCl₂

35%

cuparene

F. G. Favaloro, C. A. Goudreau, B. P. Mundy, T. Poon, S. V. Slobodzian, B. L. Jensen, *Synthetic Communications* **2001**, <u>31</u>, 1847

Me₂TiCl₂

52%

G. Haefelinger, M. Marb, *New Journal of Chemistry* **1987**, <u>11</u>, 401 (AN 1988:149977)

Me₂TiCl₂

42%
based on
recovered
starting material

B. P. Mundy, D. Wilkening, K. B. Lipkowitz, *Journal of Organic Chemistry* **1985**, <u>50</u>, 5727

Me₂Zn, TiCl₄
CH₂Cl₂, 0°C

60%

herbertene

T. Poon, B. P. Mundy, F. G. Favaloro, C. A. Goudreau, A. Greenberg, R. Sullivan, *Synthesis* **1998**, 832

Me₂TiCl₂

86%

D. J. Kerr, A. C. Willis, B. L. Flynn, *Organic Letters* **2004**, <u>6</u>, 457

# Reformatsky Reaction

## The Reaction:

## Proposed Mechanism:

## Notes:

T. Laue, A. Plagens, *Named Organic Reactions*, John Wiley and Sons, Inc., New York, 1998, pp. 224-226; M. B. Smith, J. March in *March's Advanced Organic Chemistry*, 5th ed., John Wiley and Sons, Inc., New York, 2001, p. 1212; R. L. Shiner, *Organic Reactions* **1**, 1; M. W. Rathke, *Organic Reactions* **22**, 4.

Other metals can be used for the "***Reformatsky Reaction***":

Y. S. Suh, R. D. Rieke, D. Reuben, *Tetrahedron Letters* **2004**, <u>45</u>, 180

## Examples:

K. Sakai, K. Takahashi, T. Nukano, *Tetrahedron* **1992**, <u>48</u>, 8229

J. D. Clark, G. A. Weisenburger, D. K. Anderson, P.-J. Colson, A. D. Edney, D. J. Gallagher, H. P. Kleine, C. M. Knable, M. K. Lantz, C. M. V. Moore, J. B. Murphy, T. E. Rogers, P. G. Ruminski, A. S. Shah, N. Storer, B. E. Wise, *Organic Process Research & Development* **2004**, *8*, 51

S.-i. Fukuzawa, H. Matsuzawa, S.-i. Yoshimitsu, *Journal of Organic Chemistry* **2000**, *65*, 1702

B. B. Shankar, M. P. Kirkup, S. W. McCombie, J. W. Clader, A. K. Ganguly, *Tetrahedron Letters* **1996**, *37*, 4095

| | | |
|---|---|---|
| Zn anode: | 100% | 0% |
| Zn powder: | 48% | 9% |

H. Schick, R. Ludwig, K.-H. Schwarz, K. Kleiner, A. Kunath, *Journal of Organic Chemistry* **1994**, *59*, 3161

# Reimer-Tiemann Reaction

## The Reaction:

$$\text{C}_6\text{H}_5\text{OH} \xrightarrow[\text{KOH}]{\text{CHCl}_3} \text{salicylaldehyde} + \text{H}_2\text{O} + 3 \text{ KCl}$$

## Proposed Mechanism:

## Notes:

H. Wynberg, *Chemical Reviews* **1960**, <u>60</u>, 169

M. B. Smith, J. March in *March's Advanced Organic Chemistry*, 5th ed., John Wiley and Sons, Inc., New York, 2001, p. 716; T. Laue, A. Plagens, *Named Organic Reactions,* John Wiley and Sons, Inc., New York, 1998, pp. 226-227; H. Wynberg, E. W. Meijer, *Organic Reactions* **28**, 1

## Examples:

CHCl₃, NaOH
H₂O
64%

M. E. Jung, T. I. Lazarova, *Journal of Organic Chemistry* **1997**, <u>62</u>, 1553

(yields are not generally high in these reactions)

K. M. Smith, F. W. Bobe, O. M. Minnetian, H. Hope, M. D. Yanuck, *Journal of Organic Chemistry* **1985**, <u>50</u>, 790

# Reissert-Henze Reaction

## The Reaction:

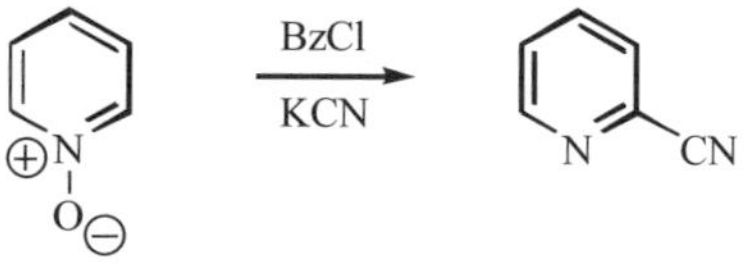

## Proposed Mechanism:

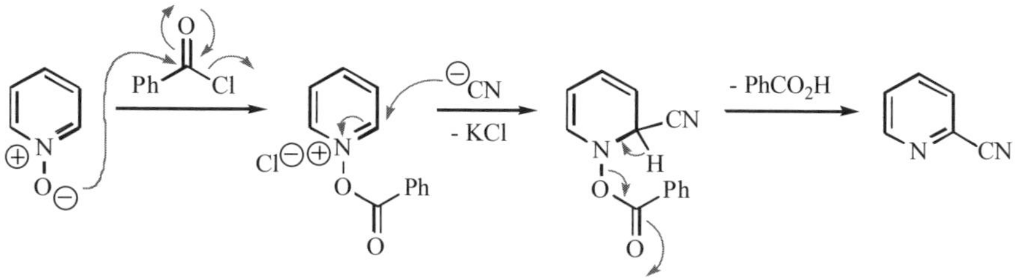

## Notes:

As will be seen in the examples, there are a number of variations on the theme to deliver the cyanide and displace the *N*-oxide oxygen. However, all follow the same general mechanistic reasoning.

## Examples:

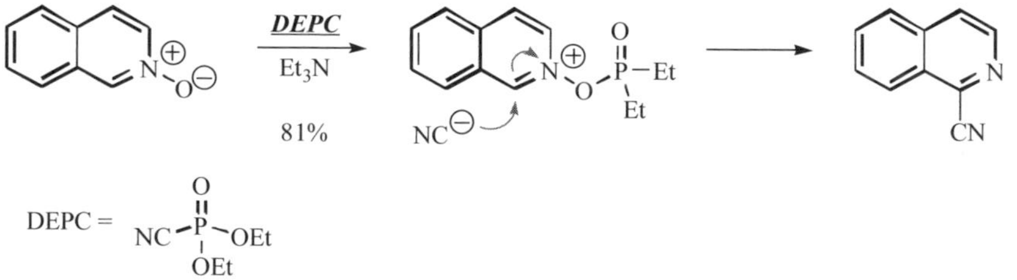

DEPC = NC–P(=O)(OEt)–OEt

Reported in: H. H. Patel, *Encyclopedia of Reagents for Organic Synthesis*, John Wiley and Sons, Inc., L. A. Paquette, Ed., New York, 1995, **3**, 1851

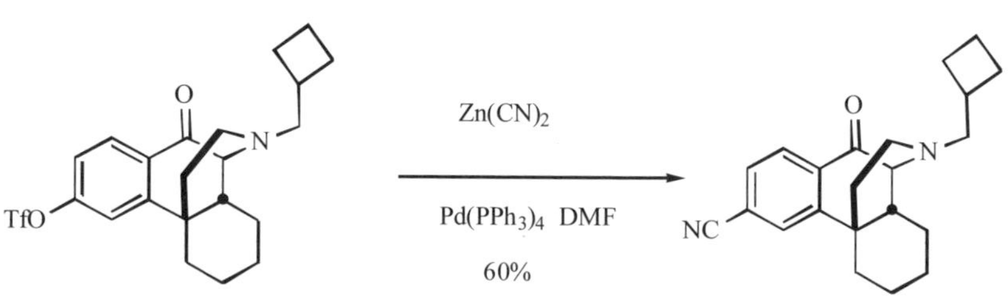

A. Zhang, J. L. Neumeyer, *Organic Letters* **2003**, <u>5</u>, 201

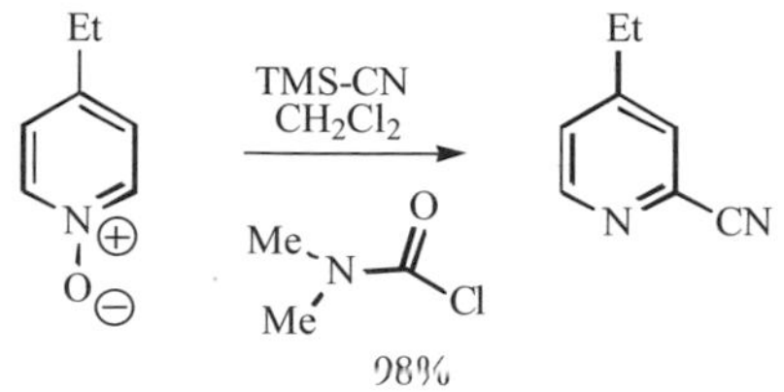

98%

R. T. Shuman, P. L. Ornstein, J. W. Paschal, P. D. Gesellchen, *Journal of Organic Chemistry* **1990**, 55, 738

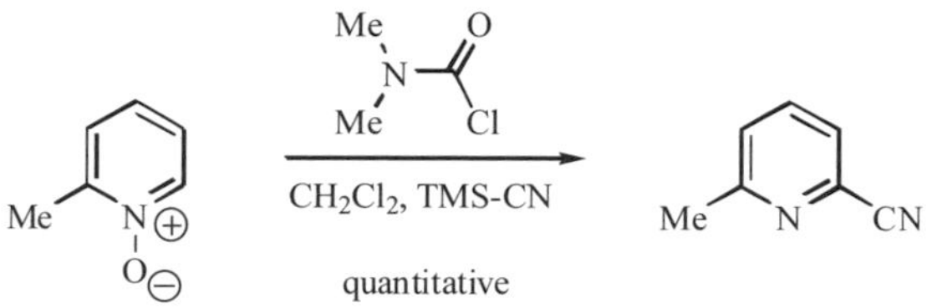

quantitative

W. K. Fife, *Journal of Organic Chemistry* **1983**, 48, 1375

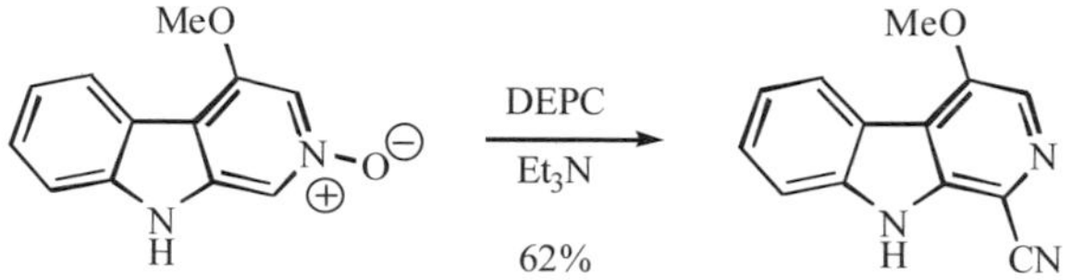

62%

H. Suzuki, C. Iwata, K. Sakurai, K. Tokumoto, H. Takahashi, M. Hanada, Y. Yokoyama, Y. Murakami, *Tetrahedron* **1997**, 53, 1593

# Reissert Indole Synthesis

## The Reaction:

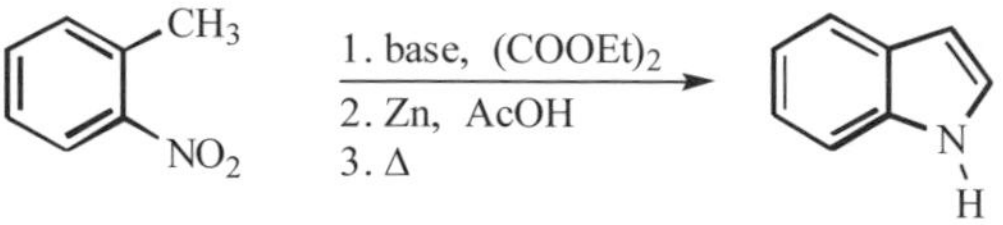

CH3
NO2
1. base, (COOEt)2
2. Zn, AcOH
3. Δ
N
H

## Proposed Mechanism:

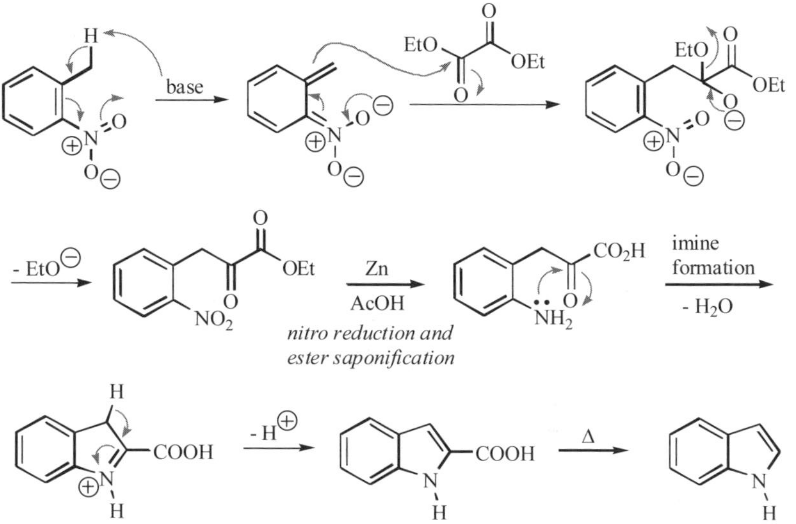

base
EtO
OEt
- EtO⊖
Zn
AcOH
nitro reduction and
ester saponification
CO2H
imine
formation
- H2O
COOH
- H⊕
Δ

## Examples:

HN
Ph
MeO2C
NO2
CO2Me
MeO
OMe
HCl
H2, 5% Pd/C
66%
NH

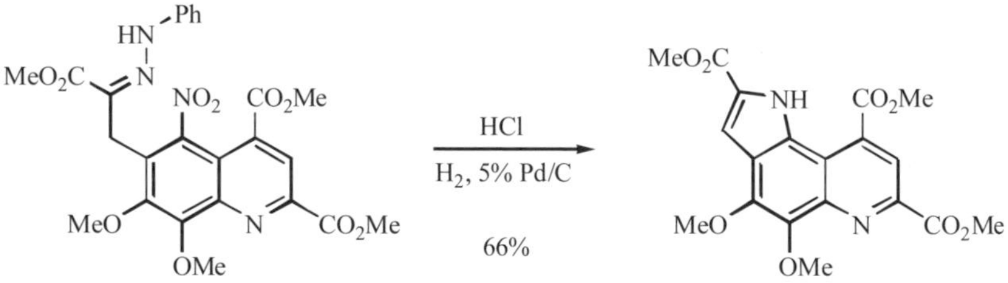

J. A. Gainor, S. M. Weinreb, *Journal of Organic Chemistry* **1982**, 47, 2833

G. Leadbetter, D. L. Fost, N. N. Ekwuribe, W. A. Remers, *Journal of Organic Chemistry* **1974**, <u>39</u>, 3580; T. Hirata, Y. Yamada, M. Matsui, *Tetrahedron Letters* **1969**, <u>10</u>, 19

# Ring Closing Metathesis (RCM)

## The Reaction:

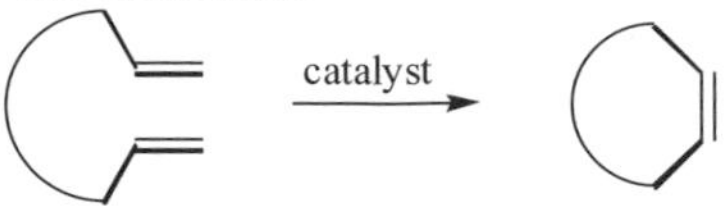

## Proposed Mechanism:

1. Generation of catalyst (one cycle).

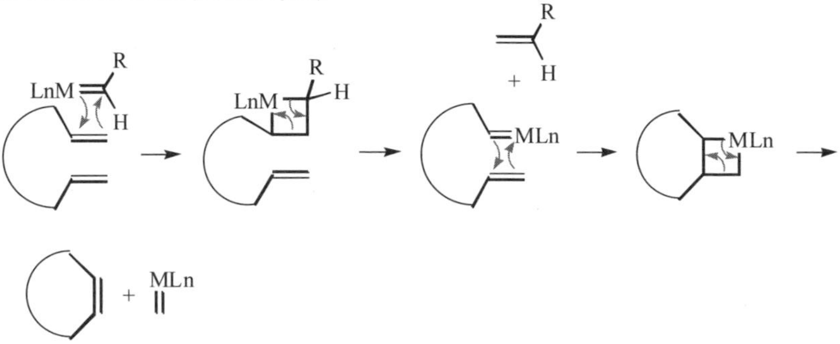

2. Catalytic cycle.

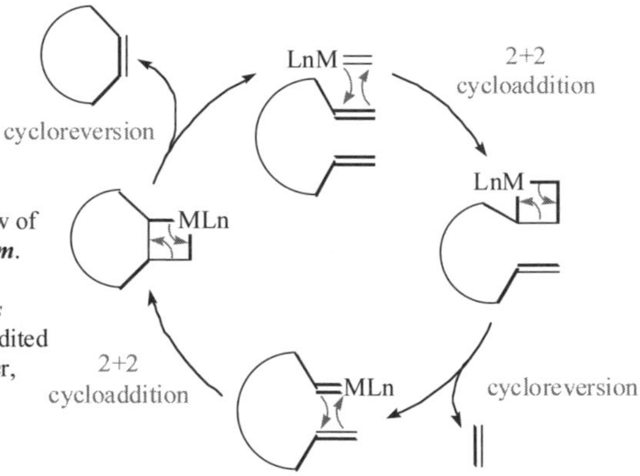

cycloreversion

2+2
cycloaddition

This is a simplified view of
the *Chauvin Mechanism*.

See: *Alkene Metathesis
in Organic Synthesis*, edited
by A. Fürstner, Springer,
Berlin, 1998

2+2
cycloaddition

cycloreversion

## Notes:

LnM═  = *Metal carbene complex*

See *Schrock* and *Grubbs* Catalysts for additional examples and information.

Variations of:

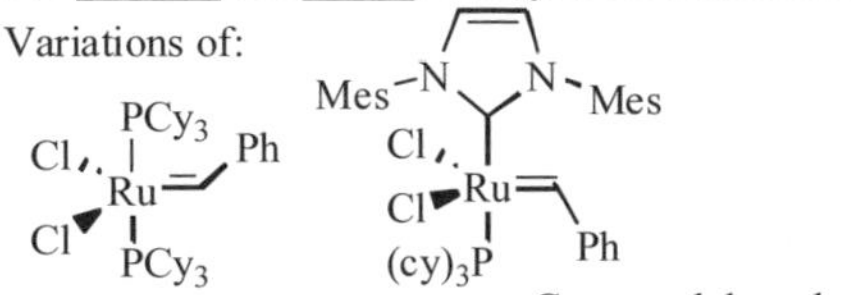

Cy = cyclohexyl
*Grubbs' catalysts*    Mes = mesitylene

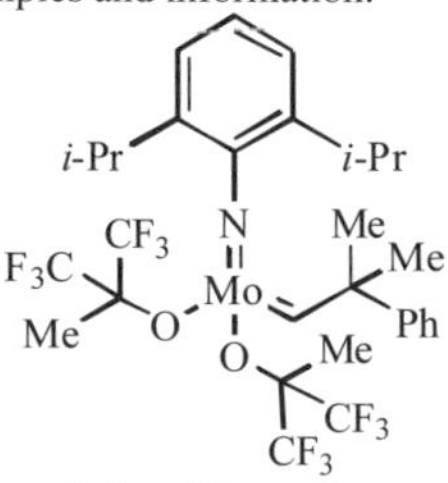

*Schrock's catalyst*

## Examples:

D. C. Harrowven, M. C. Lucas, P. D. Howes, *Tetrahedron Letters* **2000**, *41*, 8985

W. R. Zuercher, M. Scholl, R. M. Grubbs, *Journal of Organic Chemistry* **1998**, *63*, 4291

F.-D. Boyer, I. Hanna, L. Ricard, *Organic Letters* **2004**, *6*, 1817

A. K. Ghosh, C. Liu, *Journal of the American Chemical Society* **2003**, *125*, 2374

J. R. Rodrigues, L. Castedo, J. L. Mascarenas, *Organic Letters* **2000**, *2*, 3209

# Ritter Reaction

## The Reaction:

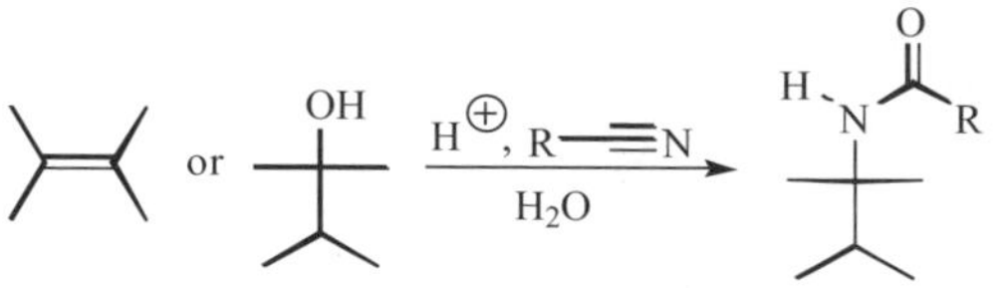

## Proposed Mechanism:

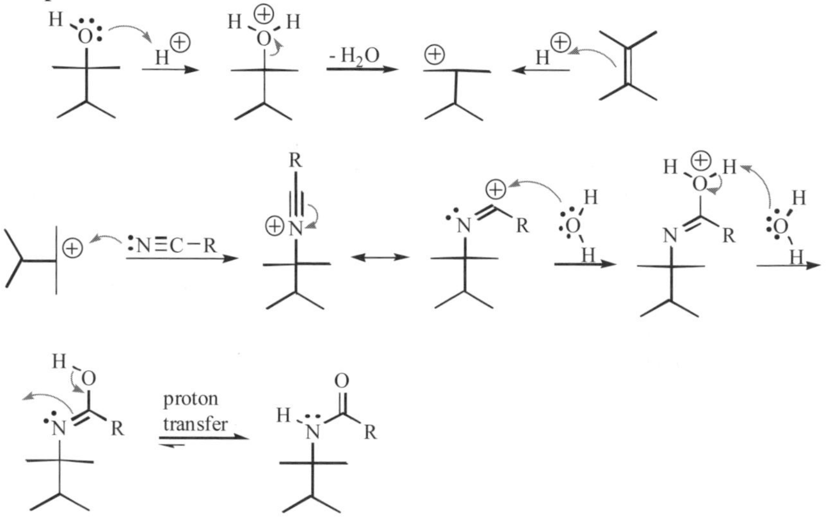

## Notes:

V. K. Ahluwalia, R. K. Parashar, *Organic Reaction Mechanisms*, Alpha Science International Ltd., Pangbourne, U.K., 2002, pp. 371-372; M. B. Smith, J. March in *March's Advanced Organic Chemistry*, 5th ed., John Wiley and Sons, Inc., New York, 2001, pp 1244-1245; L. I. Krimen, D. J. Cota, *Organic Reactions* **17**, 3

Hydrolysis of the amides provides a unique method for preparation of tertiary amines.

## Examples:

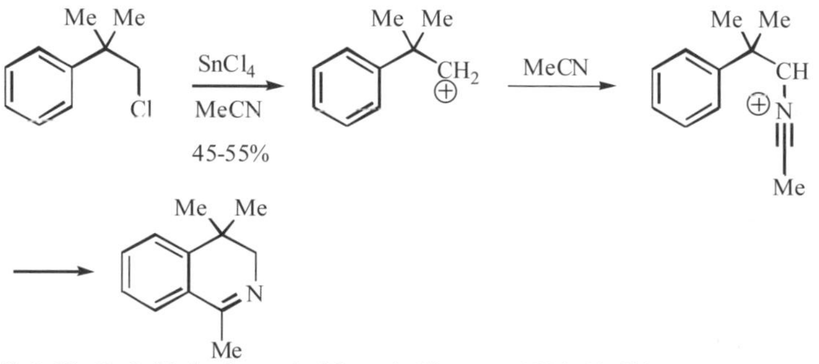

T.-L. Ho, R.-J. Chein, *Journal of Organic Chemistry* **2004**, <u>69</u>, 591

This work comments on the slow induction period of the $H_2SO_4$ reaction with acrylonitrile and alcohol, and on possible safety issues.

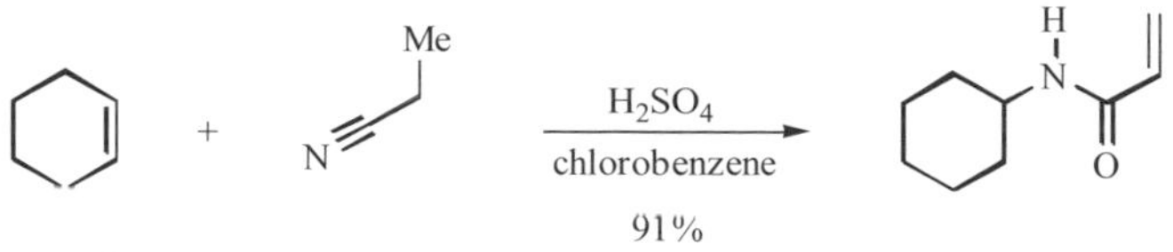

91%

S. J. Chang, *Organic Process Research and Development* **1999**, *3*, 232

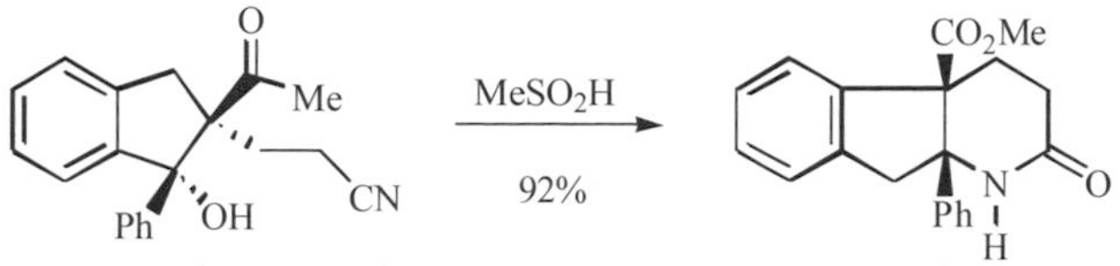

92%

K. Van Emelen, T. DeWit, G. J. Hoornaert, F. Compernolle, *Organic Letters* **2000**, *2*, 3083

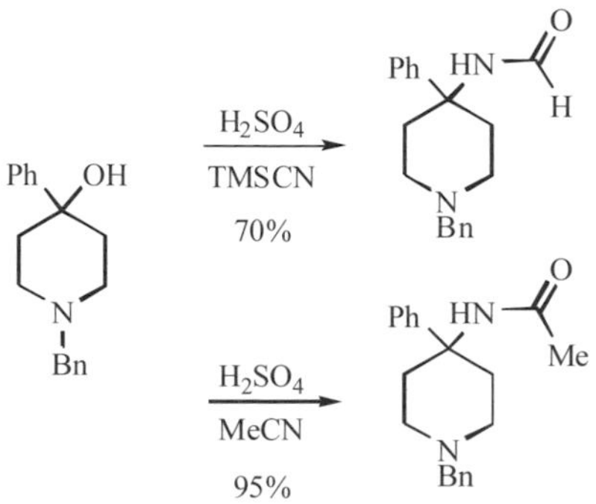

70%

95%

H. G. Chen, O. P. Goel, S. Kesten, J. Knobelsdorg, *Tetrahedron Letters* **1996**, *37*, 8129

A *modified Ritter Reaction*:

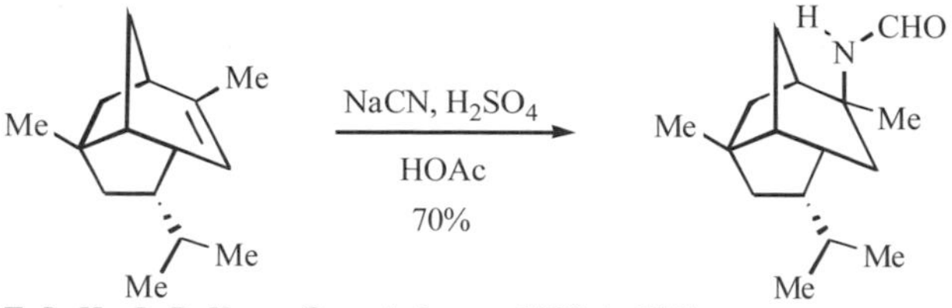

T.-L. Ho, L. R. Kung, *Organic Letters* **1999**, *1*, 1051

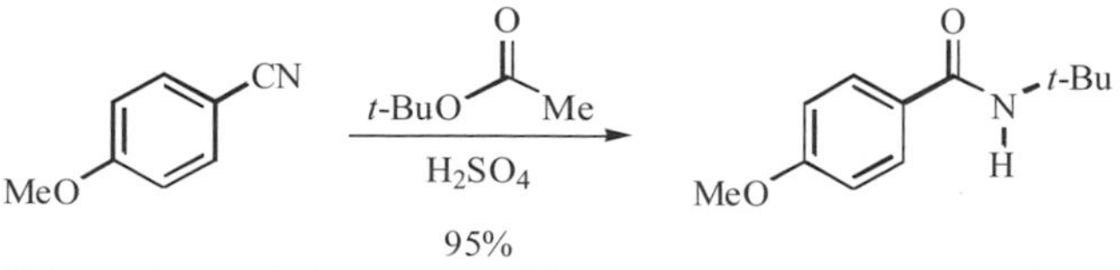

95%

K. L Reddy, *Tetrahedron Letters* **2003**, *44*, 1453

# Robinson Annulation

## The Reaction:

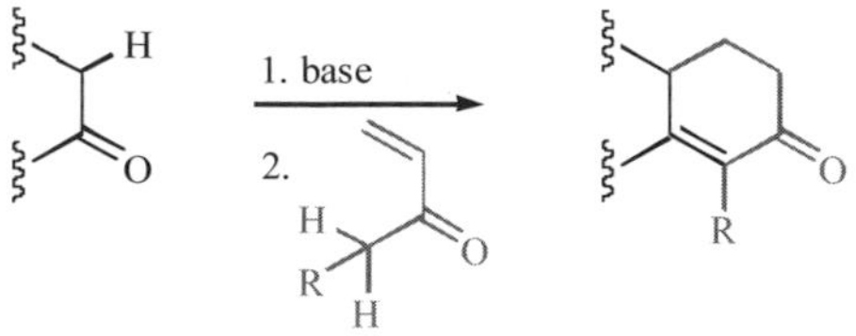

## Proposed Mechanism:

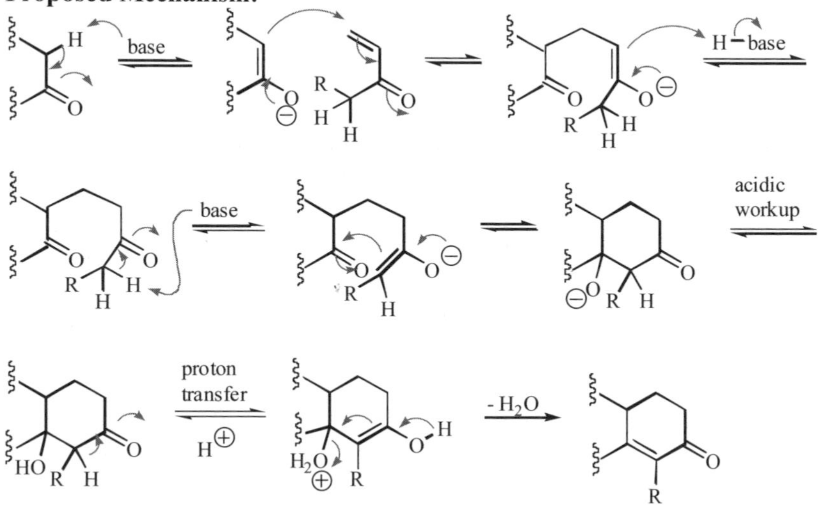

## Notes:

This is a *Michael reaction* followed by an *Aldol reaction*.

M. B. Smith, J. March in *March's Advanced Organic Chemistry*, 5th ed., John Wiley and Sons, Inc., New York, 2001, pp. 1222-1224; T. Laue, A. Plagens, *Named Organic Reactions*, John Wiley and Sons, Inc., New York, 1998, pp. 228-232

See: D. Rajagopal, R. Narayanan, S. Swaminathan, *Tetrahedron Letters* **2001**, 42, 4887 for a one-pot asymmetric annulation procedure.

The related *Stork-Jung Vinysilane annulation:*

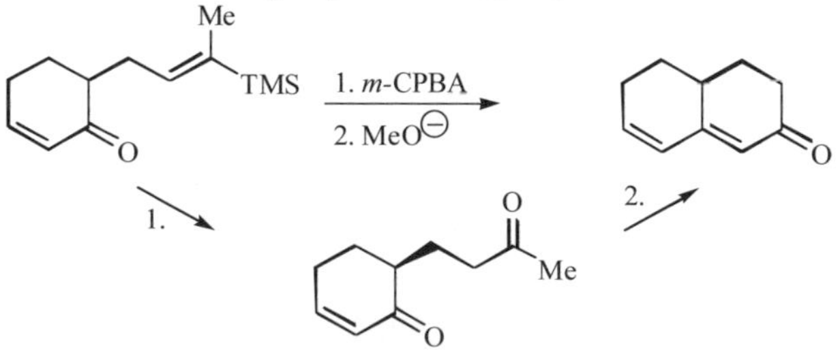

36%

B. B. Snider, B. Shi, *Tetrahedron Letters* **2001**, 42, 9123

# Examples:

A. A. Verstegen-Haaksma, H. J. Swarts, J. M. B. Jansen, A. deGroot, *Tetrahedron* **1994**, 50, 10073

R. K. Boeckman, Jr., *Tetrahedron* **1983**, 39, 925

I. Jabin, G. Revial, K. Melloul, M. Pfau, *Tetrahedron: Asymmetry* **1997**, 8, 1101

V. Zhabinskii, A. J. Minnaard, J. B. P. A. Wijnberg, A. deGroot, *Journal of Organic Chemistry* **1996**, 61, 4022

Y. Baba, T. Sakamoto, S. Soejima, K. Kanematsu, *Tetrahedron* **1994**, 50, 5645

# Robinson-Schopf Reaction

## The Reaction:

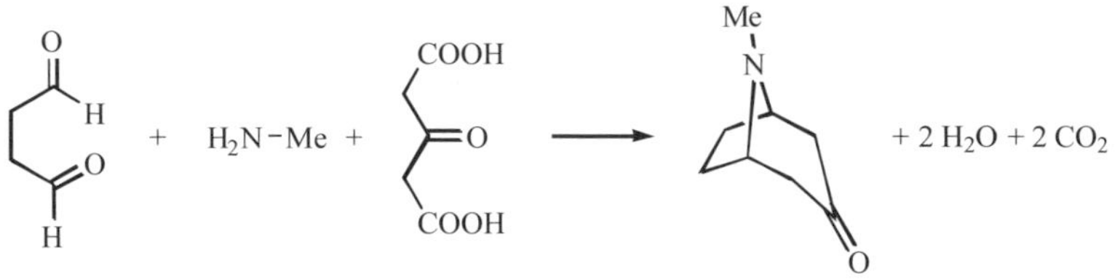

## Proposed Mechanism:

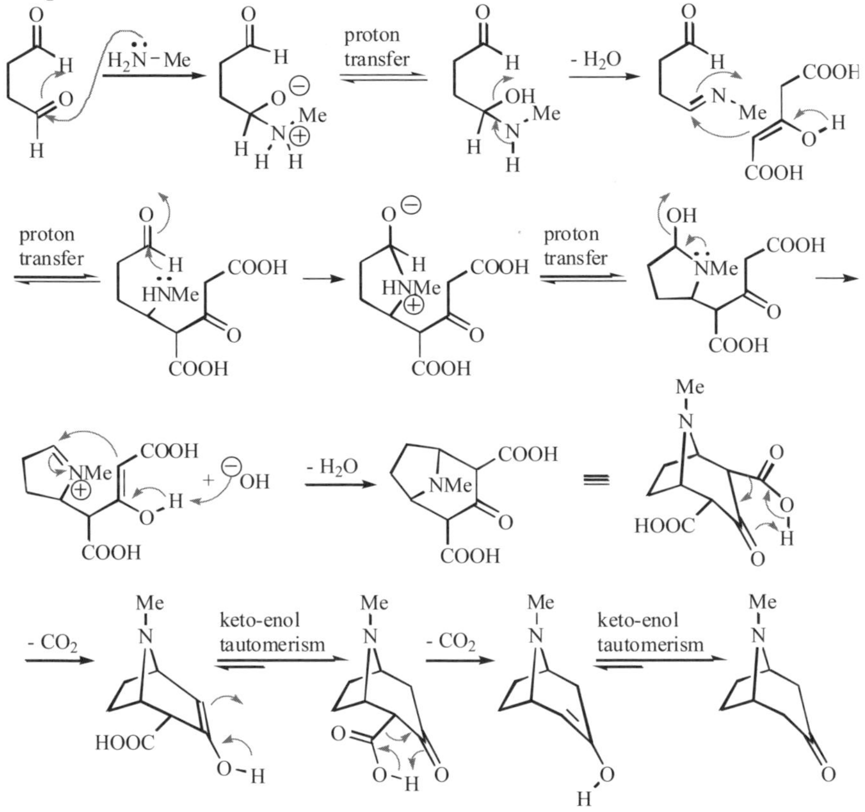

## Notes:

This reaction is a "biomimetic" approach to forming alkaloids.

## Examples:

MeOOC

Me

Me ―CHO ―CHO

HOOC〜ᴑ〜COOH

PrNH₂, THF

74%

MeOOC

Me

Me NPr ═O

T. Jarevang, H. Anke, T. Anke, G. Erkel, O. Sterner, *Acta Chemica Scandinavica* **1998**, 52, 1350 (AN 1998:770310

O ―CHO ―CHO

+ MeNH₂ +

HO₂C

O═

HO₂C

no yield given

Me N

O

O

J. Bermudez, J. A. Gregory, F. D. King, S. Starr, R. J. Summersell, *Bioorganic and Medicinal Chemistry Letters* **1992**, 2, 519

CHO CHO

HOOC〜ᴑ〜COOH O

Me-NH2

34%

Me N

O

O. L. Chapman, T. H. Koch, *Journal of Organic Chemistry* **1966**, 31, 1043

CHO CHO

HOOC〜ᴑ〜COOH O

Me-NH2

60 - 70%

Me N

O

Detailed conformational study on the product

L. A. Paquette, J. W. Heimaster, *Journal of the American Chemical Society* **1966**, 88, 763

# Rosenmund Reduction

## The Reaction:

$$\underset{R}{\overset{O}{\parallel}}\underset{Cl}{\quad} \xrightarrow[\text{Pd·BaSO}_4]{\text{H}_2} \underset{R}{\overset{O}{\parallel}}\underset{H}{\quad} + \text{HCl}$$

## Proposed Mechanism:

$$\underset{R}{\overset{O}{\parallel}}\underset{Cl}{\quad} \xrightarrow[\text{Pd·BaSO}_4]{\text{H}_2} \underset{R}{\overset{O}{\parallel}}\underset{Pd-Cl}{\quad} \longrightarrow \underset{R}{\overset{O}{\parallel}}\underset{H}{\quad}$$

J. Tsuji, K. Ohno, *Journal of the American Chemical Society*, **1968**, 90, 94

## Notes:

M. B. Smith, J. March in *March's Advanced Organic Chemistry*, 5$^{th}$ ed., John Wiley and Sons, Inc., New York, 2001, p. 532; T. Laue, A. Plagens, *Named Organic Reactions*, John Wiley and Sons, Inc., New York, 1998, pp. 232-233; E. Mosettig, R. Mozingo, *Organic Reactions* 4, 7

See: S. Siegel, *Encyclopedia of Reagents for Organic Synthesis,* John Wiley and Sons, Inc., L. A. Paquette, Ed., New York, 1995, 6, 3861

Pd-BaSO$_4$ binds less strongly to Pd-C, the product can leave the catalyst surface more readily and prevent over-reduction.

$$\underset{R}{\overset{O}{\parallel}}\underset{Cl}{\quad} \; > \; \underset{Ar}{\overset{O}{\parallel}}\underset{Cl}{\quad}$$

## Examples:

$$\underset{Cl}{\overset{O}{\parallel}}\underset{CCl_3}{\quad} \xrightarrow[\text{quinoline, petroleum ether}]{\text{H}_2, \text{Pd / C / BaSO}_4} \underset{H}{\overset{O}{\parallel}}\underset{CCl_3}{\quad}$$

50%

J. W. Sellers, W. E. Bissinger, *Journal of the American Chemical Society*, **1954**, 76, 4486

$$H_2C \diagup \diagdown (CH_2)_8 \overset{O}{\parallel} Cl \xrightarrow[\text{Pd / PPTA*}]{\text{H}_2, \text{Pd / C}} H_2C \diagup \diagdown (CH_2)_8 \overset{O}{\parallel} H$$

48%

\* PdCl$_2$ deposited on poly(*p*-phenylene terephthalamide) (PPTA)

V. G. Yadav, S. B. Chandalia, *Organic Process Research & Development* **1997**, 1, 226

*Quinoline-S* prepared according to the recipe of E. B. Hershberg, J. Cason, Method 2, *Organic Syntheses* CV 3, 626

A. I. Rachlin, H. Gurien, D. P. Wagner, *Organic Syntheses* **1971**, 51 8

E. B. Hershberg and J. Cason, Method 2, *Organic Syntheses* CV 3, 626

**Modification:**

M. Falorni, G. Giacomelli, A. Porcheddu, M. Taddei, *Journal of Organic Chemistry* **1999**, 64, 8962

# Rubottom Oxidation

## The Reaction:

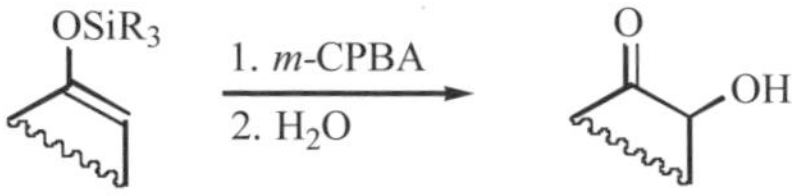

## Proposed Mechanism:

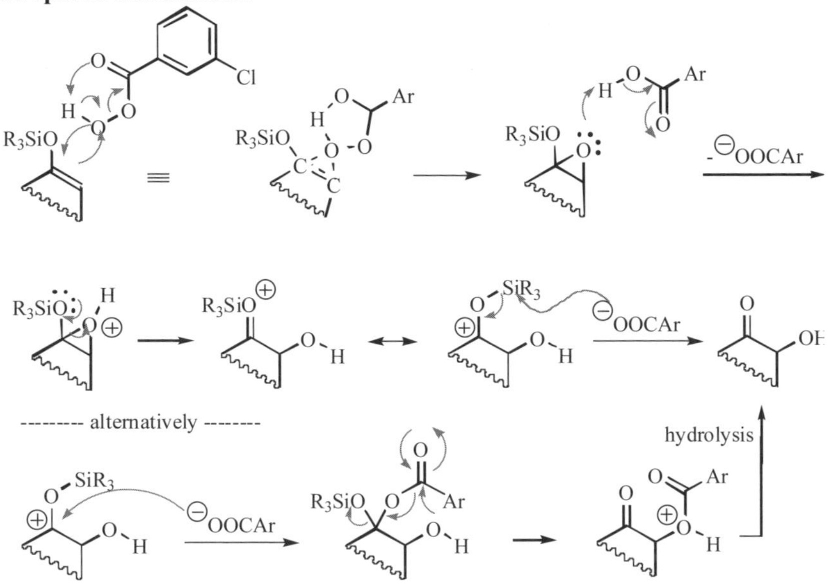

-------- alternatively -------

## Notes:

Rubottom noted that, in the absence of a hydrolytic workup, 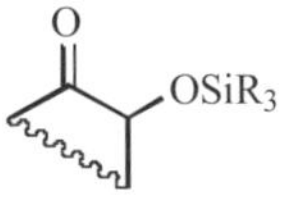 is obtained.

This is suggestive of:

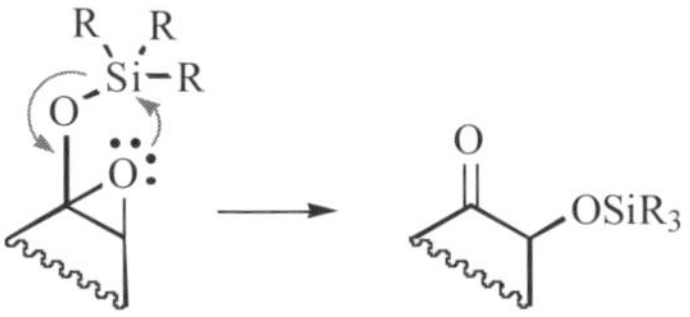

## Examples:

1. *m*-CPBA
2. hydrolysis

64%

G. M. Rubottom, M. A. Vazquez, D. R. Pelegrins, *Tetrahedron Letters* **1974**, 15, 4319

1. DDO
2. hydrolysis

76%

$$-\overset{Me}{\underset{Me}{Si}}-Et \ = TSE$$

*DDO*

J. G. Allen, S. J. Danishefsky, *Journal of the American Chemical Society* **2001**, 123, 351

1. *m*-CPBA
2. hydrolysis

55%

D. F. Taber, T. E. Christos, A. L. Rheingold, I. A. Guzei, *Journal of the American Chemical Society* **1999**, 121, 5589

*m*-CPBA

*Luche Reduction*

$NaBH_4$, $CeCl_3$

57%

Y. Xu, C. R. Johnson, *Tetrahedron Letters* **1997**, 38, 1117

1. DMDO acetone

2. DMS

80%

M. Mandel, S. Danishefsky, *Tetrahedron Letters* **2004**, 45, 3831

# Rupe Rearrangement

## The Reaction:

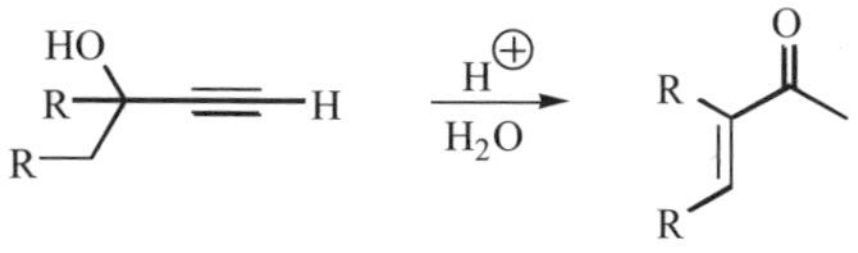

## Proposed Mechanism:

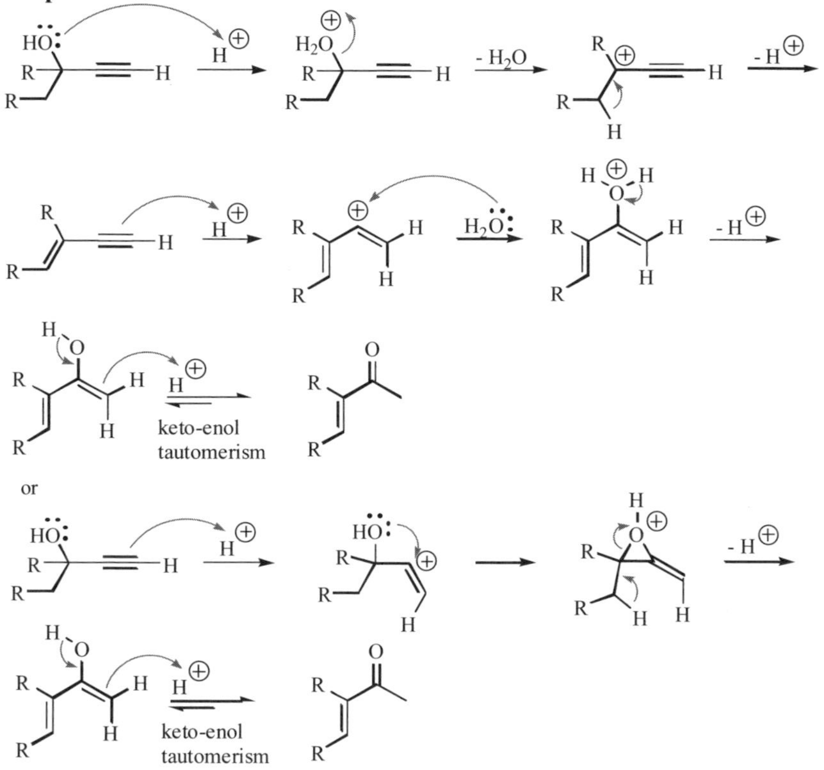

keto-enol
tautomerism

or

keto-enol
tautomerism

## Notes:

The **_Meyer-Schuster Rearrangement_** is similar to the **_Rupe Rearrangement_**.

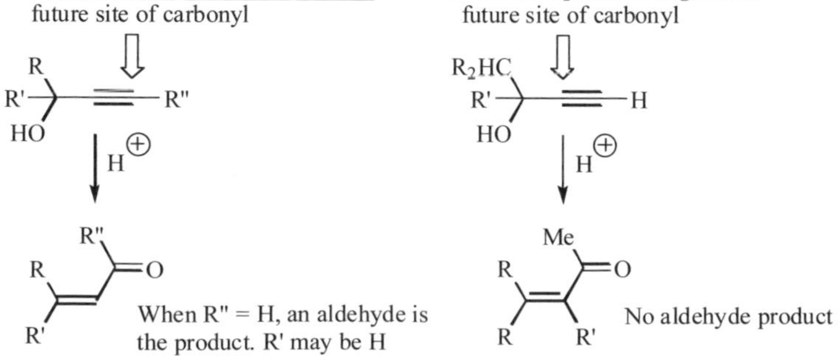

future site of carbonyl                    future site of carbonyl

When R" = H, an aldehyde is
the product. R' may be H

No aldehyde product

**Pseudo-Rupe Rearrangement**:

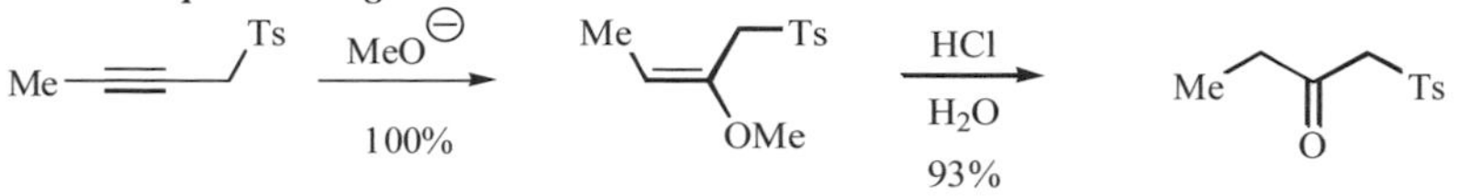

Me ≡≡ /Ts  —MeO⊖→  Me /Ts  —HCl, H₂O→  Me /Ts
                                         O
         100%           OMe
                                    93%

**Raphael Rearrangement**:

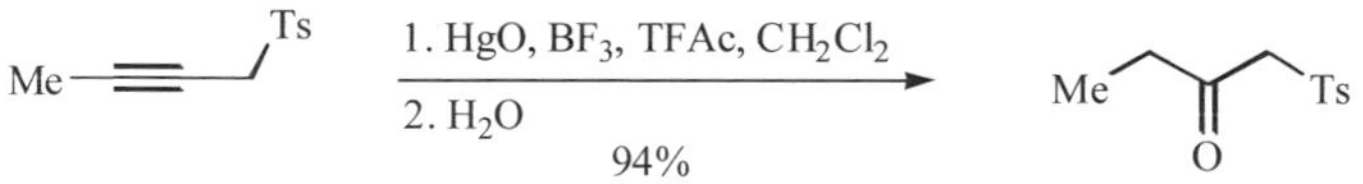

Me ≡≡ /Ts  —1. HgO, BF₃, TFAc, CH₂Cl₂ 2. H₂O→  Me /Ts
                                                    O
                    94%

V. Barre, F. Massias, D. Uguen, *Tetrahedron Letters* **1989**, 30, 7389

## Examples:

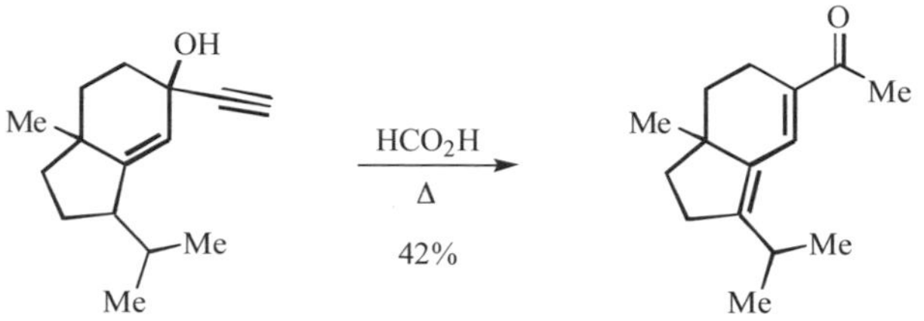

$$\xrightarrow[\Delta]{HCO_2H}$$

42%

K. Takeda, D. Nakane, M. Takeda, *Organic Letters* **2000**, 2, 1903

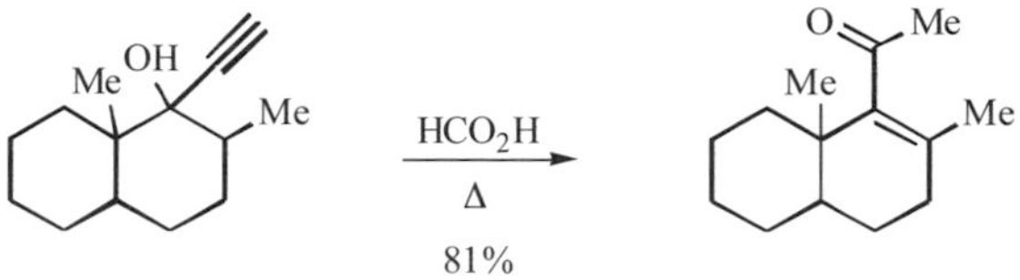

$$\xrightarrow[\Delta]{HCO_2H}$$

81%

W. S. Johnson, S. L.Gray, J. K. Crandall, D. M. Bailey, *Journal of the American Chemical Society* **1964**, 86, 1966

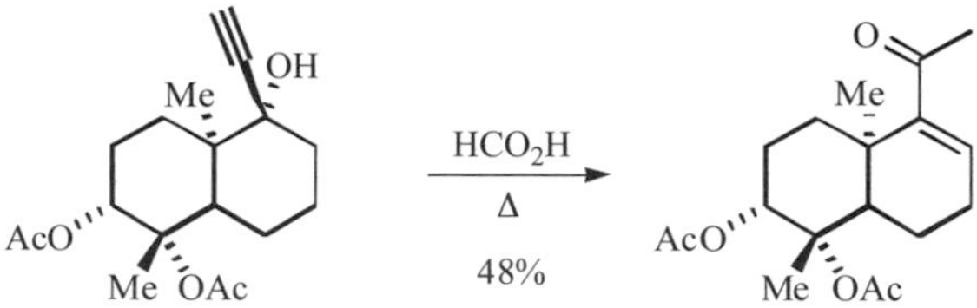

$$\xrightarrow[\Delta]{HCO_2H}$$

48%

S. W. Pelletier, S. Prabhakar, *Journal of the American Chemical Society* **1968**, 90, 5318 and earlier references.

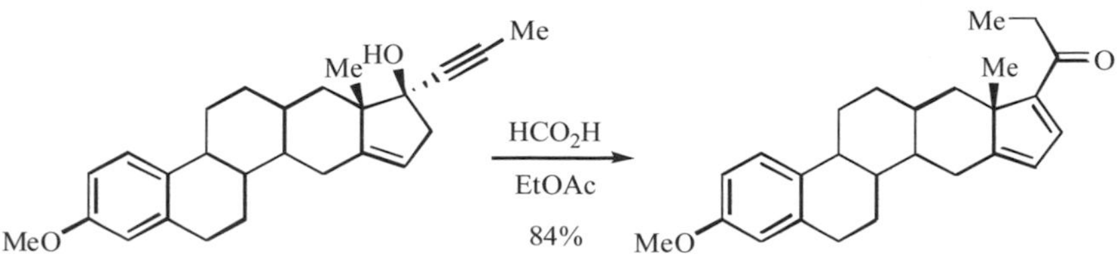

$$\xrightarrow[EtOAc]{HCO_2H}$$

84%

H. Weinman, M. Harre, H. Neh, K. Nickish, C. Skotsch, U. Tilstam, *Organic Process Research and Development* **2002**, 6, 216

# Saegusa Oxidation

## The Reaction:

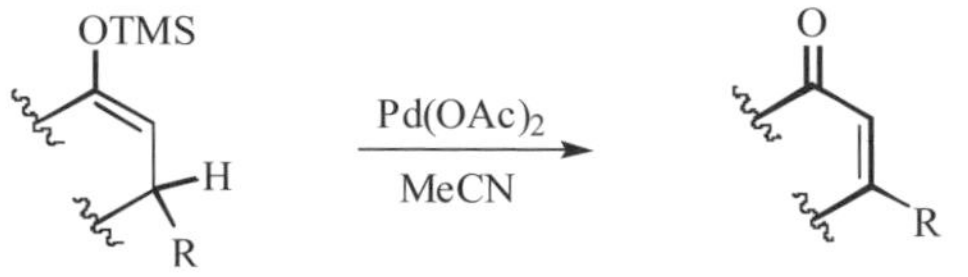

## Proposed Mechanism:

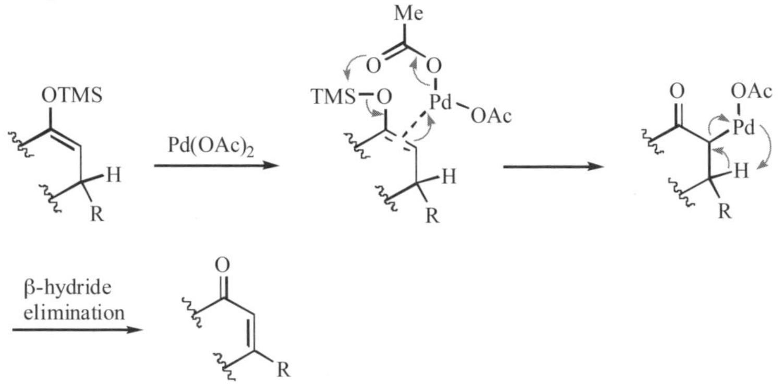

## Notes:

The original work shows that benzoquinone can be added to reoxidize the Pd reagent; however, the reaction proceeds well without it.

When applicable, there is selectivity for (*E*)-products:

only isomer formed

A Pd-containing intermediate has been isolated and fully characterized:
S. Porth, J. W. Bats, D. Trauner, G. Giester, J. Mulzer, *Angewandte Chemie, International Edition in English* **1999**, <u>38</u>, 2015

## Examples:

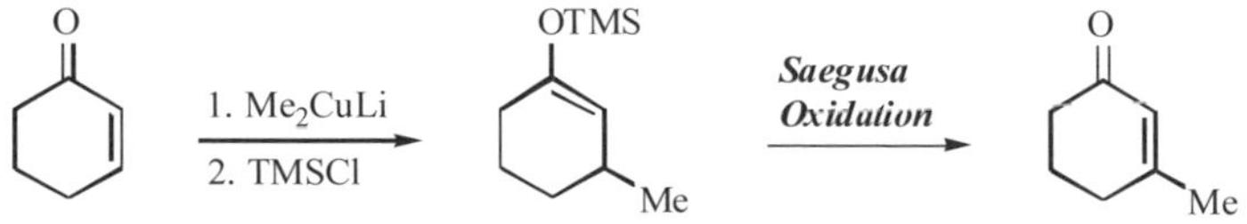

Y. Ito, T. Hirao, T. Saegusa, *Journal of Organic Chemistry* **1978**, <u>4</u>, 1011

A. M. P. Koskinen, H. Rapoport, *Journal of Medicinal Chemistry* **1985**, <u>28</u>, 1301

D. R. Williams, R. A. Turske, *Organic Letters* **2000**, <u>2</u>, 3217

S. Chi, C. H. Heathcock, *Organic Letters* **1999**, <u>1</u>, 3

# Sakurai Reaction

## The Reaction:

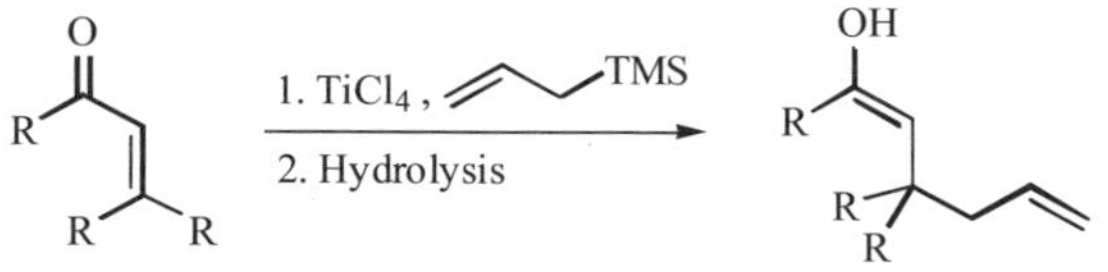

## Proposed Mechanism:

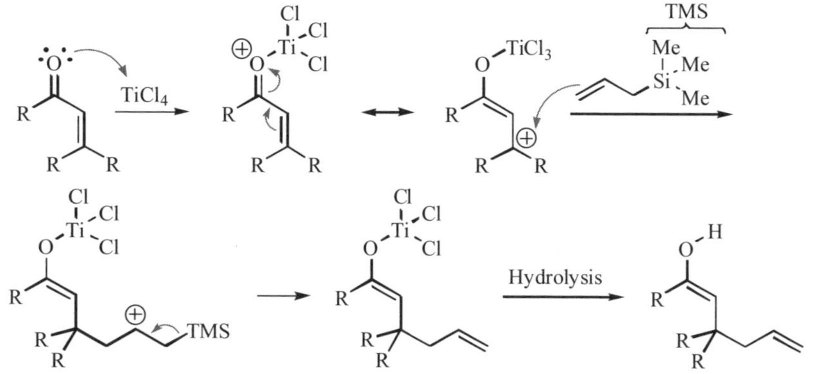

TMS/Silicon can
stabilize β carbocations.

## Notes:

This reaction is similar to the **Mukaiyama Reaction**, which employs silyl enol ethers:

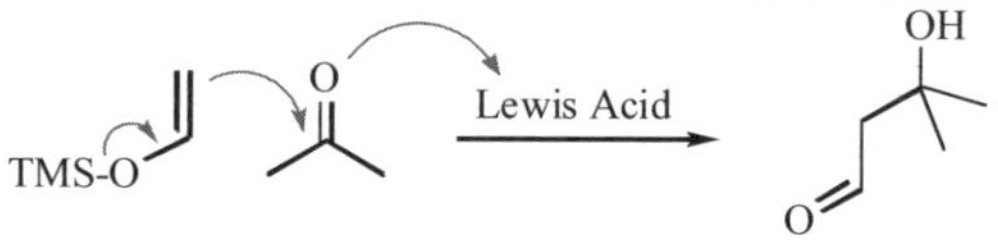

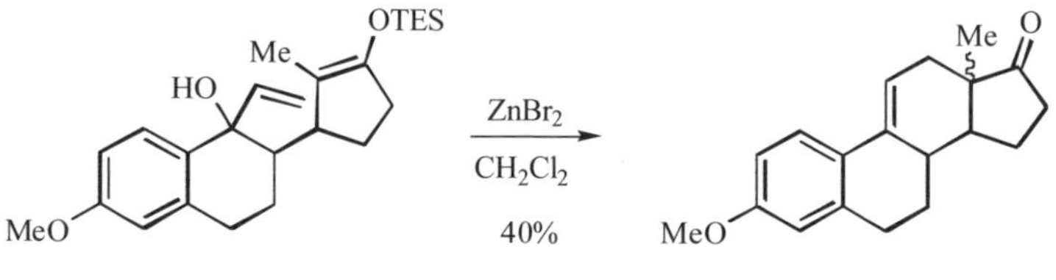

S. Dratch, T. Charnikhova, F. C. E. Saraber, B. J. M. Jansen, A. DeGroot *Tetrahedron* **2003**, 59, 4287

A useful isoprenylation reagent:

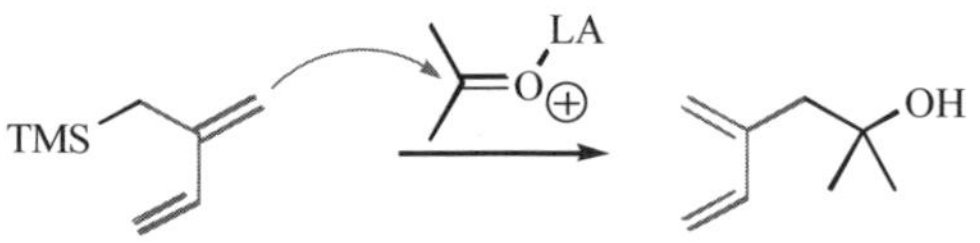

[70901-64-3]
See H. Sakurai, *Encyclopedia of Reagents for Organic Synthesis*, Ed. L. A. Paquette, John Wiley and Sons, Inc., New York, 1995, **7**, 5277

**Examples:**

G. Majetich, J. S. Song, C. Ringold, G. A. Memeth, *Tetrahedron Letters* **1990**, <u>31</u>, 2239

D. Schinzer, S. Solyon, M. Becker, *Tetrahedron Letters* **1985**, <u>26</u>, 1831

G. Majetich, K. Hull, J. Defauw, R. Desmond, *Tetrahedron Letters* **1985**, <u>26</u>, 2747

A. Hosome, H. Sakurai, *Journal of the American Chemical Society* **1977**, <u>99</u>, 1673

# Sandmeyer Reaction

## The Reaction:

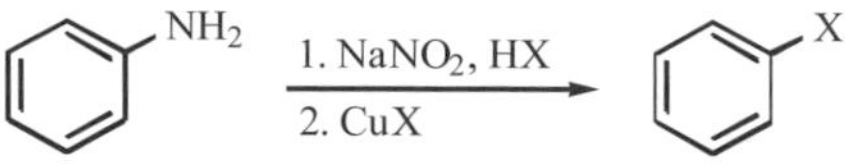

$$\text{ArNH}_2 \xrightarrow[\text{2. CuX}]{\text{1. NaNO}_2, \text{HX}} \text{ArX}$$

Works best for X = Cl, Br, and CN, however F, I, R-S, OH are also possible.

## Proposed Mechanism:

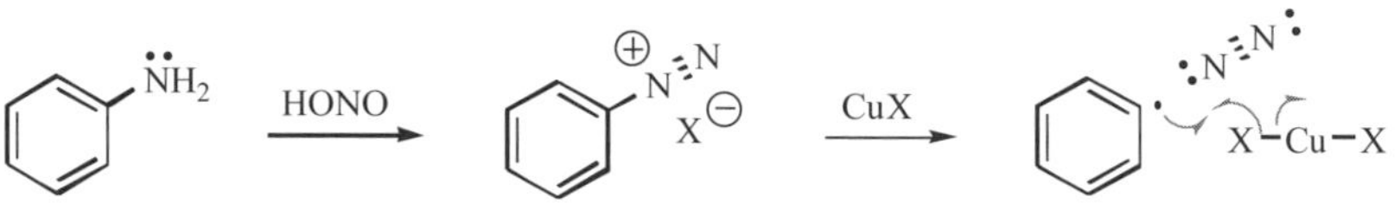

Cu donates an electron to liberate $N_2$ and is oxidized to Cu(II).

Cu is then reduced by the aryl radical and Cu(I) is regenerated.

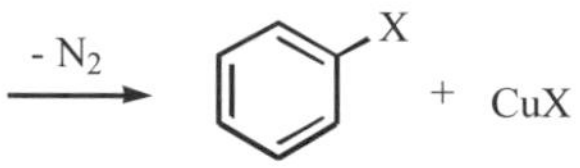

$$\xrightarrow{- N_2} \text{ArX} + \text{CuX}$$

## Notes:

For the iodination reaction, copper is not required. The iodide ion readily undergoes the necessary oxidation-reduction chemistry to push the reaction to completion:

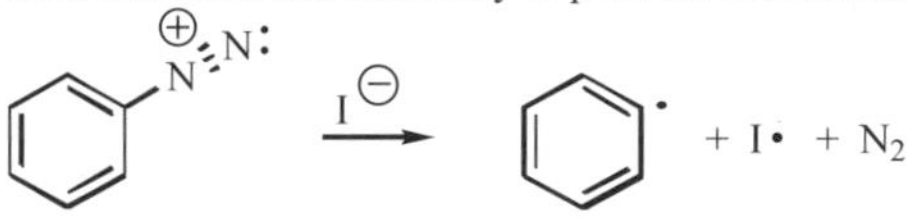

$$\xrightarrow{I^\ominus} \text{Ar}\cdot + I\cdot + N_2$$

The iodine radical combines to form $I_2$, which undergoes further chemistry with I•. The result of complex reactions with other iodine intermediates provides the net capture of an iodine radical with the aryl radical to give product.

### *Bart-Scheller Arsonylation Reaction*

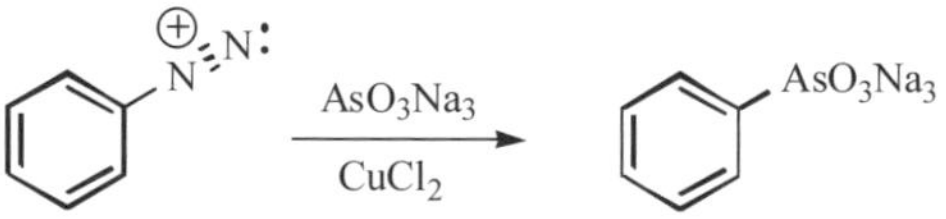

$$\xrightarrow[\text{CuCl}_2]{\text{AsO}_3\text{Na}_3} \text{Ar}-\text{AsO}_3\text{Na}_3$$

### *Körner-Contardi Reaction*

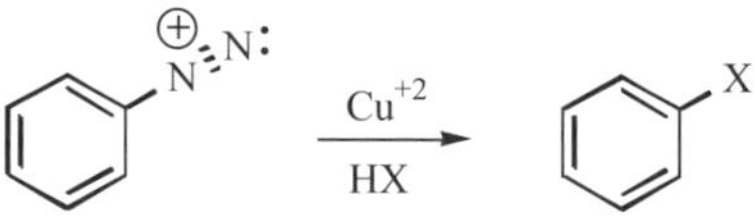

$$\xrightarrow[\text{HX}]{\text{Cu}^{+2}} \text{ArX}$$

X = Cl, Br, CN

## Examples:

C. W. Lai, C. K. Lam, H. K. Lee, T. C. W. Mak, H. N. C. Wong, *Organic Letters* **2003**, 5, 823

N. Zou, J.-F.Liu, B. Jiang, *Journal of Combinatorial Chemistry* **2003**, 5, 754

H. T. Clarke, R. R. Reed, *Organic Synthesis* **1941**, 1, 514

B. Mallesham, B. M. Rajesh, P. R. Reddy, D. Srinivas, S. Trehan, *Organic Letters* **2003**, 5, 7963

P. G. Tsoungas, M. Searcey, *Tetrahedron Letters* **2001**, 42, 6589

# Schenck Ene Reaction

## The Reaction:

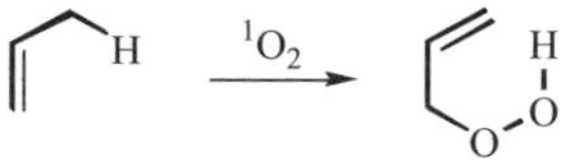

$$\text{(structure)} \xrightarrow{^1O_2} \text{(structure)}$$

## Proposed Mechanism:

$$\text{(structure)} \longrightarrow \text{(structure)}$$

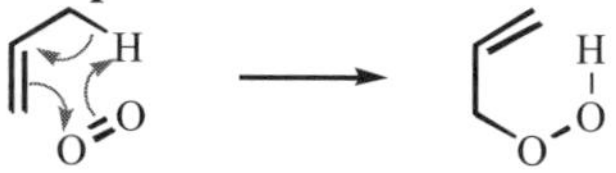

## Notes:

See H. H. Wassernan, R. W. DeSimone, *Encyclopedia of Reagents for Organic Synthesis*, Edited by L.A. Paquette, John Wiley and Sons, Inc., New York, 1995, **6**, 4478

Ways to Generate Singlet Oxygen:

1. Dye-sensitized photoexcitation (eg with TTP, below):

Photosensitizer + hv $\longrightarrow$ $^1$Photosensitizer$^*$

$^1$Photosensitizer$^*$ $\longrightarrow$ $^3$Photosensitizer$^*$

$^3$Photosensitizer$^*$ + $^3O_2$ $\longrightarrow$ Photosensitizer + $^1O_2$

2.

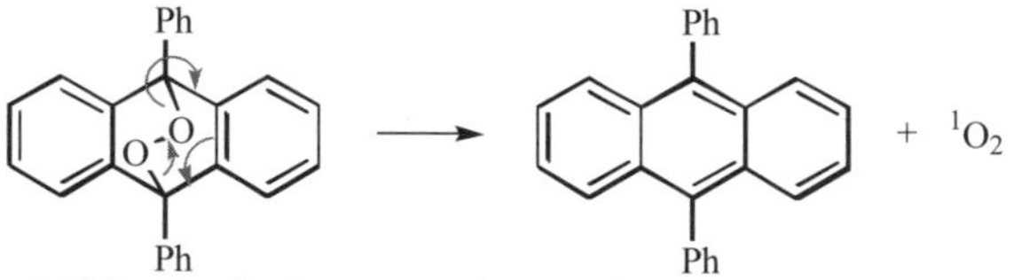

$$\text{(structure)} \longrightarrow \text{(structure)} + {}^1O_2$$

*9.10-Diphenylanthracene Endoperoxide*

3. $H_2O_2 + {}^\ominus OCl \qquad {}^1O_2 + H_2O + Cl^\ominus$

4.

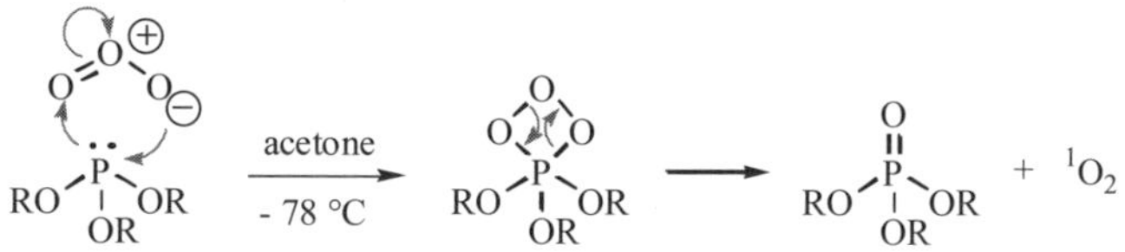

$$\text{(structure)} \xrightarrow[-78\ ^\circ C]{\text{acetone}} \text{(structure)} \longrightarrow \text{(structure)} + {}^1O_2$$

The ***Rothemund Reaction*** is useful for making substituted porphines (seen in examples on next page):

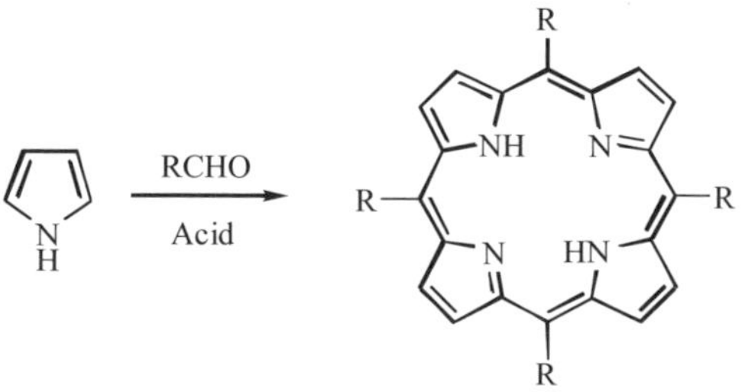

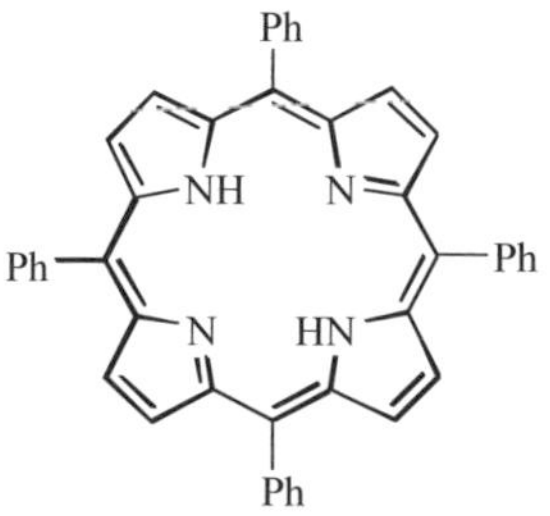

$$\text{(pyrrole)} \xrightarrow[\text{Acid}]{RCHO} \text{(porphine structure)}$$

TTP - Tetraphenylporphine

## Examples:

J.-J. Helesbeux, O. Duval, C. Dartiguelongue, D. Seraphin, J.-M. Oger, P. Richomme, *Tetrahedron* **2004**, 60, 2293

K.-Q. Ling, J.-H. Ye, X.-Y. Chen, D.-J. Ma, J.-H. Xu, *Tetrahedron* **1999**, 55, 9185

R = H,        >95%,  84 : 16 threo : erythro
R = Si(*i*-Pr)$_3$, >95% : <5  threo : erythro

W. Adam, J. Renze, T. Wirth, *Journal of Organic Chemistry* **1998**, 63, 226

89%                          88 : 12

94%                          35 : 59

T. Linker, L. Froehlich, *Journal of the American Chemical Society* **1995**, 117, 2694

# Schmidt Rearrangement, Schmidt Reaction

## The Reaction:

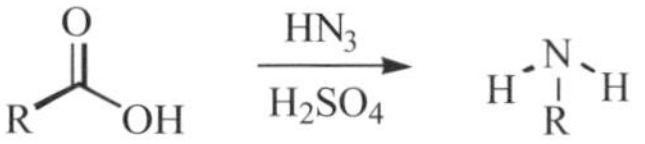

## Also classified as Schmidt Reactions:

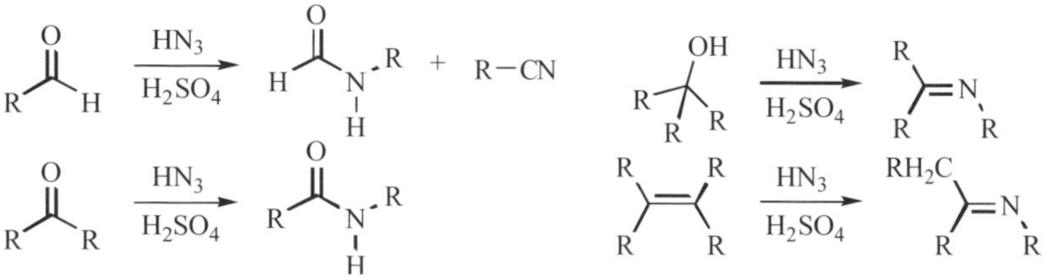

## Proposed Mechanism:

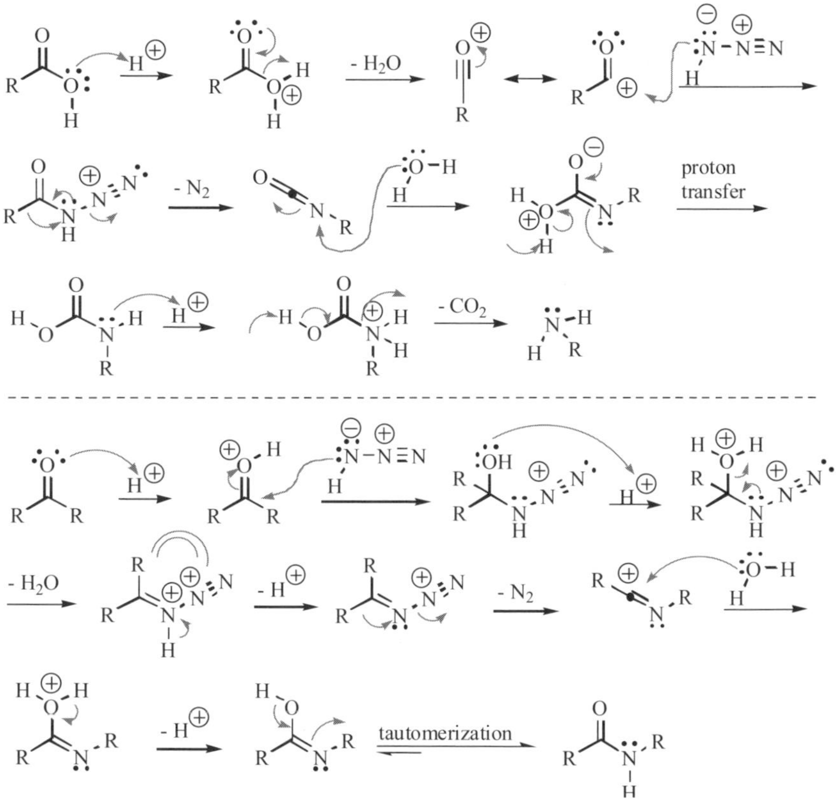

## Notes:

M. B. Smith, J. March in *March's Advanced Organic Chemistry*, 5$^{th}$ ed., John Wiley and Sons, Inc., New York, 2001, pp. 413-415; T. Laue, A. Plagens, *Named Organic Reactions*, John Wiley and Sons, Inc., New York, 1998, pp. 239-241; H. Wolff, *Organic Reactions* **3**, 8

Loss of water will generally favor the less-hindered intermediate.

As with the *Curtius*, *Hofmann* and *Lossen Rearrrangements*, there is a common isocyanate intermediate.

## Examples:

S. Arseniyadis, A. Wagner, C. Mioskowski, *Tetrahedron Letters* **2004**, <u>45</u>, 2251

R = *n*-Bu, 55%
R = *t*-Bu, 35%

M. Tanaka, M. Oba, K. Tamai, H. Suemune, *Journal of Organic Chemistry* **2001**, <u>66</u>, 2667

36 : 65

G. R. Krow, S. W. Szczepanski, J. Y. Kim, N. Liu, A. Sheikh, Y. Xiao, J. Yuan, *Journal of Organic Chemistry* **1999**, <u>64</u>, 1254

H.-J. Cristau, X. Marat, J.-P. Vors, J.-L. Pirat, *Tetrahedron Letters* **2003**, <u>44</u>, 3179

# Schmidt's Trichloroacetimidate Glycosidation Reaction

## The Reaction:

$$\text{1. NaH (cat.), Cl}_3\text{CCN}$$
$$\text{2. BF}_3\cdot\text{OEt, ROH}$$

## Proposed Mechanism:

*Therefore, only a catalytic amount of NaH is needed.*

$$\frac{\text{NaH}}{-\text{H}_2\text{ (g)}}$$

$$Cl_3C\!-\!\!\equiv\!N$$

$$\xrightarrow{\text{BF}_3\cdot\text{OEt}}$$

$$-\ F_3B\!-\!\underset{H}{N}\!\!\diagdown\!\!\overset{O}{\diagup}\!\!CCl_3$$

$$\xrightarrow{\text{HO}-\text{R}}$$

$$\xrightarrow{-\text{H}^{\oplus}}$$

## Notes:

This connection established by *Schmidt's Trichloroacetimidate Glycosidation Reaction*

K. C. Nicolaou, C. W. Hummel, M. Nakada, K. Shibayama, E. N. Pitsinos, H. Saimoto, Y. Mizuno, K. U. Baldenius, A. L. Smith, *Journal of the American Chemical Society* **1993**, <u>115</u>, 7625

**Examples:**

A. Furstner, A. F. Jeanjean, P. Razon, *Angewandte Chemie, International Edition in English* **2002**, 41, 2097

T. Ren, D. Liu, *Tetrahedron Letters* **1999**, 40, 7621

F. J. Urban, B. S. Moore, R. Breitenbach, *Tetrahedron Letters* **1990**, 31, 4421

# Scholl Reaction

## The Reaction:

$$2 \ ArH \xrightarrow{\ AlCl_3\ } Ar-Ar \ + \ H_2$$

## Proposed Mechanism:

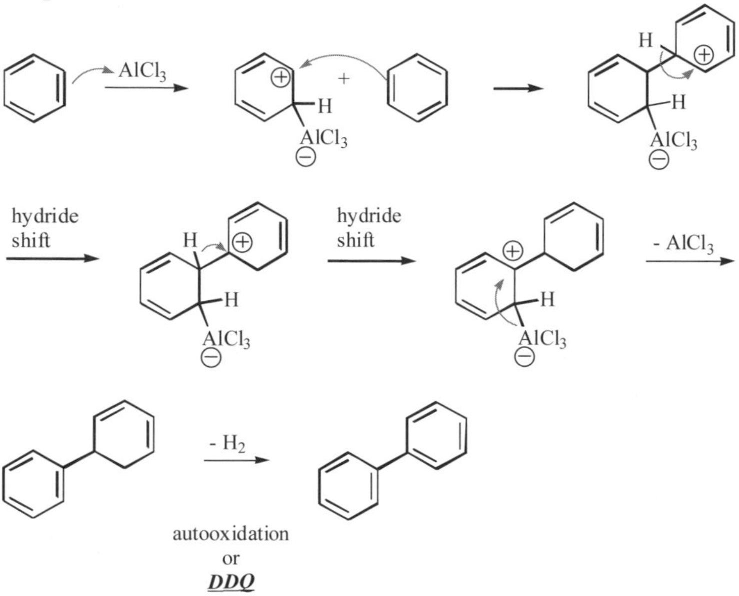

## Notes:

M. B. Smith, J. March in *March's Advanced Organic Chemistry*, 5th ed., John Wiley and Sons, Inc., New York, 2001, p. 711

## Examples:

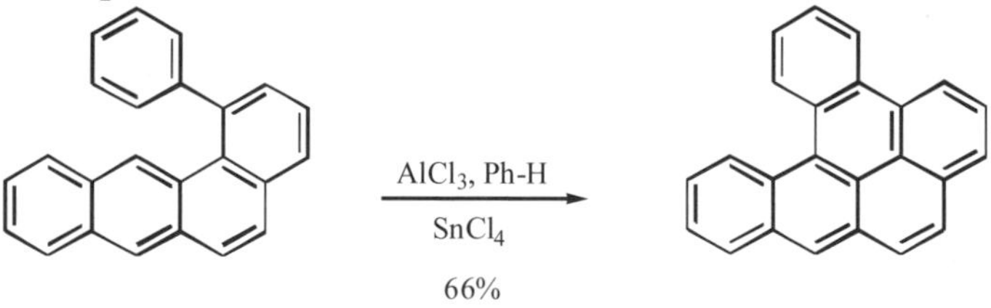

AlCl₃, Ph-H

SnCl₄

66%

F. A. Vingiello, J. Yanez, J. A. Campbell, *Journal of Organic Chemistry* **1971**, <u>36</u>, 2053

Radical-cation route:

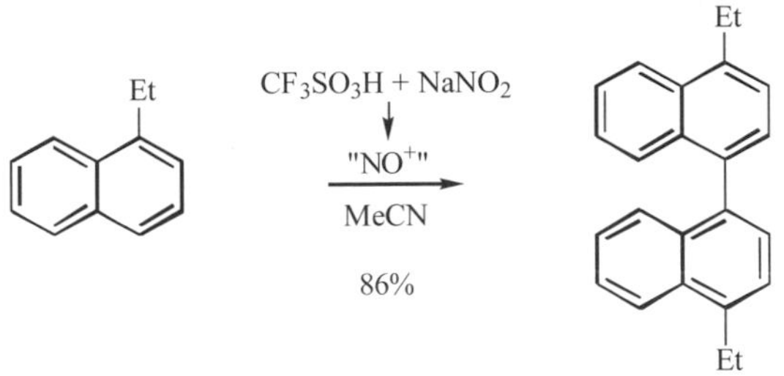

CF$_3$SO$_3$H + NaNO$_2$
↓
"NO$^+$"
MeCN

86%

M. Tanaka, H. Nakashima, M. Fujiwara, H. Ando, Y. Souma, *Journal of Organic Chemistry* **1996**, 61, 788

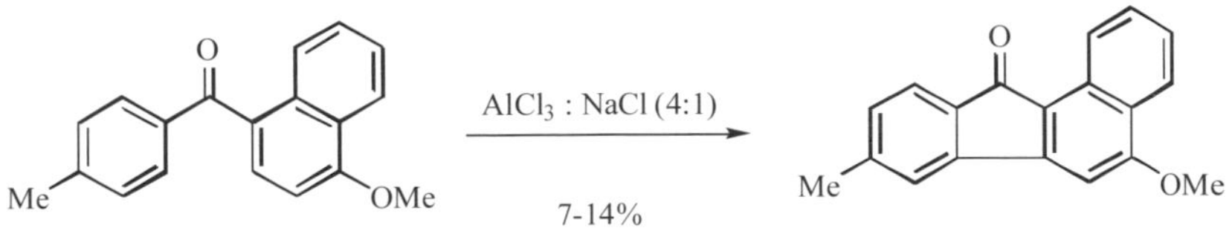

AlCl$_3$ : NaCl (4:1)

7-14%

M. E. Gross, H. P. Lankelma, *Journal of the American Chemical Society* **1951**, 73, 3439

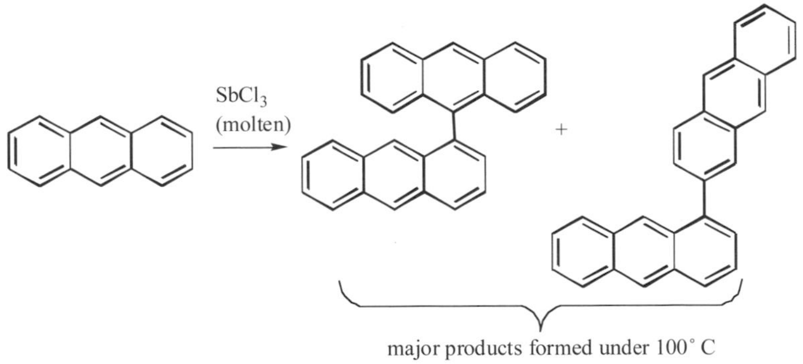

SbCl$_3$
(molten)

+

major products formed under 100° C

W. L. Poutsma, A. S. Dworkin, J. Brynestad, L. L. Brown, B. M. Bejamin, G. P. Smith, *Tetrahedron Letters* **1978**, 19, 873

# Schotten-Baumann Reaction

## The Reaction:

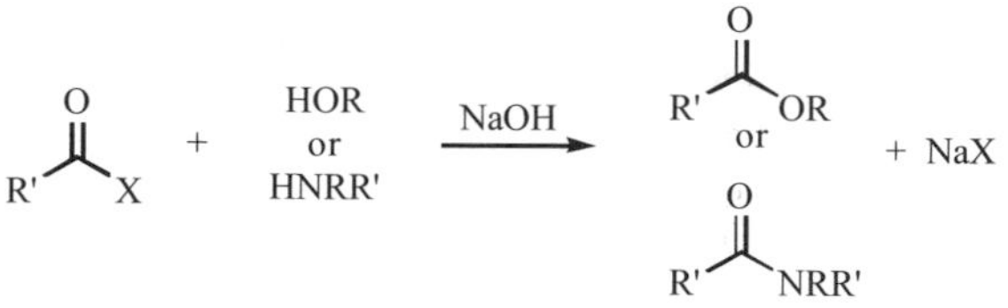

## Proposed Mechanism:

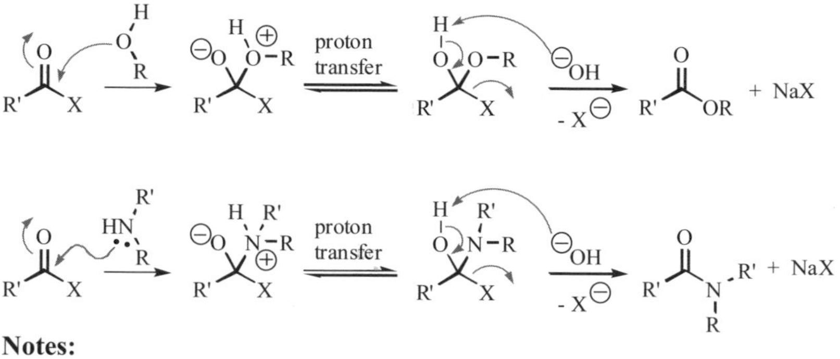

## Notes:

M. B. Smith, J. March in *March's Advanced Organic Chemistry*, 5th ed., John Wiley and Sons, Inc., New York, 2001, pp. 482, 506

Other methods for similar catalysis are known:
**Einhorn Variation**: Pyridine and DMAP (dimethylaminopyridine)
**Steglich esterification**: DMAP, trace of tosic acid, DCC (dicyclohexylcarbodiimide)
**Yamaguchi Esterification**: 2,4,6-trichlorobenzoyl chloride, (**Yamaguchii Reagent**), NEt₃, DMAP

## Examples:

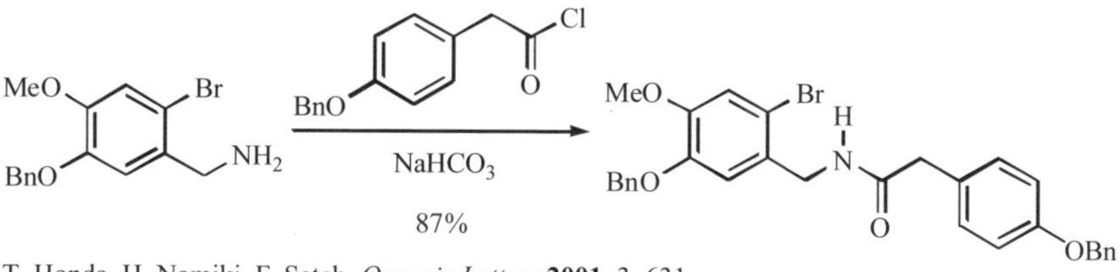

87%

T. Honda, H. Namiki, F. Satch, *Organic Letters* **2001**, *3*, 631

Y.-C. Wang, P. E. Georghiou, *Organic Letters* **2002**, 4, 2675

A. C. Hontz, E. C. Wagner, *Organic Syntheses* CV 4, 383

readily soluble in organic solvents

superior to normal Schotten-Baumann conditions for water-soluble products

J. Fitt, K. Prasad, O. Repic, T. J. Blacklock, *Tetrahedron Letters* **1998**, 39, 6991

# Schreiber Ozonolysis

## The Reaction:

A protocol that allows for the differentiation of the two ends of the alkene bond during ozonolysis.

## Proposed Mechanism:

ROH ⟶ can be ketalized with alcohol and acid

under redctive conditions, a masked aldehyde

under oxidative conditions, a masked acid

## Examples:

S. L. Schreiber, W. Liew, *Journal of the American Chemical Society* **1985**, <u>107</u>, 2980

Different possible products possible by slight variation of workup

S. L. Schreiber, R. E. Claus, J. Regan, *Tetrahedron Letters* **1982**, <u>23</u>, 3867

R = Me    68%    3.6 : 1
R = Bn    75%    4.8 : 1

D. F. Taber, K. Nakajima, *Journal of Organic Chemistry* **2001**, <u>66</u>, 2515

An anomolous ozonolysis

M. P. DeNinno, *Journal of the American Chemical Society* **1995**, <u>117</u>, 9927

# Semi-Pinacol Rearrangement

## The Reaction:
Originally:

But now commonly associated with ***Tiffeneau-Demjanov rearrangement***.
The classification now covers many reactions of the "***pinacol type***", including:

## Proposed Mechanism:

## Notes:
See ***Tiffeneau- Demjanov*** for the hydroxy-amino rearrangement: T. Laue, A. Plagens, *Named Organic Reactions*, John Wiley and Sons, Inc., New York, 1998, pp. 255-257; M. B. Smith, J. March in *March's Advanced Organic Chemistry*, 5$^{th}$ ed., John Wiley and Sons, Inc., New York, 2001, p. 1397

A complex rearrangement:

C. M. Marson, C. A. Oare, T. Walsgrove, T. J. Grinter, H. Adams. *Tetrahedron Letters* **2002**, <u>44</u>, 141

## Examples:

S.-J. Jeon, P. J. Walsh, *Journal of the American Chemical Society* **2003**, <u>125</u>, 9544

M. D. B. Fenster, G. R. Duke, *Organic Letters* **2003**, <u>5</u>, 4313

A useful methodology:

B. M. Trost, D. Keeley, M. J. Bogdanowicz, *Journal of the American Chemical Society* **1973**, <u>95</u>, 3068

K. C. Nicolaou, A. L. Smith, S. V. Wendeborn, C.-K. Whang, *Journal of the American Chemical Society* **1991**, <u>113</u>, 3106

*Reductive retro-pinacol*

T. Q. Tu, L. D. Sun, P. Z. Wang, *Journal of Organic Chemistry* **1999**, <u>64</u>, 629

# Semmler-Wolff Reaction

## The Reaction:

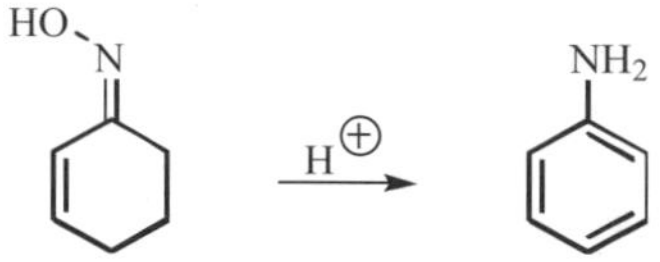

## Proposed Mechanism:

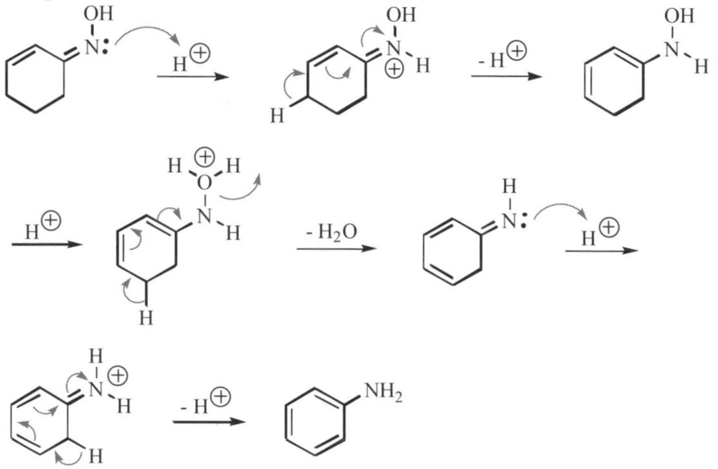

## Examples:

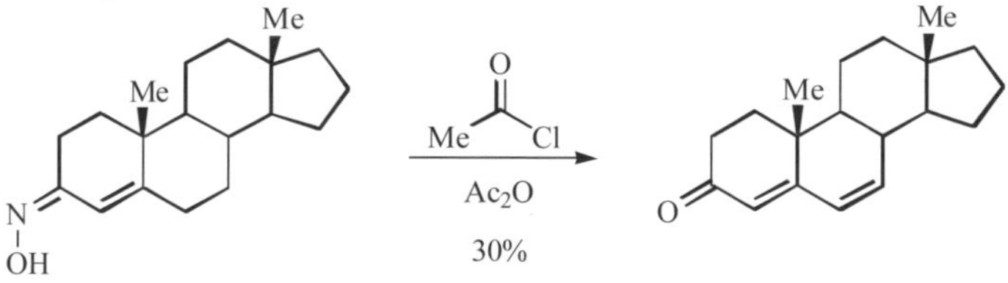

M. V. Bhatt, S. R. Raju, *Tetrahedron Letters* **1964**, <u>5</u>, 2623

A similar "special case" moted:

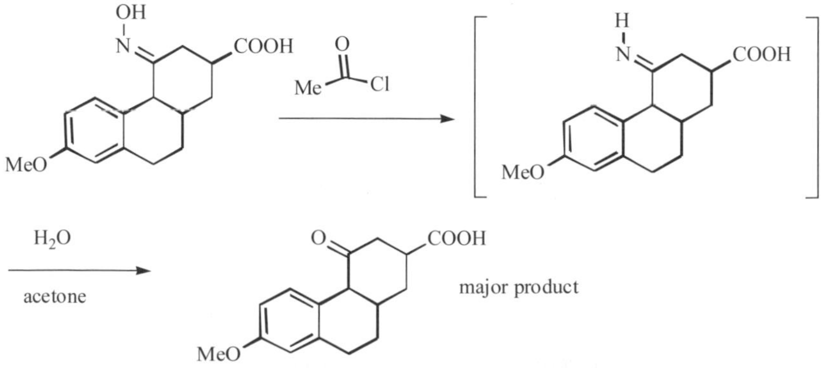

Z. G. Hajos, D. R. Parrish, M. W. Goldberg, *Journal of Organic Chemistry* **1968**, <u>33</u>, 882

Y. Kobayashi, S. Wakamatsu, *Tetrahedron* **1967**, 23, 115

J. J. Weidner, P. M. Weintraub, R. A. Schnettler, N. P. Peet, *Tetrahedron* **1997**, 53, 6303

W. K. Sprenger, J. G. Cannon, H. F. Koelling, *Journal of Organic Chemistry* **1966**, **31**, 2402

# Seyferth-Gilbert Homolgation

## The Reaction:

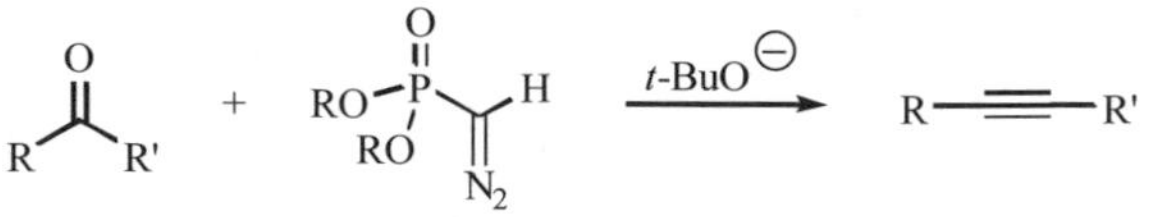

## Proposed Mechanism:

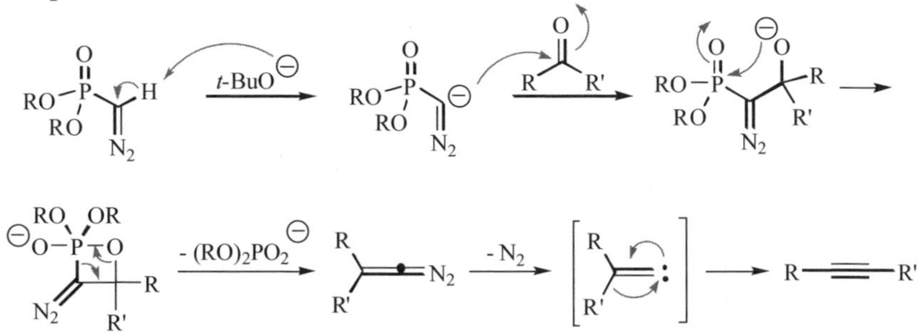

## Notes:

Starting material preparation:

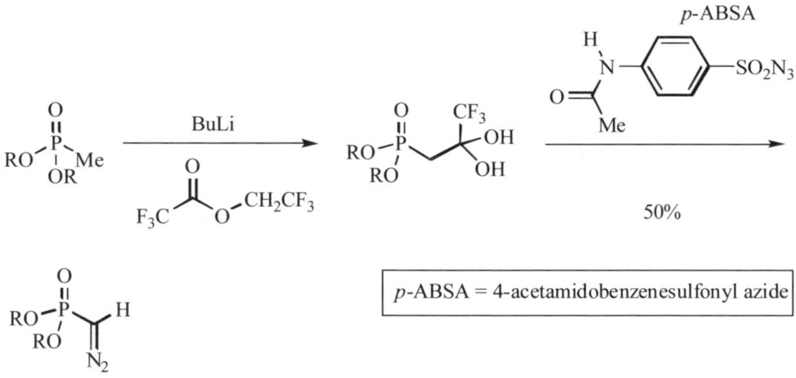

50%

| p-ABSA = 4-acetamidobenzenesulfonyl azide |

D. G. Brown, E. J. Velthuisen, J. R. Commerford, R. G. Brisbois, T. R. Hoye, *Journal of Organic Chemistry* **1996**, <u>61</u>, 2540

# Examples:

D. L. Comins, D. H. LaMunyon, X. Chen, *Journal of Organic Chemistry* **1997**, <u>62</u>, 8182

P. A. Wender, S. G. Hegde, R. D. Hubbard, L. Zhang, *Journal of the American Chemical Society* **2002**, <u>124</u>, 4956

J. B. Nerenberg, D. T. Hung, P. K. Somers, S. L. Schreiber, *Journal of the American Chemical Society* **1993**, <u>115</u>, 12621

# Shapiro Reaction

## The Reaction:

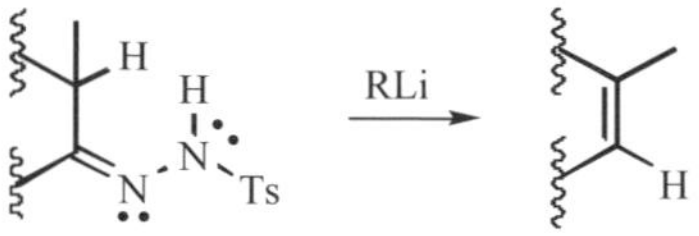

## Proposed Mechanism:

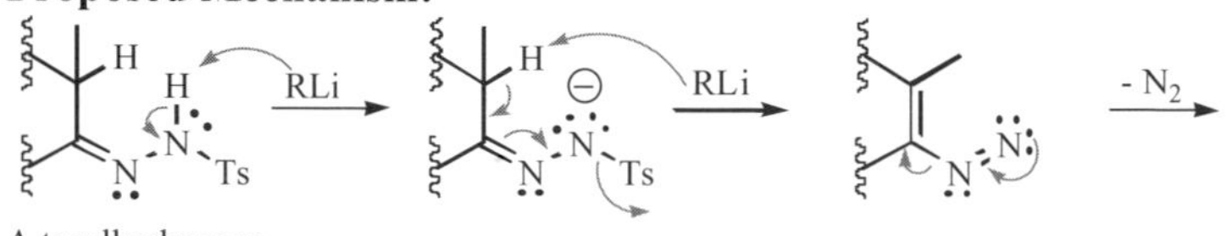

A tosylhydrazone

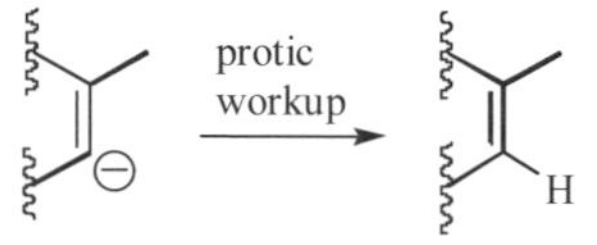

## Notes:

M. B. Smith, J. March in *March's Advanced Organic Chemistry*, 5$^{th}$ ed., John Wiley and Sons, Inc., New York, 2001, p. 1334; R. H. Shapiro, *Organic Reactions* **23**, 3; A. R. Chamberlin, S. H. Bloom, *Organic Reactions* **23**, 1

In the related **_Bamford-Stevens Reaction_**, thermodynamic bases are used and the more substituted alkene is formed.

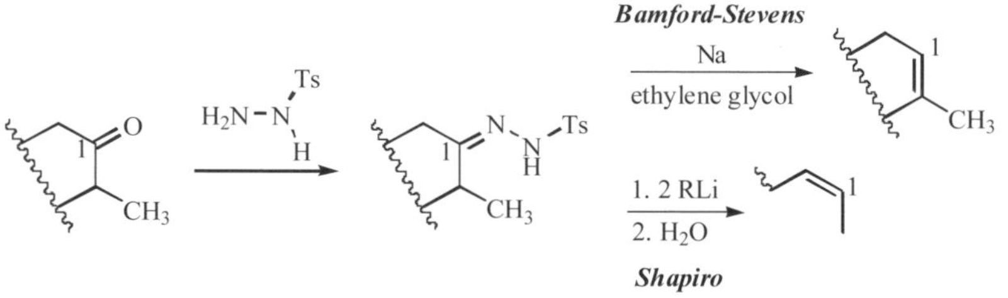

*Bamford-Stevens*

*Shapiro*

## Examples:

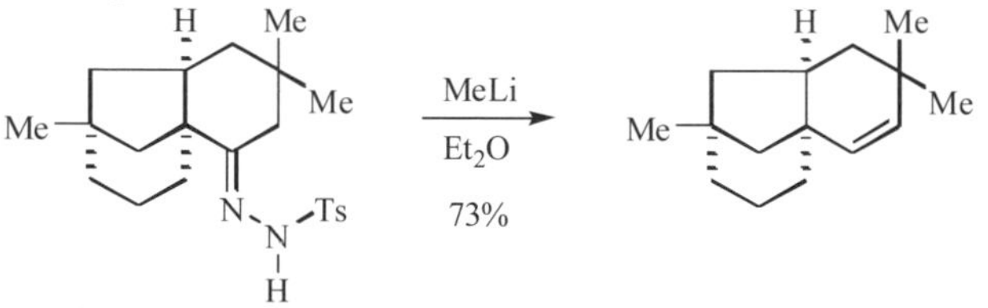

M. Ghosal, L. C. Pati, A. Roy, D. Mukherjee, *Tetrahedron* **2002**, <u>58</u>, 6179

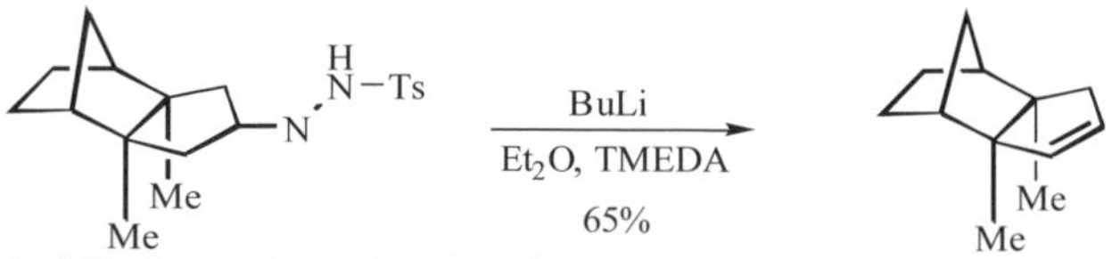

A. Srikrishna, S. Nagaraju, *Phytochemistry* **1995**, <u>40</u>, 1699

Trapping the anion formed in the reaction:

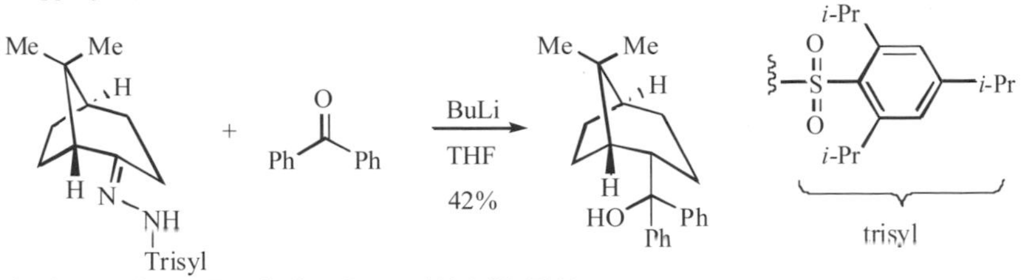

C. Liu, J.R. Sowa, *Tetrahedron Letters* **1996**, *37*, 7241

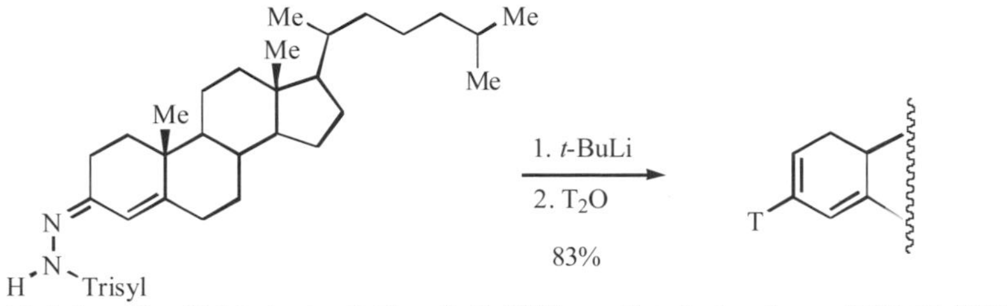

M. Saljoughian, H. Morimoto, C. Than, P. G. Williams, *Tetrahedron Letters* **1996**, *37*, 2923

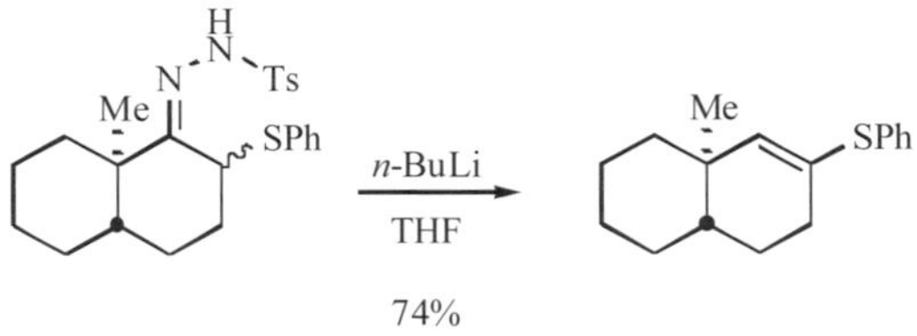

74%

D. M. Coltart, S. J. Danishefsky, *Organic Letters* **2003**, *5*, 1289

Interception of the anion intermediate in the **Shapiro Reaction** to form trisubstituted alkenes:

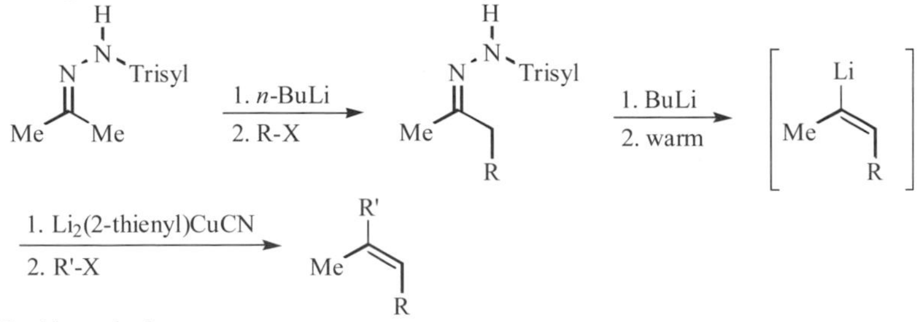

By this method:

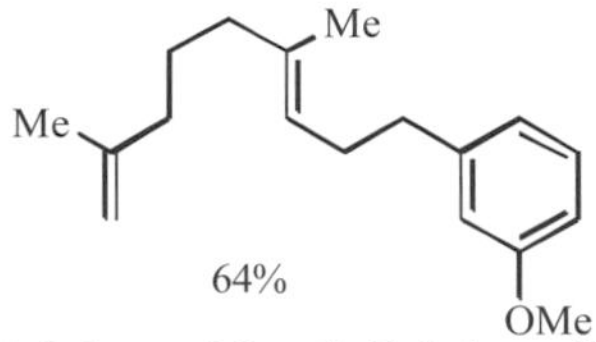

64%

E. J. Corey, J. Lee, B. E. Roberts, *Tetrahedron Letters* **1997**, *38*, 8915

# Sharpless (Catalytic Asymmetric) Aminohydroxylation
## The Reaction:

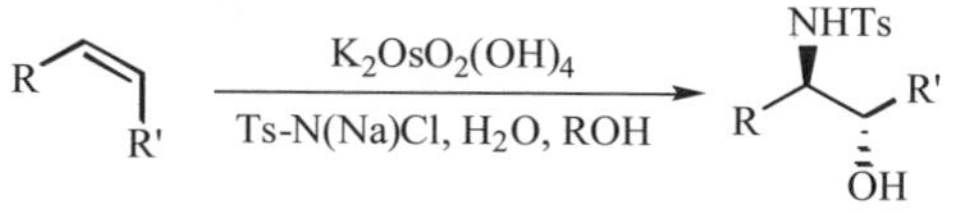

$$R \diagdown \text{(alkene with } R') \xrightarrow[\text{Ts-N(Na)Cl, H}_2\text{O, ROH}]{\text{K}_2\text{OsO}_2(\text{OH})_4} R \diagdown \text{(product with NHTs and OH)}$$

## Proposed Mechanism:

A simplified view. If L is chiral, the reaction can be rendered asymmetric:

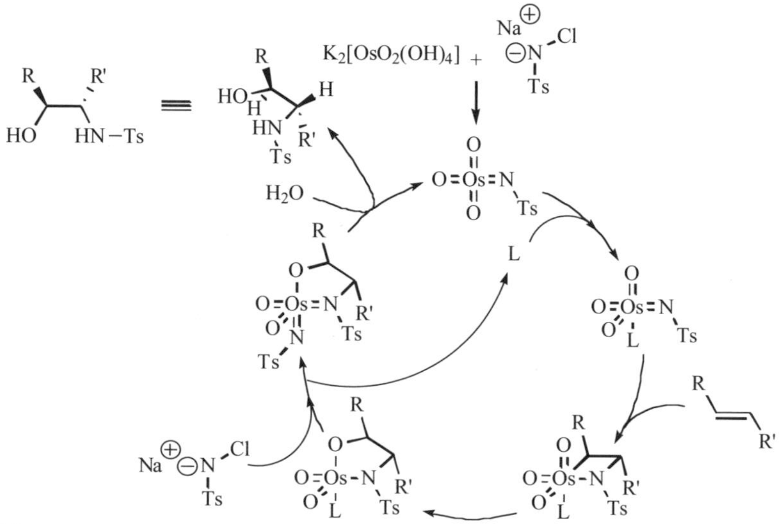

## Notes:

For recent Reviews, see: K. Muñiz, *Chemical Society Reviews* **2004**, <u>33</u>, 166; J. A. Bodkin, M. D. McLeod, *Journal of the Chemical Society, Perkin Transactions I*, **2002**, 2733

## Examples:

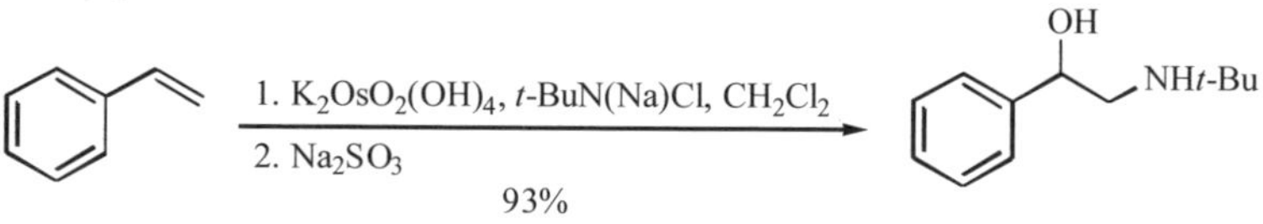

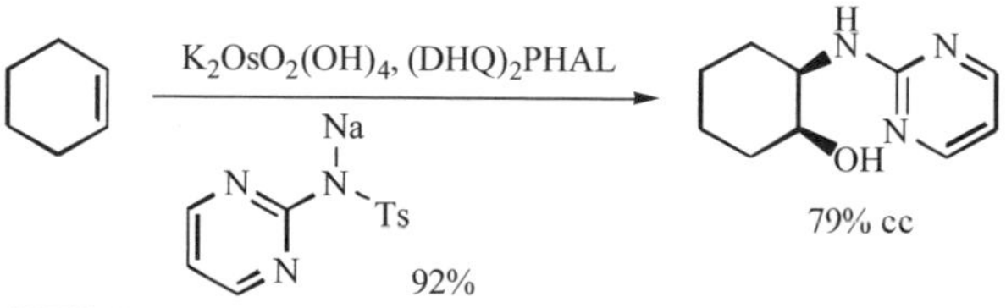

$$\xrightarrow[\text{2. Na}_2\text{SO}_3]{\text{1. K}_2\text{OsO}_2(\text{OH})_4, \ t\text{-BuN(Na)Cl, CH}_2\text{Cl}_2}$$

93%

$$\xrightarrow[\substack{\text{Na} \\ \text{N-Ts}}]{\text{K}_2\text{OsO}_2(\text{OH})_4, \ (\text{DHQ})_2\text{PHAL}}$$

92%

79% ee

(DHQ)₂PHAL: See ***Sharpless Dihydroxylation***

Reported in K. Muñiz, *Chemical Society Reviews* **2004**, <u>33</u>, 166

P. R. Blakemore, S.-K. Kim, V. K. Schulze, J. D. White, A. F. T. Yokochi, *Journal of the Chemical Society, Perkin Transactions I*, **2001**, 1831

W. Kurosawa, T. Kan, T. Fukuyama, *Journal of the American Chemical Society* **2003**, 125, 8112

L. Dong, M. J. Miller, *Journal of Organic Chemistry* **2002**, 67, 4759

# Sharpless (Asymmetric) Dihydroxylation

## The Reaction:

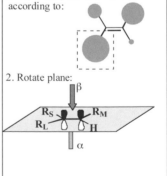

$$R_{sm}, R_{mid}, R_{lg}, H \xrightarrow[\substack{K_2CO_3, K_3Fe(CN)_6}]{K_2[OsO_2(OH)_4]} \begin{cases} \text{(DHQD)}_2\text{-PHAL} \\ \text{(see below)} \\ \text{(DHQ)}_2\text{-PHAL} \end{cases}$$

## Proposed Mechanism:

The osmium is incorporated into the optically-active AD mix, where it undergoes hydroxylation in much the same way as proposed for simple hydroxylation. See Reagent: ***Osmium tetroxide***.

## Notes:

Commercially available is AD [Asymmetric Dihydroxylation] mix (α) or (β):
α–mix contains
(DHQ)$_2$-PHAL = 1,4-bis(9-O-dihydroquinine)phtalazine

β–mix contains
(DHQD)$_2$-PHAL = 1,4-bis(9-O-dihydroquinidine)phtalazine

**How to predict products:**

1. Assign approximate "size" to groups, and place them on a plane according to:

2. Rotate plane:

A pyridazine (PYDZ) core can also be incorporated into the catalyst.

## Examples:

E. J. Corey, A. Guzman-Perez, M. C. Noe, *Journal of the American Chemical Society* **1995**, 117, 10805

A keto-hydroxylation protocol:

B. Plietker, *Organic Letters* **2004**, 6, 289

S. Lemaire-Audoire, P. Vogel, *Tetrahedron: Asymmetry* **1999**, 10, 1283

A. Guzman-Perez, E. J. Corey, *Tetrahedron Letters* **1997**, 38, 5941

B. S. J. Blagg, D. L. Boger, *Tetrahedron* **2002**, 58, 6343

# Sharpless Epoxidation

## The Reaction:

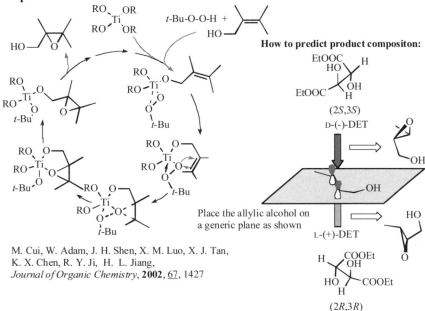

## Proposed Mechanism:

### How to predict product compositon:

M. Cui, W. Adam, J. H. Shen, X. M. Luo, X. J. Tan,
K. X. Chen, R. Y. Ji, H. L. Jiang,
*Journal of Organic Chemistry*, **2002**, <u>67</u>, 1427

(2S,3S)
D-(-)-DET

Place the allylic alcohol on
a generic plane as shown

L-(+)-DET

(2R,3R)

## Notes:

E= COOEt, R = *i*-Pr

*t*-BuOOH

C. J. Burns, C. A. Martin, K. B. Sharpless, *Journal of Organic Chemistry* **1989**, <u>54</u>, 2826

M. B. Smith, J. March in *March's Advanced Organic Chemistry*, 5<sup>th</sup> ed., John Wiley and Sons, Inc., New York, 2001, p. 1053; T. Laue, A. Plagens, *Named Organic Reactions*, John Wiley and Sons, Inc., New York, 1998, pp. 242-244

## Examples:

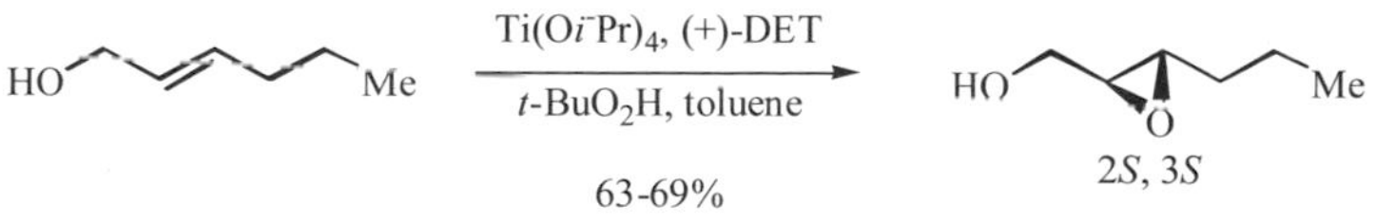

63-69%

2*S*, 3*S*

J. G. Hill, K. B. Sharpless, C. M. Exon, R. Regenye, *Organic Synthesis* CV7, 461

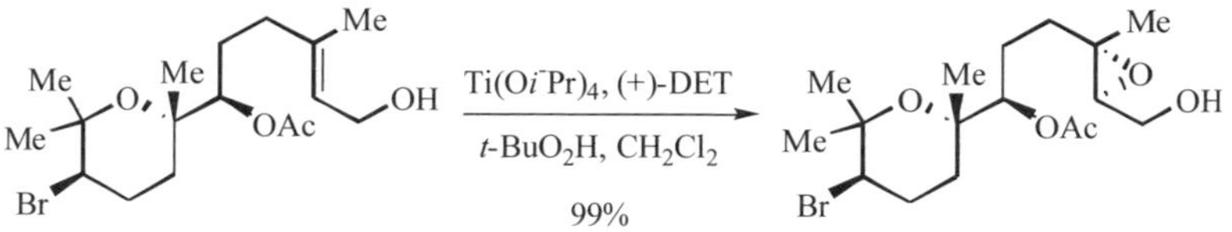

99%

F. E. McDonald, X. Wei, *Organic Letters* **2002**, 4, 593

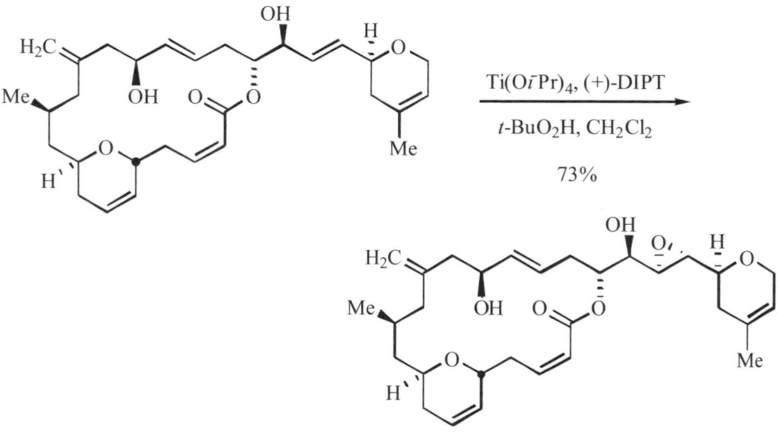

73%

I. Patterson, C. D. Savi, M. Tudge, *Organic Letters* **2001**, 3, 3149

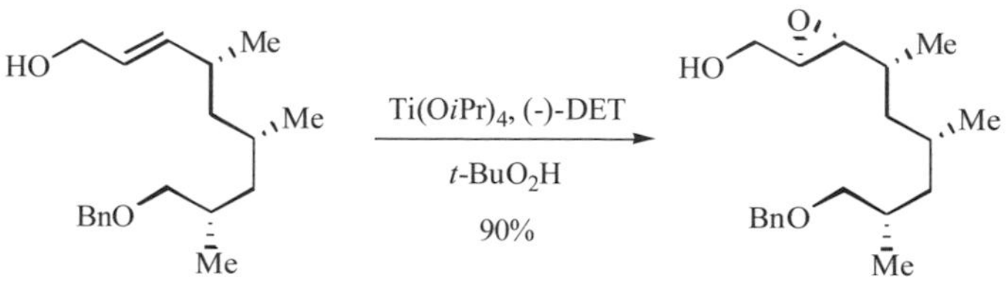

90%

A. K. Ghosh, C. Liu, *Organic Letters* **2001**, 3, 635

# Shi (Asymmetric) Epoxidation

## The Reaction:

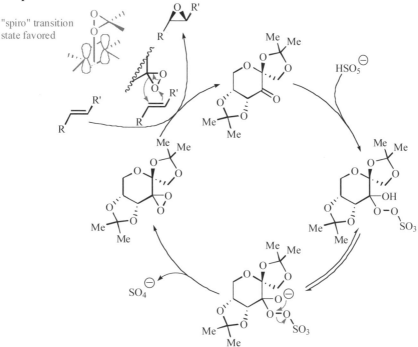

Oxone, pH 7-8
H₂O, MeCN

**_Oxone_** = 2KHSO₅ KHSO₄ K₂SO₄

## Proposed Mechanism:

"spiro" transition state favored

Y. Tu, Z.-X. Wang, Y. Shi, *Journal of the American Chemical Society* **1996**, <u>118</u>, 9806; Y. Shi, *Accounts of Chemical Research* **2004**, <u>37</u>, 488

## Notes:

*Shi Epoxidation*

81%, > 95% ee

1:1 ratio of epoxidation

*Shi Epoxidation*

90-95%

B. Olofsson, P. Somfai, *Journal of Organic Chemistry* **2003**, <u>68</u>, 2514

## Examples:

D. W. Hoard, E .D. Moher, M. J. Martinelli, B. H. Norman, *Organic Letters* **2002**, *4*, 1813

F. E. McDonald, X. Wei, *Organic Letters* **2002**, *4*, 593

A two-phase system has been reported:

N. Hashimoto, A. Kanda, *Organic Process Research & Development* **2002**, *6*, 405

*m*CPBA ~2 : 1
*Shi*    ~3.5 : 1

D. W. Hoard, E. D. Moher, M. J. Martinelli, B. H. Norman, *Organic Letters* **2002**, *4*, 1813

# Simmons-Smith Cyclopropanation

## The Reaction:

$$\text{alkene} \xrightarrow[\text{[Zn-Cu]}]{CH_2I_2} \text{cyclopropane} + ZnI_2 + Cu$$

## Proposed Mechanism:

$$CH_2I_2 + Zn\text{-}Cu$$

$$\downarrow \text{oxidative addition}$$

## Notes:

M. B. Smith, J. March in *March's Advanced Organic Chemistry*, 5th ed., John Wiley and Sons, Inc., New York, 2001, p. 1088; T. Laue, A. Plagens, *Named Organic Reactions*, John Wiley and Sons, Inc., New York, 1998, pp. 244-246; A. B. Charette, A. Beauchemin, *Organic Reactions* **58**, 1; H. E. Simmons, T. L. Cairns, S. A. Vladuchick, C. M. Hoiness, *Organic Reactions* **20**, 1

The cyclopropanation is sensitive to steric effects, adding from the less-hindered face. Having a neighboring hydroxyl group will generally accelerate the reaction, and will direct the cyclopropanation *syn* to the hydroxyl group; even into sterically congested alkenes.

W.G. Dauben, A.C. Ashcraft, *Journal of the American Chemical Society* **1963**, <u>85</u>, 3673

There are a number of modifications of this reaction; all providing the same product:

*Furukawa's Reagent*        $Et_2Zn + ICH_2I$ (1:1)
*Sawada Reagent*            $EtZnI + CH_2I_2$
*Kobayashi Variation*       $I_2 + Et_2Zn$

A. K. Ghosh, C. Liu, *Organic Letters* **2001**, <u>3</u>, 635

## Examples:

O. Cheng, T. Kreethadumrongdat, T. Cohen, *Organic Letters* **2001**, <u>3</u>, 2121

M. Harmata, P. Rashatasakhon, *Organic Letters* **2000**, <u>2</u>, 2913

D. A. Evans, J. D. Burch, *Organic Letters* **2001**, <u>3</u>, 503

B. R. Aavula, Q. Cui, E. A. Mash, *Tetrahedron: Asymmetry* **2000**, <u>11</u>, 4681

T. Onoda, R. Shirai, Y. Koiso, S. Iwasaki, *Tetrahedron Letters* **1996**, <u>37</u>, 4397

# Skraup Reaction

## The Reaction:

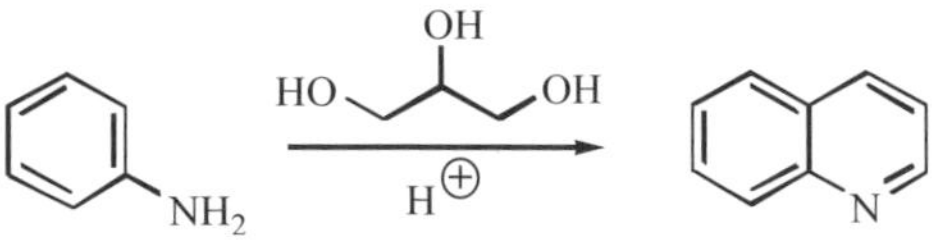

## Proposed Mechanism:

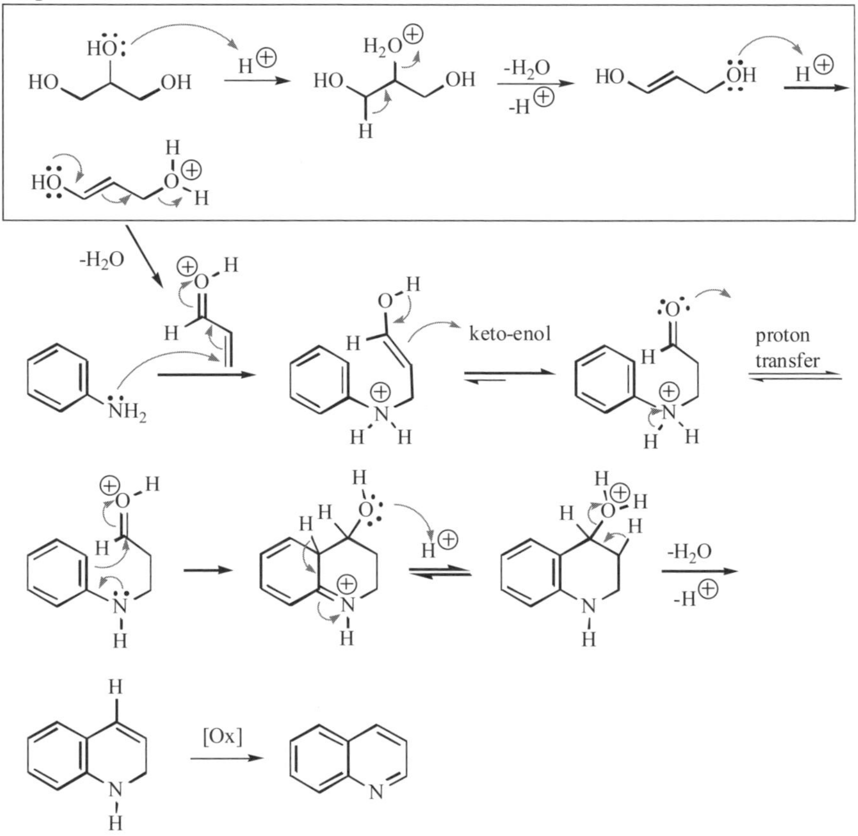

## Notes:

T. Laue, A. Plagens, *Named Organic Reactions*, John Wiley and Sons, Inc., New York, 1998, pp. 246-248; R. H. F. Manske, M. Kulka, *Organic Reactions* **7**, 2

See also the very similar ***Deobner-von Miller reaction***.

## Examples:

A "modified" *Skraup*:

M.-E. Theoditou, L. A. Robinson, *Tetrahedron Letters* **2002**, <u>43</u>, 3907

P. M. O'Neill, R. C. Storr, B. K. Park, *Tetrahedron* **1998**, <u>54</u>, 4615

Taken directly to next step

H.-Y. Choi, D. Y. Choi, *Journal of the American Chemical Society* **2001**, <u>123</u>, 9202

A "modified" *Skraup*:

Y.-Y Ku, T. Grieme, P. Raje, P. Sharma, H. E. Morton, M. Rozema, S. A. King, *Journal of Organic Chemistry* **2003**, <u>68</u>, 3238

# Smiles Rearrangement

## The Reaction:

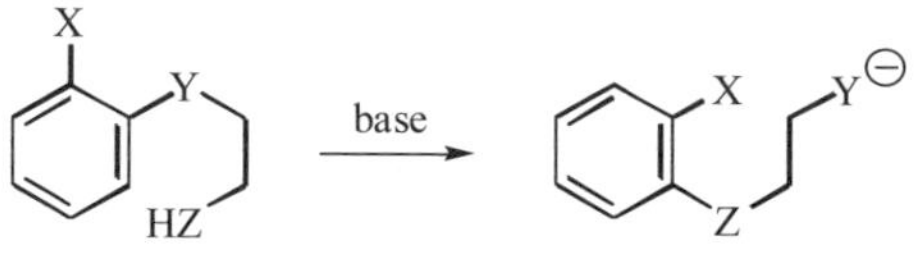

X = NO$_2$, SO$_2$R
Y = S, SO, SO$_2$, O, COO$^{\ominus}$
ZH = OH, NHR, SH, CH$_2$R, CONHR

## Proposed Mechanism:

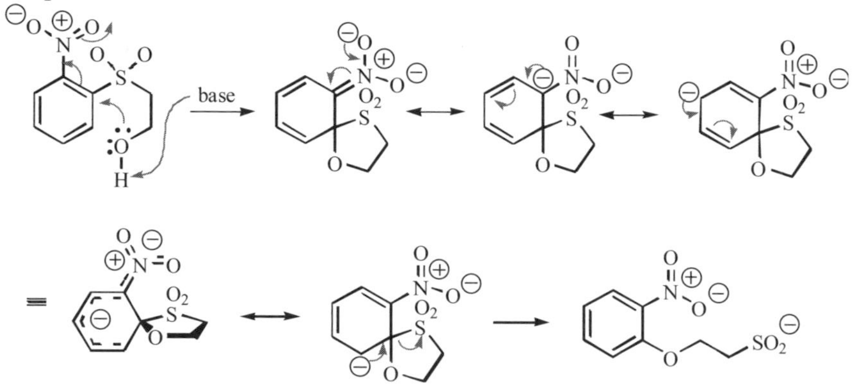

## Notes:

M. B. Smith, J. March in *March's Advanced Organic Chemistry*, 5$^{th}$ ed., John Wiley and Sons, Inc., New York, 2001, p. 879; W. E. Truce, E. M. Kreider, W. W. Brand, *Organic Reactions* **18**, 2

## Examples:

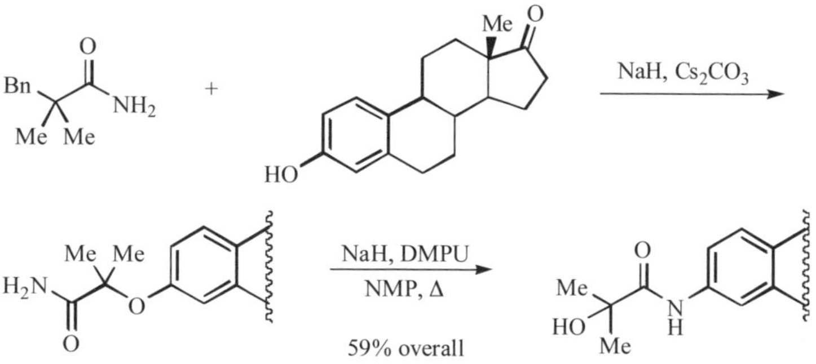

DMPU = N,N'-Dimethylpropyleneura; NMP = *N*-methylpyrrolidine

J. J. Weidner, P. M. Weintraub, R. A. Schnettler, N. P. Peet, *Tetrahedron* **1997**, *53*, 6303

C. Bonini, M. Funicello, R. Scialpib, P. Spagnolob, *Tetrahedron* **2003**, *5*, 7515

DNBSA = 2,4-Dinitrobenzenesulfonamide =

TEA = triethylamine = Et₃N

via:

V. J. Huber, R. A. Bartsch, *Tetrahedron* **1998**, *54*, 9281

# Smith-Tietze Coupling

## The Reaction:

## Proposed Mechanism:

## Notes:

a) A. B. Smith, III, A. M. Boldi, *Journal of the American Chemical Society* **1997**, <u>119</u>, 6925; (b) A. B. Smith, III, L. Zhuang, C. S. Brook, Q. Lin, W. H. Moser, R. E. L. Trout, A. M. Boldi *Tetrahedron Letters* **1997**, <u>38</u>, 8671; (c) L. F. Tietze, H. Geissler, J. A. Gewert, U. Jakobi, *Synlett* **1994**, 51

For the very first report of a *C*- to *O*-silyl rearrangement occurring after epoxide opening with 2-TMS-1,3-dithiane, see: P. F. Jones, M. F. Lappert, A. C. Szary, *Journal of the Chemical Society, Perkin Transactions 1* **1973**, 2272

For a recent variation of this theme:

A. B. Smith, III, D.-S. Kim, *Organic Letters* **2004**, <u>6</u>, 1493

## Examples:

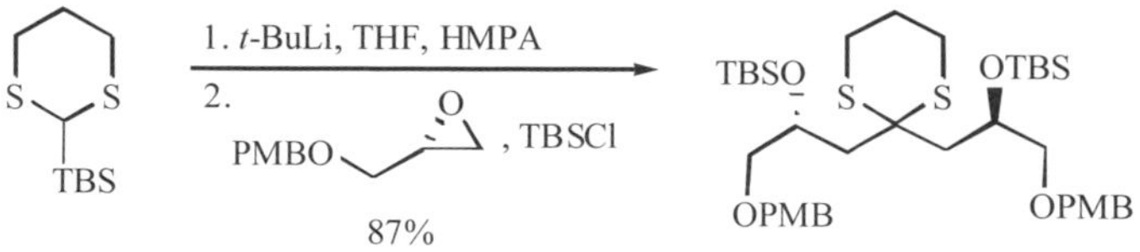

87%

K.J. Hale, M. G. Hummersone, and G. S. Bhatia, *Organic Letters* **2000**, <u>2</u>, 2189; K. J. Hale, M. G. Hummersone, J. Cai, S. Manaviazar, G. S. Bhatia, J. A. Lennon, M. Frigerio, V. M. Delisser, A. Chumnongsaksarp, N. Jogiya, A. Lemaitre, *Pure and Applied Chemistry* **2000**, <u>72</u>, 1659

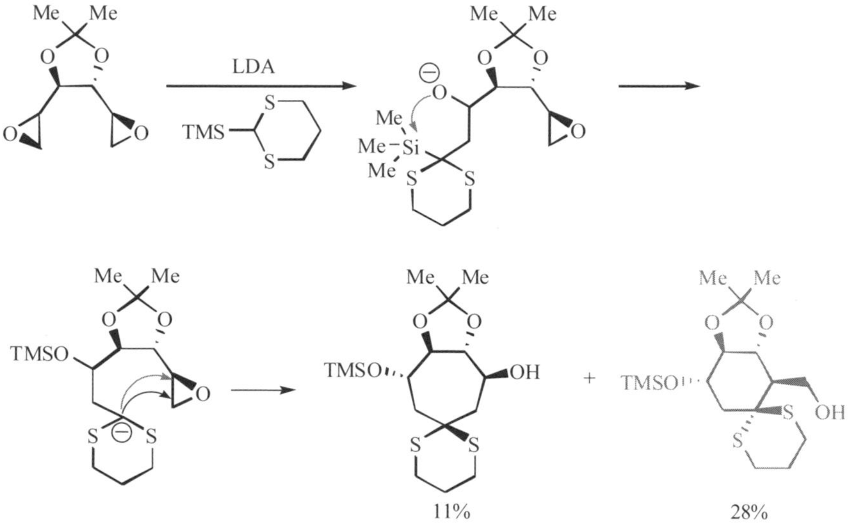

11%						28%

C. Gravier-Pelletier, W. Maton, T. Dintinger, C. Tellier, Y. Le Merrera, *Tetrahedron* **2003**, <u>59</u>, 8705

## Sommelet-Hauser Rearrangement

### The Reaction:

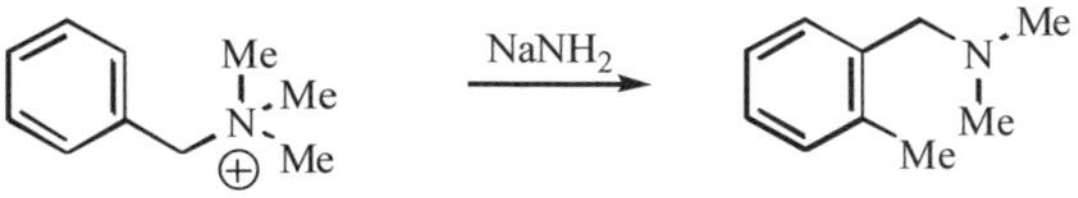

### Proposed Mechanism:

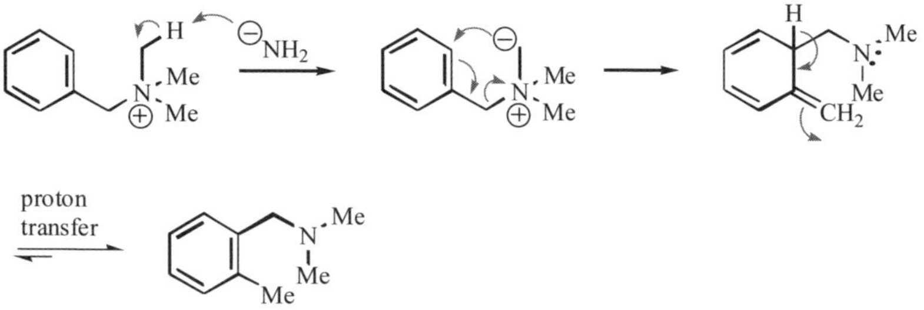

### Notes:

There can be competition between **Sommelet-Hauser** and **_Stevens Rearrangement_** mechanisms:

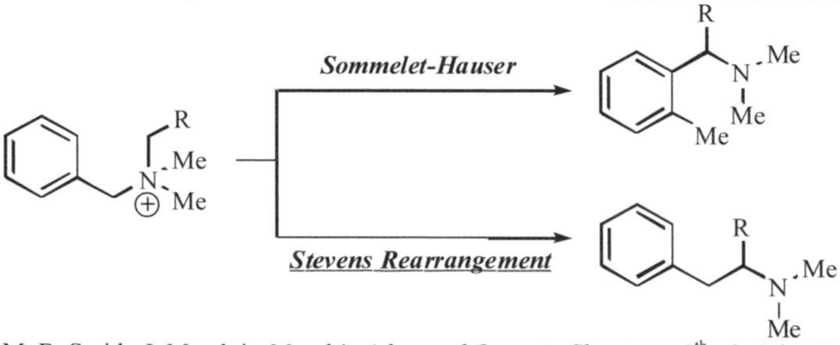

M. B. Smith, J. March in *March's Advanced Organic Chemistry*, 5th ed., John Wiley and Sons, Inc., New York, 2001, pp. 877, 1420, 1455; S. H. Pine, *Organic Reactions* **18**, 4

### Examples:

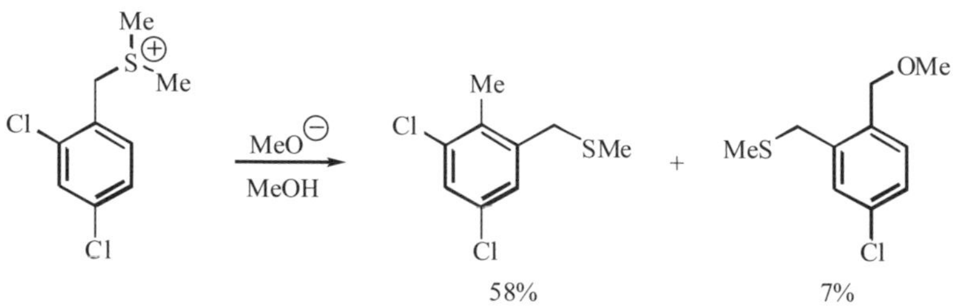

58%                                              7%

T.-J. Lee, W. J. Holtz, *Tetrahedron Letters* **1983**, <u>24</u>, 2071

When G = -OMe, this intermediate product could be isolated as the only product of the reaction.

proton transfer

N. Shirai, Y. Watanabe, Y. Sato, *Journal of Organic Chemistry* **1990**, <u>55</u>, 2767

CsF
HMPA

89%                                                        82 : 18

A. Sakuragi, N. Shirai, Y. Sato, Y. Kurono, K. Hatano, *Journal of Organic Chemistry* **1994**, <u>59</u>, 148

MeLi

60%                                                        3 : 2

Y. Maeda, Y. Sato, *Journal of Organic Chemistry* **1966**, <u>61</u>, 5188

NaOH
benzene

87%                                                        92 : 8

A. Jonczyk, D. Lipiak, *Journal of Organic Chemistry* **1991**, <u>56</u>, 6933

# Sommelet Oxidation

## The Reaction:

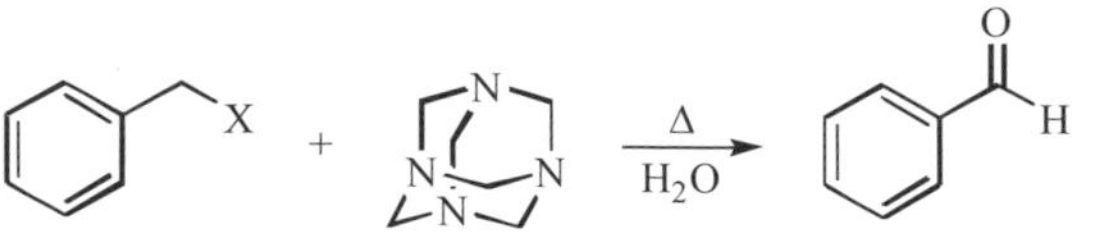

HMTA
Hexamethylenetetramine

## Proposed Mechanism:

The mechanism is not clear, except for the first steps. Workup appears to play an important role in the product evolution: J. D. Hayler, S. L. B. Howie, R. G. Giles, A. Negus, P. W. Oxley, T. C. Walsgrove, M. Whiter, *Organic Process Research & Development* **1998**, 2, 3

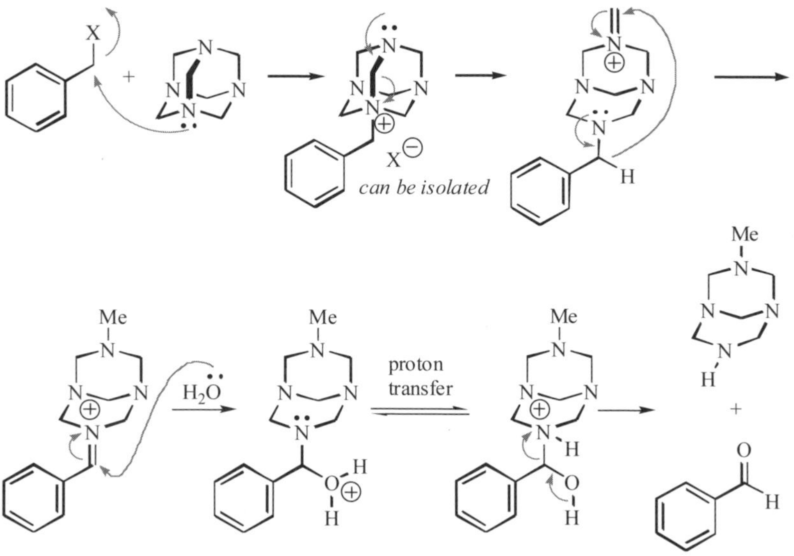

can be isolated

proton
transfer

## Notes:

M. B. Smith, J. March in *March's Advanced Organic Chemistry*, 5th ed., John Wiley and Sons, Inc., New York, 2001, p.1536; S. J. Angyal, *Organic Reactions* **8**, 4

Other products from the reaction can be explained by different hydride transfers and cleavages:

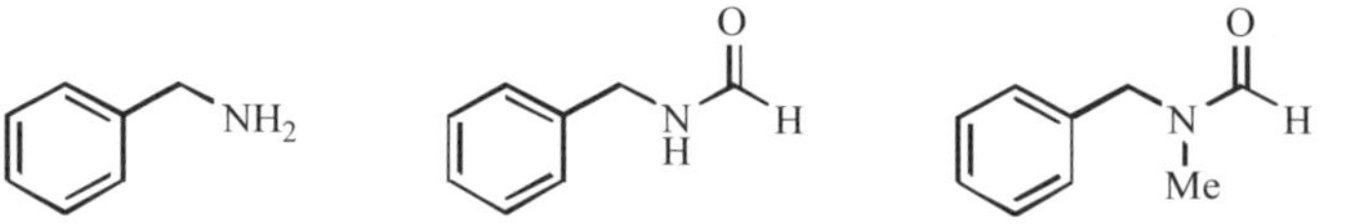

## Examples:

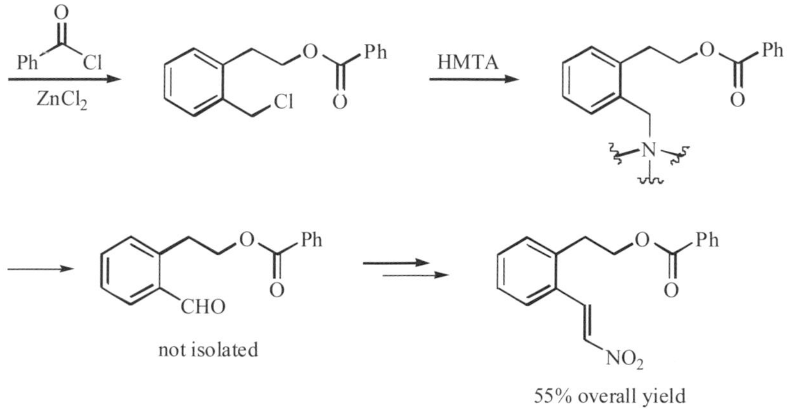

Ph
Cl
O
ZnCl2
O
Ph
Cl
O
HMTA
O
Ph
O
N
O
Ph
O
CHO
not isolated
O
Ph
O
NO2
55% overall yield


J. D. Hayler, S. L. B. Howie, R. G. Giles, A. Negus, P. W. Oxley, T. C. Walsgrove, M. Whiter, *Organic Process Research & Development* **1998**, 2, 3

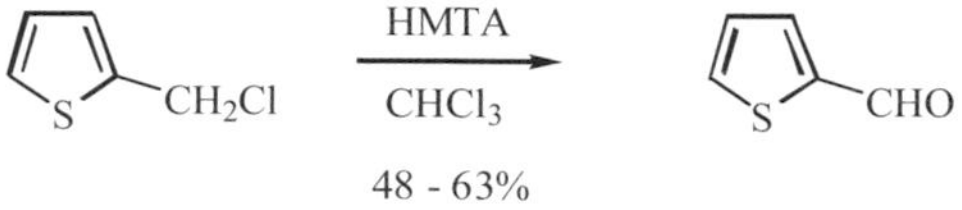

S
CH2Cl
HMTA
CHCl3
48 - 63%
S
CHO


K. B. Wiberg, *Organic Syntheses*, CV 3, 811

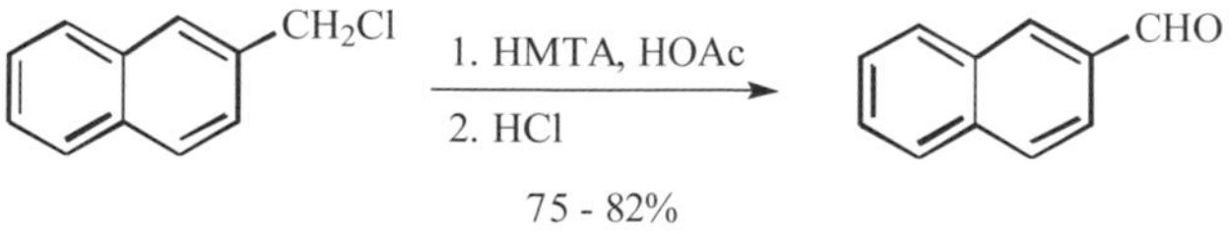

CH2Cl
1. HMTA, HOAc
2. HCl
75 - 82%
CHO


S. J. Angyal, J. R. Tetaz, J. G. Wilson, *Organic Syntheses* CV 4, 690

# Sonogashira Coupling

## The Reaction:

$$R-X \ + \ H\text{———}R' \ \xrightarrow[\text{CuI, NEt}_3]{\text{PdCl}_2(\text{PPh}_3)_2} \ R\text{———}R'$$

## Proposed Mechanism:

R = aryl, vinyl, or pyridinyl
X = halogen

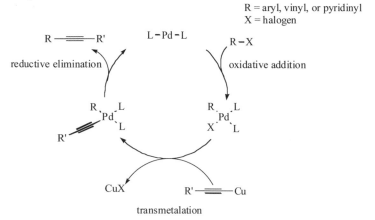

reductive elimination

oxidative addition

transmetalation

## Notes:

For a recent summary of the reaction see:
R. R. Tykwinski, *Angewandte Chemie International Edition in English* **2003**, <u>42</u>, 1566

## Examples:

A useful selection containing a number of functional groups is presented.

$$\xrightarrow[\text{microwave}]{\substack{\text{PdCl}_2(\text{PPh}_3)_2 \\ \text{CuI, Et}_2\text{N, DMF,}}}$$

89-98%

M. Erdelyi, A. Gogol, *Journal of Organic Chemistry* **2003**, <u>68</u>, 6431

In a number of examples with high yields. P(*t*-Bu)$_3$ is useful for helping the reaction take place at room temperature.

$$\xrightarrow[\text{2\% CuI, } i\text{-Pr}_2\text{N, dioxane}]{3\% \text{ PdCl}_2(\text{PhCN})_2, \ 6\% \text{ P}(t\text{-Bu})_3}$$

T. Hundertmark, A. F. Littke, S. L. Buchwald, G. C. Fu, *Organic Letters* **2000**, <u>2</u>, 1729

P. A. Jacobi, H. Lui, *Journal of Organic Chemistry* **1999**, <u>64</u>, 1778

H.-F. Chow, C.-W. Wan, K.-H. Low, Y.-Y. Yeung, *Journal of Organic Chemistry* **2001**, <u>66</u>, 1910

W. R. Goundry, J. E. Baldwin, V. Lee, *Tetrahedron* **2003**, <u>59</u>, 1719

D. L. Comins, A. L. Williams, *Organic Letters* **2001**, <u>3</u>, 3217

M. Feuerstein, H. Doucet, M. Santelli *Tetrahedron Letters* **2004**, <u>45</u>, 1603

# Staudinger Reaction

## The Reaction:

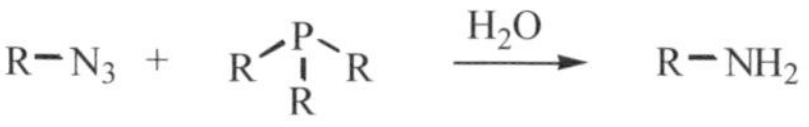

R−N₃ + R−P(R)R → (H₂O) → R−NH₂

## Proposed Mechanism:

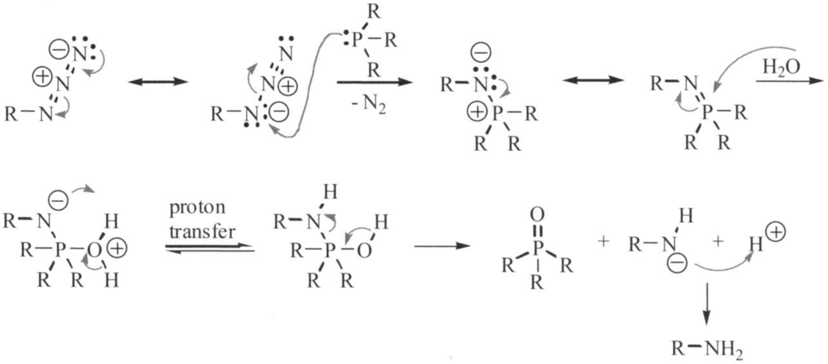

## Notes:

Other reducing agents can be used including LiAlH₄, Fe, and Na₂S₂O₄.

A unique tandem **Staudinger-Wittig reaction** finds application in lactam synthesis:

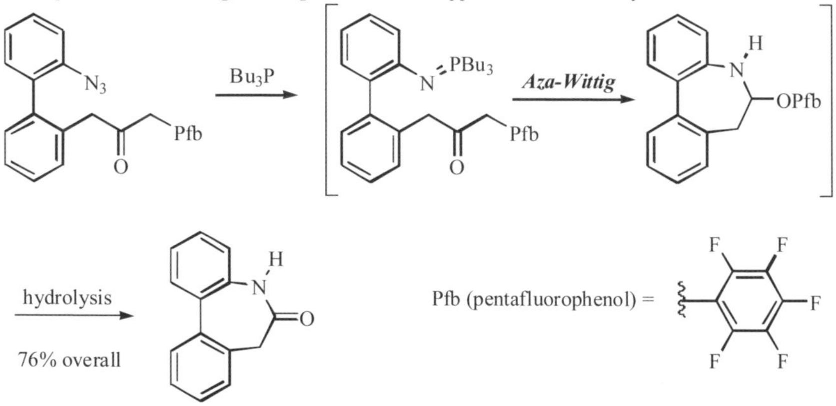

hydrolysis

76% overall

Pfb (pentafluorophenol) =

H. Fuwa, Y. Okamura, Y. Morohashi, T. Tomita, T. Iwatsubo, T. Kan, T. Fukuyam, H. Natsugari, *Tetrahedron Letters* **2004**, <u>45</u>, 2320

## Examples:

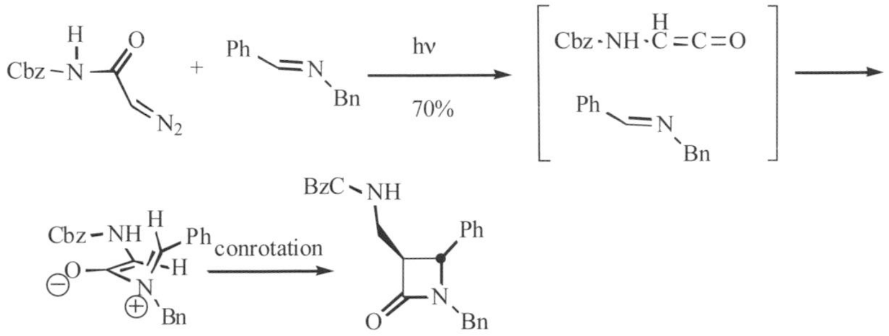

Y. Liang, L. Jiao, S. Zhang, J. Xu, *Journal of Organic Chemistry* **2005**, <u>70</u>, 334

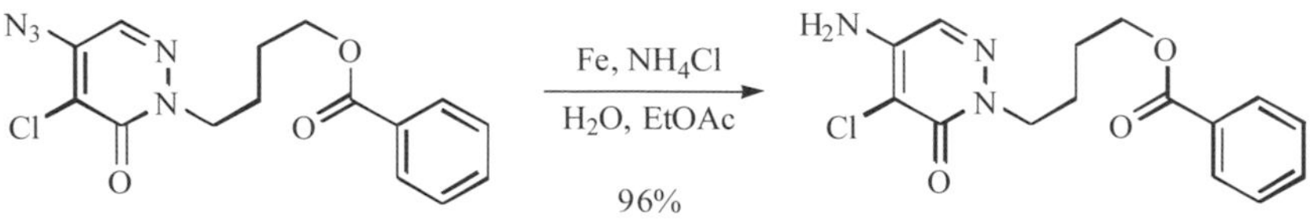

96%

S.-D. Cho, W.-Y. Choi, S.-G. Lee, Y.-J. Yoon, S.-C. Shin, *Tetrahedron Letters* **1996**, <u>37</u>, 7059

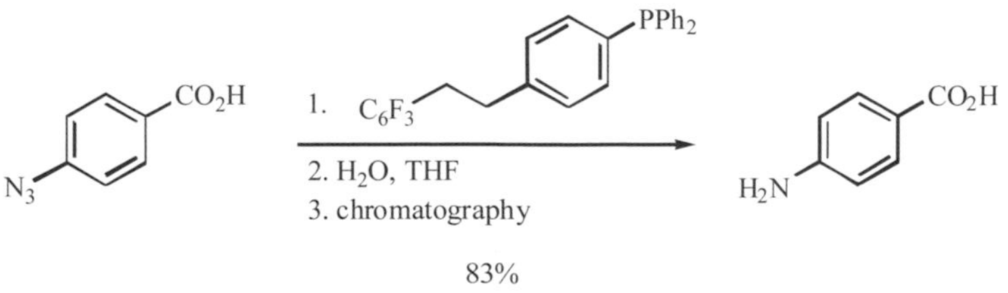

1. C₆F₃

2. H₂O, THF

3. chromatography

83%

C. W. Lindsley, Z. Zhao, R. C. Newton, W. H. Leister, K. A. Strauss, *Tetrahedron Letters* **2002**, <u>43</u>, 4467

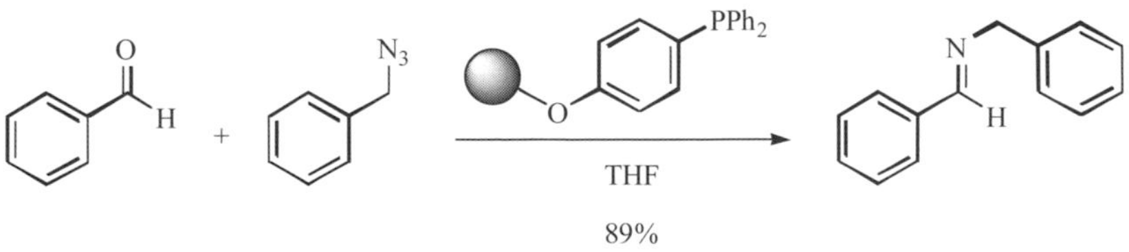

THF

89%

A. B. Charette, A. A. Boezio, M. K. Janes, *Organic Letters* **2000**, <u>2</u>, 3777

# Stetter Reaction (Stetter 1,4-Dicarbonyl Synthesis)

## The Reaction:

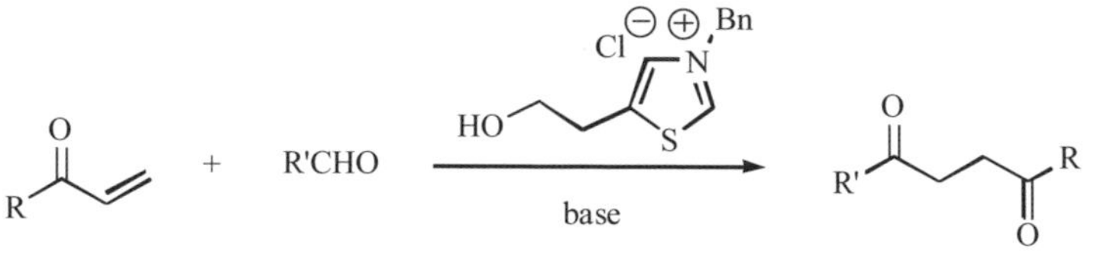

## Proposed Mechanism:

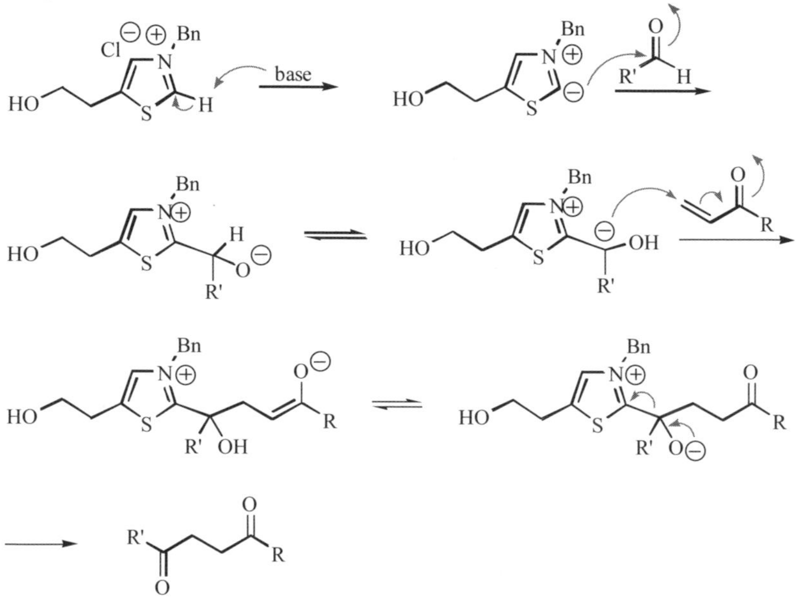

## Notes:

H. Stetter, H. Kuhlmann, *Organic Reactions* **40**, 4

This reaction bears much mechanistic similarity to a vinylogous ***Benzoin Condensation***.

The 1,4-dicarbonyl compounds provide access to ***Paal-Knorr*** sequences. A one-pot, three-step reaction sequence has been reported:

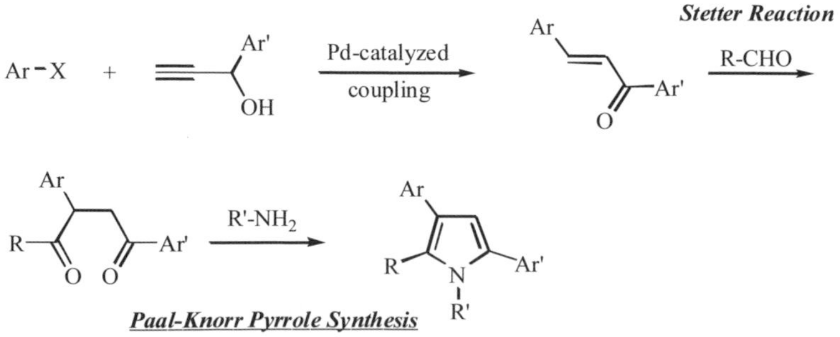

***Paal-Knorr Pyrrole Synthesis***

R. U. Braun, K. Zeitler, T. J. J. Müller *Organic Letters* **2001**, *3*, 3297

## Examples:

PBu₃ is used in place of 3-benzyl-5-(hydroxyethyl)-4-methylthiazolium chloride in this example.

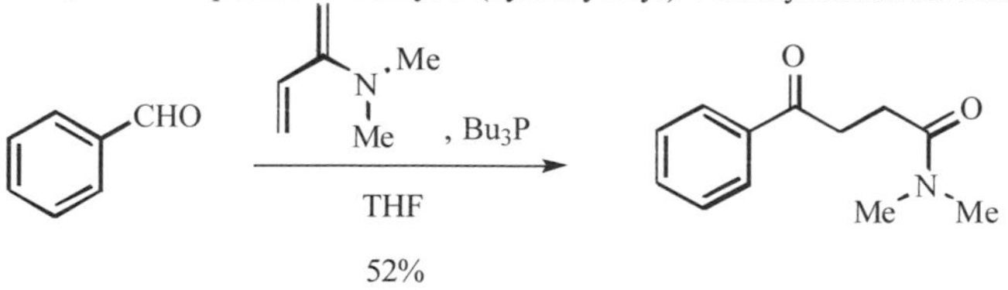

52%

J. H. Gong, Y. J. Im, K. Y. Lee, J. N. Kim, *Tetrahedron Letters* **2002**, <u>43</u>,1247

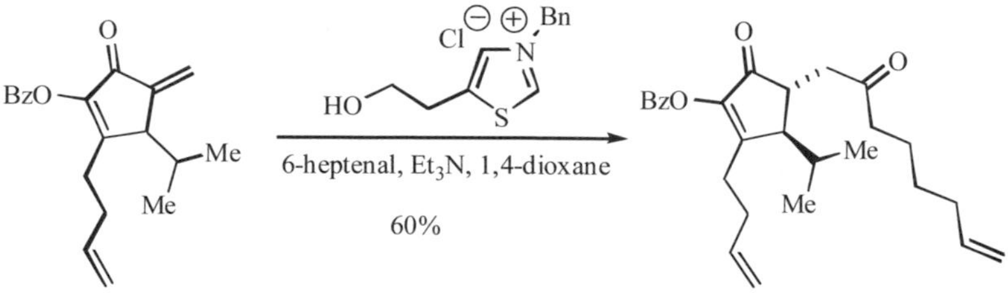

60%

P. E. Harrington, M. A. Tius, *Organic Letters* **1999**, <u>1</u>, 649

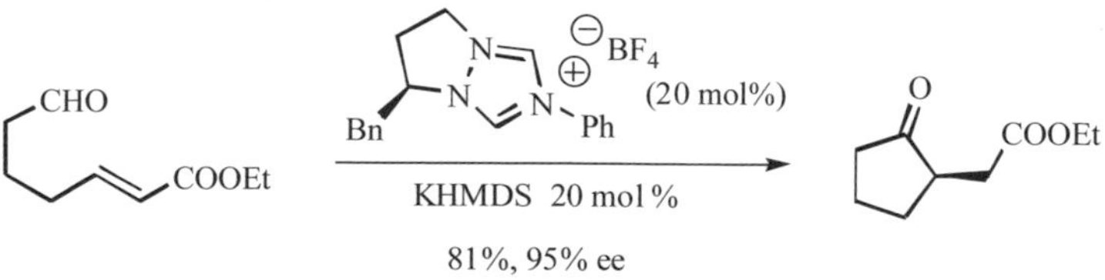

81%, 95% ee

M. S. Kerr, J. R. de Alaniz, T. Rovis, *Journal of the American Chemical Society* **2002**, <u>124</u>, 10298

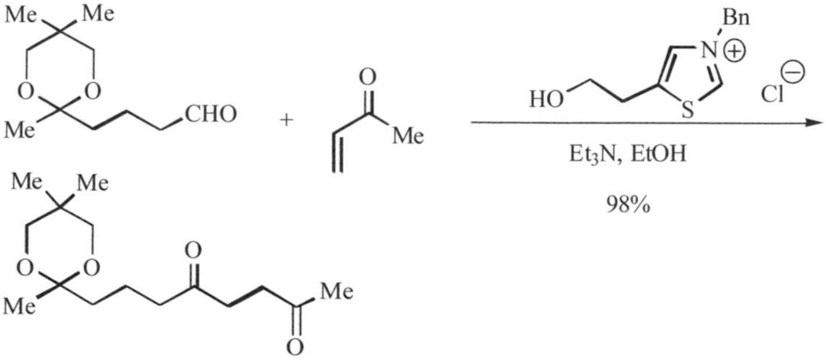

98%

M. P. Sant, W.B. Smith, *Journal of Organic Chemistry* **1993**, <u>58</u>, 5479

# Stevens Rearrangement

## The Reaction:

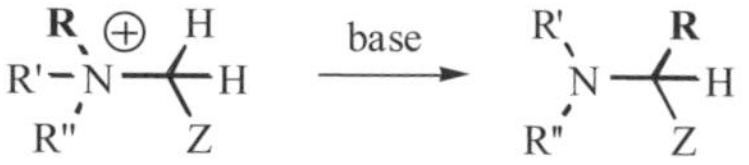

$$R\text{'}-\overset{\overset{\displaystyle R}{\oplus}}{\underset{R\text{''}}{N}}\overset{H}{\underset{Z}{\diagdown}}H \quad \xrightarrow{\text{base}} \quad \overset{R\text{'}}{\underset{R\text{''}}{N}}\overset{R}{\diagdown}\overset{H}{\underset{Z}{\diagdown}}H$$

Z = ketone, ester, aryl

R (with approximate migratory aptitude) = propargyl > allyl > benzyl > alkyl (S. H. Pine, *Organic Reactions* **18**, 4)

## Proposed Mechanism:

A much-used mechanism:

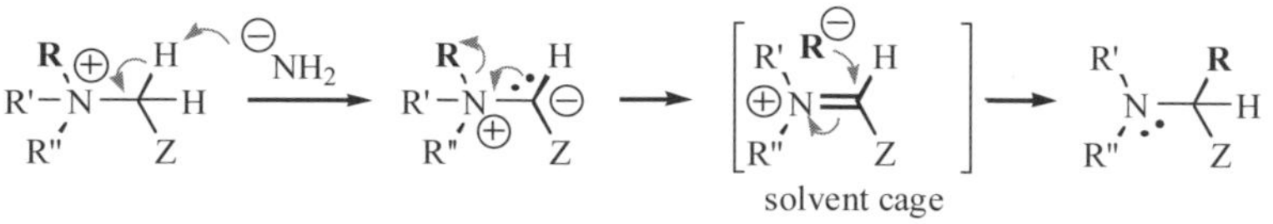

Evidence also favors:

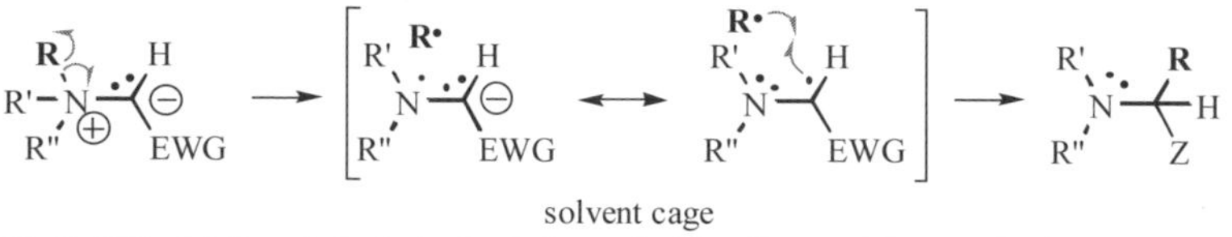

solvent cage

W. D. Ollis, M. Rey, I. O. Sutherland, *Journal of the Chemical Society, Perkin Transaction I* **1983**, 1009

## Notes:

M. B. Smith, J. March in *March's Advanced Organic Chemistry*, 5$^{th}$ ed., John Wiley and Sons, Inc., New York, 2001, p. 1419; T. Laue, A. Plagens, *Named Organic Reactions*, John Wiley and Sons, Inc., New York, 1998, pp. 248-250; S. H. Pine, *Organic Reactions* **18**, 4

The reaction is often associated with migrations on sulfur:

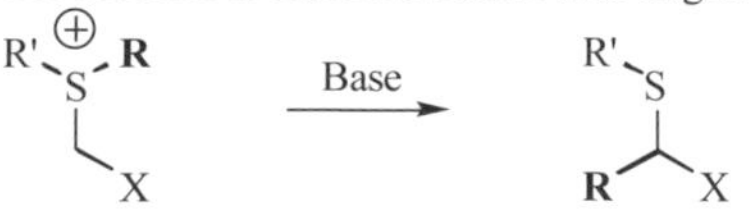

## Examples:

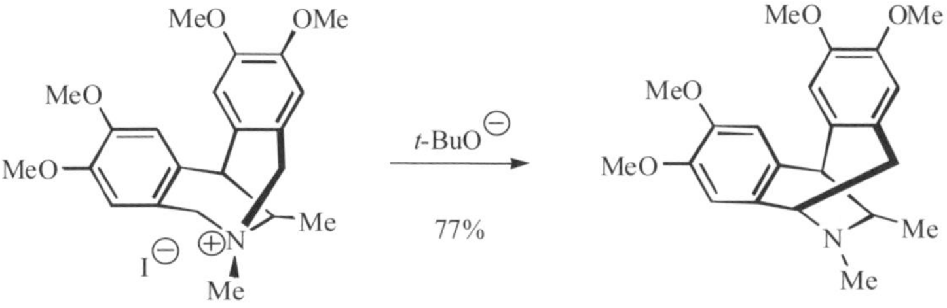

77%

S. Hanessian, M. Mauduit, *Angewandte Chemie, International Edition in English* **2001**, _40_, 3810

F. P. Marmsaeter, G. K. Murphy, F. G. West, *Journal of the American Chemical Society* **2003**, <u>125</u>, 14724

S. Knapp, G. J. Morriello, G. A. Doss, *Tetrahedron Letters* **2002**, <u>43</u>, 5797

J. A. Vanecko, J. F. G. West, *Organic Letters* **2002**, <u>4</u>, 2813

# Stille Coupling

## The Reaction:

$$R-X \ + \ R'-Sn(R'')_3 \ \xrightarrow[\text{THF or DMF}]{\text{Pd(0)}} \ R-R'$$

## Proposed Mechanism:

Catalytic Cycle:

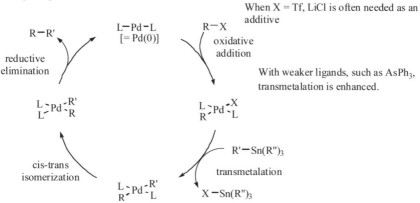

When X = Tf, LiCl is often needed as an additive

With weaker ligands, such as AsPh$_3$, transmetalation is enhanced.

## Notes:

In many respects this is a very general and useful reaction. It is compatible with a variety of functional groups and is run under relatively neutral conditions. The organotin intermediates are easy to prepare; however, toxicity of the tin reagents makes this method more of a problem than some of the other coupling protocols.

M. B. Smith, J. March in *March's Advanced Organic Chemistry*, 5$^{th}$ ed., John Wiley and Sons, Inc., New York, 2001, p. 931; V. Farina, V. Krishnamurthy, W. J. Scott, *Organic Reactions* **50**.

See also the book based on the *Organic Reactions* article: V. Farina, V. Krishnamurthy, W. J. Scott, *The Stille Reaction*, John Wiley and Sons, Inc., New York, 1998,

## Examples:

Pd$_2$(dba)$_3$, AsPh$_3$

cyclohexane

38%

C. Boden, G. Pattenden, *Synlett* **1994**, 181 (AN 1995:6867)

S. T. Handy, X. Zhang, *Organic Letters* **2001**, *3*, 233

R. E. Maleczka, Jr., W. P. Gallagher, I. Terstiegee, *Journal of the American Chemical Society* **2000**, 122, 384

C. M. Rayner, P. C. Astles, L. A. Paquette, *Journal of Organic Chemistry* **1991**, *56*, 1489

# Stille Carbonylative Coupling

## The Reaction:

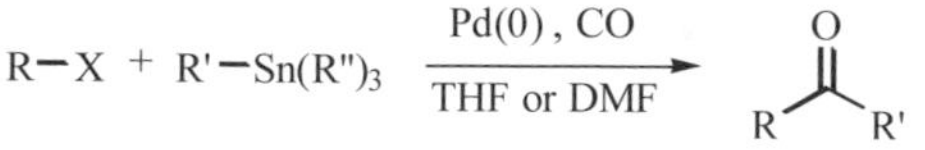

$$R-X + R'-Sn(R'')_3 \xrightarrow[\text{THF or DMF}]{\text{Pd(0), CO}} \underset{R \quad R'}{\overset{O}{\Vert}}$$

## Proposed Mechanism:

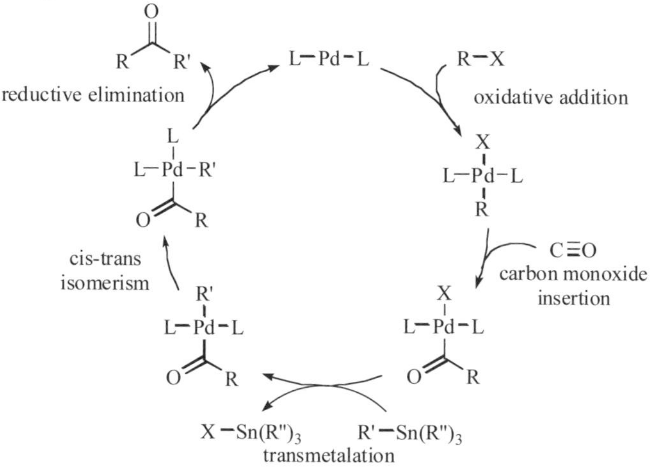

## Notes:

See also the book based on the *Organic Reactions* article: V. Farina, V. Krishnamurthy, W. J. Scott, The *Stille Reaction*, John Wiley and Sons, Inc., New York, 1998,

## Examples:

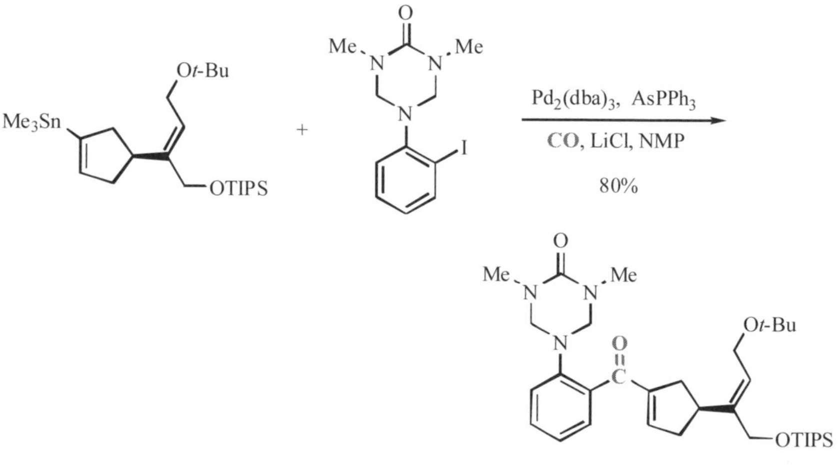

S. D. Kight, L. E. Overman, G. Pairaudeau, *Journal of the American Chemical Society* **1993**, <u>115</u>, 9293

S. Ceccarelli, U. Piarulli, C. Gennart, *Journal of Organic Chemistry* **2000**, <u>65</u>, 6254

|              |     |
| ------------ | --- |
| No added CuI | 38% |
| added CuI    | 63% |

R. D. Mazzola, Jr., S. Giese, C. L. Benson, F. G. West, *Journal of Organic Chemistry* **2004**, <u>69</u>, 220

# Stille-Kelly Reaction

## The Reaction:

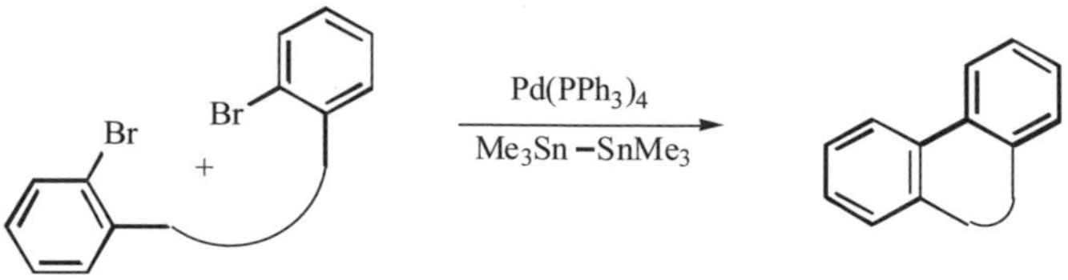

## Proposed Mechanism:

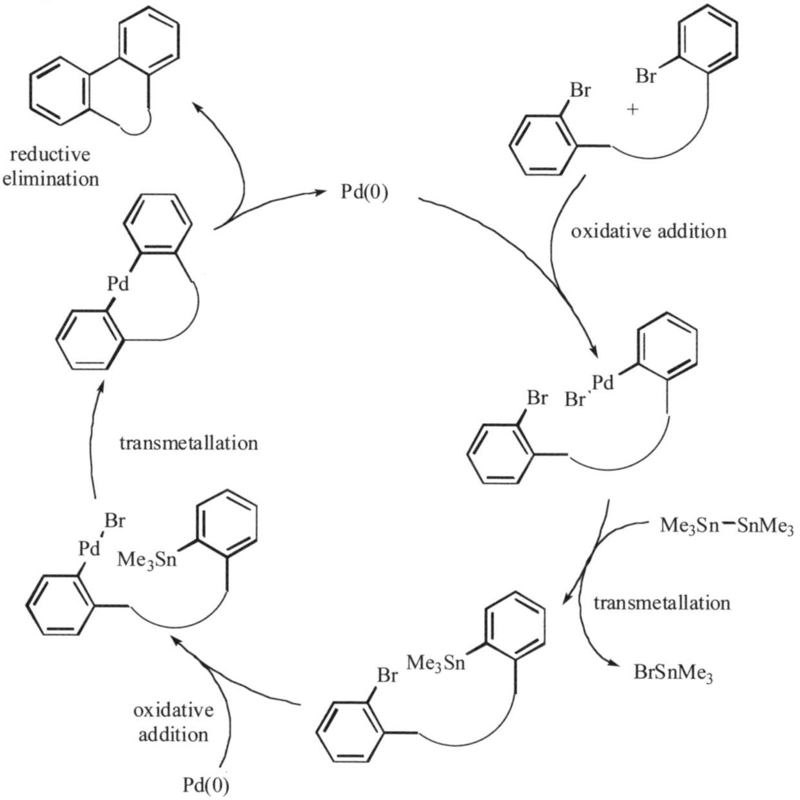

## Examples:

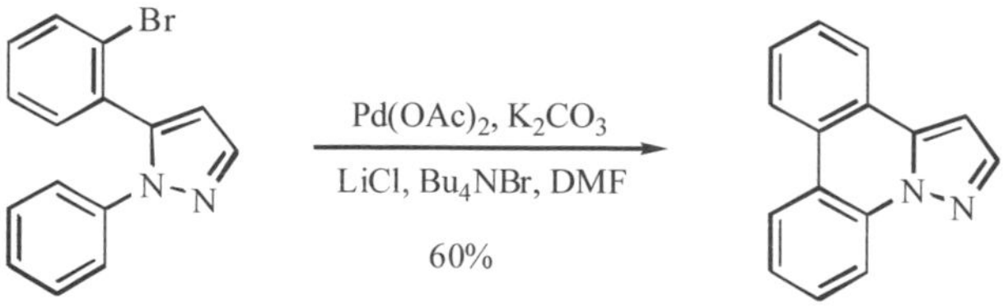

Pd(OAc)$_2$, K$_2$CO$_3$

LiCl, Bu$_4$NBr, DMF

60%

S. Hernandez, R. SanMartin, I. Tellitu, E. Dominguez, *Organic Letters* **2003**, *5*, 1095

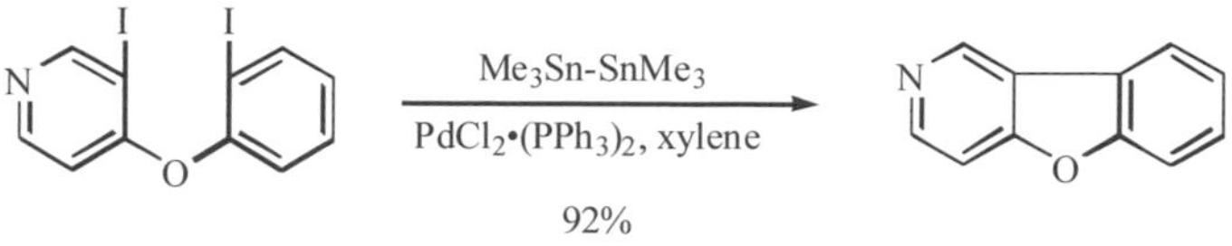

Me$_3$Sn-SnMe$_3$

PdCl$_2$•(PPh$_3$)$_2$, xylene

92%

W. S. Yue, J. J. Li, *Organic Letters* **2002**, *4*, 2201

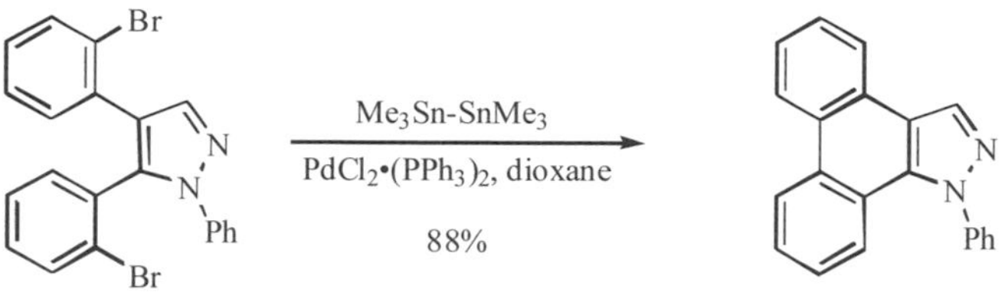

Me$_3$Sn-SnMe$_3$

PdCl$_2$•(PPh$_3$)$_2$, dioxane

88%

R.Olivera, R. SanMartin, E, Dominguez, *Journal of Organic Chemistry* **2000**, *65*, 7010

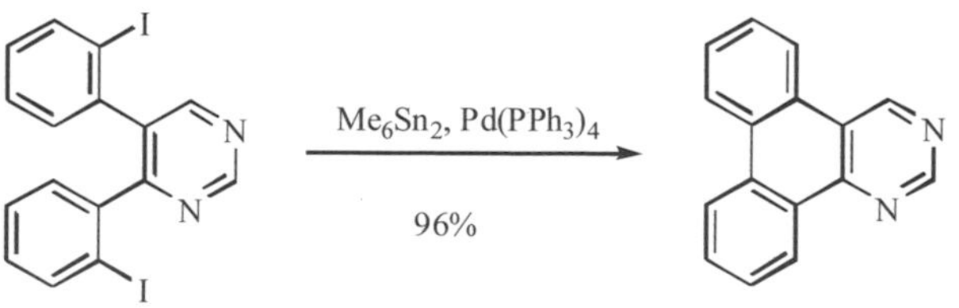

Me$_6$Sn$_2$, Pd(PPh$_3$)$_4$

96%

R. Olivera, R. SanMartin, I. Tellitu, E. Domínguez, *Tetrahedron* **2002**, *58*, 3021

# Stobbe Condensation

## The Reaction:

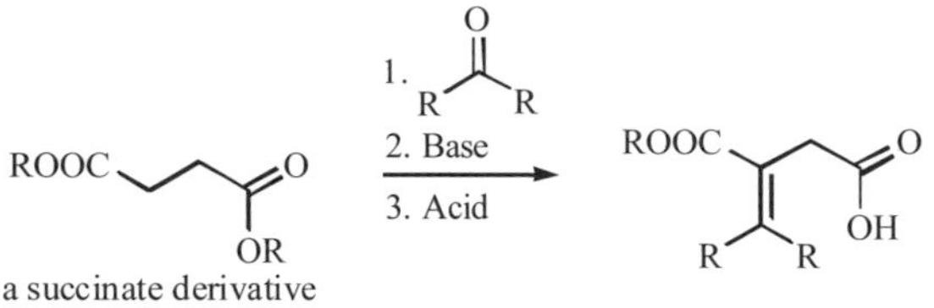

1.
2. Base
3. Acid
ROOC
OR
a succinate derivative
OH

## Proposed Mechanism:

Base
ROOC
OR
Base

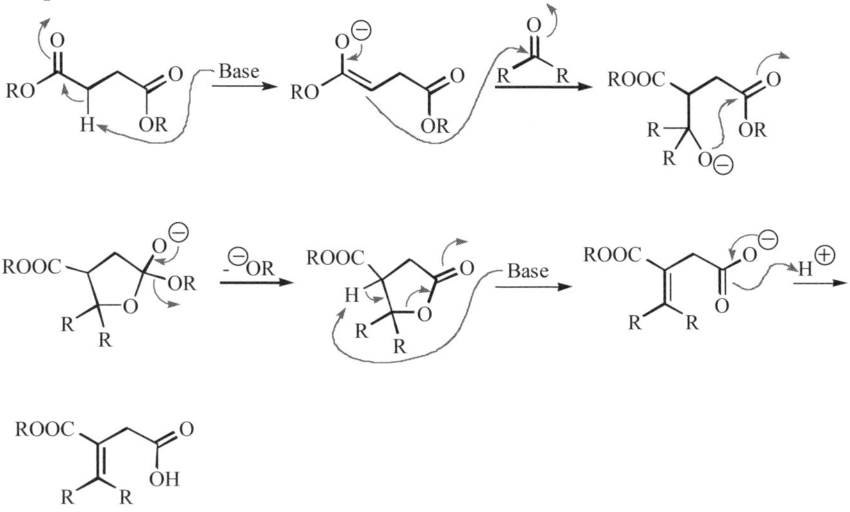

## Notes:

M. B. Smith, J. March in *March's Advanced Organic Chemistry*, 5th ed., John Wiley and Sons, Inc., New York, 2001, p. 1224; W. S. Johnson, G. H. Daub, *Organic Reactions* **6**, 1

## Examples:

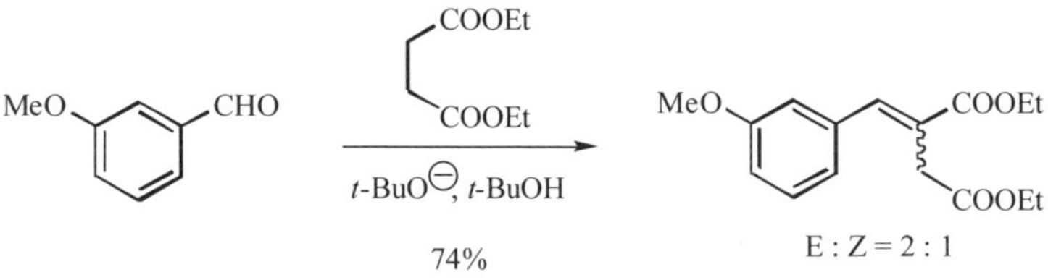

MeO
CHO
COOEt
COOEt
t-BuO⊖, t-BuOH
74%
E : Z = 2 : 1

D. L. Boger, J. A. McKie, H. Cai, B. Cacciari, P. G. Baraldi, *Journal of Organic Chemistry* **1996**, 61, 1710

J. Liu, N. R. Brooks, *Organic Letters* **2002**, *4*, 3521

A. Piettre, E. Chevenier, C. Massardier, Y. Gimbert, A. E. Green, *Organic Letters* **2002**, *4*, 3139

J. D. White, P. Hrnclar, F. Stappenbeck, *Journal of Organic Chemistry* **1999**, *64*, 7871

R. Baker, P. H. Briner, D. A. Evans, *Chemical Communications* **1978**, 410

# Stork Enamine Synthesis

## The Reaction:

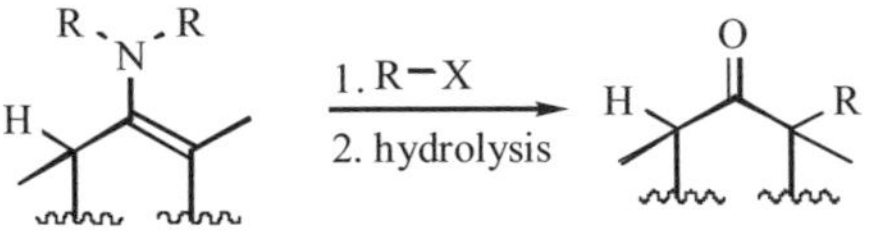

## Proposed Mechanism:

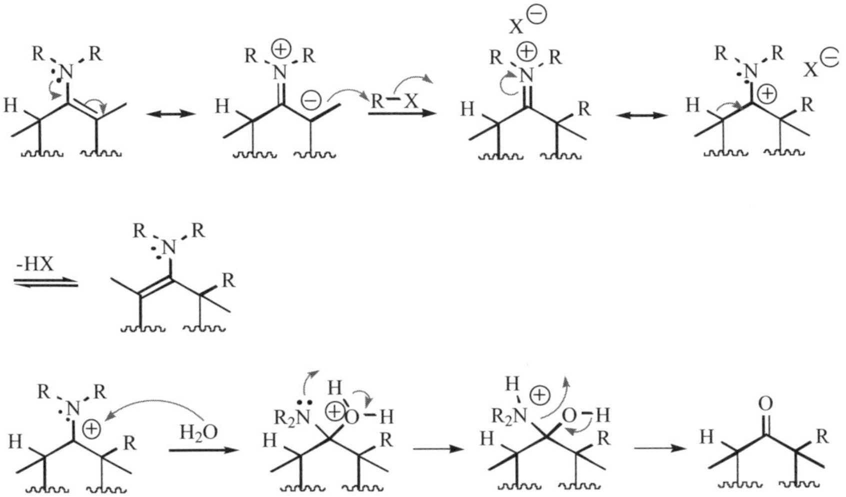

## Notes:

M. B. Smith, J. March in *March's Advanced Organic Chemistry*, 5th ed., John Wiley and Sons, Inc., New York, 2001, pp. 555, 787; T. Laue, A. Plagens, *Named Organic Reactions*, John Wiley and Sons, Inc., New York, 1998, pp. 250-253

Starting material preparation:

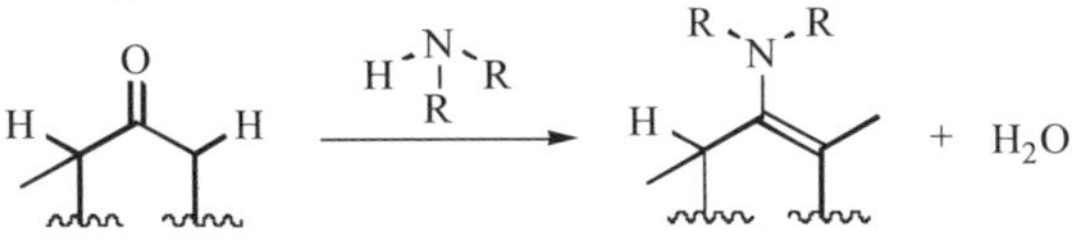

The amine used to make enamine is usually pyrrolidine, morpholine, piperidine or diethylamine.

Besides alkyl halides, the electrophile can be activated aryl halides, epoxides, anhydrides and *Michael additions* to activated alkenes.

## Examples:

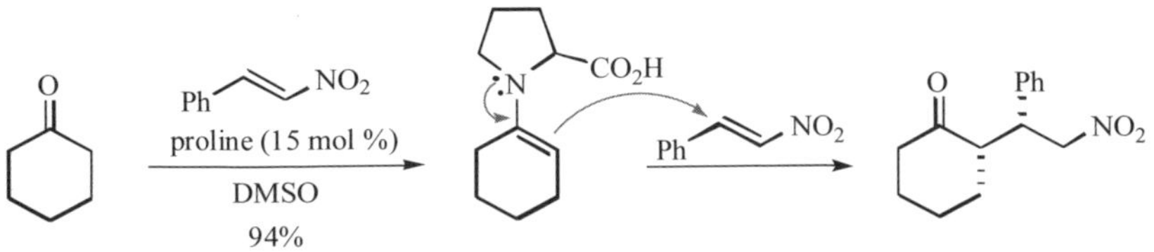

B. Kist, P. Pojarliev, H. J. Martin, *Organic Letters* **2001**, *3*, 2423

Stork
enamine synthesis
Pd(OAc)$_2$, PPh$_3$
Trost's TMM Reaction
45%

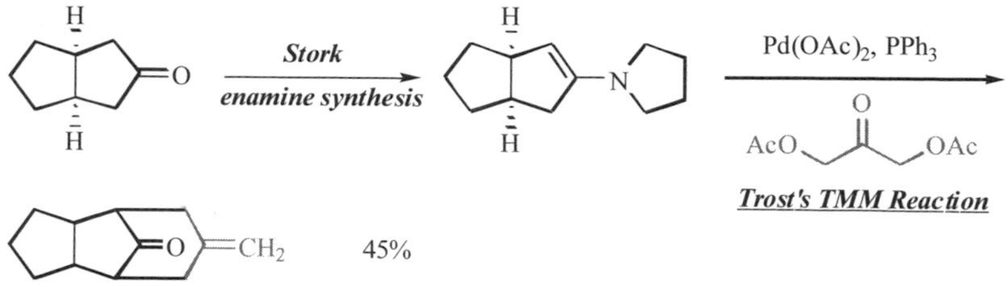

F. Bouno, A. Tenaglia, *Journal of Organic Chemistry* **2000**, 65, 3869

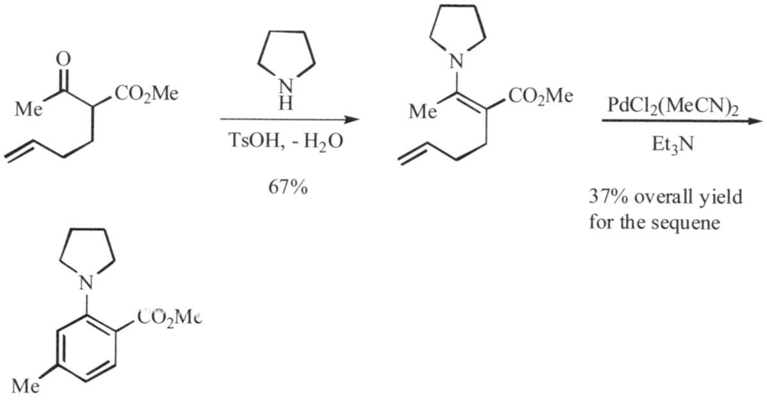

TsOH, - H$_2$O
67%
PdCl$_2$(MeCN)$_2$
Et$_3$N
37% overall yield
for the sequene

Method of: S. Hunig, E. Lücke, W. Brenninger, *Organic Syntheses*, CV 5, 808
in: T. Ishikawa, E. Vedo, R. Tani, S. Saito, *Journal of Organic Chemistry* **2001**, 66, 186

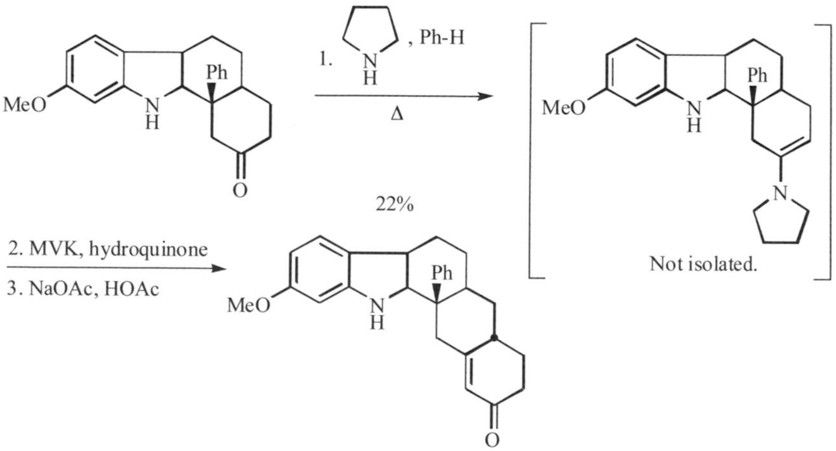

1. , Ph-H
Δ
22%
2. MVK, hydroquinone
3. NaOAc, HOAc
Not isolated.

J. J. Li, B. K. Trivedi, J. R. Rubin, B. D. Roth, *Tetrahedron Letters* **1998**, 39, 6111

# Stork-Wittig Olefination

## The Reaction:

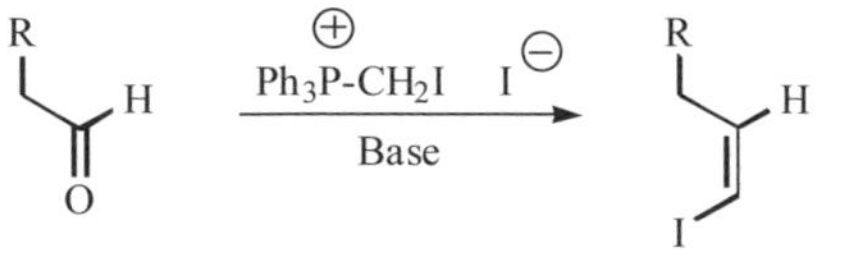

## Proposed Mechanism:

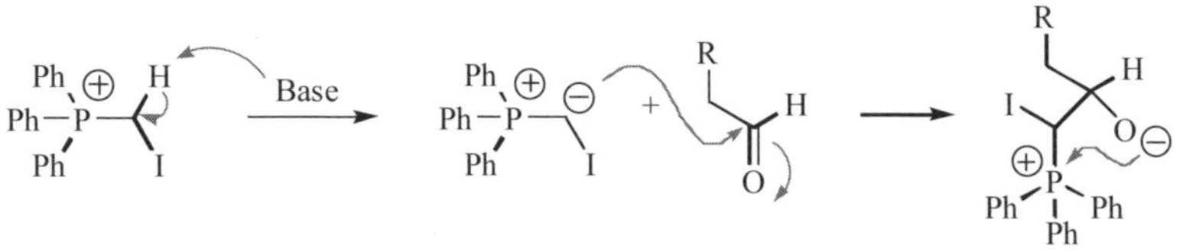

See: **_Wittig Reaction_**

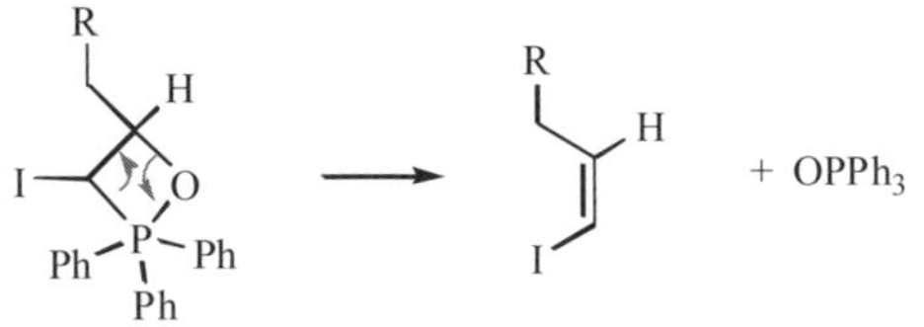

## Notes:
The **_Takai Reaction_** provides the same transformation, but arrives at the E- configuration rather than the Z-.

## Examples:

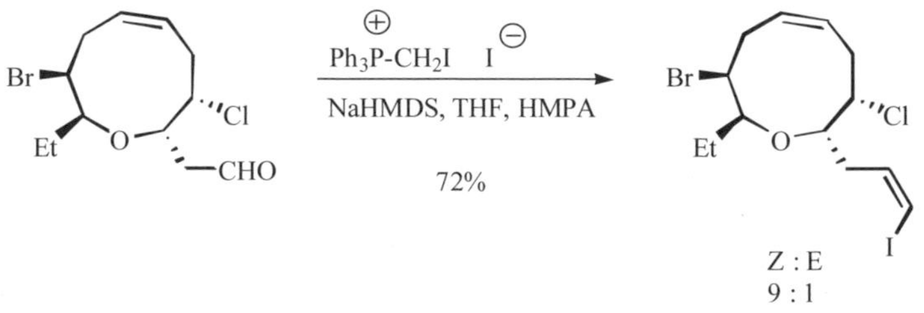

$$Z : E$$
$$9 : 1$$

M. T. Crimmins, M. T. Powell, *Journal of the American Chemical Society* **2003**, <u>125</u>, 7592

K. Lee, J. K. Cha, *Journal of the American Chemical Society* **2001**, <u>123</u>, 559

S. S. Harried, C. P. Lee, G. Yang, T. I. H. Lee, D. C. Myles, *Journal of Organic Chemistry* **2003**, <u>68</u>, 6646

J. E. Davies, A. B. Holmes, J. P. Adams, *Journal of the American Chemical Society* **1999**, <u>121</u>, 4900

# Strecker Amino Acid Synthesis

## The Reaction:

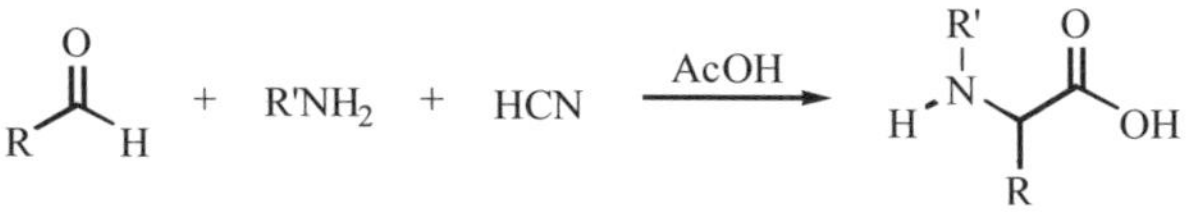

## Proposed Mechanism:

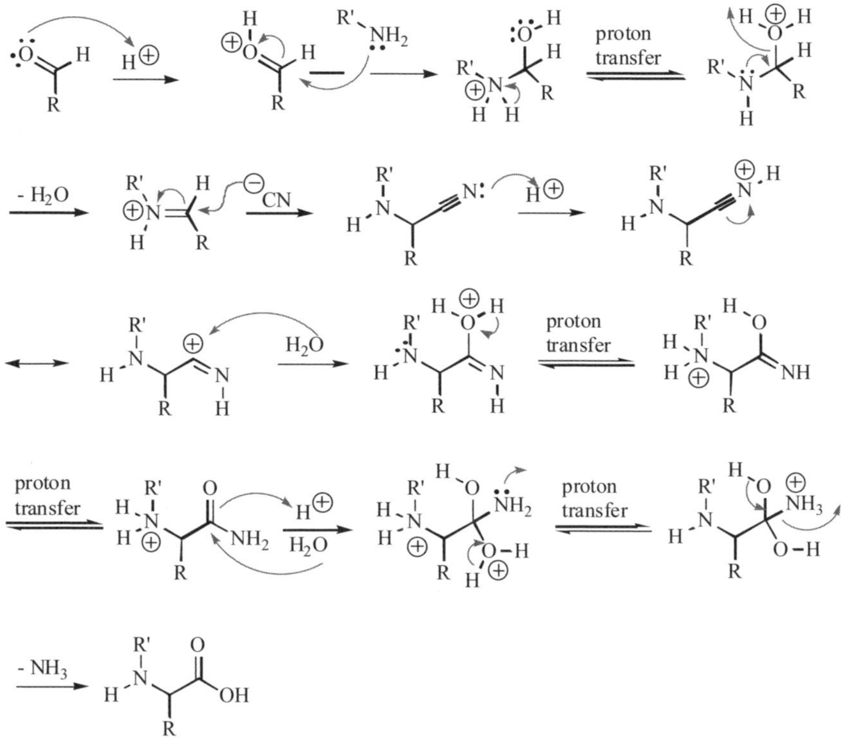

## Notes:

T. Laue, A. Plagens, *Named Organic Reactions*, John Wiley and Sons, Inc., New York, 1998, pp. 253-254

## Examples:

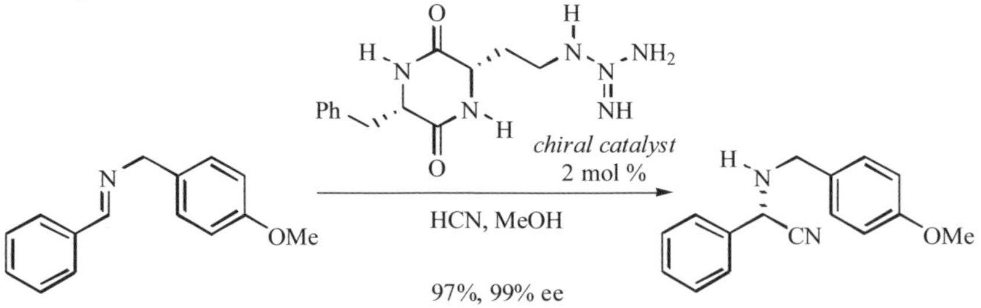

97%, 99% ee

M. S. Iyer, K. M. Gigstad, N. D. Namdev, M. Lipton, *Journal of the American Chemical Society* **1996**, 118, 4910

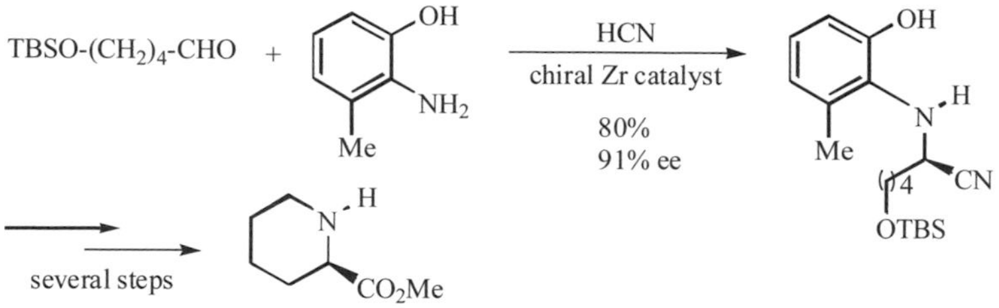

TBSO-(CH$_2$)$_4$-CHO + [structure: 2-amino-3-methylphenol with OH and NH$_2$, Me] $\xrightarrow[\substack{\text{chiral Zr catalyst} \\ 80\% \\ 91\% \text{ ee}}]{\text{HCN}}$ [structure: product with OH, N-H, Me, (CH$_2$)$_4$ CN, OTBS]

$\xrightarrow{\text{several steps}}$ [structure: piperidine with N-H and CO$_2$Me]

H. Ishitani, S. Komiyama, Y. Hasegawa, S. Kobayashi, *Journal of the American Chemical Society* **2000**, <u>122</u>, 762

An anomolous reduction of a **Strecker nitrile**:

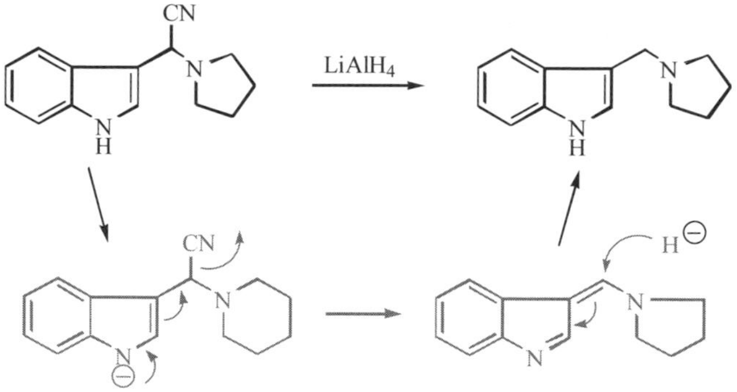

[structure: indole-3-yl with CN and pyrrolidine substituent] $\xrightarrow{\text{LiAlH}_4}$ [structure: indol-3-ylmethyl-pyrrolidine]

[structure with CN and piperidine, indolyl anion] $\longrightarrow$ [structure: exocyclic methylene indolenine with pyrrolidine, H$^{\ominus}$]

P. Rajagopalan, B. G. Advani, *Tetrahedron Letters* **1965**, <u>6</u>, 2197

# Strecker Degradation

## The Reaction:

$$+ NH_3 + CO_2$$

## Proposed Mechanism:
G. P. Rizzi, *Journal of Organic Chemistry* **1969**, <u>34</u>, 2002

## Notes:
The formation of aldehydes (flavorings) in roasting of cocoa beans, for example, is caused by *Strecker degradation* of amino acids.

A. Arnoldi, C. Arnoldi, O. Baldi, A. Griffini, *Journal of Agricultural and Food Chemistry* **1987**, <u>35</u>, 1035

Other degradation reactions:
*Bergmann Degradation*

*Darapsky Degradation (Procedure)*

## *von Braun (Amide) Degradation / Reaction*

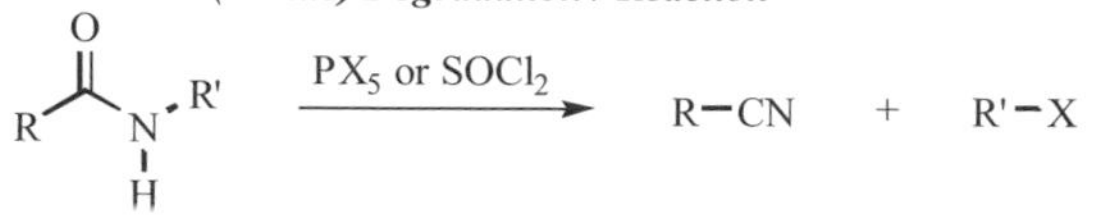

$$R-CN \quad + \quad R'-X$$

## Examples:

This was used as a visual test for amino acids, where alloxane reacted with the amino acid to produce murexide, a colored compound:

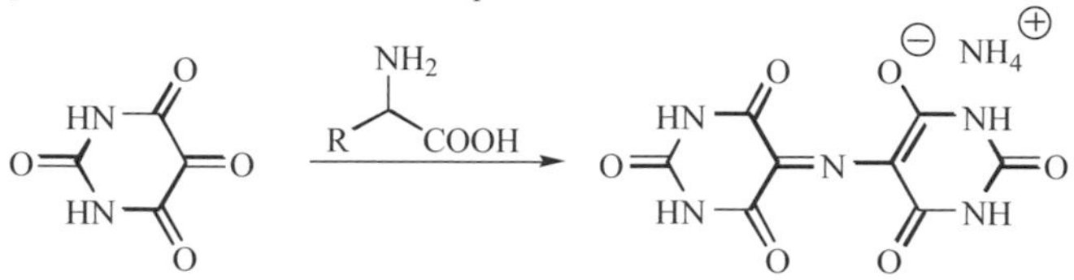

M. F. Aly, G. M. El-Nagger, T. I. El-Emary, R. Grigg, S. A. M. Metwally, S. Sivagnnam, *Tetrahedron* **1994**, 50, 895

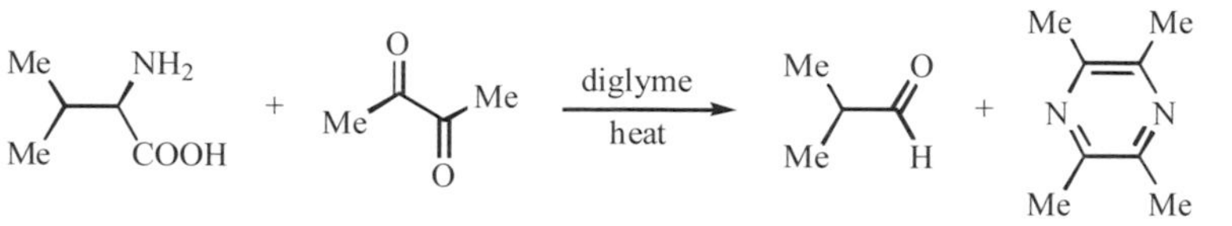

low yields of isolated products, but rapid loss of the butane dione.

G. P. Rizzi, *Journal of Organic Chemistry* **1969**, 34, 2002

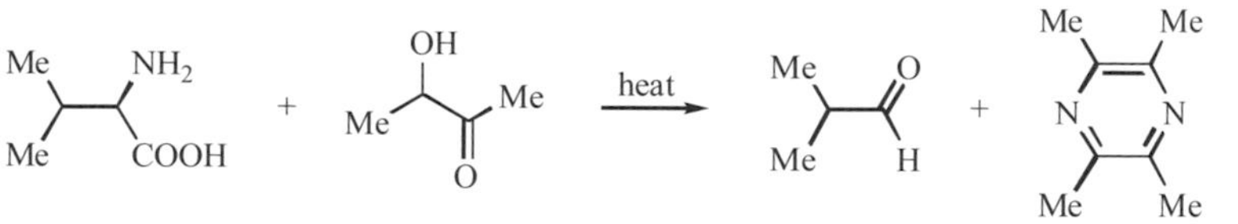

GC-MS analysis of head space

C.-K. Shu, *Journal of Agricultural and Food Chemistry* **1998**, 46, 1515

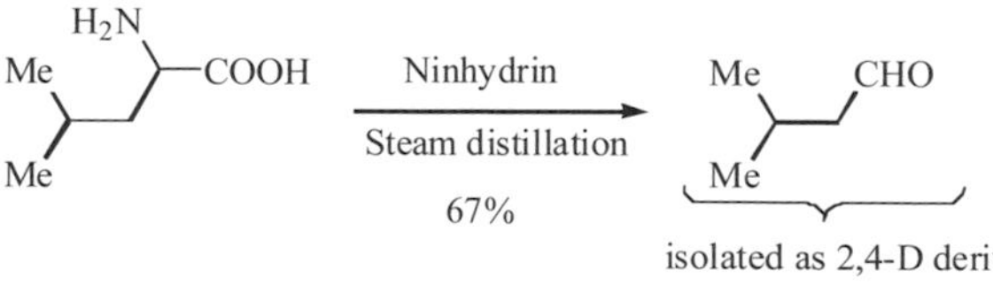

isolated as 2,4-D derivative

W. S. Fones, *Journal of the American Chemical Society* **1954**, 76, 1377

# Suzuki Coupling

## The Reaction:

N. Miyauri, A. Suzuki, *Chemical Reviews* **1995**, <u>95</u>, 2457

$$R-X \quad + \quad \underset{R'}{\overset{H}{\underset{}{\bigg\rangle}}}=\underset{R''}{\overset{B-OR^*}{\underset{}{\bigg\langle}}} \quad \xrightarrow[\text{NaOR or NaOH}]{Pd(0)} \quad \underset{R'}{\overset{H}{\underset{}{\bigg\rangle}}}=\underset{R''}{\overset{R}{\underset{}{\bigg\langle}}}$$

Where X = I >> Br > OTf >> Cl.

See: A useful review: S. R. Chemler, D. Trauner, S. J. Danishefsky, *Angewandte Chemie International Edition in English* **2001**, <u>40</u>, 4544

## Proposed Mechanism:

$$\underset{L}{\overset{L}{\underset{}{}}}\text{Pd}\underset{L}{\overset{L}{\underset{}{}}}$$

$$\downarrow - 2L$$

L—Pd—L
[= Pd(0)]

reductive elimination

cis-trans isomerism

R—X
oxidative addition

$$\underset{R}{\overset{L}{}}\text{Pd}\underset{L}{\overset{X}{}}$$

NaOR

NaX

$$\underset{R}{\overset{L}{}}\text{Pd}\underset{L}{\overset{OR}{}}$$

transmetalation

$$RO-B\underset{OR^*}{\overset{OR^*}{}}$$

## Notes:

General sources of common boron reagents:

$$R\text{-Li} \; + \; B(R'O)_3 \; \Longleftarrow \; R-B\underset{OR'}{\overset{OR'}{}} \; \xrightarrow{\text{Hydrolysis}} \; R-B\underset{OH}{\overset{OH}{}}$$

$$R\diagdown=\diagup \Longrightarrow R-\!\!\!\equiv\!\!\!- \;+\; HB\underset{O}{\overset{O}{}}$$

$$\underset{}{\overset{R}{\underset{}{\overset{|}{B}}}}$$

$$BCl_3 \; + \; R_3SiH$$

$$\downarrow$$

$$R-\!\!\!\equiv\!\!\!-H \; \xrightarrow{HBCl_2} \; \underset{H}{\overset{R}{\underset{}{\bigg\rangle}}}=\underset{Cl}{\overset{B-Cl}{\underset{}{\bigg\langle}}} \; \xrightarrow{R'OH} \; \underset{H}{\overset{R}{\underset{}{\bigg\rangle}}}=\underset{R'O}{\overset{B-OR'}{\underset{}{\bigg\langle}}}$$

## Examples:

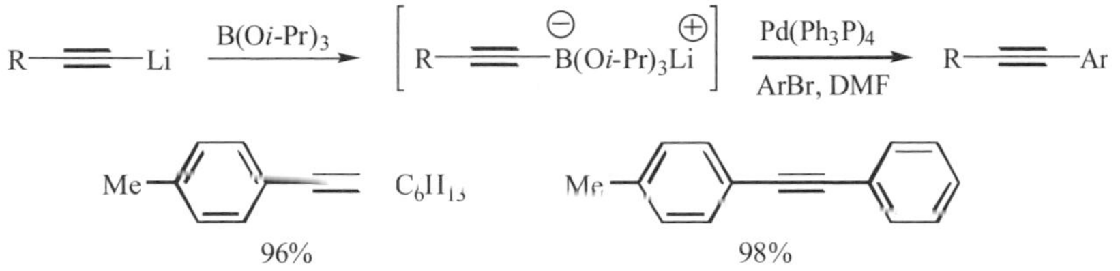

R——≡——Li  $\xrightarrow{\text{B(O}i\text{-Pr)}_3}$  [ R——≡——$\overset{\ominus}{\text{B}}$(O$i$-Pr)$_3\overset{\oplus}{\text{Li}}$ ]  $\xrightarrow[\text{ArBr, DMF}]{\text{Pd(Ph}_3\text{P)}_4}$  R——≡——Ar

Me——⟨ ⟩——≡——C$_6$H$_{13}$        Me——⟨ ⟩——≡——⟨ ⟩

96%                                      98%

A.-S. Castanet, F. Colobert, T. Schlama, *Organic Letters*, **2000**, <u>2</u>, 3559.

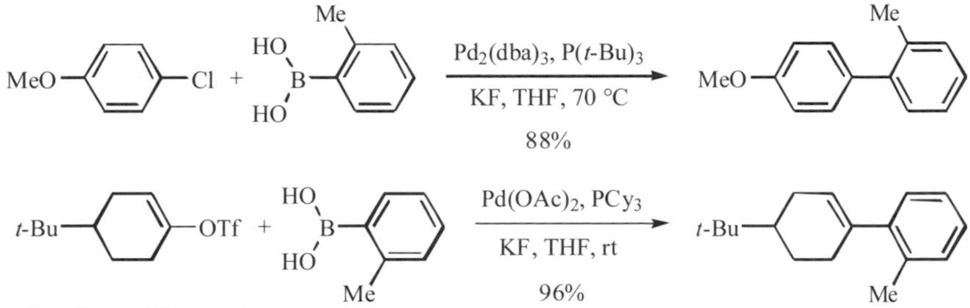

MeO——⟨ ⟩——Cl  +  HO—B—OH (2-Me-phenyl)  $\xrightarrow[\text{KF, THF, 70 °C}]{\text{Pd}_2(\text{dba})_3, \text{P}(t\text{-Bu})_3}$  MeO——⟨ ⟩——⟨ ⟩ (2-Me)

88%

$t$-Bu——⟨cyclohexenyl⟩——OTf  +  HO—B—OH (2-Me-phenyl)  $\xrightarrow[\text{KF, THF, rt}]{\text{Pd(OAc)}_2, \text{PCy}_3}$  $t$-Bu——⟨cyclohexenyl⟩——⟨ ⟩ (2-Me)

96%

dba: dibenzylidene acetone

A.F. Littke, C. Dai, G.C. Fu, *Journal of the American Chemical Society* **2000**, <u>127</u>, 4020

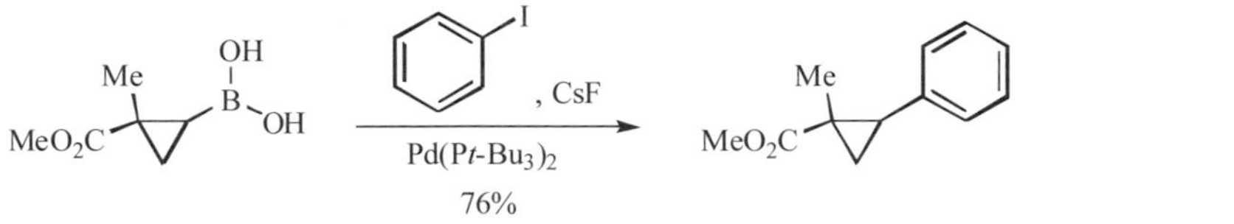

MeO$_2$C—(cyclopropane, Me)—B(OH)$_2$  +  ⟨ ⟩—I , CsF  $\xrightarrow[\text{Pd(P}t\text{-Bu}_3)_2]{}$  MeO$_2$C—(cyclopropane, Me)—Ph

76%

M. Rubina, M. Rubin, V. Geforgyan, *Journal of the American Chemical Society* **2003**, <u>125</u>, 7198

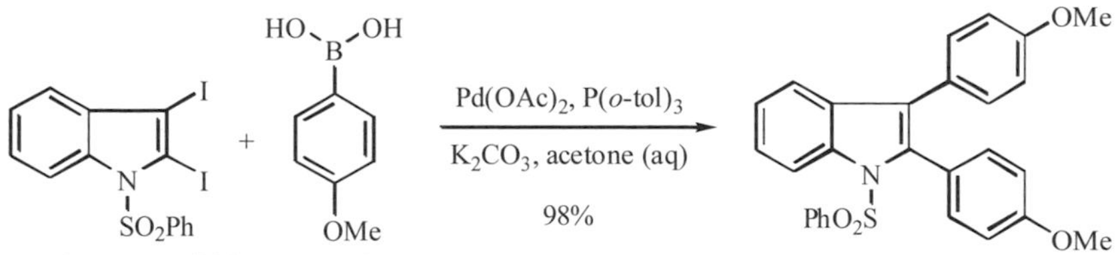

(3-I-2-I-indole, N-SO$_2$Ph)  +  HO—B—OH (4-OMe-phenyl)  $\xrightarrow[\text{K}_2\text{CO}_3, \text{acetone (aq)}]{\text{Pd(OAc)}_2, \text{P}(o\text{-tol})_3}$  (indole with 3-(4-OMe-phenyl), 2-(4-OMe-phenyl), N-SO$_2$Ph)

98%

Y. Liu, G.W. Gribble, *Tetrahedron Letters* **2000**, <u>41</u>, 8717

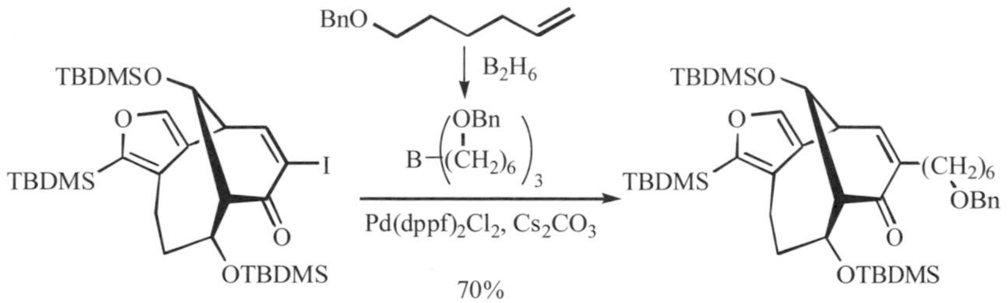

BnO⌒⌒⌒⌒═  $\xrightarrow{\text{B}_2\text{H}_6}$  B—[(CH$_2$)$_6$—OBn]$_3$

(TBDMS-bicyclic-I structure)  $\xrightarrow[\text{Pd(dppf)}_2\text{Cl}_2, \text{Cs}_2\text{CO}_3]{}$  (TBDMS-bicyclic-(CH$_2$)$_6$OBn structure)

70%

D. Meng, S. J. Danishfesky, *Angewandte Chemie, International Edition in English* **1999**, <u>38</u>, 1485

# Swern Oxidation

## The Reaction:

DMSO - Dimethylsulfoxide

## Proposed Mechanism:

*A sulfur ylide is formed with base which then abstracts the
α proton, generating the carbonyl and dimethylsulfide.*

## Notes:

M. B. Smith, J. March in *March's Advanced Organic Chemistry*, 5th ed., John Wiley and Sons, Inc.,
New York, 2001, p. 1516; T. T. Tidwell, *Organic Reactions* **39**, 3

Triflouroacetic anhydride was used before oxalyl chloride and is also known as the *Swern
Oxidation*. Acetic anhydride can also be used. (*Albright-Goldman Oxidation*)

An alternative to the *Swern Oxidation*:

Extremely mild, easy workup.

90%                                    74%                                    90%

L. DeLuca, G. Giacomelli, A. Porcheddu, *Journal of Organic Chemistry* **2001**, <u>66</u>, 7807

## Examples:

X. Fang, U. K. Gandarange, T. Wang, J. D. Sol, D. S. Garvey, *Journal of Organic Chemistry* **2001**, <u>66</u>, 4019

M. K. W. Choi, P. H. Toy, *Tetrahedron* **2004**, <u>60</u>, 2875

M. F. Semmelhack, Y. Jujang, D. Ho, *Organic Letters* **2001**, <u>3</u>, 2403

(taken directly into a ***Wittig reaction***

58% overall

T. J. Speed, D. M. Thamattoor, *Tetrahedron Letters* **2002**, <u>43</u>, 367

# Takai Reaction

## The Reaction:

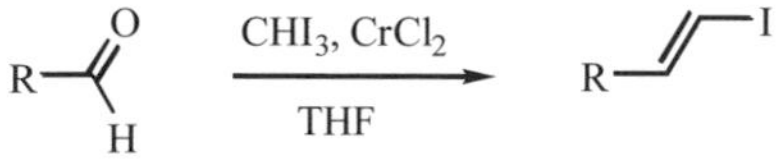

## Proposed Mechanism:

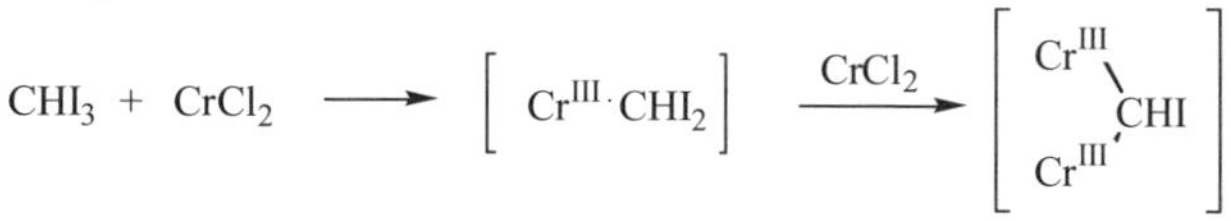

The two Cr-containing intermediates can distribute as shown below:

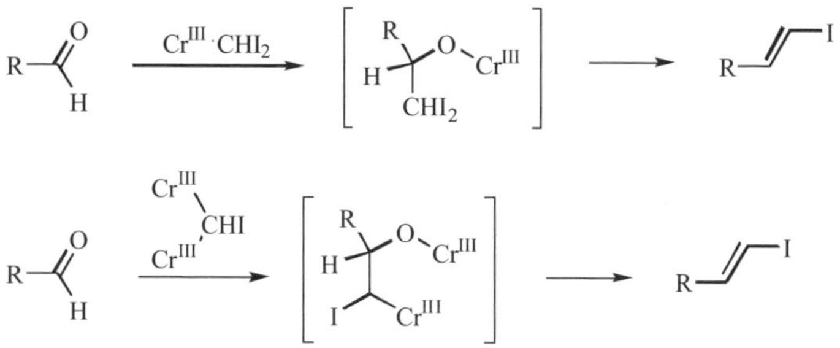

## Notes:

There is also a **_Takai Coupling protocol_**:

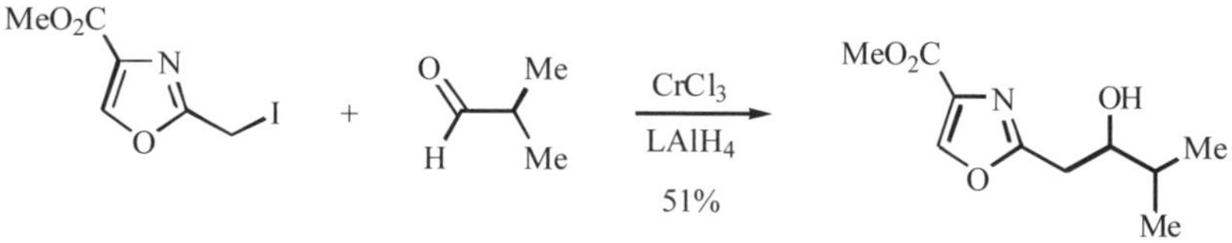

P. Breuilles, D. Uquen, *Tetrahedron Letters* **1998**, <u>39</u>, 3149

See: **_Stork-Wittig Olefination_** for a related procedure:

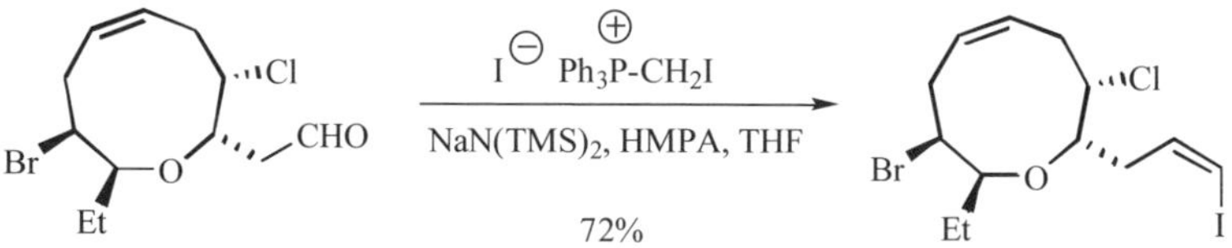

M. T. Crimmins, M. T. Powell, *Journal of the American Chemical Society* **2003**, <u>125</u>, 7592

# Examples:

K. Takai, K. Nitta, K. Utimoto, *Journal of the American Chemical Society* **1986**, <u>108</u>, 7408

S. Chackalamannil, R. Davies, A. T. McPhail, *Organic Letters* **2001**, <u>3</u>, 1427

J. Cossy, D. Bauer, V. Bellosta, *Tetrahedron* **2002**, <u>58</u>, 5909

I. C. Gonzalez, C. J. Forsyth, *Journal of the American Chemical Society* **2000**, <u>122</u>, 9099

M. E. Jung, B. T. Fahr, D. C. D'Amico, *Journal of Organic Chemistry* **1998**, <u>63</u>, 2982

R. Garg, R. S. Coleman, *Organic Letters* **2001**, <u>3</u>, 3487

# Tebbe Reagent / Olefination

## The Reaction:

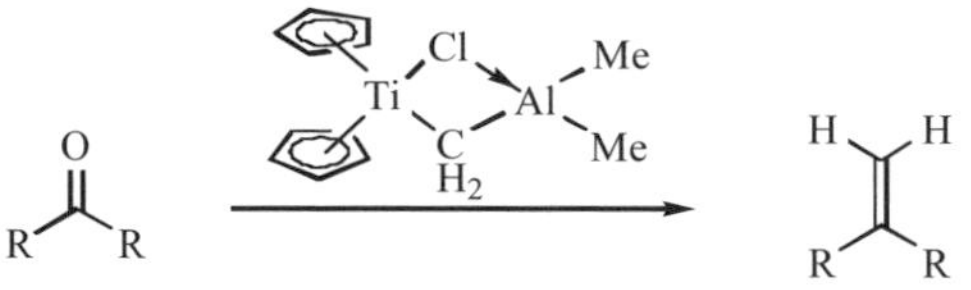

## Proposed Mechanism:

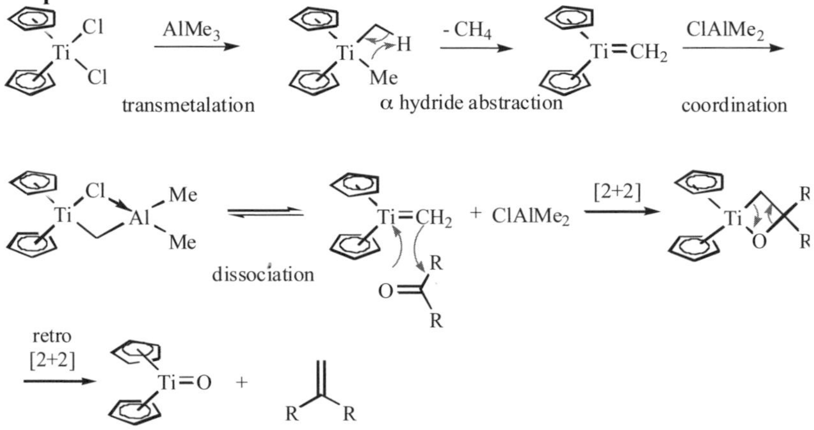

AlMe3
transmetalation
- CH4
α hydride abstraction
ClAlMe2
coordination
dissociation
+ ClAlMe2
[2+2]
retro
[2+2]

## Notes:

See ***Tebbe Reagent***

***Tebbe reactions*** on beads:

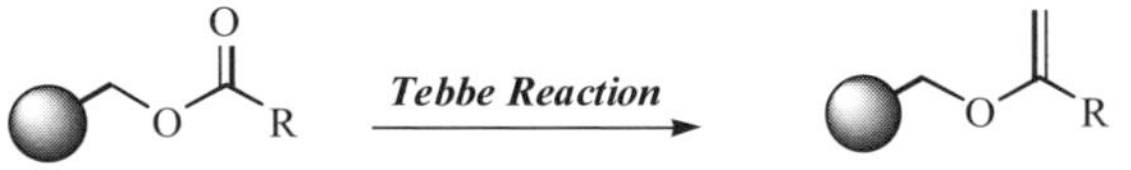

Tebbe Reaction

A. G. M. Barrett, P. A. Procopiou, U. Voigtmann, *Organic Letters* **2001**, 3, 3165

The ***Oshima modification*** provides another entry into methylenation:

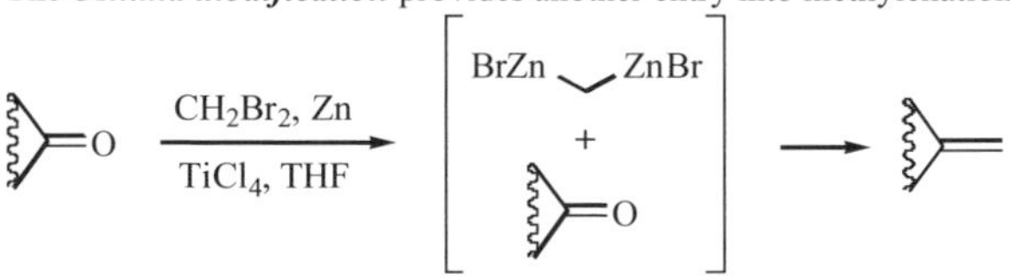

CH2Br2, Zn
TiCl4, THF
BrZn ZnBr
+

For details of competitive regiochemical analysis with a number of reagents, see:

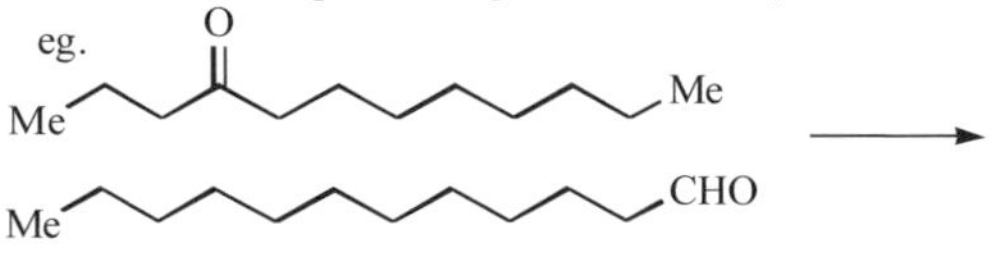

eg.

T. Okazoe, J.-I. Hibino, K. Takai, H. Nozakil, *Tetrahedron Letters* **1985**, 26, 5581

## Examples:

M. E. Jung, J. Pontillo, *Tetrahedron* **2003**, 59, 2729

K. C. Nicolaou, M. H. D. Postema, C. F. Claiborne, *Journal of the American Chemical Society* **1996**, 118, 1565

M. Cortés, J. A. Valderrama, M. Cuellar, V. Armstrong, M. Preite, *Journal of Natural Products* **2001**, 64, 348

N. A. Petasis, M. A. Patane, *Tetrahedron Letters* **1990**, 31, 6799

# Thiele-Winter Reaction (Acetoxylation) (Thiele Reaction)

## The Reaction:

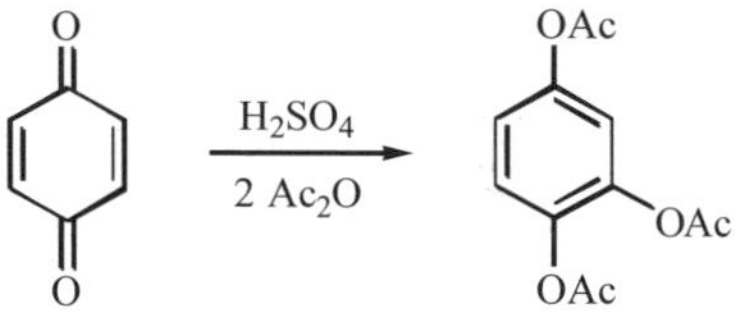

## Proposed Mechanism:

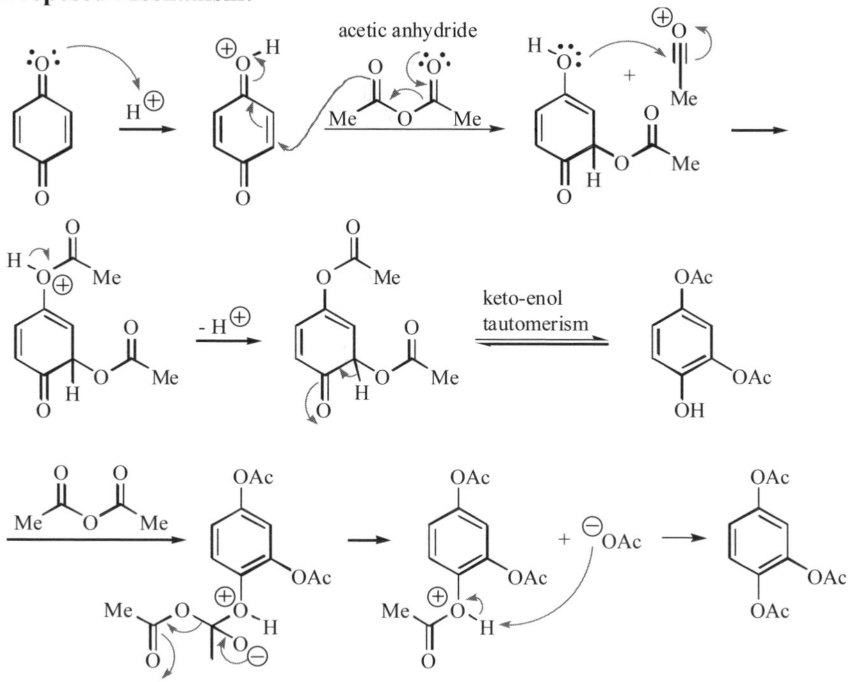

## Notes:

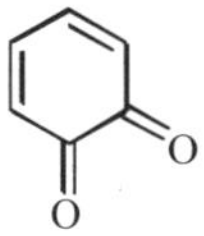

*Ortho*-quinones will give the same product.

## Examples:

D. Villemin, N. Bar, M. Hammadi, *Tetrahedron Letters* **1997**, <u>38</u>, 4777

P. A. Reddy, C. D. Gutsche, *Journal of Organic Chemistry* **1993**, <u>58</u>, 3245

W. M. Mclamore, *Journal of the American Chemical Society* **1951**, <u>73</u>, 2225

S. Spyroudis, N. Xanthopoulou, *Journal of Organic Chemistry* **2002**, <u>67</u>, 4612

# Thorpe Reaction (If intermolcular, known as the Thorpe-Ziegler Reaction.)

## The Reaction:

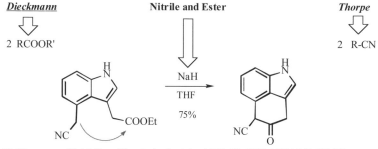

## Proposed Mechanism:

## Notes:

M. B. Smith, J. March in *March's Advanced Organic Chemistry*, 5th ed., John Wiley and Sons, Inc., New York, 2001, pp. 1219, 1238; for ***Thorpe-Ziegler***, p. 1239.

The ***Thorpe reaction*** is often better than the ***Diekmann Cyclization*** for ring sizes > 7.

What of the reaction of an ester and a nitrile?

| ***Diekmann*** | **Nitrile and Ester** | *Thorpe* |
|---|---|---|
| 2 RCOOR' | | 2 R-CN |

NaH
THF
75%

H. Plienenger, W. Muller, *Chemische Berichte* **1960**, <u>93</u>, 2029 (AN 1961:13347)

## Examples:

1. *t*-BuO⊖, *t*-BuOH
2. AcOH, H₃PO₄
68%

A. Toro, P. Nowak, P. Deslongchamps, *Journal of the American Chemical Society* **2000**, <u>122</u>, 4526

L. Lu, R. K. Shoemaker, D. M. S. Wheeler, *Tetrahedron Letters* **1989**, 30, 6993

A. Hashimoto, A. K. Przybyl, J. T. M. Linders, S. Kodato, X. Tian, J. R. Deschamps, C. George, J. L. Flippen-Anderson, A. E. Jacobson, K. C. Rice, *Journal of Organic Chemistry* **2004**, 69, 5322

G. Seitz, H. Monnighoff, *Tetrahedron Letters* **1971**, 12, 4889

J. J. Bloomfield, P. V. Fennessey, *Tetrahedron Letters* **1964**, 5, 2273

N. E. Kayaleh, R. C. Gupta, F. Johnson, *Journal of Organic Chemistry* **2000**, 65, 4515

# Tiffeneau-Demjanov Rearrangement

## The Reaction:

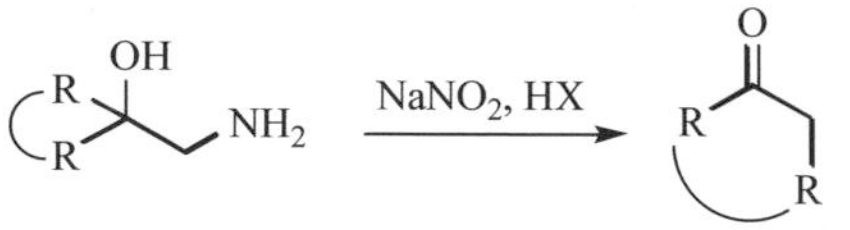

## Proposed Mechanism:

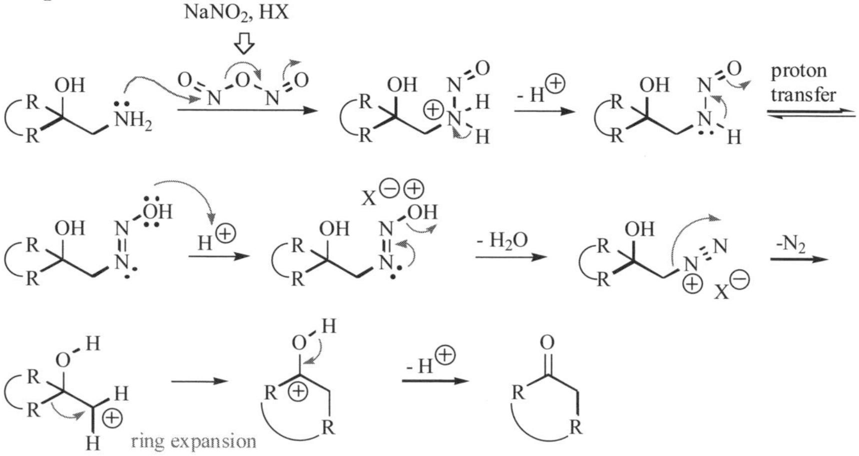

## Notes:

M. B. Smith, J. March in *March's Advanced Organic Chemistry*, 5$^{th}$ ed., John Wiley and Sons, Inc., New York, 2001, p. 1399

Possible sources of starting material:

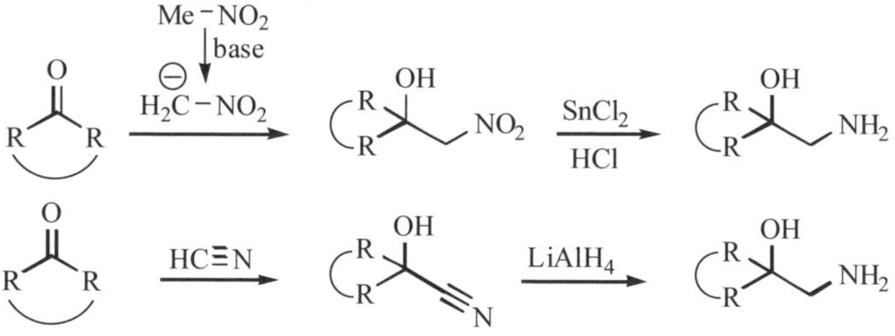

*Takuchi Modification*

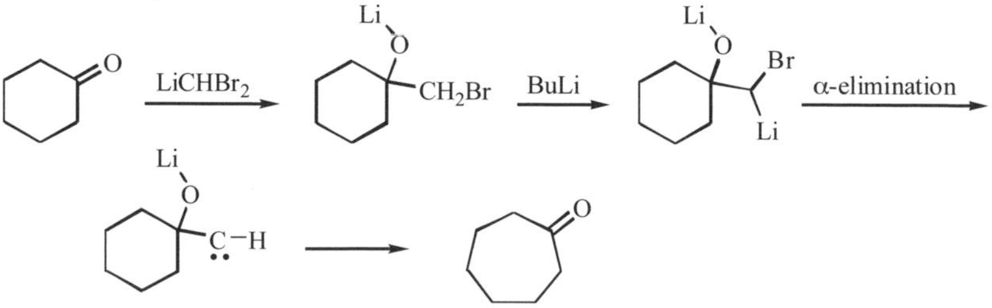

## Examples:

Z. Wang, D. Yang, A. K. Mohanakrishnan, E. Hamel, M. Cushman, *Journal of Medicinal Chemistry* **2000**, **43**, 2419

S. Kim, R. Bishop, D. C. Craig, I. G. Dance, M. L. Scudder, *Journal of Organic Chemistry* **2002**, **67**, 3221

D. Fattori, S. Henry, P. Vogel, *Tetrahedron* **1993**, **49**, 1649

J. T. Lumb, G. H. Whitham, *Tetrahedron* **1965**, **21**, 499

R. B. Woodward, J. Gostel, I. Ernest, R. J. Friary, G. Nestler, H. Raman, R. Sitrin, C. Suter, J. K. Whitesell, *Journal of the American Chemical Society* **1973**, **95**, 6853

# Tischenko Reaction

## The Reaction:

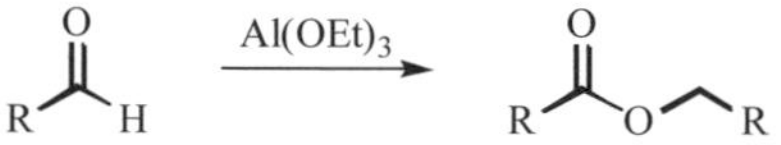

$$\underset{R}{\overset{O}{\|}}\underset{H}{\overset{}{\|}} \xrightarrow{\text{Al(OEt)}_3} \underset{R}{\overset{O}{\|}}\overset{}{O}\overset{}{R}$$

## Proposed Mechanism:

This alcoholate serves as catalyst.

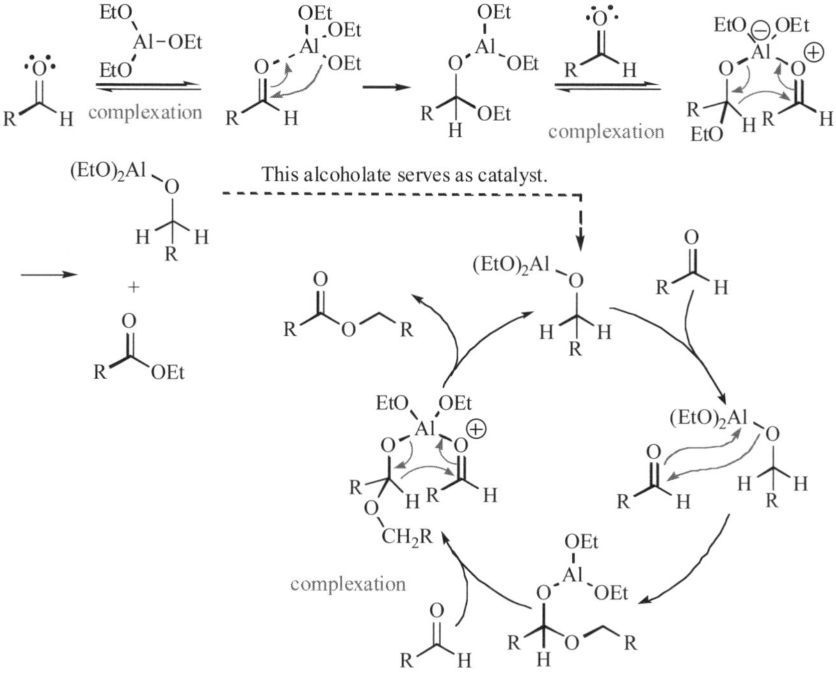

A simpler view:

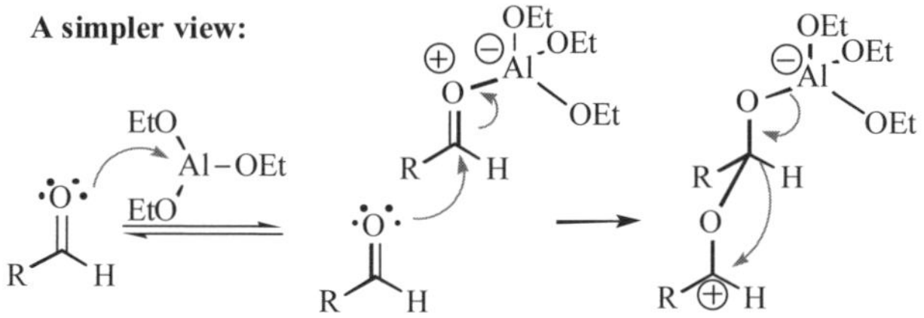

## Notes:

An "*Aldol-Tischenko reaction*" is sometimes observed as a byproduct of *Aldol condensation*:

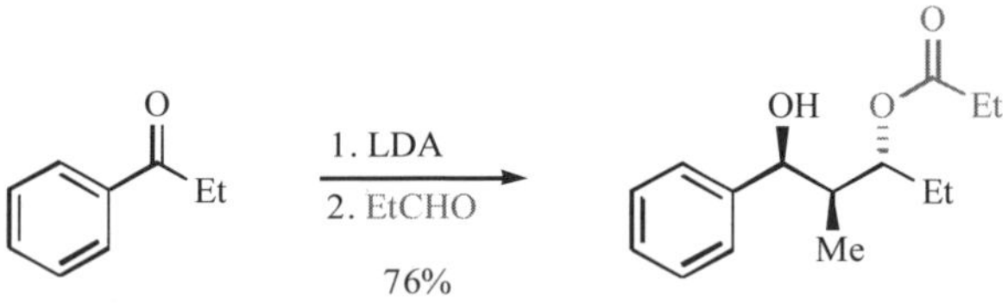

1. LDA
2. EtCHO

76%

P. M. Bodnar, J. T. Shaw, K. A. Woerpel, *Journal of Organic Chemistry* **1997**, <u>62</u>, 5674 via:

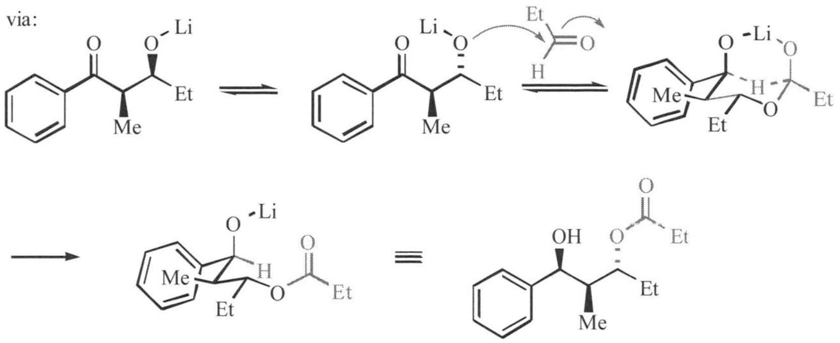

**Examples:**

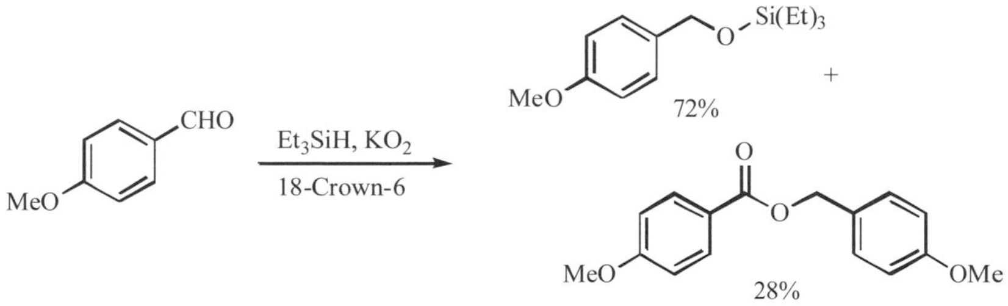

72%

+

28%

F. LeBideau, T. Coradin, D. Gourier, J. Henique, E. Samuel, *Tetrahedron Letters* **2000**, <u>41</u>, 5215

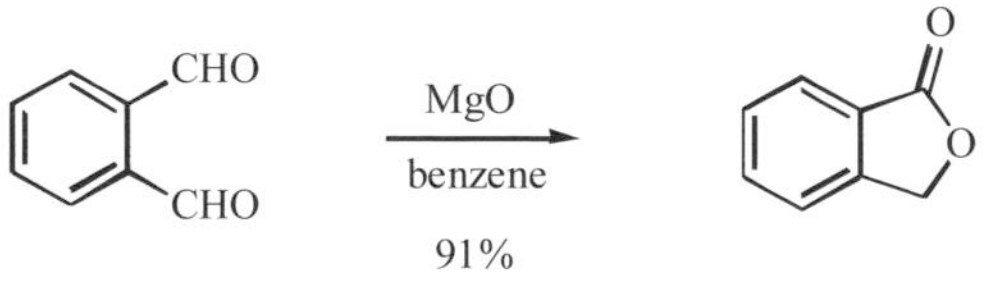

MgO

benzene

91%

T. Sek, H. Tachikawa, T. Tamada, H. Hattori, *Journal of Catalysis* **2003**, <u>217</u>, 117

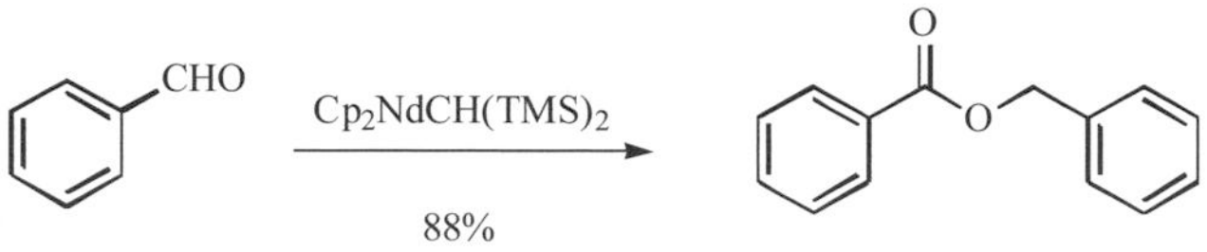

Cp₂NdCH(TMS)₂

88%

S.-Y. Onozawa, T. Sakakura, M. Tanaka, M. Shiro, *Tetrahedron* **1996**, <u>52</u>, 4291

This reaction will take place slowly by storing an aqueous solution of the aldehyde at 60 °C.

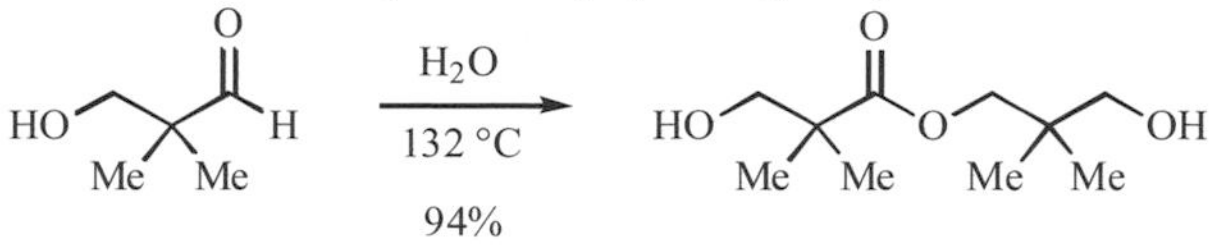

$H_2O$

132 °C

94%

G. K. Finch, *Journal of Organic Chemistry* **1960**, <u>25</u>, 2219

# Trost's TMM (trimethylenemethane) Cycloaddition

## The Reaction:

## Proposed Mechanism:

EWG can stabilize
the anion and allow
for bond rotation.

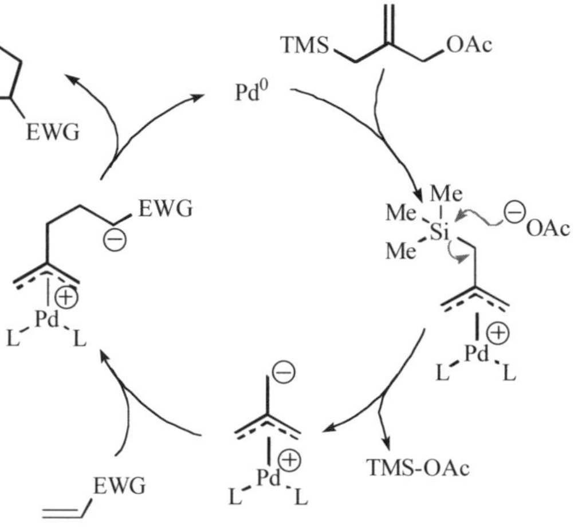

## Notes:

An analysis of the use of orbital theory to rationalize product formation from the diradical TMM
analog formed by nitrogen extrusion:

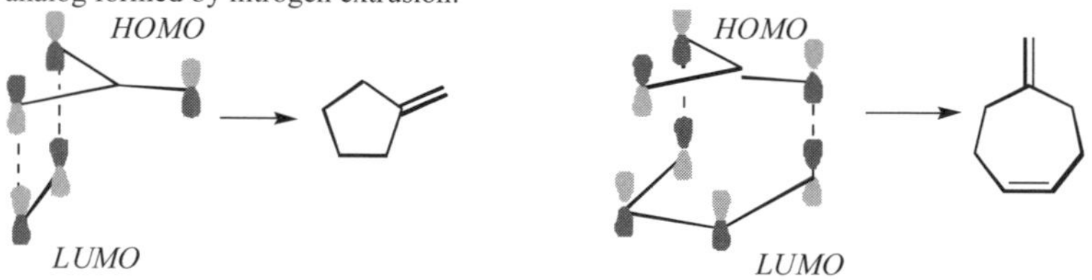

*HOMO*

*LUMO*

*HOMO*

*LUMO*

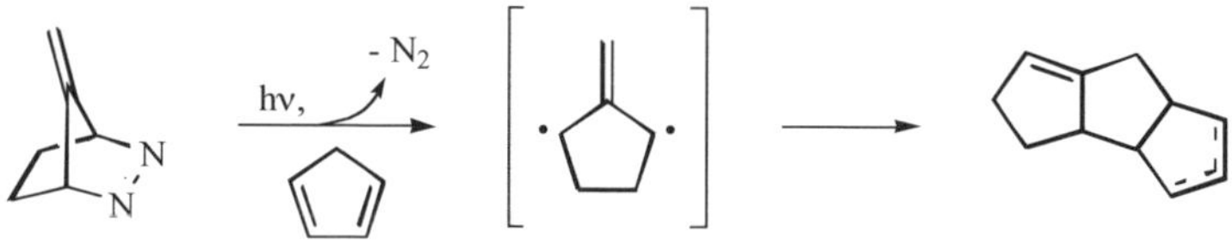

A series of products of which these are the major components.

R. K. Siemionke, J. A. Berson, *Journal of the American Chemical Society* **1980**, 102, 3870

**Examples:**

AcO—/—TMS

$\xrightarrow[\text{Pd(OAc)}_2, \text{P(O}i\text{-Pr})_3, \text{toluene}]{}$

60%

J. Cossy, D. Belotti, J. P. Pete, *Tetrahedron* **1990**, <u>46</u>, 1859

AcO—/—TMS

$\xrightarrow[\text{Pd(PPh}_3)_4, \text{DPPE, THF}]{\text{CO}_2\text{Me} / \text{CO}_2\text{Me}}$

50%

MeO$_2$C      CO$_2$Me

25:1 / trans: cis    toluene
1.3 / 1 trans: cis    THF

B. M. Trost, D. M. T. Chan, *Journal of the American Chemical Society* **1983**, <u>105</u>, 2315

$\xrightarrow[\text{Pd(OAc)}_2, \text{P}(i\text{-Pr})_3, \text{toluene}]{\text{AcO} \diagup \text{TMS}}$

$\cdots$CO$_2$Me

50%

$\cdots$CO$_2$Me

A. Heumann, S. Kadly, A. Tenagli, *Tetrahedron* **1994**, <u>50</u>, 539

$\xrightarrow[\text{Pd(OAc)}_2, \text{P(OEt)}_3, \text{THF}]{\text{AcO} \diagup \text{TMS}}$

98%

MeO$_2$C

L. A. Paquette, D. R. Sauer, D. G. Cleary, M. A. Kinsella, C. M. Blackwell, L. G. Anderson, *Journal of the American Chemical Society* **1992**, <u>114</u>, 7375

# Tscherniac-Einhorn Reaction

## The Reaction:

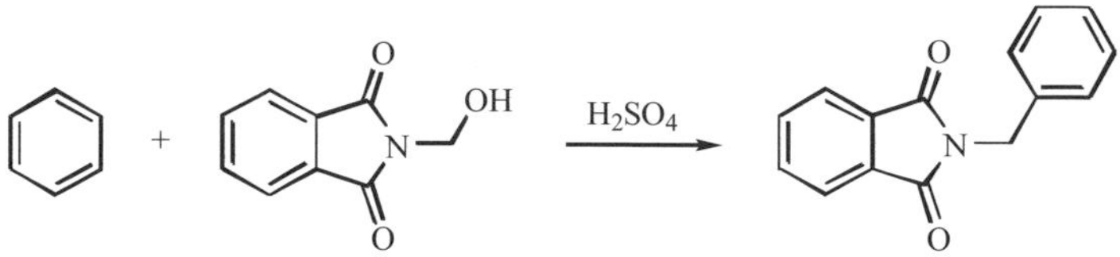

## Proposed Mechanism:

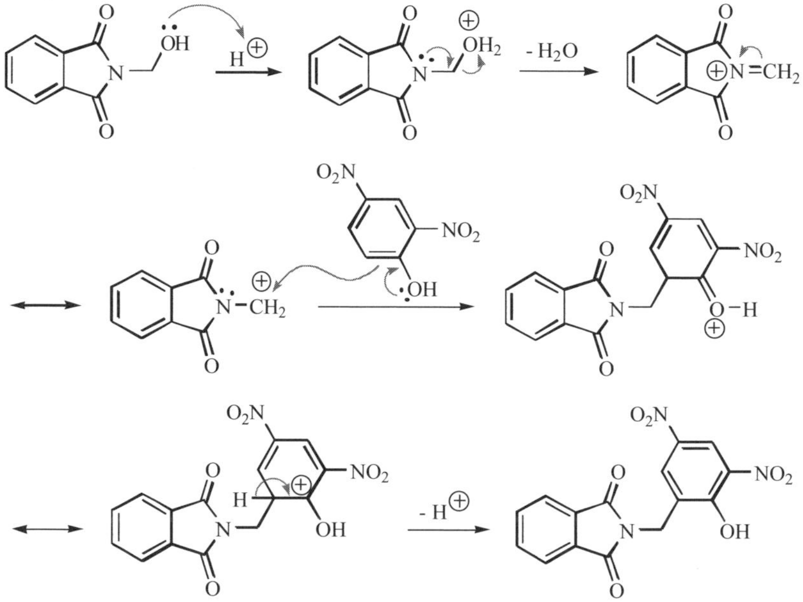

4 total resonance structures

H. E. Zaugg, R. W. DeNet, J. E. Fraser, A. M. Kotre, *Journal of Organic Chemistry* **1969**, <u>34</u>, 14

## Notes:

Sometimes the reaction can continue with hydrolysis of the phthalimide group:

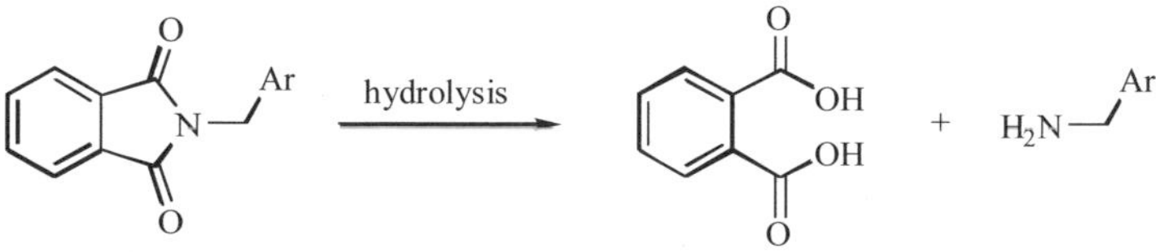

## Examples:

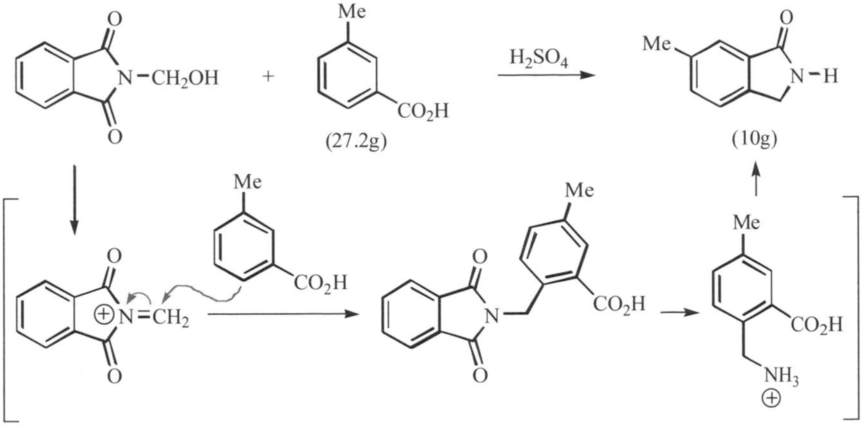

(27.2g)                                                                                    (10g)

P. S. Anderson, M. E. Christy, C. D. Colton, W. Halczenko, G. S. Ponticello, K. L. Shepard, *Journal of Organic Chemistry* **1979**, <u>44</u>, 1519

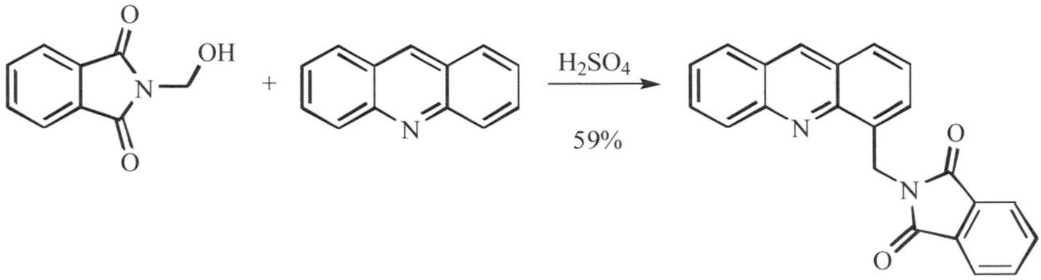

59%

F. Hess, E. Cullen, K. Grozinger, *Tetrahedron Letters* **1971**, <u>12</u>, 2591

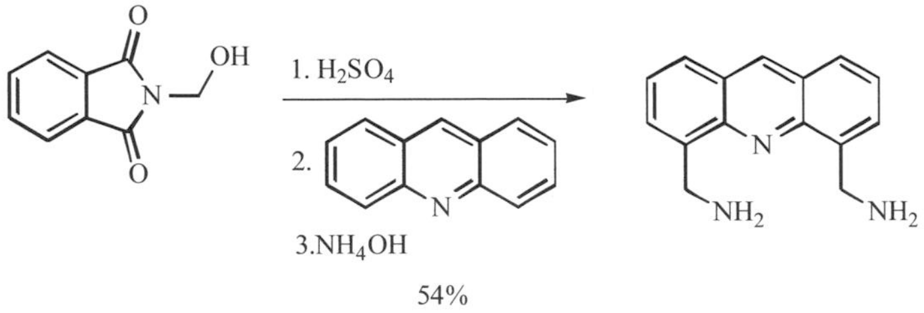

1. H₂SO₄

2.

3. NH₄OH

54%

F. K. Hess, P. B. Stewart, *Journal of Medicinal Chemistry* **1975**, <u>18</u>, 320

# Tsuji-Trost Reaction

## The Reaction:

$$\xrightarrow[\text{THF, 60-70 °C}]{\text{Pd}_2(\text{dba})_3, \text{dppb, Nu}}$$

X = leaving group

J. Tsuji, H. Takahashi, M. Morikawa, *Tetrahedron Letters* **1965**, <u>6</u>, 4387
B. M. Trost, T. J. Fullerton, *Journal of the American Chemical Society* **1973**, <u>95</u>, 292

## Proposed Mechanism:

$$\xrightarrow{\text{Pd(0)}}$$

:Nu

## Notes:

major product
when: R' = H                   R' = SiR$_3$

L = ligand

V. Branchadell, M. Moreno-Manas, R. Pleixats, S. Thorimbert, C. Commandeur, C. Boglio, M. Malacria, *Journal of Organometallic Chemistry* **2003**, <u>687</u>, 337

## Examples:

1. NaH, THF, DMSO
2. Pd$_2$(dba)$_3$, (*i*-PrO)$_3$P, CHCl$_3$
3.

55 - 67%

J. H. Hong, M. J. Shim, B. O. Ro, O. H. Ko, *Journal of Organic Chemistry* **2002**, <u>67</u>, 6837

A water-mediated, transition metal free reaction:

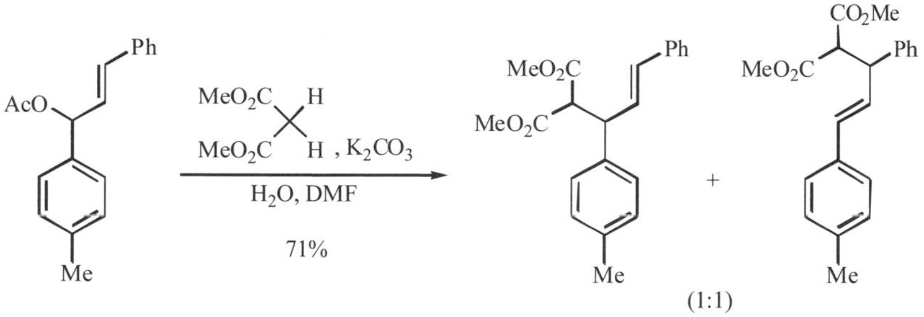

71%

(1:1)

C. Chevrin, J. Le Bras, F. Henin, J. Muzart, *Tetrahedron Letters* **2003**, <u>44</u>, 8099

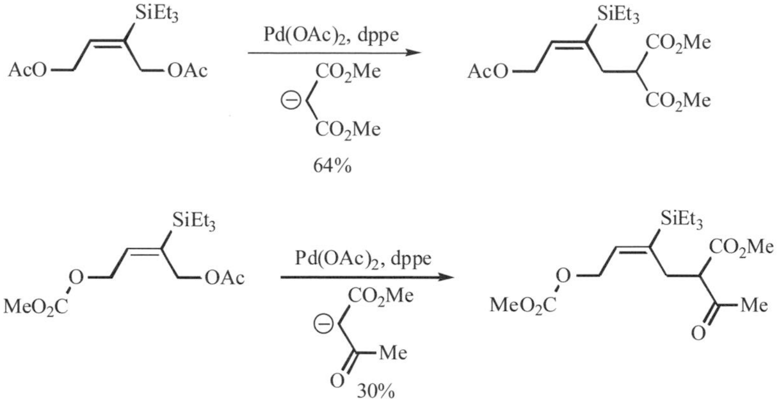

64%

30%

V. Branchadell, M. Moreno-Manas, R. Pleixats, S. Thorimbert, C. Commandeur, C. Boglio, M. Malacria, *Journal of Organometallic Chemistry* **2003**, <u>687</u>, 337

Catalytic allylation of aldehydes:

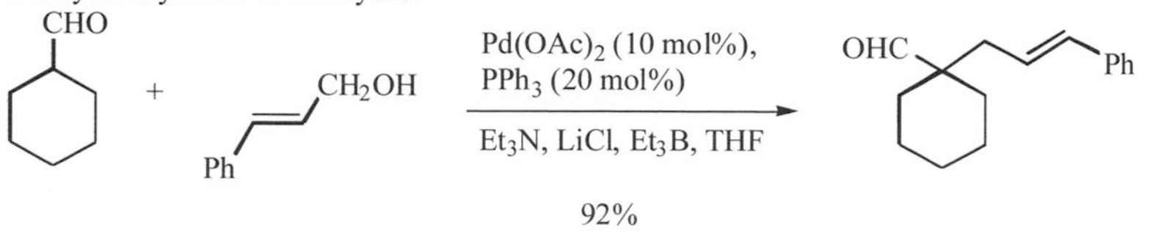

92%

M. Kimura, Y. Horino, R. Mukai, S. Tanaka, Y. Tamaru, *Journal of the American Chemical Society* **2001**, <u>123</u>, 10401

# Ueno-Stork Cyclization

## The Reaction:

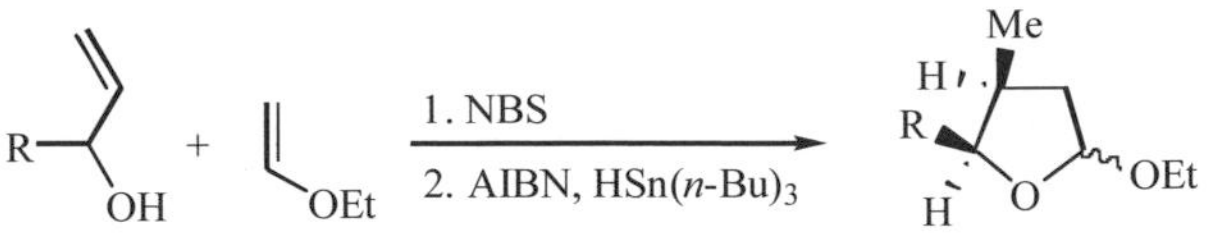

## Proposed Mechanism:

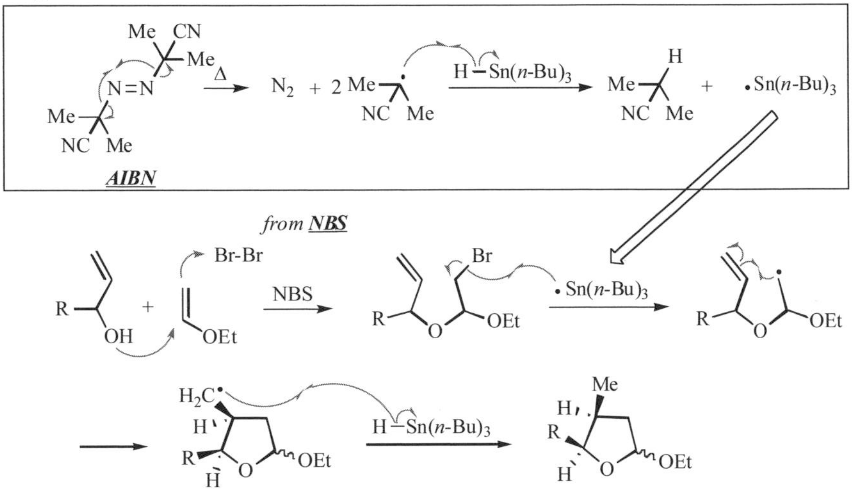

*from **NBS***

## Examples:

50 - 55%

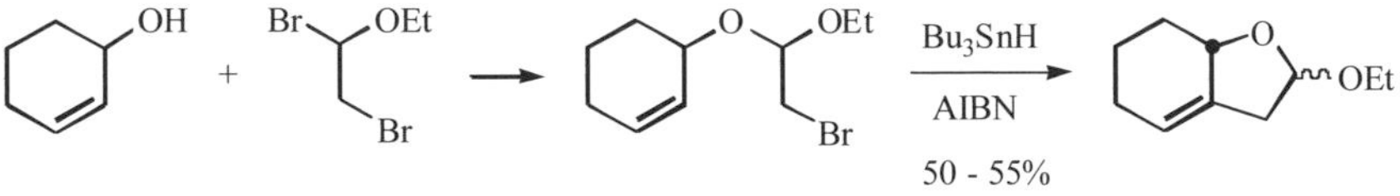

G. Stork, R. Mook, Jr., S. A. Biller, S. D. Rychnovsky, *Journal of the American Chemical Society* **1983**, <u>105</u>, 3741

F. Villar, O. Equey, P. Renaud, *Organic Letters* **2000**, 2, 1061

F. Villar, O. Equey, P. Renaud, *Organic Letters* **2000**, 2, 1061

1 : 1 mixture

F. Villar, P. Renaud, *Tetrahedron Letters* **1998**, 39, 8655

# Ugi Reaction

## The Reaction:

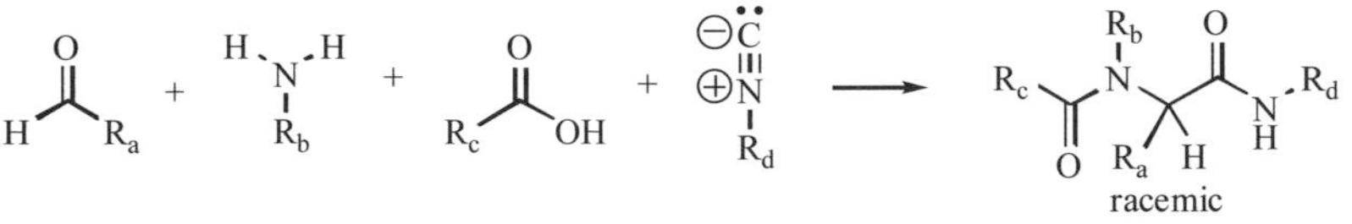

racemic

## Proposed Mechanism:

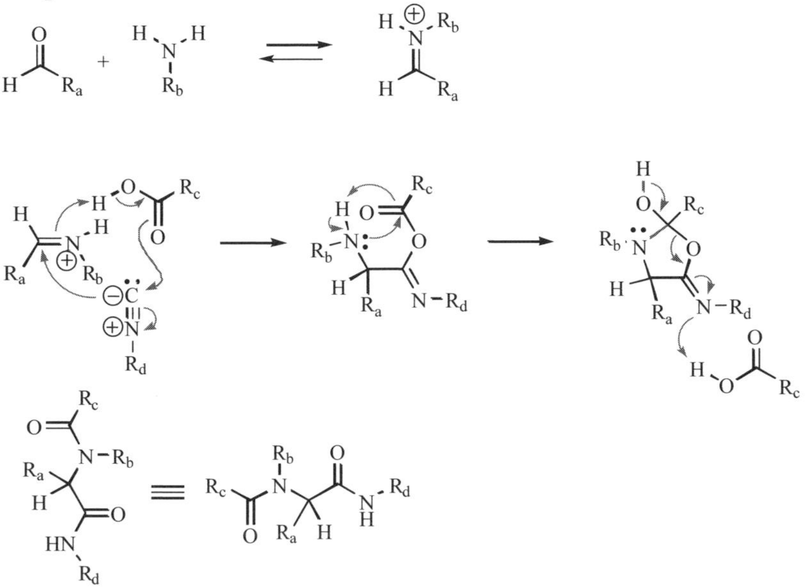

## Notes:

This reaction is similar to the ***Passerini reaction***. It is characterized by the four components going into the reaction mix:

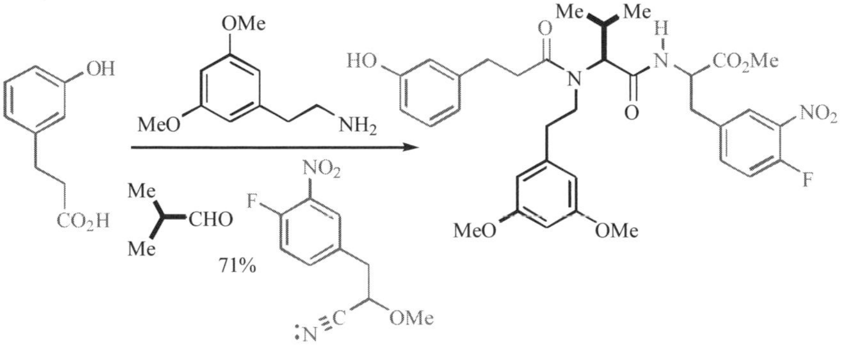

71%

P. Cristau, J.-P. Vors, J. Zhu, *Organic Letters* **2001**, <u>3</u>, 4079

## Examples:

A chemical library developed

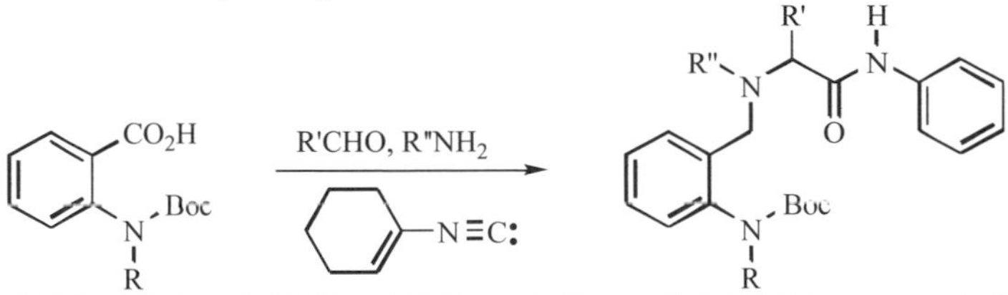

C. Hulme, J. Peng, S.-Y. Tang, C.J. Burns, I. Morize, R. Labaudiniere, *Journal of Organic Chemistry* **1998**, <u>63</u>, 8021

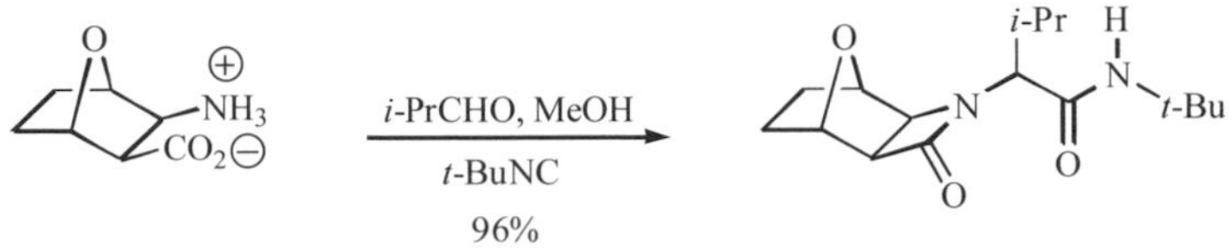

96%

A. Basso, L. Banfi, R. Riva, G. Guanti, *Tetrahedron Letters* **2004**, <u>45</u>, 587

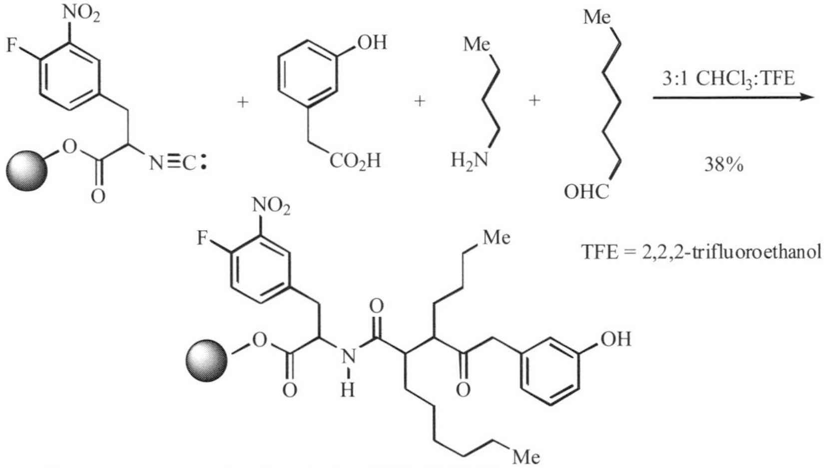

TFE = 2,2,2-trifluoroethanol

P. Cristau, J.-P. Vors, J. Zhu, *Tetrahedron* **2003**, <u>59</u>, 7859

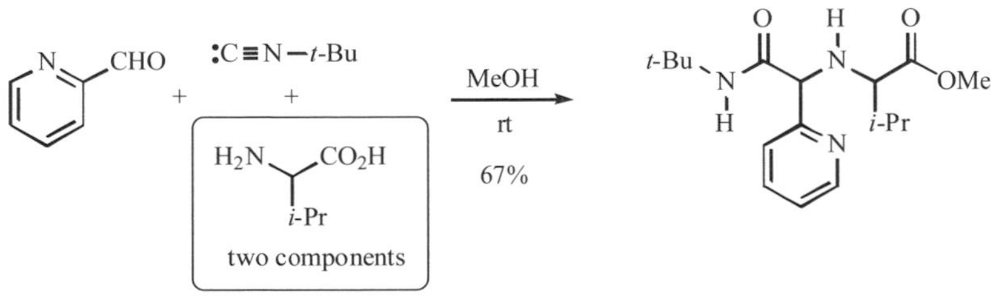

two components

67%

G. Dyker, K. Breitenstein, G. Henkel, *Tetrahedron: Asymmetry* **2002**, <u>13</u>, 1929

# Ullmann Coupling Reaction

## The Reaction:

$$2 \ Ar{-}I \quad \xrightarrow[\Delta]{Cu} \quad Ar{-}Ar \ + \ CuI_2$$

## Proposed Mechanism:

Ar-I + Cu $\xrightarrow{\text{heat}}$ Ar$-$Cu$-$I $\xrightleftharpoons{\;\;\;}$ Ar-Cu $\xrightarrow{\text{Ar-I}}$ 

oxidative addition

oxidative addition

$\xrightarrow{\hspace{2cm}}$ Ar-Ar

reductive elimination

## Notes:

M. B. Smith, J. March in *March's Advanced Organic Chemistry*, 5<sup>th</sup> ed., John Wiley and Sons, Inc., New York, 2001, p. 870.

Problem with mixed coupling:

$$Ar{-}I \ + \ Ar'{-}I \quad \xrightarrow[\Delta]{Cu} \quad \begin{array}{c} Ar{-}Ar \\ + \\ Ar{-}Ar' \\ + \\ Ar'{-}Ar' \end{array} \ + \ CuI_2$$

## Examples:

1. Cu, Δ
2. KCN, NH$_3$

90%

G. Vlád, I. Horváth, *Journal of Organic Chemistry* **2002**, <u>67</u>, 6550

Cu
Δ, DMF

51%

C. W. Lai, C. K. Lam, H. K. Lee, T. C. W. Mak, H. N. C. Wong, *Organic Letters* **2003**, <u>5</u>, 823

$$\text{Ar-X} \xrightarrow[\text{H}_2\text{O, CO}_2 \text{ (l)}]{\text{Pd/C, Zn}} \text{Ar-Ar}$$

X = I, Br, Cl

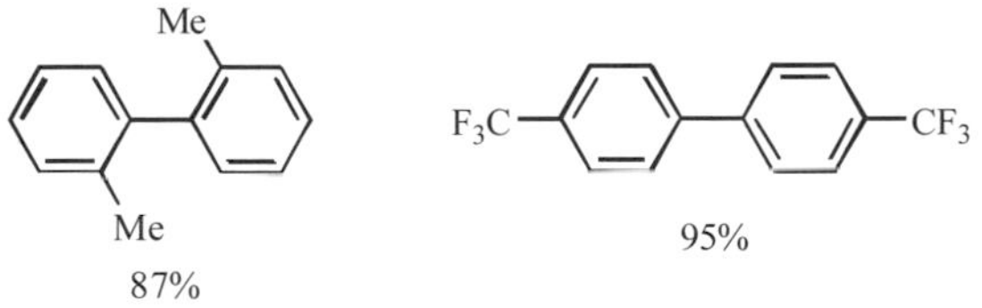

87%

95%

J.-H. Li, Y.-X. Xie, D.-L. Yin, *Journal of Organic Chemistry* **2003**, <u>68</u>, 9867

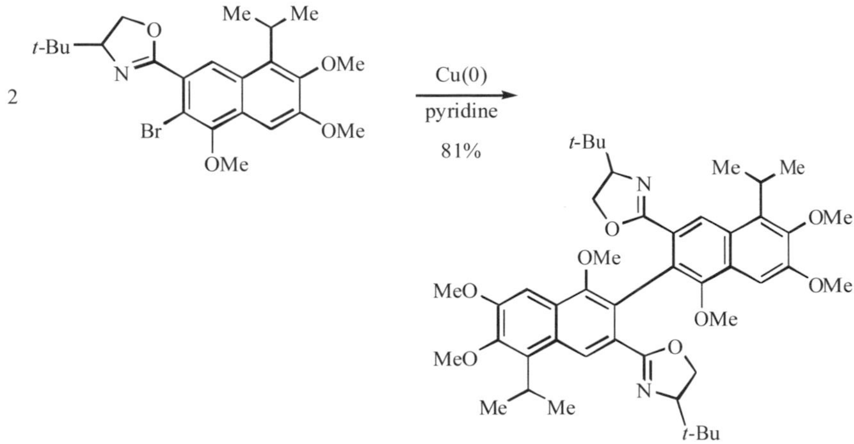

A. I. Meyers, J. J. Willemsen, *Tetrahedron Letters* **1996**, <u>37</u>, 791

# Ullmann Ether Synthesis

## The Reaction:

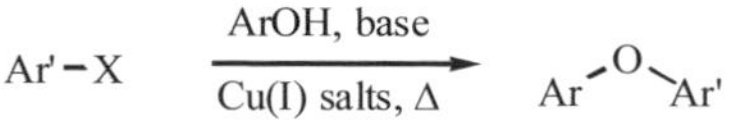

$$Ar'-X \xrightarrow[\text{Cu(I) salts, } \Delta]{\text{ArOH, base}} Ar^{-O}{}_{\diagdown}Ar'$$

X includes halides, $-NO_2$, $-NR_3{}^+$, $-OR$, or $-OH$

## Proposed Mechanism:

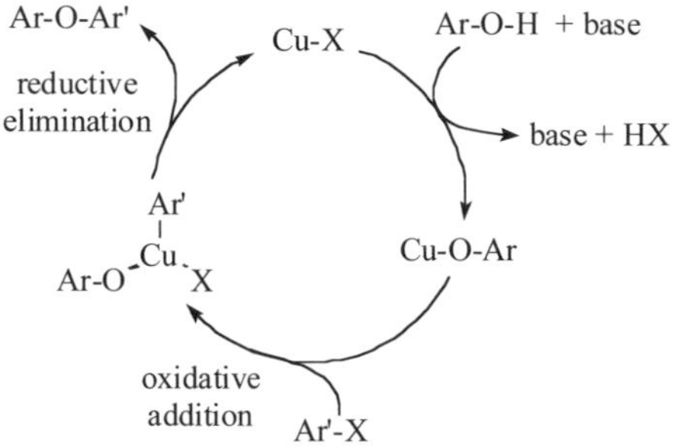

Ar-O-Ar'

reductive
elimination

Cu-X

Ar-O-H + base

base + HX

$\overset{Ar'}{\underset{}{\mid}}$
$Ar-O^{-}\overset{Cu}{\diagdown}X$

Cu-O-Ar

oxidative
addition   Ar'-X

## Notes:

***Ullmann-type reactions*** can replace Ar-O with Nu-H.

The copper-catalyzed addition of amines and amides (***Goldberg Reaction***) are placed in this section:

## Goldberg Reaction

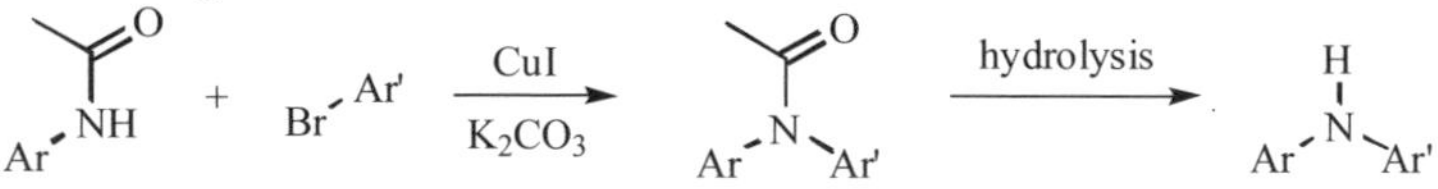

$$\underset{Ar}{\overset{O}{\diagup\diagdown}}{}_{\diagdown NH} \; + \; Br^{-}Ar' \xrightarrow[K_2CO_3]{CuI} \underset{Ar}{\overset{O}{\diagup\diagdown}}{}_{\diagdown N\diagdown Ar'} \xrightarrow{\text{hydrolysis}} \underset{Ar}{\overset{H}{\underset{}{\mid}}}{}_{N\diagdown Ar'}$$

## Ullmann Reaction (Jourdan-Ullmann Synthesis)

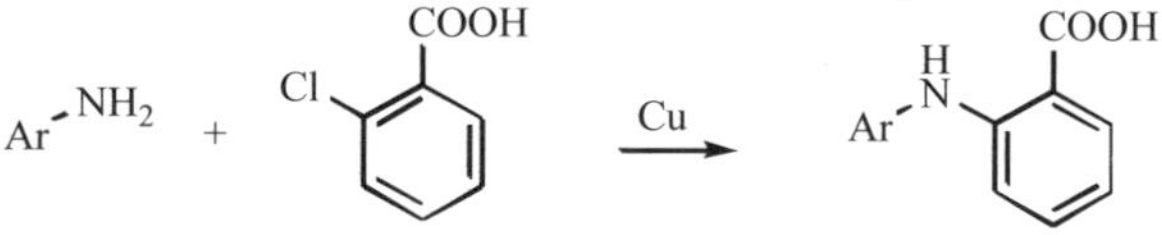

$$Ar^{-NH_2} \; + \; \underset{Cl}{\overset{COOH}{\bigcirc}} \xrightarrow{Cu} Ar^{-\overset{H}{N}}\overset{COOH}{\bigcirc}$$

There is also the related **Jourdan Synthesis**:

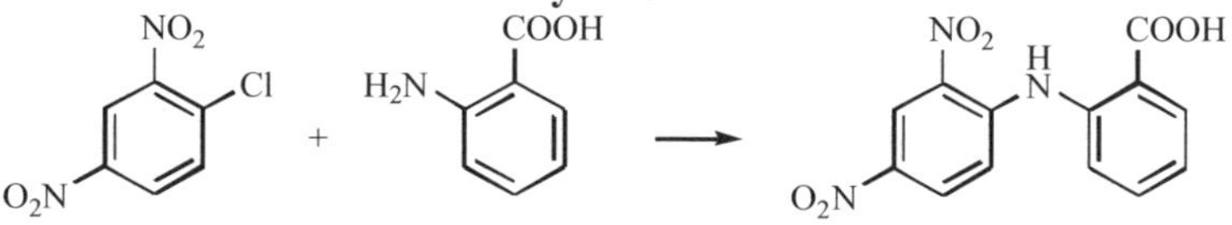

$$\underset{O_2N}{\overset{NO_2}{\bigcirc}}{}_{Cl} \; + \; H_2N^{-}\overset{COOH}{\bigcirc} \longrightarrow \underset{O_2N}{\overset{NO_2}{\bigcirc}}{}_{\overset{H}{N}}\overset{COOH}{\bigcirc}$$

## Examples:

$$\overset{OH}{\bigcirc} \; + \; I^{-}\underset{t\text{-Bu}}{\bigcirc} \xrightarrow[\substack{\text{NMP, microwave} \\ 90\%}]{CuI, Cs_2CO_3} \bigcirc^{-O}{}_{\diagdown}\underset{t\text{-Bu}}{\bigcirc}$$

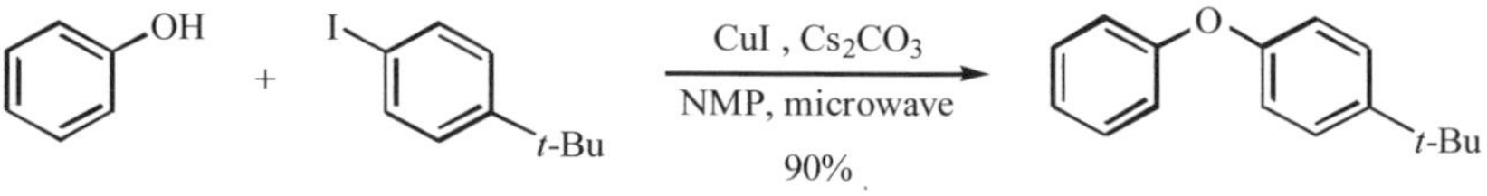

Y.-J. Wu, H. He, *Tetrahedron Letters* **2003**, <u>44</u>, 3445

D. Ma, Q. Cai, *Organic Letters* **2003**, <u>5</u>, 3799

A. V. R. Rao, J. K. Chakraborty, K. L. Reddy, A. S. Rao, *Tetrahedron Letters* **1992**, <u>33</u>, 4799

M. Wolter, G. Nordmann, G. E. Job, S. L. Buchwald, *Organic Letters* **2002**, <u>4</u>, 973

J. H. M. Lange, L. J. F. Hofmeyer, F. A. S. Hout, S. J. M. Osnabrug, P. C. Verveer, C. G. Kruse, R. W. Feenstra, *Tetrahedron Letters* **2002**, <u>43</u>, 1101

J. A. Ragen, B. P. Jones, M. J. Castaldi, P. D. Hill, T. W. Makowski, *Organic Syntheses*, <u>CV10</u>, 418

# Upjohn Dihydroxylation Protocol

## The Reaction:

$$\text{OsO}_4, \text{NMO}$$

$$t\text{-BuOH}, \text{H}_2\text{O}$$

HO   OH

V. VanRheenen, R. C. Kelly, D. Y. Cha, *Tetrahedron Letters* **1976**, <u>17</u>, 1973

## Proposed Mechanism:

OsO₄

OsO₃

HO   OH

Me
|
N

**NMM**
*N*-Methyl morpholine

Me   O⁻
N⊕

**NMO**
*N*-Methyl morpholine *N*-oxide

## Notes:

A procedure for the catalytic use of OsO₄:

$$\text{OsO}_4 \text{ (cat.)}, \text{NMO}$$

$$t\text{-BuOH}, \text{H}_2\text{O}$$

HO   OH

91%                          79%                                79%

V. VanRheenen, R. C. Kelly, D. Y. Cha, *Tetrahedron Letters* **1976**, <u>17</u>, 1973

## Examples:

Modification making $H_2O_2$ the terminal oxidant. general scheme: example substrates below.

cat. $OsO_4$, cat. NMM, cat. flavin, cat. TEAA

$H_2O_2$, solvent

OH OH

Me ~~~~~~~~~~~ Me

95%

91%

Ph ~~~~

Ph~O~~~~

91%

95%

TEAA = $Et_4N^+AcO^-$ = tetraethylammonium acetate

= flavin analog

S. Y. Jonsson, K. Farnegård, J.-E. Backvall *Journal of the American Chemical Society* **2001**, <u>123</u>, 1365

DMAP-$OsO_4$

[Bmim]$PF_6$

73-99%

OH OH

Modification using

DMAP-$OsO_4$

recyclable - reuseable

[Bmim]$PF_6$ = 1-butyl-3-methylimidazolium hexafluorophosphate

Q. Yao *Organic Letters* **2002**, <u>4</u>, 2197

$OsO_4$, NMO, quinuclidine

76%

68 : 32

$OsO_4$, TMEDA, $CH_2Cl_2$

82%

99 : 1

PMB = *p*-methoxybenzyl

T. J. Donohoe, J. W. Fisher, P. J. Edwards *Organic Letters* **2004**, <u>6</u>, 465

# Vilsmeier-Haack Reaction

## The Reaction:

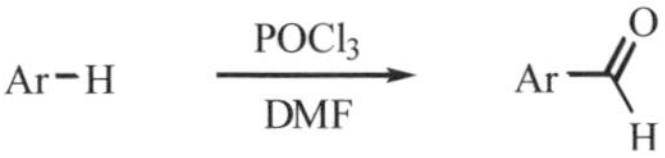

$$Ar-H \xrightarrow[\text{DMF}]{\text{POCl}_3} Ar-\overset{O}{\underset{H}{\overset{\|}{C}}}$$

## Proposed Mechanism:

**Vilsmeier Reagent**

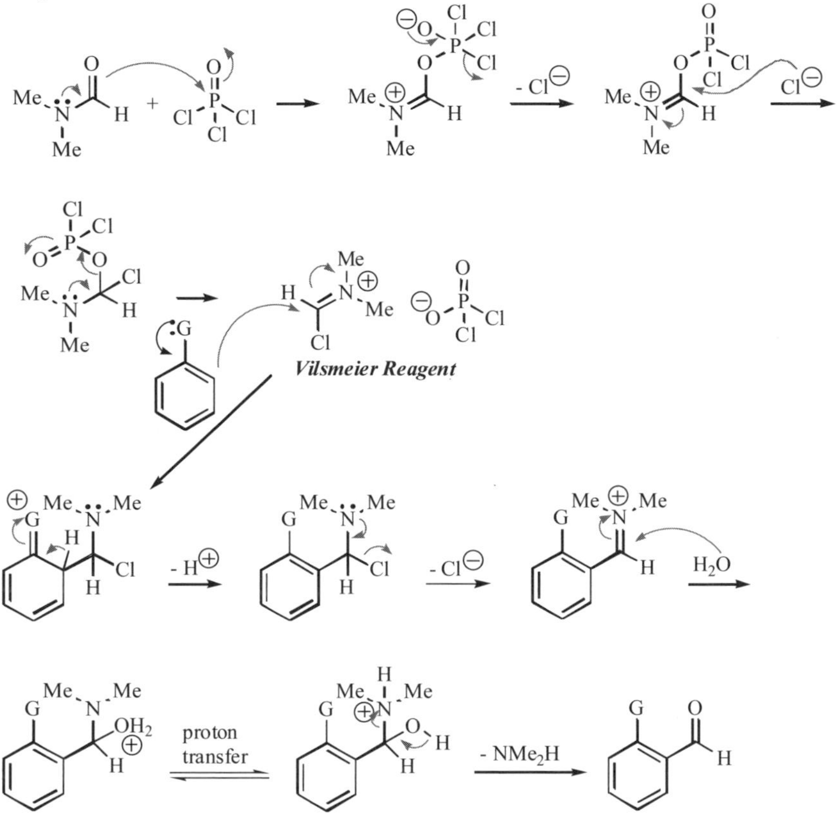

## Notes:

A substituted phenyl group is shown for illustration. Heterocycles are also common substrates.

The reaction is more efficient when the aromatic ring has electron-donating groups as the G-substituent in the scheme above.

*Vilsmeier* chemistry can also be carried out on alkenes.

## Examples:

S. Hesse, G. Kirsch, *Tetrahedron Letters* **2002**, <u>43</u>, 1213

P. G. M. Wutz, J. M. Northuis, T. A. Kwan, *Journal of Organic Chemistry* **2000**, <u>65</u>, 9223

R. A. Aungst, Jr., C. Chan, R. L. Funk, *Organic Letters* **2001**, <u>3</u>, 2611

D. L. Comins, A. L. Williams, *Organic Letters* **2001**, <u>3</u>, 3217

# Vinylcyclopropane-Cyclopentene Rearrangement

## The Reaction:

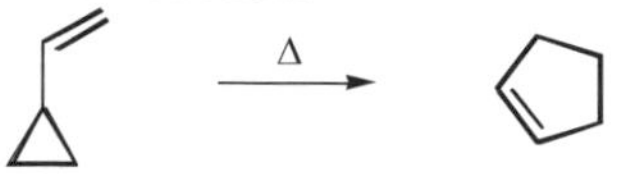
Δ

## Proposed Mechanism:

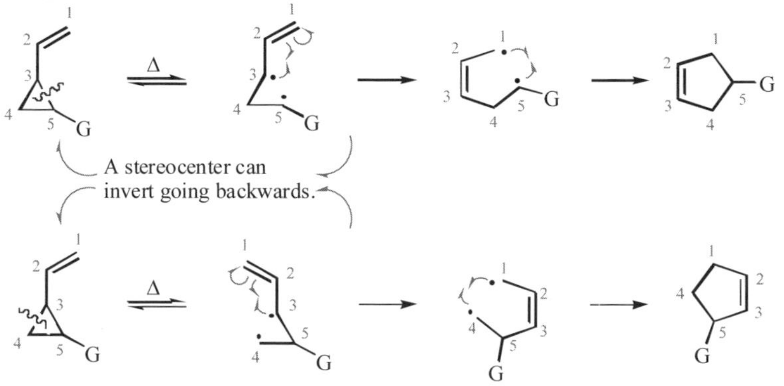
Δ
G
A stereocenter can
invert going backwards.

## Examples:

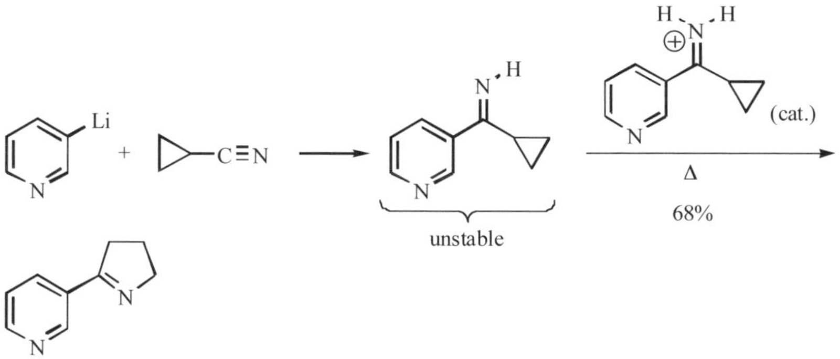
Li
C≡N
N
H
(cat.)
Δ
68%
unstable

R. V. Stevens, M. C. Ellis, M. P. Wentland, *Journal of the American Chemical Society* **1968**, 90, 5576

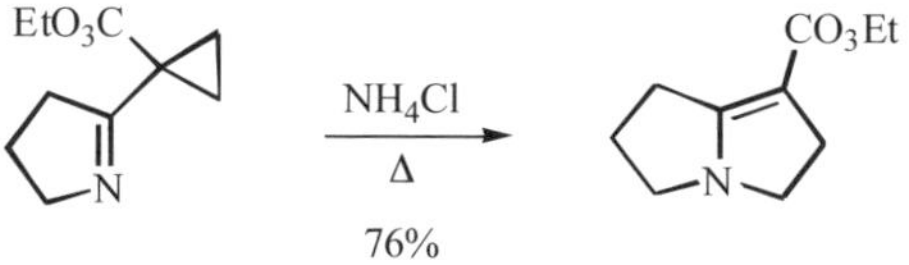
$EtO_3C$
$CO_3Et$
$NH_4Cl$
Δ
76%

H. W. Pinnick, Y.-H. Chang, *Tetrahedron Letters* **1979**, 20, 837

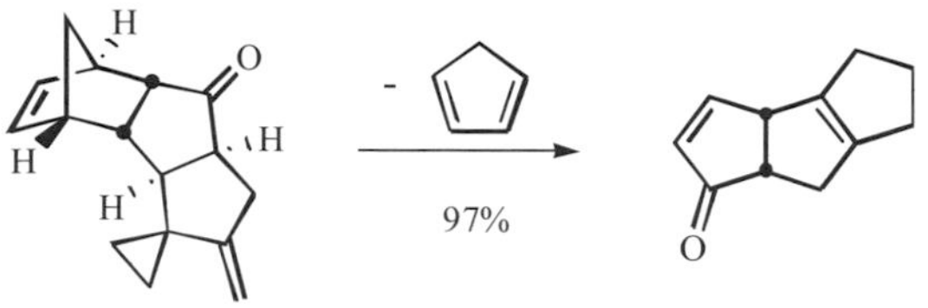

97%

B. M. Trost, J. R. Paraquette, *Journal of Organic Chemistry* **1994**, 59, 7568

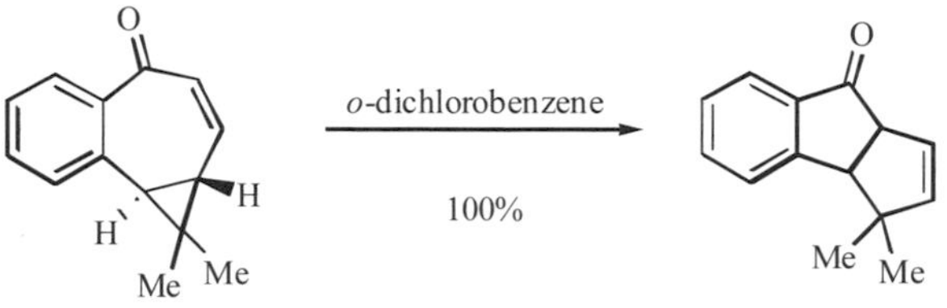

*o*-dichlorobenzene

100%

J. L. Wood, A. B. Smith, III, *Journal of the American Chemical Society* **1992**, 114, 10075

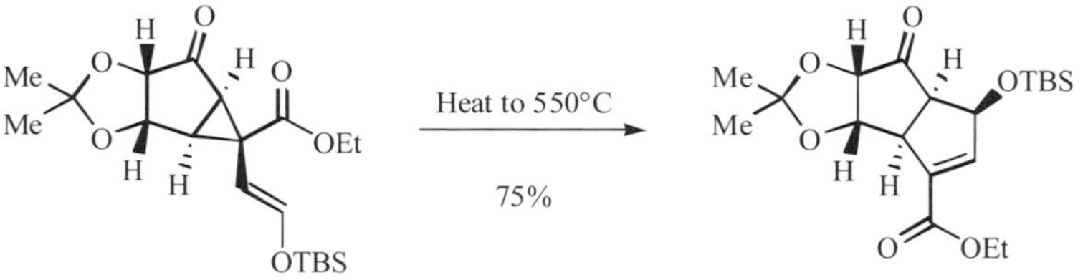

Heat to 550°C

75%

T. Hudlicky, A. Fleming, L. Radesca, *Journal of the American Chemical Society* **1989**, 111, 6691

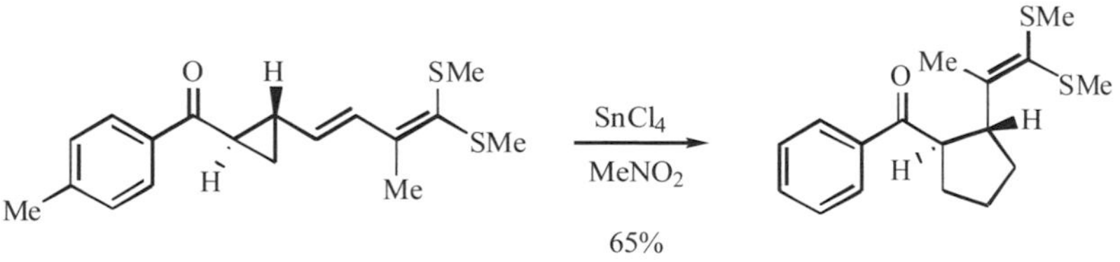

SnCl₄

MeNO₂

65%

J. Satyanarayana, M. V. B. Rao, H. Ila, H. Junjappa, *Tetrahedron Letters* **1996**, 37, 3565

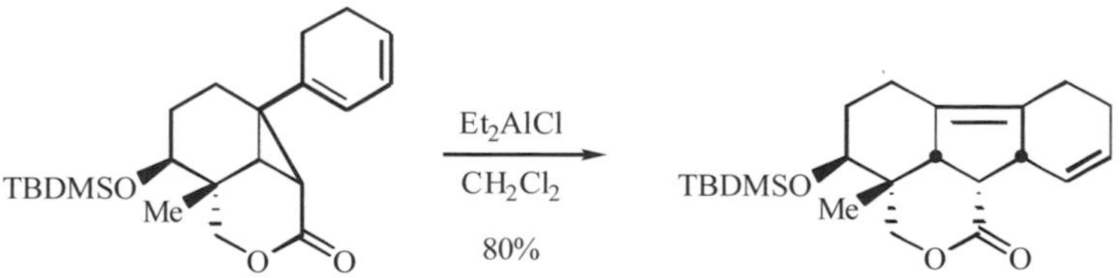

Et₂AlCl

CH₂Cl₂

80%

E. J. Corey, A. G. Myers, *Journal of the American Chemical Society* **1985**, 107, 5574

# von Braun Reaction

## The Reaction:

$$\begin{matrix} R \\ \diagdown \\ N-R' \\ \diagup \\ R'' \end{matrix} \ + \ Br-C\equiv N \ \longrightarrow \ \begin{matrix} R \\ \diagdown \\ N-C\equiv N \\ \diagup \\ R'' \end{matrix} \ + \ Br-R'$$

## Proposed Mechanism:

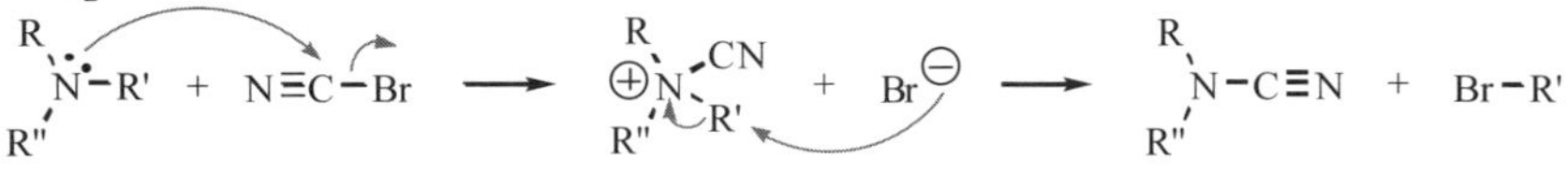

$$\begin{matrix} R \\ \diagdown \\ N-R' \\ \diagup \\ R'' \end{matrix} \ + \ N\equiv C-Br \ \longrightarrow \ \begin{matrix} R \\ \diagdown \oplus \quad CN \\ N \\ \diagup \diagdown \\ R'' \quad R' \end{matrix} \ + \ Br^{\ominus} \ \longrightarrow \ \begin{matrix} R \\ \diagdown \\ N-C\equiv N \\ \diagup \\ R'' \end{matrix} \ + \ Br-R'$$

## Notes:

Cyanogen bromide has been classified as a "counter-attack" reagent.
See: J. R. Hwu, B. A. Gilbert, *Tetrahedron* **1989**, <u>45</u>, 1233

After attack at the cyano group, the released bromide counter-attacks at the least hindered position.

## Examples:

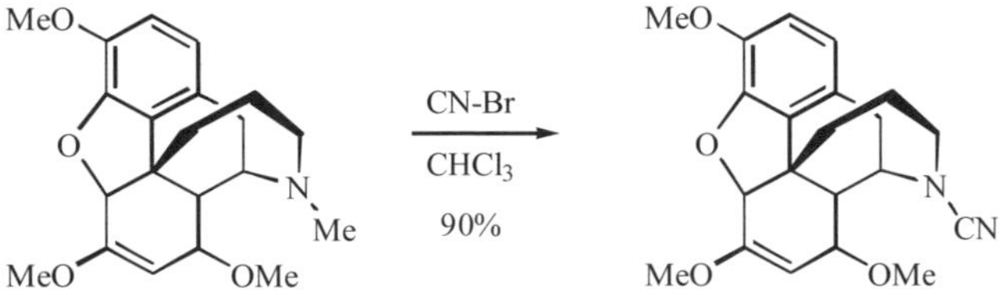

H. Rapoport, C. H. Lovell, H. R. Reist, M. E. Warren, Jr., *Journal of the American Chemical Society* **1967**, <u>89</u>, 1942

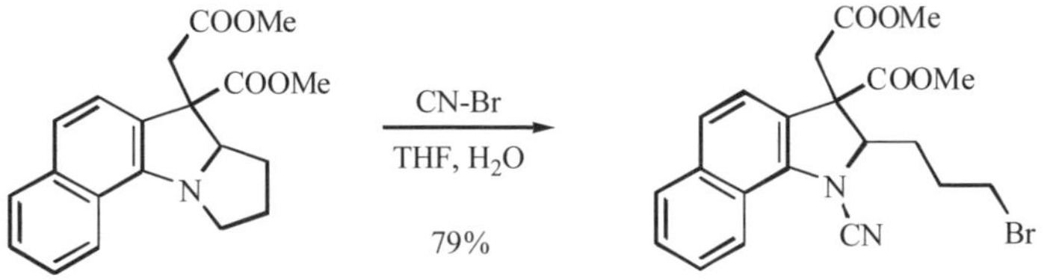

W. Verboom, G. W. Visser, D. N. Reinhoudt, *Tetrahedron* **1982**, <u>38</u>, 1831

S. S.Lee, Y. J. Lin, M. Z. Chen, Y. C. Wu, C. H. Chen, *Tetrahedron Letters* **1992**, <u>33</u>, 6309

Y. Nakahara, T. Niwaguchi, H. Ishii, *Tetrahedron* **1977**, <u>33</u>, 1591

H. Niwa, M. Toda, S. Ishimaru, Y. Hirata, S. Yamamura, *Tetrahedron* **1974**, <u>30</u>, 3031

This work was a kinetic study showing the reaction is extremely fast.

No yield given

G. Fodor, S. Abidi, *Tetrahedron Letters* **1971**, <u>12</u>, 1369

# von Richter Reaction

**The Reaction:**

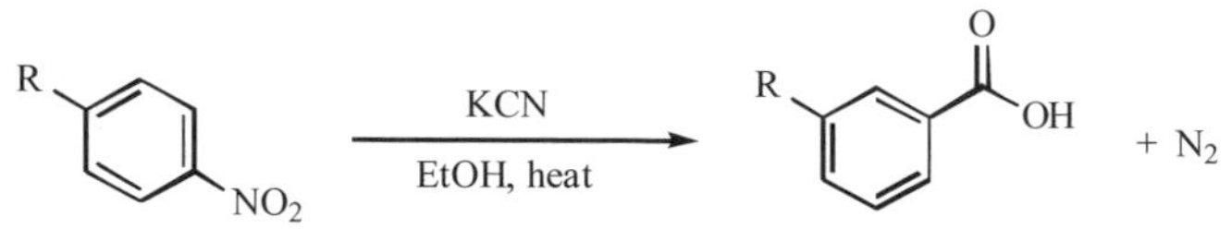

**Proposed Mechanism:**

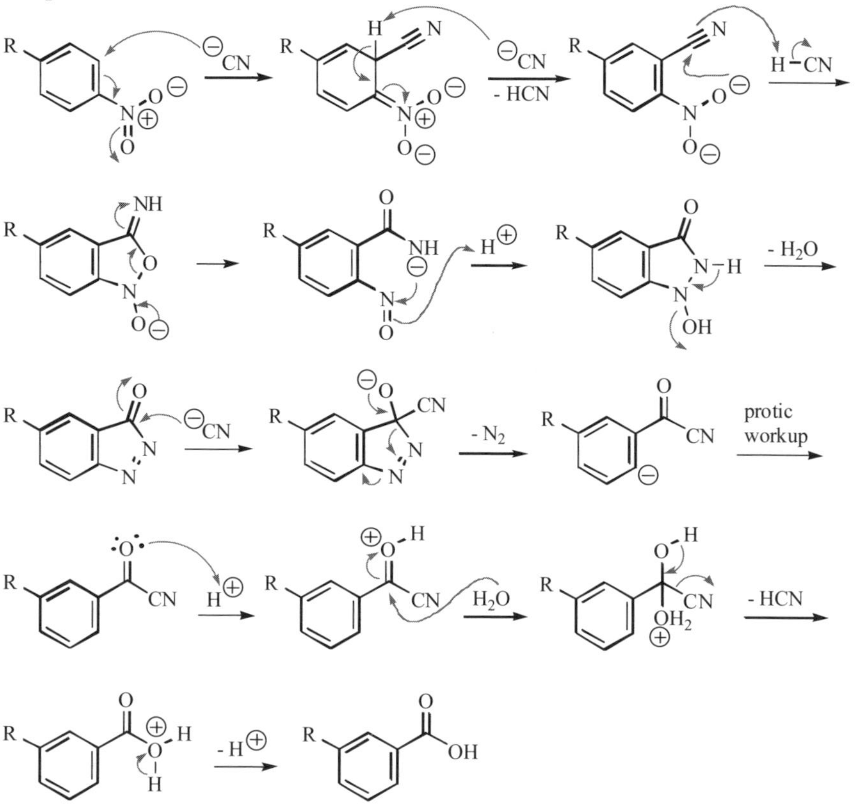

M. Rosenblum, *Journal of the American Chemical Society* **1960**, <u>82</u>, 3796

## Examples:

Note solvent change:

G. T. Rogers, T. L. V. Ulbricht, *Tetrahedron Letters* **1968**, <u>9</u>, 1029

This substrate was subjected to the reaction conditions to test the validity of it being a proposed intermediate.

K. M. Ibne-Rasa, E. Koubak, *Journal of Organic Chemistry* **1963**, <u>28</u>, 3240

J. F. Bunnett, M. Rauhut, M. D. Knutson, G. E. Bussell, *Journal of the American Chemical Society* **1954**, <u>76</u>, 5755

# Wacker Oxidation Reaction

## The Reaction:

## Proposed Mechanism:

See O. Hamed, C. Thompson, P.M. Henry, *Journal of Organic Chemistry* **1997**, <u>62</u>, 7082 for useful mechanistic discussion.

## Notes:

Attack of water is always at the more substituted carbon of the alkene. Given competition between a terminal alkene and a more substituted one, preference will be for the terminal bond.

### *Catalytic Asymmetric Wacker-Type Cyclization*

Y. Uozumi, K. Kato, T. Hayashi, *Journal of the American Chemical Society* **1997**, <u>119</u>, 5063

## Examples:

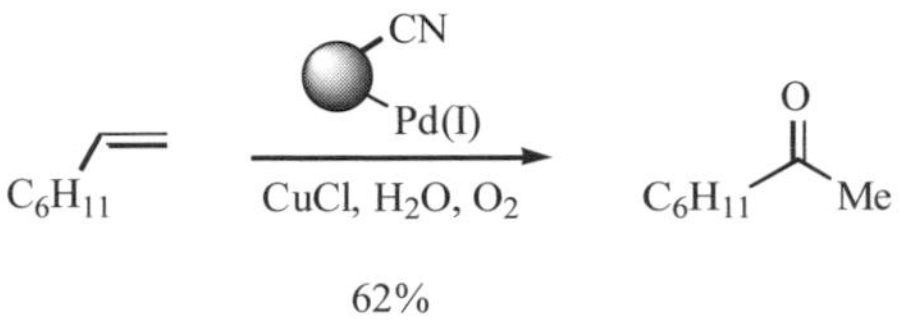

62%

J.-H. Ahn, D. C. Sherrington, *Macromolecules* **1996**, 29, 4164

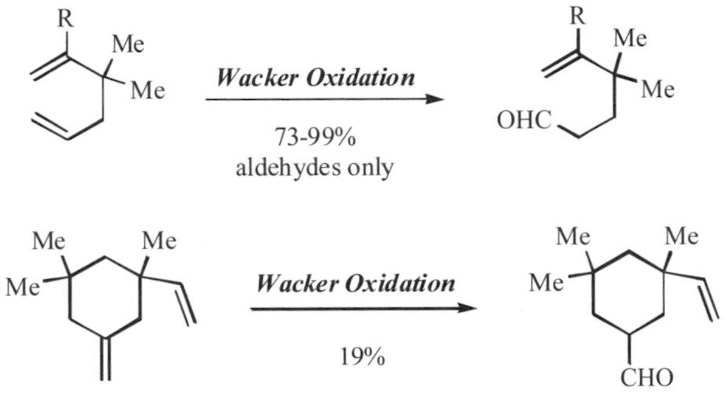

73-99%
aldehydes only

19%

T.-L. Ho, M. H.Chang, C. Chen, *Tetrahedron Letters* **2003**, 44, 6955

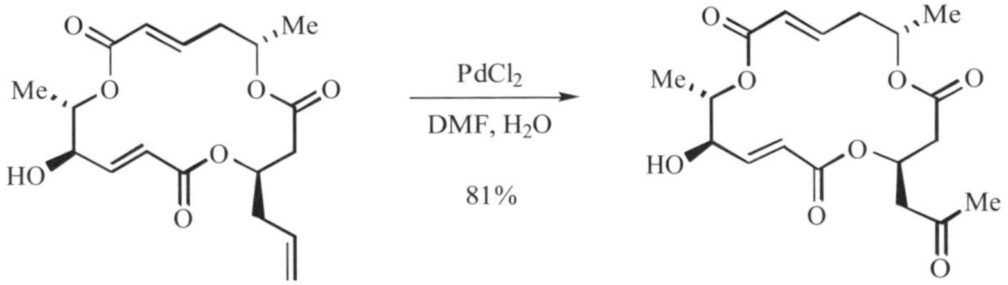

PdCl₂

DMF, H₂O

81%

Y. Kobayashi, Y.-G. Wang, *Tetrahedron Letters* **2002**, 43, 4381

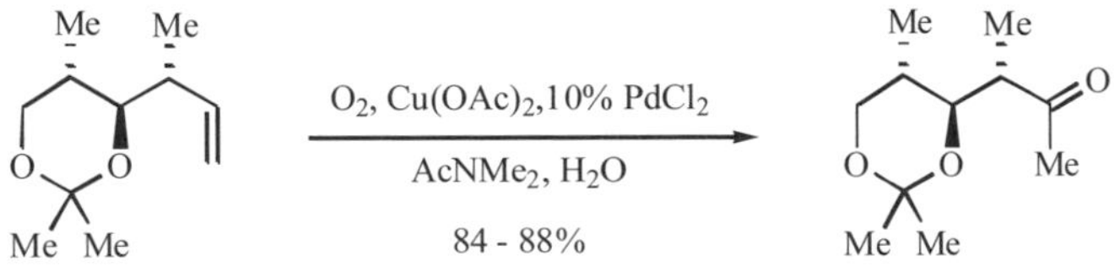

O₂, Cu(OAc)₂,10% PdCl₂

AcNMe₂, H₂O

84 - 88%

A. B. Smith, III, Y. S. Cho, G. K. Friestad, *Tetrahedron Letters* **1998**, 39, 8765

# Wagner-Meerwein Rearrangement

See: L. Birladeanu, "The Story of the Wagner-Meerwein Rearrangement", *Journal of Chemical Education* **2000**, *77*, 858

## The Reaction:

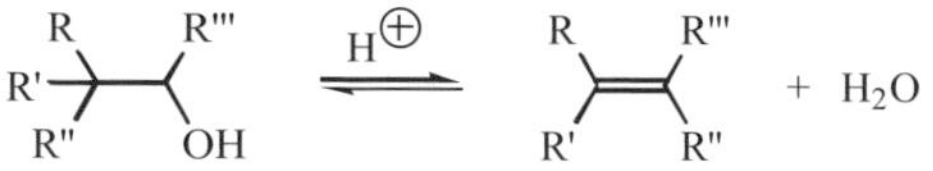

## Proposed Mechanism:

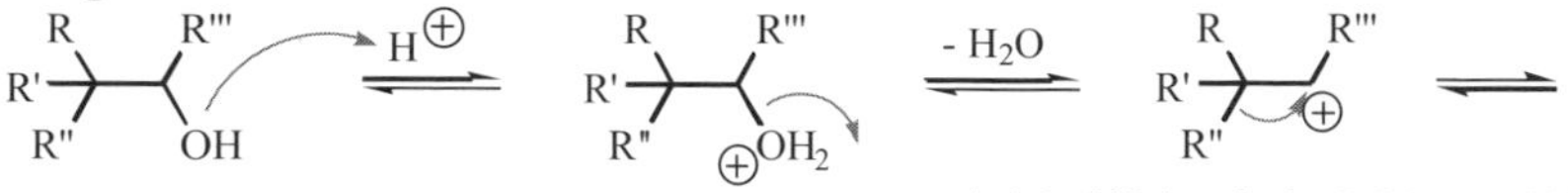

A 1,2-shift is called a ***Whitmore Shift***.

## Notes:

M. B. Smith, J. March *in March's Advanced Organic Chemistry*, 5ᵗʰ ed., John Wiley and Sons, Inc., New York, 2001, p. 1393

The rearrangement of camphene hydrochloride is called the ***Nametkin Rearrangement***

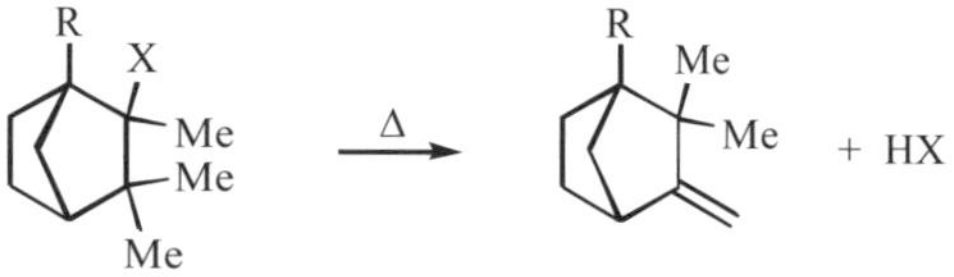

The "classical" - "non-classical" carbocation controversy concerned the ***Wagner-Meerwein rearrangement*** of norbornyl systems:

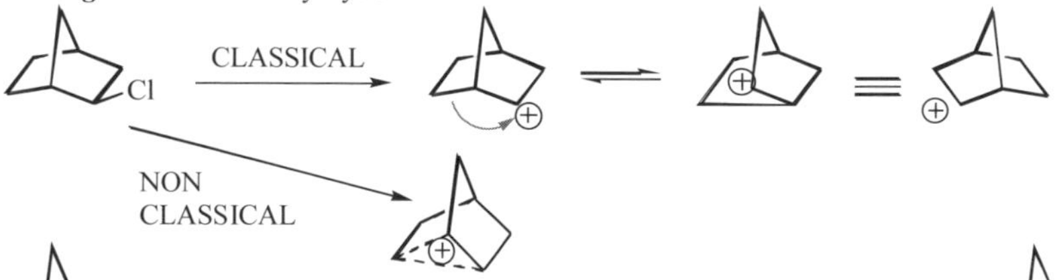

undergoes solvolysis reaction significantly faster than the endo isomer:

The reaction is also regulated by stereoelectronic factors:

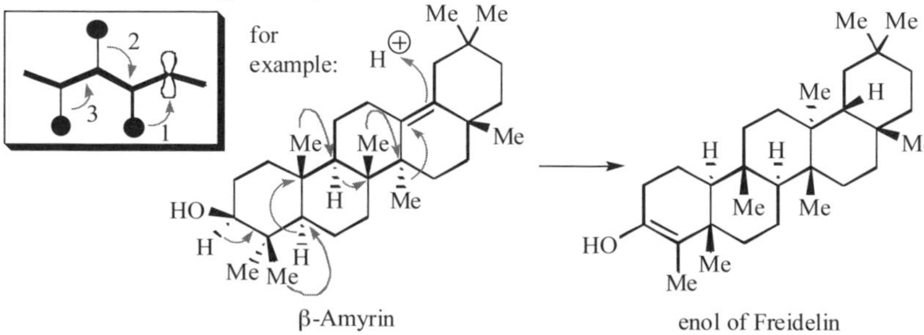

β-Amyrin                          enol of Freidelin

E. J. Corey, J. J. Ursprung, *Journal of the American Chemical Society* **1956**, *78*, 5041

## Examples:

## Demjanov (Demyanov) Rearrangement

## The Reaction:

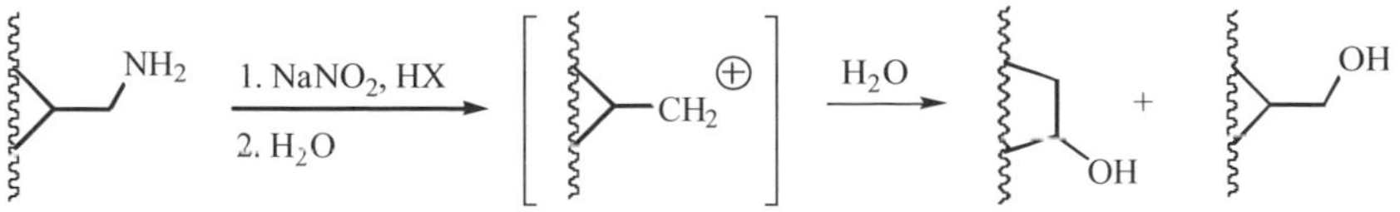

## Examples:

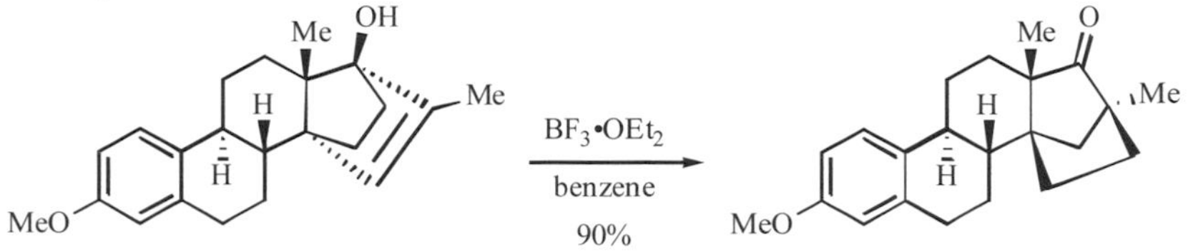

J. R. Bull, K. Bischofberger, R. I. Thomson, J. L. M. Dillen, P. H. Van Rooyen, *Journal of the Chemical Society, Perkin Transactions 1* **1992**, 2545

A detailed biosynthetic pathway analysis:

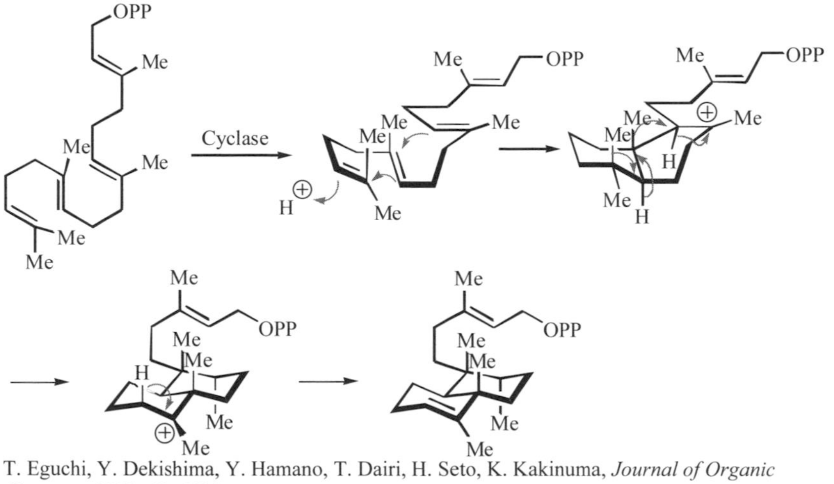

T. Eguchi, Y. Dekishima, Y. Hamano, T. Dairi, H. Seto, K. Kakinuma, *Journal of Organic Chemistry* **2003**, *68*, 5433

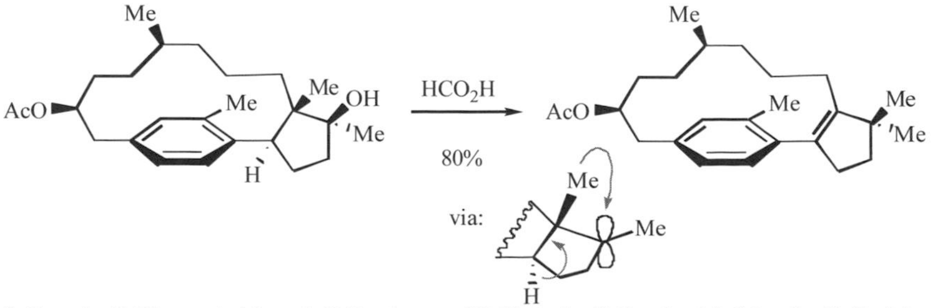

S. Baeurle, T. Blume, A. Mengel, C. Parchmann, W. Skuballa, S. Baesler, M. Schaefer, D. Suelzle, H.-P. Wrona-Metzinger, *Angewandte Chemie, International Edition in English* **2003**, *42*, 3961

# Watanabe-Conlon Transvinylation

## The Reaction:

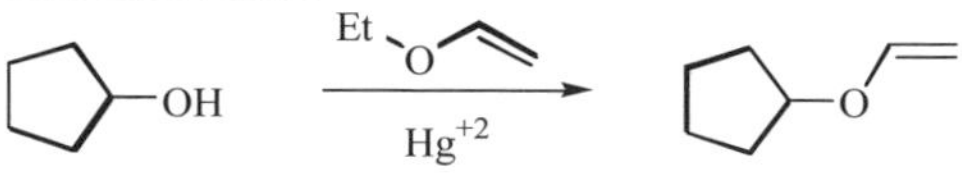

## Proposed Mechanism:

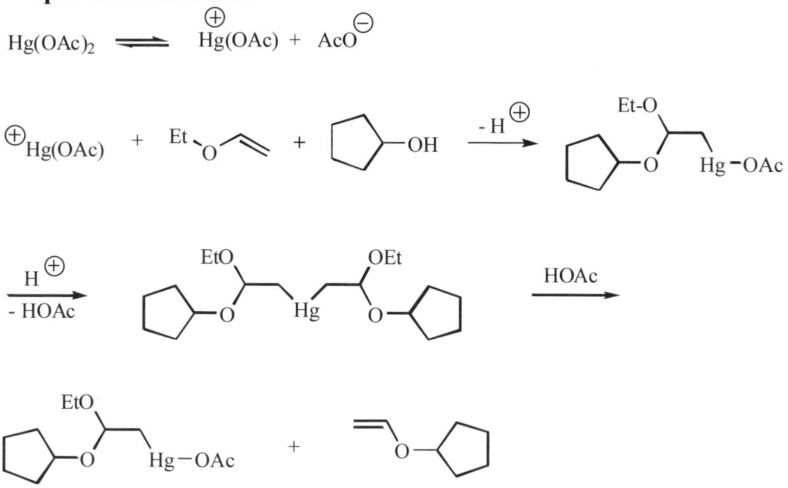

$$Hg(OAc)_2 \rightleftharpoons \overset{\oplus}{Hg(OAc)} + Ac\overset{\ominus}{O}$$

## Notes:

This reaction generally uses about 10% molar equivalent of $Hg(OAc)_2$.

This reaction is generally part of a sequence followed by ***Claisen rearrangement*** to generate a remote functional group (aldehyde).

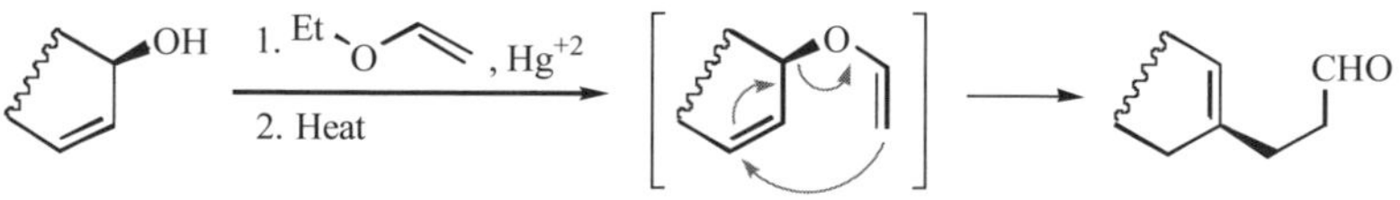

This approach was developed by Burgstahler for:

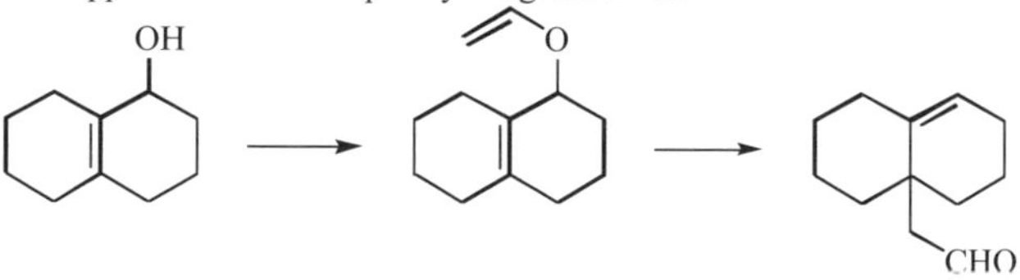

A. W. Burgstahler, I. C. Nordin, *Journal of the American Chemical Society* **1959**, <u>81</u>, 3151

# Examples:

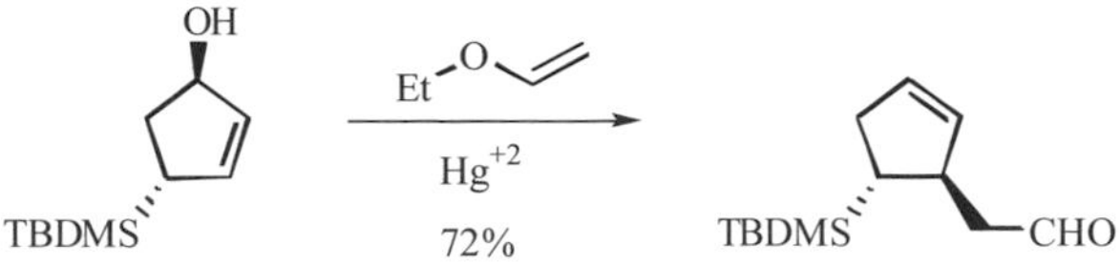

D. L. J. Clive, H. W. Manning, *Journal of the Chemical Society, Chemical Communications.* **1993**, 666

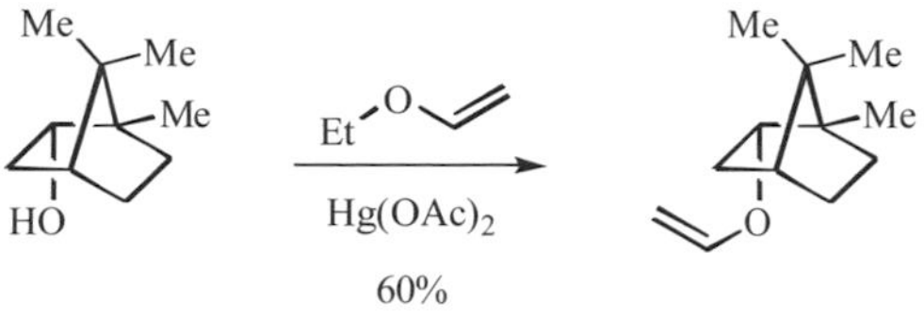

60%

V. Godebout, S. Leconte, F. Levasseur, L. Duhamel, *Tetrahedron Letters* **1996**, 37, 7255

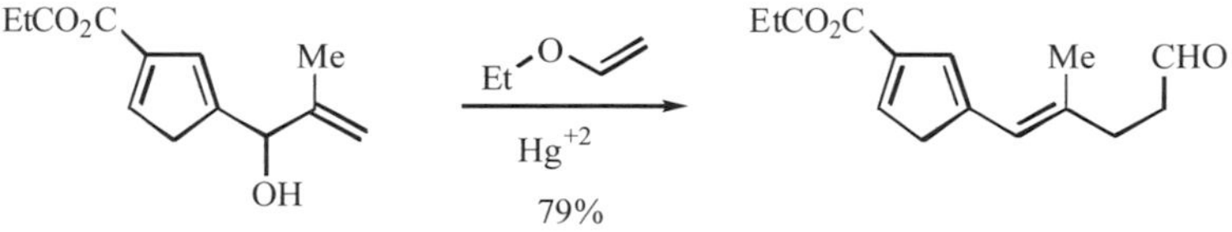

79%

D. H. Williams, D. J. Faulkner, *Tetrahedron* **1996**, 52, 4245

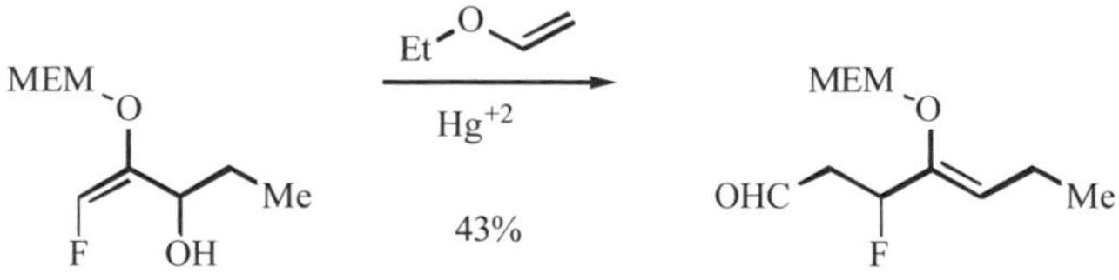

43%

S. T. Patel, J. M. Percy, S. D. Wilkes, *Tetrahedron* **1995**, 51, 11327

# Weiss Reaction

## The Reaction:

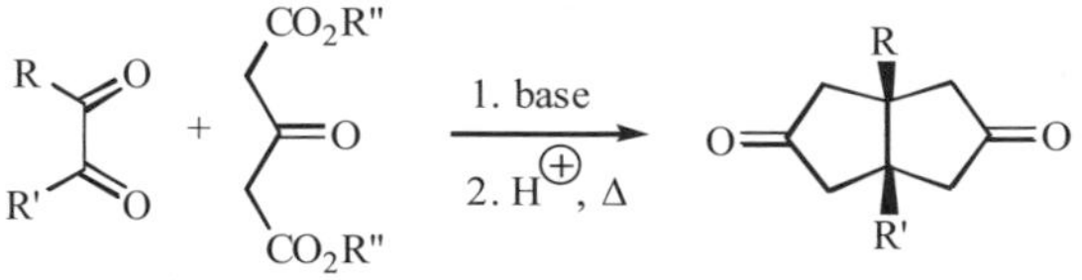

R's = H, alkyl, aryl or could be a cyclobutane ring or larger

## Proposed Mechanism:

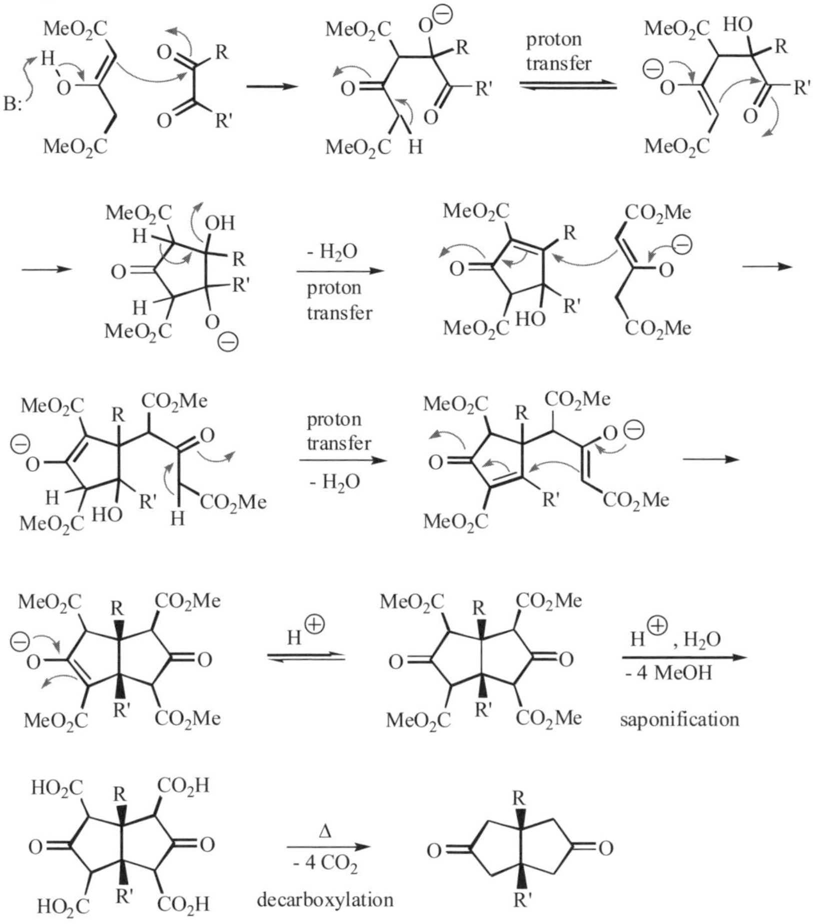

- H$_2$O
proton transfer

proton transfer
- H$_2$O

H$^{\oplus}$

H$^{\oplus}$ , H$_2$O
- 4 MeOH
saponification

$\Delta$
- 4 CO$_2$
decarboxylation

## Notes:

A number of examples are reported in:

U. Weiss, J. M. Edwards, *Tetrahedron Letters* **1968**, 9, 4885

**Examples:**

1. MeO$_2$C$\quad$CO$_2$Me, 2% HCO$_3^{\ominus}$
2. HCl, HOAc

>70%

G. Kubiak, X. Fu, K. Bupta, J. M. Cook, *Tetrahedron Letters* **1990**, 31, 4285

1. MeO$_2$C$\quad$CO$_2$Me, pH 5.6
2. 10% HCl, HOAc

85%

(dashed bonds meant to idicate a mixture of isomers)

L. A. Paquette, M. A. Kesselmayer, G. E. Underiner, S. D. House, R. D. Rogers, K. Meerholz, J. Heinze, *Journal of the American Chemical Society* **1992**, 114, 2652

MeO$_2$C$\quad$CO$_2$Me

K$_2$CO$_3$, MeOH

48 - 50%

$t$-BuO$_2$C$\quad$CO$_2t$-Bu

$t$-BuO$_2$C$\quad$CO$_2t$-Bu

A. K. Gupta, J. M. Cook, U. Weiss, *Tetrahedron Letters* **1988**, 29, 2535

COOCH$_3$

+

COOCH$_3$

1. NaOH    58 - 63%
2. H$^{\oplus}$, Δ  88 - 90%

S. H. Bertz, J. M. Cook, A. Gawish, U. Weiss, *Organic Synthesis* CV 7, 50

# Wharton Olefination

## The Reaction:

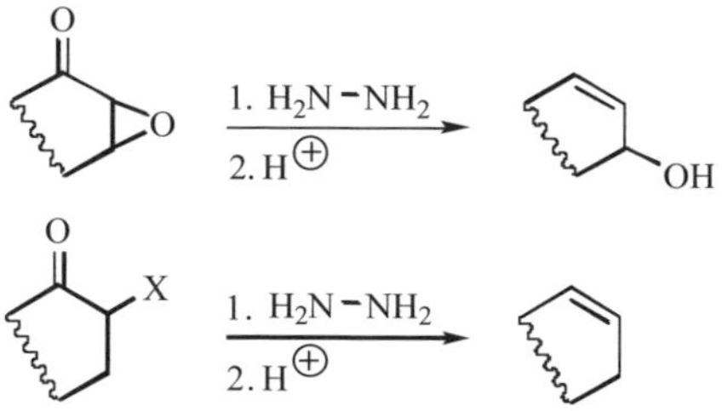

## Proposed Mechanism:

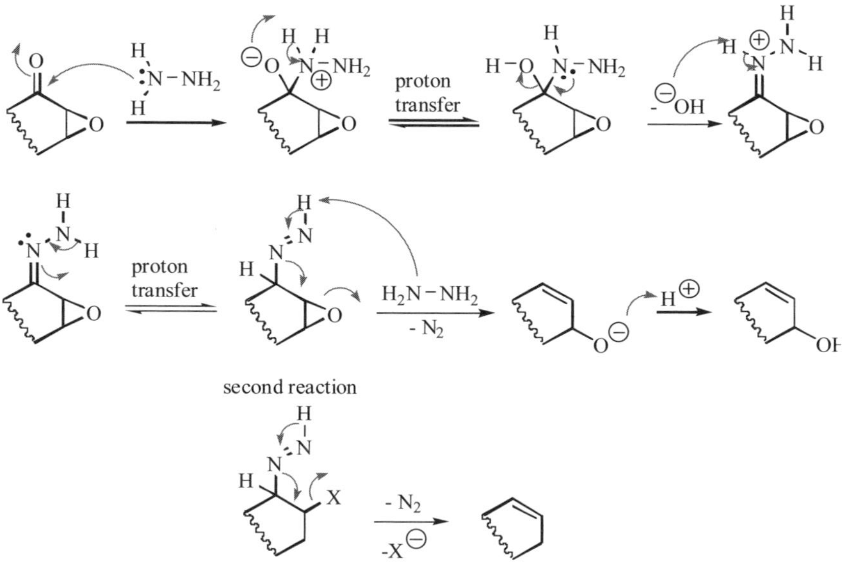

second reaction

## Examples:

50%

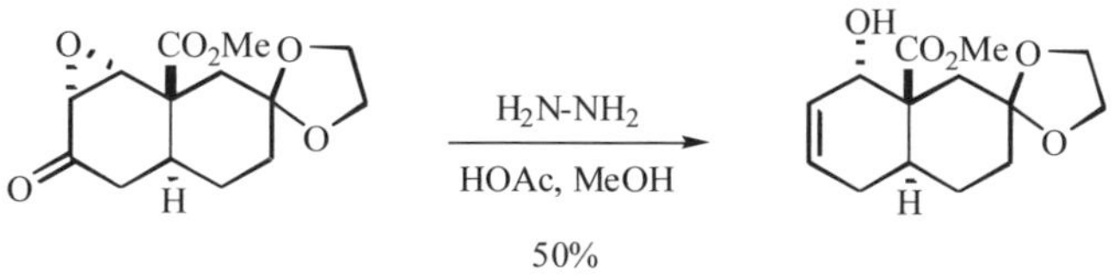

P. A. Zoretic, R. J. Chambers, G. D. Marbury, A. A. Riebiro, *Journal of Organic Chemistry* **1985**, 50, 2981

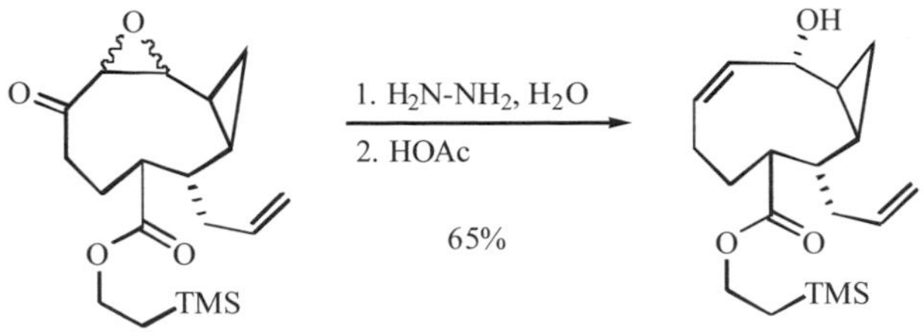

1. H$_2$N-NH$_2$, H$_2$O

2. HOAc

65%

G. Kim, M. Y. Chu-Moyer, S. J. Danishefsky, G. K. Schulte, *Journal of the American Chemical Society* **1993**, <u>115</u>, 30

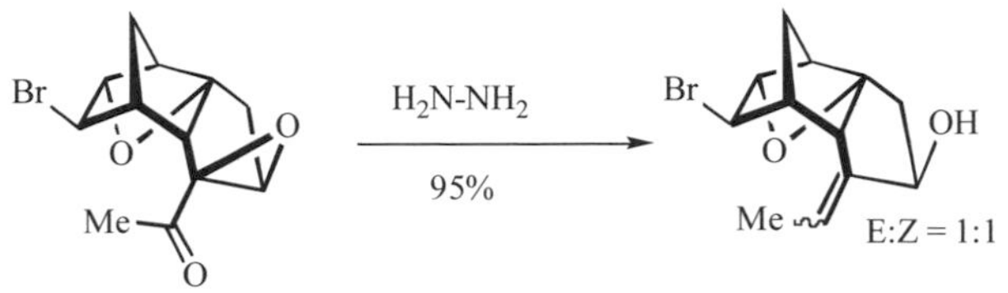

H$_2$N-NH$_2$

95%

E:Z = 1:1

T. Sugahara, H. Fukuda, Y. Iwabuchi, *Journal of Organic Chemistry* **2004**, <u>69</u>, 1744

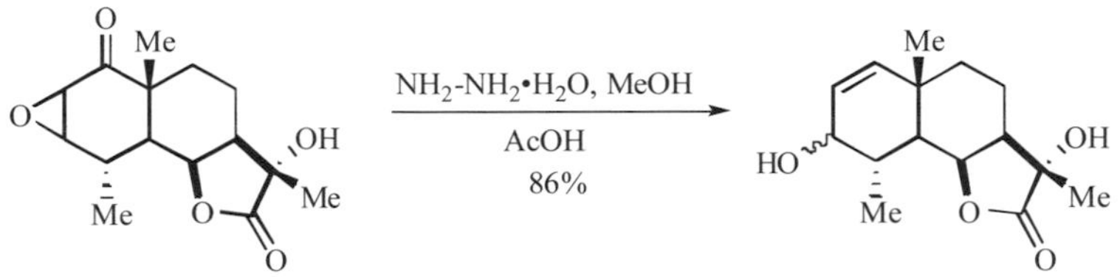

NH$_2$-NH$_2$•H$_2$O, MeOH

AcOH

86%

F. J. Moreno-Dorado, F. M. Guerra, F. J. Aladro, J. M. Bustamante, Z. D. Jorge, G. M. Massanet, *Journal of Natural Products* **2000**, <u>63</u>, 934

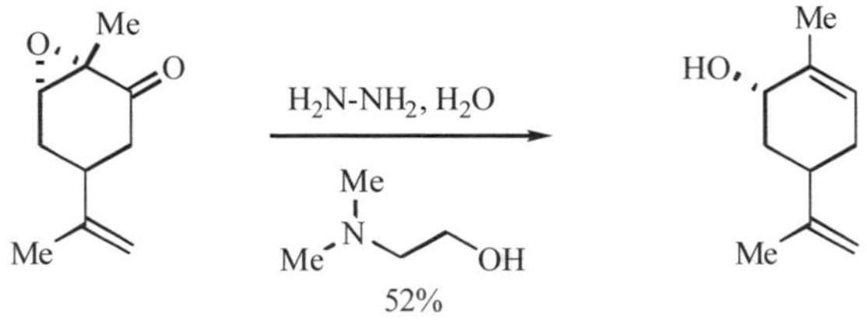

H$_2$N-NH$_2$, H$_2$O

52%

L. Castedo, J. L. Mascarnenas. M. Mourino, *Tetrahedron Letters* **1987**, <u>28</u>, 2099

# Wichterle Reaction

## The Reaction:

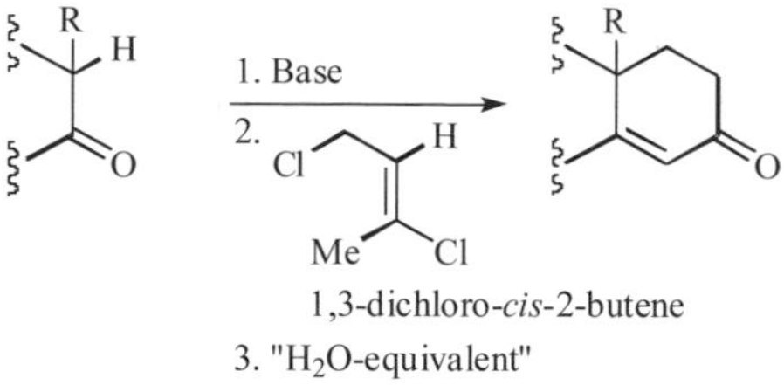

1,3-dichloro-*cis*-2-butene

## Proposed Mechanism:

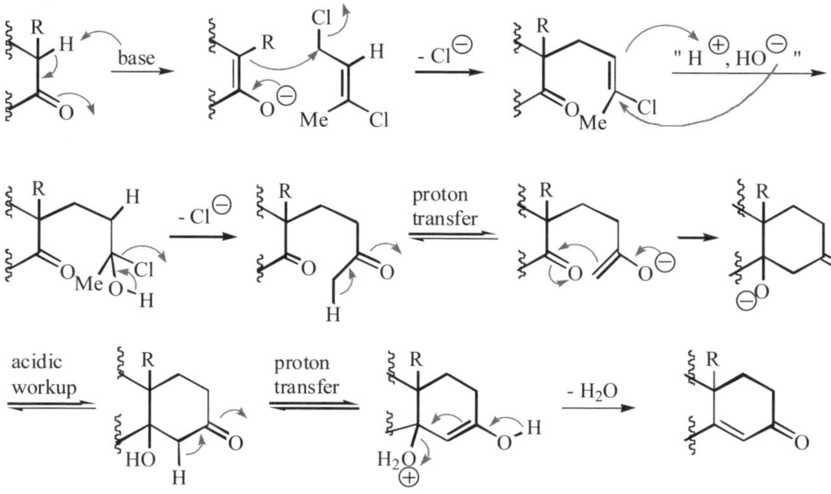

## Notes:

This is a modification of the **_Robinson annulation_** in which 1,3-dichloro-*cis*-2-butene is used in place of methyl vinyl ketone.

A number of "$H^+$ $OH^-$" equivalents are used: **_Oxymercuration_** conditions or formic acid and protic acid, followed by hydrolysis of the formate ester are two common approaches.

A variation of this approach:

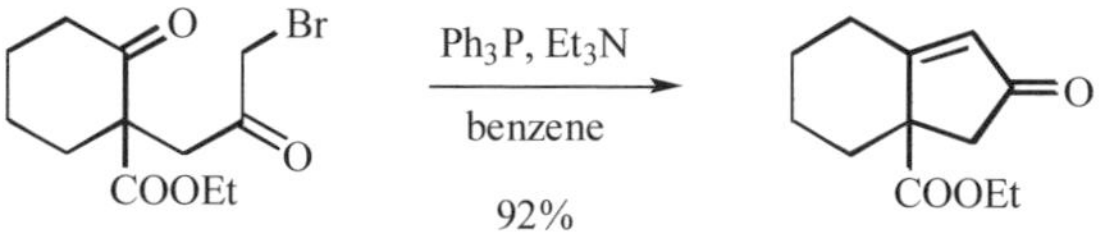

M. P. VanBrunt, R. O. Ambenge, S. W. Weinreb, *Journal of Organic Chemistry* **2003**, <u>68</u>, 3323

## Examples:

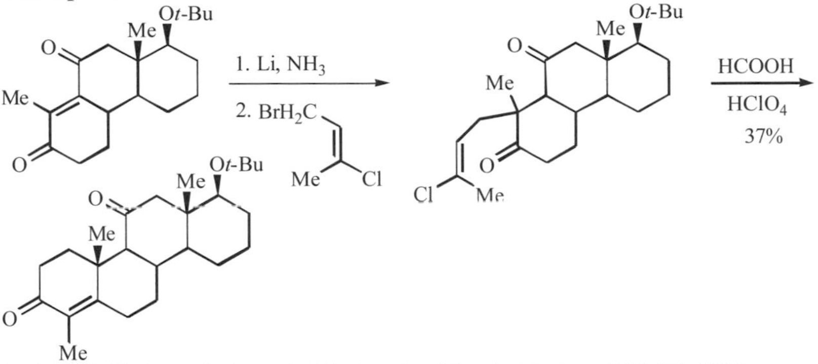

G. Stork, E. W. Logusch, *Journal of the American Chemical Society*, **1980**, <u>102</u>, 1219

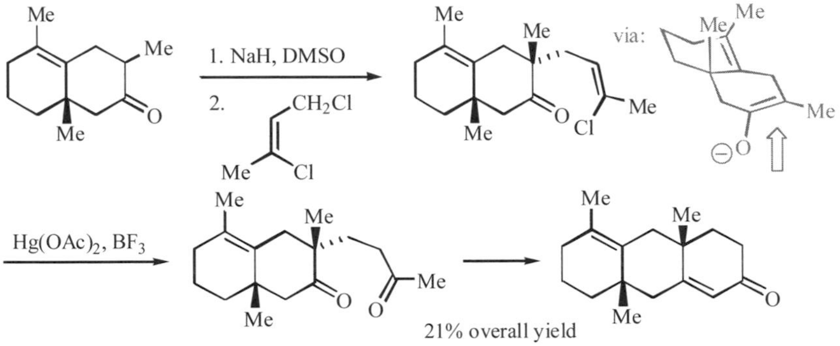

21% overall yield

L. A. Paquette, D. T. Belmont, Y.-L. Hsu, *Journal of Organic Chemistry* **1985**, <u>50</u>, 4667

## Fujimoto-Belleau Reaction

**The Reaction:**

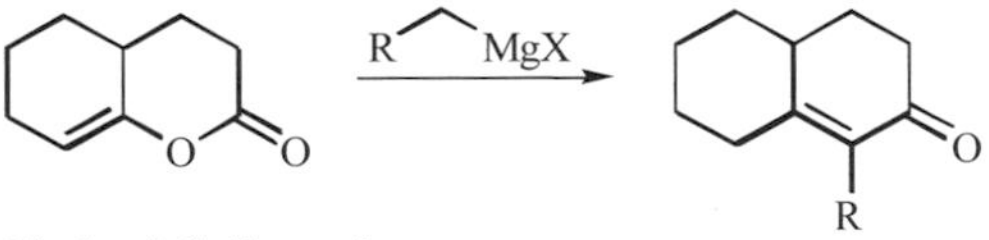

**Mechanistic Example:**

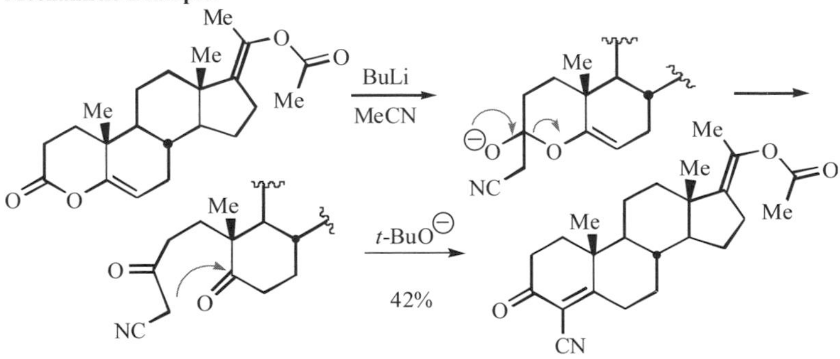

42%

M. Haase-Held, M. Hatzis, J. Mann, *Journal of the Chemical Society Perkin Transactions 1* **1993**, 2907

# Widman-Stoermer Synthesis

## The Reaction:

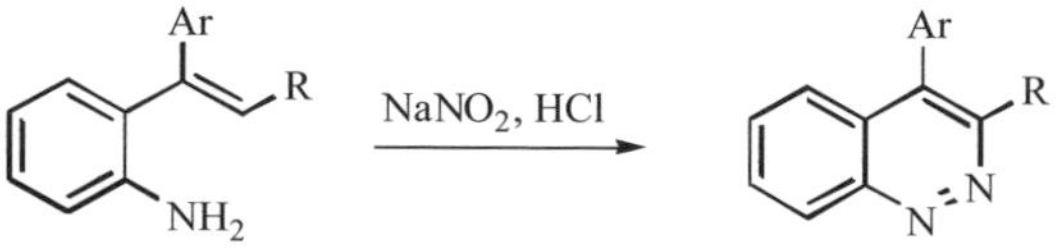

## Proposed Mechanism:

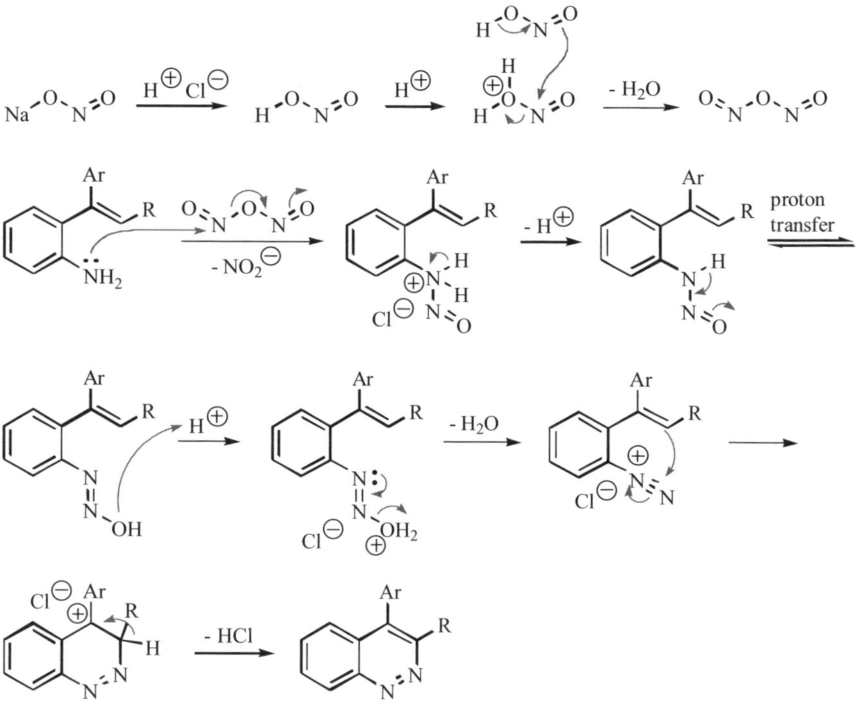

## Notes:

See the ***Borsche Cinnoline Synthesis*** and the ***von Richter Cinnoline Synthesis*** for other preparations of cinnolines.

## Examples:

J. W. Barton, N. D. Pearson, *Journal of the Chemical Society, Perkin Transactions 1,* **1987**, 1541

B. S. Ross, R. A. Wiley, *Journal of Medicinal Chemistry* **1985**, 28, 870

M. H. Palmer, P. S. McIntyre, *Tetrahedron* **1971**, 27, 2913

# Willgerodt-Kindler Reaction

## The Reaction:

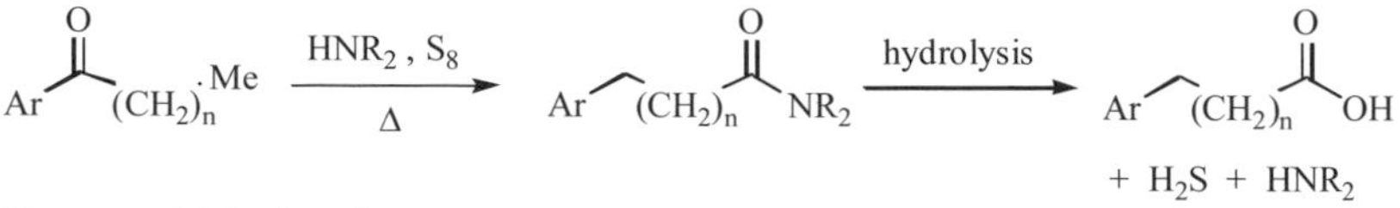

$$\underset{Ar}{\overset{O}{\|}}\underset{(CH_2)_n}{\bigwedge}\cdot Me \xrightarrow[\Delta]{HNR_2,\ S_8} \underset{Ar}{\overset{O}{\|}}\underset{(CH_2)_n}{\bigwedge}NR_2 \xrightarrow{hydrolysis} \underset{Ar}{\overset{O}{\|}}\underset{(CH_2)_n}{\bigwedge}OH$$

$$+\ H_2S\ +\ HNR_2$$

## Proposed Mechanism:

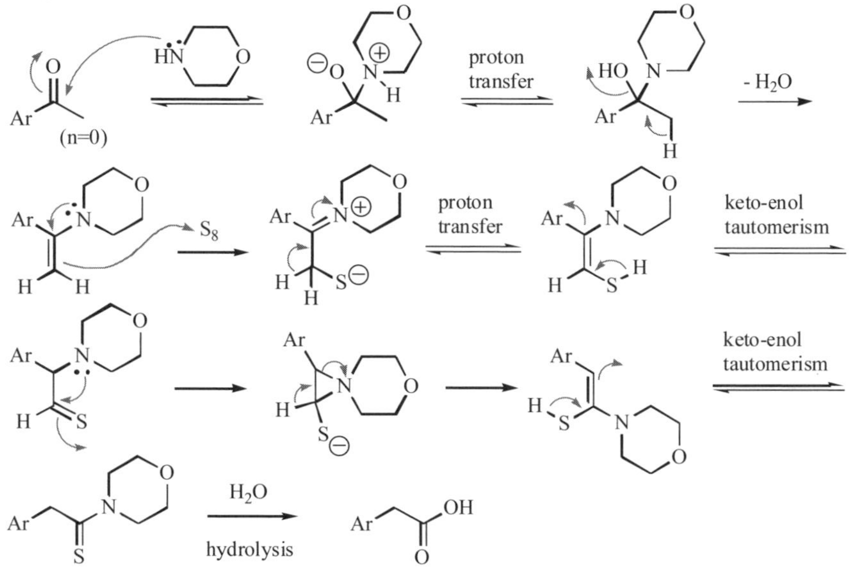

## Notes:

M. B. Smith, J. March in *March's Advanced Organic Chemistry*, 5th ed., John Wiley and Sons, Inc., New York, 2001, p. 1567; T. Laue, A. Plagens, *Named Organic Reactions*, John Wiley and Sons, Inc., New York, 1998, pp. 267-269.

The original *Willgerodt Reaction* conditions required high temperature and pressure, with use of ammonium polysulfide $(NH_4)_2S_x$ and $H_2O$ to give either an amide or the ammonium salt of the corresponding acid. *Kindler's modification*, shown above, eliminated these problems and substituted $S_8$ and a dry amine, most commonly morpholine.

The reaction will introduce the acid at the terminal carbon, no matter where the carbonyl position is occupied in the starting material:

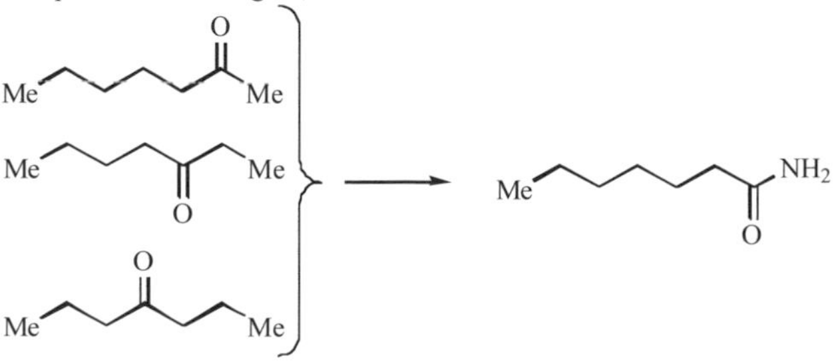

L. Cavalieri, D. B. Pattison, M. Carmack, *Journal of the American Chemical Society* **1945**, <u>67</u>, 1783

## Examples:

$S_8$, morpholine; NaOH; MeSO$_2$-C$_6$H$_4$-CH$_2$-CO$_2$H reaction scheme, 57%

I. W. Davies, J.-F. Marcoux, E. G. Corley, J. Journet, D.-W Cai, M Palucki, J. Su, R. D. Larsen, K. Rossen, P. J. Pye, L. DeMichele, P. Dormer, P. J. Reider, *Journal of Organic Chemistry* **2000**, 65, 8415

1. $S_8$, morpholine, Ph-SO$_3$H
2. Br-CH$_2$-CO$_2$H
3. H$_2$S
24%

G. Levesque, P. Arsene, V. Fanneau-Bellenger, T.-N. Pham, *Biomacromolecules* **2000**, 1, 387

$S_8$, morpholine, microwave, 81%

$S_8$, morpholine, microwave, 55 - 74%

G = Cl, Me, NH$_2$, OMe

M. Mooshabadi, K. Aghapoor, H. R. Darabi, M. M. Mujtahedi, *Tetrahedron Letters* **1999**, 40, 7549

$S_8$, morpholine, 60%

T. Bacchetti, A. Alemagna, B. Daniel, *Tetrahedron Letters* **1965**, 6, 2001

# Williamson Ether Synthesis

## The Reaction:

$$R\text{-}O\text{-}H \xrightarrow[\text{2. R'-L}]{\text{1. base}} R\text{-}O\text{-}R'$$

L = leaving group = -X, -OTs, -OMs, O-SO$_3$R'

## Proposed Mechanism:

$$R\text{-}O\text{-}H \xrightarrow{\text{base}} R\text{-}O^{\ominus} \quad R'\text{-}L \xrightarrow{-L^{\ominus}} R\text{-}O\text{-}R'$$

Most *Williamson Ether Syntheses* proceed by an S$_N$2 mechanism. Stereochemical inversions can be expected, where appropriate, as a result.

## Notes:

Secondary R groups usually give low yields.

Tertiary R groups are typically not successful due to elimination:

$$\xrightarrow{\text{base}} \quad + \quad HOR$$

## Examples:

62%

S. Hecht, J. M. Frechet, *Journal of the American Chemical Society* **1999**, 121, 4084

75:25

M. Attolini, T. Boxus, S. Biltresse, J. Marchand-Brynaert, *Tetrahedron Letters* **2002**, 43, 1187

92%

E. E. Dueno, F. Chu, S.-I. Kim, J. W. Jung, *Tetrahedron Letters* **1999**, 40, 1843

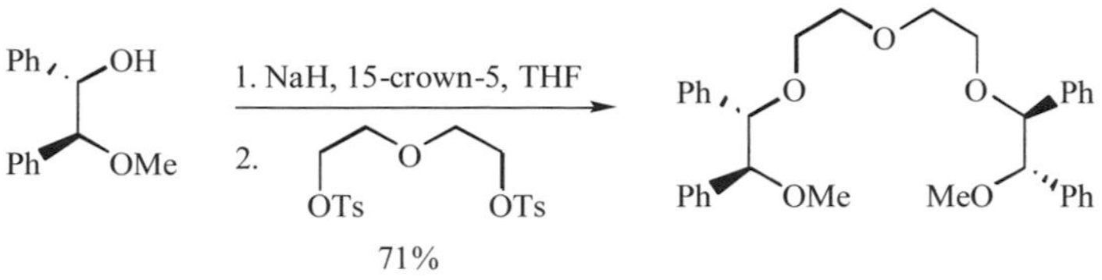

71%

H. C. Aspinall, N. Greeves, W.-M. Lee, E. G. McIver, P. M. Smith, *Tetrahedron Letters* **1997**, <u>38</u>, 4679

This work discusses the question of substitution vs. elimination, and has useful commentary from a synthetic point of view.

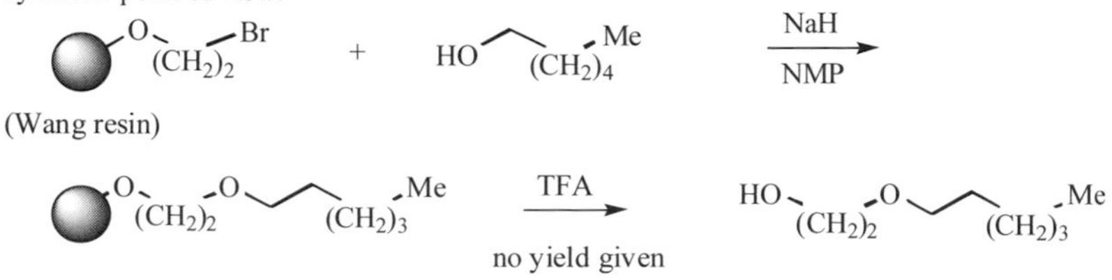

(Wang resin)

no yield given

A. Weissberg, A. Dahan, M. Portnoy, *Journal of Combinatorial Chemistry* **2001**, <u>3</u>, 154

# Wittig Indole Synthesis

## The Reaction:

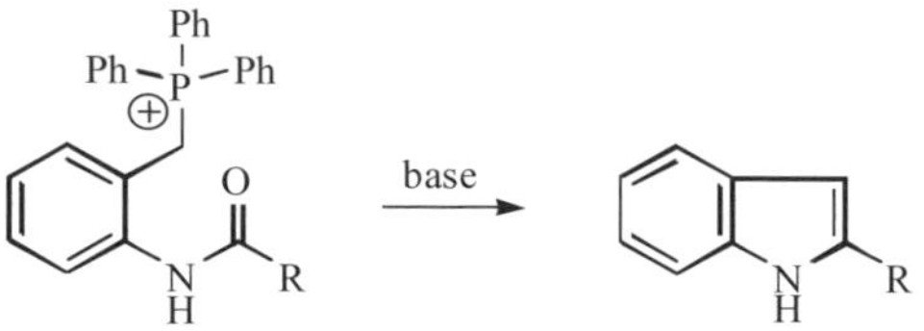

## Proposed Mechanism:

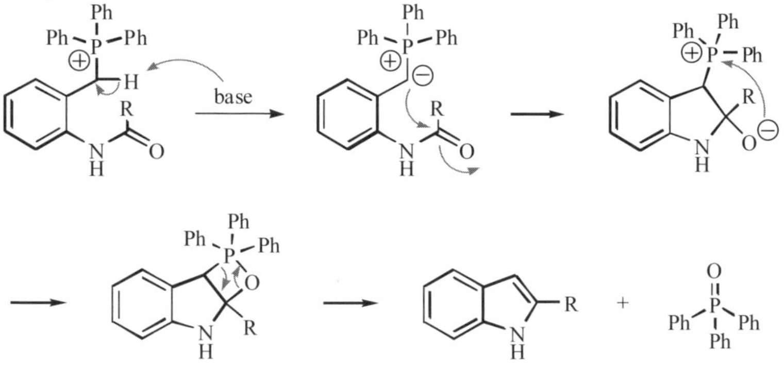

## Examples:

K. Miyashita, K. Kondoh, K. Tsuchiya, H. Miyabe, T. Imanishi, *Journal of the Chemical Society: Perkin Transactions 1* **1996**, 1261

B. Danieli, G. Lesma, G. Palmisano, D. Passarella, A. Silvani, *Tetrahedron* **1994**, <u>50</u>, 6941

M. Le Corre, Y. Le Stane, A. Hercouet, H. Le Brown, *Tetrahedron* **1985**, <u>41</u>, 5313

# Wittig Reaction (Wittig Olefination Reaction)

## The Reaction:

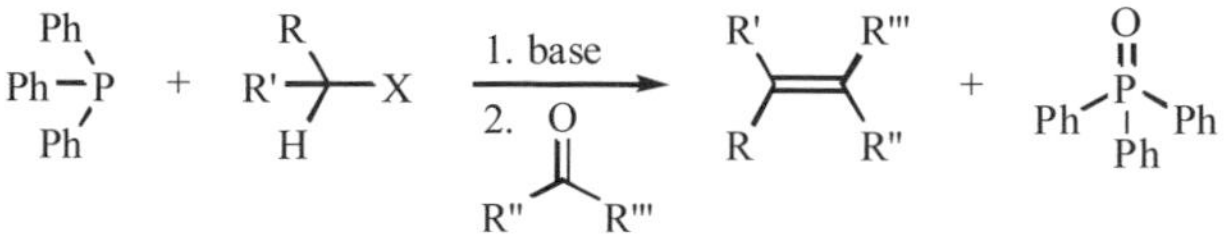

## Proposed Mechanism:

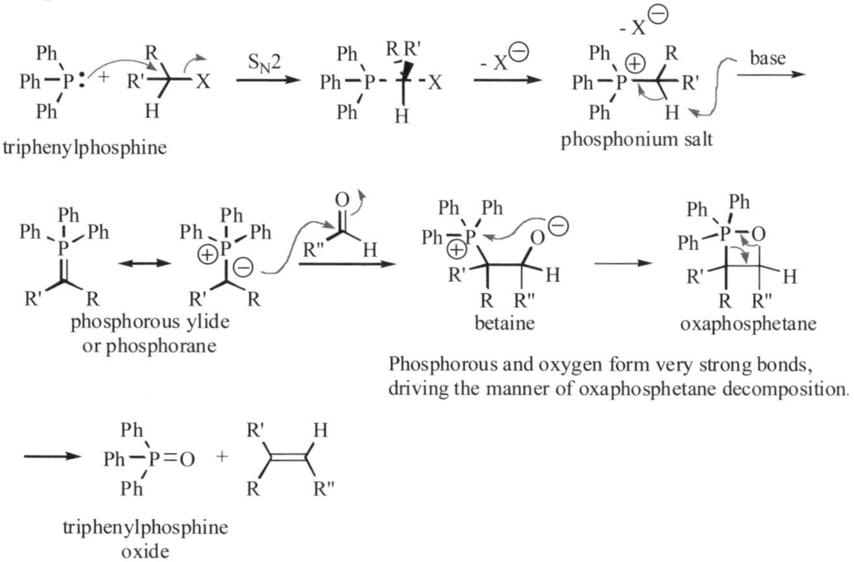

triphenylphosphine

phosphonium salt

phosphorous ylide
or phosphorane

betaine

oxaphosphetane

Phosphorous and oxygen form very strong bonds,
driving the manner of oxaphosphetane decomposition.

triphenylphosphine
oxide

## Notes:

Other phosphines may be used for this reaction, but the choice should not contain a protons that
could be abstracted as is the proton on the halide coupling partner, as a mixture of desired and
undesired ylides would be formed.

Usually, one uses a strong base such as BuLi, NaNH$_2$ / NH$_3$, NaH or NaOR.

Preferred anti attack of ylide,       Bond rotation follows to       As a result, this reaction often
minimizing steric interactions.       form the betaine.              gives the cis / Z - alkene.

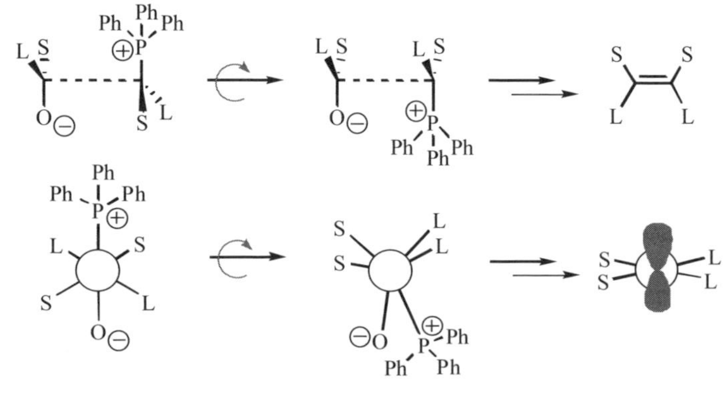

If the halide contains an electron withdrawing group, the negative charge in the ylide is delocalized, decreasing its nucleophilicity and reactivity. Aldehydes may still react, but ketones most likely will not.

*Bestman's Protocol*: The intermediate ylide can be cleaved by ozone. By careful addition one can carry out a unique *Wittig reaction*:

## Examples:

72%

S. P. Chavan, R. K. Kharul, R. R. Kale, D. A. Khobragade, *Tetrahedron* **2003**, *59*, 2737

65%

R. K. Boeckman, Jr., T. R. Aless, *Journal of the American Chemical Society* **1982**, *104*, 3216

1. NaN₃
2. Bu₃P, toluene

- N₂,

15 - 25%

Reaction with polymer-supported PhP₃ rather than Bu₃P gave:

43%

B. J. Neubert, B. B. Snider, *Organic Letters* **2003**, *5*, 765

# [1,2]-Wittig Rearrangement

## The Reaction:

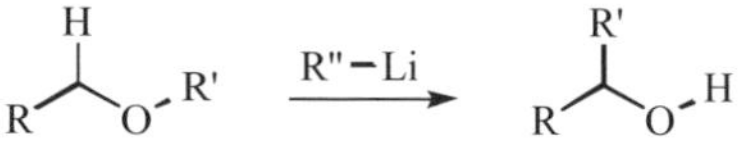

$$R = H, Alkl, Aryl, Alkenyl, Alkynyl, -COOR, -COOM$$

R = H, Alkl, Aryl, Alkenyl, Alkynyl, -COOR, -COOM
R' = Alkyl, Ally, Benzyl, Aryl

## Proposed Mechanism:

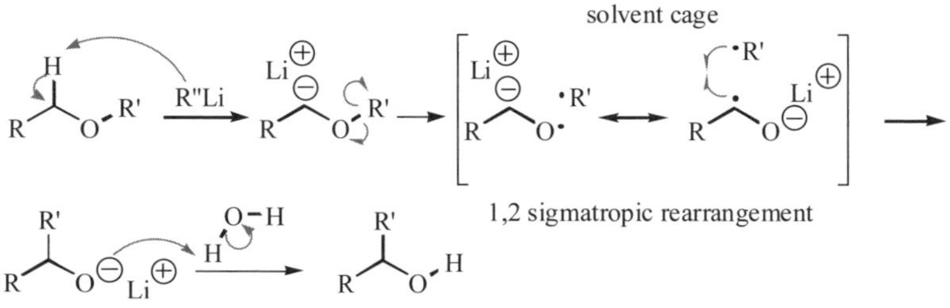

solvent cage

1,2 sigmatropic rearrangement

## Notes:

The oxygen can be replaced by nitrogen and then the reaction is known as *[1,2]-Aza-Wittig Rearrangement*:

## Examples:

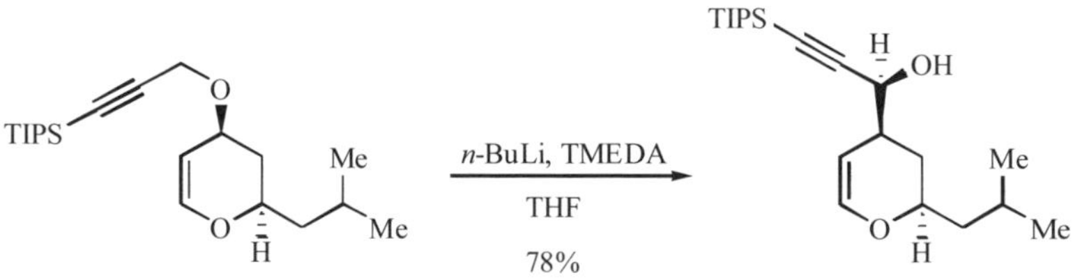

*n*-BuLi, TMEDA

THF

78%

P. Wipf, T. H. Graham, *Journal of Organic Chemistry* **2003**, <u>68</u>, 8798

R. E. Maleczka, Jr., F. Geng, *Journal of the American Chemical Society* **1998**, <u>120</u>, 8551

K. Tomooka, H. Yamamoto, T. Nakai, *Journal of the American Chemical Society* **1996**, <u>118</u>, 3317

A. Garbi, L Allain, F. Chorki, M. Ourevitch, B. Crousse, D. Bonnet-Delpon, T. Nakai, J.-P. Begue, *Organic Lett*ers **2001**, <u>3</u>, 2529

Via dehydration of major product:

L. Lemiegre, T. Regnier, J.-C. Combret, J. Maddaluno, *Tetrahedron Letters* **2003**, <u>44</u>, 373

# [2,3]-Wittig Rearrangement

## The Reaction:

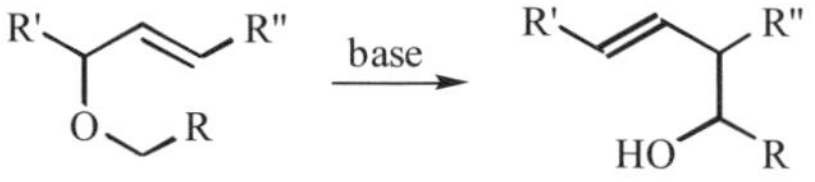

## Proposed Mechanism:

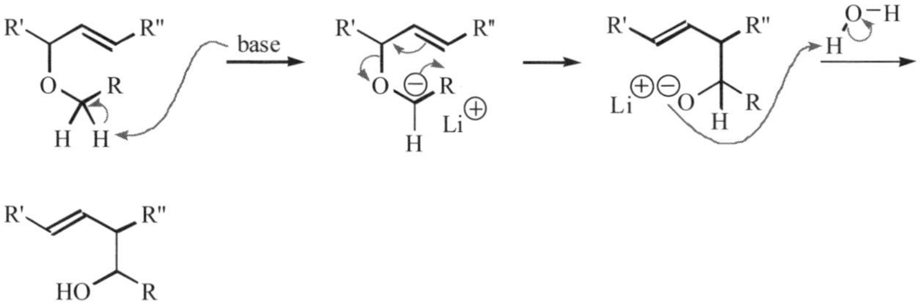

## Examples:

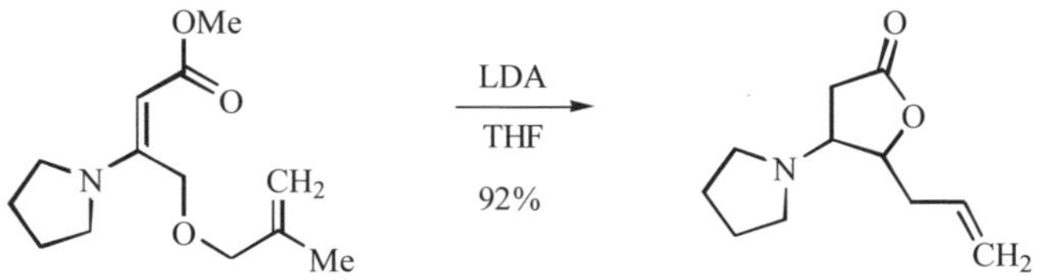

Y. J. Li, P.-T. Lee, C.-M. Yang, Y.-K. Chang, Y.-C. Weng, Y.-H. Liu, *Tetrahedron Letters* **2004**, <u>45</u>, 1865

A. S. Balnaves, G. McGowan, P. D. P. Shapland, E. J. Thomas, *Tetrahedron Letters* **2003**, <u>44</u>, 2713

(24:76)

K. Tomooka, T. Igarashi, N. Kishi, T. Nakai, *Tetrahedron Letters* **1999**, <u>40</u>, 6257

M. Tsubuki, K. Takahashi, T. Honda, *Journal of Organic Chemistry* **2003**, <u>68</u>, 10183

# Wolff Rearrangement

## The Reaction:

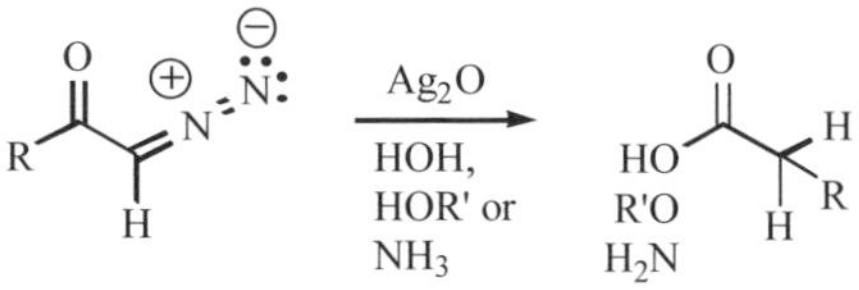

## Proposed Mechanism:

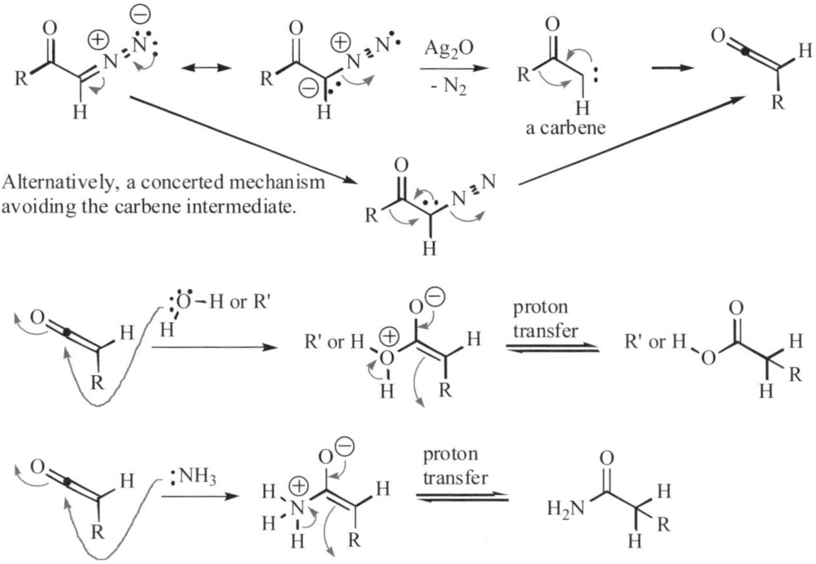

Alternatively, a concerted mechanism avoiding the carbene intermediate.

## Notes:

M. B. Smith, J. March in *March's Advanced Organic Chemistry*, 5th ed., John Wiley and Sons, Inc., New York, 2001, p. 1405

### Smith's Vinylogous Wolff Rearrangement:

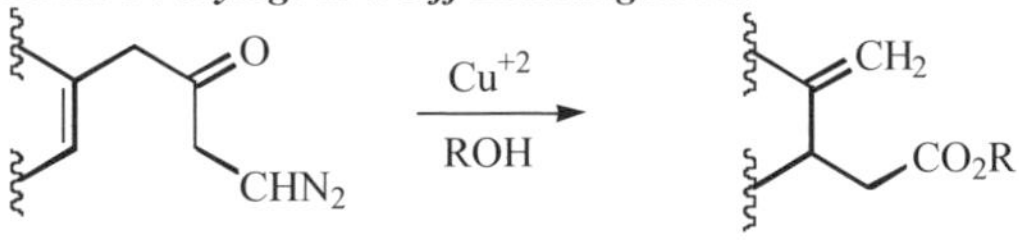

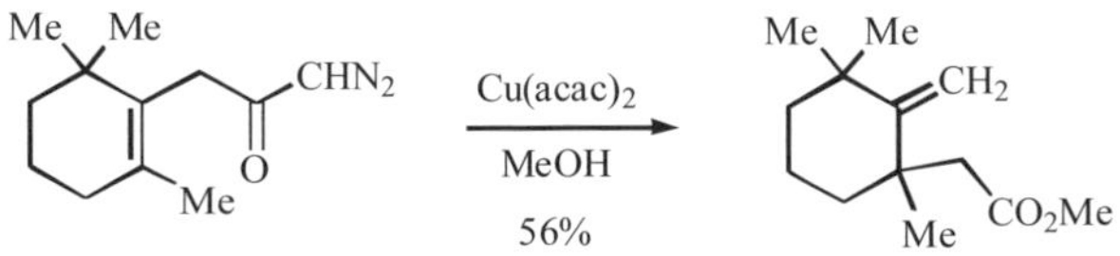

56%

A. B. Smith, III, B. H. Toder, S. J. Branca, *Journal of the American Chemical Society* **1985**, <u>106</u>, 1995

Ag$_2$O,        56 : 42
CuSO$_4$, Δ    0 : 85
hv             92 : 8

Smith, et. al. (continued)

92%

S. G. Sudrik, S. P. Chavan, K. R. S. Chandrakumar, S. Pal, S. K. Date, S. P. Chavan, H. R. Sonawane, *Journal of Organic Chemistry* **2002**, 67, 1574

1. NaH, H-CO$_2$Et
2. Ts, N$_3$
3. hv

48%

D. P. Walker, P. A. Grieco, *Journal of the American Chemical Society* **1999**, 121, 9891

hv

MeOH

48%

H. Yang, K. Foster, C. R. J. Stephenson, W. Brown, E. Roberts, *Organic Letters* **2000**, 2, 2177

Cu(acac)$_2$

MeOH, Δ

50%

B. Saha, G. Bhattacharjee, U. R.Ghatak, *Tetrahedron Letters* **1986**, 27, 3913

# Wolff-Kishner Reduction

## The Reaction:

$$\text{R}\overset{\text{O}}{\underset{}{\parallel}}\text{R'} \xrightarrow[\text{KOH, } \Delta]{\text{H}_2\text{N}-\text{NH}_2 \cdot \text{H}_2\text{O}} \text{R}\overset{\text{H}}{\underset{\text{R'}}{\parallel}}\text{H}$$

## Proposed Mechanism:

hydrazine

proton transfer

$- \text{H}_2\text{O}$

$- \text{N}_2$

M. B. Smith, J. March in *March's Advanced Organic Chemistry*, 5$^{th}$ ed., John Wiley and Sons, Inc., New York, 2001, p. 1548

## Notes:

See *catecholborane* for a mild and selective alternative to the *Wolff-Kishner reduction*.

$$\xrightarrow{\text{Ts-NH-NH}_2}$$

Enone systems can undergo rearrangement

The commonly used approach (above) is better recognized as the *Huang-Minlon modification*.

Under *Wolff-Kishner* conditions, a primary amine can be converted to an alcohol:

$$\xrightarrow[\text{diethylene glycol, 210°C}]{\text{hydrazine, KOH}}$$

54%

S. M. A. Rahman, H. Ohno, N. Maezaki, C. Iwata, T. Tanaka, *Organic Letters* **2000**, _2_, 2893

## Examples:

M. Harmata, P. Rashatasakhon, *Organic Letters* **2001**, <u>3</u>, 2533

H. Nagata, N. Miyazawa, K. Ogasawara, *Organic Letters* **2001**, <u>3</u>, 1737

J. P. Marino, M. B. Rubio, G. Cao, A. de Dios, *Journal of the American Chemical Society* **2002**, <u>124</u>, 13398

A. Srikrishna, K. Anebouselvy, *Journal of Organic Chemistry* **2001**, <u>66</u>, 7102

# Woodward Modification of the Prevost Reaction

## The Reaction:

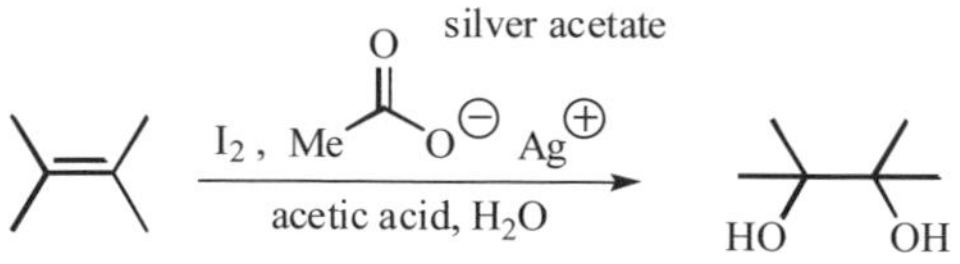

*cis*

## Proposed Mechanism:

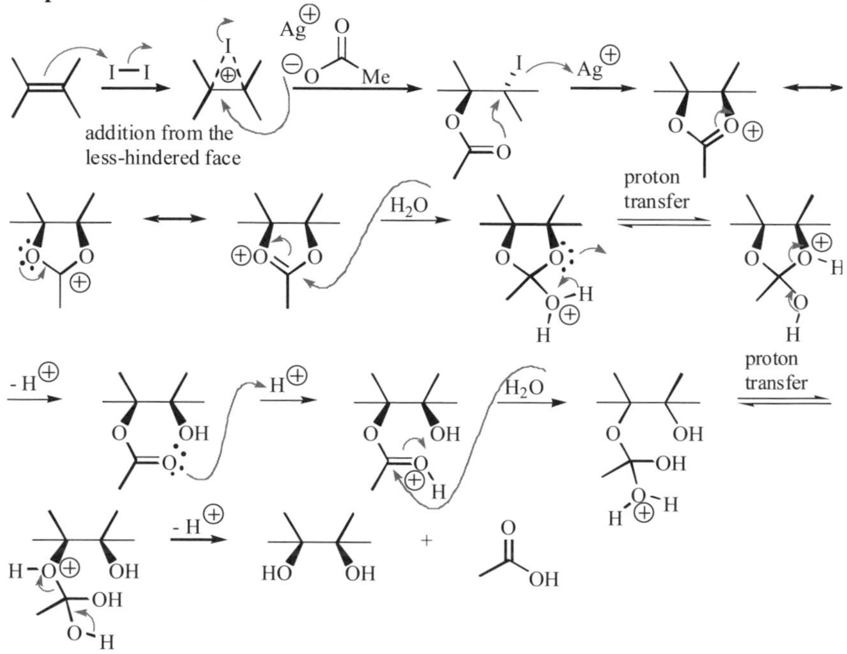

## Notes:

The *Prevost Reaction*:

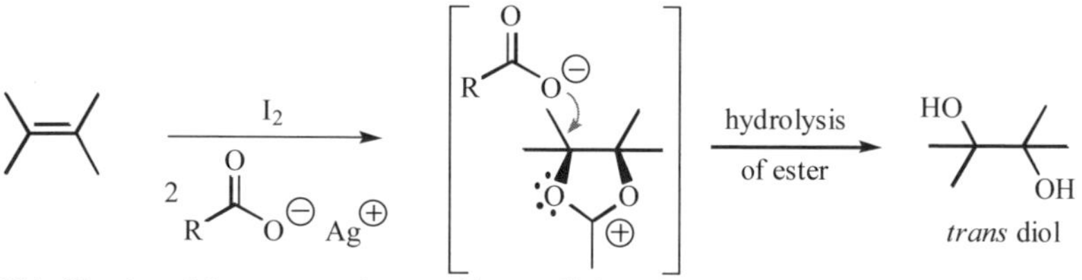

*trans* diol

This *Woodward-Prevost reaction* provides *cis* diols at the more hindered face. This is due to the first step, the addition of iodine from the less-hindered face.

Contrast this reaction with OsO₄ or MnO₄⁻

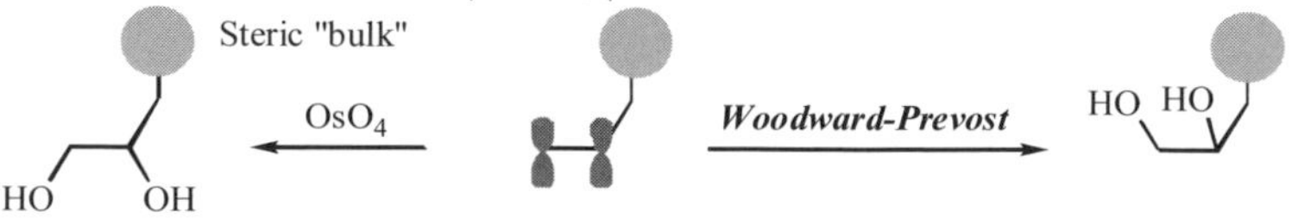

## Examples:

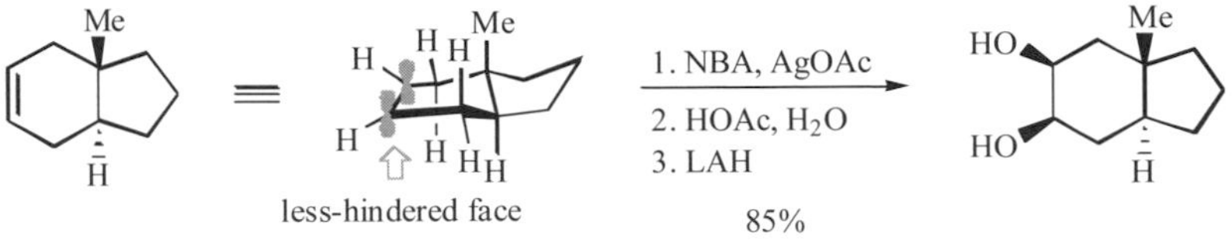

1. NBA, AgOAc
2. HOAc, H$_2$O
3. LAH

less-hindered face

85%

D. Jasserand, J. P. Girard, J. C. Rossi, R. Granger *Tetrahedron Letters* **1976**, 17, 1581

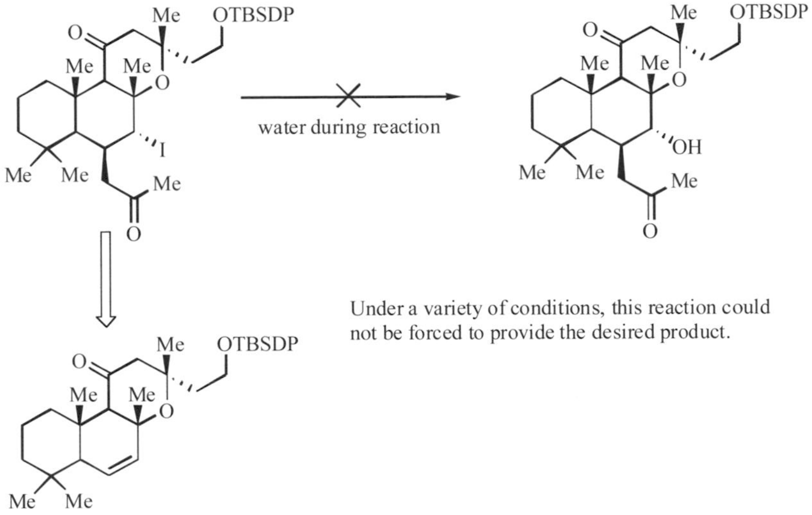

water during reaction

Under a variety of conditions, this reaction could not be forced to provide the desired product.

S. Hamm, L. Hennig. M, Findeisen, D. Muller, P. Weizel, *Tetrahedron* **2000**, 56, 1345

Examples are available to show that show the necessary acetate participation does not always take place:

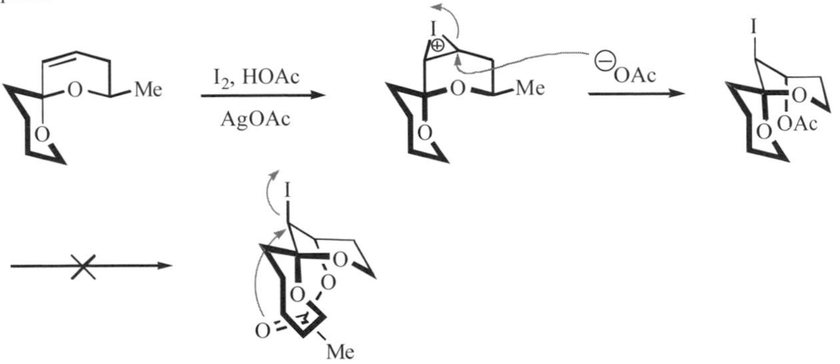

I$_2$, HOAc

AgOAc

M. A. Brimble, M. R. Nairn, *Journal of Organic Chemistry* **1996**, 61, 4801

# Wurtz (Coupling) Reaction and Related Reactions

In the **Wurtz Reaction**, both halides are alkyl. For the **Wurtz-Fittig Reaction**, there is one alkyl and one aryl group, while in the **Fittig Reaction**, both coupling partners are aryl halides.

## Wurtz (Coupling) Reaction:

$$2 \ R-X \xrightarrow{\text{2 Na}} R-R \ + \ 2 \ NaX$$

## Fittig Reaction

$$2 \ Ar-X \xrightarrow{\text{2 Na}} Ar-Ar \ + \ 2 \ NaX$$

## Wurtz-Fittig Reaction

$$Ar-X \ + \ R-X \xrightarrow{\text{2 Na}} Ar-R \ + \ 2 \ NaX$$

## Proposed Mechanism:

Much is not known about the details of the reaction.
It could be a simple $S_N1$ or $S_N2$:

$$R-X \xrightarrow{\text{2 Na}} Na-X \ + \ R^{\ominus}Na^{\oplus}$$

$$R^{\ominus}Na^{\oplus} \ + \ R-X \begin{cases} R^{\ominus}Na^{\oplus} + R^{\oplus} \longrightarrow R-R \\ R^{\ominus}Na^{\oplus} + R-X \longrightarrow R-R \end{cases}$$

or a radical pathway could be followed:

$$R-X \ + \ R-X \xrightarrow{\text{2 Na}} 2 \ NaX \ + \left[ R \cdot \quad \cdot R \right] \longrightarrow R-R$$
$$\text{solvent cage}$$

## Notes:

M. B. Smith, J. March in *March's Advanced Organic Chemistry*, 5th ed., John Wiley and Sons, Inc., New York, 2001, p. 1452; T. Laue, A. Plagens, *Named Organic Reactions*, John Wiley and Sons, Inc., New York, 1998, p. 282

See: T. D. Lash, D. Berry, *Journal of Chemical Education* **1985**, 62, 85

## Examples:

J. W. Morzycki, S. Kalinowski, Z. Lotowski, J. Rabiczko, *Tetrahedron* **1997**, 53, 10579

Br—⟨◇⟩—Cl $\xrightarrow[\text{dioxane}]{\text{2 Na}}$ ⟨◇⟩

78 - 94%

G. M. Lampman, J. C. Aumiller, *Organic Syntheses* CV 6, 133

TMSCl + (Pr)(I)(Pr) $\xrightarrow[\substack{\text{EtO}_2 \\ 63\%}]{\text{Na wire}}$ (Pr)(TMS)(Pr)

TMSCl + (Hex)(I) $\xrightarrow[\substack{\text{EtO}_2 \\ 64\%}]{\text{Na wire}}$ (Hex)(TMS)

P. F. Hudrlik, A. K. Kulkarni, S. Jain, A. M. Hudrlik, *Tetrahedron* **1983**, *39*, 877

(Me)(Br)(Me)(Me) + CD₃I $\xrightarrow[\substack{\text{cyclohexane} \\ 40\%}]{\text{Na}}$ (Me)(CD₃)(Me)(Me)

T. L. Kwa, C. Boelhouwer, *Tetrahedron* **1969**, *25*, 5771

(Ph)(OMe)(OMe)(OMe) + *t*-Bu—Cl $\xrightarrow[35\%]{\text{Na (molten)}}$ (Ph)(OMe)(OMe)(*t*-Bu)

C. C. Chappelow, Jr., R. L. Elliott, J. T. Goodwin, Jr., *Journal of Organic Chemistry* **1962**, *27*, 1409

Ph—(Cl)(Cl) $\xrightarrow[60\%]{\substack{\text{CuCl} \\ \text{DMSO}}}$ Ph—(Cl)(Cl)—Ph

(Ph)(Cl) $\xrightarrow[27\%]{\substack{\text{CuCl} \\ \text{DMSO}}}$ (Ph)(Ph)

H. Nozaki, T. Shirafuji, Y. Yamamoto, *Tetrahedron* **1969**, *25*, 3461

⟨⟩—I $\xrightarrow[\substack{\text{H}_2\text{O, N}_2 \\ 81\%}]{\text{Mn, CuCl}_2}$ ⟨⟩—⟨⟩

J. Ma, T.-H. Chan, *Tetrahedron Letters* **1998**, *39*, 2499

# Yamaguchi Esterification / Reagent

## The Reaction:

**2,4,6-trichlorobenzoyl chloride**
(*Yamaguchi reagent*)

## Proposed Mechanism:

DMAP = *N*,*N*-Dimethylaminopyridine

4 resonance structures

## Examples:

R. Nakamura, K. Tanino, M. Miyashita, *Organic Letters* **2003**, <u>5</u>, 3583

M. Berger, J. Mulzer, *Journal of the American Chemical Society* **1999**, <u>121</u>, 8393

P. A. Wender, J. L. Baryza, C. E. Bennett, F. C. Bi, S. E. Brenner, M. O. Clarke, J. C. Horan, C. Kan, E. Lacôte, B. Lippa, P. G. Nell, T. M. Turner, *Journal of the American Chemical Society* **2001**, <u>124</u>, 13684

# Yamamoto Esterification

## The Reaction:

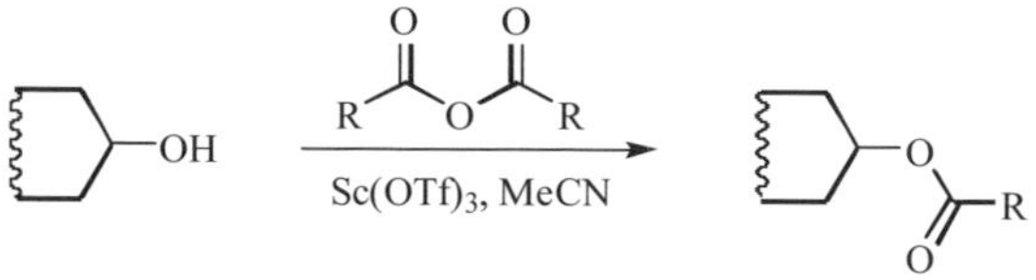

## Proposed Mechanism:

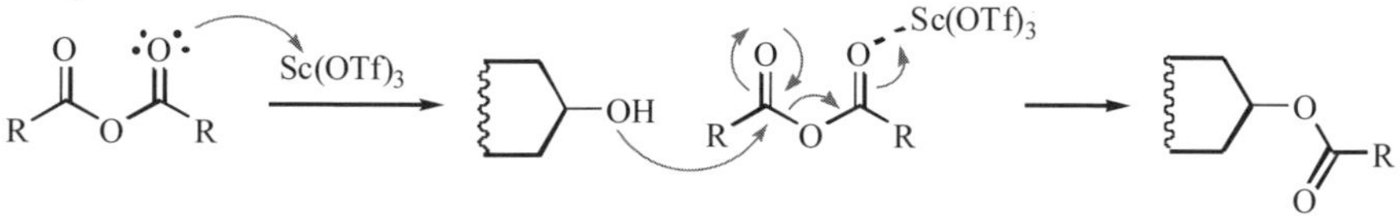

## Notes:

Sc(OTf)$_3$ is commercially available. This reaction provides good yields and is able to esterify relatively hindered alcohols:

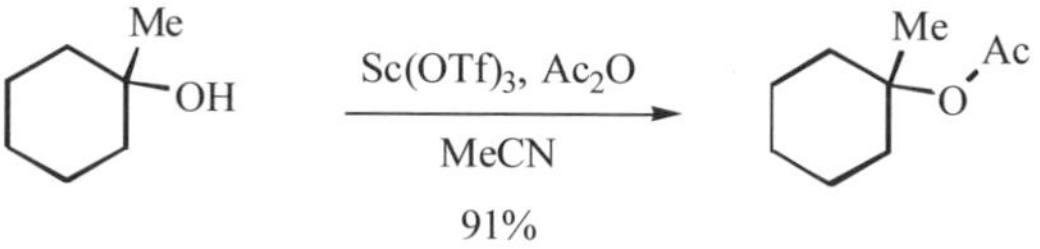

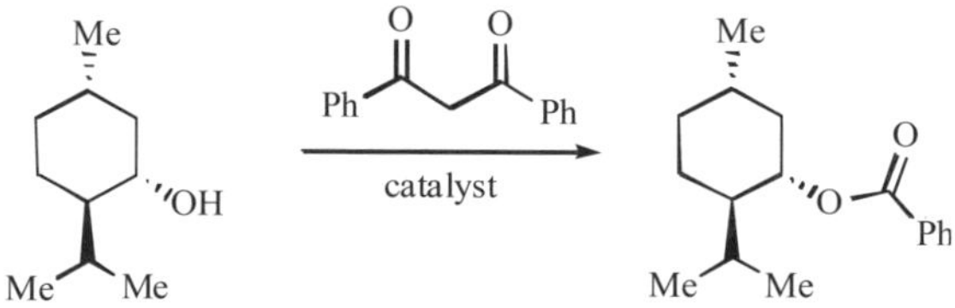

| Catalyst | Yield |
|----------|-------|
| DMAP, Et$_3$N | ` 75% |
| Sc(OTf)$_3$ | > 95% |

K. Ishihara, M. Kubota, H. Kurihara, H. Yamamoto, *Journal of Organic Chemistry* **1996**, <u>61</u>, 4560

Alcohols react preferentially over phenols; an observation not common to other methods.

The reaction is readily extended to lactone synthesis:

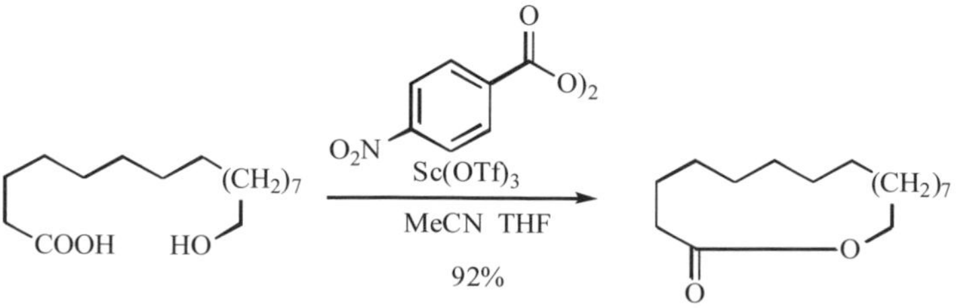

K. Ishihara, M. Kubota, H. Kurihara, H. Yamamoto, *Journal of Organic Chemistry* **1996**, <u>61</u>, 4560

## Examples:

Yamamoto has examined a number of catalyst systems for the reaction. A simple process using HfCl₄ in a soxhlet extractor has shown useful characteristics:

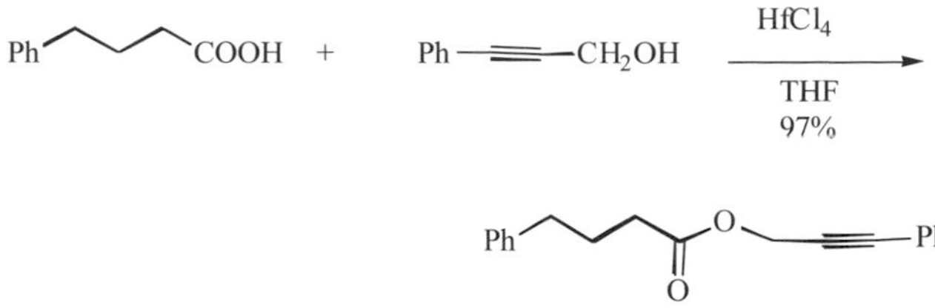

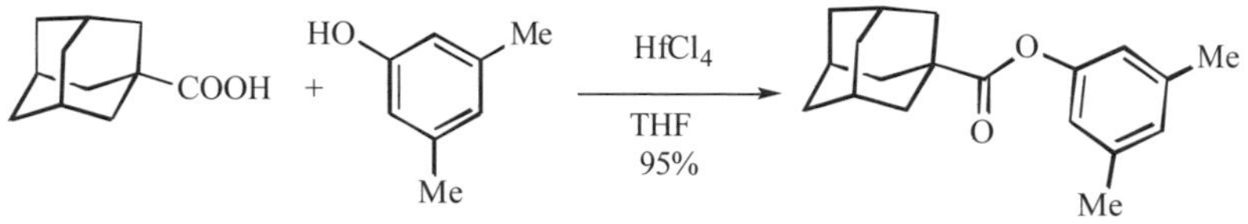

K. Ishihara, S. Ohara, H. Yamamoto, *Science* **2000**, <u>290</u>, 1140

# NAME REAGENTS and ACRONYMS

In this section we provide a summary of Name Reagents and Acronyms. A few additional reagents are included. We have tried to provide a simple summary of the chemistry within the format:

## Name or Acronym
### Structure
Chemical Name
[CAS Number]
### Commercially available or Preparation:

> Reference to, *Encyclopedia of Reagents for Organic Synthesis*, John Wiley and Sons, Inc., L. A. Paquette, Ed., New York, 1995, **Volume number** page

### Notes:

### Examples:

# Adam's Catalyst
## PtO₂
Platinum (IV) Oxide
[1314-15-4]
## Commercially available

A. O. King, I. Shinkai, *Encyclopedia of Reagents for Organic Synthesis*, John Wiley and Sons, Inc., L. A. Paquette, Ed., New York, 1995, **6**, 4162

## Notes:

**1**. PtO₂ is a pre-catalyst. In the presence of hydrogen gas and solvent, the oxide is reduced to a fresh platinum surface; the active catalyst.

**2**. The **Brown procedure**[2] for generating a fresh platinum catalyst uses NaBH₄ to reduce the Pt.[1]

## Examples:

1.[2]

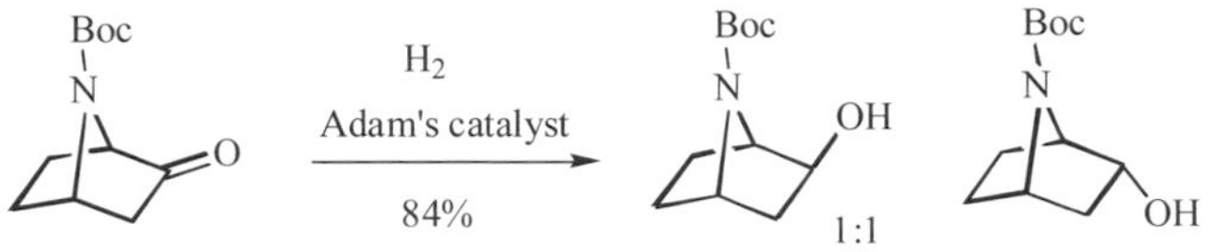

Boc
H₂
Adam's catalyst
84%

Boc
OH

Boc
OH
1:1

2.[3]

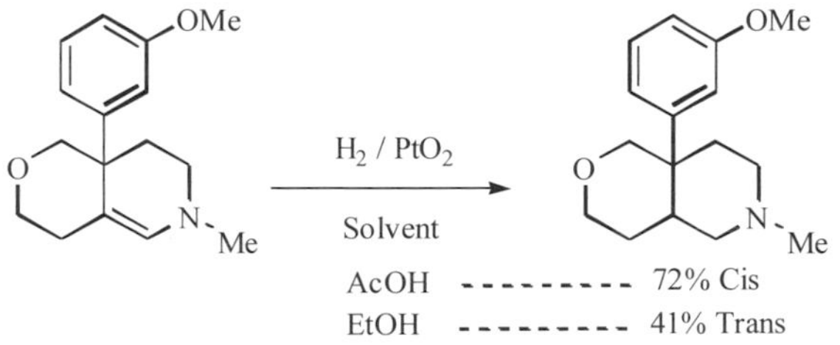

OMe

O
N-Me

H₂ / PtO₂
Solvent
AcOH  - - - - - - - - -  72% Cis
EtOH  - - - - - - - - -  41% Trans

OMe

O
N-Me

3.[4]

O
Me
Me
CH₂
Me

Catalyst
H₂
Pd / C
Adam's

O
Me
Me
Me
Me

7        :

O
Me
Me
Me
Me

only
3

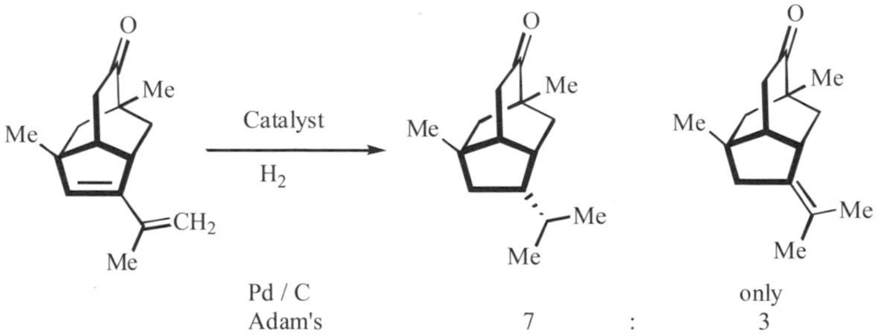

[1] H. C. Brown, C. A. Brown, *Journal of the American Chemical Society* **1962**, 84, 1493

[2] K. C. V. Ramanaia, N. Zhu, C. Klein-Stevens, M. L. Trudell, *Organic Letters* **1999**, 1, 1439

[3] Reported in: A. O. King, I. Shinkai, *Encyclopedia of Reagents for Organic Synthesis*, John Wiley and Sons, Inc., L. A. Paquette, Ed., New York, 1995, **6**, 4162

[4] G. A. Schiehser, J. D. White, *Journal of Organic Chemistry* **1980**, 45, 1864

# Adams Reagent

**Zn(CN)₂**

Zinc cyanide

[557-21-1]

## Commercially available

H. Heaney, *Encyclopedia of Reagents for Organic Synthesis*, John Wiley and Sons, Inc., L. A. Paquette, Ed., New York, 1995, **8**, 5564

**Notes:** A useful source of HCN for the *Gattermann-type formylation* reaction.

## Examples:

1.[1]

1. $Zn(CN)_2$ , HCl

2. $AlCl_3$, HCl

3. $H_2O$, Heat

100%

2.[2]

1. $Zn(CN)_2$ / HCl
   KCl / $Et_2O$

2. EtOH / $H_2O$

51%          8%

3.[3]

$Zn(CN)_2$ / $AlCl_3$ / HCl

65%

4.[4]

1. $Zn(CN)_2$, $C_2H_2Cl_2$

HCl, $AlCl_3$

2. HCl / $H_2O$

75 - 81%

[1] E. Montgomery, *Journal of the American Chemical Society* **1924**, <u>46</u>, 1518

[2] Z. Yang, H. B. Liu, C. M. Lee, H. Mou Chang, H. N. C. Wong, *Journal of Organic Chemistry* **1992**, <u>57</u>, 7248

[3] A. V. R. Rao, N. Sreenivasan, D. Reddeppa, N. Reddy, V. H. Deshpande, *Tetrahedron Letters* **1987**, <u>28</u>, 455

[4] R. C. Fuson, E. C. Horning, S. P. Rowland, M. L. Ward, *Organic Syntheses* <u>CV3</u>, 549

# AIBN

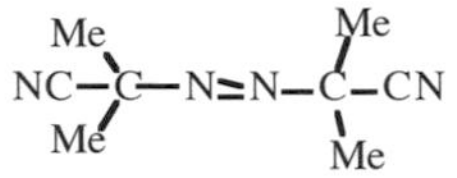

$$NC-\underset{\underset{Me}{|}}{\overset{\overset{Me}{|}}{C}}-N=N-\underset{\underset{Me}{|}}{\overset{\overset{Me}{|}}{C}}-CN$$

2,2'-Azobisisobutyronitrile
[78-67-1]

**Commercially available.** This and related reagents can be prepared by the general scheme

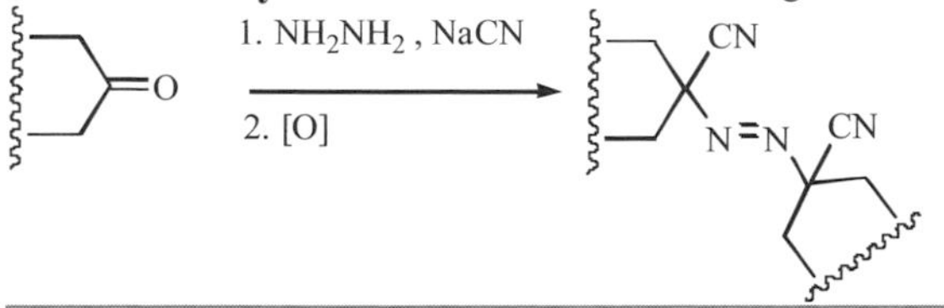

1. NH$_2$NH$_2$ , NaCN

2. [O]

N. S. Simpkins, *Encyclopedia of Reagents for Organic Synthesis*, John Wiley and Sons, Inc., L. A. Paquette, Ed., New York, 1995, **1**, 229

## Notes:

This reagent is useful for initiating radical reactions. Since it is a radical initiator, only small amounts are needed. For reactions at higher temperature, ACN (1,1'-azobis-1-cyclohexanenitrile [2094-98-6]) has been suggested (see Example 2).[1]

Initiation:

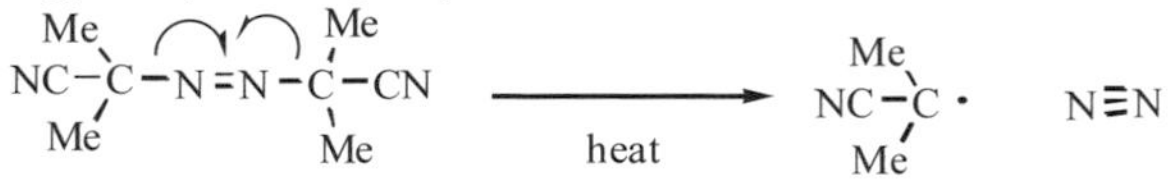

$$NC-\overset{\overset{Me}{|}}{\underset{\underset{Me}{|}}{C}}-N=N-\overset{\overset{Me}{|}}{\underset{\underset{Me}{|}}{C}}-CN \xrightarrow{heat} NC-\overset{\overset{Me}{|}}{\underset{\underset{Me}{|}}{C}}\cdot \quad N\equiv N$$

$$Bu-\underset{\underset{Bu}{|}}{\overset{\overset{Bu}{|}}{Sn}}-H \quad \cdot\overset{\overset{Me}{|}}{\underset{\underset{Me}{|}}{C}}-CN \longrightarrow Bu-\underset{\underset{Bu}{|}}{\overset{\overset{Bu}{|}}{Sn}}\cdot$$

## Example

1.[2]

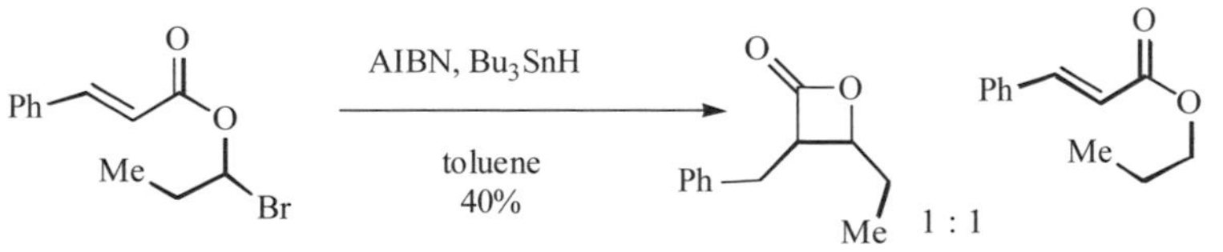

AIBN, Bu$_3$SnH

toluene
40%

1 : 1

2.[3]

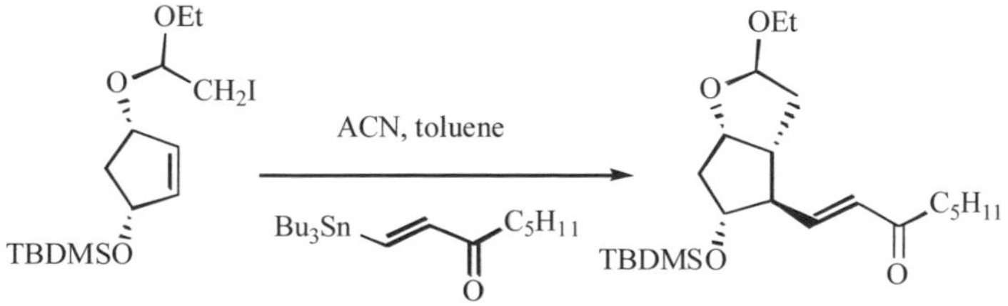

ACN, toluene

Bu$_3$Sn〜C$_5$H$_{11}$

AIBN gave lower yield    72%    ACN = Azobiscyclohexylnitrile

[1] G. E. Keck, D. A. Burnett, *Journal of Organic Chemistry* **1987**, <u>52</u>, 2958
[2] K. Castle, C.-S. Hau, J. B. Sweeney, C. Tindall, *Organic Letters* **2003**, <u>5</u>, 757
[3] G. E. Keck, D. A. Burnett, *Journal of Organic Chemistry* **1987**, <u>52</u>, 2958

# Albright-Goldman Reagent[1]

## DMSO-Ac₂O

[DMSO]  [Acetic anhydride]

**Commercially available.  Both DMSO and Ac₂O are readily available.**

T. T. Tidwell, *Encyclopedia of Reagents for Organic Synthesis*, John Wiley and Sons, Inc., L. A. Paquette, Ed., New York, 1995, **3**, 2145

**Notes:**  The mechanism of this oxidation is suggested to be:

See ***DMSO-based Oxidations***, in *Aldehyde Syntheses*

## Examples:

1.[2]

DMSO / Ac₂O

89%

2.[3]

80 - 85%

3.[4]

Bisulfite complex

DMSO -Ac₂O

90%

[1] J. D. Albright, L. Goldman, *Journal of the American Chemical Society* **1967**, <u>89</u>, 2416

[2] E. Montgomery, *Journal of the American Chemical Society* **1924**, <u>46</u>, 1518

[3] (a) J. D. Albright, L. Goldman, *Journal of American Chemical Society* **1965**, <u>87</u>, 4214
(b) *Journal of Organic Chemistry* **1965**, <u>30</u>, 1107

[4] P. G. M. Wuts, C. L. Bergh, *Tetrahedron Letters* **1986**, <u>27</u>, 3995

# Allyl Boron Reagents

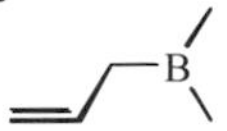

## Notes:

The reagents described are used to generate chiral allyl groups. In reaction with aldehydes:

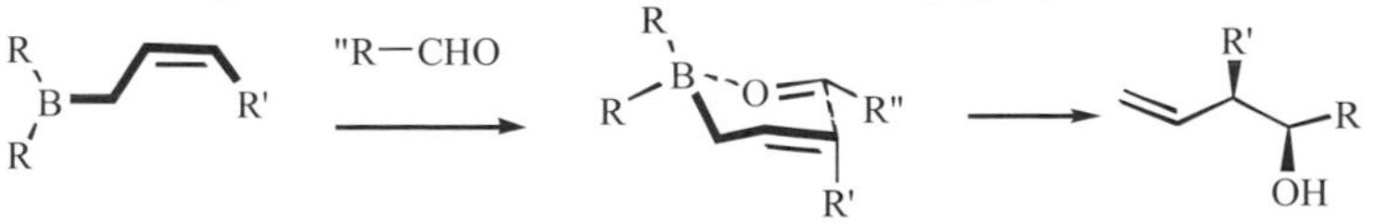

Examples: of chiral allyl-boron reagents:

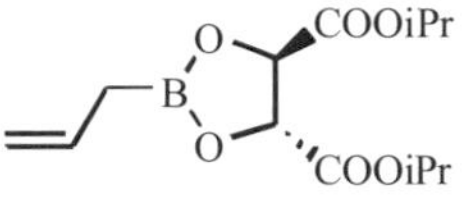

**Rousch Reagent**

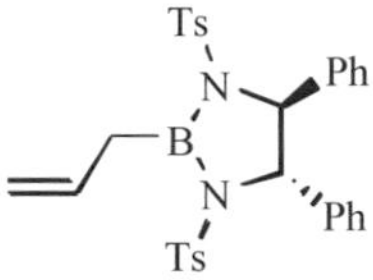

**Corey Reagent**

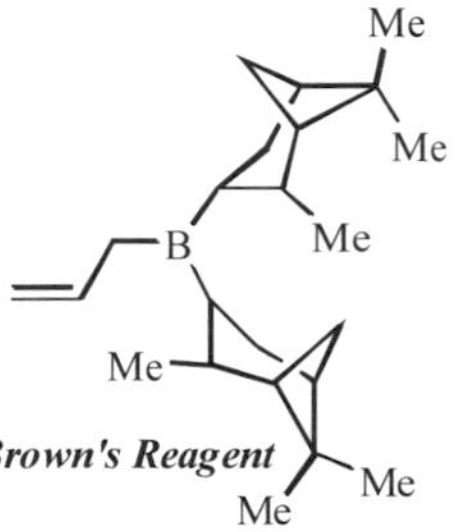

**Brown's Reagent**

## Examples:

1.[1]

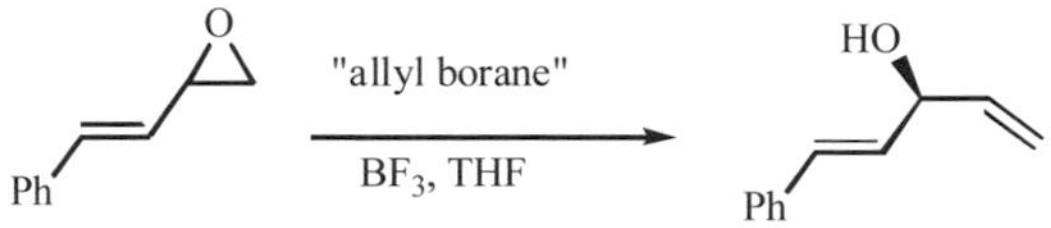

"allyl borane"

BF$_3$, THF

| | *Yield* | *Percent ee* |
|---|---|---|
| Rousch | 70 | 47 |
| Corey | 33 | 37 |
| Brown | 80 | 96 |

2.[2]

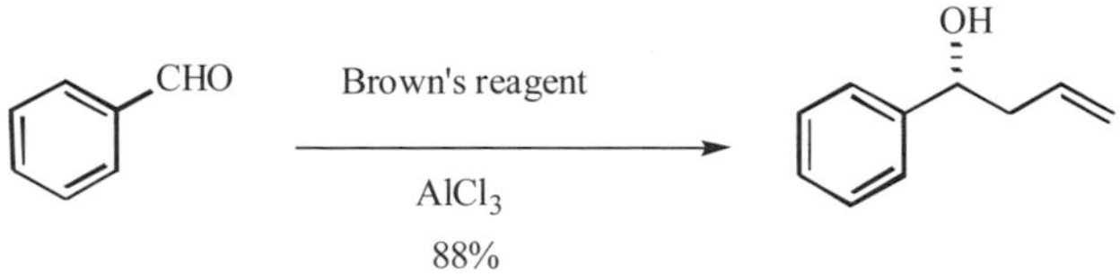

Brown's reagent

AlCl$_3$

88%

[1] M. Lautens, M. L. Madess, E. L. O. Sauer, S. G. Ouellet, *Organic Letters* **2002**, 4, 83
[2] T. Ishiyama, T. Ahiko, N. Miyaura, *Journal of the American Chemical Society* **2002**, 124, 12414

# Attenburrow's Oxide[1,2]

## MnO₂

Manganese dioxide

[1313-13-9]

## Commercially available

G. Cahiez, M. Alami, *Encyclopedia of Reagents for Organic Synthesis*, John Wiley and Sons, Inc.,
L. A. Paquette, Ed., New York, 1995, **5**, 3229

## Notes:

This is a selective oxidizing reagent for sensitive substrates, particularly alcohols adjacent to pi-
systems. The preparation of the active reagent is critical to its success.

## Examples:

1.[3]

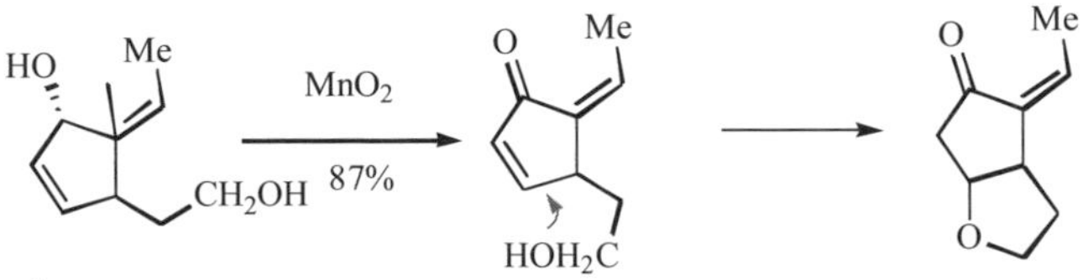

2.[4]

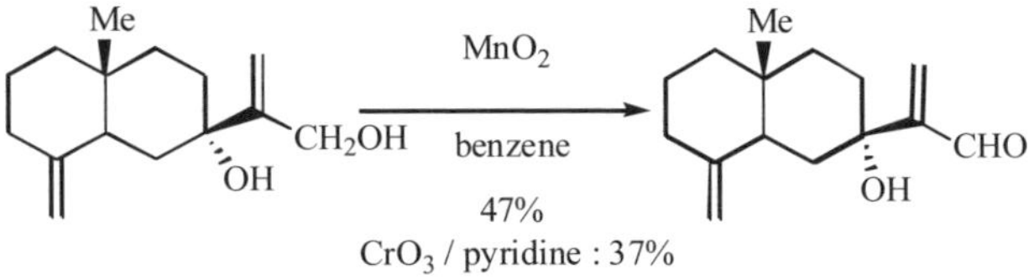

3.[5]

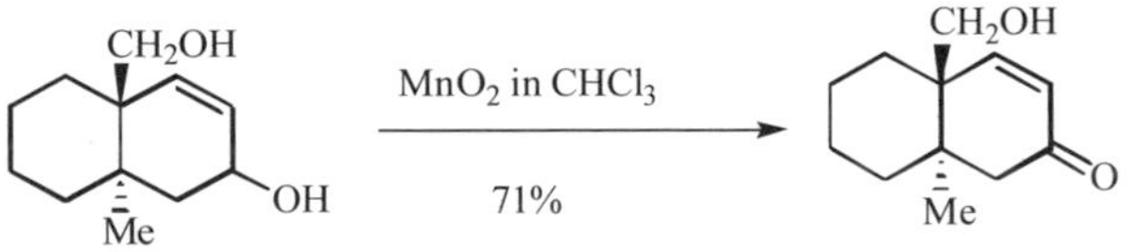

4.[6]

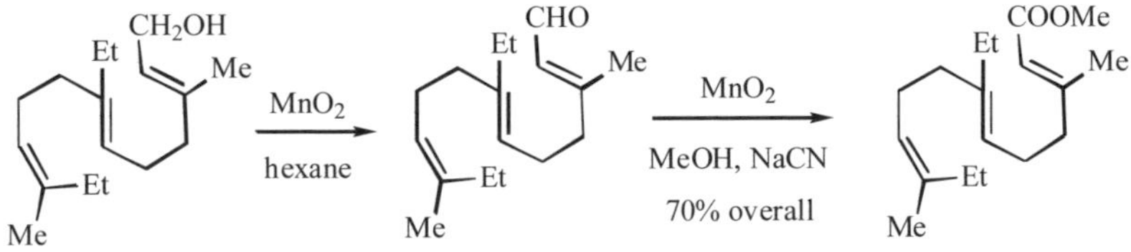

---

[1] A. J. Fatiadi, *Synthesis* **1976**, <u>65</u>, 133

[2] J. Attenburrow, A. F. B. Cameron, J. H. Chapman, R. M. Evans, B. A. Hems, A. B. A. Jansen, T.
Walker, *Journal of the Chemical Society* **1952**, 1094

[3] K. M. Brummond, P. C. Still, H. Chen, *Organic Letters* **2004**, <u>6</u>, 149

[4] J. Cuomo, *Journal of Agricultural and Food Chemistry* **1985**, <u>33</u>, 717

[5] P. C. Mukharji, A. N. Ganguly, *Tetrahedron* **1969**, <u>25</u>, 5281

[6] E. J. Corey, J. A. Katzenellenbogen, N. W. Gilman, S. A. Roman, B. W. Erickson, *Journal of the
American Chemical Society* **1968**, <u>90</u>, 5618

# 9-BBN

9-Borobicyclo[3.3.1]nonane
[280-64-8]
[21205-91-4] Crystalline dimer

**Commercially available.** Can be prepared from the reaction of 1,5-cyclooctadiene with diborane.

| J. A. Soderquist, A. Negron, *Encyclopedia of Reagents for Organic Synthesis*, John Wiley and Sons, Inc., L. A. Paquette, Ed., New York, 1995, **1**, 622 |
| --- |

## Notes:

As with borane, this reagent exists as a dimer. Unlike diborane, the solid dimer has limited stability to the atmosphere, but should be stored in an inert environment. ***Commercially available solutions should be maintained under inert atmosphere and anhydrous conditions***. Due to its steric bulk, 9-BBN is seen to be more selective than diborane. This is even observed in reactions where the 9-BBN group is a non-reacting center.[1]

$BR_2 = BMe_2$

$BR_2 = 9\text{-}BBN$

|  | 1 | : | 1 |
| --- | --- | --- | --- |
|  | 98 | : | 2 |

## Examples:

1.[2]

9-BBN

THF

This intermediate is taken on to the next step, a ***Suzuki reaction***.

2.[3]

9-BBN

THF

After mild oxidation to form the alcohol (see ***Brown's Hydroboration***) the ***Swern oxidation*** formed the aldehyde in overall 98% yield

---

[1] S. C. Pellegrinet, M. A. Silva, J. M. Goodman, *Journal of the American Chemical Society* **2001**, 123, 8832
[2] A. Kamatani, L. E. Overman, *Organic Letters* **2001**, 3, 1229
[3] N. C. Kallan, R. L. Halcomb, *Organic Letters* **2000**, 2, 2687

# Belleau's Reagent

$$Ph-O-\phantom{}\!\!\!\!\!\!\langle\text{ring}\rangle\!\!-\!\!P\!\!\stackrel{S}{\underset{S}{\diagdown}}\!S\!\!\diagup\!P\!\!\stackrel{S}{\underset{S}{\diagdown}}\!-\!\langle\text{ring}\rangle\!-\!O-Ph$$

1,3,2,4-Dithiadiphosphetane-2,4-bis(4-phenoxyphenyl)-2,4-disulfide
[88816-02-8]

**Preparation:**[1] Details of the synthesis are not reported except that the procedure followed H. Z. Lecher, R. A. Greenwood, K. C. Whitehouse, T. H. Chao, *Journal of the American Chemical Society* **1956**, <u>78</u>, 5018

**Notes:**
Similar to **_Lawesson's_** and **_Davy's reagents_**. Used as a mild thionation reagent for esters and lactones.

**Examples:**

1.[2]

2.[3]

3.[4]

---

[1] G. Lajoie, F. Lépine, L. Maziak, B. Belleau, *Tetrahedron Letters* **1983**, <u>24</u>, 3815
[2] A.G. M. Barrett, A. C. Lee, *Journal of Organic Chemistry* **1992**, <u>57</u>, 2818
[3] A. P. Degnan, A. I. Meyers, *Journal of American Chemical Society* **1999**, <u>121</u>, 2762
[4] S. V. Ley, A. Priour, C. Heusser, *Organic Letters* **2002**, <u>4</u>, 711

# BINAP

2,2'-Bis(diphenylphosphino)-1,1'-binaphththyl
[76189-55-4] R-(+)
[76189-56-5] S-(-)

## Commercially available

K. Kitamura, R. Noyori, *Encyclopedia of Reagents for Organic Synthesis*, John Wiley and Sons, Inc., L. A. Paquette, Ed., New York, 1995, **1**, 509

**Notes:**  A chiral diphosphine ligand to be used with transition metal catalyzed reactions.

## Examples:

1.[1]

(R)  91% ee

(S)  85% ee

2.[2]

*Geraniol*

*(R)-Citronellol*

Ru(OAc)₂  (R)-BINAP

*Nerol*

*(S)-Citronellol*

3.[3]

Rh(acac)(C₂H₄)₂
R-BINAP   dioxane/ water
83%

---

[1] T. Ohta, H. M. Kitamura, K. Nagai, R. Noyori, *Journal of Organic Chemistry* **1987**, <u>52</u>, 3174

[2] H. Takaya, T. Ohta, S. Inoue, M. Tokunaga, M. Kitamura, R. Noyori, *Organic Syntheses* **1994**, <u>72</u>, 74

[3] T. Hayashi, M. Takahashi, Y. Takaya, M. Ogasawara, *Organic Syntheses* **2001**, <u>79</u>, 84

# BINOL

(R or S)-1,1'-bi-2,2'-naphthol
[18531-94-7]

## Commercially available

K. Mikama, Y. Motoyama, *Encyclopedia of Reagents for Organic Synthesis*, John Wiley and Sons, Inc., L. A. Paquette, Ed., New York, 1995, **1**, 397

**Notes:** As a reagent for introducing chirality, a BINOL-based crown ether has been reported:[1]

## Examples:

1.[2]

100%

2.[3]

92% ee

84%

[1] R. C. Helgeson, J. M. Timko, P. Moreau, S. C. Peacock, J. M. Mayer, D. J. Cram, *Journal of the American Chemical Society* **1974**, 96, 6762
[2] K. Ishihara, S. Nakamura, M. Kaneeda, H. Yamamoto, *Journal of the American Chemical Society* **1996**, 118, 12854
[3] Z.-B. Li, L. Pu, *Organic Letters* **2004**, 6, 1065

# BITIP

### Ti(*i*-PrO)₄ / BINOL

Binol/Titanium isopropoxide

## Preparation:[1]

## Notes:

A useful chiral catalyst.

## Examples:

1.[2]

Intermediates of this type are useful for a pyran annulation protocol:

2.[3]

For an example of other uses of chiral Lewis acids (*CLA*) to synthesis:[4]

[1] G. E. Keck, K. H. Tarbet, L. S. Geraci, *Journal of the American Chemical Society* **1993**, 115, 8467

[2] G. E. Keck, J. A. Covel, T. Schiff, T. Yu, *Organic Letters* **2002**, 4, 1189[2]

[3] G. E. Keck, X.-Y. Li, D. Khrishnamurthy, *Journal of Organic Chemistry* **1995**, 60, 5998

[4] C. Wolf, Z. Fadul, P. A. Hawes, E. C. Volpe, *Tetrahedron Asymmetry* **2004**, 15, 1987

# BMDA

Me
Me—⟨
      N⊖  ⊕ MgBr
Me—⟨
      Me

Bromomagnesium diisopropylamide
[50715-01-0]

**Preparation:** From the reaction of RMgX with diisopropylamine.

Me
Me—⟨
      N–H    Et₂O or THF
Me—⟨                    ⟶
      Me    R = Me or Et

R—MgBr

Me
Me—⟨
      N⊖  ⊕ MgBr
Me—⟨
      Me

R. H. Erickson, *Encyclopedia of Reagents for Organic Synthesis*, John Wiley and Sons, Inc., L. A. Paquette, Ed., New York, 1995, **1**, 740

**Notes:** Shown to be an extremely useful base for generating thermodynamic enolates.[1] Useful for *aldol condensations*.

**Examples:**

1.[2]

|            | 1 | : | 99 |
|------------|---|---|----|
| LDA, DME, TMSCl | 1 | : | 99 |
| BMDA,Et₃N, TMSCl | 97 | : | 3 |

2.[3]

BMDA, THF   71%

Stereochemistry is highly dependent on reaction conditions

[1] M. E. Krafft, R. A. Holton, *Tetrahedron Letters* **1983**, 24, 1345

[2] A. Yanagisawa, T. Watanabe, T. Kikuchi, H. Yamamoto, *Journal of Organic Chemistry* **2000**, 65, 2979

[3] T. Fukuyama, K. Akasaka, D. S. Karanewsky, C.-L. J. Wang, G. Schmid, Y. Kishi, *Journal of the American Chemical Society* **1979**, 101, 262

# BMS

BH$_3$-Me$_2$S
Borane Dimethylsulfide
[13292-87-0]

## Commercially available

M. Zaidlewics, *Encyclopedia of Reagents for Organic Synthesis*, John Wiley and Sons, Inc., L. A. Paquette, Ed., New York, 1995, **1**, 634

**Notes:** 1. Use as other *hydroborating reagents*. Its value is in the increased reagent stability and solvent solubility. Like other hydroborating agents, it is stable to an array of functional groups. It is useful for the reduction of ozonides.
2. A recent report describes replacing the dimethyl sulfide with larger alkyl-substituted sulfides to reduce stench.[1]

## Examples:

1.[2]

78% with this stereochemistry

2.[3]

73%

3.[4]

2. BMS

100%

[1] M. Zaidlewicz, J. V. B. Kanth, H. C. Brown, *Journal of Organic Chemistry* **2000**, <u>65</u>, 6697
[2] D. A. Evans, J. Bartroli, T. Godel, *Tetrahedron Letters* **1982**, <u>23</u>, 4577
[3] R. M. Bannister, M. H. Brookes, G. R. Evans, R. B. Katz, N. D. Tyrrell, *Organic Process Research & Development* **2000**, <u>4</u>, 467
[4] L. A. Flippin, D. W. Gallagher, K. Jalali- Araghi, *Journal of Organic Chemistry* **1989**, <u>54</u>, 1430

# Boc-Cl

$$Cl-\underset{\underset{O}{\|}}{C}-O-\underset{\underset{Me}{|}}{\overset{\overset{Me}{|}}{C}}\sim Me$$

*t*-Butoxycarbonylchloride
*t*-Butyl chloroformate
[24608-52-4]

**Preparation:** Prepared from *t*-butyl alcohol (or K⁺ *t*-BuO⁻) and phosgene.

G. Sennyey, *Encyclopedia of Reagents for Organic Synthesis*, John Wiley and Sons, Inc., L. A. Paquette, Ed., New York, 1995, **2**, 859

## Notes:
Reagent used for the protection of amino groups. Because of the steric influence of the Boc protecting group, it is quite stable to base hydrolysis and catalytic reduction with hydrogen.

See 1-(*t*-butoxycarbonyl)imidazole [49761-82-2] (I. Grapsas, S. Mobashery, *Encyclopedia of Reagents for Organic Synthesis*, John Wiley and Sons, Inc., L. A. Paquette, Ed., New York, 1995, **2**, 835) and di-*t*-butyl dicarbonate (Boc₂O) [24424-99-5](M. Wakselman, *Encyclopedia of Reagents for Organic Synthesis*, John Wiley and Sons, Inc., L. A. Paquette, Ed., New York, 1995, **3**, 1602 for similar roles as providing Boc protecting groups.

## Examples:
1.¹

2.²

3.³

---

[1] R. C. F. Jones, A. K. Crockett, *Tetrahedron Letters* **1993**, _34_, 7459
[2] Y. Basel, A. Hassner, *Journal of Organic Chemistry* **2000**, _65_, 6368
[3] D. K. Mohapatra, A. Datta, *Journal of Organic Chemistry* **1999**, _64_, 6879

# BOMCl

—CH₂-O-CH₂Cl

Benzyl chloromethyl ether
Source of the BOM (**Benzyloxymethyl**) protecting group
[3587-60-8]

## Commercially available

H. W. Pinnick, *Encyclopedia of Reagents for Organic Synthesis*, John Wiley and Sons, Inc., L. A. Paquette, Ed., New York, 1995, **1**, 327.

**Notes:** This is often used as a protecting group for alcohols, where it is observed that primary alcohols form more readily than secondary hydroxy groups, which in turn are more reactive than tertiary alcohols. As with most benzylic ethers, this protecting group can be removed by hydrogenolysis over Pd or by metal-ammonia reduction.[1]

## Examples:

1.[2]

2.[3]

3.[4]

---

[1] See: T. W. Greene, P. G. M. Wuts, *Protective Groups in Organic Synthesis*, Wiley-Interscience, New York. Third Edition, 1999, p 36

[2] W. R. Roush B. B. Brown, *Journal of Organic Chemistry* **1993**, <u>58</u>, 2162

[3] K. C. Nicolaou, D. A. Claremon, W. E. Barnette, *Journal of the American Chemical Society* **1980**, <u>102</u>, 6611

[4] S. M. A. Rahman, H. Ohno, T. Murata, H. Yoshino, N. Satoh, K. Murakami, D. Patra, C. Iwata, N. Maezaki, T. Tanaka, *Organic Letters* **2001**, <u>3</u>, 619

# Brassard's Diene

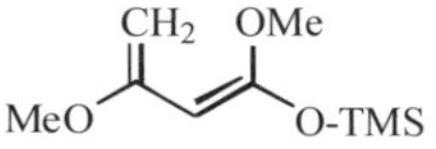

Silan, [(1,3-dimethoxy-1,3-butadienyl)oxy]trimethyl-,(*E*)-
[90857-62-8]

## Preparation:[1]

The general preparation was along the lines:

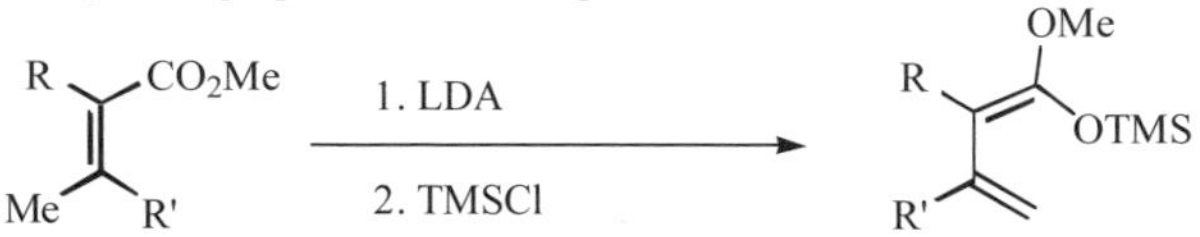

**Notes:**  See similarity with ***Chan's***,[2] ***Danishefsky's*** and ***Rawal's Diene***.
For chemistry of related dienes, see: S. Danishefsky, *Accounts of Chemical Research* **1981**, <u>14</u>, 400.

## Examples:

1.[3]

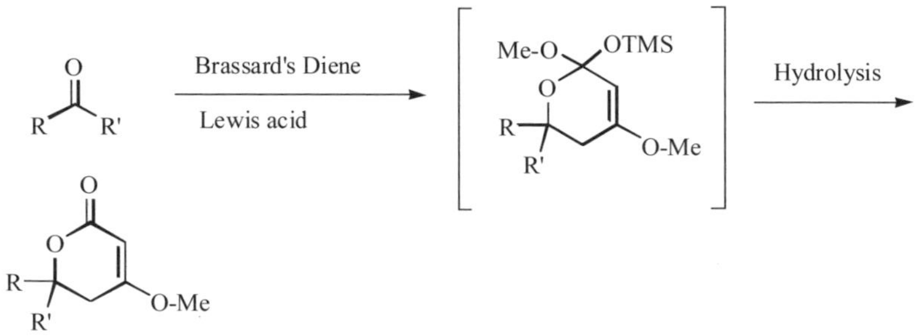

2.[4]

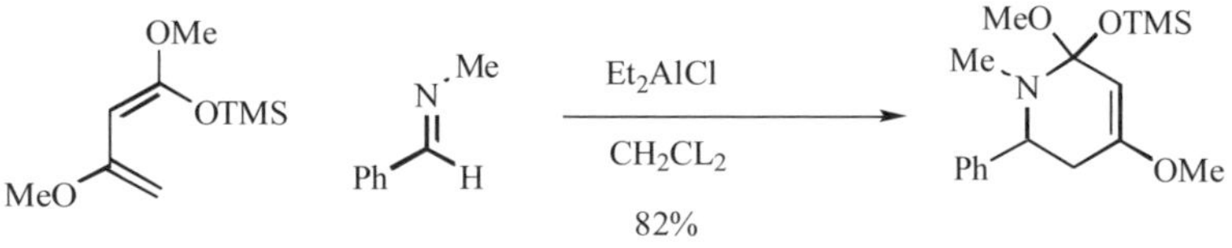

[1] P. Brassard, J. Savard, *Tetrahedron Letters* **1979**, <u>21</u>, 4911
[2] P. Brownbridge, T. H. Chan, M. A. Brook, G. J. Kang, *Canadian Journal of Chemistry* **1983**, <u>61</u>, 688
[3] M. M. Midland, R. S. Graham, *Journal of the American Chemical Society* **1984**, <u>106</u>, 4294
[4] M. M. Midland, J. I. McLaughlin, *Tetrahedron Letters* **1988**, <u>29</u>, 4653

# Bredereck's Reagent

Me—N—Me  Me
|       |
H—C—O—C—Me
|       |
N.      Me
Me  Me

*t*-Butoxybis(dimethylamino)methane
[5815-08-7]

## Preparation:

$$H-\overset{NMe_2}{\underset{NMe_2}{\overset{(+)}{C}}} \xrightarrow{\text{t-BuO}^{\ominus}} [\quad] \xleftarrow{\text{t-BuO}^{\ominus}} NC-\overset{NMe_2}{\underset{NMe_2}{C}}$$

## Commercially available

> W. Kantlehner, *Encyclopedia of Reagents for Organic Synthesis*, John Wiley and Sons, Inc., L. A. Paquette, Ed., New York, 1995, **2**, 828

**Notes:** This reagent places an enamine adjacent to a carbonyl group, in essence, providing a mild formylation protocol.

## Examples:

1.[1]

Bredereck's reagent

quantitative

2.[2]

Bredereck's Reagent

69%

3.[3]

Bredereck's Reagent

88%

[1] F. E. Ziegler, J.-M. Fang, *Journal of Organic Chemistry* **1981**, <u>46</u>, 827

[2] R. Jakse, S. Recnik, J. Svete, A. Golobic, L. Golic, B. Stanovnik, *Tetrahedron* **2001**, <u>57</u>, 839

[3] E. Morera, F. Pinnen, G. Lucente, *Organic Letters* **2002**, <u>4</u>, 1139

# Burgess Reagent

$$Et-\underset{Et}{\overset{Et}{N}}\overset{\oplus}{\underset{}{}}-\underset{O}{\overset{O}{S}}-N\diagdown{}^{\ominus}_{COOMe}$$

Methyl, *N*-(triethylammoniosulfonyl)carbamate
[29864-56-8]

## Commercially available

P. Taibi, S. Mobushery, *Encyclopedia of Reagents for Organic Synthesis*, John Wiley and Sons, Inc., L. A. Paquette, Ed., New York, 1995, **5**, 3345

**Notes:** Efficient catalyst for the stereospecific *cis*-dehydration of 2 and 3 alcohols. Also useful for the preparation of oxazolines, a "***benzyl Burgess reagent***" has been reported[1], as has been a polymer-supported reagent.[2]

## Examples:

1.[3]

2.[4]

3.[5]

---

[1] M. R. Wood, J. Y. Kim, K. M. Books, *Tetrahedron Letters* **2002**, *43*, 3887

[2] P. Wipf, S. Venkatraman, *Tetrahedron Letters* **1996**, *37*, 4659

[3] D. J. Goldsmith, H. S. Kezar, III, *Tetrahedron Letters* **1980**, *21*, 3543

[4] P. Wipf, S. Venkatraman, *Tetrahedron Letters* **1996**, *37*, 4659

[5] P. Wipf, W. Xu, *Journal of Organic Chemistry* **1996**, *61*, 6556

# *n*-Butyl Tin Hydride (TBTH)

(*n*-Bu)₃SnH
tri-*n*-butylstannane
[688-73-3]

## Commercially available

T. V. RajanBabu, *Encyclopedia of Reagents for Organic Synthesis*, John Wiley and Sons, Inc.,
L. A. Paquette, Ed., New York, 1995, **7**, 5016

## Notes:

Reactions with TBTH are often initiated with ***AIBN***. Useful for reduction of the C-X bond to C-H.

$$Bu_3SnH + AIBN \longrightarrow Bu_3Sn$$
$$Bu_3Sn + R\text{-}X \longrightarrow R + Bu_3SnX$$
$$R + Bu_3SnH \longrightarrow R\text{-}H + Bu_3Sn$$

The reagent will selectively remove I, Br, or Cl in the presence of other functional groups.

Since this is a source of radicals, the reagent can find use in ***Barton-McCombie Reaction*** and ***Dowd-Beckwith Ring Expansion***, as well as radical cyclization reactions.

## Examples:

1.[1]

2.[2]

3. Heathcock has found that a Zn-Ag reagent is extremely useful for the conversion:[3]

---

[1] D. P. Curran, D. M. Rakiewicz, *Tetrahedron* **1985**, <u>41</u>, 3943
[2] P. Dowd, S. C. Choi, *Journal of the American Chemical Society* **1987**, <u>109</u>, 6548
[3] R. D. Clark, C. H. Heathcock, *Journal of Organic Chemistry* **1972**, <u>88</u>, 3657

# CAN

## Ce(NH$_4$)$_2$(NO$_3$)$_6$
Cerium(IV) ammonium nitrate, Ceric ammonium nitrate
[16774-21-3]

## Commercially available

T.-L. Ho, *Encyclopedia of Reagents for Organic Synthesis*, John Wiley and Sons, Inc., L. A. Paquette, Ed., New York, 1995, **2**, 1029

**Notes:** A very strong one-electron oxidizing agent. See also *CAS*.

R = H, or Me

## Examples:

1.[1]

2.[2]

3.[3]

Rapid deproptection, and can be carried out in the presence of a number of other functional groups.

4.[4] Can be used to convert a hydrazone to an ester.

[1] K. Pachamuthu, Y. D. Vankar, *Journal of Organic Chemistry* **2001**, <u>66</u>, 7511

[2] A. J. Clark, C. P. Dell, J. M. McDonagh, J. Geden, P. Mawdsley, *Organic Letters* **2003**, <u>5</u>, 2063

[3] A. Ates, A. Gautier, B. Leroy, J.-M. Planncher, Y. Quesnel, I. E. Marko, *Tetrahedron Letters* **1999**, <u>40</u>, 1799

[4] B. Stefane, M. Kocevar, S. Polanc, *Tetrahedron Letters* **1999**, <u>40</u>, 4429

# CAS

### Ce(NH₄)₄(SO₄)₄

Cerium(IV) ammonium sulfate
Ceric ammonium sulfate
[7637-03-8] hydrate
[10378-47-9] anhydrous

## Commercially available

M. Periasamy, U. Radhakrishnan *Encyclopedia of Reagents for Organic Synthesis*, John Wiley and Sons, Inc., L. A. Paquette, Ed., New York, 1995, **2**, 1029

**Notes:** Strong one-electron oxidizing agent. See also **CAN**.

## Examples:

1.[1]

CAS, MeCN

H₂O   78%

Note that the major product from the CAS reaction was the minor product from the peracid-initiated Baeyer-Villiger oxygen insertion.

2.[2]

CAS

1,2-shift

[1] G. Mehta, P. N. Pandeyl, T.-L. Ho, *Journal of Organic Chemistry* **1976**, 41, 953
[2] M. V. Brett, M. Periasany, *Tetrahedron* **1994**, 50, 3515

# Catecholborane

[274-07-7]

## Commercially available

**Preparation:** A new convenient preparation has been reported.[1]

B$_2$H$_6$

triglyme

85%

MeO$\sim$O$\sim$O$\sim$OMe

[112-49-2]  bp 216

triglyme

---

M. S. VanNieuwenhze, *Encyclopedia of Reagents for Organic Synthesis*, John Wiley and Sons, Inc., L. A. Paquette, Ed., New York, 1995, **2**, 1017

**Notes:**

A special reducing agent for:[2]

providing an alternative to the ***Wolff-Kishner reduction***.

A number of functional groups do not react with catecholborane: These include: organohalogen, alcohols and thiols, ethers, amides, nitro groups, and sulfones. Slowly reducing groups include acid chlorides, esters and nitriles. Aldehydes, ketones, imines, and sulfoxides are reduced. A hydroborating agent.

## Examples:

1.[3]

1.

CHCl$_3$

2. NaOAc
83-88%

2.[4]

Bu$\longequal$H

78-87%

C$_4$H$_9$

---

[1] J. V. B. Kanth, M. Periasamy, H. C. Brown, *Organic Process Research & Development* **2000**, <u>4</u>, 550

[2] G. W. Kabalka, J. D. Baker, Jr., *Journal of Organic Chemistry* **1975**, <u>40</u>, 1834

[3] G. W. Kabalka, R. Hutchins, N. R. Natale, D. T. C. Yang, V. Broach, *Organic Syntheses* <u>CV6</u>, 293

[4] N. Miyaura, A. Suzuki, *Organic Syntheses* <u>CV8</u>, 532

# Cbz-Cl

Benzyl chloroformate. source of the carbobenzyloxy (Cbz-) protecting group.
[501-53-1]

## Commercially available

P. Sampson, *Encyclopedia of Reagents for Organic Synthesis*, John Wiley and Sons, Inc., L. A. Paquette, Ed., New York, 1995, **1**, 323

**Notes:** An important functional group for protecting amines as carbamates and alcohols as benzyl carbonates. Notable for the protection of amino acids during peptide synthesis.

## Examples:

1.[1]

2.[2]

3.[3]

98-88%

---

[1] D. L. Comins, C. A. Brooks, R. S. Al-awar, R. R. Goehring, *Organic Letters* **1999**, <u>1</u>, 229
[2] W. R. Baker, J. D. Clark, R. L. Stephens, K. H. Kim, *Journal of Organic Chemistry* **1988**, <u>53</u>, 2340
[3] M. Carrasco, R. V. Jones, S. Kamel, H. Rapoport, T. Truong, *Organic Syntheses* <u>CV9</u>, 63

# Chan's Diene

TMSO    OTMS          TMSO    OMe

⟍⟍⟋⟍⟋OMe          ⟍⟍⟋⟍OTMS

[81114-98-9]                [74590-73-1]

1,3-Bis-(trimethylsilyloxy)-1-methoxy-buta-1,3-diene
3,7-dioxa-2,6-disilanon-4-ene

## Preparation:[1]

> T. Kitahara, *Encyclopedia of Reagents for Organic Synthesis*, John Wiley and Sons, Inc., L. A. Paquette, Ed., New York, 1995, **5**, 3330

**Notes:**  See ***Brassard's***, ***Danishefsky's*** and ***Rawal's Diene*** and for similar reactions; different substitution patterns.

## Examples:

1.[2]

2.[3]

ca. 30%

3.[4]

[1] P. Brownbridge, T. H. Chan, *Tetrahedron Letters* **1980**, 21, 3423
[2] D. A. Evans, E. Hu, J. D. Burch, G. Jaeschke, *Journal of the American Chemical Society* **2002**, 124, 5654
[3] G. E. Keck, T. Yu, *Organic Letters* **1999**, 1, 289
[4] I. Paterson, R. D. M. Davies, A. C. Heimann, R. Marquez, A. Meyer, *Organic Letters* **2003**, 5, 4477

# Collins and Sarrett Reagents[1]

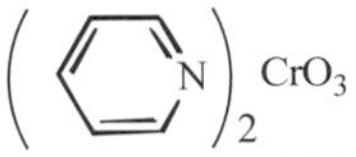

Chromium trioxide bispyridine complex [26412-88-4]
[1333-82-0]    [110-86-1]

**Preparation:** Chromium trioxide is added to pyridine, with cooling. See: *Organic Reactions*, **1998**, <u>53</u>, 15 for details of preparation. Order of mixing is critical. Addition of pyridine to chromium trioxide may result in fire.

F. Freeman, *Encyclopedia of Reagents for Organic Synthesis*, John Wiley and Sons, Inc., L. A. Paquette, Ed., New York, 1995, **4**, 2272

**Notes:**

1.  A non-acidic oxidizing reagent. It is quite tolerant of other functional groups.
2.  Similar to the ***Sarrett Reagent***. The major difference appears that the classical ***Collins Reagent*** is filtered while the ***Sarrett*** is used in pyridine.
3.  Allylic oxidations may also be carried out with the related ***Chromium(VI) Oxide-3,5-Dimethylpyrazole*** (DMP) complex.[2]

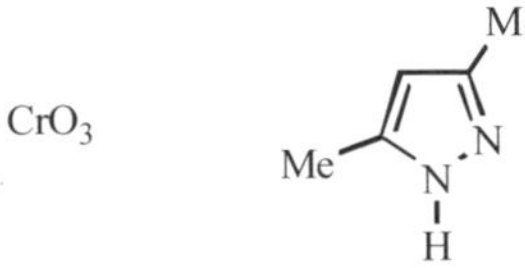

4.  The ***Ratcliffe Reagent*** is prepared in situ in dichloromethane.[3]
5.  Largely replaced by ***PCC*** and ***PDC***.
6.  See ***Chromium-based Oxidizing agents***.

**Examples:**

1.[4]

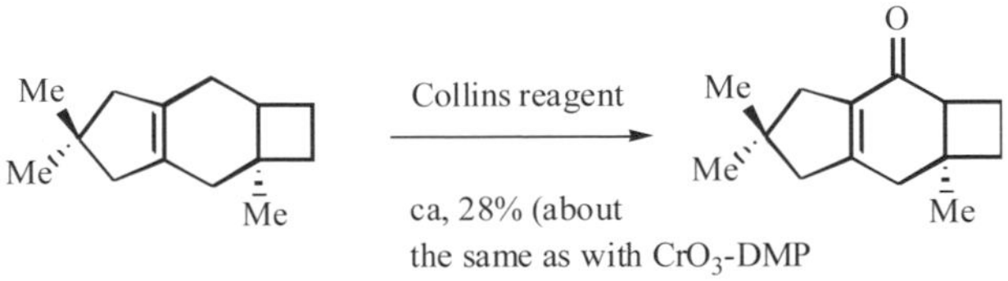

ca, 28% (about
the same as with $CrO_3$-DMP

2.[5]

60%

---

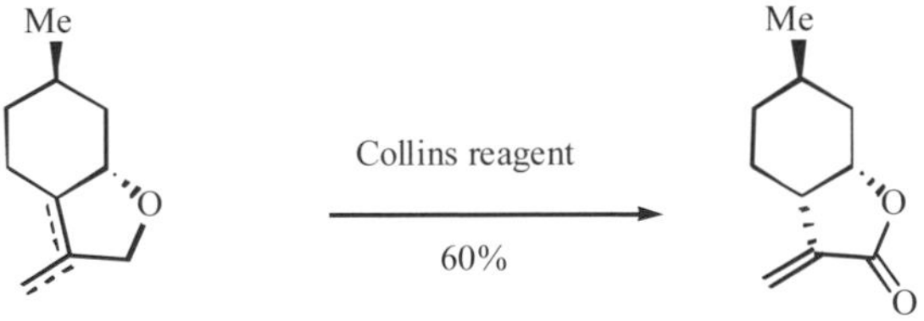

[1] J. C. Collins, W. W. Hess, F. J. Frank, *Tetrahedron Letters* **1968**, <u>9</u>, 3363
[2] E. J. Corey, G. W. J. Fleet, *Tetrahedron Letters* **1973**, <u>14</u>, 4459
[3] R. W. Ratcliff, *Organic Syntheses* <u>CV 6</u>, 373
[4] M. Harmata, G. J. Bohnert, *Organic Letters* **2003**, <u>5</u>, 59
[5] H. M. C. Ferraz, M. V. A. Grazini, C. M. R. Ribeiro, U. Brocksom, T. J. Brocksom, *Journal of Organic Chemistry* **2000**, <u>65</u>, 2606

# Collman's Reagent[1]
## $Na_2Fe(CO)_4$
Disodium tetracarbonylferrate
[14878-31-0]
[59733-73-2] dioxane complex
## Commercially available as dioxane complex

R. D. Pike, *Encyclopedia of Reagents for Organic Synthesis*, John Wiley and Sons, Inc., L. A. Paquette, Ed., New York, 1995, **4**, 2299

## Notes:
A general scheme:

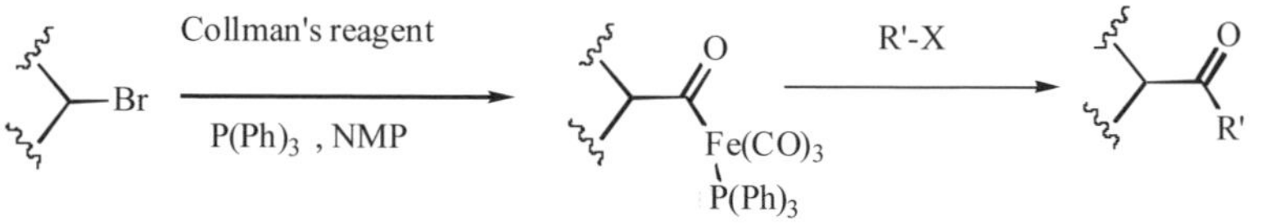

## Examples:
## Isomerization catalyst
1.[2]

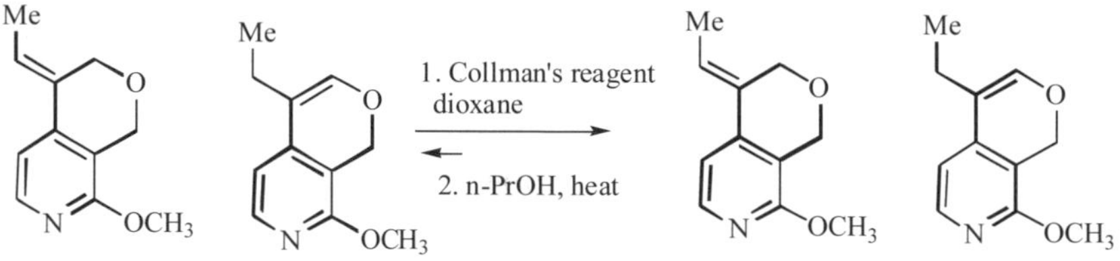

| | | | |
|---|---|---|---|
| 1 | : | 2.3 | |

2.[3]

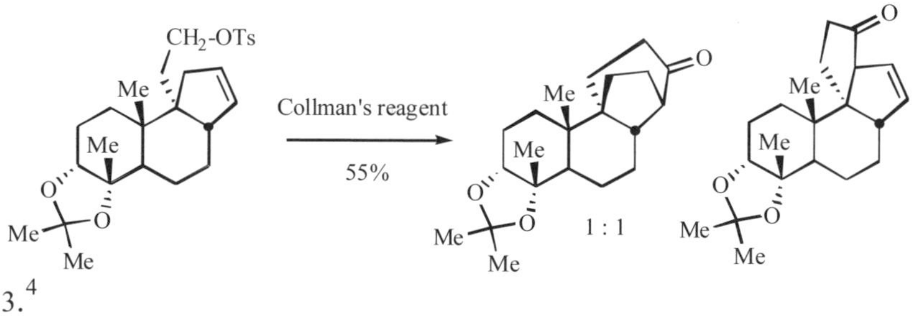

3.[4]

---

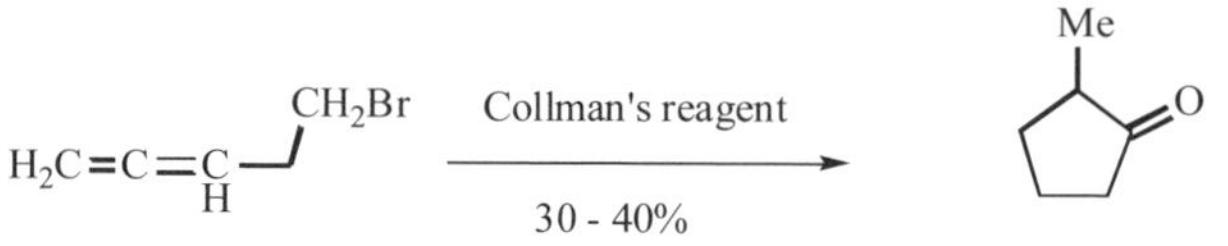

[1] a) J. P. Collman, *Accounts of Chemical Research* **1975**, <u>8</u>, 342 (b) J. P. Collman, R. G. Finke, J. N. Cawse, J. I. Brauman, *Journal of the American Chemical Society*, **1977**, <u>99</u>, 2515 (c) J. P. Collman, R. G. Finke, J. N. Cawse, J. I. Brauman, *Journal of the American Chemical Society* **1978**, <u>100</u>, 4766

[2] D. Bankston, F. Fang, E. Huie, S. Xie, *Journal of Organic Chemistry* **1999**, <u>64</u>, 3461

[3] J. E. McMurry, A. Andrcjs, G. M. Ksander, J. H. Musser, M. A. Johnson, *Tetrahedron* **1981**, <u>37</u>, 319

[4] J. Y. Mérour, J. L. Roustan, C. Charrier, J. Benaim, J. Collin, P. Cadiot, *Journal of Organometallic Chemistry* **1979**, <u>168</u>, 337

# Comins' Reagent

2-[N,N-Bis(trifluoromethylsulfonyl)amino]-5-chloropyridine
[145100-51-2]

**Commercially available**

**Preparation:**[1]

**Notes:** Reagent used for the conversion of carbonyl groups to –OTf derivatives.

**Examples:**

1.[2]

Comin's reagent

75%
Only isomer

2.[3]

Comin's reagent

75%

3.[4]

KHMDS

Comin's reagent

94%

[1] D. L. Comins, A. Dehghani, C. J. Foti, S. P. Joseph, *Organic Syntheses* CV9, 165
[2] S. M. Ceccarelli, U. Piarulli, J. Telsera, C. Gennaria, *Tetrahedron Letters* **2001**, 42, 7421
[3] J. A. Marshall, E. A. Van Devender, *Journal of Organic Chemistry* **2001**, 66, 8037
[4] K. Yamamoto, C. H. Heathcock, *Organic Letters* **2000**, 2, 1709

# Copper-Organometallics

### *Gilman Cuprates*

$$2 \text{ R-Li} \xrightarrow{\text{CuI}} R_2\text{CuLi}$$

### *Normant Cuprates*

$$2 \text{ R-MgX} \xrightarrow{\text{CuI (cat)}} R_2\text{CuMgX}$$

### *Knochel Cuprates*

$$2 \text{ R-ZnX} \xrightarrow[\text{LiCl}]{\text{CuCN}} RCu(CN)ZnX$$

The ***Knochel cuprates*** are able to tolerate a large variety of other functional groups imbedded in "R".

Reactivity differences are readily seen in the 1,4-addition to enones:

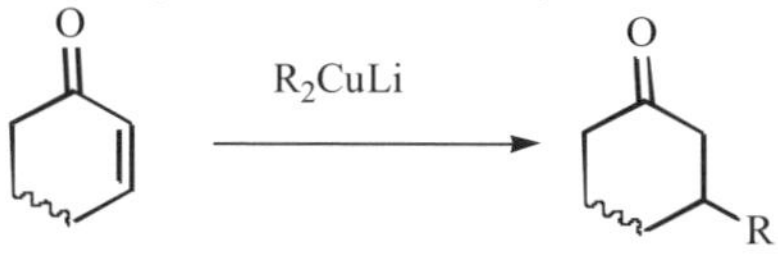

The ***Normant variation*** adds the same way. Since the CuI is only catalytic, it becomes obvious that the reagent is much more reactive than the ***Grignard*** starting material.

For the ***Knochel reaction*** an added Lewis acid is needed to enhance the 1,4-addition; otherwise 1,2-addition would take place. A major advantage of the ***Knochel reagent*** is that other functional groups (ester, carbonyl, nitrile, Cl, sulfoxide and terminal alkyne) are possible with the organo-zinc starting material.

# Corey-Chaykovsky Reagent [1,2]

$$H_2C^{\ominus} \quad H_3C-\overset{\oplus}{\underset{H_3C}{S}}-O^{\ominus} \quad X^{\ominus}$$

Trimethylsulfoxonium iodide
[1774-47-6]

## Commercially available. Preparation:[3]

J. S. Ng, C. Liu, *Encyclopedia of Reagents for Organic Synthesis*, John Wiley and Sons, Inc., L. A. Paquette, Ed., New York, 1995, **7**, 5335

**Notes:** This reagent finds use in epoxidation of carbonyl groups. Can also cyclopropanate enone systems.

## Examples:

1.[4]

2.[5]

3.[6]

4.[7]

---

[1] Review: Y. G. Gololobov, A. N. Nesmeyanov, V. P. Lysenco, I. E. Boldeskul, *Tetrahedron* **1987**, 43, 2609

[2] E. J. Corey, M. Chaykovsky, *Journal of the American Chemical Society* **1962**, 84, 867

[3] Details of the preparation and use of the reagent are provided: E. J. Corey, M. Chaykovsky, *Journal of the American Chemical Society* **1965**, 87, 1350

[4] C. F. D. Amigo, I. G. Collado, J. R. Hanson, R. Hernandez-Galan, P. B. Hitchcock, A. J. Macías-Sánchez, D. J. Mobbs, *Journal of Organic Chemistry* **2001**, 66, 4327

[5] J. S. Ng, *Synthetic Communications* **1990**, 20, 1193

[6] C. Mahaim, L. Schwager, P.-A. Carrupt, P. Vogel, *Tetrahedron Letters* **1983**, 24, 3603

[7] E. J. Corey, M. Chaykovsky, *Organic Syntheses* CV5, 755

# Corey-Kim Reagent

*N*-Chlorosuccinimide-dimethyl sulfide
[39095-38-0]

**Preparation:** By reaction of **_NCS_** with dimethylsulfide. Prepared *in situ* in solvents such as dichloromethane, THF or toluene.

R. C. Kelly, *Encyclopedia of Reagents for Organic Synthesis*, John Wiley and Sons, Inc., L. A. Paquette, Ed., New York, 1995, **2**, 1208

**Notes:** Mild oxidizing agent (*See **DMSO Oxidations in Aldehyde Syntheses***). A source of the S-Me electrophile. For an odorless protocol:[1] Instead of DMSO, dodecyl methyl sulfide is used in the reaction; thus, eliminating the dimethylsulfide by-product.

**Examples:**

1.[2]

2.[3]

No yield provided; however, it was reported that this process was not as effective as the **_Swern Oxidation_**.

3.[4]

90 - 93%

In this report a number of additional examples of primary and secondary alcohol oxidations are provided.

[1] K. Nishide, S. Ohsugi, M. Fudesaka, S. Kodama, M. Node, *Tetrahedron Letters* **2002**, *43*, 5177
[2] M. Kawahara, A. Nishida, M. Nakagawa, *Organic Letters* **2000**, *2*, 675
[3] J. Z. Ho, R. M. Mohareb, J. H. Ahn, T. B. Sim, H. Rapoport, *Journal of Organic Chemistry* **2003**, *68*, 109
[4] E. J. Corey, C. U. Kim, P. F. Misco, *Organic Syntheses* CV 6, 220

# Corey-Seebach Reagent

2-Lithio-1,3-dithiane     [36049-90-8]

## Preparation:

M. Kolb, *Encyclopedia of Reagents for Organic Synthesis*, John Wiley and Sons, Inc., L. A. Paquette, Ed., New York, 1995, **5**, 2983

**Notes:** An "umpolung" reagent.

The partial positive charge on carbon          Negative charge

## Examples:

1.[1]

96%

2.[2]

TBDMS-O     87%     TBDMS-O

3.[3]

BuLi / THF

81%

TBDMSO

4.[4]

1. BuLi

2. C$_{14}$H$_{29}$Br

HgCl$_2$

HgO
MeOH/H$_2$O

C$_{14}$H$_{29}$-CHO

47-55%

[1] P. Gros, P. Hansen, P. Caubere, *Tetrahedron* **1996**, 52, 15147

[2] P. G. Steel, E.J. Thomas, *Journal of the Chemical Society, Perkin Transaction 1* **1997**, 371

[3] B. G. Hazra, S. Basu, B. B. Bahule, V. S. Pore, B. N. Vyas, V. M. Ramraj, *Tetrahedron* **1997**, 53, 4909

[4] D. Seebach, A. K. Beck, *Organic Syntheses* CV6, 869

# Corey's Reagent
## PCC

Pyridinium Chlorochromate
[26299-14-9]

**Commercially available.** An improved preparation has been reported.[1]

---

G. Piancatelli, *Encyclopedia of Reagents for Organic Synthesis*, John Wiley and Sons, Inc., L. A. Paquette, Ed., New York, 1995, **6**, 4356

## Notes:

PCC can be used under solvent-free conditions.[2] Mechanistic studies.[3]
PCC/$I_2$ is useful for the conversion:[4]

## Examples:
1.[5]

2.[6]

---

[1] S. Agarwal, H. P. Tiwari, J. P. Sharma, *Tetrahedron* **1990**, _46_, 4417
[2] P. Salehi, H. Firouzabadi, A. Farrokhi, M. Gholizadeh, *Synthesis* **2001**, 2237
[3] S. Agarwal, H. P. Tiwari, J. P. Sharma, *Tetrahedron* **1990**, _46_, 1963
[4] G. Piancatelli, *Encyclopedia of Reagents for Organic Synthesis*, John Wiley and Sons, Inc., L. A. Paquette, Ed., New York, 1995, **6**, 4356
[5] G. I. Hwang, J.-H. Chung, W. K. Lee, *Tetrahedron* **1996**, _52_, 12111
[6] J. Cossy, S. BouzBouz, M. Laghgar, B. Tabyaoui, *Tetrahedron Letters* **2002**, _43_, 823

## Cornforth Reagent

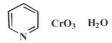 CrO₃   H₂O

Chromium trioxide / Pyridine / Water
[110-86-1]
[1333-82-0]

## Fieser's Reagent

**CrO₃   HOAc**
Chromium trioxide  Acetic Acid
[1333-82-0]           [64-19-7]

F. Freeman, *Encyclopedia of Reagents for Organic Synthesis*, John Wiley and Sons, Inc., L. A. Paquette, Ed., New York, 1995, **2**, 1273

# Criegee Reagent
## $Pb(OAc)_4$
Lead tetraacetate (LTA)
[546-67-8]

## Commercially available

M. L. Mihailovic, Z. Cekovic, *Encyclopedia of Reagents for Organic Synthesis*, John Wiley and Sons, Inc., L. A. Paquette, Ed., New York, 1995, **5**, 2949

## Notes:

The "named reagent" is most closely associated with the cleavage of 1,2-diols.
*Cis*-diols are cleaved more readily than *trans* diols. Different mechanistic interpretations are invoked for the two processes:

**Reported[1]**

Examples:

1.[2]

2.[3]

---

[1] M. L. Mihailovic, Z. Cekovic, *Encyclopedia of Reagents for Organic Synthesis*, John Wiley and Sons, Inc., L. A. Paquette, Ed., New York, 1995, **5**, 2949

[2] J. Xia, Y.-Z. Hui, *Tetrahedron: Asymmetry* **1997**, 8, 451

[3] M. G. Banwell, G. S. Forman, *Journal of the Chemical Society, Perkin Transactions 1* **1996**, 2565

# Crown Ether

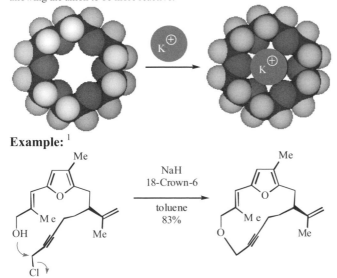

12-Crown-4          15-Crown-5          18-Crown-6

21-Crown-7              24-Crown-8

## n-Crown-m

Where **n** gives ring size and **m**- gives number of heteroatoms

## Commercially available

These materials, generally used as cosolvents, are useful in holding cations in their cavaties; thus, allowing the anion to be more reactive.

K⊕ →

## Example: [1]

NaH
18-Crown-6

toluene
83%

---

[1] J. A. Marshall, G. S. Bartley, E. M. Wallace, *Journal of Organic Chemistry* **1996**, <u>61</u>, 5729

# CSA

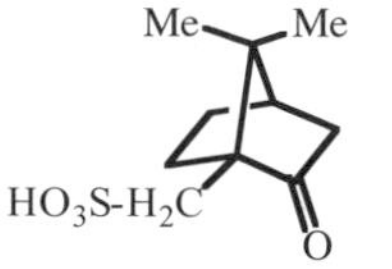

Camphorsulfonic Acid
[5872-08-2]

## Commercially available

E. M. Leahy, *Encyclopedia of Reagents for Organic Synthesis*, John Wiley and Sons, Inc., L. A.
Paquette, Ed., New York, 1995, **2**, 969

**Notes:**  A useful acid catalyst.

**Examples:**

1.[1]

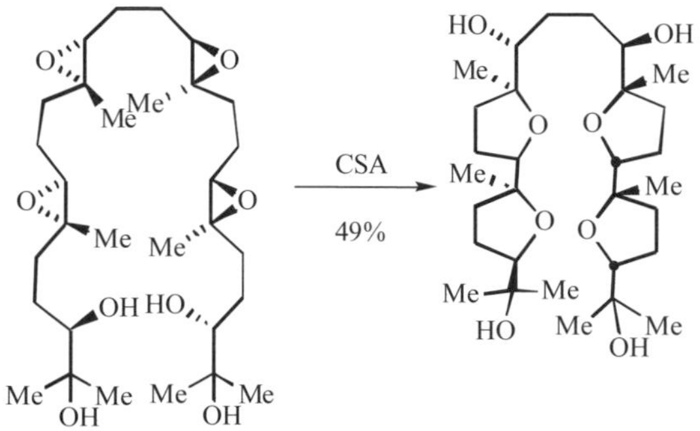

2.[2]

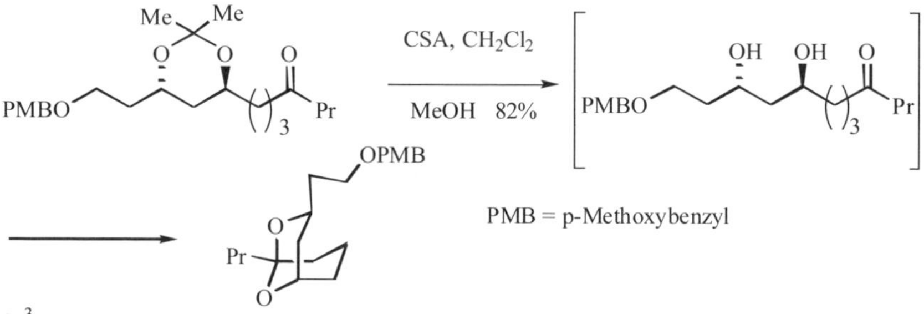

PMB = p-Methoxybenzyl

3.[3]

96%

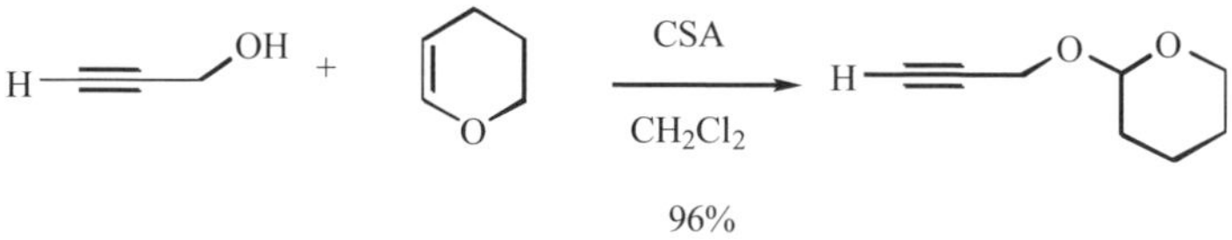

[1] X. Xiong, E. J. Corey, *Journal of the American Chemical Society* **2000**, <u>122</u>, 9338
[2] J. Aiguade, J. Hao, C. J. Forsyth, *Organic Letters* **2001**, <u>3</u>, 979
[3] A. G. Myers, B. Zheng, *Organic Syntheses* <u>CV10</u>, 2350

# DABCO

[280-57-9]
1,4-Diazabicylo[2.2.2]octane, triethylenediamine, TED

## Commercially available

U. V. Mallavadhani, *Encyclopedia of Reagents for Organic Synthesis*, John Wiley and Sons, Inc., L. A. Paquette, Ed., New York, 1995, **2**, 1494

**Notes:** Useful for organometallic complexation. Selective base used for the ***Bayliss-Hillman reaction***. Has been used for the ***Haller-Bauer*** cleavage reaction. Streitwieser[1] has examined a new scale of amine basicity ( $K_{ip}$ ), where the *lower* the $pK_{ip}$ values the stronger the base. In this analysis:

| Proton Sponge | ***DABCO*** | ***DMAP*** | Quinuclidine | ***DBU*** |
|---|---|---|---|---|
| $pK_{ip}$    2.15 | 0.80 | 0.61 | 0.15 | -3.78 |

## Examples:

1.[2]

TMS≡COOMe + PhCHO →(DABCO, PhH) 73%

(74) / (26)

2.[3]

TsCl, DABCO
TEA
91%

3.[4]

Me–CHO + =COOMe →(DABCO / MeOH) 79%

---

[1] A. Streitieser, Y.-J. Kim, *Journal of the American Chemical Society* **2000**, <u>48</u>, 11783
[2] Y. Matsuya, K. Hayashi, H. Nemoto, *Journal of the American Chemical Society* **2003**, <u>125</u>, 646
[3] Z. Wang, S. Campagna, K. Yang, G. Xu, M. E. Pierce, J. M. Fortunak, P. N. Confalone, *Journal of Organic Chemistry* **2000**, <u>65</u>, 1889
[4] C. Behrens, L. Paquette, *Organic Syntheses* <u>CV10</u>, 2316

# Dane's Diene

MeO
1-Vinyl-6-methoxy-3,4-dihydronaphthalene
[2811-50-9]

## Notes:

This is a diene useful for the construction of steroid skeletal.

## Examples:

1.[1]

Me___CHO

EtOOC

CH₂Cl₂   MeO
92%
Chiral catalyst

94% ee

Chiral catalyst

2.[2]

MeO

H₂O₂

38% overall

3.[3]

MeO

TiCl₂(O-iPr)₂
80%

MeO

---

[1] Q.-Y. Hu, P. D. Rege, E. J. Corey, *Journal of the American Chemical Society* **2004**, <u>126</u>, 5984
[2] S. Woskiand, M. Koreeda, *Journal of Organic Chemistry* **1992**, <u>57</u>, 5736
[3] G. Quinkert, M. del Grosso, A. Bucher, M. Bauch, W. Döring, J. W. Bats, G. Dürner, *Tetrahedron Letters* **1992**, <u>33</u>, 3617

# Danishefsky's Diene

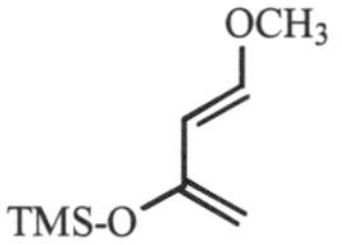

OCH₃

TMS-O

*trans*-1-methoxy-3-trimethylsiloxy-1,3-butadiene
[59414-23-21]

T. Kitahara, *Encyclopedia of Reagents for Organic Synthesis*, John Wiley and Sons, Inc., L. A. Paquette, Ed., New York, 1995, **5**, 3395

## Examples:

1.[1]

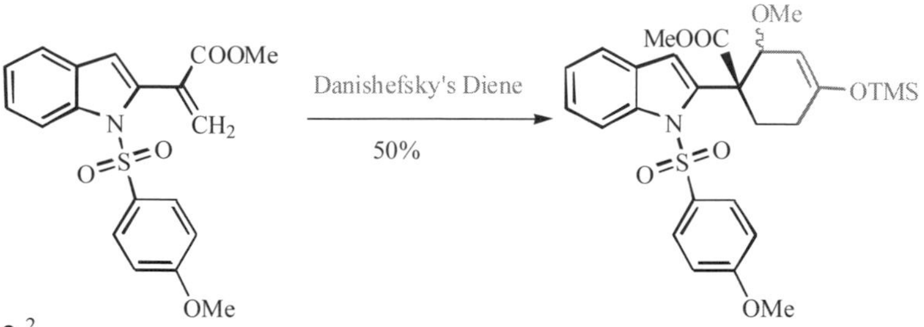

2.[2]

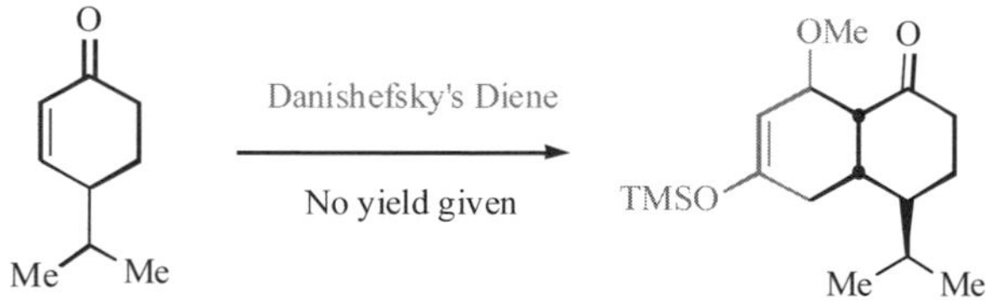

3.[3]

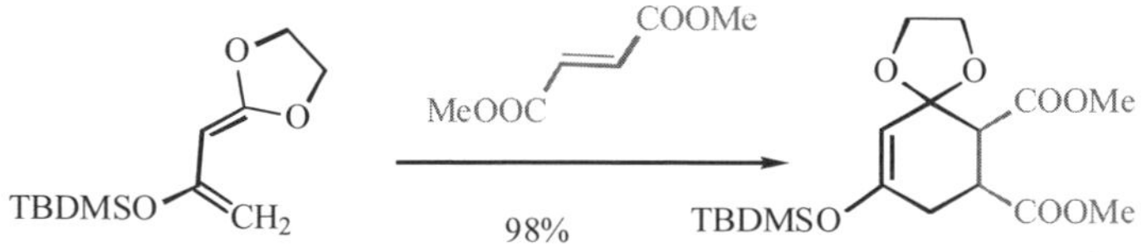

modified reagent

[1] B. Danieli, G. Lesma, M. Luzzani, D. Passarella, A. Silvani, *Tetrahedron* **1996**, <u>52</u>, 11291
[2] B. Ye, H. Nakamura, A. Murai, *Tetrahedron* **1996**, <u>52</u>, 6361
[3] J. P. Konopelski, R.A. Kalsar, *Tetrahedron Letters* **1993**, <u>34</u>, 4587

# DAST

Et
  \
   N—SF$_3$
  /
Et

(Diethylamino)sulfur trifluoride
[38078-09-0]

## Commercially available

A. H. Fauq, *Encyclopedia of Reagents for Organic Synthesis*, John Wiley and Sons, Inc., L. A.
Paquette, Ed., New York, 1995, **3**, 1787

## Notes:

A useful fluorinating reagent.  Readily converts the –OH
group of an alcohol to –F.  Deoxy-Fluor Reagent® [bis-(2-
methoxyethyl) amino sulfur trifluoride] has been found to be
more stable.[1]

## Examples:

1.[1]

Deoxo-Fluor Reagent

CH$_2$Cl$_2$, SbCl$_3$

95%

2.[2]

iPr$_2$NET,  CH$_2$Cl$_2$

91%

a *Weinreb amide*

3.[3]

DAST

EtOAc

99%  Note rearrangement

4.[4]

Et$_2$NSF$_3$

CH$_2$Cl$_2$

67%

[1] G. S. Lal, E. Lobach, A. Evans, *Journal of Organic Chemistry* **2000**, <u>65</u>, 4830
[2] A. R. Tunoori, J. M. White, G. I. George, *Organic Letters* **2000**, <u>2</u>, 4091
[3] D. J. Hallett, U. Gerhard, S. C. Goodacre, L. Hitzel, T. J. Sparey, S. Thomas, M. Rowley, R. G.
Ball, *Journal of Organic Chemistry* **2000**, <u>65</u>, 4984
[4] W. J. Middleton, E.M. Bingham, *Organic Syntheses* <u>CV6</u>, 835

# Davis's Oxaziridine

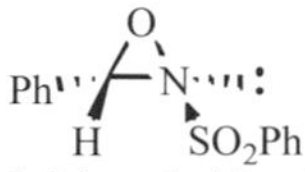

2-(phenylsulfonyl)-3-phenyloxaziridine
[63160-13-4]]

## Preparation:

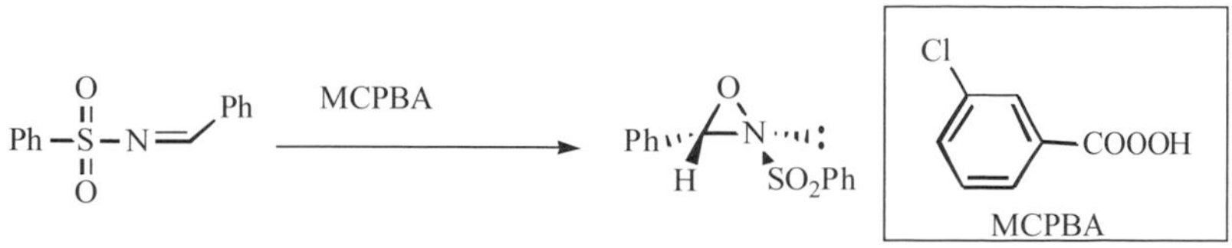

$$Ph-\overset{\overset{O}{\|}}{\underset{\underset{O}{\|}}{S}}-N=\diagup Ph \xrightarrow{\text{MCPBA}}$$

B.-C. Chen, F.A. Davis, *Encyclopedia of Reagents for Organic Synthesis*, John Wiley and Sons, Inc., L. A. Paquette, Ed., New York, 1995, **6**, 4054

## Notes:

A useful oxidizing agent. Will convert organic sulfides to sulfoxides without overoxidation and disubstituted enamines to α-aminoketones. Most useful is the oxidation of carbanions to hydroxyl groups.

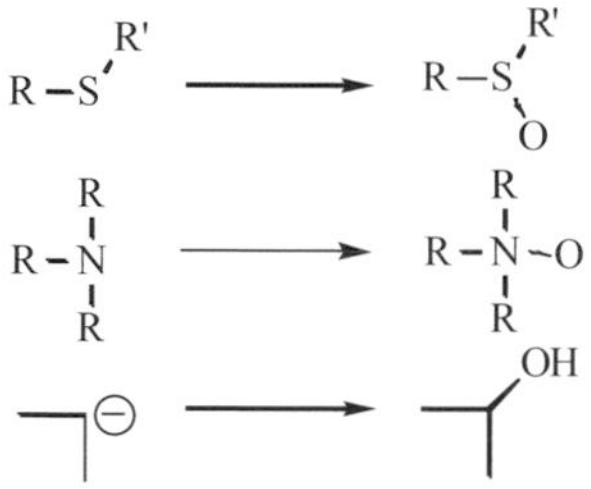

$$R-\overset{R'}{\underset{}{S}} \longrightarrow R-\overset{R'}{\underset{\underset{O}{}}{S}}$$

$$R-\overset{R}{\underset{R}{N}} \longrightarrow R-\overset{R}{\underset{R}{N}}-O$$

## Examples:
1.[1]

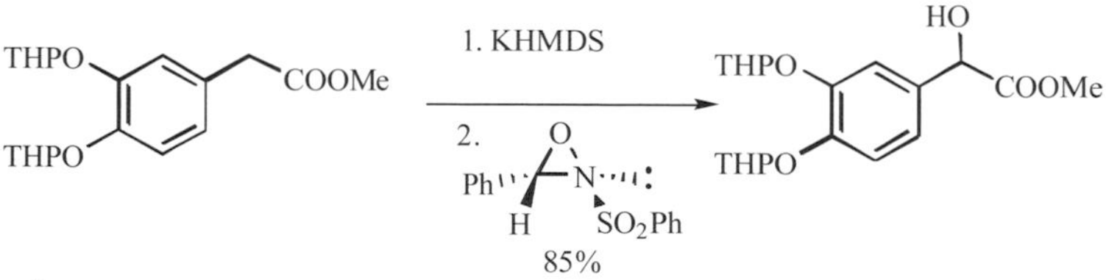

1. KHMDS

2. 85%

2.[2]

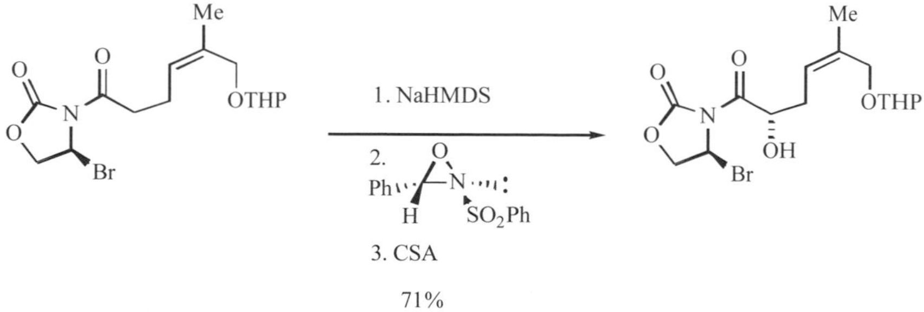

1. NaHMDS

2.

3. CSA

71%

[1] J. Narayanan, Y. Hayakawa, J. Fan, K. L. Kirk, *Bioorganic Chemistry* **2003**, <u>31</u>, 191
[2] J. D. White, R. G. Carter, K. F. Sundermann, *Journal of Organic Chemistry* **1999**, <u>64</u>, 684

# Davy's Reagent

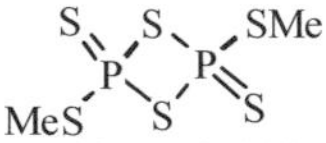

2,4-bis(methylthio)-1,3,2,4-dithiadiphosphetane-2,4-disulfide
[82737-61-9]

## Preparation:

$$MeOH + P_4S_{10} \longrightarrow$$

J. Voss, *Encyclopedia of Reagents for Organic Synthesis*, John Wiley and Sons, Inc., L. A. Paquette, Ed., New York, 1995, **1**, 535

## Notes:

More reactive than ***Lawesson's Reagent***. Can convert a carboxylic acid directly to a dithioester. See also, ***Belleau's Reagent***.

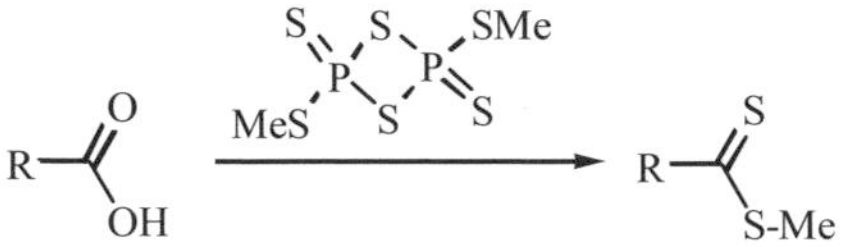

Is also useful for the general conversion of a carbonyl group to a thiacarbonyl.

## Examples:

1.[1]

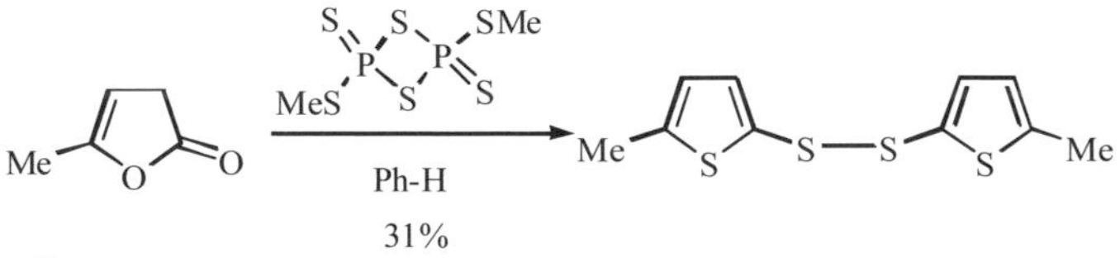

Ph-H

31%

2.[2]

dioxane

2. heat
37%

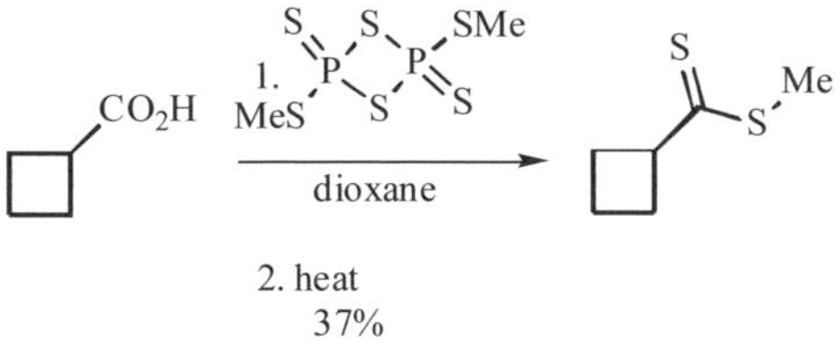

3.[3]

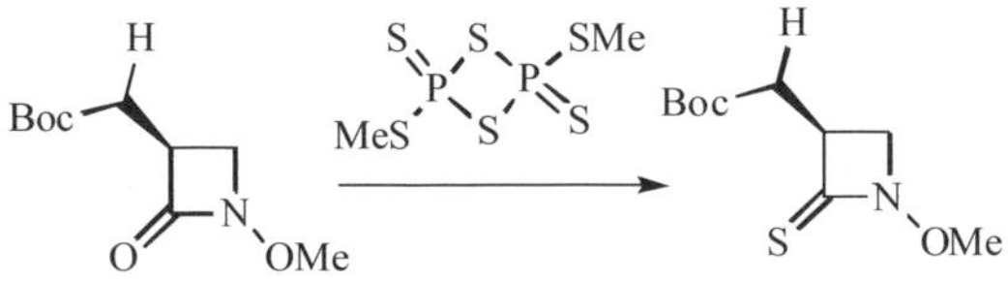

[1] N. M. Yousif, U. Pedersen, B. Yde, S.-O. Lawesson, *Tetrahedron* **1984**, 40, 2663
[2] Y. Vallee, S. Masson, J.-L, Ripoli, *Tetrahedron* **1990**, 46, 3928
[3] J. Nieschalk, E. Schaumann, *Liebigs Annelen* **1996**, 141 (AN 1995: 48264)

# DBN

1,5-Diazabicyclo[4.3.0]non-5-ene
[3001-72-7]

## Commercially available

A. C. Savoco, *Encyclopedia of Reagents for Organic Synthesis*, John Wiley and Sons, Inc., L. A. Paquette, Ed., New York, 1995, **2**, 1491

## Notes:

An organic base, soluble in a variety of common organic solvents. Useful for equilibrations and elimination reactions.

## Examples:

1.[1]

DBN, Ph-H

95%

2.[2]

DBN

| G = H | 77 | 23 |
| G = COOMe | 21 | 79 |

3.[3]

DBN

toluene

75%

4.[4]

DBN

THF
100%

[1] E. Piers, M. Gilbert, K. L. Cook, *Organic Letters* **2000**, *2*, 1407
[2] R. C. Mease, J. A. Hirsch, *Journal of Organic Chemistry* **1984**, *49*, 2927
[3] G.-C. Wei, T. J. Chow, Y.-P. Yang, Y.-J. Chen, *Tetrahedron* **1993**, *49*, 2201
[4] F. Berree, E. Marchand, G. Morel, *Tetrahedron Letters* **1992**, *33*, 6155

# DBU

[6674-22-2]
1,5-Diazabicyclo[5.4.0]undec-7-ene
## Commercially available

A. C. Savaca, *Encyclopedia of Reagents for Organic Synthesis*, John Wiley and Sons, Inc., L. A. Paquette, Ed., New York, 1995, **2**, 1497

**Notes:** Streitwieser[1] has examined a new scale of amine basicity ($K_{ip}$), where the *lower* the $pK_{ip}$ values the stronger the base. In this analysis:

| Proton Sponge | *DABCO* | *DMAP* | Quinuclidine | *DBU* |
|---|---|---|---|---|

| $pK_{ip}$ | 2.15 | 0.80 | 0.61 | 0.15 | -3.78 |

## Examples:

1.[2]

2.[3]

3.[4]

[1] A. Streitieser, Y.-J. Kim, *Journal of the American Chemical Society* **2000**, 48, 11783
[2] W.-C. Shieh, S. Dell, O. Repic, *Organic Letters* **2001**, 3, 4279
[3] F. A. Luzzio, D. Y. Duveau, *Tetrahedron: Asymmetry* **2002**, 13, 117
[4] P. Magnus, I. K. Sebhat, *Tetrahedron* **1998**, 54, 15509

# DCC

1,3-Dicyclohexylcarbodiimide
[538-75-0]

## Commercially available

J. S. Albert, A. D. Hamilton, *Encyclopedia of Reagents for Organic Synthesis*, John Wiley and
Sons, Inc., L. A. Paquette, Ed., New York, 1995, **3**, 1751

**Notes:** Dehydrating agent often used to form esters, amides or anhydrides.

$$Z = —OR$$

$$—N\diagdown$$

$$—O\diagdown_C^{\parallel}_O-R'$$

With **_DMSO_**, provides a mild oxidizing agent (**_Pfitzner-Moffatt Oxidation_**). The reagent has been
useful in forming the peptide linkage from amino acids.
With **_DMAP_** is used for the **_Steglich Esterification_**.

## Examples:

1.[1]

2.[2]

---

[1] J. R. P. Cetusic, F. R. Greene, III, P. R. Graupner, M. P. Oliver, *Organic Letters* **2002**, <u>4</u>, 1307
[2] D. Goubet, P. Meric, J.-R. Dormoy, P. Moreau *Journal of Organic Chemistry* **1999**, <u>64</u>, 4516

# DDO

Me
O
O
Me

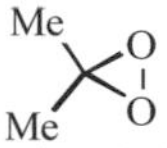

Dimethyldioxirane
[74087-85-7]

## Preparation:[1,2]

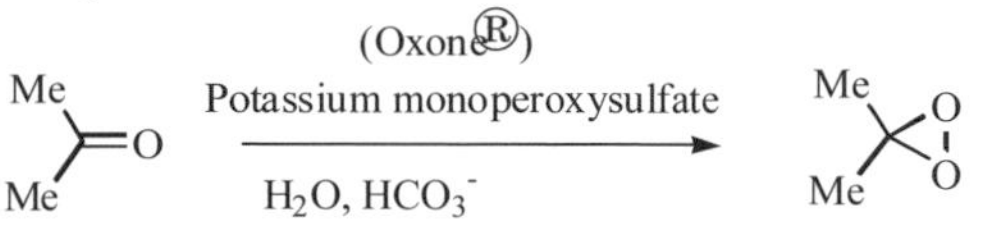

Me
O
Me
(Oxone®)
Potassium monoperoxysulfate
$H_2O$, $HCO_3^-$
Me
O
O
Me

***Oxone*** is commercially available

| |
|---|
| J. A. Crandall, *Encyclopedia of Reagents for Organic Synthesis*, John Wiley and Sons, Inc., L. A. Paquette, Ed., New York, 1995, **3**, 206 |

## Notes:

A selective and reactive oxidizing agent. Will epoxidize α,β–unsaturated carbonyl compounds. In epoxidation reactions, there is a strong steric influence directing the facial selectivity:

R
O
Me
O
Me
R
O

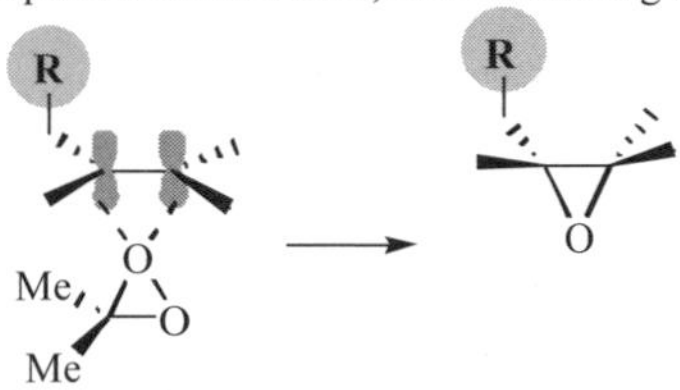

The amount of *trans* isomer increases with increasing steric bulk of the R- group

## Examples:

1.[3]

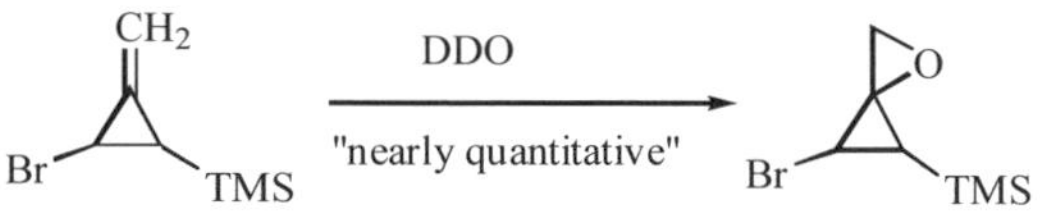

CH2
Br
TMS
DDO
"nearly quantitative"
O
Br
TMS

2.[4]

O
O
O
Me
Me
Me
Me
Me
DDO
96%
O
O
O
O
Me
Me
Me
Me
Me

3.[5]

O
O
O
O
tBu
1. DDO
2. TsOH, H2O
94%
O
O
O
O
OH
OH
tBu

---

[1] $2KHSO_3 \cdot KHSO_4 \cdot K_2SO_4$ is commercially available as ***Oxone***®.
[2] See: R. W. Murray, M. Singh, *Organic Syntheses* CV9, 288 for details of preparation.
[3] W. E. Billups, V. A. Litosh, R. K. Saini, A. D. Daniels, *Organic Letters* **1999**, 1, 115
[4] B. C. Raimundo, C. H. Heathcock, *Organic Letters* **2000**, 2, 27
[5] M. T. Crimmins, J. M. Pace, P. G. Nantermet, A. S. Kim-Meade, J. B. Thomas, S. H. Watterson, A. S. Wagman, *Journal of the American Chemical Society* **1999**, 121, 10249

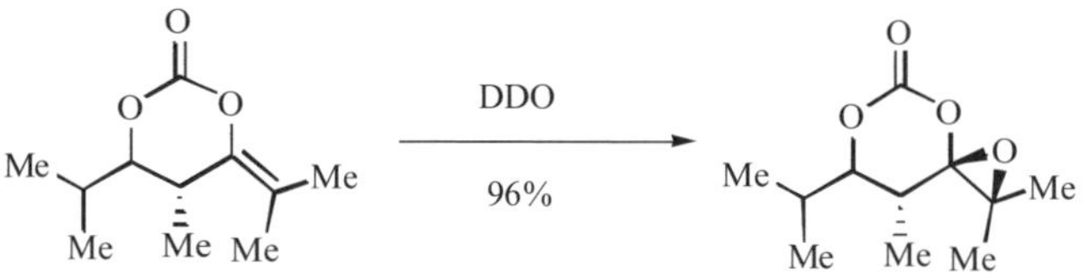

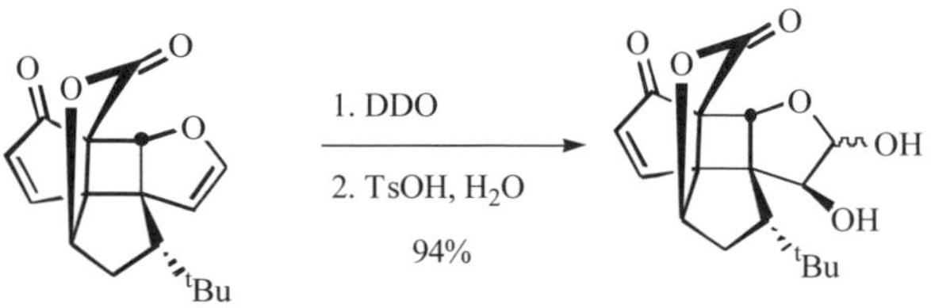

# DDQ

2,3-Dichloro-5,6-dicyano-1,4-benzoquinone
[84-58-2]

## Commercially available

D. R. Buckle, *Encyclopedia of Reagents for Organic Synthesis*, John Wiley and Sons, Inc., L. A. Paquette, Ed., New York, 1995, **3**, 1699

**Notes:** An oxidizing agent useful for dehydrogenations; particularly those resulting in aromatization, extended conjugation from aromatic systems, and the formation of enones. Oxidation of phenol provides quinones.

## Examples:

1.[1]

DDQ, CHCl$_3$

100%

2.[2]

DDQ

dioxane

83 - 85%

3.[3]

DDQ

dioxane

81%

4.[4] Rearrangement may accompany dehydrogenation:

DDQ, Ph-H

78%

1 S. Cossu, O. DeLucchi, *Tetrahedron* **1996**, 52, 14247

2 J. W. A. Findlay, A. B. Turner, *Organic Syntheses* CV 5, 428

3 S. R. Cheruka, M. P. Padmanilayam, J. L. Vennerstrom, *Tetrahedron Letters* **2003**, 44, 3701

4 Reported by D.R. Buckle, *Encyclopedia of Reagents for Organic Synthesis,* John Wiley and Sons, Inc., L. A. Paquette, Ed., New York, 1995, **3**, 1700 (E. A. Braude, L. M. Jackman, R. P. Linstead, G. Lowe, *Journal of the Chemical Society* **1960**, 3123)

# DEAD

EtOOC − N ≡ N − COOEt

Diethyl Azodicarboxylate
[1972-28-7]

## Commercially available

E. J. Stoner, *Encyclopedia of Reagents for Organic Synthesis*, John Wiley and Sons, Inc., L. A.
Paquette, Ed., New York, 1995, **3**, 1790

## Notes:

An oxidizing agent. Useful for the dealkylation of amines; and the conversion of pyrimidines to
purines. It is most often associated with triphenylphosphine (***TPP***) in the ***Mitsunobu reaction***.

## Examples:

1.[1]

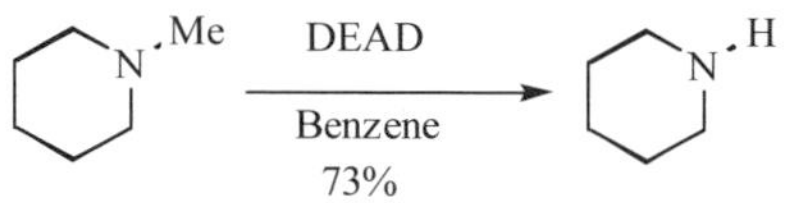

2.[2]

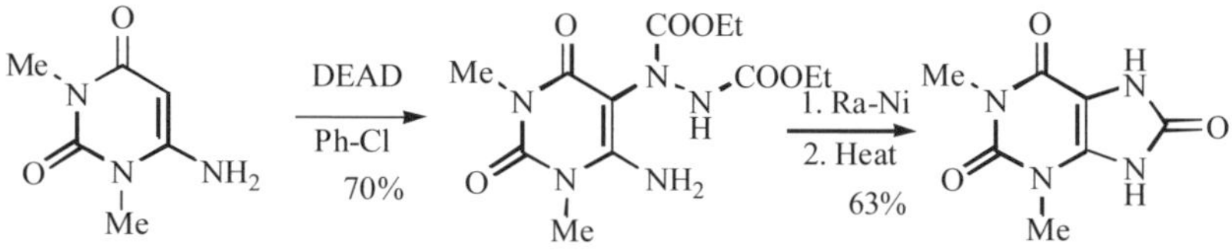

3.[3]

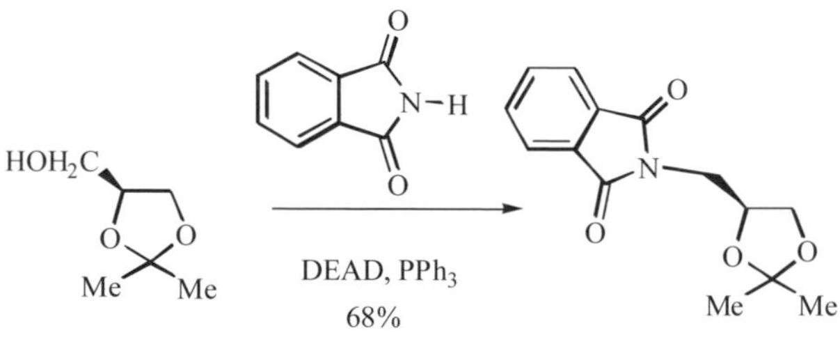

4.[4]

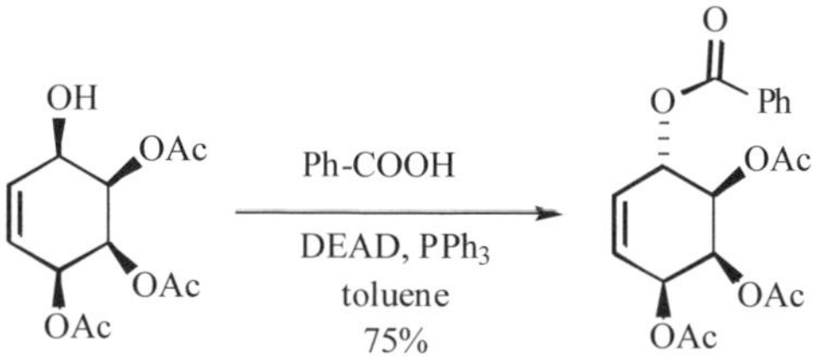

---

[1] E. M. Smissman, A. Makriyannis, *Journal of Organic Chemistry* **1973**, <u>38</u>, 1652

[2] E. C. Taylor, F. Sowinski, *Journal of Organic Chemistry* **1974** <u>39</u> 907

[3] B. B. Lohray, A. S. Reddy, V. Bhushan, *Tetrahedron: Asymmetry* **1996**, <u>7</u>, 2411

[4] A. Patti, C. SanFilippo, M. Piattelli, G. Nicolosi, *Tetrahedron: Asymmetry* **1996**, <u>7</u>, 2665

# Dess-Martin Reagent

1,1-Triacetoxy-1,1-dihydro-1,2-benziodoxol-3(1H)-one
periodinane
[87413-09-0]

## Preparation:

Robert J. Boeckman, Jr., *Encyclopedia of Reagents for Organic Synthesis*, John Wiley and Sons, Inc., L. A. Paquette, Ed., New York, 1995, **7**, 4982

**Notes:** Selective oxidizing agent. Since a mechanistic requirement of this reagent functioning ability is a complexation with the substrate, it is exceedingly important that only the –OH group binds efficiently. A useful advantage of this oxidation protocol is that it takes place under essentially neutral conditions. See: ***Dess-Martin Oxidation***.

## Examples:

1.[1] Note the preferential oxidation of the primary alcohol

2.[2]

---

[1] D. A. Evans, J. R. Gage, J. L. Leighton, *Journal of the American Chemical Society* **1992**, 114, 9434
[2] A. G. Myers, P. S. Dragovich, *Journal of the American Chemical Society* **1992**, 114, 5859

# Diazald⊕

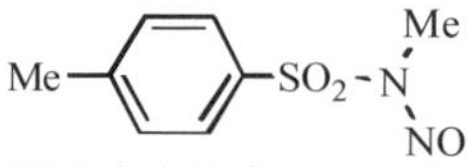

N-Methyl-N-nitroso-p-toluenesulfonamide
[80-11-5]

## Commercially available

Y. Terao, M. Sekiya, *Encyclopedia of Reagents for Organic Synthesis*, John Wiley and Sons, Inc., L. A. Paquette, Ed., New York, 1995, **5**, 3555

## Notes:

Precursor of diazomethane, a highly toxic and unpredictably explosive gas. Possibly a carcinogen. Read details carefully before preparation. Reactions with and preparation of diazomethane should be carried out in a fume hood and behind a sturdy safety shield. Rough glass surfaces and strong sunlight are known to initiate detonation.

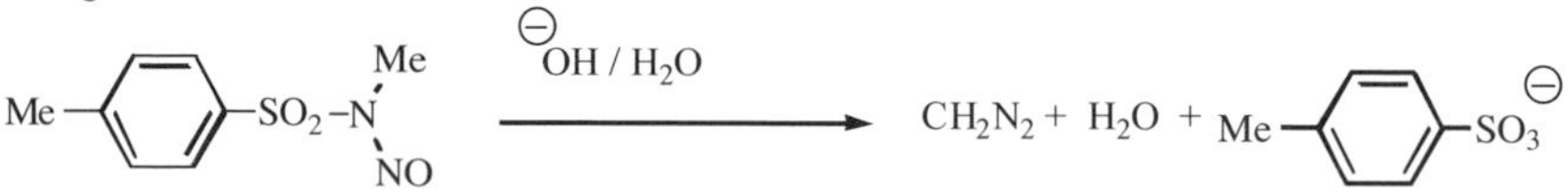

## Mechanism

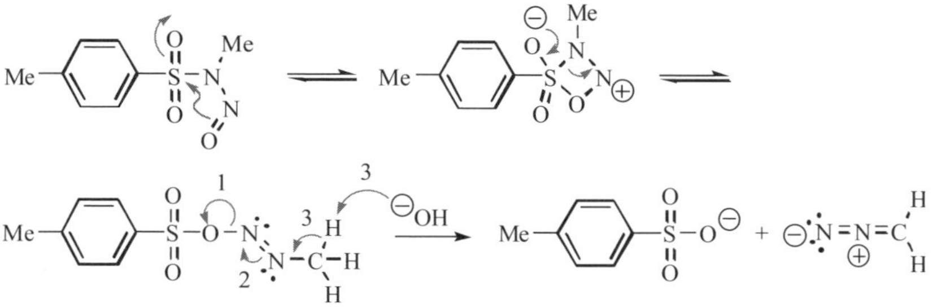

Example of using diazomethane to prepare methyl ester:[1]

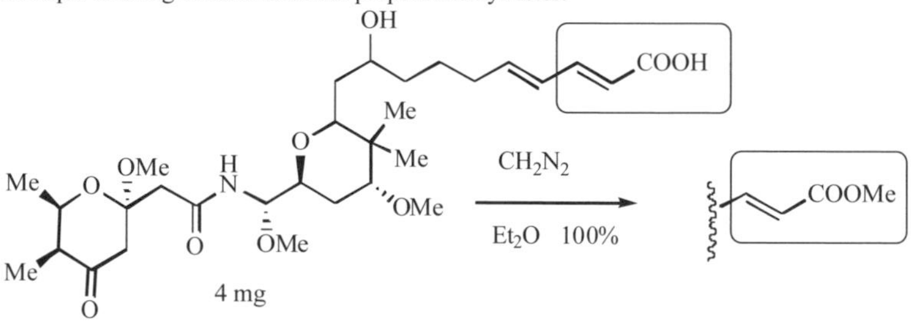

Example of using diazomethane to prepare methyl ether:[2]

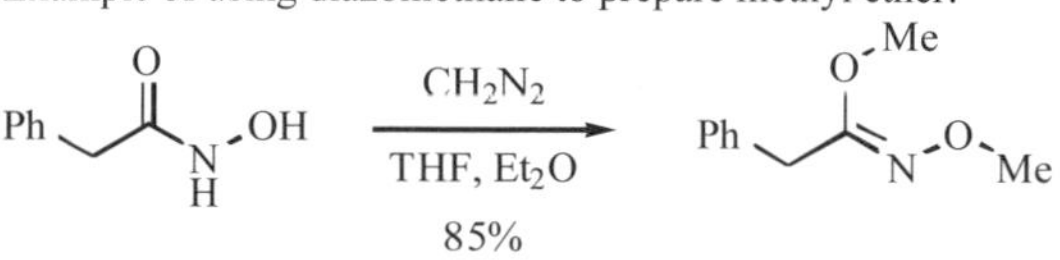

---

[1] D. Vuong, R. J. Capon, E. Lacey, J. H. Gill, K. Heiland, T. Friedel, *Journal of Natural Products* **2001**, 64, 640

[2] A. Leggio, A. Liguori, A. Napoli, C. Siciliano, G. Sindona, *Journal of Organic Chemistry* **2001**, 66, 2246

# Diazomethane

$$H_2C\overset{\oplus}{=}N\overset{\ominus}{=}N\text{:} \longleftrightarrow \overset{\ominus}{H_2C}-\overset{\oplus}{N}\equiv N\text{:}$$

[334-88-3]

## Preparation:

Major sources are the basic hydrolysis of *N*-methyl-*N*-nitrosourea (or *N*-methyl-*N*-nitroso-*p*-toluenesulfonamide (***Diazald*®**) or 1-methyl-3-nitro-1-nitrosoguanidine (**MNNG**).

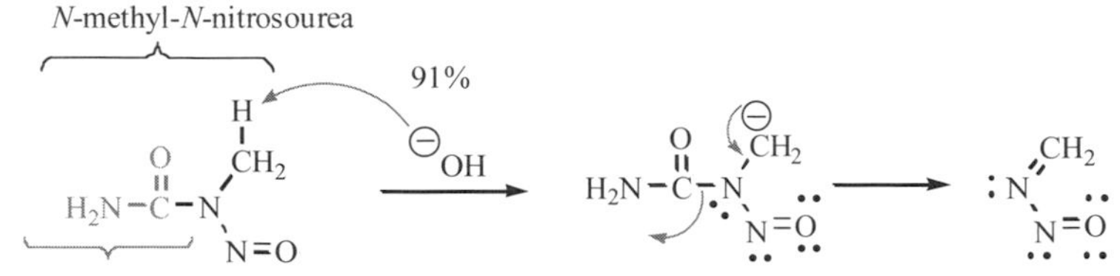

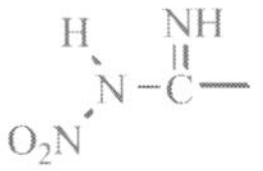

*N*-methyl-*N*-nitrosourea

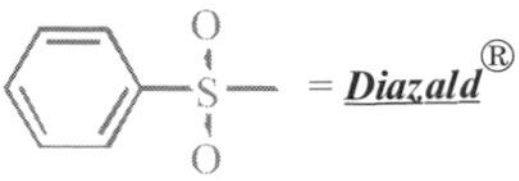

= ***Diazald*®**

1-Methyl-3-nitro-1-nitrosoguanidine (MNNG)

---

T. Sammakia, *Encyclopedia of Reagents for Organic Synthesis*, John Wiley and Sons, Inc., L. A. Paquette, Ed., New York, 1995, **2**, 1512

---

## Notes:

Diazomethane is prepared immediately prior to using it. It is a powerful methylating agent, particularly useful for mild preparation of methyl esters of acids.

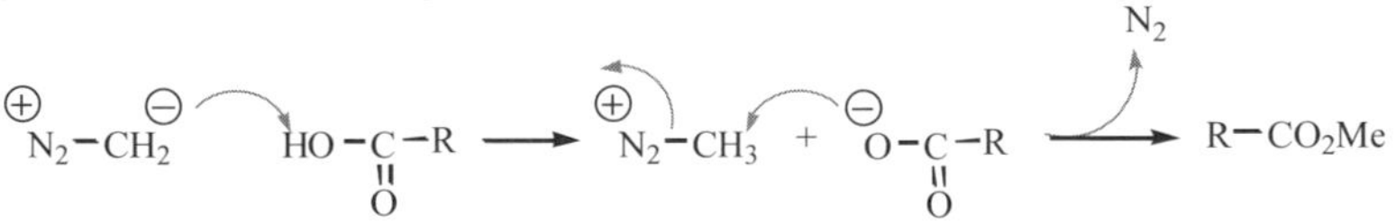

Diazomethane is also useful for cyclopropanation reactions and for reaction with acid chlorides to produce diazo ketones. [*See **Wolff Rearrangement***] It can be used to generate methyl ethers and *N*-methylations. Reaction with ketones can provide ring expansions.[1]

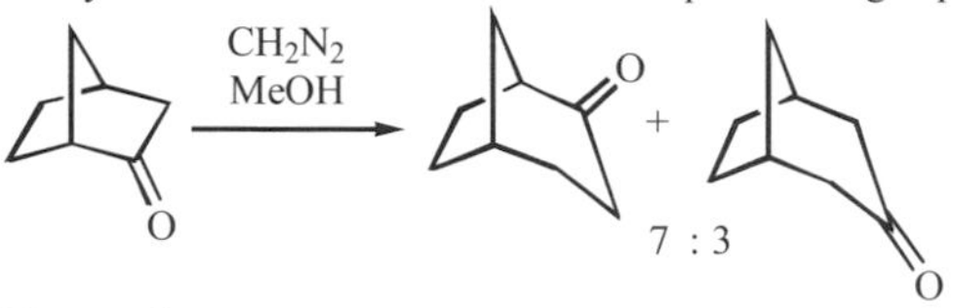

7 : 3

## Examples:

1.[2]

100%

---

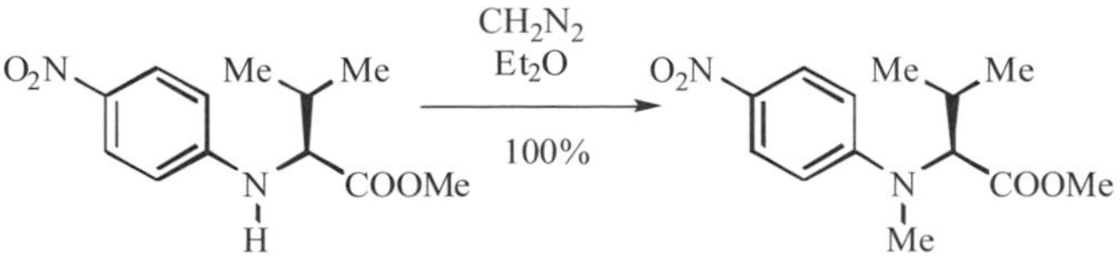

[1] Y. Auberson, R. Mampuya Bimwala, P. Vogel, *Tetrahedron Letters* **1991**, _32_, 1637
[2] M. L. Di Gioia, A. Leggio, A. Le Pera, A. Liguori, A. Napoli, C. Siciliano, G. Sindona, *Journal of Organic Chemistry* **2003**, _68_, 7416

# DIBAL (DIBAL-H, DIBAH)

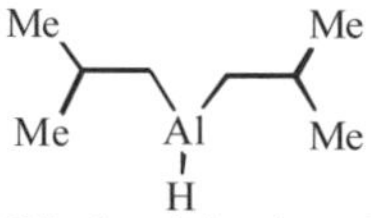

Diisubutylaluminum hydride
[1191-15-7]

## Commercially available

P. Galatsis, *Encyclopedia of Reagents for Organic Synthesis*, John Wiley and Sons, Inc., L. A. Paquette, Ed., New York, 1995, **3**, 1908

**Notes:** A reducing agent. Alcohols are generated from aldehydes, ketones, esters and acid chlorides. Nitriles can be converted to aldehydes. Tosylates will be replaced by –H; halides are inert. Amides are reduced to amines. Reduction of lactones can provide a useful synthetic strategy:

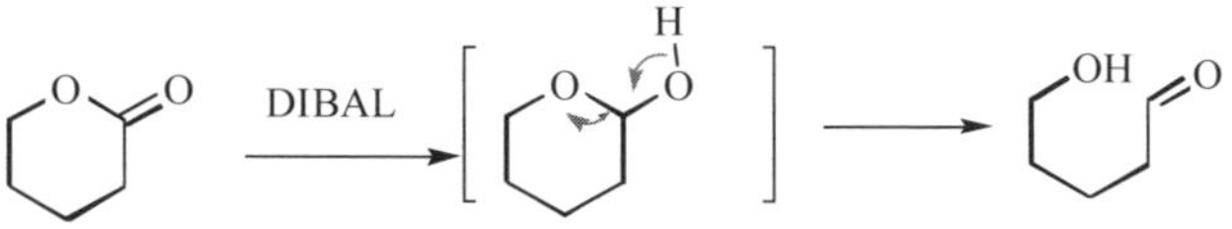

## Examples:

1.[1]

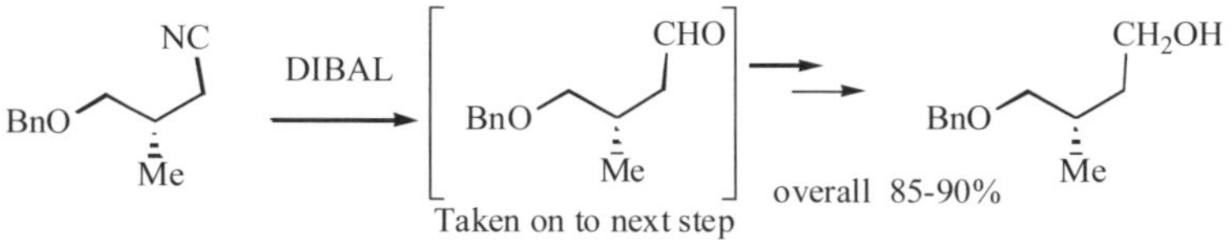

Taken on to next step     overall 85-90%

2.[2]

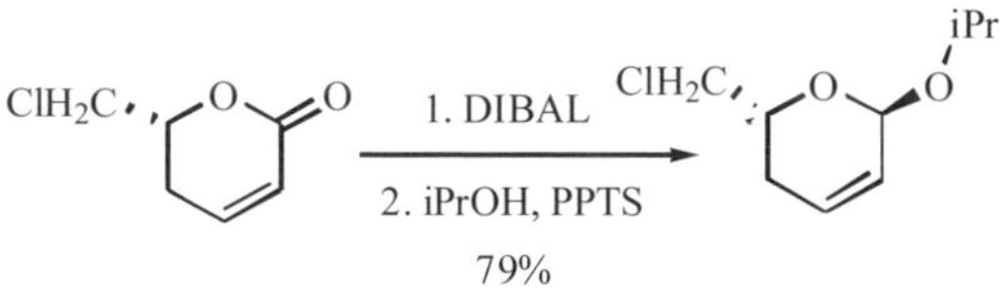

1. DIBAL

2. iPrOH, PPTS

79%

3.[3]

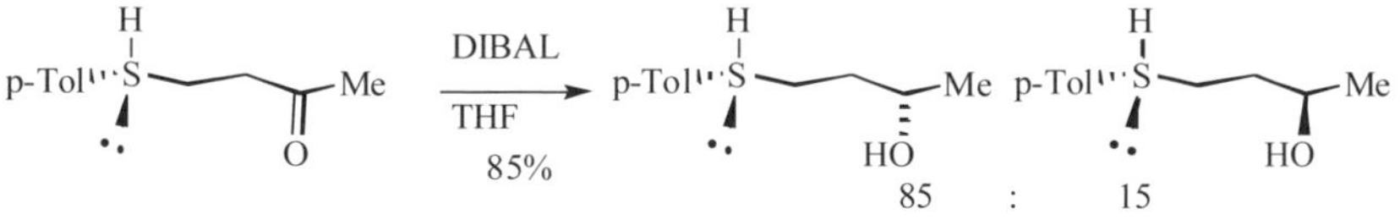

DIBAL

THF

85%

85   :   15

4.[4]

DIBAH

benzene

80%

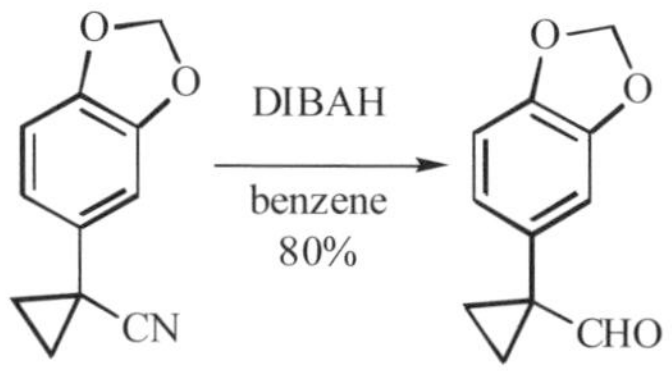

[1] A. K. Ghosh, C. Liu, *Organic Letters* **2001**, <u>3</u>, 635
[2] J. L. Vicario, A. Job, M. Wolberg, M. Muller, D. Enders, *Organic Letters* **2002**, <u>4</u>, 1023
[3] G. Solladié, F. Colobert, F. Somny, *Tetrahedron Letters* **1999**, <u>40</u>, 1227
[4] R. V. Stevens, L. E. Dupree, P. L. Lowenstein, *Journal of Organic Chemistry* **1972**, <u>37</u>, 977

# Diimide

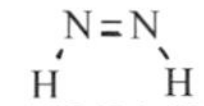

N=N
H     H
[15626-42-3]

## Preparation:

H   H          Oxidation
 \ /
  N–N          ──────────────→          N=N
 / \                                    H   H
H   H

---

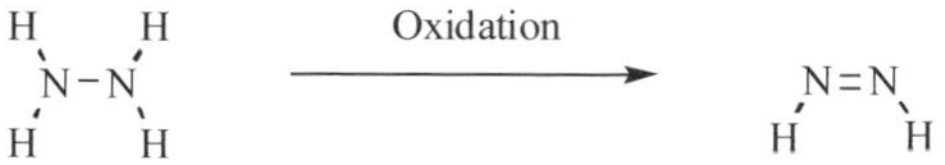

D. J. Pasto, *Encyclopedia of Reagents for Organic Synthesis*, John Wiley and Sons, Inc., L. A. Paquette, Ed., New York, 1995, **3**, 1892

---

## Notes:

The reagent can be prepared in situ, by the copper-catalyzed oxidation of hydrazine. Hydrogen peroxide or oxygen are often the oxidants. The acid-catalyzed decomposition of potassium azodicaroylate provides a useful source of diimide:

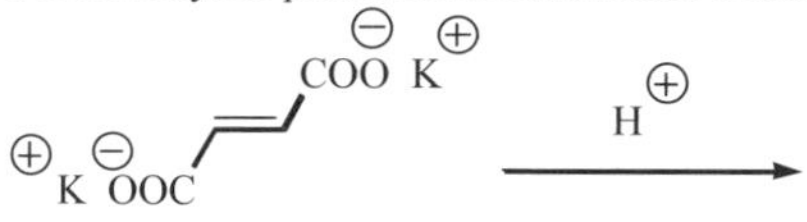

as is the decomposition of sulfonylhydrazines:

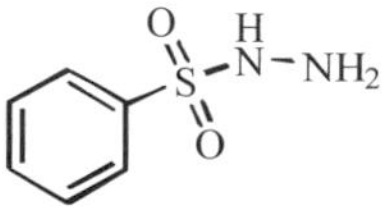

This reduction finds unique application because:

1. The reagent is extremely tolerant of other functional groups.
2. Reduction is *cis*, from the less-hindered face.

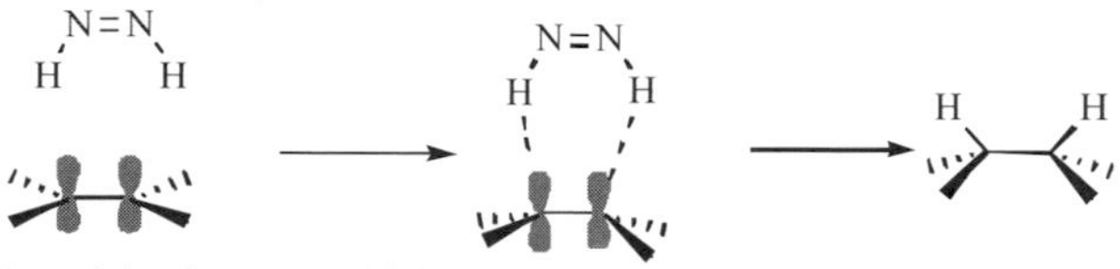

3. Reactivity decreases with increasing substitution about the alkene bond
4. Alkynes reduce faster than alkenes
5. Alkenes with electron-withdrawing groups react faster than those with electron-donating groups.
6. Use of deuterium or tritium-labeled diimde affords a method of *cis* reduction without scrambling of label.

## Examples:

1.[1]

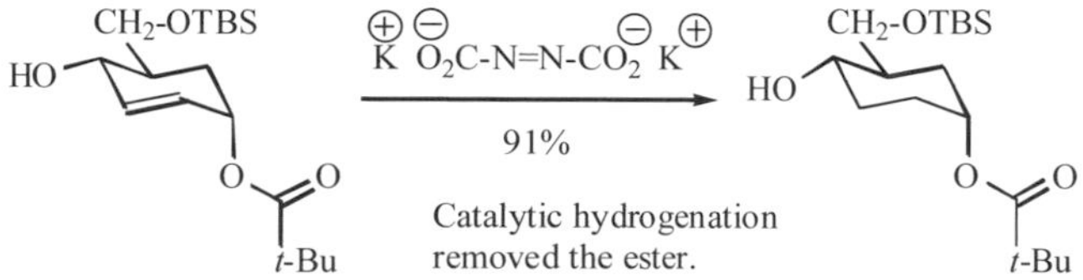

91%

Catalytic hydrogenation removed the ester.

2.[2]

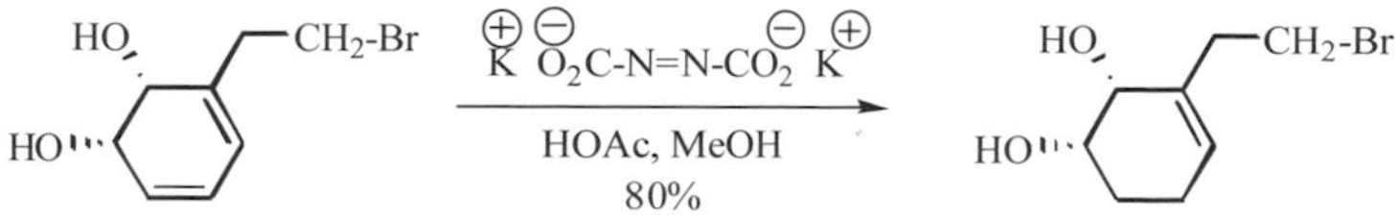

HOAc, MeOH
80%

---

[1] M. H. Haukaas, G. A. O'Doherty, *Organic Letters* **2002**, 4, 1771
[2] D. A. Frey, C. Duan, T. Hudlicky, *Organic Letters* **1999**, 1, 2085

# Diisopinocampheylborane
## (Ipc₂BH)

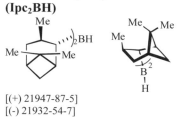

[(+) 21947-87-5]
[(-) 21932-54-7]

## Preparation:

$BH_3$- DMS

Either (+) or (-)

R. K. Dhar, *Encyclopedia of Reagents for Organic Synthesis*, John Wiley and Sons, Inc., L. A. Paquette, Ed., New York, 1995, **3**, 1914

## Notes:
A useful reagent for asymmetric hydroboration. See ***Brown's Hydroboration Reaction***.

## Examples:
1.[1]

1. (+) Ipc₂BH, THF

2. HO⁻, $H_2O_2$

73%

92% ee

2.[2]

1. Ipc₂BH

2. [O]

83%

3.[3]

40%

[1] A. E. Greene, M.-J. Luche, A. A. Serra, *Journal of Organic Chemistry* **1985**, 50, 3957
[2] H. M. Hess, H. C. Brown, *Journal of Organic Chemistry* **1967**, 32, 4138
[3] E. M. Flamme, W. R. Roush, *Journal of the American Chemical Society* **2002**, 124, 13644

# Dimsylate

$$\text{Me}-\overset{\overset{\displaystyle O}{\|}}{\underset{\displaystyle \cdot\cdot}{S}}-CH_2\text{-}\xi$$

Methylsulfinylmethide

**Preparation:** The "dimsyl anion" is generated from dimethylsulfoxide (DMSO) by use of base. The resulting lithio- or sodio- derivative is generally used in the DMSO solvent.

> M. Harmata, *Encyclopedia of Reagents for Organic Synthesis*, John Wiley and Sons, Inc., L. A. Paquette, Ed., New York, 1995, **7**, 4596.

**Notes:** This is generally found as a base or nucleophile.

### Generally used as M Dimsylate:

M= Na [15590-23-5]; K [15590-26-8]; Li [57741-62-5]

Harmata suggests that the use of K*t*-BuO/DMSO as a basic mixture most likely involves *t*-BuO as the actual base. This is seen in Example 2, where addition of Crown ether doesn't influence the already high degree of dissociation for the active base, *t*-BuO⁻.

## Examples:

1.[1]

2.[2]

| | | | |
|---|---|---|---|
| | 20 | : | 80 |
| | 48 | : | 52 |

---

[1] W.-C. Cheng, C.-C. Lin, M. J. Kurth, *Tetrahedron Letters* **2002**, <u>43</u>, 2967

[2] R. D. Bach, J. W. Knight, *Tetrahedron Letters* **1979**, <u>20</u>, 3815

# DMAP

4-(Dimethylamino)pyridine
[1122-58-3]
[82942-26-5] Polymer-bound

## Commercially available

A. Hassner, *Encyclopedia of Reagents for Organic Synthesis*, John Wiley and Sons, Inc., L. A. Paquette, Ed., New York, 1995, **3**, 2022

**Notes:** A basic catalyst for amine and alcohol acylations, (particularly useful for hindered alcohols), macrocyclization, and for ***Steglich esterification***. ***DMAP/PCC*** provides a useful oxidation protocol.[1]

## Examples:

1.[2]

2.[3]

3.[4]

---

[1] F. S. Guziec, Jr., *Encyclopedia of Reagents for Organic Synthesis,* John Wiley and Sons, Inc., L. A. Paquette, Ed., New York, 1995, **3**, 2024

[2] D. Bonafoux, I. Ojima, *Organic Letters* **2001**, <u>3</u>, 2333

[3] P. G. M. Wuts, S. W. Ashford, A. M. Anderson, J. R. Atkins, *Organic Letters* **2003**, <u>5</u>, 1483

[4] B. Liang, D. J. Richard, P. S. Portonovo, M. M. Joullié, *Journal of the American Chemical Society* **2001**, <u>123</u>, 4469

# DMPU

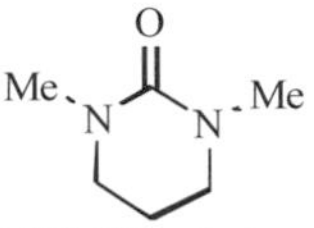

*N*,*N*'-Dimethylpropyleneurea
[7226-23-5]

## Commercially available

A. K. Beck, D. Seebach, *Encyclopedia of Reagents for Organic Synthesis*, John Wiley and Sons, Inc., L. A. Paquette, Ed., New York, 1995, **3**, 2123

## Notes:

A co-solvent with properties and reaction enhancements similar to **_HMPA_**. It is a dipolar aprotic solvent, miscible in water and most organic solvents. Can be cooled to dry ice temperature.

## Examples:

1.[1]

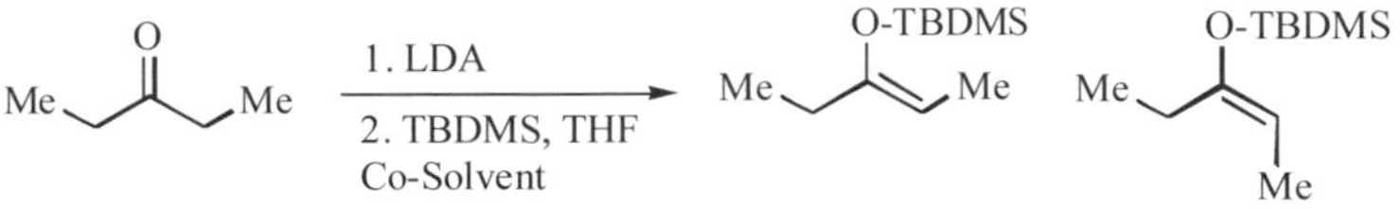

| Co-Solvent | Ratio | | |
|---|---|---|---|
| None | 30 | : | 70 |
| **_HMPA_** | 95 | : | 5 |
| **_DMPU_** | 94 | : | 6 |

2.[2]

3.[3]

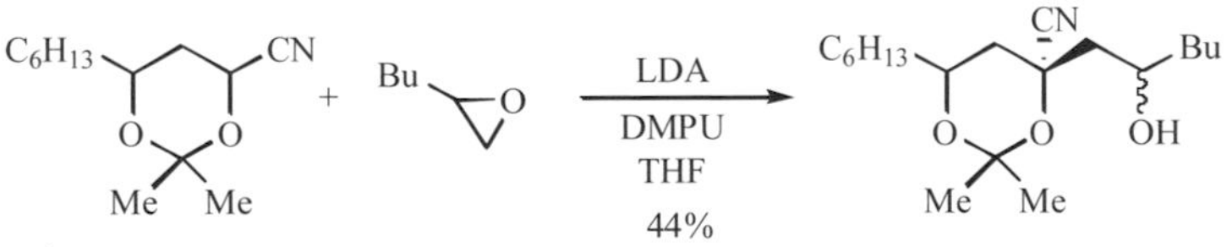

4.[4]

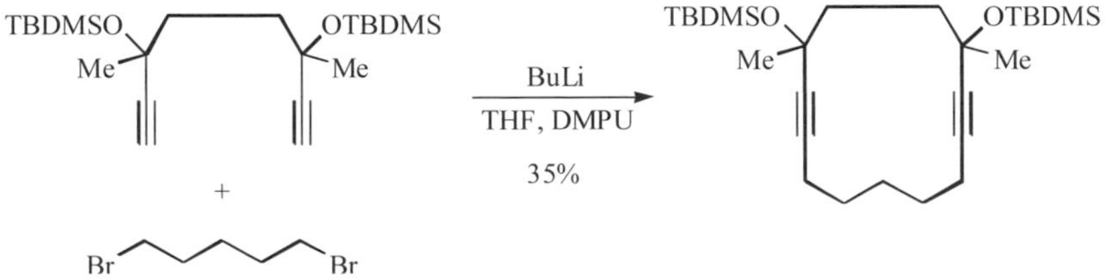

---

[1] T. Mukhopadhyay, D. Seebach, *Helvetica Chemica Acta*, **1982**, _65_, 385; Reported by A. K. Beck, D. Seebach, *Encyclopedia of Reagents for Organic Synthesis, John Wiley and Sons, Inc.*, L. A. Paquette, Ed., New York, 1995, _3_, 2123

[2] S. Poulain, N. Nairet, H. Patin, *Tetrahedron Letters* **1996**, _37_, 7703

[3] S. D. Rychnovsky, S.S. Swenson, *Journal of Organic Chemistry* **1997**, _62_, 1333

[4] C. Boss, R. Keese, *Tetrahedron* **1997**, _53_, 3111

# DMSO

$$O^-\overset{\overset{\bullet}{\underset{Me}{S}}}{S}{}^{'Me} \qquad \underset{:\overset{..}{O}:}{\overset{Me\;\cdot\cdot\;Me}{\underset{\uparrow}{S}}} \quad \text{and by formal charge} \qquad \underset{O}{\overset{Me\diagdown\;\diagup Me}{\underset{\parallel}{S}}}$$

Dimethylsulfoxide
[67-68-5]
## Commercially available

A. P. Krapcho, *Encyclopedia of Reagents for Organic Synthesis*, John Wiley and Sons, Inc., L. A. Paquette, Ed., New York, 1995, **3**, 2141

**Notes:** A polar, aprotic solvent; miscible in water and many organic solvents. Efficiently solvates cations. Can greatly enhance the rates of nucleophilic displacement reactions.

Will reduce the order of reactivity for the halide ions

| **Protic Solvents** | **DMSO** |
|---|---|
| I > Br > Cl > F | F > Cl > Br > I |

A number of alcohol oxidations are based on the general scheme: **See DMSO based Oxidations under _Aldehyde Syntheses_**

See: **Dimsyl group**

$$Me-\overset{\overset{O}{\parallel}}{\underset{..}{S}}-CH_2-\}$$

*Methylsulfinylmethide*

**Preparation** The "dimsyl anion" is generated from dimethylsulfoxide (DMSO) by use of base. The resulting lithio- or sodio- derivative is generally used in the DMSO solvent.

$$H_3C-\overset{\overset{O}{\parallel}}{S}-CH_3 \quad \xrightarrow{\text{Base}} \quad H_3C-\overset{\overset{O}{\parallel}}{S}-CH_2{}^{\ominus}\;M^{\oplus}$$

Sodium dimsylate (M. Harmata, *Encyclopedia of Reagents for Organic Synthesis*, *John Wiley and Sons, Inc.*, L. A. Paquette, Ed., New York, 1995, **7**, 4596.) is often used to initiate ***Wittig reactions***.[1]

See use in the ***Krapcho Dealkoxycarbonylation*** reaction..

---

[1] J. A. Deyrup, M. F. Betkouski, *Journal of Organic Chemistry* **1975**, <u>40</u>, 284

# Dppe (DIPHOS)

Ph—P        P—Ph
    |        |
    Ph      Ph
1,2-Bis(diphenylphosphino)ethane
[1663-45-2]

## Commercially available

G. T. Whiteker, *Encyclopedia of Reagents for Organic Synthesis*, John Wiley and Sons, Inc., L. A. Paquette, Ed., New York, 1995, **1**, 515

**Notes:** Ligand. Used for Pd-catalyzed nucleophilic reactions at allylic positions. See also, ***Dppp***. With Br$_2$ forms a useful brominating agent, *1,2-bis(diphenylphosphino)ethane tetrabromide* [7726-95-6].[1] This reagent is useful for the conversion:

R-OTHP $\longrightarrow$ R-Br

Ph$_2$P    PPh$_2$ + Br$_2$ $\longrightarrow$ BrPh$_2$P$^{\oplus}$    $^{\oplus}$PPh$_2$Br
                                         Br$^{\ominus}$        Br$^{\ominus}$

## Examples:

1.[2]

2.[3]

dppe-2Br$_2$
CH$_2$Cl$_2$
63%

3.[4]

Ni(dppe)Br$_2$ / Zn
THF

84%        79%

---

[1] Reported by L. A. Paquette, *Encyclopedia of Reagents for Organic Synthesis,* John Wiley and Sons, Inc., L. A. Paquette, Ed., New York, 1995, **1**, 517

[2] J.-C. Fiaud, J.-L. Malleron, *Chemical Communications* **1981**, 1159

[3] C. M. Rayner, P. C. Astles, L. A. Paquette, *Journal of the American Chemical Society* **1992**, <u>114</u>, 3925

[4] Y.-C. Huang, K. K. Majumdar, C.-H. Cheng, *Journal of Organic Chemistry* **2002**, <u>67</u>, 1682

# Dppp

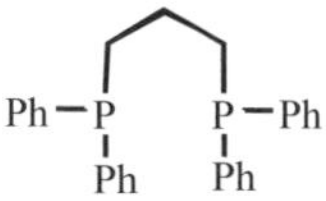

1,2-Bis(diphenylphosphino)propane
[6737-42-4]

## Commercially available

G. T. Whiteker, *Encyclopedia of Reagents for Organic Synthesis*, John Wiley and Sons, Inc., L. A. Paquette, Ed., New York, 1995, **1**, 521

## Notes:

Ligand. Used for Pd-catalyzed reactions of aryl and vinyl halides and triflates. Also for *Kumuda coupling* reactions. See *Dppe* for a similar reagent.

## Examples:

1.[1]

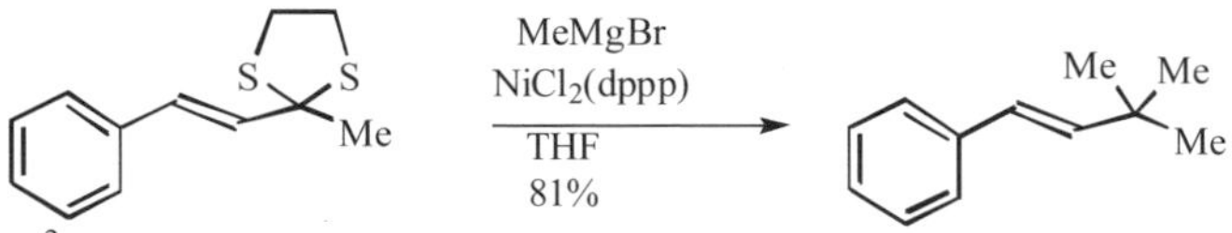

2.[2]

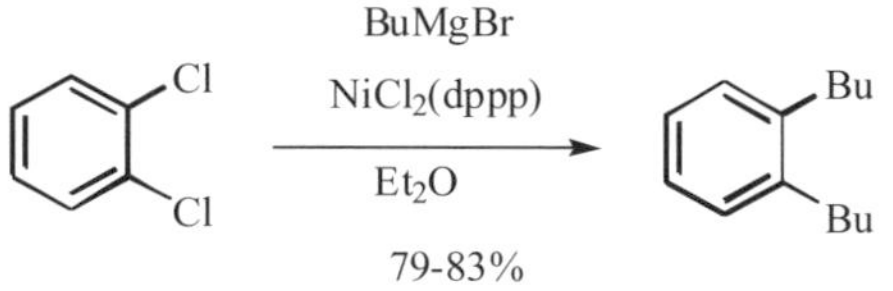

3.[3]

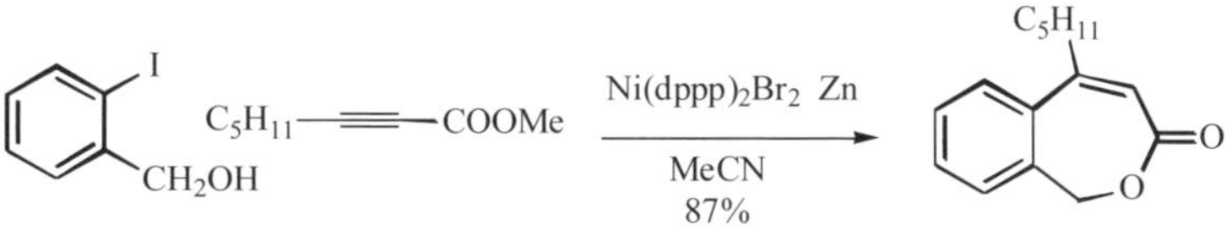

4.[4]

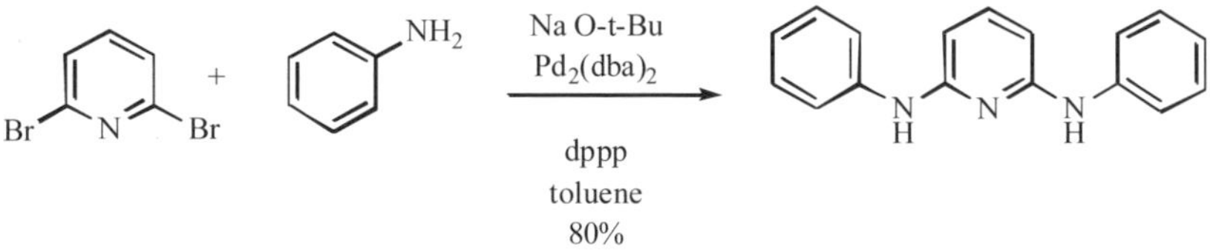

[1] T.-M. Yuan, T.-Y. Luh, *Organic Syntheses* CV 9, 649
[2] M. Kumada, K. Tamao, K. Sumitani, *Organic Syntheses* CV 6, 407
[3] D. K. Rayabarapu, C.-H. Cheng, *Journal of the American Chemical Society* **2002**, 124, 5630
[4] S. Wagaw, S. L. Buchwald, *Journal of Organic Chemistry* **1996**, 61, 7240

# Eaton's Reagent
### P$_2$O$_5$ / MeSO$_3$H
[39394-84-8]

**Preparation:** Add MeSO$_3$H to P$_2$O$_5$, (10:1, m/m) stir until P$_2$O$_5$ is dissolved. **Commercially available**

L. A. Dixon, *Encyclopedia of Reagents for Organic Synthesis*, John Wiley and Sons, Inc., L. A. Paquette, Ed., New York, 1995, **6**, 4129

**Notes:** A useful alternative to PPA (Polyphosphoric acid)

## Examples:
1.[1]

P$_2$O$_5$ , MeSO$_3$H

94%

2.[2]

Eaton's reagent

92%

3.[3]

P$_2$O$_5$ , MeSO$_3$H

no yield given

91%

9%

4.[4]

P$_2$O$_5$  MeSO$_3$H

92%

[1] P. W. Jeffs, G. Molina, N. A. Cortese, P. R. Hauck, J. Wolfram, *Journal of Organic Chemistry* **1982**, 47, 3876
[2] X.-J. Hao, M. Node, K. Fuji, *Journal of the Chemical Society, Perkin Transaction 1* **1992**, 1505
[3] F. E. Ziegler, J.-M. Fang, C. C. Tam, *Journal of the American Chemical Society* **1982**, 104, 7174
[4] P. E. Eaton, G. R. Carlson, J. T. Lee, *Journal of Organic Chemistry* **1973**, 38, 4071

# Ender's Reagent (SAMP, RAMP)

| SAMP:<br>(S)-1-Amino-2-methoxymethylpyrrolidine | RAMP:<br>(R)-1-Amino-2-methoxymethylpyrrolidine |
|---|---|
| | |

[59983-39-0]

## Commercially available

D. Enders, M. Klatt, *Encyclopedia of Reagents for Organic Synthesis*, John Wiley and Sons, Inc.,
L. A. Paquette, Ed., New York, 1995, **1**, 178

**Notes:** Chiral directing group for alkylation of carbonyl

## Examples:

1.[1]

2.[2]

---

[1] A. B. Smith, III, H. Ishiyama, Y. S. Cho, K. Ohmoto, *Organic Letters* **2001**, <u>3</u>, 3967
[2] A. Toro, P. Nowak, P. Deslongchamps, *Journal of the American Chemical Society* **2000** <u>122</u>, 4526

# Eschenmoser's Salt

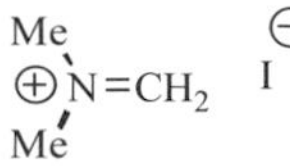

Me
$\overset{\oplus}{N}$=CH$_2$   I$^\ominus$
Me

Dimethyl(methylene)ammonium iodide
[33627-00-6]
The chloride salt is known as **Böhme's salt**

## Commercially available

E. F. Kleinman, *Encyclopedia of Reagents for Organic Synthesis*, John Wiley and Sons, Inc., L. A. Paquette, Ed., New York, 1995, **3**, 2090

**Notes:** Useful reagents for **Mannich reactions** with active methylene compounds. Particularly useful for the preparation of exo methylene groups.

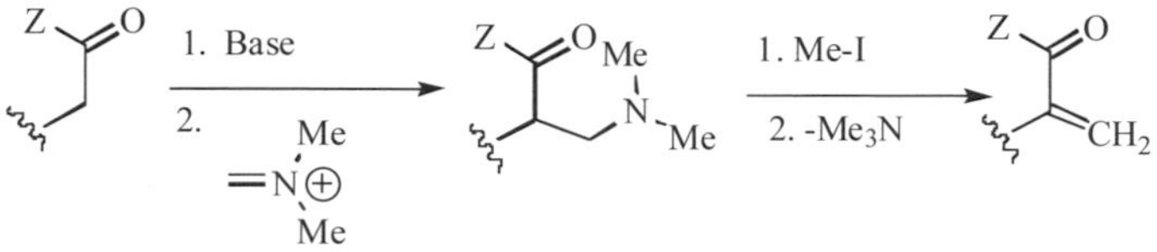

Also for the conversion:

## Examples:

1.[1]

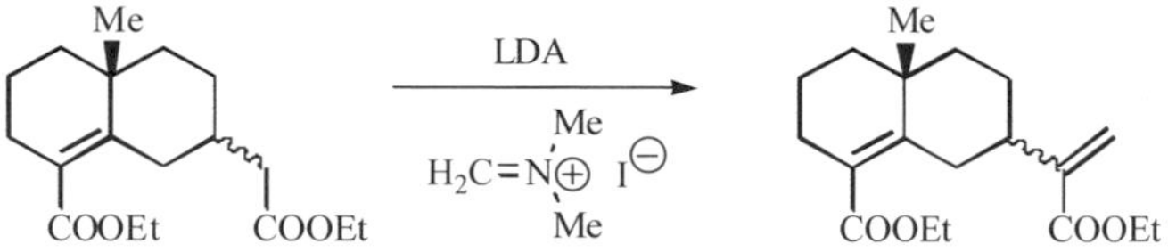

2.[2]

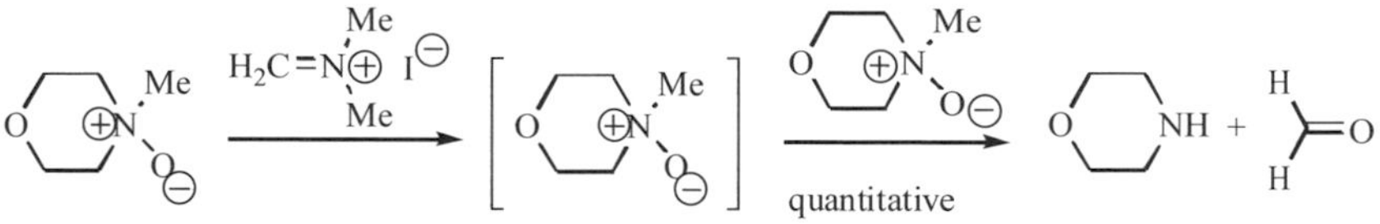

quantitative

3.[3]

62%

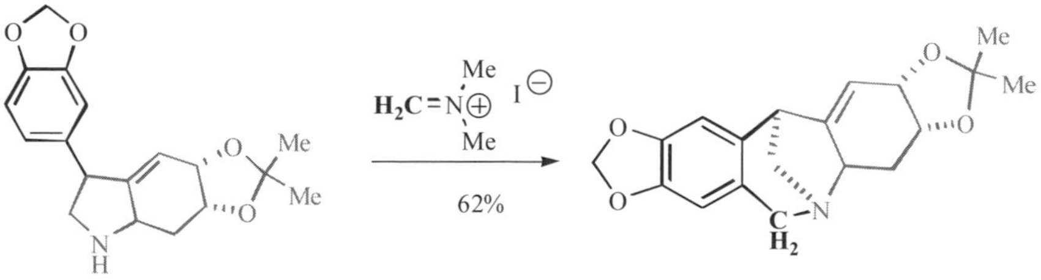

[1] F. Bohlmann, H. Suding, *Liebigs Annalen der Chemie* **1985**, _1_, 160 (AN 1985:166963)
[2] T. Rosenau, A. Potthast, P. Kosma, C.-L. Chen, J. S. Gratzl, *Journal of Organic Chemistry* **1999**, _64_, 2166
[3] C.-K. Sha, A-W. Hong, C.-M. Huang, *Organic Letters* **2001**, _3_, 2177

# Fetizon's Reagent

Ag$_2$CO$_3$ / Celite

[534-16-7] / [61790-53-2]

## Commercially available

M. Fetizon, *Encyclopedia of Reagents for Organic Synthesis*, John Wiley and Sons, Inc., L. A. Paquette, Ed., New York, 1995, **6**, 4448

## Notes:

Celite™ is a commercially available inorganic material; mostly SiO$_2$. Generally finds use as a filtering aid.

This is a mild oxidizing agent for conversion of primary alcohols to aldehydes and secondary alcohols to ketones.

## Examples:

1.[1] One of the earliest reports of using silver as an oxidizing agent:

2.[2]

3.[3]

4.[4]

near quantitative yield

---

[1] H. Rapoport, H. N. Reist, *Journal of the American Chemical Society* **1955**, 77, 490

[2] R. L. Funk, G. L. Bolton, J. U. Daggett, M. M. Hansen, L. H. M. Horcher, *Tetrahedron* **1985**, 41, 3479

[3] Y. R. Lee, J. Y. Suk, B. S. Kim, *Organic Letters* **2000**, 2, 1387

[4] D. P. Walker, P. A. Grieco, *Journal of the American Chemical Society* **1999**, 121, 9891

# Fmoc-Cl

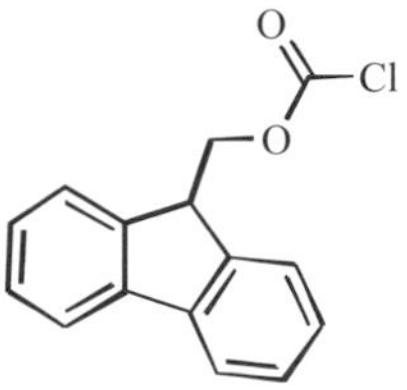

9-Fluorenylmethyl chloroformate
[28920-43-6]

## Commercially Available.

R. L. Polt, *Encyclopedia of Reagents for Organic Synthesis*, John Wiley and Sons, Inc., L. A. Paquette, Ed., New York, 1995, **4**, 2545

## Notes:

A useful protecting group for nitrogen.

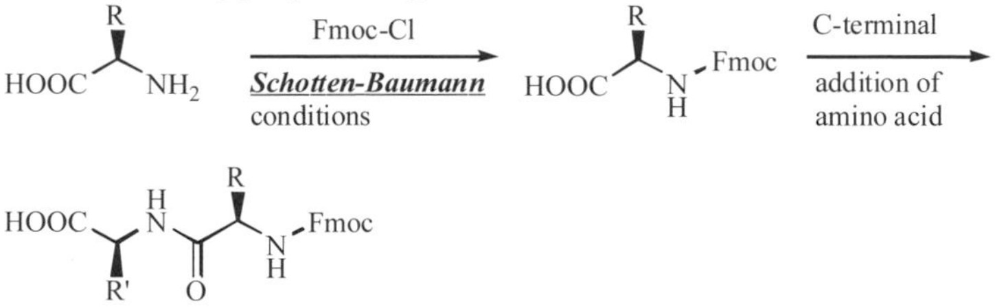

## Examples:

1.[1]

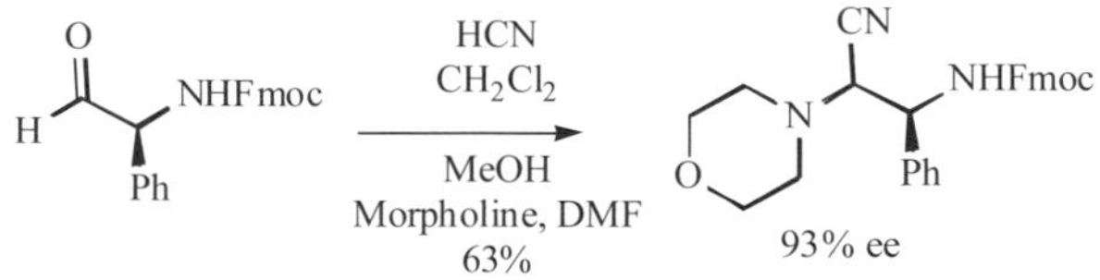

2.[2]

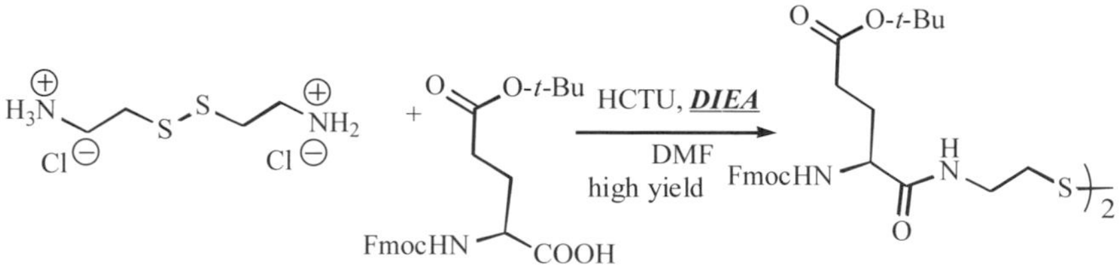

HCTU = 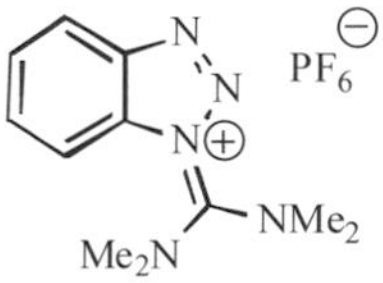 [490019-20-0] a strong condensing agent

[1] A. G. Myers, B. Zhong, D. W. Kung, M. Movassaghi, B. A. Lanman, S. Kwon, *Organic Letters* **2000**, 2, 3337
[2] A. Ortiz-Acevedo, G. R. Dieckmann, *Tetrahedron Letters* **2004**, 45, 6795

# Furukawa's Cyclopropanation reagent[1]

### I-CH₂ZnEt

## Preparation:
   Et₂Zn + I-CH₂-I

---

P. Knochel, *Encyclopedia of Reagents for Organic Synthesis*, John Wiley and Sons, Inc., L. A.
Paquette, Ed., New York, 1995, **3**, 1861

---

## Notes:
Cyclopropanation reagent. A similar reagent, prepared from EtZnI + CH₂I₂, [I-CH2-Zn-Et]
[33598-72-0] is known as the ***Sawada Reagent***. [P. Knochel, *Encyclopedia of Reagents for Organic
Synthesis*, John Wiley and Sons, Inc., L. A. Paquette, Ed., New York, 1995, **4** 2473]
Advantage over ***Simmons-Smith*** is the homogeneity of the reaction; and the high yields with enol
ethers. (Not true for ***Sawada Reagent***.)

## Examples:

1.[2]

2.[3]

3.[4]

---

[1] J. Furukawa, N. Kawabata, J. Nishimura, *Tetrahedron Letters* **1968**, <u>9</u>, 3495
[2] A. J. Blake, A. J. Highton, T. N. Majid, N. S. Simpkins, *Organic Letters* **1999**, <u>1</u>, 1787
[3] D. A. Evans, J. D. Burch, *Organic Letters* **2001**, <u>3</u>, 503
[4] R. Hilgenkamp, C. K. Zercher, *Organic Letters* **2001**, <u>3</u>, 3037

# Furukawa's reagent

DMAP

+ CH$_3$-SO$_2$Cl

Mesyl chloride
Methanesulfonyl chloride

[124-63-0, 1122-58-3]

**Preparation:** Stir the alcohol with MsCl, ***DMAP***, H$_2$O, and CH$_2$Cl$_2$ at rt.

V. Vaillancourt, M. M. Cudahy, *Encyclopedia of Reagents for Organic Synthesis*, John Wiley and Sons, Inc., L. A. Paquette, Ed., New York, 1995, **5**, 3311

## Notes:
Dehydrating agent

## Examples:

1.[1]

2.[2]

---

[1] D. L. Comins, C. A. Brooks, R. S. Al-war, R. R. Goehring, *Organic Letters* **1999**, <u>1</u>, 229
[2] A. Nakazato, T. Kumagai, T. Okubo, H. Tanaka, S. Chaki, S. Okuyama, K. Tomisawa, *Bioorganic & Medicinal Chemistry* **2000**, <u>8</u> 1183

# Garner's Aldehyde

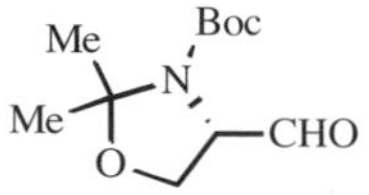

3-(*tert*-butoxycarbonyl)-2,2-dimethyl-4-formyloxazolidin
        [102308-32-7]

## Preparation

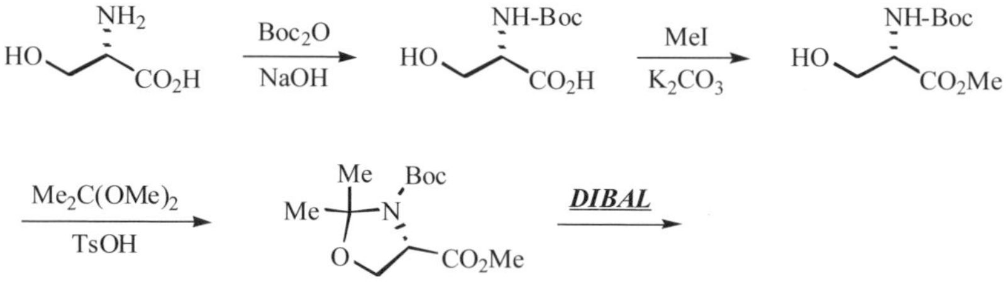

**Notes:** An extremely useful chiral reagent.
**Examples:**
1.[1]

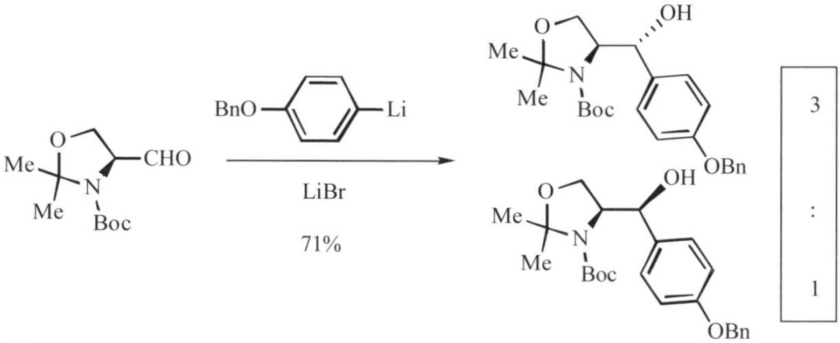

2.[2]

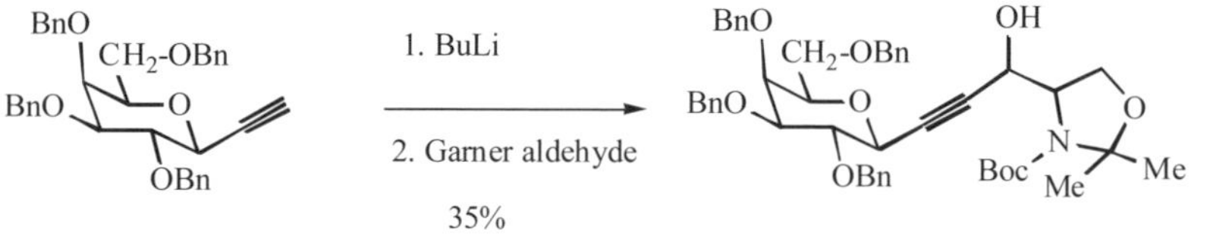

3.[3]

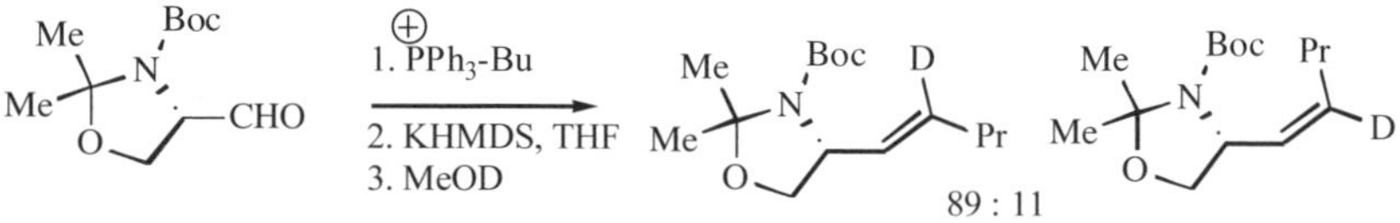

[1] N. Okamoto, O. Hara, K. Makino, Y. Hamada, *Journal of Organic Chemistry* **2002**, <u>67</u>, 9210
[2] A. Dondoni, G. Mariotti, A. Marra, *Journal of Organic Chemistry* **2002**, <u>67</u>, 4475
[3] J. S. Oh, B. H. Kim, Y. G. Kim, *Tetrahedron Letters* **2004**, <u>45</u>, 3925

# Grubb's Reagent

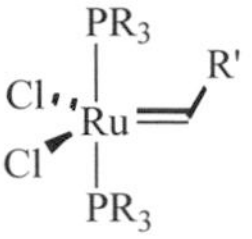

## Commercially Available

## Notes:

The labs of R.M. Grubbs has prepared and examined a number of catalysts along the general structure shown above. See **_RCM Reaction_** for additional examples. These catalysts continue to evolve, but bear the same general characteristics. The catalysts are generally tolerant to an array of ot\her functional groups. The catalysts are generally more stable than the **_Schrock catalysts_**. Examples include:

1

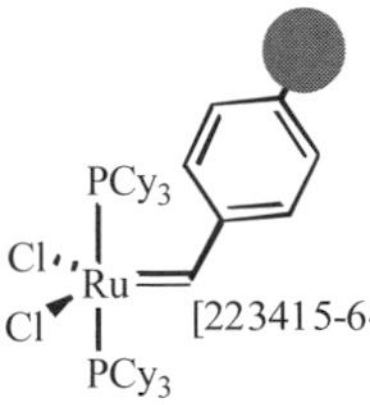

[223415-64-3]

Catalyst readily recovered for reuse.

2.

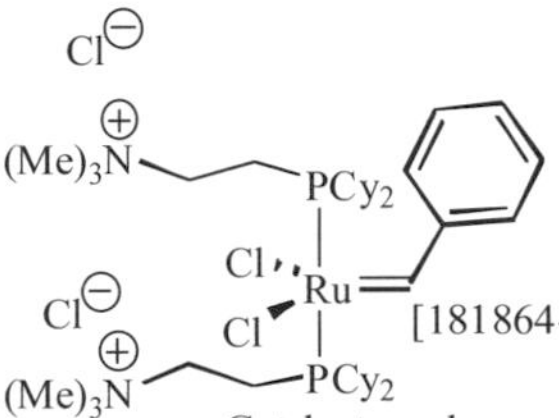

[181864-83-5]

Catalyst can be used in protic solvents, air-sensitive.

3.

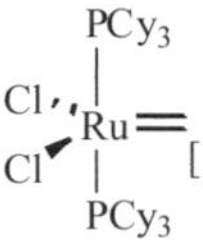

[171368-36-8]

4.

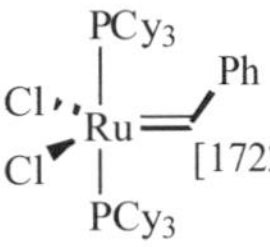

[172222-30-9]

air-sensitive.

5.

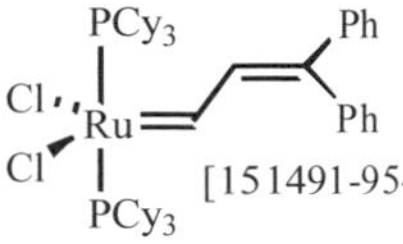

[151491-95-1]

6.

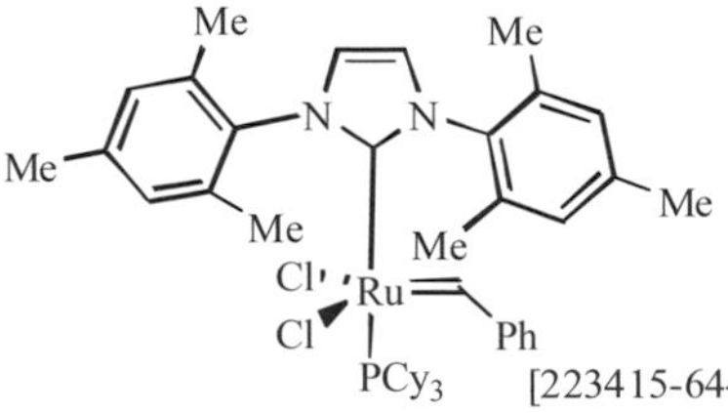

[223415-64-3]

**_Nolan's Catalyst_**

7.

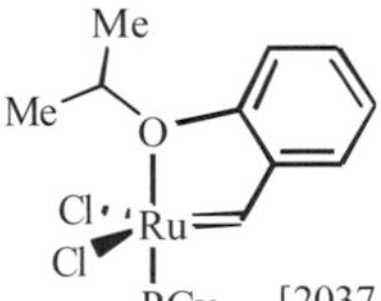

[203714-71-0]

**_Hoveyda-Grubbs Catalyst_**

# Grundmann's Ketone
## *Windaus-Grundmann Ketone*

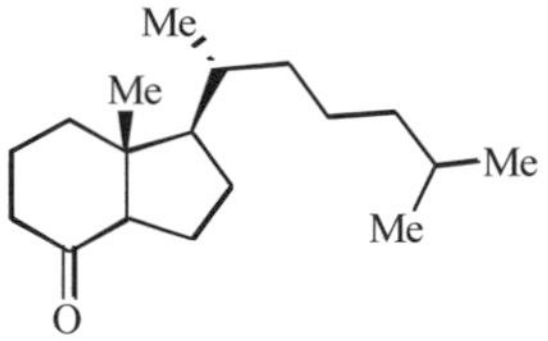

[66251-18-1 ]
(1*R*,7a*R*)-7a-methyl-1-((*R*)-6-methylheptan-2-yl)-octahydroinden-4-one

# Examples:

1.[1]

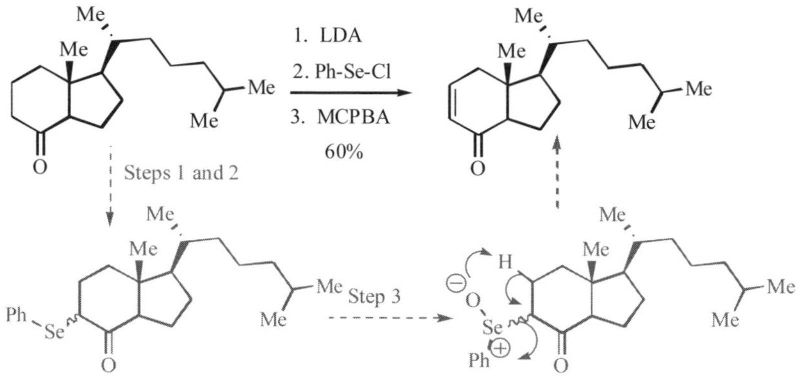

2.[2]

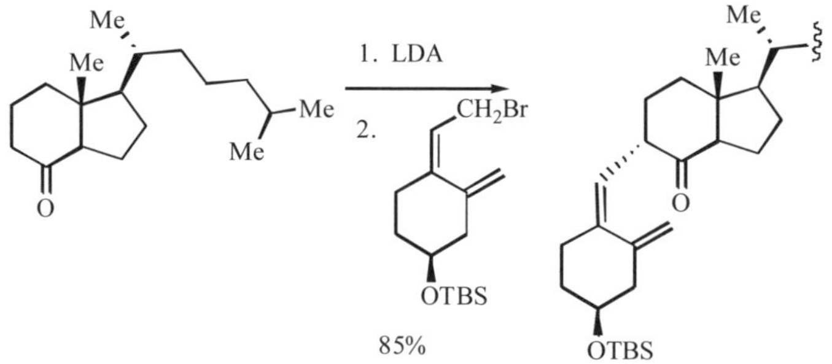

[1] W. H. Okamura, G.-D. Zhu, D. K. Hill, R. J. Thomas, K. Ringe, D. B. Borchard, A. W. Norman, L.
J. Mueller *Journal of Organic Chemistry* **2002**, <u>67</u>, 1637
[2] E. M. Codesido, L. Castedo, J. R. Granja, *Organic Letters* **2001**, <u>3</u>, 1483

# HMPA, HMPT

Me   Me
  N
  |
O=P—N   Me
  |       Me
  N
Me   Me

Hexamethylphosphoric Triamide
[680-31-9]

## Commercially available.

R. R. Dykstra, *Encyclopedia of Reagents for Organic Synthesis*, John Wiley and Sons, Inc., L. A. Paquette, Ed., New York, 1995, **4**, 2668

**Notes:** A powerful Lewis base able to form strong cation complexes. This attribute is often associated with greatly increased reactivities when used in solvent. This is particularly true when used with organolithium reagents. Soluble in most solvents.

## Examples:

1.[1]

BuLi

HMPA / THF
81%

2.[2]

Bu₃SnH / AIBN

HMPA

69%

3.[3]

+ C₁₂H₂₃-CH₂I

BuLi

HMPA, Et₂O

60%

[1] S. SanKaranarayanan, A. Sharma, S. Chattopadhyay, *Tetrahedron: Asymmetry* **1996**, *7*, 2639
[2] E. J. Enholm, P. E. Whitley, Y. Xie, *Journal of Organic Chemistry* **1996**, *61*, 5384
[3] J. R. Al Dulayymi, M. S. Baird, M. J. Simpson, G. R. Port, *Tetrahedron* **1996**, *52*, 12509

# HMTA

Hexamethylenetetramine
[100-97-0]

## Commercially Available

S. N. Kilényi, *Encyclopedia of Reagents for Organic Synthesis*, John Wiley and Sons, Inc., L. A. Paquette, Ed., New York, 1995, **4**, 2666

**Notes:**  HMTA finds use for:

Oxidation of benzylic halides (See: ***Sommelet Reaction***)

$$Ar\text{-}CH_2\text{-}X \;+\; \text{HMTA} \quad\longrightarrow\quad \xrightarrow[\Delta]{\text{dil } H^{\oplus}} \quad Ar\text{-}CHO$$

Ammonolysis of alkyl halides (See: ***Delépine Reaction***)

$$R\text{-}CH_2\text{-}X \;+\; \text{HMTA} \quad\longrightarrow\quad \xrightarrow[\Delta]{\text{dil } H^{\oplus}} \quad R\text{-}CH_2\text{-}NH_2$$

*hexamethylenetetramine*

Electrophilic formylation of aromatics (See: ***Duff Reaction***)

$$EDG\text{-}C_6H_5 \quad\xrightarrow[\text{Aq. HOAc}]{\text{(HMTA)}}\quad EDG\text{-}C_6H_4\text{-}CHO$$

Provides imidazoles from 1,2-diketones

$$\text{(1,2-diketone)} \quad\xrightarrow[\text{NH}_4\text{OAc, HOAc}]{}\quad \text{(imidazole)}$$

# Hünig's Base, DIPEA, DIEA

Diisopropylethylamine
[7087-68-5]

## Commercially available

K. L. Sorgi, *Encyclopedia of Reagents for Organic Synthesis*, John Wiley and Sons, Inc., L.A. Paquette, Ed., New York, 1955, **3**, 1933

**Notes:** Because of the severe steric constraints of this amine, it serves well as a proton scavenger and is quite resistant to quaternization.

## Examples:

1.[1]

Hunig's base, PMBM-Cl

$CH_2Cl_2$

84%

Hunig's base, PMBM-Cl

$CH_2Cl_2$

100%

PMBM-Cl = *p*-methoxybenzylchloromethyl ether

2.[2]

1. Hunig's base, $SiH_2I_2$, $Cl_2Cl_2$

2. $Ph\text{-}CH_2\text{-}NH_2$

89%

3.[3]

Pd/C ,

*t*-BuOH , Hunig's base

70%

[1] A. P. Kozikowski, J.-P. Wu, *Tetrahedron Letters* **1987**, 28 5125
[2] S. Gastaldi, S. M. Weinreb, D. Stien, *Journal of Organic Chemistry* **2000**, 65, 3239
[3] S. Boisnard, A.-C. Carbonnelle, J. Zhu, *Organic Letters* **2001**, 3, 2061

# Hydroboration Reagents[1]

See *Brown's Hydroboration Reaction*
## *Diborane*

Easily prepared in laboratory quantities by reaction of NaBH$_4$ and BF$_3$ in THF <u>under dry, inert atmosphere.</u>

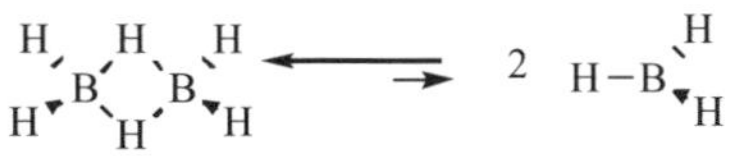

In the absence of stabilization boron, borane
tends to maintain the dimeric form.

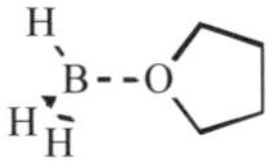

**BH$_3$-THF**

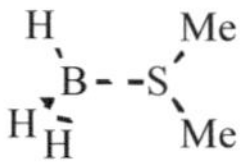

**BH$_3$-DMS**
See ***BMS***

## Alkyl Derivatives:

## Disiamylborane

Prepared by addition of 2-methyl-2-butene to diborane

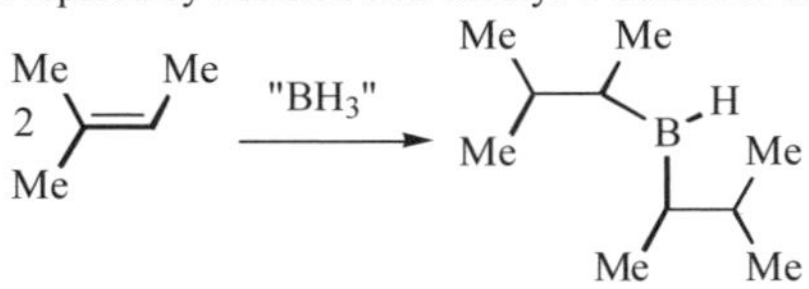

A hindered borane providing increased reaction selectivity:

6  94  Borane

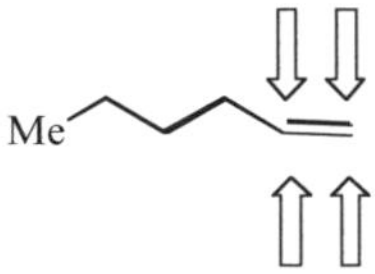

Ref 1, page 282

1  99  Disiamylborane

## Thexylborane

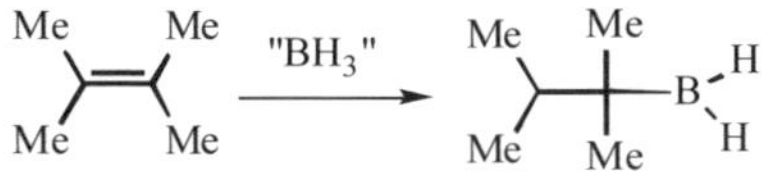

A hindered borane particularly useful for reaction with dienes:

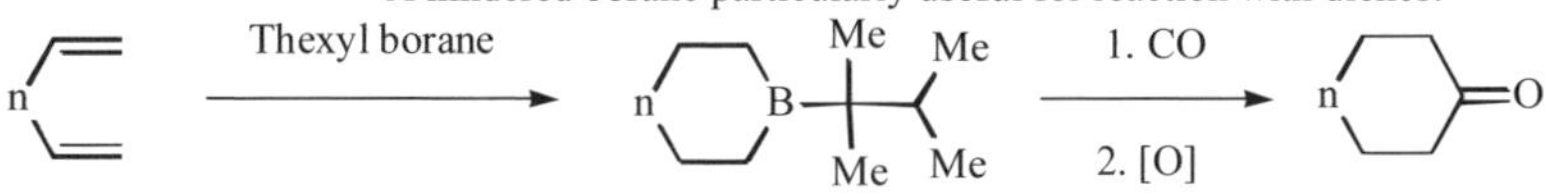

Thexyl borane                                    1. CO

                                                 2. [O]

---

[1] H. C. Brown, *Boranes in Organic Chemistry*, Cornell University Press, Ithaca, 1972. This is a dated but extremely interesting perspective written by the major driving force in the development of the extremely useful chemistry of boron.

### *9-BBN*

A useful borane derivative

### *Diisopinocampheylborane (Ipc₂BH)*

A chiral hydroborating agent

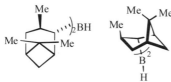

**See also:** *Catecholborane*

# Hydrogenation Catalysts

## Ni

### Ni₂B
### Nickel Boride
[12007-01-1]
T. J. Caggiano, *Encyclopedia of Reagents for Organic Synthesis*, John Wiley and Sons, Inc., L. A. Paquette, Ed., New York, 1995, **6**, 3694

### Ni Catalysts
[7440-02-0]
C. R. Sarko, M. DiMare, *Encyclopedia of Reagents for Organic Synthesis*, John Wiley and Sons, Inc., L. A. Paquette, Ed., New York, 1995, **6**, 3701

### *Raney Nickel*

## Pd
### *Pt/BaSO₄*
####     *Rosenmund Catalyst*

### *Pd/CaCO₃ / Pb-poisoned*
####     *Lindlar Catalyst*

### Pd/C
[7440-05-3]
A. O. King, I. Shinka, *Encyclopedia of Reagents for Organic Synthesis*, John Wiley and Sons, Inc., L. A. Paquette, Ed., New York, 1995, **6**, 3867

### Pd/Gr (graphite)
[59873-73-3]
E. M. Leahy, *Encyclopedia of Reagents for Organic Synthesis*, John Wiley and Sons, Inc., L. A. Paquette, Ed., New York, 1995, **6**, 3887

### *Pd(OH)₂*
####     *Pearlman's Catalyst*

## Pt

### Pt/Al₂O₃
[7440-06-4]
A. O. King, I. Shinka, *Encyclopedia of Reagents for Organic Synthesis*, John Wiley and Sons, Inc., L. A. Paquette, Ed., New York, 1995, **6**, 4159

### Pt/C
#### Heyn's Catalyst [7440-06-4]
A. O. King, I. Shinka, *Encyclopedia of Reagents for Organic Synthesis*, John Wiley and Sons, Inc., L. A. Paquette, Ed., New York, 1995, **6**, 4160

***<u>$PtO_2$</u>***

***<u>Adam's Catalyst</u>***

**Rh**

**$Rh/Al_2O_3$**

[7440-16-6]

S. Siegel, *Encyclopedia of Reagents for Organic Synthesis*, John Wiley and Sons, Inc., L. A. Paquette, Ed., New York, 1995, **6**, 4405

**$Rh(PPh_3)_3Cl$**

***<u>Wilkinson's Catalyst</u>***

# Ru

**Ru Catalysts**

S. Siegel, *Encyclopedia of Reagents for Organic Synthesis*, John Wiley and Sons, Inc., L. A. Paquette, Ed., New York, 1995, **6**, 4410

# Jacques Reagent

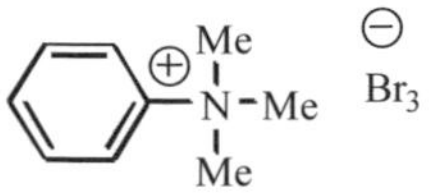

Phenyltrimethylammonium perbromide
**PTT, PTAB**
[4207-56-1]

## Preparation:

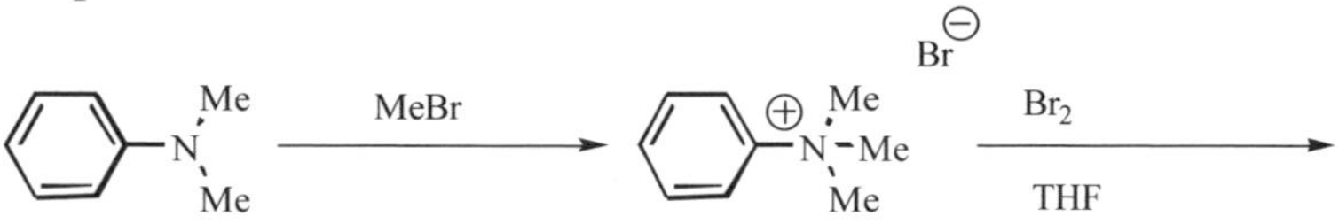

N. DeKimpe, *Encyclopedia of Reagents for Organic Synthesis*, John Wiley and Sons, Inc., L. A. Paquette, Ed., New York, 1995, **6**, 4098

**Notes:** A brominating agent

## Examples:

1.[1]

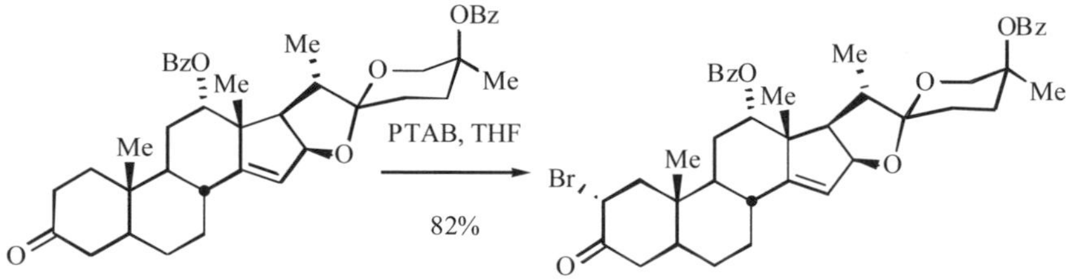

2.[2]

3.[3]

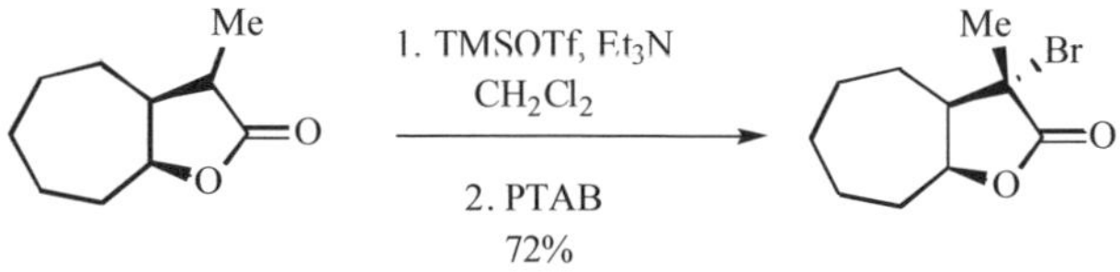

[1] S. Lee, P. L. Fuchs, *Organic Letters* **2002**, *4*, 317
[2] S. K. Burt, A. Padwa, *Organic Letters* **2002**, *4*, 4135
[3] Y. Higuchi, F. Shimoma, M. Ando, *Journal of Natural Products* **2003**, *66*, 810

# Jones Reagent

$$HO-\overset{\overset{\displaystyle O}{\|}}{\underset{\underset{\displaystyle O}{\|}}{Cr}}-OH \quad + H_2SO_4 + H_2O$$

**Preparation:** CrO$_3$ + water, then <u>carefully</u> sulfuric acid or CrO$_3$ + sulfuric acid, then diluted with water to a specific volume. CAUTION: *READ DIRECTIONS FOR PREPARATION*

> F. Freeman, *Encyclopedia of Reagents for Organic Synthesis*, John Wiley and Sons, Inc., L. A. Paquette, Ed., New York, 1995, **2**, 1261

**Notes:** A useful reagent in that one can titrate the oxidation. Primary alcohols are converted to acids; secondary alcohols to ketones; alkenes and alkynes are resistant. Related reagents:

| *Cornforth Reagent* | *Fieser's Reagent* |
|---|---|
| Pyridine / Chromic acid | Acetic Acid / Chromic Acid |

## Examples:

1.[1]

No yield provided; product taken directly to a diazomethane esterification. Overall yield OK.

2.[2]

89%

75%

3.[3]

70%

4.[4]

79-88%

---

[1] J.-H. Tai, M.-Y. Chang, A.-Y. Lee, N.-C. Chang, *Journal of Organic Chemistry* **1999**, <u>64</u>, 659
[2] J. R. Henry, S. M. Weinreb, *Journal of Organic Chemistry* **1993**, <u>58</u>, 4145
[3] A. K. Sharma, P. J. Hergenrother, *Organic Letters* **2003**, <u>5</u>, 2107
[4] J. Meinwald, J. Crandall, W. E. Hymans, *Organic Syntheses* <u>CV5</u>, 866

# KHMDS

### KN(SiMe₃)₂

Potassium hexamethyldisilazide, potassium bis(trimethylsilyl)amide
[40949-94-8]

## Commercially available

Can be prepared from hexamethyldisilane and potassium hydride:

B. T. Watson, *Encyclopedia of Reagents for Organic Synthesis*, John Wiley and Sons, Inc., L. A. Paquette, Ed., New York, 1995, **6**, 4234

## Notes:

A strong, sterically demanding, base useful for preparation of kinetic enolates. Soluble in THF, ether and toluene.

## Examples:

1.[1]

95% cis ring juncture

95% trans ring juncture

# LHMDS

[4039-32-1]
Lithium hexamethyldisilazide, LiHMDS

## Preparation:

M. Gray, V. Snieckus, *Encyclopedia of Reagents for Organic Synthesis*, John Wiley and Sons, Inc., L.A. Paquette, Ed., New York, 1995, **5**, 3127

**Notes:** For preparation of kinetic enolates.

---

[1] (a) G. Stork, J. O. Gardner, R. K. Boeckman, Jr., K. A. Parker, *Journal of the American Chemical Society* **1973**, 95, 2014 (b) G. Stork, R. K. Boeckman, Jr. *Journal of the American Chemical Society* **1973**, 95, 2016

# Koser's Reagent

HTIB Hydroxy(tosyloxy)iodobenzene
[12415-97-5]

## Commercially available

G. F. Koser , *Encyclopedia of Reagents for Organic Synthesis*, John Wiley and Sons, Inc., L. A. Paquette, Ed., New York, 1995, **4**, 2782

## Examples:

1.[1]

93%

2.[2]

45%

3.[3]

NaCl, CH$_2$Cl$_2$

95%

4.[4]

CF$_3$-CH$_2$-I-OH   OTs

58%

A fluoro analog of the *Koser's reagent*.

[1] T. Nabana, H. Togo, *Journal of Organic Chemistry* **2002**, <u>67</u>, 4362
[2] A. Kirschning, *Journal of Organic Chemistry* **1995**, <u>60</u>, 1228
[3] P. Bovonsombat, E. Djuardi, E. McNelis, *Tetrahedron Letters* **1994**, <u>35</u>, 2841
[4] V. V. Zhdankin, C.J. Kuehl, A. J. Simonsen, *Journal of Organic Chemistry* **1996**, <u>61</u>, 8272

# Kulinkovich Reagent

Bis(2-propanolato)[(1,2-h)-1-propene]titanium
[169123-69-7]

## Preparation:

A. J. Phillips, *e-Encyclopedia of Reagents for Organic Synthesis*, L.A. Paquette, Ed., John Wiley
& Sons, Inc., online reference available at *http://www.interscience.wiley.com.*

**Notes:**  A reagent that allows for a unique preparation of cyclopropanol derivatives.

## Examples:

1.[1]

2.[2]

3.[3]

[1] J. Lee, H. Kim, J. K. Cha, *Journal of the American Chemical Society* **1995**, 117, 9919
[2] E. J. Corey, S. A. Rao, M. C. Noe, *Journal of the American Chemical Society* **1994**, 116, 9345
[3] J. Lee, H. Kim, J. K. Cha, *Journal of the American Chemical Society* **1996**, 118, 4198

# L-Selectride®
## LiBH(s-Bu)3
Lithium Tri-s-butylborohydride
[38721-52-7]

## Commercially available

J. L. Hubbard , *Encyclopedia of Reagents for Organic Synthesis*, John Wiley and Sons, Inc., L. A. Paquette, Ed., New York, 1995, **5**, 3172

**Notes:**  Selective reducing agent.  The potassium and sodium Selectrides are also available. Will convert C-X to C-H, with I>Br>Cl.

## Examples:

1.[1]

2.[2]

3.[3]

[1] E. Piers, M. A. Romero, *Tetrahedron* **1993**, <u>49</u>, 5791
[2] J. Nokami, M. Ohkura, Y. Dan-Oh,Y. Sakamoto, *Tetrahedron Letters* **1991**, <u>32</u>, 2409
[3] W. Oppolzer, C. Chapuis, G. Bernardinelli, *Tetrahedron Letters* **1984**, <u>25</u>, 5885

# Lawesson's Reagent

2,4-bis(4-methoxyphenyl)-1,3-dithia-2,4-diphosphetane-2,4-disulfide   [19172-47-5]

## Commercially available

## Preparation:

J. Voss, *Encyclopedia of Reagents for Organic Synthesis*, John Wiley and Sons, Inc., L. A. Paquette, Ed., New York, 1995, **1**, 530

**Notes:** A combination of phosphorus pentasulfide and hexamethyldisiloxane can be used to prepare thionoesters and thionolactones in comparable yields and with a simplified workup.[1] A solvent-free conversion of ketones, lactones, esters and amides, to the corresponding thio analogs with *Lawesson's Reagent*.[2]
See also: *Belleau's* and *Davy's Reagents*.

## Examples:
1.[3]

2.[4]

[1] T. J. Curphey, *Tetrahedron Letters* **2002**, 43, 371
[2] R. S. Varma, D. Kumar, *Organic Letters* **1999**, 1, 697
[3] V. M. Sonpatki, M. R. Herbert, L. M. Sandvoss, A. J. Seed, *Journal of Organic Chemistry* **2001**, 66, 7283
[4] A. Padwa, M. D. Danca, *Organic Letters* **2002**, 4, 715

# LDA

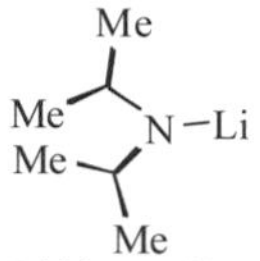

Lithium Diisopropylamide
[4111-54-0]

## Commercially available

## Preparation:

W. I. Iwema Bakker, P. L. Wong & V. Snieckus, *Encyclopedia of Reagents for Organic Synthesis*, John Wiley and Sons, Inc., L. A. Paquette, Ed., New York, 1995, **5**, 3096

**Notes:** An extremely common and widely used base. Often used to remove an acidic proton leading to a "kinetic enolate". When carried out with TMS-Cl, the process allows for the isolation of a kinetic silyl enol ether.

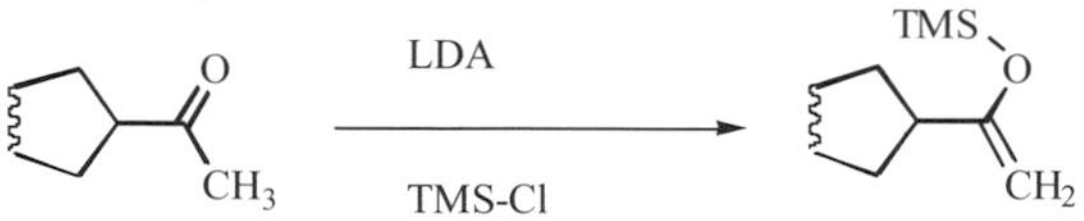

Trapping the kinetic enolate of a methyl ketone with diethyl phosphochloridate provides an enol phosphate, that can, in turn, be converted to an alkyne:

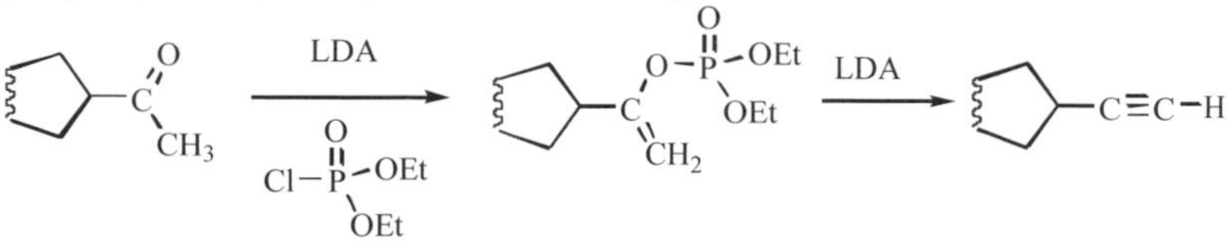

The reagent is extremely useful for the preparation of cross-conjugated enolates of α,β-unsaturated systems:

## Example:[1]

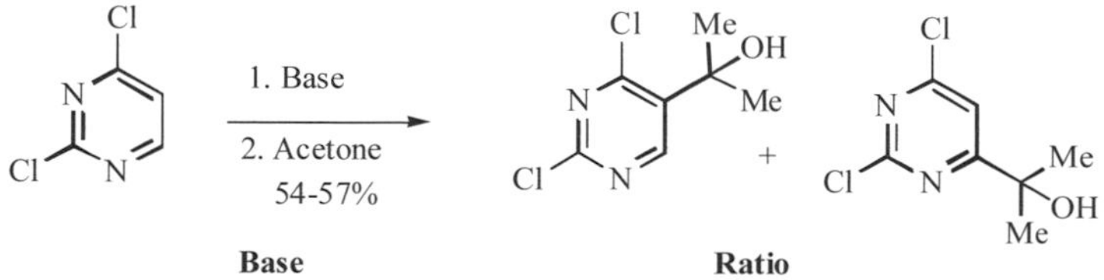

| Base | Ratio | |
|------|-------|---|
| LDA | 100 | 0 |
| LTMP | 35 | 65 |

---

[1] N. Plé, A. Turck, P. Martin, S. Barbey, G. Quéguiner, *Tetrahedron Letters* **1993**, <u>34</u>, 1605

# Lemieux-Johnson Reagent
## NaIO$_4$-OsO$_4$
[7790-28-5, 20816-12-0]
## Preparation:

A. G. Wee, B. Liu, *Encyclopedia of Reagents for Organic Synthesis*, John Wiley and Sons, Inc., L. A. Paquette, Ed., New York, 1995, **7**, 4616

## Examples:
1.[1]

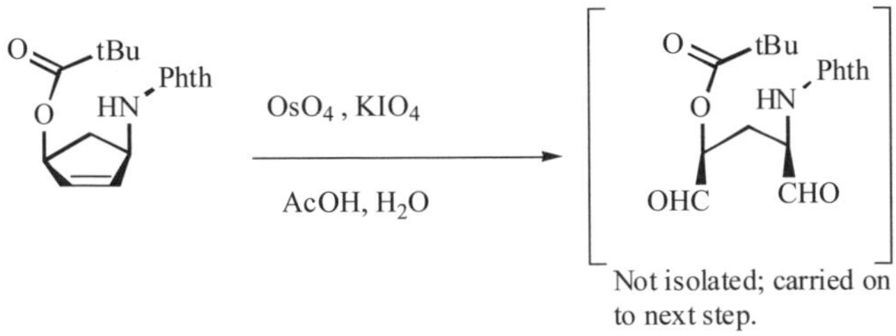

Not isolated; carried on to next step.

2.[2]

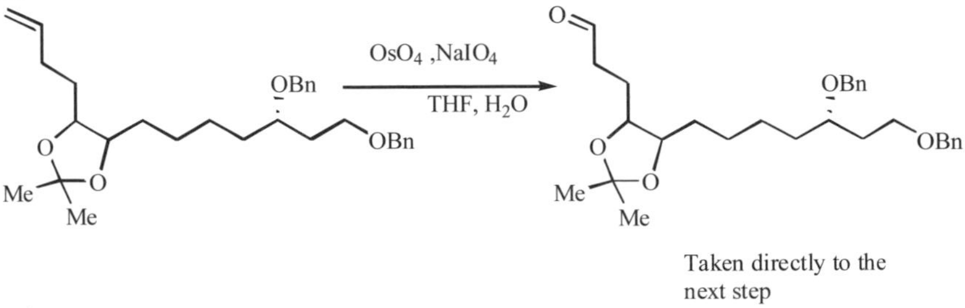

Taken directly to the next step

3.[3]

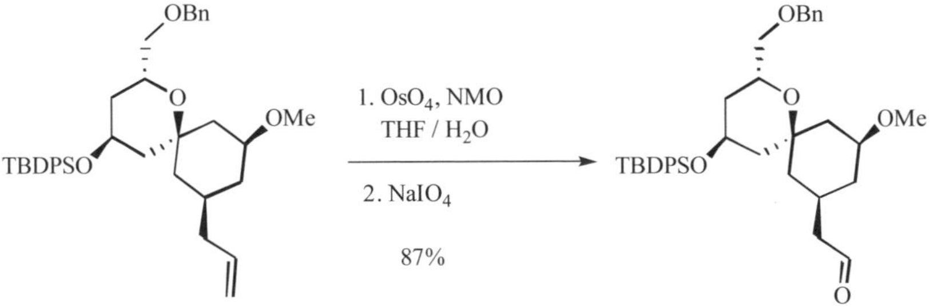

[1] F. A. Luzzio, A. V. Mayrov, W. D. Figg, *Tetrahedron Letters* **2000**, <u>41</u>, 2275
[2] S. Takahashi, A. Kubota, T. Nakata, *Organic Letters* **2003**, <u>5</u>, 1353
[3] D. Zuev, L. A. Paquette, *Organic Letters* **2000**, <u>2</u>, 679

# Lemieux-von Rudloff Reagent
## NaIO$_4$ / KMnO$_4$
[7790-28-5, 7722-64-7]

**Preparation:**

An aqueous solution (often with a miscible organic cosolvent) of NaIO$_4$ and KMnO$_4$ with strong stirring.

A. G. Wee, B. Liu , *Encyclopedia of Reagents for Organic Synthesis*, John Wiley and Sons, Inc., L. A. Paquette, Ed., New York, 1995, **7**, 4620

**Notes:**

Since KMnO$_4$ is a strong oxidizing agent, one expects the following reaction generalization:

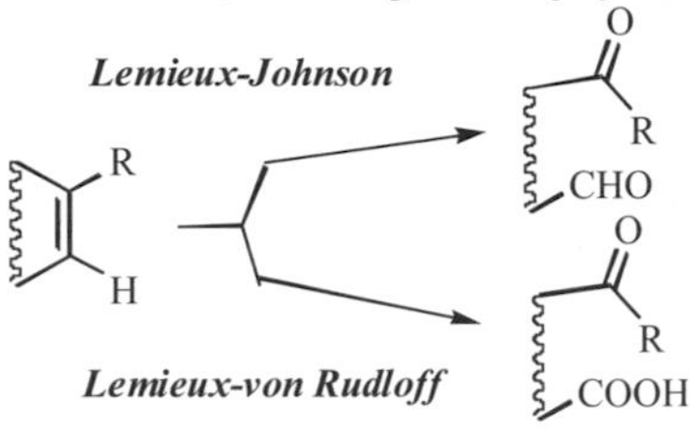

**Examples:**

1.[1]

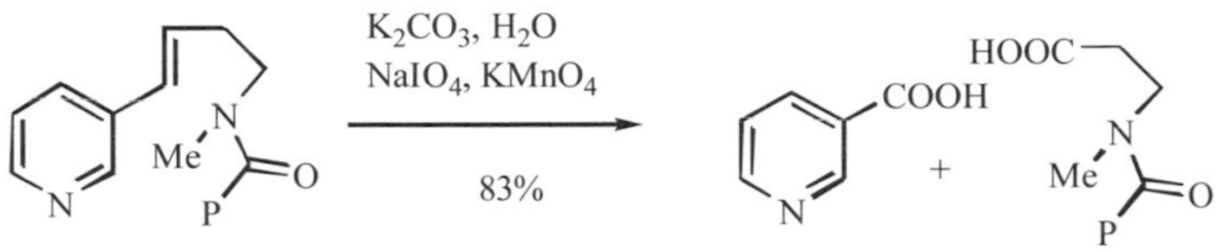

2.[2]

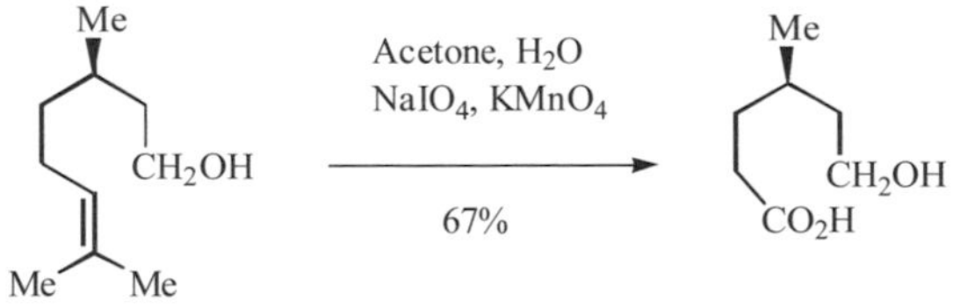

---

[1] A. A. Liebman, B. P. Mundy, H. Rapoport, *Journal of the American Chemical Society* **1967**, <u>89</u>, 664

[2] C. G. Overberger, H. Kayel, *Journal of the American Chemical Society* **1967**, <u>89</u>, 5640

# Lindlar's Catalyst
## Pd / CaCO₃ / PbO
[7440-05-3], [471-34-1], [1317-36-8]

### Preparation:

A. O. King, I. Shinkai, *Encyclopedia of Reagents for Organic Synthesis*, John Wiley and Sons, Inc., L. A. Paquette, Ed., New York, 1995, **6**, 3866

**Notes:** Major use is the *cis* reduction of an alkyne to an alkene.

### Examples:
1.[1]

2.[2]

3.[3]

4.[4]

---

[1] A. Tai, F. Matsumura, H. C. Coppel, *Journal of Organic Chemistry* **1969**, <u>34</u>, 2180
[2] A. Fulrstner, T. Dierkes, *Organic Letters* **2000**, <u>2</u>, 2463  Quinoline was added to the reduction mixture
[3] T. Itoh, N. Yamazaki, C. Kibayashi, *Organic Letters* **2002**, <u>4</u>, 2469
[4] T. Lindel, M. Hochgörtel, *Journal of Organic Chemistry* **2000**, <u>65</u>, 2806

# Lombardo Reagent
### $CH_2Br_2$, $TiCl_4$, Zn

**Preparation:**

$$Zn\ dust \xrightarrow[\text{slowly add } TiCl_4]{CH_2Br_2,\ THF}$$

N. A. Petasis, J. P. Staszewshi, *Encyclopedia of Reagents for Organic Synthesis*, John Wiley and Sons, Inc., L. A. Paquette, Ed., New York, 1995, **3**, 1565

**Notes:** Original report: L. Lombardo, *Tetrahedron Letters* **1982**, _23_, 4293 Sometimes called the *Oshima-Lombardo Reagent*.

**Examples:**

1.[1]

Lombardo reagent

$CH_2Cl_2$

87%

2.[2]

Bn   Br

Zn, $TiCl_4$

80%

3.[3]

Zn

$TiCl_4$, $CH_2Cl_2$

81%

4.[4]

Lombardo reagent

70%

[1] M. Furber, L. N. Mander, *Journal of the American Chemical Society* **1988**, _110_, 4084

[2] H. Huang, C. J. Forsyth, *Journal of Organic Chemistry* **1995**, _60_, 5746

[3] L. Plamondon, J. D. Wuest, *Journal of Organic Chemistry* **1991**, _56_, 2066

[4] P. Magnus, B. Mugrage, M. DeLuca, G. A. Cain, *Journal of the American Chemical Society* **1989**, _111_, 786

# LTMP (LiTMP)

Lithium 2,2,6,6-tetramethylpiperidide
[38227-87-1]

## Preparation:

M. Campbell, V. Snieckus, *Encyclopedia of Reagents for Organic Synthesis*, John Wiley and Sons, Inc., L. A. Paquette, Ed., New York, 1995, **5**, 3166

**Notes:**  Hindered base with low nucleophilicity.  Is stable with TMS-Cl; thus reactions can be carried out under equilibrium-controlled conditions, where substrate, TMS-Cl and *LTMP* are present at the same time.

See also:  *LDA*, *NHMDS*, *KHMDS*, *LHMDS* for other strong bases

As a generalization,

## Examples:

1.[1]

1. LTMP, HMPA

2. Ph-Se-Cl
   40%

Reaction failed with LDA

2.[2]

LTMP

90 - 92%

---

[1] A. B. Smith, III, R. E. Richmond, *Journal of Organic Chemistry* **1981**, 46, 4814
[2] C. S. Shiner, A. H. Berks, A. M. Fisher, *Journal of the American Chemical Society* **1988**, 110, 957

# Luche Reagent[1]
### NaBH₄ + CeCl₃
[1191-15-7]

**Preparation:** Equal amounts of NaBH$_4$ and CeCl$_3$ (7 H$_2$O) in methanol

L. A. Paquette, *Encyclopedia of Reagents for Organic Synthesis*, John Wiley and Sons, Inc., L. A. Paquette, Ed., New York, 1995, **2**, 1031

**Notes:** Specifically reduces enones to allylic alcohols via 1,2-addition. Sometimes stereochemistry is different from NaBH$_4$ alone.

## Examples:
1.[2]

2.[3]

3.[4]

4.[5]

[1] J.-L. Luche, A. L. Gemal, *Journal of the Chemical Society, Chemical Communications* **1978**, 976
[2] D. L. J. Clive, S. Sun, V. Gagliardini, M. K. Sano, *Tetrahedron Letters* **2000**, <u>41</u>, 6259
[3] F. J. Moreno-Dorado, F. M. Guerra, F. J. Aladro, J. M. Bustamanta, Z. D. Jorge, G. M. Massanet, *Tetrahedron* **1999**, <u>55</u>, 6997
[4] C.-K. Sha, A.-W. Hong, C.-M. Huang, *Organic Letters* **2001**, <u>3</u>, 2177
[5] K. Takao, G. Watanabe, Y. Yasui, K. Tadano, *Organic Letters* **2002**, <u>4</u>, 2941

# MAD

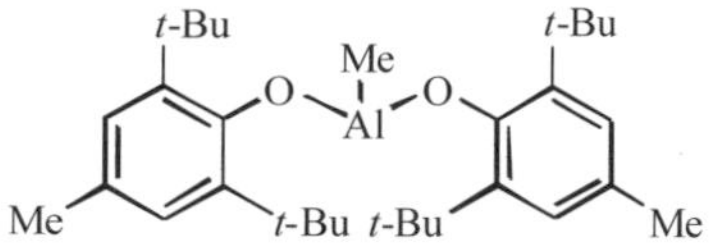

Methylaluminum bis(2,6-di-t-butyl-4-methylphenoxide)
[65260-44-8]

**Preparation:**[1] prepared by reaction of trimethylaluminum with 2 equiv of 2,6-di-t-butyl-4-methylphenol in toluene or CH$_2$Cl$_2$.

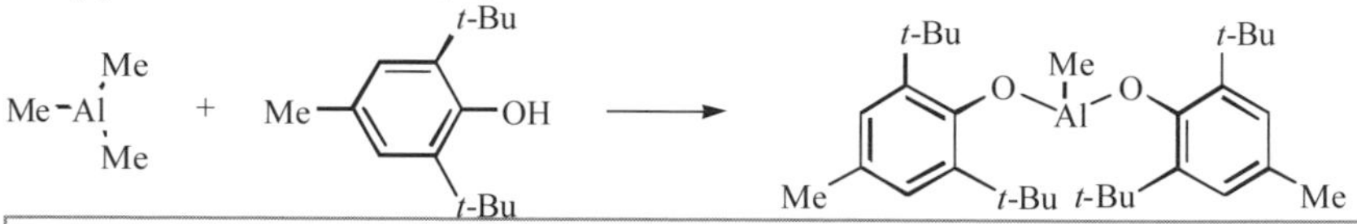

K. Maruoka, H. Yamamoto, *Encyclopedia of Reagents for Organic Synthesis*, John Wiley and Sons, Inc., L. A. Paquette, Ed., New York, 1995, **5**, 3415

**Notes:** For a similar reagent, see **DAD** (Dimethylaluminum 2,6-Di-t-butyl-4-methylphenoxide) [86803-85-2]

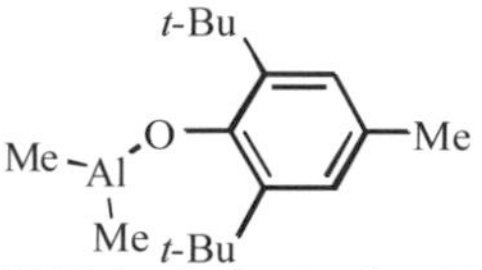

MAD is a bulky Lewis acid used to complex with functional groups to provide a steric discrimination to a number of organic transformations.

## Examples:

**1.**[2] **Selectivity in carbonyl reaction**. Here we see that sterically less-hindered ketones are preferentially complexed, leaving the more-hindered ketone to react.

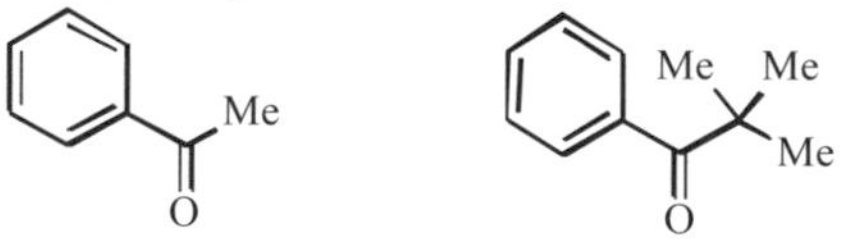

| Ratio of Reduction | 1 | 16 | with 2 equiv. of MAD |

**2.**[1] Increased carbonyl discrimination:

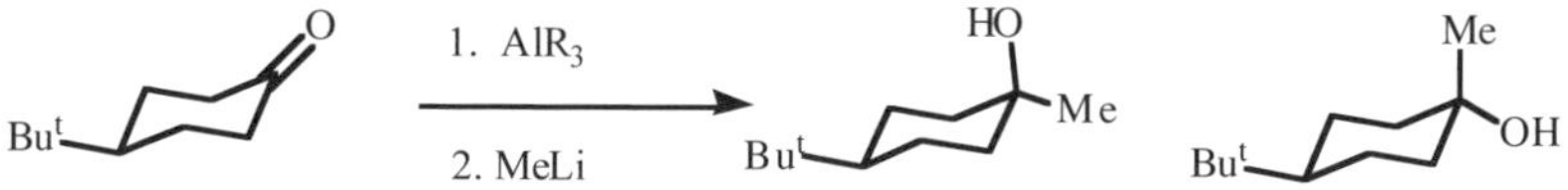

| AlR$_3$ | Ratio |
|---------|-------|
| None | 79 : 21 |
| MAD | 1 : 99 |

---

[1] K. Maruoka, T. Itoh, M. Sakurai, K. Nonoshita, H. Yamamoto, *Journal of the American Chemical Society* **1988**, 110, 3588. This article provides a wealth of tabulated data for a number of cyclic systems.
[2] K. Maruoka, Y. Araki, H. Yamamoto, *Journal of the American Chemical Society* **1988**, 110, 2650

| AlR₃ | Yield | Ratio |
|------|-------|-------|
| None | 77% | 75 : 25 |
| MAD | 82% | 1 : 99 |

## 2. [1] Oxygen Complexation. 

Here we see that sterically less-hindered ethers are preferentially complexed.

| Lewis Acid | Rel. Complexation | | |
|------------|-------------------|---|---|
| BF₃ etherate | NONE | | |
| SnCl₄ | 2 | : | 1 |
| MAD | 99 | : | 1 |

## 3. [2]

A reagent has been developed that will preferentially complex with aldehydes, so that a ketone can be reacted in the presence of an aldehyde.

| | | | | |
|---|---|---|---|---|
| MAD / MeMgI | 9 | : | 1 | |
| Me₂AlNMePh / MeLi | 1 | : | 6 | |

[1] S. Sato, H. Yamamoto, *Journal of the Chemical Society, Chemical Communications* **1997**, 1585
[2] K. Maruoka, Y. Araki, H. Yamamoto, *Tetrahedron Letters* **1988**, _29_, 3101

# Magic Methyl and Related Reagents

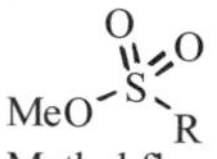

$$MeO-\overset{\overset{O}{\underset{\parallel}{}}}{\underset{\underset{O}{\parallel}}{S}}-R$$

Methyl fluorosulfonate, R = F
Methyl triflate, R = CF$_3$

## Commercially available

R.W. Alder, J.G.E. Phillips, *Encyclopedia of Reagents for Organic Synthesis*, John Wiley and Sons, Inc., L. A. Paquette, Ed., New York, 1995, **5**, 3617

## Notes:

Both reagents are powerful methylating agents, reacting with O, N, and S. Read precautions carefully. The ***Methyl fluorosulfonate (Magic Methyl)*** is no longer commercially available. Methyl triflate is extremely (sometimes violently) reactive towards amines. Except for extremely hindered or unactivated amines, not generally required for the task.

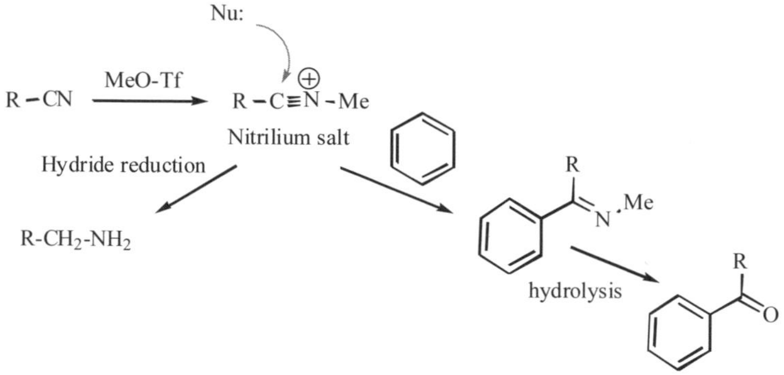

## Examples:

1.[1]

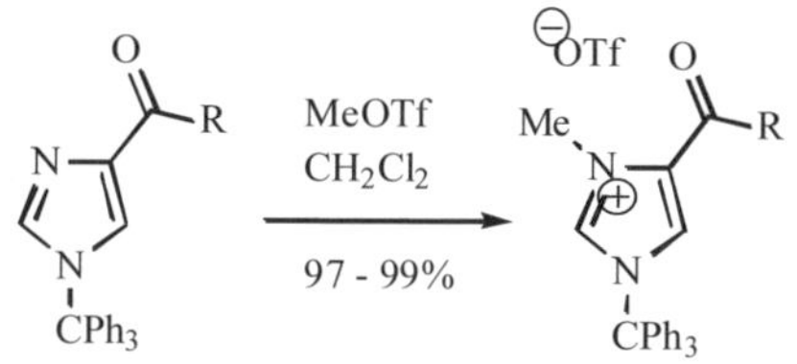

2.[2]

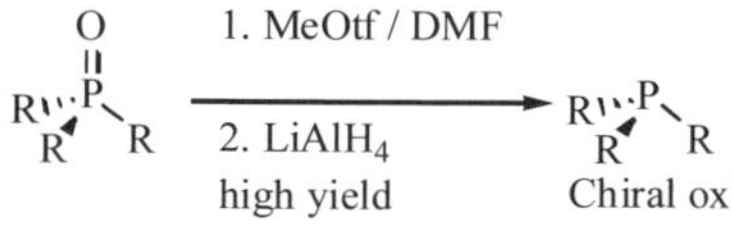

Chiral oxides are reduced with inversion at P.

---

[1] B.-C. Chen, A. P. Skoumbourdis, J. E. Sundeen, G. C. Rovnyak, S. C. Traeger, *Organic Process Research & Development* **2000**, 4, 613
[2] T. Imamoto, S. Kikuchi, T. Miura, Y. Wada, *Organic Letters*, **2001**, 3, 87

# Mander's Reagent
### NC-COOMe
[17640-15-2]
### Commercially available.

L. N. Mander, *Encyclopedia of Reagents for Organic Synthesis*, John Wiley and Sons, Inc., L. A. Paquette, Ed., New York, 1995, **5**, 3466

**Notes:** Used for the α-carbomethoxylation of carbonyl compounds.

## Examples:
1.[1]

2.[2]

3.[3]

4.[4]

81-84% overall yield

[1] I. Efremov, L. A. Paquette, *Journal of the American Chemical Society* **2000**, *122*, 9324
[2] C. Chen, M.E. Layton, S. M. Sheehan, M. D. Shair, *Journal of the American Chemical Society* **2000**, *122*, 7424.
[3] J. D. Winkler, M. B. Rouse, M. F. Greaney, S. J. Harrison, Y. T. Jeon, *Journal of the American Chemical Society* **2002**, *124*, 9726
[4] S. R. Crabtree, L. N. Mander, S. P. Sethi, *Organic Syntheses* CV9, 619

# Markiewicz Reagent

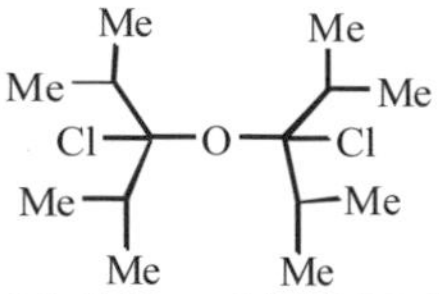

1,3-dichloro-1,1,3,3-tetraisopropyl disiloxane [TIPSCl]
[69304-37-6]

## Commercially available

J. Slade, *Encyclopedia of Reagents for Organic Synthesis*, John Wiley and Sons, Inc., L. A. Paquette, Ed., New York, 1995, **3**, 1730.

**Notes:** A useful protecting group; primary alcohols react about $10^3$ faster than secondary alcohols.

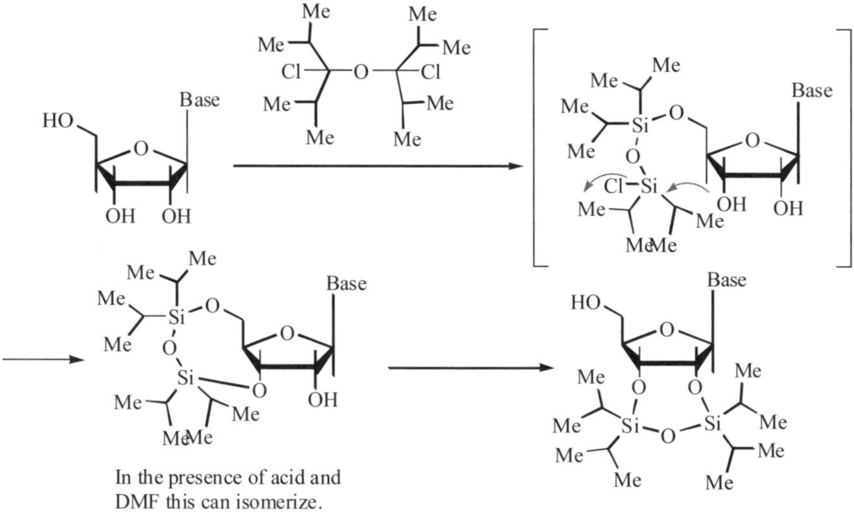

In the presence of acid and DMF this can isomerize.

## Examples:
1.[1]

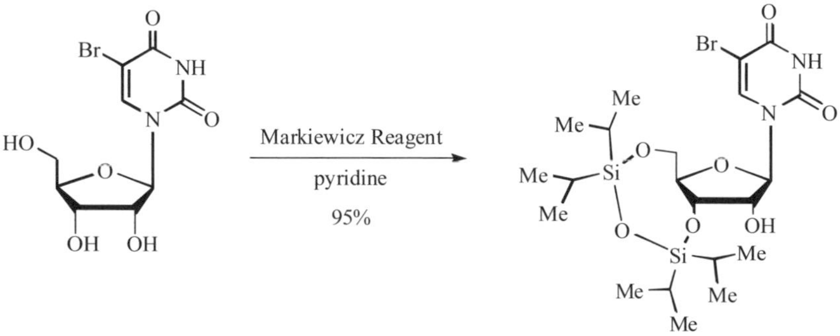

Markiewicz Reagent

pyridine

95%

---

[1] S.C. Holmes, A.A. Arzumanov, M.J. Gait, *Nucleic Acid Research* **2003**, <u>31</u>, 2759.

# Martin sulfurane

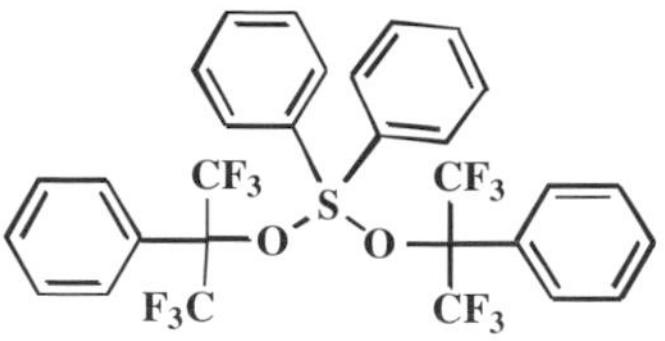

[32133-82-7]
Diphenylbis(1,1,1,3,3,3-hexafluoro-2-phenyl-2-propoxy)sulfurane

## Commercially available

B. A. Rodin, *Encyclopedia of Reagents for Organic Synthesis*, John Wiley and Sons, Inc., L. A. Paquette, Ed., New York, 1995, **4**, 2201

## Notes:

Dehydrating agent; tertiary alcohols react instantly. Primary alcohols are often unreactive. Pinacols are often converted to epoxidea:

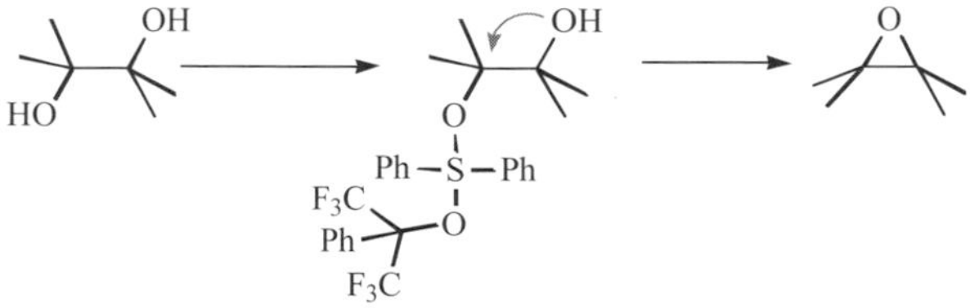

## Examples:

1[1].

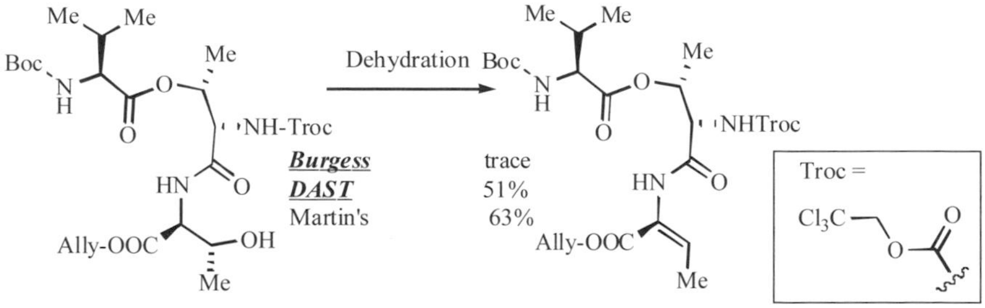

2.[2]

| | | | | |
|---|---|---|---|---|
| ***Burgess*** | No yield given | 1 | : | 4 |
| Martin | | 92% | | |

---

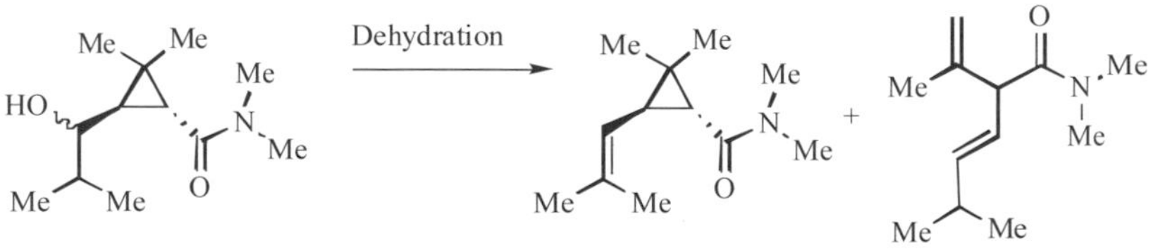

[1] F. Yokokawa, T. Shioiri, *Tetrahedron Letters* **2002**, 43, 8673
[2] M. Majewski, V. Snieckus, *Tetrahedron Letters* **1982**, 23, 1343

# Matteson's Reagent

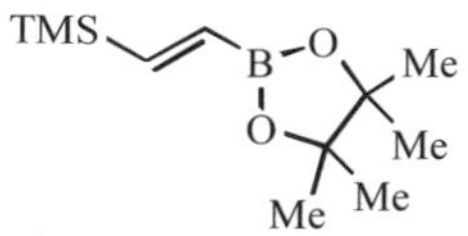

Pinacol *E*-l-trimethylsilyl-1-propene-3-boronate
[126688-99-1]

**Preparation:** D. J. S. Tsai, D. S. Matteson, *Tetrahedron Letters* **1981**, 22, 2751

## Examples:

1.[1]

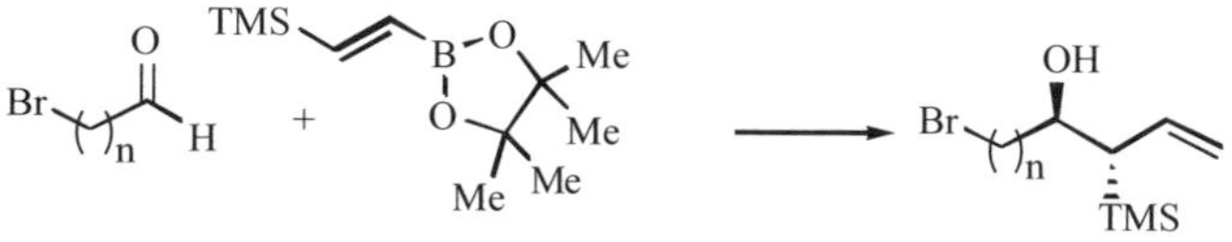

2.[2]

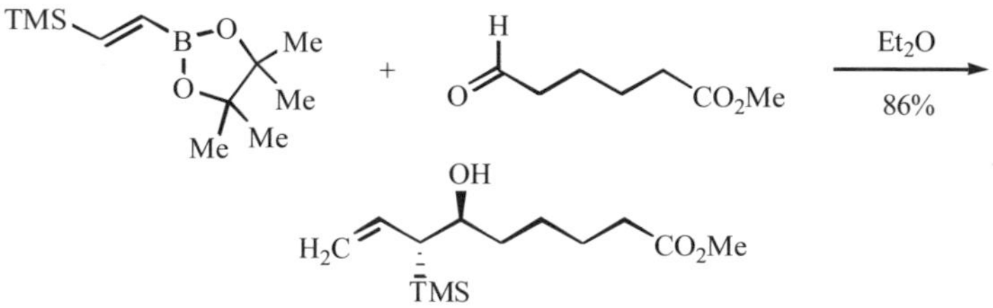

3.[3]

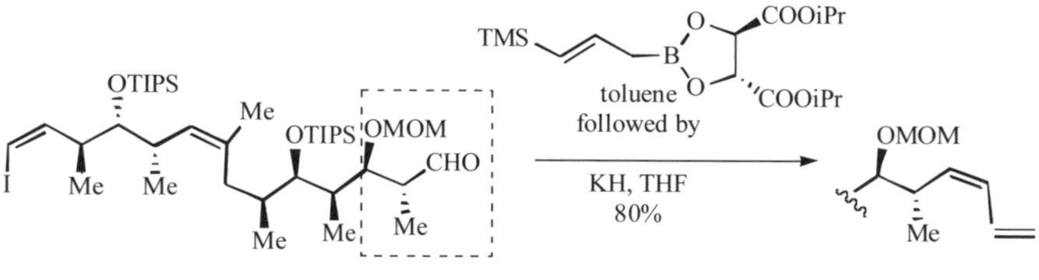

This sequence does not use the Matteson boronate intermediate; however, was referenced as the method for the conversion.

---

[1] B. P. Hoag, D. L. Gin, *Macromolecules* **2000**, 33, 8549
[2] D. A. Smith, K. N. Houk, *Tetrahedron Letters* **1991**, 32, 1549
[3] S. S. Harried, C. P. Lee, G. Yang, T. I. H. Lee, D. C. Myles, *Journal of Organic Chemistry* **2003**, 68, 6646

# McMurry's Reagent[1]
## TiCl₄ / reducing agent

**Preparation:** Essentially this reaction involves the preparation of a low-valent titanium reagent that then couples carbonyl groups, including esters to aldehydes/ketones. Generally, $TiCl_4$ is reduced with some reducing agent ($LiAlH_4$, Zn, Mg).

**Notes:** See *McMurry Olefination*.

**Examples:**

1.[2]

$TiCl_3$ / Zn/Cu

DME
90%

2.[3]

Ti-graphite / DME

60-82%

3.[4]

TiCl₄ / Zn

CH₂Cl₂   82%

4.[5]

TiCl₄ / Zn

78%

[1] J. E. McMurry, T. Lectka, J. G. Rico, *Journal of Organic Chemistry* **1989**, 54, 3748
[2] J. E. McMurry, G. J. Haley, J. R. Matz, J. C. Clardy, J. Mitchell, *Journal of the American Chemical Society* **1986**, 108, 515
[3] A. Fürstner, O. R. Thiel, N. Kindler, B. Bartkowska, *Journal of Organic Chemisty* **2000**, 65, 7990
[4] J. Nakayama, H. Machida, R. Satio, M. Hoshino, *Tetrahedron Letters* **1985**, 26, 1983
[5] F. B. Mallory, K. E. Butler, A. Bérubé, E. D. Luzik, Jr., C. W. Mallory, E. J. Brondyke, R. Hiremath, P. Ngo, P. Carroll, *Tetrahedron* **2001**, 57, 3715

# Meldrum's Lactone

2,2-Dimethyl-1,3-dioxane-4,6-dione
[2033-24-1]

## Commercially available

O. Yonemitsu, K. Hirao, *Encyclopedia of Reagents for Organic Synthesis*, John Wiley and Sons, Inc., L. A. Paquette, Ed., New York, 1995, **3**, 2056

## Notes:

Useful for the general preparation of β-keto esters[1]

## Examples:

1.[2]

2.[3]

---

[1] Y. Oikawa, K. Sugano, O. Yonemitsu, *Journal of Organic Chemistry* **1978**, <u>43</u>, 2087
[2] B. Hin, P. Majer, T. Tsukamoto, *Journal of Organic Chemistry* **2002**, <u>67</u>, 7365
[3] R. Bruns, A. Wernicke, P. Köll, *Tetrahedron* **1999**, <u>55</u>, 9793

# MEMCl

MeO—O—Cl

**Source of the 2-methoxyethylmethyl (MEM) group**
2-Methoxyethoxymethyl chloride
[3970-21-6]

## Commercially available

P. G. M. Wuts, *Encyclopedia of Reagents for Organic Synthesis*, John Wiley and Sons, Inc., L. A. Paquette, Ed., New York, 1995, **5**, 3351

**Notes:** This alcohol protecting is easily attached and readily removed by Lewis acids such as zinc bromide and titanium tetrachloride. Phenols can be protected (reaction of the sodium salt with MEMCl) and deprotected with TFA. More easily removed than the MOM group.

## Examples:

1.[1] Example of a removal of an enol-MEM group

Me₃SiCl

MeOH

53%

2.[2]

3.[3]

An analysis of how product ratios depended on the stereochemistry in the starting material.

---

[1] G. A. DeBoos, J. J. Fullbrook, J. M. Percy, *Organic Letters* **2001**, 3, 2859
[2] J. R. Donaubauer, T.C. McMorris, *Tetrahedron Letters* **1980**, 21, 2771
[3] T. A. Blumenkopf, G. C. Look, L. E. Overman *Journal of the American Chemical Society*, **1990**, 112, 4399

# Meyer's Reagent

5,6-Dihydro-2,4,4,6-tetramethyl-1,3(4*H*)-oxazine
[26939-18-4]

## Commercially available

## Preparation: *Ritter reaction*

T. D. Nelson, A.I. Meyers, *Encyclopedia of Reagents for Organic Synthesis,* John Wiley and Sons, Inc., L. A. Paquette, Ed., New York, 1995, **3**, 888

**Notes:** Useful reagent for the preparation of aldehydes and ketones

1. Base
2. R-X
3. NaBH₄ / HOAc
4. Hydrolysis

1. Base
2. RX then MeI
3. **R'MgX**
4. H⊕, H₂O

, via

# Meyer's Reagent

Tetrahydro-3-isopropyl-7a-methylpyrrolol[2,1-b]-5(6*H*)-one
(3*S,cis*) [98203-44-2]; (3*R,cis*) [123808-97-9]

## Commercially available

T. D. Nelson, A.I. Meyers, *Encyclopedia of Reagents for Organic Synthesis,* John Wiley and Sons, Inc., L. A. Paquette, Ed., New York, 1995, **7**, 4756

**Notes:**

Chiral alkylation can be performed

# Nafion-H
## Reagent

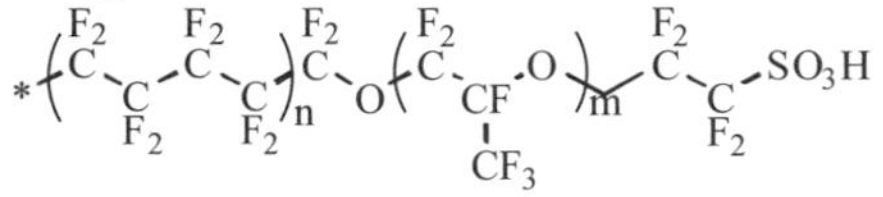

$$*\left(C\begin{smallmatrix}F_2\\C\end{smallmatrix}-C\begin{smallmatrix}\\F_2\end{smallmatrix}-C\begin{smallmatrix}F_2\\C\end{smallmatrix}\right)_n O\left(C\begin{smallmatrix}F_2\\CF\\\\CF_3\end{smallmatrix}O\right)_m C\begin{smallmatrix}F_2\\C\end{smallmatrix}-C\begin{smallmatrix}\\F_2\end{smallmatrix}-SO_3H$$

Nafion-H
[63937-00-8]

## Commercially available

Y. El-Kattan, J. McAtee, *Encyclopedia of Reagents for Organic Synthesis,* John Wiley and Sons, Inc., L. A. Paquette, Ed., New York, 1995, **6**, 3677

## Notes:

An extremely useful acid catalyst with the acidity of concentrated sulfuric acid. Has been proposed as a replacement for mineral acids in the instructional organic laboratory for dehydration and esterification reactions.[1]

## Examples:
1.[2]

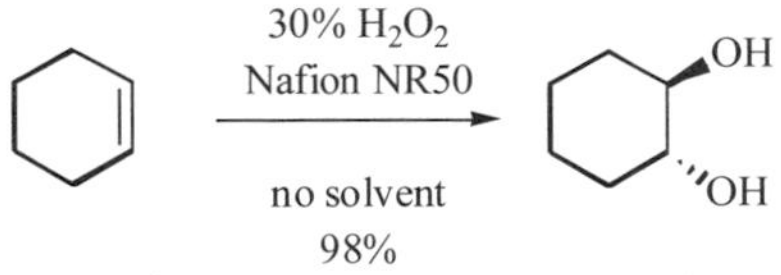

30% $H_2O_2$
Nafion NR50

no solvent
98%

The resin could be reused; after 10 cycles almost no loss in yield.

2.[3]

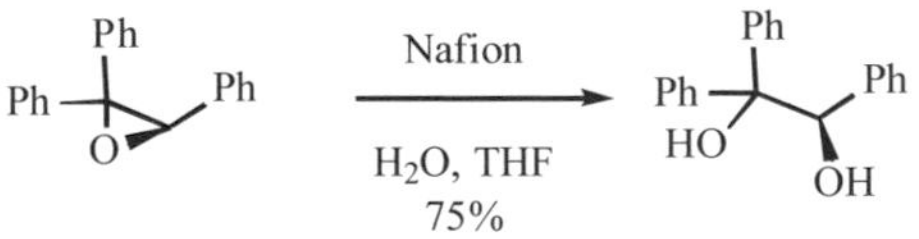

Nafion

$H_2O$, THF
75%

3.[4]

Nafion H

4-t-butyl-o-xylene

90%

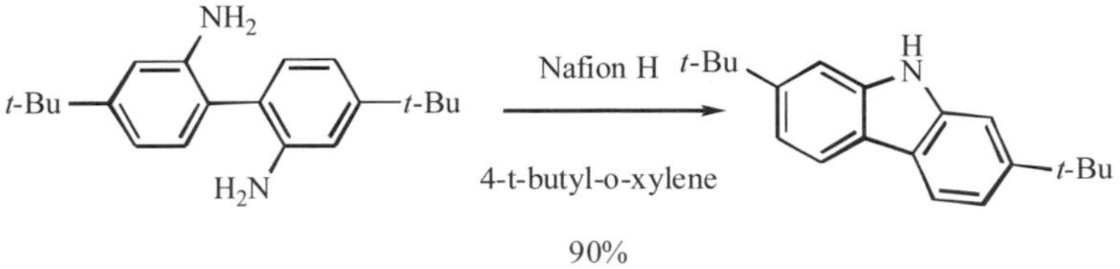

[1] M. P. Doyle, B. F. Plummer, *Journal of Chemical Education* **1993**, 70, 493
[2] Y. Usui, K. Sato, M. Tanaka, *Angewandte Chemie International Edition in English* **2003**, 42, 5623
[3] B. D. Brandes, E. N. Jacobsen, *Journal of Organic Chemistry* **1994**, 59, 4378
[4] T. Yamato, C. Hideshima, K. Suehiro, M. Tashiro, G. K. S. Prokash, G. A. Olah, *Journal of Organic Chemistry* **1991**, 56, 6248

# *N,N'*-Carbonyldiimidazole
## CDI

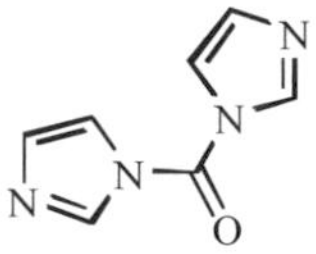

[530-62-1]

## Commercially available

A. Armstrong, *Encyclopedia of Reagents for Organic Synthesis,* John Wiley and Sons, Inc., L. A. Paquette, Ed., New York, 1995, **2**, 1006

## Notes:

Useful for activation of carboxylic acids..

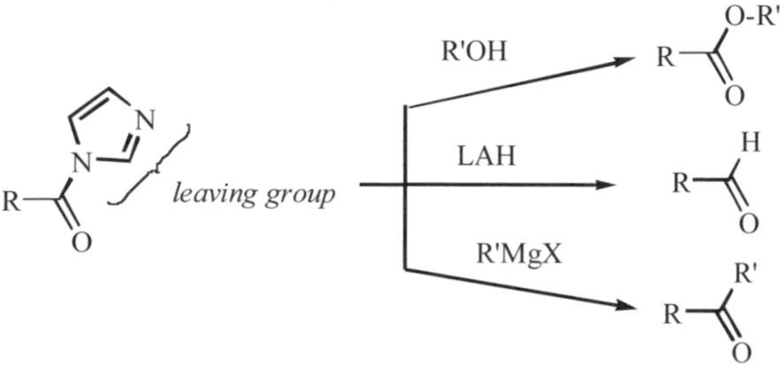

## Examples:

1.[1]

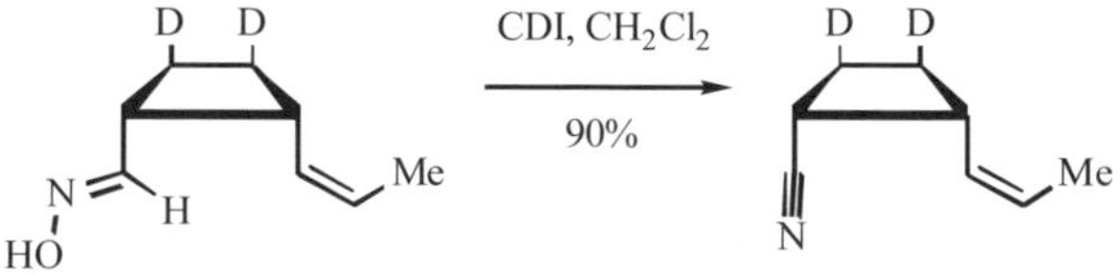

2[2]

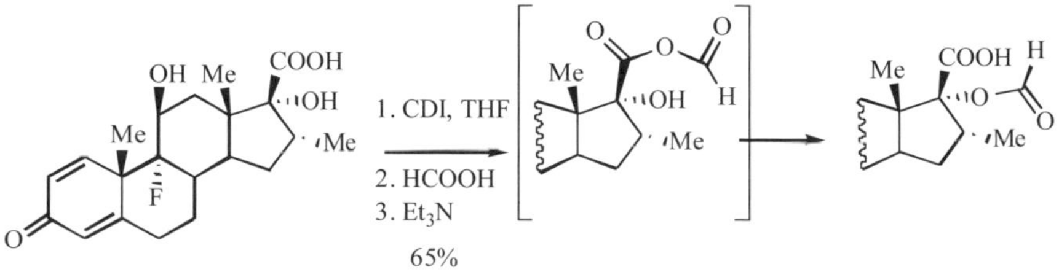

---

[1] R. E. Conrow, G. W. Dillow, L. Baker, L. Xue, O. Papadopoulou, J. K. Baker, B. S. Scott, *Journal of Organic Chemistry* **2002**, <u>67</u>, 6835

[2] W. von E. Doering, X. Cheng, K. Lee, Z. Lin, *Journal of the American Chemical Society* **2002**, <u>124</u>, 11642

# NaHMDS, NaN(SiMe₃)₂

Sodium Hexamethyldisilazide, NaHMDS
[1070-89-9]

## Commercially available

B. T. Watson, *Encyclopedia of Reagents for Organic Synthesis*, John Wiley and Sons, Inc., L. A. Paquette, Ed., New York, 1995, **7**, 4564

**Notes:** Strong, sterically-hindered base, soluble in THF, ether, and toluene. The Li and K analogs are also available. The reagent also can act as a nucleophile under certain conditions.

## Example

1.[1]

Use of NaH resulted in significant *N*-alkylation.

2.[2]

3.[3] The lithio reagent serves much the same function:

---

[1] S. Chow, K. Wen, Y. S. Sanghvib, E. A. Theodorakisa, *Bioorganic & Medicinal Chemistry Letters* **2002**, *13*, 1631
[2] X. Li, D. Lantrip, P. L. Fuchs, *Journal of the American Chemical Society* **2003**, *125*, 14262
[3] S. B. Hoyt, L. E. Overman, *Organic Letters* **2000**, *2*, 3241

# NBS

N-Bromosuccinimide
[128-08-5]

## Commercially available

S. C. Virgil, *Encyclopedia of Reagents for Organic Synthesis*, John Wiley and Sons, Inc., L. A. Paquette, Ed., New York, 1995, **1**, 768

**Notes:** Used for allylic and benzylic brominations (***Wohl-Ziegler Reaction***). With moist DMSO the reagent is useful for bromohydrin formation, providing trans addition of bromine and water. Can brominate alpha to carbonyl in carbonyl (carboxyl)-containing compounds. With DMF useful for aromatic bromination of activated aromatic rings, such as phenols, aromatic ethers, aniline derivatives and activated heterocyclic compounds. For similar chemistry, see also ***NBA, N-Bromoacetamide***.

reactivity differences

In allylic bromination, the radical nature of the reaction can lead to rearranged products.

## Examples:

1.[1]

NBS / LIGHT
AIBN
91%

2.[2]

NBS
Benzoyl peroxide
93%

3.[3]

NBS MeOH
80%

4.[4] Useful for the conversion:

β-cyclodextrin
NBS, H₂O

$$R \underset{(H)R'}{\overset{H}{\diagdown}} \text{—OTHP} \quad \xrightarrow[\text{NBS, H}_2\text{O}]{\beta\text{-cyclodextrin}} \quad R \underset{(H)R'}{\diagdown} \text{O}$$

[1] T. R. Kelly, D. Xu, G. Martınez, H. Wang, *Organic Letters* **2002**, *4*, 1527
[2] H. Khatuya, *Tetrahedron Letters* **2002**, *42*, 2643
[3] D. L. Boger, S. Ichikawa, H. Jiang, *Journal of the American Chemical Society* **2000**, *122*, 12169
[4] M. Narender, M. S. Reddy, K. R. Rao, *Synthesis* **2004**, 1741

# NCS

N-Chlorosuccinimide
[128-09-6]

## Commercially available

S. C. Virgil, *Encyclopedia of Reagents for Organic Synthesis*, John Wiley and Sons, Inc., L. A. Paquette, Ed., New York, 1995, **2**, 1205

**Notes:** Used for the a-chlorination of carbonyl compounds, sulfoxides and sulfides (See: *Ramberg-Bäcklund rearrangement*). Also used for preparing *N*-chloroamines.

## Examples:

1.[1]

2.[2]

3.[3]

---

[1] V. K. Aggarwal, G. Boccardo, J. M. Worrall, H. Adams, R. Alexander, *Journal of the Chemical Society Perkin Transactions 1* **1997**, 11
[2] N. DeKimpe, C. Stevens, M. Virag, *Tetrahedron* **1996**, <u>52</u>, 3303
[3] B. B. Snider, T. Liu, *Journal of Organic Chemistry* **1997**, <u>62</u>, 5630

# Ni(acac)₂

Nickel(II) acetylacetonate
[3264-82-2]

## Commercially available

J. Doyon, *Encyclopedia of Reagents for Organic Synthesis,* John Wiley and Sons, Inc., L. A. Paquette, Ed., New York, 1995, **6**, 3689

**Notes:** Catalyst for polymerization and dimerization of alkynes and alkenes. Has been used for catalyzing conjugate additions to enones. Useful as a Lewis acid.

## Examples:

1.[1]

2.[2]

3.[3]

[1] K. Shibata, M. Kimura, M. Shimizu, Y. Tamaru, *Organic Letters* **2001**, *3*, 2181
[2] M. Lozanov, J. Montgomery, *Journal of the American Chemical Society* **2002**, *124*, 2106
[3] A. E. Greene, J. P. Lansard, J. L. Luche, C. Petrier, *Journal of Organic Chemistry* **1984**, *49*, 931

# Ni(cod)₂

Bis(1,5-cyclooctadiene)nickel
[1295-35-8]

## Commercially available

P. A. Wender, T.E. Smith, *Encyclopedia of Reagents for Organic Synthesis*, John Wiley and Sons, Inc., L. A. Paquette, Ed., New York, 1995, **1**, 450

## Notes:

Useful source of nickel(0).

## Examples:

1.[1]

2.[2]

3.[3]

4.[4]

[1] M. F. Semmelhack, L.S. Ryono, *Journal of the American Chemical Society* **1975**, 97, 3873
[2] P. A. Wender, N. C. Ihle, C. R. D. Correia, *Journal of the American Chemical Society* **1988**, 110, 5904
[3] S. Arai, K.Tokumaru, T. Aoyama, *Tetrahedron Letters* **2004**, 45, 1845
[4] M. Lautens, S. Ma, *Journal of Organic Chemistry* **1996**, 61, 7246

# NMO

*N*-Methlymorpholine-*N*-oxide  [7529-22-8]

## Commercially available

M. R. Sivik, *Encyclopedia of Reagents for Organic Synthesis*, John Wiley and Sons, Inc., L. A. Paquette, Ed., New York, 1995, **5**, 3545

**Notes:**  A useful and relatively mild oxidizing agent.  Most current use attempts to use catalytic amounts.of OsO$_4$; thus NMO is used as a co-oxidant with OsO$_4$ for hydroxylation of alkenes. A method has been developed that allows catalytic amounts of NMM (*N*-methylmorpholine) to be used. [1]

## Examples:

1. [2]

2. [3]

*NMO* can carry out the conversion of an activated halogen compound to an aldehyde

3. [4]

$(6 \quad : \quad 1)$

---

[1] K. Bergstad, J. J. N. Piet, J. E. Backvall, *Journal of Organic Chemistry* **1999**, <u>64</u>, 2545

2. J. A. Marshall, S. Beaudoin, K. Lewinskit, *Journal of Organic Chemistry* **1993**, <u>58</u>, 5876

[3] W. P. Griffith, J. M. Jolliffe, S. V. Ley, K. F. Springhorn, P. D. Tiffin, *Synthetic Communications* **1992**, <u>22</u>, 1967 (AN 1992:489712)

[4] J. K. Cha, W. J. Christ, Y. Kishi, *Tetrahedron Letters* **1983**, <u>24</u>, 3943

# Noyori's Reagent

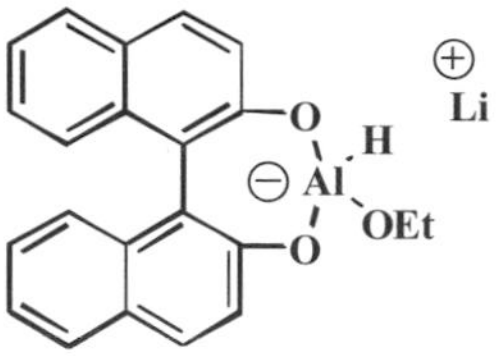

[16853-85-3] ·Lithium aluminum hydride-2,2´-dihydroxy-1,1´-binaphthyl ((*R*)-BINAL)
[18531-94-7] ·Lithium aluminum hydride-2,2´-dihydroxy-1,1´-binaphthyl ((*S*)-BINAL)

## Preparation:

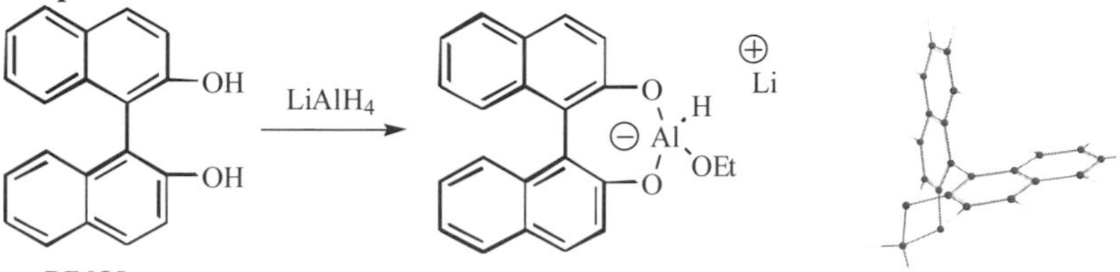

BINOL

## Commercially available

A. S. Gopalan, H. K. Jacobs, *Encyclopedia of Reagents for Organic Synthesis*, John Wiley and Sons, L. A. Paquette, Ed., New York, 1995, **5**, 3022

**Notes:** Used for enantioselective reduction of carbonyl groups. The reducing agent (for the appropriate enantiomer) can be designated in carton style to show the twist in the ring systems as:

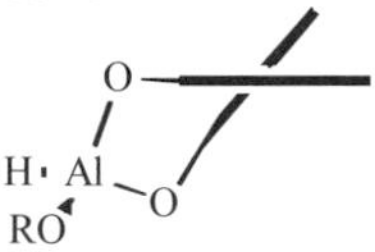

## Examples:

1.[1]

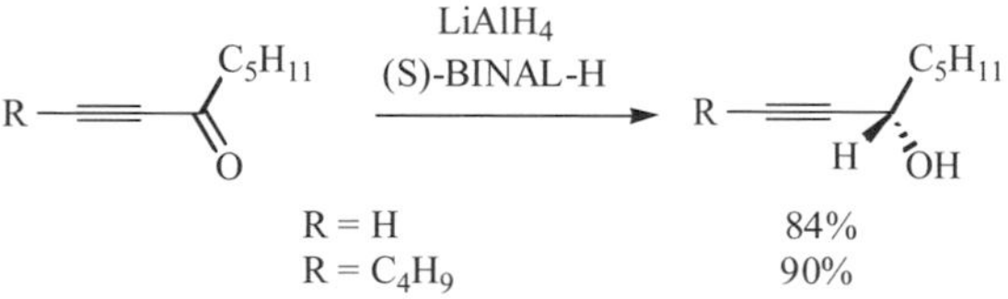

| | |
|---|---|
| R = H | 84% |
| R = C4H9 | 90% |

2.[2]

---

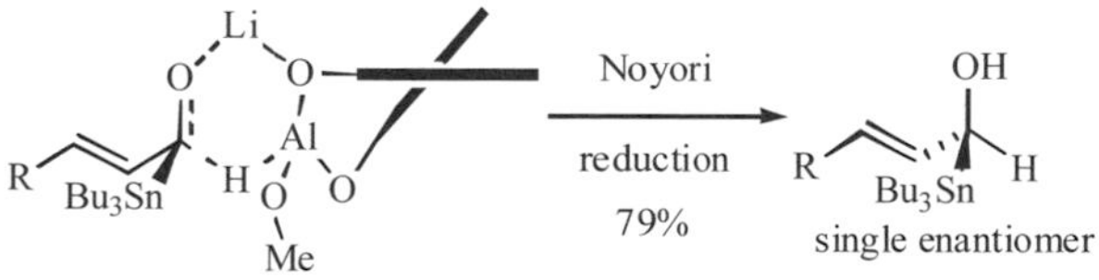

[1] M. Nishizawa, M. Yamada, R. Noyori, *Tetrahedron Letters* **1981**, <u>22</u>, 247
[2] J. A. Marshall, W. Y. Gung, *Tetrahedron Letters* **1998**, <u>29</u>, 1657

# Nystead Reagent

I-Zn-CH$_2$-Zn-I

Bis(iodozincio) methane
[31729-70-1]
Similar reagent

[41114-59-4]
Most likely a complex mixture
resulting from Schlenk equilibrium.

## Commercially available

S. Matsubara, *e-Encyclopedia of Reagents for Organic Synthesis*, L.A. Paquette, Ed.,, John Wiley
& Sons, Inc.,  online reference available at *http://www.interscience.wiley.com.*

**Notes:**  The bromo- variation, due to poor solubility is of limited use.  Useful for the BF$_3$-
catalyzed methylenation of aldehydes in the presence of ketones.

## Examples:

1.[1]

2.[2]

very little attack here.

3.[3]

via:

---

[1] S. Sugihara, M. Utimoto, *Synlett*, **1998**, 313; from S. Matsubara, *e-Encyclopedia of Reagents for Organic Synthesis*, John Wiley and Sons, Ltd., 2003

[2] K. Ukai, A. Arioka, H. Yoshino, H. Fushimi, K. Oshima, K. Utimoto, S. Matsubara, *Synlett* **2001**, 513; from S. Matsubara, *e-Encyclopedia of Reagents for Organic Synthesis*, John Wiley and Sons, Ltd., 2003

[3] S. Matsubara, K. Ukai, H. Fushimi, Y. Yokota, H. Yoshino, K. Oshima, K. Omoto, A. Ogawa, Y. Hiokib, H. Fujimotob, *Tetrahedron* **2002**, <u>58</u>, 8255

# Osmium tetroxide

## OsO₄

$OsO_4$

[20816-12-0]

### Commercially available

Y. Gao, *Encyclopedia of Reagents for Organic Synthesis*, John Wiley and Sons, Inc., L. A. Paquette, l York, 1995, **6**, 3801

**Notes:** Useful for the *cis*-dihydroxylation of alkenes. Attacks alkenes from the less-hindered face.

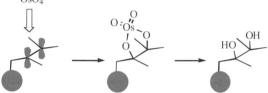

In examples containing an allylic hydroxyl (or alkoxy) group, the newly formed *cis*-diol will be anti to the existing group. This appears to override other steric factors (see footnote 1).

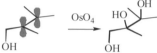

Early experiments used $OsO_4$ in stoichiometric amounts, more modern use reoxides the $OsO_2$ formed in the reaction: Common reagents for this include ***TBHP*** and ***NMO***

For studies on a polymer-bound $OsO_4$ reagent, see: S. Kobayashi, M. Endo, S. Nagayama, *Journal of the American Chemical Society* **1999**, <u>121</u>, 11229

A recoverable, microencapsulated $OsO_4$ has been described: S. Nagayama, M. Endo, S. Kobayashi, *Journal of Organic Chemistry* **1998**, <u>63</u>, 6094

### Examples:

1.[1]

catalytic
OsO₄, acetone
water; NMO,
tBuOH

No yield reported

This was a detailed study of the stereochemistry of the osmylation reaction.

6     :     1

2.[2]

t-BuOOH
t-BuOH

Cat. OsO₄

73%

---

[1] J. K. Cha, W. J. Christ, Y. Kishi, *Tetrahedron* **1984**, <u>40</u>, 2247

[2] K. B. Sharpless, K. Akashi, *Journal of the American Chemical Society* **1976**, <u>98</u>, 1986

# Oxone[®][1]

## (2KHSO₅.KHSO₄.K₂SO₄)

A "triple salt", providing a convenient souce of
potassium monoperoxysulfate (potassium hydrogen persulfate)
[37222-66-5]

## Commercially available

J. M. Crandall, *Encyclopedia of Reagents for Organic Synthesis*, John Wiley and Sons, Inc., L. A.
Paquette, Ed., New York, 1995, **6**, 4265

**Notes:** This reagent is a useful oxidizing agent.

The reagent can provide for ***Baeyer-Villiger*** conditions and reacts with nitrogen and sulfur to form
*N*-oxides and sulfones, respectively.

As a reagent, it is commonly found in the preparation of dimethyldioxirane (***DDO***), itself a useful
oxidizing agent.

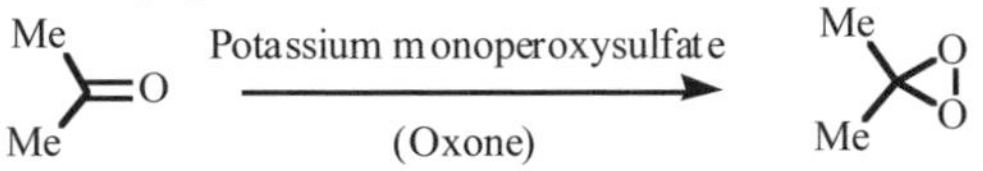

The oxidation of aldehydes to acids might follow the following mechanism:

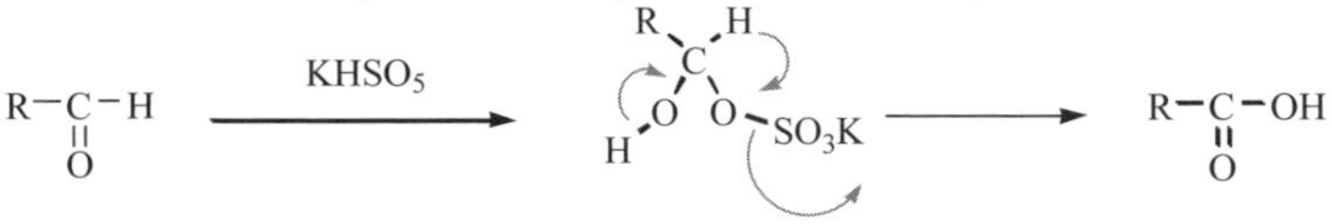

## Examples:

1.[2]

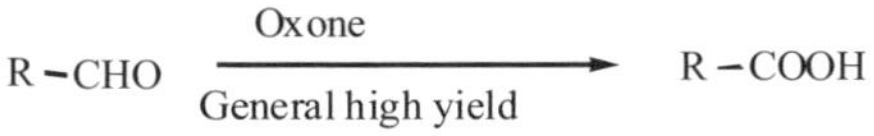

R—CHO  $\xrightarrow{\text{Oxone}}$  R—COOH
      General high yield

2.[3]

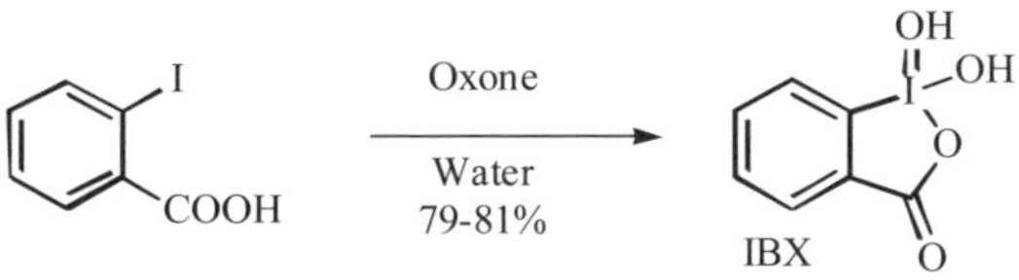

3.[4] Use of Oxone for conversion of C-B to C-OH bonds in the hydroboration reaction.

4.[5]

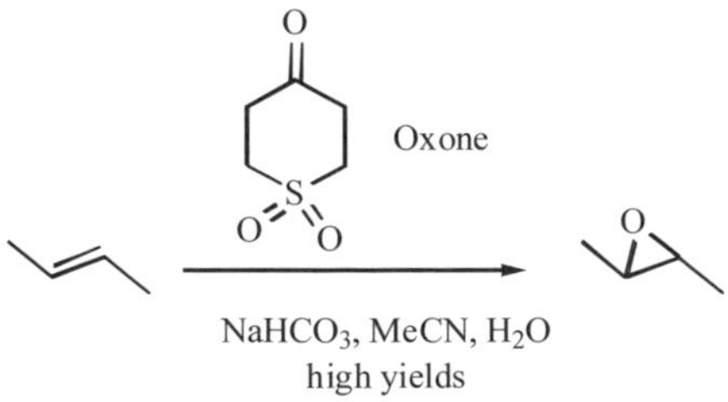

NaHCO₃, MeCN, H₂O
high yields

---

[1] Also commercially available as Curox® and Caroat®
[2] B. R. Travis, M. Sivakumar, G. O. Hollist, B. Borhan, *Organic Letters* **2003**, 5, 1031
[3] M. Frigerio, M. Santagostino, S. Sputore, *Journal of Organic Chemistry* **1999**, 64, 4537
[4] D. H. B. Ripin, W. Cai, S. J. Brenek, *Tetrahedron Letters* **2000**, 41, 5817
[5] D. Yang, Y.-C. Yip, G.-S. Jiao, M.-K. Wong, *Journal of Organic Chemistry* **1998**, 63, 8952

# Ozone

$$\ddot{O}=\overset{\cdot\cdot}{\underset{\cdot\cdot}{O}}\overset{\cdot\cdot\overset{\cdot\cdot}{O}\cdot}{}$$

O₃
[10028-15-6]

**Preparation:** generated by passing dry oxygen through two electrodes connected to AC.

R. A. Berglund, *Encyclopedia of Reagents for Organic Synthesis*, John Wiley and Sons, Inc., L. A. Paquette, Ed., New York, 1995, **6**, 3837

**Notes:** For organic chemists the major use of this oxidizing agent is in the cleavage of alkene bonds. This can be followed by either oxidative or reductive workups. The reaction can be run in a number of ***common solvents***.

Among the reducing agents for the reductive workup, Me₂S is one of the most common. Hydrogen peroxide is a common oxidative workup.

*Criegee Mechanism for Ozonolysis*

| 1,3-dipolar cycloaddition | 1,2,3-trioxolane | carbonyl oxide | 1,2,4-trioxolane (ozonide) |

*Griesbaum Coozonoysis Reaction[1]*

**Examples:**

1.[2]

---

[1] Y. Tang, Y. Dong, J. M. Karle, C.A. DiTusa, J. L. Vennerstrom, *Journal of Organic Chemistry* **2004**, <u>69</u>, 6470

[2] P. H. Dussault, J. M. Raible, *Organic Letters* **2000**, <u>2</u>, 3377.

# Pd(dba)₂

Bis(dibenzylideneacetone)palladium(0)
[32005-36-0]

## Commercially available

J. R. Stille, *Encyclopedia of Reagents for Organic Synthesis*, John Wiley and Sons, Inc.,
L. A. Paquette, Ed., New York, 1995, **1**, 482

## Notes:

Catalytic amounts of Pd(dba)₂ will activate allylic acetates to nucleophilic attack.

Triphenylphosphine or *1,2-bis(diphenylphosphino)ethane (dppe)* are often part of the reaction
mixture.

## Examples:

1.[1]

2.[2]

3.[3]

4.[4]

---

[1] K. H. Shaughnessy, B. C. Harmann, J. F. Hartwig, *Journal of Organic Chemistry* **1998**, _63_, 6546
[2] S. E. Denmark, J. Y. Choi, *Journal of the American Chemical Society* **1999**, _121_, 5821
[3] S. Cacchi, G. Fabrizi, A. Goggiamani, G. Zappia, *Organic Letters* **2001**, _3_, 2539
[4] K. Yamazaki, Y. Kondo, *Journal of Combinatorial Chemistry* **2002**, _4_, 191

# Pearlman's Catalyst
## Pd(OH)$_2$ / C
Palladium(II) hydroxide on carbon
[7440-05-3]
## Commercially available

A. O. King, I. Shinkai, *Encyclopedia of Reagents for Organic Synthesis*, John Wiley and Sons, Inc., L. A. Paquette, Ed., New York, 1995, **6**, 3888

**Uses:** A catalyst that finds use in the removal of benyzyl groups under hydrogenolysis conditions.

## Examples:

1.[1]

2.[2]

3.[3]

[1] Y. Al-Abed, N. Naz, D. Mootoo, W. Voelter, *Tetrahedron Letters* **1996**, 37, 8641
[2] W. Notz, C. Hartel, B. Waldscheck, R. R. Schmidt, *Journal of Organic Chemistry* **2001**, 66, 4250
[3] D. E. DeMong, R. M. Williams, *Journal of the American Chemical Society* **2003** 125, 8561

# Phase-transfer Catalysts
## Commercially available
**Notes:**  A phase transfer reagent is a compound soluble in both an aqueous and organic phase.

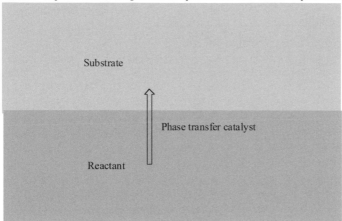

Generally the phase transfer catalyst will be either a salt (ex, KMnO4, NaOCl, etc) where the anion portion of the reactant is transported to the substrate by way of a cationic material with solubility in both phases.

We also include crown ethers as phase transfer reagents.  In these case the cation of the reactant is most likely coordinated.

## Examples:

Triethylbenzylammonium bromide *[TEBA]*
[56-37-1], P.B. Savage, *Encyclopedia of Reagents for Organic Synthesis*, John Wiley and Sons, Inc.,
L. A. Paquette, Ed., New York, 1995, **1**, 376

$Bu_4N^+ Br^-$
Tetrabutylammonium bromide
[1643-19-2], A.B. Charette, *Encyclopedia of Reagents for Organic Synthesis*, John Wiley and Sons,
Inc., L. A. Paquette, Ed., New York, 1995, **7**, 3632

$(Me(CH_2)_7]_3N-Me)^+ Cl^-$
Methyltrioctylammonium chloride  [Aliquat]®
[5137-55-3]. M.E. Bos, *Encyclopedia of Reagents for Organic Synthesis*, John Wiley and Sons, Inc.,
L. A. Paquette, Ed., New York, 1995, **7**, 3632

# PPA
## Unspecified structure
Polyphosphoric acid
[8017-16-1]
### Commercially available

J. H. Dodd, *Encyclopedia of Reagents for Organic Synthesis*, John Wiley and Sons, Inc., L. A. Paquette, Ed., New York, 1995, **6**, 4169

**Notes:** A strong mineral acid of unspecified structure. It contains a mixture of phosphoric acids and derivatives, with a high content of $P_2O_5$. Does chemistry similar to that found with ***Eaton's reagent***.

Useful for ***Friedel-Crafts acylation*** reactions:

for ring sizes of 5 - 7

## Examples:

1.[1]

82%

2.[2]

HOAC
PPA
96%

3.[3]

PPA
50%

---

[1] A. Jones, P. U. Civcir, *Tetrahedron* **1997**, <u>53</u>, 11529
[2] A. P. Venkov, I. I. Ivanov, *Tetrahedron* **1996**, <u>52</u>, 12299
[3] M. M. Ramana, P. V. Potnis, *Synthesis* **1996**, 1090

# PPTS

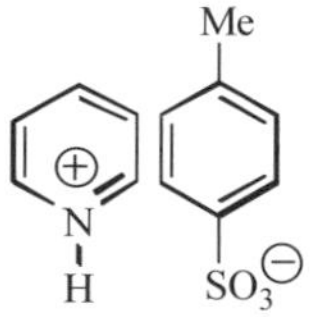

Pyridinium *p*-toluenesulfonate
[24057-28-1]

## Commercially available

A. A. Galan, *Encyclopedia of Reagents for Organic Synthesis*, John Wiley and Sons, Inc., L. A. Paquette, Ed., New York, 1995, **6**, 4375

**Notes:** A weak acid, useful for substrates with acid-sensitive functional groups. Has been used to place and remove alcohol protecting groups:

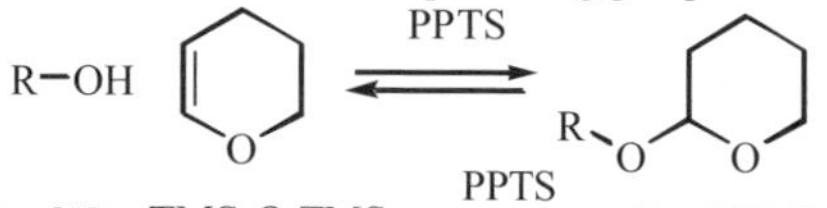

R—OH + TMS-O-TMS $\xrightleftharpoons{\text{PPTS}}$ R—OTMS

## Examples:

1.[1]

Selective removal of protecting groups:

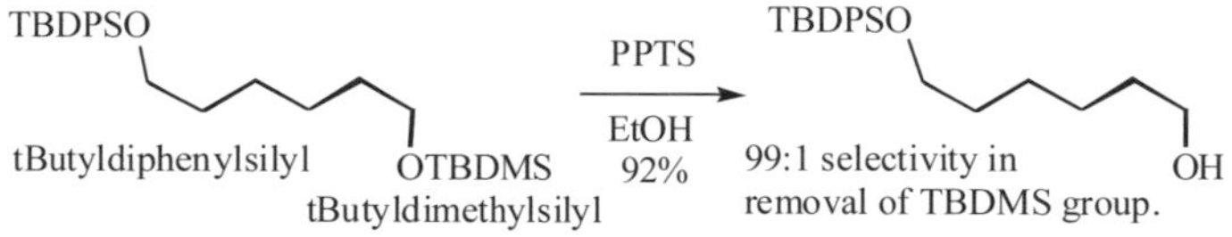

tButyldiphenylsilyl          OTBDMS    92%    99:1 selectivity in
                     tButyldimethylsilyl              removal of TBDMS group.

2.[2]

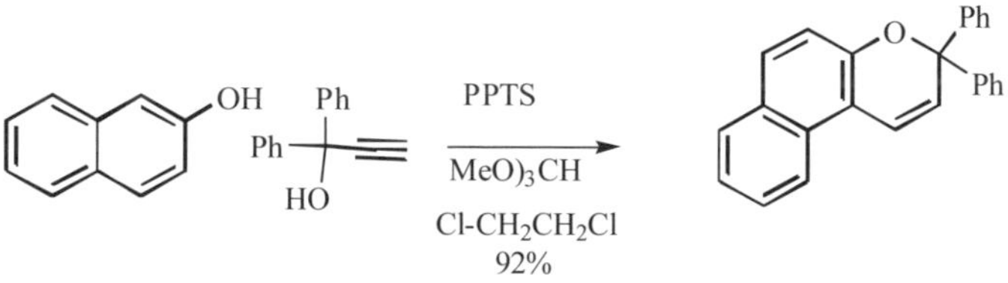

3.[3]

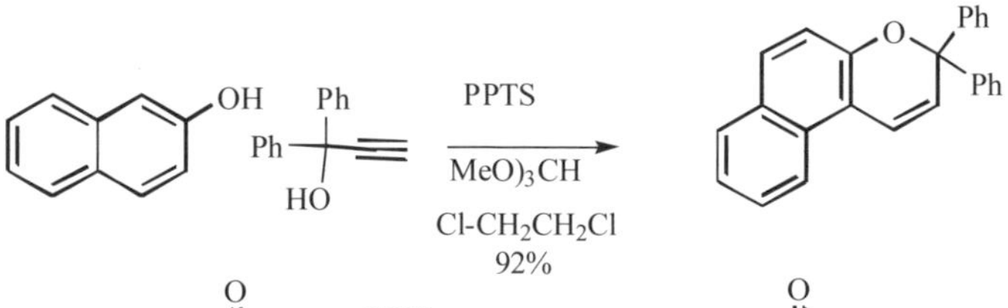

4.[4]

---

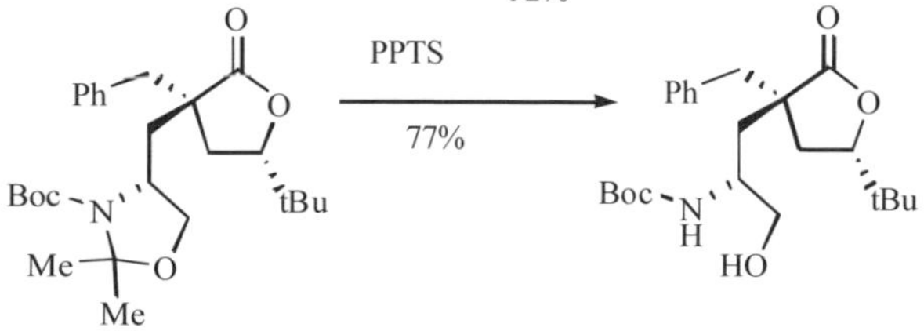

[1] C. Prakash, S. Saleh, I. A. Blair, *Tetrahedron Letters* **1989**, 30, 19

[2] W. Zhao, E. M. Carreira, *Organic Letters* **2003** 5, 4153

[3] H. Choo, Y. Chong, C. K. Chu, *Organic Letters* **2001**, 3, 1471

[4] K. Nakamura, T. J. Baker, M. Goodman, *Organic Letters* **2000**, 2, 2967

# Proton Sponge

Proton Sponge

[20734-58-1]
1,8-bis(dimethylamino)naphthalene

**Commercially available.** Can be prepared:[1]

Me-SO$_4$-Me , NaH

B. A. Barner, *Encyclopedia of Reagents for Organic Synthesis*, John Wiley and Sons, Inc., L. A. Paquette, Ed., New York, 1995, **1**, 494

**Notes:** A very strong base (pKa = 21.1) with poor nucleophilicity due to steric effects. This makes it a useful organic base for reactions sensitive to most other Lewis bases.

**Examples:**

1.[2]

POCl$_3$, proton sponge

PO(OMe)$_3$

Other organic bases were not effective

2.[3]

proton sponge

dioxane

92%

3.[4]

proton sponge

CH$_2$Cl$_2$

81%

[1] H. Quast, W. Risler, G. Döllscher, *Synthesis* **1972**, 558
[2] T. Kovács, L. Ötvös, *Tetrahedron Letters* **1988**, 29, 4525
[3] R. Lavilla, T. Gotsens, M. Guerrero, J. Bosch, *Synthesis* **1995**, 382
[4] S. T. Sigurdsson. B. Seeger, U. Kutzke, F. Eckstein, *Journal of Organic Chemistry* **1996**, 61, 3883

# PTSA

Me—⬡—SO₃H

[104-15-4]
Tosic Acid

## Commercially available.

G. S. Hamilton, *Encyclopedia of Reagents for Organic Synthesis*, John Wiley and Sons, Inc., L. A. Paquette, Ed., New York, 1995, **7**, 4941.

## Examples:

1.[1]

MeO)₃CH

PTSA

Quant

2.[2]

HO    OH

TsOH, PhH

3.[3]

PTSA / H₂O

4.[4]

PTSA

CH₂Cl₂

51%

[1] G. A. Kraus, I. Kim, *Organic Letters* **2003**, <u>5</u>, 1191
[2] K. M. Brummond, J. Lu, *Organic Letters* **2001**, <u>3</u>, 1347
[3] S. B. Singh, *Tetrahedron Letters* **2000** <u>41</u>, 6973
[4] D. Dugat, A. Chiaroni, C. Riche, J. Royer, H.-P. Husson, *Tetrahedron Letters* **1997**, <u>38</u>, 5801

# Raney Nickel
## Ni-Al
[106-51-4]
## Commercially Available

T.-K. Yang, D.-S. Lee, *Encyclopedia of Reagents for Organic Synthesis*, John Wiley and Sons, Inc., L. A. Paquette, Ed., New York, 1995, **6**, 4401

## Notes:

1. Many forms, based on ratio of NaOH to aluminum
2. Particularly useful for reducing C-S bonds.[1]

## Examples:

1.[2]

2.[3]

3.[4] A method for making peptides:

5.[5]

[1] See:H. Atkins, *Organic Reactions*, Vol 8, Chapter 1; G . R. Pettit, E. E. vanTamelen, *Organic Reactions*, Vol 12, Chapter 5.

[2] S.-M. Yang, S. K. Nandy, A. R. Selvakumar, J.-M. Fang, *Organic Letters* **2000**, 2, 3719

[3] M. Couturier, J. L. Tucker, B. M. Andresen, P. Dubé, J. T. Negri, *Organic Letters* **2001**, 3, 465

[4] Z. Y. Liang, P. E. Dawson, *Journal of the American Chemical Society* **2001**, 123, 526

[5] B. Klenke, I. H. Gilbert, *Journal of Organic Chemistry* **2001**, 66, 2480

# Rawal's Diene

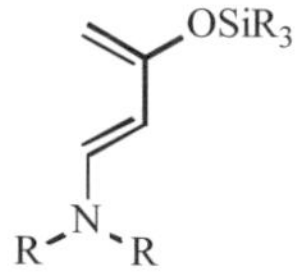

## Preparation:[1]

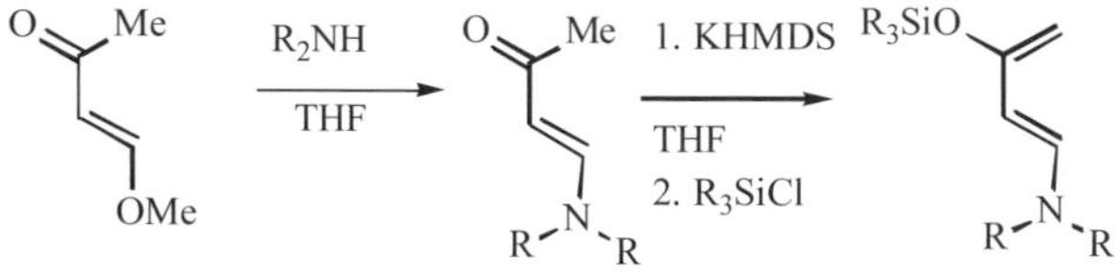

## Notes:

See: **Brassard's**, **Chan's**, and *Danishefsky's dienes*.

*Rawal's diene* has been shown to be about 25 times more reactive than ***Danishefsky's diene***.[1]

See: ***Diels-Alder Reaction***.

## Examples:

1.[1]

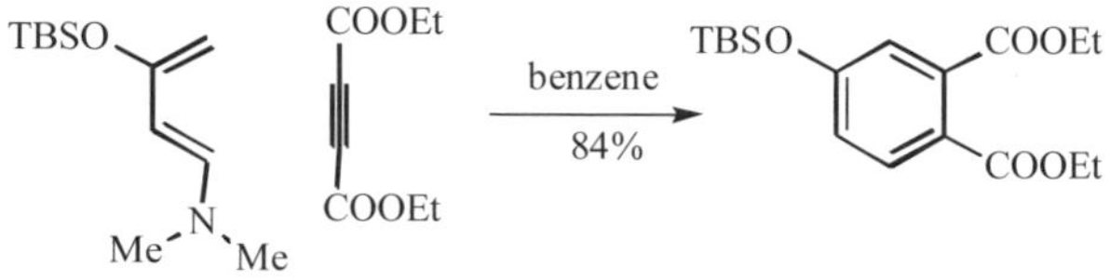

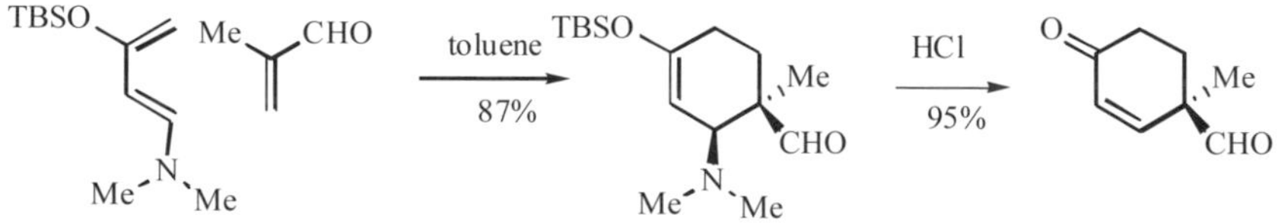

2.[2]

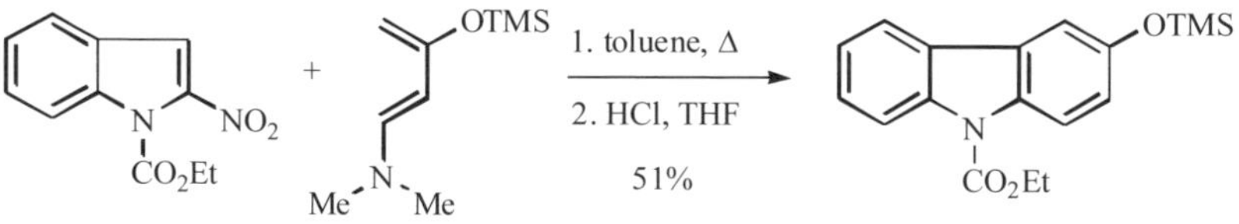

---

[1] S. A. Kozmin, V. H. Rawal, *Journal of Organic Chemistry* **1997**, <u>62</u>, 5252

[2] T. L. S. Kishbaugh, G. W. Gribble, *Tetrahedron Letters* **2001**, <u>42</u>, 4783

# Rosenmund Catalyst
## Pd / BaSO₄
[7440-05-3]
## Commercially available

S. Siegel,*Encyclopedia of Reagents for Organic Synthesis*, John Wiley and Sons, Inc., L. A.
Paquette, Ed., New York, 1995 **6**, 3861

**Notes:** Used for the conversion of an acid chloride to an aldehyde. Quinoline is often added to
further reduce catalyst activity. See: ***Rosenmund Reduction***.

An improved procedure uses THF as solvent with 2,6-dimethylpyridine as base.[1]

## Examples:
1.[2]

$H_2$ / Pd / BaSO₄

toluene, heat

95%

2.[3]

$H_2$ / Pd-C

2,6-lutidine
Yield not provided, however, from thedieater
the yield was 74%

R = (+) menthyl

3.[4]

$H_2$ / Rosenmund Catalyst

toluene   98%

4.[5] A modified ***Rosenmund Reduction***:

Pd-C

"Quinoline S"

toluene, NaOAc

64-83%

[1] A. W. Burgstahler, L. O. Weigel, C. G. Shaefer, *Synthesis* **1976**, 767
[2] S. Danishefsky, M. Hirama, K. Gombatz, T. Harayama, E. Berman, P.F. Schuda, *Journal of the American Chemical Society* **1979**, 101, 7020
[3] M. O. Duffey, A. LeTiran, J. P. Morken, *Journal of the American Chemical Society* **2003**, 125, 1458
[4] S. Chimichi, M. Boccalini, B. Cosimelli, *Tetrahedron* **2002**, 58, 4851
[5] A. I. Rachlin, H. Gurien, D. P. Wagner, *Organic Syntheses* CV6, 1007

# Schlosser's Base
## (Sometimes referred to as the *Lochmann-Schlosser Base*)
### *n*-BuLi / *t*-BuO⁻ K⁺

**Preparation:** BuLi, after removal of hexane, is taken up in cooled (-90 C) THF. At reduced temperature, (-50 C) is added the *t*-BuOK and material from which a proton is to be abstracted.[1,2]

| X. Xia, *Encyclopedia of Reagents for Organic Synthesis*, John Wiley and Sons, Inc., L.A.Paquette, Ed., New York, 1995, **2**, 923 |
| --- |

**Notes:** The combination of *n*-BuLi / K⁺*t*-BuO⁻ is considerably more basic than *n*-BuLi alone. It can exchange C-H bonds of $pK_a$ 35-50.[3]

## Examples:

1.[4] After removal of an allylic proton, a useful contathermodynamic isomerization:

This procedure provided a simple method for converting α- to β-pinene in high yield and enantiomeric purity.

2.[5]

3.[6]

---

[1] a.) M. Schlosser (Editor), in *Organometallics in Synthesis. A Manual*, John Wiley and Sons, New York, 1994, Chapter 1.; b.) See M. Schlosser, O. Desponds, R. Lehmann, E. Moret, G. Rauchschwalbe, *Tetrahedron* **1993**, <u>49</u>, 10175.

[2] Y. Naruse, Y. Ito, S. Inagaki, *Journal of Organic Chemistry* **1999**, <u>64</u>, 639 examine a number of different base combinations in a study on ethylations.

[3] M. Schlosser, S. Strunk, *Tetrahedron Letters* **1984**, <u>25</u>, 741

[4] H. C. Brown, M. Zaidlewicz, K. S. Bhat, *Journal of Organic Chemistry* **1989**, <u>54</u>, 1764

[5] F. Cominetti, A. Deagostino, C. Prandi, P. Venturello, *Tetrahedron* **1998**, <u>54</u>, 14603

[6] S. V. Kessar, P. Singh, K. N. Singh, S. K. Singh, *Chemical Communications* **1999**, 1927

# Schwartz Reagent

Chlorobis(cyclopentadienyl)hydridozirconium
[37342-97-5]

## Commercially available

## Preparation:[1]

---

| T. Takahashi, N. Suzuki, *Encyclopedia of Reagents for Organic Synthesis*, John Wiley and Sons, Inc, New York, edited by L.A. Paquette, 1995, **2**, 1082 |

## Notes:

Lipshutz[2] notes that the shelf-life of the reagent suggests that a simple in-situ preparation serves well:

$$Cp_2ZrCl_2 \xrightarrow[\text{THF}]{\text{LiEt}_3\text{BH}} \left[ Cp_2Zr(H)Cl \right] \xrightarrow{R\text{—}\equiv} $$

| $\overset{\oplus}{E}$ source | Alkyne | $\overset{\oplus}{E}$ | Percent |
|---|---|---|---|
| $I_2$ | (structure, O–C(=O)Ph) | I | 80 |
| $H_2O$ | (structure, S–Ph) | H | 88 |

The hydrozirconation of alkenes will generally result in positional isomerization of the Zr to the least-hindered position. The Zr-intermediate can be replaced with Br ($Br_2$, NBS), I ($I_2$), Cl (**NCS**), OH (MCPBA, or basic aq. hydrogen peroxide).

## Examples:

1.[3]

BnO... TMS

1. $Cp_2Zr(H)Cl$

2. $I_2$, $CH_2Cl_2$

50%

30 : 1

2.[4] Also converts **Weinreb amides** to aldehydes

$$R\text{—}C(=O)\text{—}N(Me)(OMe) \xrightarrow[\text{THF}]{Cp_2Zr(D)Cl} R\text{—}C(=O)\text{—}D$$

---

[1] S. L. Buchwald, S. J. LaMaire, R. B. Nielsen, B. T. Watson, S. M. King, *Organic Syntheses* CV9, 162

[2] B. H. Lipshutz, R. Keil, E. L. Ellsworth, *Tetrahedron Letters* **1990**, 31, 7257

[3] A. Arefolov, N. F. Langille, J. S. Panek, *Organic Letters* **2001**, 3, 3281

[4] J. T. Spletstoser, J. M. White, G. I. George, *Tetrahedron Letters* **2004**, 45, 2787

# Schweizer's Reagent

Vinyltriphenylphosphonium bromide
[5044-52-0]

## Commercially available

Edward E. Schweizer, *Encyclopedia of Reagents for Organic Synthesis*, John Wiley and Sons, Inc, New York, edited by L.A. Paquette, 1995, **7**, 5508

**Notes:**  Useful reagent for generating functionalized ***Wittig intermediates***.

Unique application for non-conjugated dienes[1]

1.[2]

2.[3]

Overall yield for this three-component, one step reaction was 70%

---

[1] B. O'Connor, G. Just, *Tetrahedron Letters* **1985**, <u>26</u>, 1799.
[2] E. E. Schweizer, *Journal of the American Chemical Society* **1964**, <u>86</u>, 2744
[3] H. H. Posner, S. B. Lu, *Journal of the American Chemical Society* **1985**, <u>107</u>, 1424

# Schwesinger P4 base

Phosphorimidic triamide, (1,1-dimethylethyl)tris[tris(dimethylamino)phosphoranylidene]
Phosphazene Base P4-*t*-Bu
[111324-04-0]

## Commercially available

Reinhard Schwesinger, *Encyclopedia of Reagents for Organic Synthesis*, John Wiley and Sons,
Inc, New York, edited by L.A. Paquette, 1995, **6**, 4110

**Notes:**

Extremely strong base. Has excellent solublizing properties, and complex cation so that the anion is
more reactive.

## Examples:

1.[1]

No conversion found with NaOH, *DABCO*, *DMAP*, *DBU*.

2.[2]

3.[3]

| | | | | |
|---|---|---|---|---|
| KHMDS | 20% | 1 | | 99 |
| P - 4/ THF | 78% | 5 | | 95 |

[1] G. Jenner, *Tetrahedron* **2002**, <u>58</u>, 4311

[2] T. Imahor1, Y. Kondo, *Journal of the American Chemical Society* **2003**, <u>125</u>, 8082

[3] D. A. Alonso, C. Najera, M. Varea, *Tetrahedron Letters* **2004**, <u>45</u>, 573

# Selenium Dioxide
### SeO$_2$
Selenium (IV) Oxide
[7446-08-4]
## Commercially available

> William J. Hoekstra, *Encyclopedia of Reagents for Organic Synthesis*, John Wiley and Sons, Inc., L. A. Paquette, Ed., New York, 1995, **6**, 4437

## Notes:
Primarily noted for its ability to oxidize (hydroxylate) saturated allylic, propargylic or cabon atom alpha to a carbonyl group.

**General observations:** See reference 2 for additional comments.

CH$_2$ > CH$_3$ > CH

most favored ◄-- least favored

Ring oxidation preferred; and α to the more substituted alkene position.

Oxidation α to a carbonyl group is called a ***Riley oxidation***.

For mechanistic studies see: E. J. Corey, J. P. Schaefer, *Journal of the American Chemical Society* **1960**, <u>82</u>, 918

## Examples:
2.[1]

SeO$_2$, EtOH
90%

3.[2]

SeO$_2$ oxidation
>98% this isomer

This was a detailed study of the regiochemistry of SeO$_2$ allylic oxidation.

---

[1] R. W. Curley, Jr., C. J. Ticoras, *Journal of Organic Chemistry* **1986**, <u>51</u>, 256
[2] H. Rapoport, U. T. Bhalerao, *Journal of the American Chemical Society* **1971**, <u>93</u>, 4835

# Schrock's Catalyst

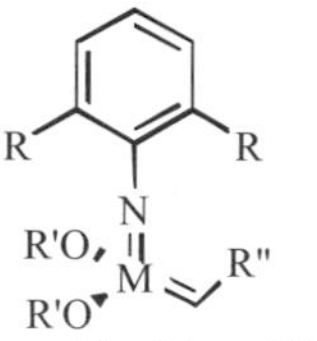

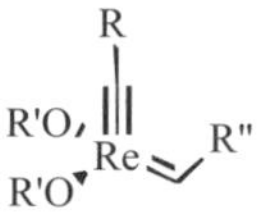

M = Mo or W

These catalysts generally contain W, Mo or Re, and have a general structure:

## Commercially available
## Notes:
The Schrock catalysts are very reactive and subtle changes in substitution can modify reactivity. The catalysts are generally tolerant to other functional groups. Tend to be less stable to air, water or heat than *Grubbs catalysts*. Useful for *RCM reactions*.

1.

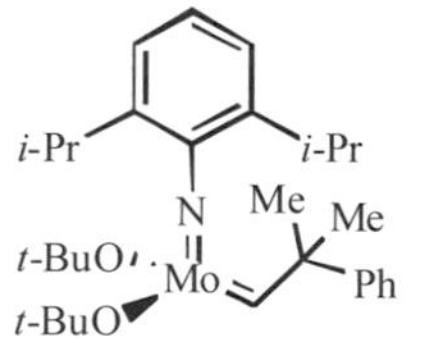

[126949-65-3]

Reactions need to carried out in Schlenk lines.

2.

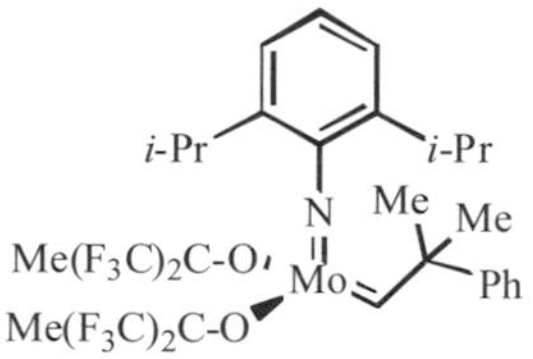

[136070-71-8]

Reactions need to carried out in Schlenk lines.

3.

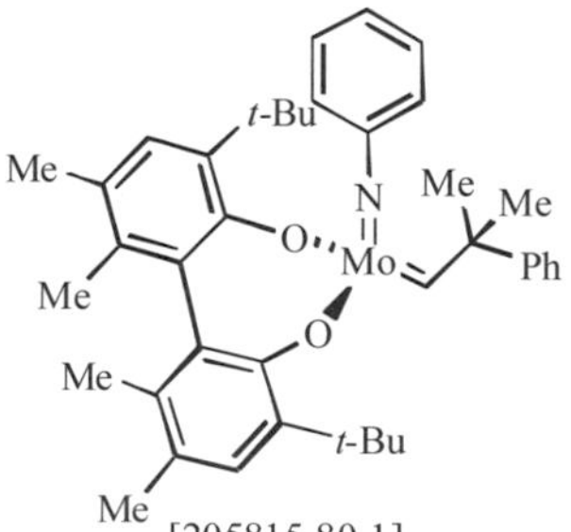

[205815-80-1]

*Schrock-Hoveyda catalyst*

4.

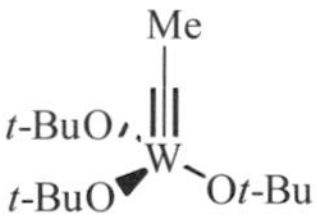

[78234-36-3]

# SMEAH

$$\underset{\text{Na}}{\overset{\oplus}{}} \quad \text{OCH}_2\text{CH}_2\text{OMe}$$
$$\text{H} - \underset{\underset{\ominus}{|}}{\overset{|}{\text{Al}}} - \text{OCH}_2\text{CH}_2\text{OMe}$$
$$\underset{\text{H}}{}$$

Sodium bis(2-methoxyethoxy)aluminum hydride
[22722-98-1]

## Commercially available.

M. Gugelchuk, *Encyclopedia of Reagents for Organic Synthesis*, John Wiley and Sons, Inc., L. A. Paquette, Ed., New York, 1995, **7**, 4518

**Notes:** Selective reducing agent.

## Examples:

1.[1]

| | | | |
|---|---|---|---|
| LiAlH$_4$ | 19 | 49 | 32 |
| SMEAH | 100 | 0 | 0 |

2.[2]

3.[3]

[1] S. E. Denmark, T. K. Jones, *Journal of Organic Chemistry* **1982**, <u>47</u>, 4595
[2] M. Ishizaki, O. Hoshino, Y. Iitaka, *Journal of Organic Chemistry* **1992**, <u>57</u>, 7285
[3] R. B. Gammill, L. T. Bell, S. A. Mizsak, *Tetrahedron Letters* **1990**, <u>31</u>, 5301

# Sodium Nitrite
## NaNO$_2$
[6732-00-0]
**Commercially available**. Less expensive than Li or K counterparts; thus more commonly used.

K. J. McCullough, *Encyclopedia of Reagents for Organic Synthesis*, John Wiley and Sons, Inc., L. A. Paquette, Ed., New York, 1995, **7** 4604

## Notes:
General source of the unstable nitrous acid; most often formed *in-situ*.

For synthetic organic chemistry the reagent finds use for preparing diazo compounds:

$$ArNH_2 \ + \ NaNO_2 \ + 2 \, HX \ \longrightarrow \ ArN_2{}^+X^- \ + \ NaX \ + 2 \, H_2O$$

**Proposed Mechanism:**

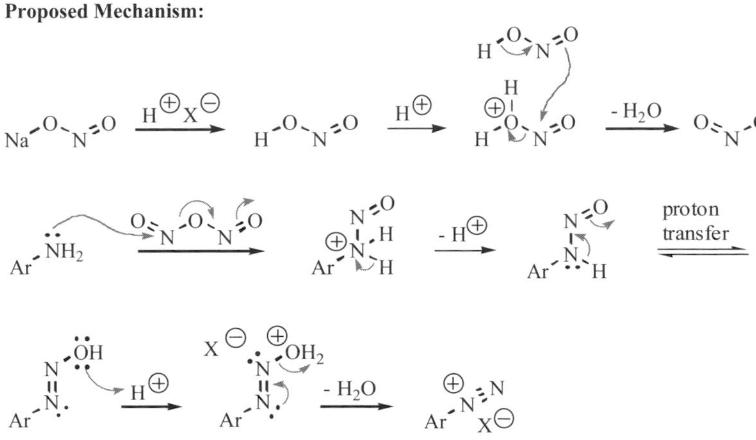

Other variations of the reaction:

**Griess (Diazotization) Reaction**

$$2 \, ArNH_2 \ + \ N_2O_3 \ + \ 2 \, HNO_3 \ + \ H_2O \ \longrightarrow \ 2 \, ArN_2{}^+NO_3^- \ + \ 4 \, H_2O$$

**Knoevenagel (Diazotization) Method**

$$ArNH_2 \ + \ RONO \ + \ HX \ \longrightarrow \ ArN_2{}^+X^- \ + \ ROH \ + \ H_2O$$

**Witt (Diazotization) Method**

$$2 \, ArNH_2 \ + \ Na_2S_2O_5 \ + \ 4 \, HNO_3 \ \longrightarrow \ 2 \, ArN_2{}^+NO_3^- \ + \ Na_2S_2O_7 \ + \ 4 \, H_2O$$

A selection of reactions using this general approach include:

*Bart Reaction; Borsche (Cinnoline) Synthesis; Demjanov Rearrangment; Diazo Reaction; Gattermann Reaction; Gattermann Method; Pschorr Arylation; Sandmeyer Reaction; Schiemann Reaction (Balz-Schiemann Rxn); Tiffeneau-Demjanov Reaction; Widman-Stoermer (Cinnoline) Synthesis*

# Solvents
## Benzene-based solvents

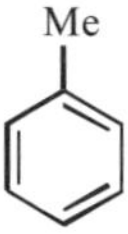

*Toluene*

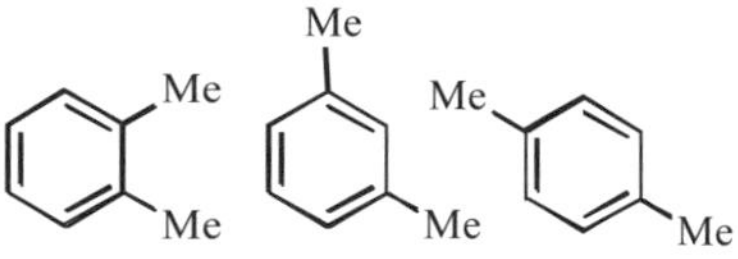

*Xylenes*

o-          m-          p-

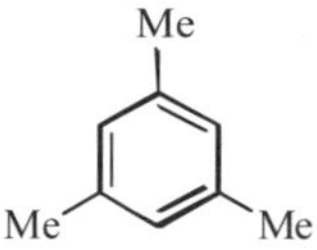

**Mesitylene** 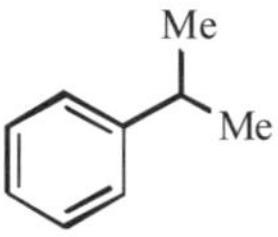          **Cumene**

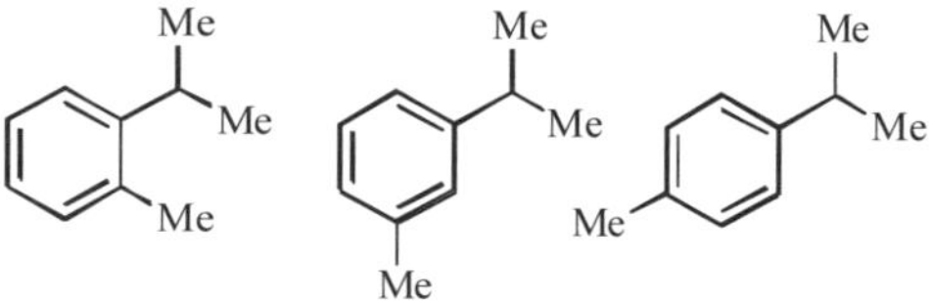

*Cymene*

o-          m-          p-

## Ether-based solvents

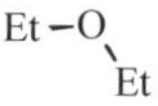

*Et₂O*          *MTBE*
Diethyl ether          Methyl *t*-butyl ether

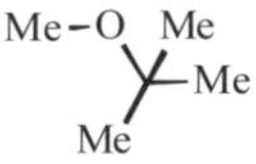

*THF*
Tetrahydrofuran  Dioxane

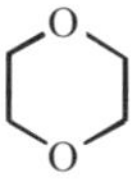

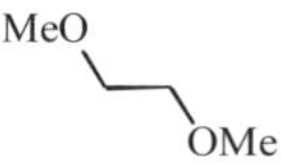

Glyme          Diglyme
**DME**
Dimethoxyethane

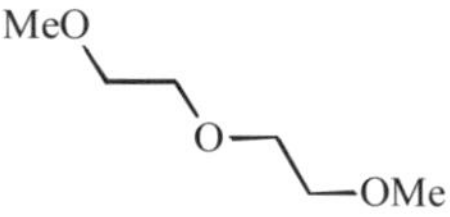

Ethers can stabilize cations; diglyme can provide properties similar to those found in ***crown ethers***.

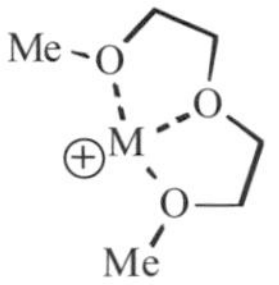

## Pyridine-based solvents

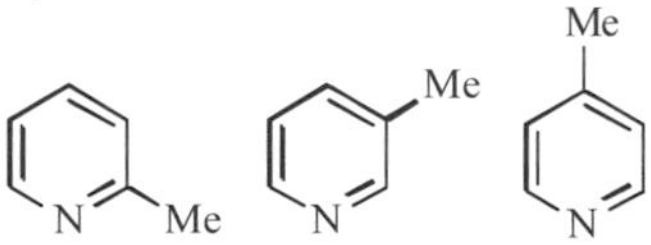

*Picolines*

| 2-, | 3-, | and | 4- picoline |
| 2-, | 3-, | and | 4-methylpyridine |

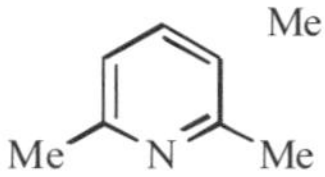

**Lutidine**     **Lupetidine**
  2,6-

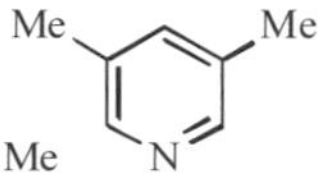

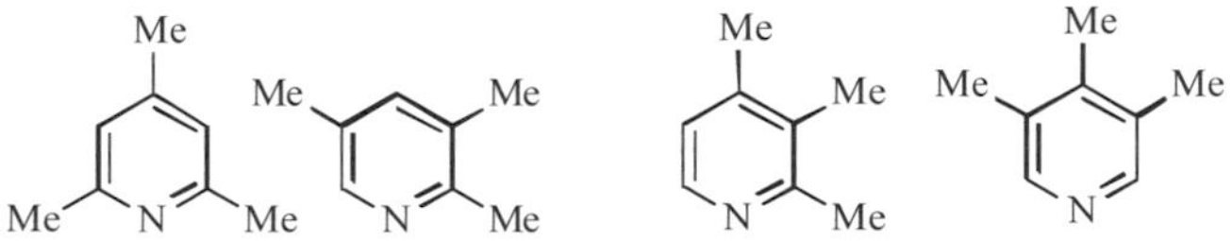

**Collidines**

| 2,4,6- | 2,3,5- | 2,3,4- | 3,4,5- |

## Additional Common Solvents and Additives

**Acetone:**

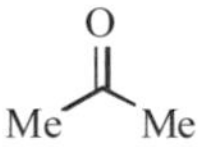

**Acetonitrile: MeCN**

**Carbon tetrachloride: CCl$_4$**

**Chloroform: CHCl$_3$**

**Dichloromethane: CH$_2$Cl$_2$**

**Dimethylformamide, DMF:**

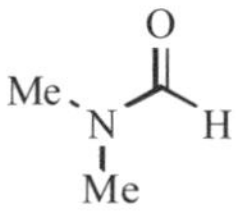

**Dimethylsulfoxide, _DMSO_:**

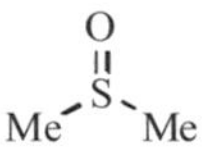

**Hexamethylphosphoramide, _HMPA_, _HMPT_:**

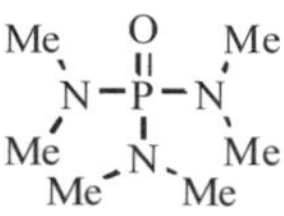

# Stiles Reagent

MeOMgOCO$_2$Me
Methyl magnesium carbonate (MMC)
[4861-79-4]

**Preparation:** Bubble carbon dioxide into a solution of magnesium methoxide. Upon heating, a series of equilibrium reactions forms the reagent:

CO$_2$ + Mg(OMe)$_2$ ⇌ MeO-MgO-CO$_2$Me x CO$_2$ ⇌ (MeO-CO$_2^-$)$_2$Mg
  Magnesium methoxide [109-88-6]

D. Caine, *Encyclopedia of Reagents for Organic Synthesis*, John Wiley and Sons, Inc., L. A.
Paquette, Ed., New York, 1995, **5**, 3204

L. N. Mander, *e-Encyclopedia of Reagents for Organic Synthesis* L.A. Paquette, Ed.,, John Wiley
& Sons, Inc., online reference available at *http://www.interscience.wiley.com.*

## Notes:

1. Used for placing a carboxyl group adjacent to an existing carbonyl.
2. A useful method of incorporating an alkyl group with the carboxylation has been reported.[1]

## Examples:

1.[2]

2.[3]

3.[4]

Me−NO$_2$  →[Stiles Reagent]  MeOOC-CH$_2$-NO$_2$

---

[1] E. S. Hand, S. C. Johnson, D. C. Baker, *Journal of Organic Chemistry* **1997**, <u>62</u>, 1348
[2] A. S. Kende, J. Chen, *Journal of the American Chemical Society* **1985**, <u>107</u>, 7184
[3] F. G. Favaloro, Jr., T. Honda, Y. Honda, G. W. Gribble, N. Suh, R. Risingsong, M. B. Sporn,
*Journal of Medicinal Chemistry* **2002**, <u>45</u>, 4801
[4] Reported in K. B. G. Torssell, K.V. Gothelf, *Encyclopedia of Reagents for Organic Synthesis*,
John Wiley and Sons, Inc., L. A.Paquette, Ed., New York, 1995, **6**, 3745

# Stryker's Reagent

## (Ph₃PCuH)₆

hexa-*m*-hydrohexakis(triphenylphosphine)hexacopper
[33636-93-0]

## Commercially available:

**Preparation:** An expedient preparation of *Stryker's reagent*.[1]

J. F. Daeuble, J. M. Stryker, *Encyclopedia of Reagents for Organic Synthesis*, John Wiley and Sons, Inc., L. A. Paquette, Ed., New York, 1995, **4**, 2651

## Notes:

Conjugate addition of hydride. The reagent will not react with isolated double bonds, carbonyl groups. Many functional groups resistant to reaction.

## Examples:

1.[2]

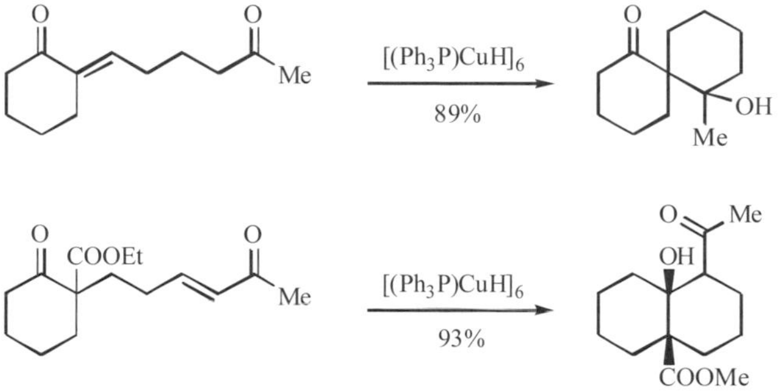

2.[3]

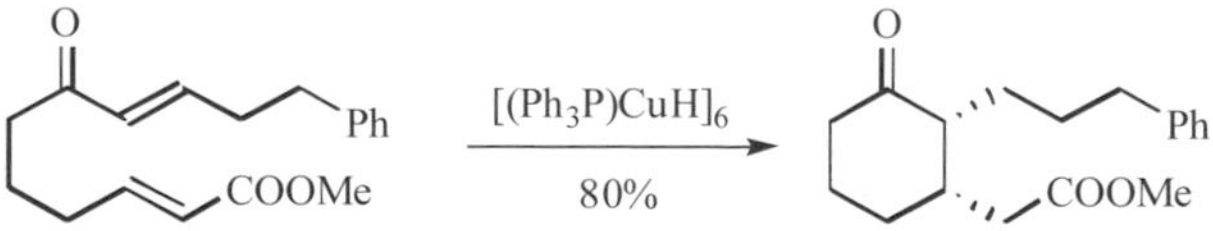

3.[4]

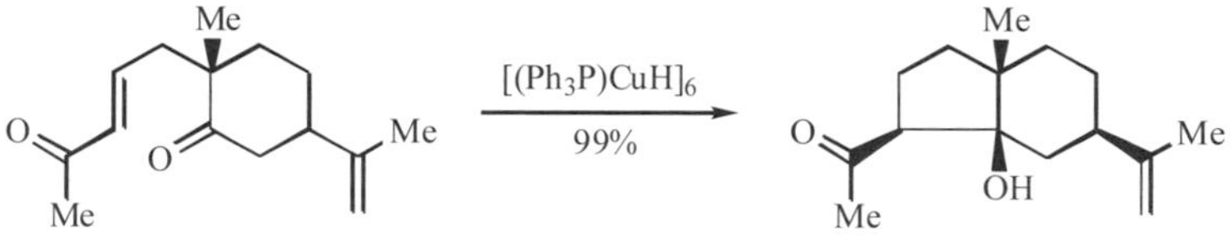

4.[5]

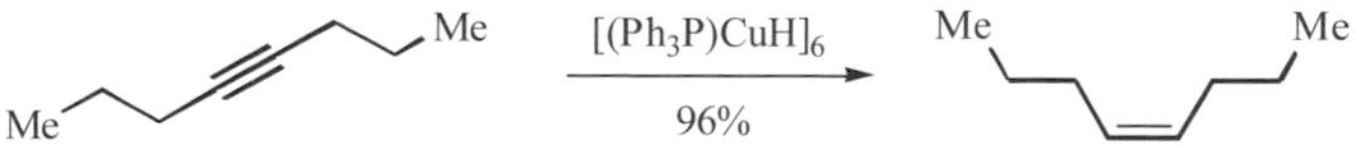

[1] P. Chiu, Z. Li, K. C. M. Fung *Tetrahedron Letters*, **2003**, 44, 455.
[2] P. Chiu, C.-P. Szeto, Z. Geng, K.-F. Cheng, *Organic Letters*, **2001**, 3, 1901
[3] T. M. Kamenecka, L. E. Overman, S. K. Ly Sakata, *Organic Letters* **2002**, 4, 79
[4] P. Chiu, C. P. Szeto, Z. Geng, K. F. Cheng, *Tetrahedron Letters* **2001**, 42, 4091
[5] J. F. Daeuble, C. McGettigan, J.M. Stryker, *Tetrahedron Letters* **1990**, 31, 2397

# Super hydride

$$\left[ \begin{array}{c} Et \\ | \\ Et-B-H \\ | \\ Et \end{array} \right]^{\ominus} \quad Li^{\oplus}$$

Lithium triethylborohydride
[22560-16-3]

## Commercially available

M. Zaidlewicz, H. C. Brown, *Encyclopedia of Reagents for Organic Synthesis*, John Wiley and Sons, Inc., L. A. Paquette, Ed., New York, 1995, **5**, 3180

**Notes:** An extremely powerful source of hydride. If one assigns a relative value of 1 for the nucleophilic character of hydride from NaBH$_4$, the hydride from LiAlH$_4$ is about 250 times more reactive and that from Super hydride is about 10,000 times more reactive.

The reagent is particularly useful for the reaction C-X -> C-H, where X = I > Br > Cl > F. The replacement of tosylate and mesylate are also readily carried out.

## Examples:
1.[1]

| Hydride Source | Yield | Ratio | | |
|---|---|---|---|---|
| *DIBAL* | 64 | 25 | : | 75 |
| Super hydride | 83 | 85 | : | 15 |
| *L-Selectride* | 65 | 75 | : | 25 |

2.[2]

Super hydride

THF

97%

3.[3]

1. Super hydride
2. *Swern oxidation* then *Wittig-Horner*

90%

---

[1] E. A. Bercot, D. E. Kindrachuk, T. Rovis, *Organic Letters* **200X**, XX, XXX; see a similar study: N. Pourahmady, E. J. Eisenbraun *Journal of Organic Chemistry* **1983**, 48, 3067
[2] Naoki Toyooka, Maiko Okumura, Hideo Nemoto, *Journal of Organic Chemistry* **2002**, 67, 6078
[3] N. Toyooka, A. Fukutome, H. Nemoto, J. W. Daly, T. F. Spande, H. M. Garraffo, T. Kaneko, *Organic Letters* **2002**, 4, 1715

# TBAF

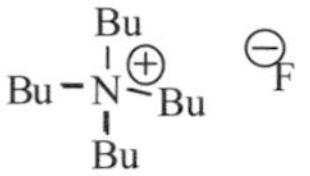

Tetrabutylammonium fluoride
[100-85-6]

## Commercially available

H.-Y. Li, *Encyclopedia of Reagents for Organic Synthesis*, John Wiley and Sons, Inc., L. A. Paquette, Ed., New York, 1995, **7**, 4728

**Notes:**  Useful for the removal of silyl groups:

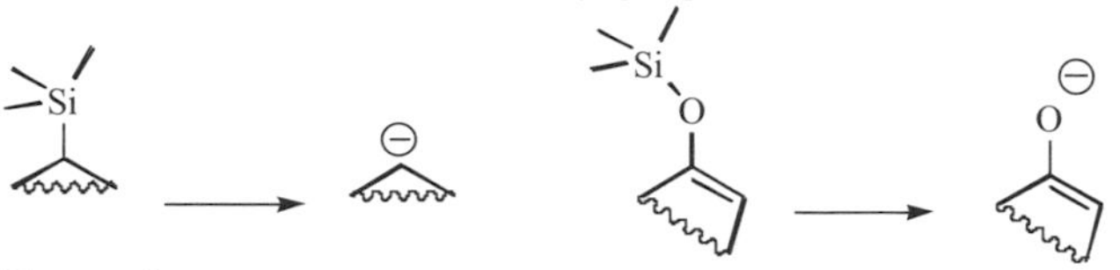

## Examples:

1.[1]

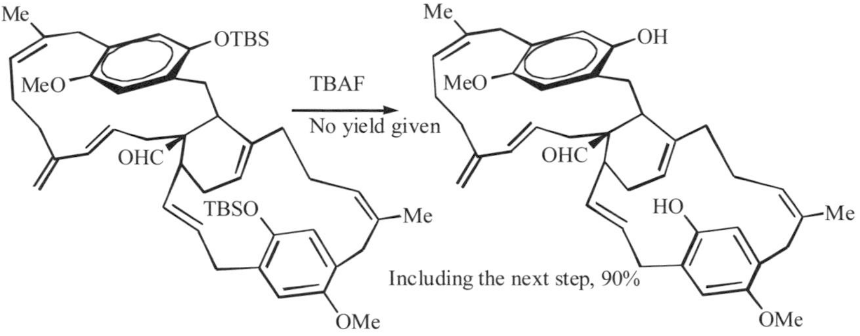

TBAF
No yield given

Including the next step, 90%

2.[2]

Promotion of S$_N$Ar reactions:

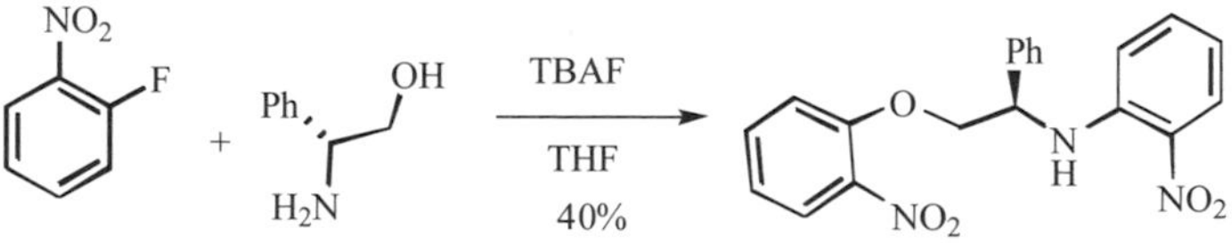

TBAF

THF

40%

3.[3]

Note reloaction of acyl group
during the deprotection.

TBAF

70%

---

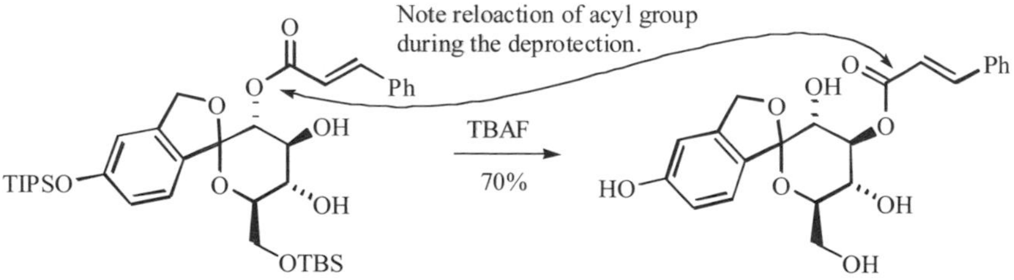

[1] M. E. Layton, C. A. Morales, M. D. Shair, *Journal of the American Chemical Society* **2002**, <u>124</u>, 773

[2] T. Temal-Laib, J. Chastanet, J. Zhu, *Journal of the American Chemical Society* **2002**, <u>124</u>, 583

[3] C. Hamdouchia, C. Jaramillo, J. Lopez-Pradosb, A. Rubioa, *Tetrahedron Letters* **2002**, <u>43</u>, 3875

# TBHP

Me
   \  O
Me —<  |  ‵OH
   /
Me   Me

*t*-Butyl hydroperoxide
[75-91-2]

## Commercially available.

K. Jones, T. E. Wilson, S. S. Nikam, *Encyclopedia of Reagents for Organic Synthesis*, John
Wiley and Sons, Inc., L. A. Paquette, Ed., New York, 1995, **2**, 880

**Notes:**  Oxidizing agent.  See also ***THP*** (triphenylmethyl hydroperoxide), [4198-93-0].  Is useful
with OsO$_4$ for catalytic hydroxylation reactions,

## Examples:

1.[1]

Pr —≡— Et   —CuCl$_2$—→   Pr —≡— Et
              TBHP / tBuO            ‖
                 91%                  O

2.[2]

TBDPSO ⌁ CH$_2$OH   —(+) DET, Ti(O*i*-Pr)$_4$—→   TBDPSO ⌁ O ‵H
        |                      TBHP                      |   H   CH$_2$OH
       Me                       90%                     Me

3.[3]  A method for removing thioketals:

⌁ ⟨S–S⟩   —TBHO—→   ⌁ ⟨ ⟩=O
          refluxing MeOH

[1] P. Li, W. M. Fong, L. C. F. Chao, S. H. C. Fung, I. D. Williams, *Journal of Organic Chemistry*
**2001**, _66_, 4087
[2] R. M. Garbaccio, S. J. Stachel, D. K. Baeschlin, S. J. Danishefsky, *Journal of the American
Chemical Society* **2001**, _123_, 10903
[3] N. B. Barhate, P. D. Shinde, V. A. Mahajan, R. D. Wakharkar, *Tetrahedron Letters* **2002**, _43_, 6031

# TEBA (TEBAC)

Benzyltriethylammonium chloride
[56-37-1]

## Commercially available

P. B. Savage, *Encyclopedia of Reagents for Organic Synthesis*, John Wiley and Sons, Inc., L. A. Paquette, Ed., New York, 1995, **1**, 376

**Notes:**  Useful phase transfer reagent.

## Examples:

1.[1]

CHCl$_3$,  NaOH
TEBA,  H$_2$O
80%

2.[2]

A phase transfer approach to substituted ureas via a ***Curtius Rearrangement***

NaN$_3$, TEBA
toluene
84%

*t*-BuNH$_2$

[1] M. Makosza, M. Wawrzyniewicz, *Tetrahedron Letters* **1969**, <u>10</u>, 4659
[2] G. Groszek, *Organic Process Research & Development* **2002**, <u>6</u>, 759

# Tebbe Reagent

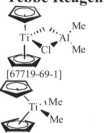

[67719-69-1]

See: The *Petasis modification* of the *Tebbe Reagent*.[1]  N. A. Petasis, *Encyclopedia of Reagents for Organic Synthesis*, John Wiley and Sons, Inc., L. A. Paquette, Ed., New York, 1995 **1**, 470 [1271-66-5]

## Commercially available

D. A. Strauss, *Encyclopedia of Reagents for Organic Synthesis*, John Wiley and Sons, Inc., L. A. Paquette, Ed., New York, 1995, **2**, 1078

## Notes:
1.   Particularly useful for conversion of lactones and esters to enol ethers
2.   Can serve as alternative to Wittig reagent for the methylenation of hindered ketones.
3.   See *Tebbe Reaction*.

## Examples:
1.[2]

Me, R / O O  →(Tebbe Reagent, 87 - 92% overall)→  Me, R / O CH₂ (taken directly to next step)  →(Claisen Rearrangement)→  product with R, R, Me, O, Me

2.[3]

Me / O C(=O) Me  →(Tebbe Reagent, 85%)→  Me / O C(=CH₂) Me  →(Heat)→

Me / Me C=O  →(Tebbe Reagent, 90%)→  Me / Me C=CH₂

---

[1] This reagent is useful due to its low acidity and basicity, allowing it to be used in the presence of easily enolizable carbonyl compounds.  The reagent will selectively react with a simple carbonyl in the presence of a carboxyl-type carbonyl group.
[2] W. A. Kinney, M. J. Coghlan, L. A. Paquette, *Journal of the American Chemical Society* **1986**, 106, 6868
[3] J. W. S. Stevenson, T. A. Bryson, *Tetrahedron Letters* **1982**, 23, 3143

# TEMPO

[2564-83-2]
2,2,6,6-Tetramethylpiperidin-1-oxyl

## Commercially available

F. Montanari, S. Quici, *Encyclopedia of Reagents for Organic Synthesis*, John Wiley and Sons, Inc., L. A. Paquette, Ed., New York, 1995, **7**, 4821

**Notes:** In conjunction with other oxidizing agents, this reagent provides mild conditions for oxidations, for example:[1]

## Examples:

1.[2]

2.[3]

3.[4]

*Trichloroisocyanuric acid*

---

[1] P. L. Anelli, F. Montanari, S. Quici, *Organic Syntheses* <u>CV8</u>, 367. A number of other primary alcohol oxidations are reported.
[2] K. Sakuratani, H. Toga, *Synthesis* **2003**, 21
[3] Jeffrey B. Epp, Theodore S. Widlanski, *Journal of Organic Chemistry* **1999**, <u>64</u>, 293
[4] L. De Luca, G. Giacomelli, A. Porcheddu, *Organic Letters* **2001**, <u>3</u>, 3041

## 1,1,3,3-Tetramethyldisilazane

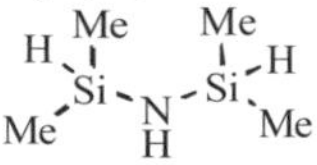

[15933-59-2]

### Commercially available

K. Tamao, *Encyclopedia of Reagents for Organic Synthesis*, John Wiley and Sons, Inc., L. A. Paquette, Ed., New York, **1995**, **7**, 4809

### Notes:

Tamoa notes how this is particularly useful for polyol synthesis:

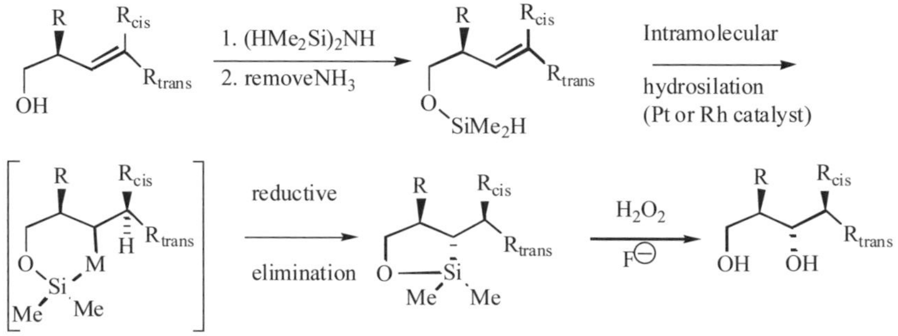

### Examples
1.[1]

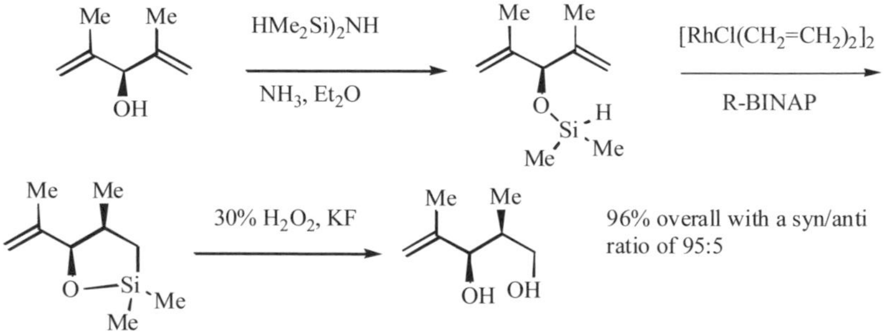

96% overall with a syn/anti ratio of 95:5

2.[2]

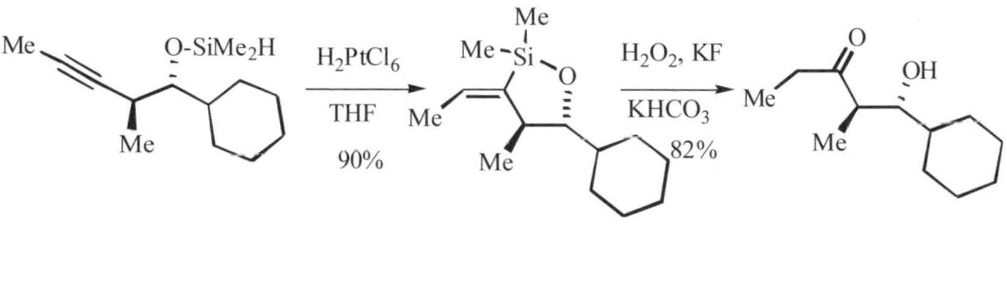

---

[1] K. Tamao, T. Tanaka, T. Nakajima, R. Sumiya, H. Arai, Y. Ito, *Tetrahedron Letters* **1990**, <u>31</u>,7333
[2] J. A. Marshall, M. M. Yanik, *Organic Letters* **2000**, <u>2</u>, 2173

# TMEDA

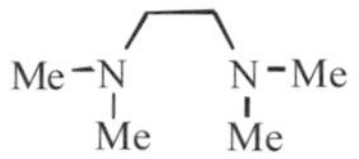

Me−N        N−Me
    |        |
   Me      Me

*N,N,N',N'*-Tetramethylethylenediamine
[110-18-9]

## Commercially available

R. K. Haynes, S. C. Vonwiller, *Encyclopedia of Reagents for Organic Synthesis*, John Wiley and Sons, Inc., L. A. Paquette, Ed., New York, 1995, **7**, 4811

## Notes:

Enhances reactivity of organolithium reagents.
Note the influence of TMEDA on the lithiation of naphthyl methyl ether: (D.A. Shirley, C.F. Cheng, *Journal of Organometallic Chemistry* **1969**, 20, 251.)

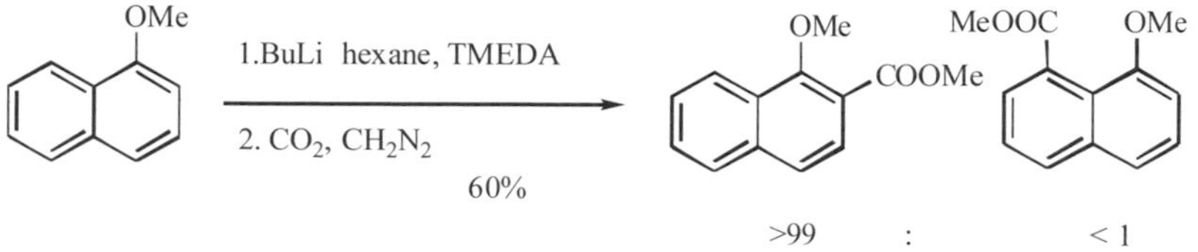

60%

>99    :    < 1

## Examples:

1.[1]

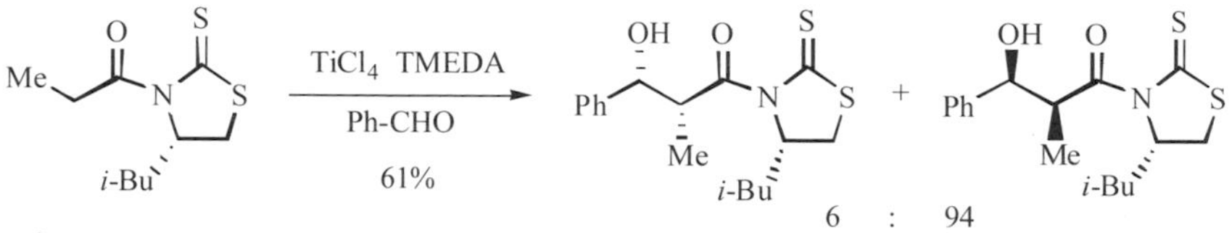

TiCl$_4$ TMEDA

Ph-CHO

61%

6 : 94

2.[2]

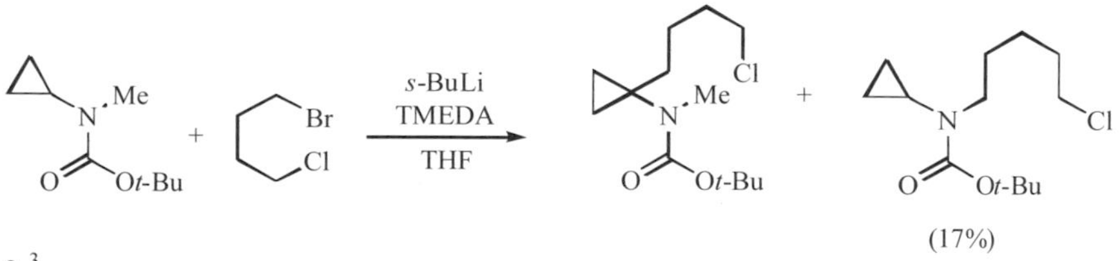

*s*-BuLi
TMEDA

THF

(17%)

3.[3]

CuI, TMEDA

TMS-Cl, THF

78%

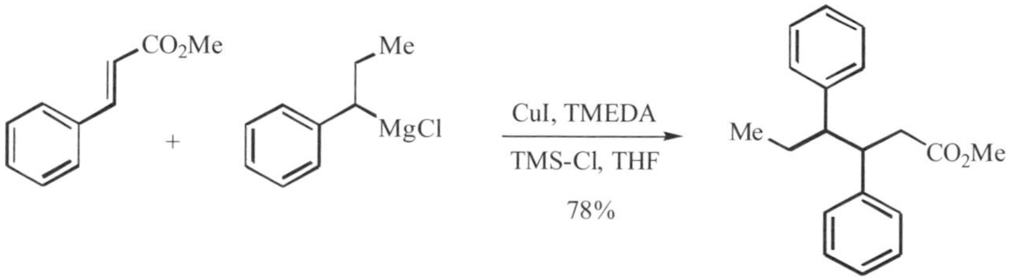

[1] M. T. Crimmins, K. Chaudhary, *Organic Letters* **2000**, 2, 775
[2] Y. S. Park, P. Beak, *Tetrahedron*, **1996**, 52, 12333
[3] P. S. Van Heerden, B. C. B. Bezuidenhoudt, D. Ferreira, *Tetrahedron* **1996**, 52, 12313

# TMSOTf

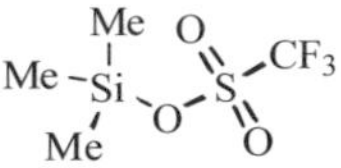

Trimethylsilyl trifluoromethanesulfonate
[88248-68-4]

## Commercially available

JJ. Sweeny, G. Perkins, *Encyclopedia of Reagents for Organic Synthesis*, John Wiley and Sons, Inc., L. A. Paquette, Ed., New York, 1995, **7**, 5315

**Notes:** A very efficient Lewis acid. Useful for the conversion of ketones to their corresponding TMS enol ethers:

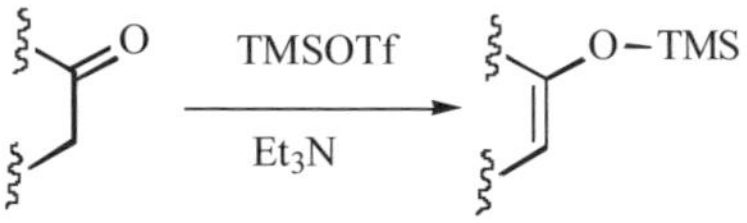

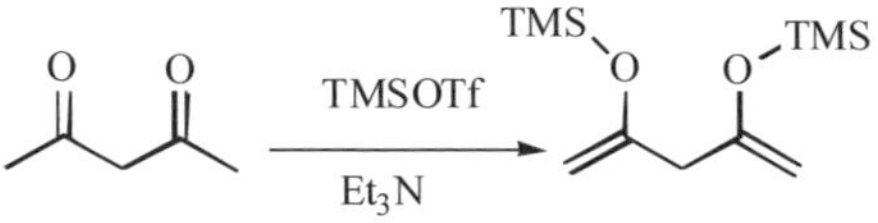

Can serve as catalyst for the selective ketalization of multicarbonyl-containing compounds. Sterically-hindered or conjugated carbonyl groups will react much more slowly to the reaction:

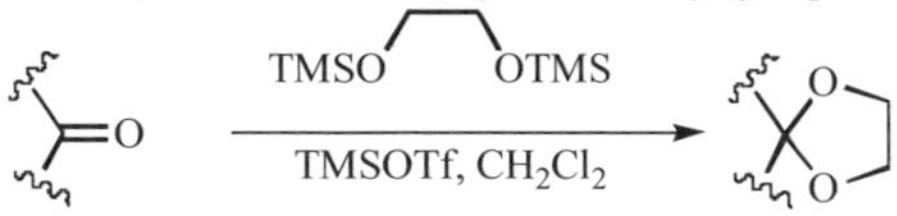

## Examples:

1.[1]

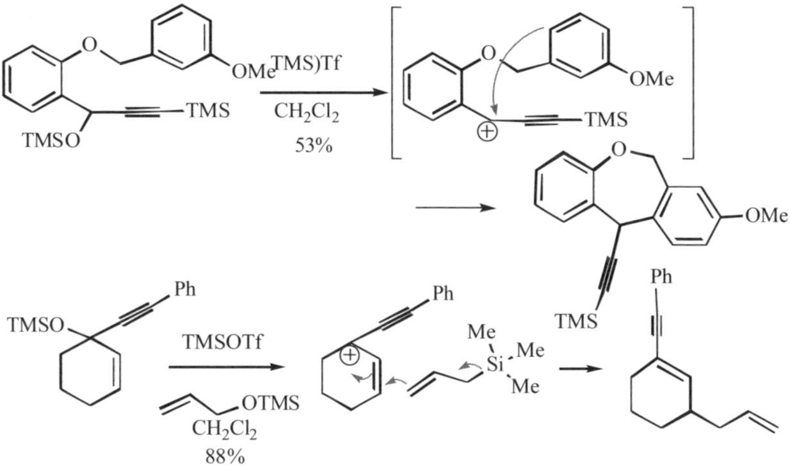

2.[2]

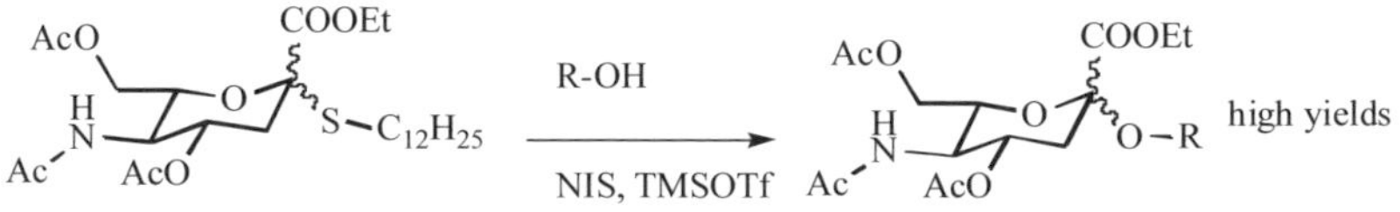

high yields

[1] T. Ishikawa, M. Okano, T. Aikawa, S. Saito, *Journal of Organic Chemistry* **2001**, <u>66</u>, 4635

[2] K. Matsuoka, T. Onaga ,T. Mori, J.-I. Sakamoto,T. Koyama N. Sakairi, K. Hatanoa, D. Terunuma, *Tetrahedron Letters* **2004**, <u>45</u>, 9384

3.[1]

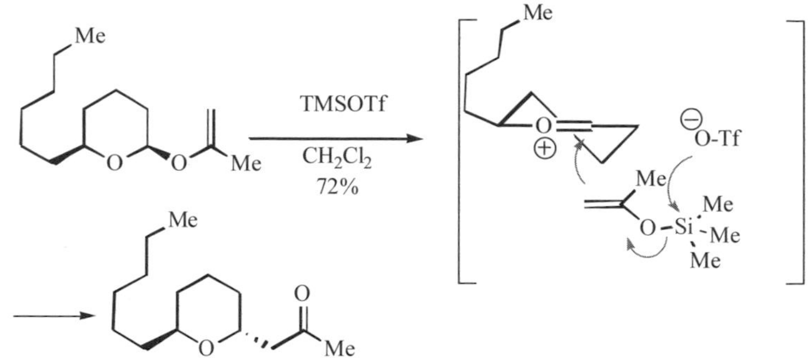

4.[2]

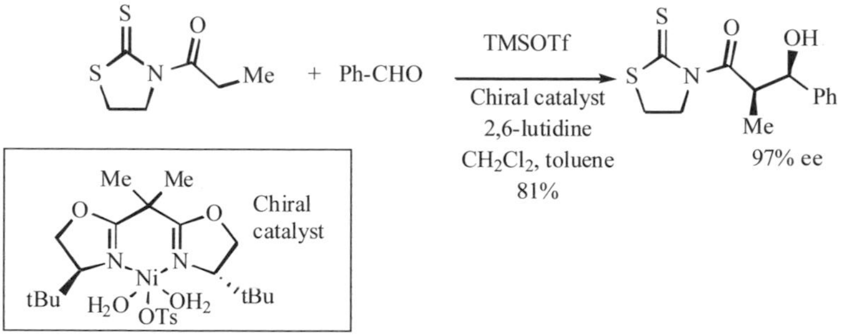

[1] D. J. Dixon, S. V. Ley, E. W. Tate, *Journal of the Chemical Soc*iety, *Perkin Transaction 1* **1999**, 2665

[2] D. A. Evans, C. W. Downey, J. L. Hubbs, *Journal of the American Chemical Society* **2003**, 125, 8706

# TPAP
## Pr₄N⁺RuO₄⁻
Tetra-*n*-propylammonium perruthenate
[114615-82-6]

## Commercially available

S. V. Ley, J. Norman, *Encyclopedia of Reagents for Organic Synthesis*, John Wiley and Sons, Inc.,
L. A. Paquette, Ed., New York, 1995, **7**, 4827

**Notes:** Mild oxidizing agent. Generally carried out in the presence of a co-oxidant, often **_NMO_**.

## Examples:
1.[1]

2.[2]

93% overall

3.[3]

[1] T. Hu, N. Takanaka, J. S. Panek, *Journal of the American Chemical Society* **1999**, 121, 9229
[2] L. A. Paquette, H.-J. Kang, C. S. Ra, *Journal of the American Chemical Society* **1992**, 114, 7387
[3] H. Miyaoka, M. Yamanishi, Y. Kajiwara, Y. Yamada, *Journal of Organic Chemistry* **2003**, 68,
3476

# Triton-B

Me
|  .Me
N⊕
|
Me    OH ⊖

Benzyltriethylammonium hydroxide
[100-85-6]

## Commercially available

M. E. Bos, *Encyclopedia of Reagents for Organic Synthesis*, John Wiley and Sons, Inc., L. A.
Paquette, Ed., New York, 1995, **1**, 382

**Notes:** This reagent (see structural similarity with ***TEBA***) serves as a phase transfer catalyst and a
base.

## Examples:

1.[1]

2.[2]

3.[3]

---

[1] N. Yoshida, K. Ogasawara, *Organic Letters* **2000**, <u>2</u>, 1461
[2] F. Wendling, M. Miesch, *Organic Letters* **2001**, <u>3</u>, 2689
[3] M. R. Younes, M. M. Chaabounib, A. Bakloutia, *Tetrahedron Letters* **2001**, <u>42</u>, 3167

# Vedejs Reagent

O
O\ ‖ -O
⎮-Mo-⎮
O-/ \-O
Pyd  HMPA

**MoOPH**
[23319-63-3]
Oxodiperoxymolybdenum(pyridine)(hexamethyphospotriamide)

## Preparation:[1]

1. 30% $H_2O_2$

MoO₃  ⟶  2. HMPA

3. Dry, then Py / THF

O
O\ ‖ -O
⎮-Mo-⎮
O-/ \-O
Pyd  HMPA

---

E. Vedejs, *Encyclopedia of Reagents for Organic Synthesis*, John Wiley and Sons, Inc., L. A. Paquette, Ed., New York, 1995, **6**, 3825

---

## Notes:

The reagent is useful for the hydroxylation of enolate anions.

⊖ ⊕
    O  M            MoOPH            HO    O
  ⟶

## Examples:

1.[2]

BTSO⟍ ⟋ ⟍ O        1. LDA        BTSO⟍ ⟋ ⟍ O
                                              HO
         2. MoOPH
                    66%
But                                     But

2.[3]

                H                    1. LDA                        H
  ⟍ ⟍ ⟍ N ⟍ O                                      ⟍ ⟍ N ⟍ O
        ⎮                            2. MoOPH              HO ⎮
        O   Ph                               77%              O   Ph

---

[1] E. Vedejs, S. Larsen, *Organic Synthesis* **1985**, <u>64</u>, 127
[2] K.Takeda, Y. Sawada, K. Sumi, *Organic Letters* **2002**, <u>4</u>, 1031
[3] S. Makino, K. Shintani, T. Yamatake, O. Hara, K. Hatano, Y. Hamata, *Tetrahedron* **2002**, <u>58</u>, 9737

# Verkade's Superbase

2,8,9-trialkyl-2,5,8,9-tetraaza-1-phosphabicyclo[3.3.3]undecane
R = Me, *i*-Pr, *i*-Bu

## Commercially available

## Notes:
A strong base, with calculated pKa of 29.[1]
See: **_Schwesinger P4 base_** for similar non-ionic bases.

Ready reaction on P due to:[2]

## Examples:

1.[2]

(*E*)-selective

---

[1] B. Kovacevic, D. Baric, Z. B. Maksic, *New Journal of Chemistry* **2004**, <u>28</u>, 284 (AN 2004:87563)
[2] Z. Wang, J. G. Varkade, *Tetrahedron Letters* **1998**, <u>39</u>, 9331

# Weinreb Amide[1]

**Preparation:** These intermediates are readily prepared from an activated acid (acid chloride) and the methoxymethyl amine.

## Notes:

1.    This useful functional group is particularly well-suited to react with hydride or organometallic reagents.

2. For application on solid support.[2]

## Examples:

1.[3]

2.[4]

77%

---

[1] S. Nahm, S. M. Weinreb, *Tetrahedron Letters* **1981**, <u>22</u>, 3815

[2] T. Q. Dinh, R. W. Armstrong, *Tetrahedron Letters* **1996**, <u>37</u>, 1161

[3] A. Gomtsyan, R.J. Koenig, C.-H. Lee, *Journal of Organic Chemistry* **2001**, <u>66</u>, 3613

[4] Y.-G Suh, J.-K. Jung, S.-Y. Seo, K.-H. Min, D.-Y. Shin, Y.-S. Lee, S.-H. Kim, H.-J. Park, *Journal of Organic Chemistry* **2002**, <u>67</u>, 4127

# Wieland-Miescher Ketone

8a Methyl-3,3,8,8a-tetrahydro-2*H*,7*H*-naphthalene-1,6-dione
[20007-72-1]

## Commercially available

## Preparation:[1]

**Notes:**  Chiral preparations include the proline-catalyzed reactions[2] and recently an aldolase antibody 38C2 method has been reported.[3]  See also:[4]

## Examples:
1.[5]

2.[6]

[1] S. Ramachandran, M. S. Newman, *Organic Syntheses*, CV5, 486
[2] G. Zhong, T. Hoffmann, R. A. Lerner, S. Danishefsky, C. F. Barbas, III, *Journal of the American Chemical Society* **1997**, 119, 8131
[3] B. List, R. A. Lerner, C. F. Barbas III, *Organic Letters* **1999**, 1, 59
[4] D. Rajagopal, R. Narayanan, S. Swaminathan, *Tetrahedron Letters* **2001**, 42, 4887
[5] S. Karimi, *Journal of Natural Products* **2001**, 64, 406
[6] K. Park, W. J. Scott, D. F. Wiemer, *Journal of Organic Chemistry* **1994**, 59, 6313

# Wilkinson's Catalyst

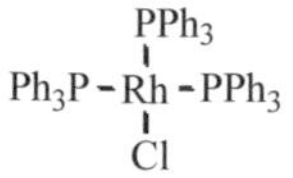

$$PPh_3$$
$$Ph_3P - Rh - PPh_3$$
$$Cl$$

Chlorotris(triphenylphosphine)rhodium (I)
[14694-95-2]

## Commercially available

K. Burgess, W. A. van derDonk, *Encyclopedia of Reagents for Organic Synthesis*, John Wiley and Sons, Inc., L. A. Paquette, Ed., New York, 1995, **2**, 1253

## Notes:

Useful for homogeneous reduction of alkenes. As a consequence of the reagent bulk, it is understandable that the reactivity of alkene reduction is dependent on substitution; the less-substituted alkenes react faster. Also, reduction occurs from the less-hindered face in a *cis*-stereochemistry. Many other functional groups are tolerated by conditions.

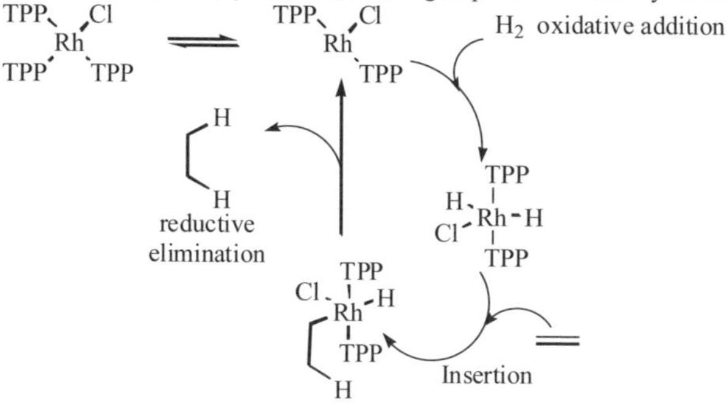

## Examples:
1.[1]

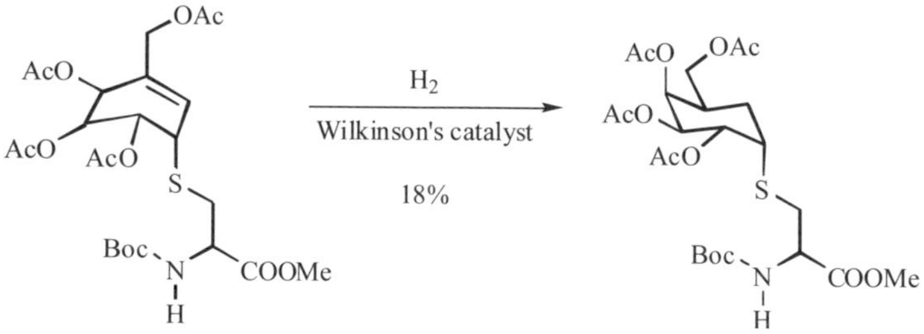

$$H_2$$

Wilkinson's catalyst

18%

[1] L. J. Whalen, R. L. Halcomb, *Organic Letters* **2004**, <u>6</u>, 3221

2.[1] Use in hydroformylation:

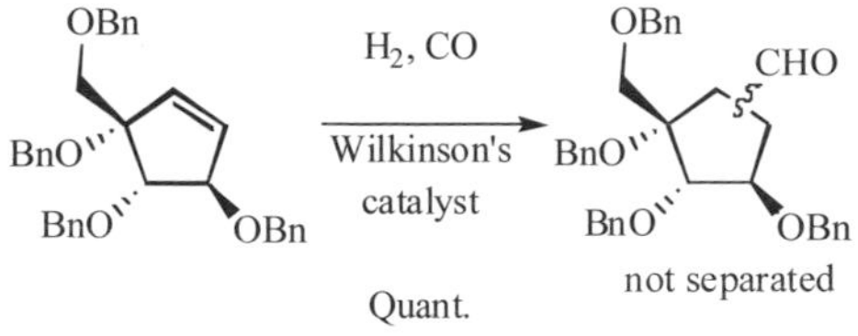

OBn
BnO
BnO
OBn
$H_2$, CO
Wilkinson's catalyst
Quant.
OBn
CHO
BnO
BnO
OBn
not separated

3.[2]

t-BuO
O
Et
O
$H_2$, $Rh(PPH_3)_3Cl$
EtOH
95%
t-BuO
O
Et
O

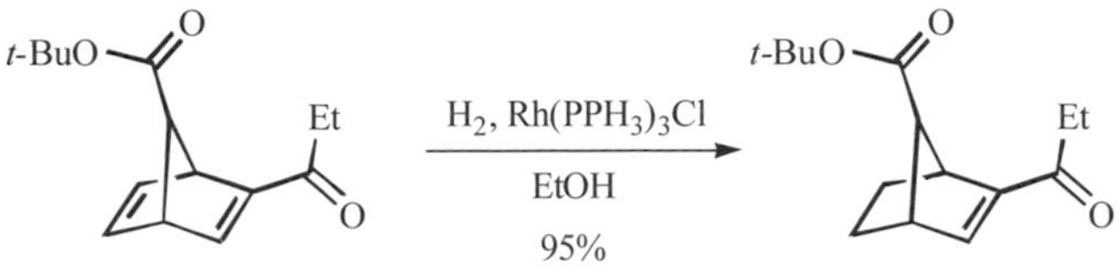

[1] M. Seepersauda, M. Kettunenb, A. S. Abu-Surrahc, T. Repob, W. Voelterd, Y. Al-Abed, *Tetrahedron Letters* **2002**, 43, 1793

[2] H. M. L. Davies, E. Saikali, N. J. S. Huby, V. J. Gilliatt, J. J. Matasi, T. Sexton, S. R. Childers, *Journal of Medicinal Chemistry* **1994**, 37, 1262

# Yamada's Reagent

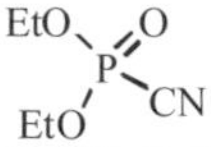

DEPC, diethyl phosphorocyanidate, diethyl phosphoryl cyanide, diethyl cyanophosphonate
[2942-58-7]

**Commercially available**. Prepared by the reaction of triethylphosphite with cyanogen bromide

H. H. Patel, *Encyclopedia of Reagents for Organic Synthesis*, John Wiley and Sons, Inc., L. A. Paquette, Ed., New York, 1995, **3**, 1851

**Notes:** Useful for carboxylic acids activation towards the preparations of esters, thioesters and amides:

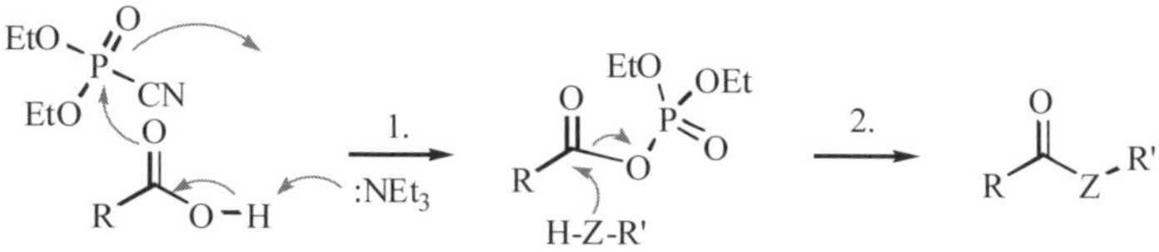

The reagent also finds application in the conversion:

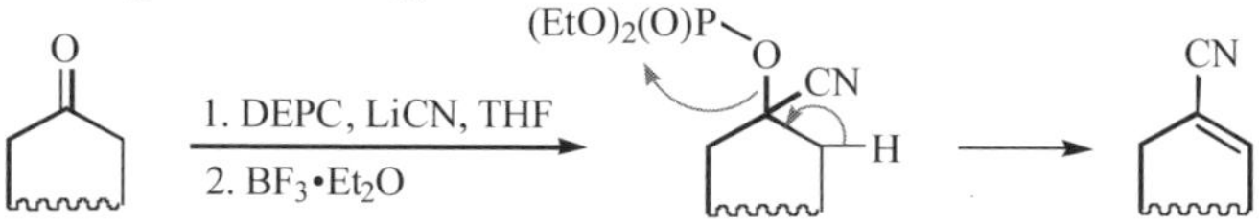

**Examples:**

1.[1]

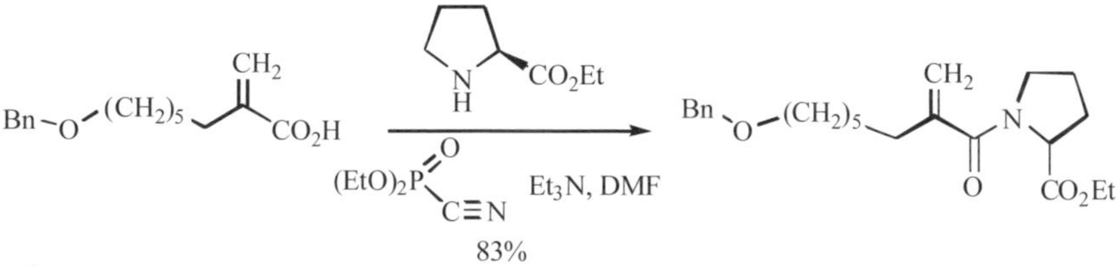

2.[2]

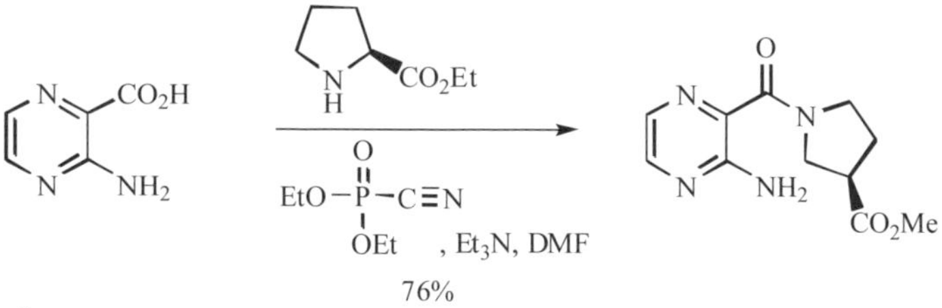

3.[3]

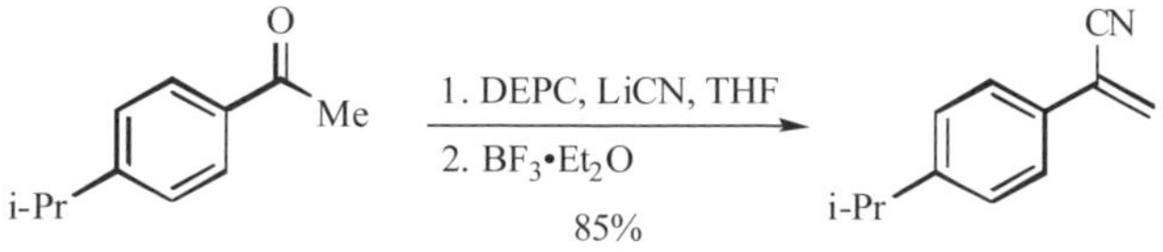

---

[1] J. Jew, J. Kim, B. Jeong. H. Park, *Tetrahedron: Asymmetry* **1997**, <u>8</u>, 1187
[2] T. Okawa, S. Eguchi, *Tetrahedron Letters* **1996**, <u>37</u>, 81
[3] S. Harusawa, R. Yoneda, T. Kurihara, Y. Hamada, T. Shioin, *Tetrahedron Letters* **1984**, <u>25</u>, 427

# Zeise's Dimer

Di-μ-chlorodichlorobis(η2-ethene) diplatinum
[12073-36-8]

## Commercially available

A. J. Phillips, *e-Encyclopedia of Reagents for Organic Synthesis*, L.A. Paquette, Ed.,, John Wiley & Sons, Inc., online reference available at *http://www.interscience.wiley.com.*

## Notes:
Useful for opening cyclopropane rings and hydrating alkynes.

## Examples:

1.[1]

2.[2]

3.[3]

R = H, Me    77 - 80%    18 - 20%

4.[4]

[1] J. Beyer, P. R.Skaanderup, R. Madsen, *Journal of the American Chemical Society* **2000**, 122, 9575
[2] K. Ikura, I. Ryu, N. Kambe, N. Sonoda, *Journal of the American Chemical Society* **1992**, 114, 1520
[3] Y. Chen, J. K. Snyder, *Journal of Organic Chemistry* **2001**, 66, 6943
[4] Z. Ye, M. Dimke, P. W. Jennings, *Organometallics* **1993**, 12, 1026

# INDEX